# Basic
# Circuit
# Theory

# Basic Circuit Theory

**Charles A. Desoer**

and

**Ernest S. Kuh**

Department of Electrical Engineering
and Computer Sciences
University of California, Berkeley

**McGraw-Hill Book Company**

New York   St. Louis   San Francisco
London   Sydney   Toronto
Mexico   Panama

*To the University of California*
*on Its Centennial*

**Basic Circuit Theory**

*Library of Congress Catalog Card Number* 68-9551

16575

1 2 3 4 5 6 7 8 9 0    M A M M    7 6 5 4 3 2 1 6 9

# Preface

This book is an outgrowth of a course intended for upper-division students in electrical engineering. The course is 20 weeks long and consists of three lectures and one two-hour recitation per week. Included in the book are enough additional materials to accommodate a full year's course. We assume that students have completed work in basic physics and mathematics, including some introduction to differential equations and some acquaintance with matrices and determinants. Previous exposure to some circuit theory and to electronic circuits is helpful, but not necessary.

**Conceptual basis**  Although this book covers most of the material taught in conventional circuit courses, the point of view from which it is considered is significantly different. The most important feature of this book is a novel formulation of lumped-circuit theory which accommodates linear and nonlinear, time-invariant and time-varying, and passive and active circuits. In this way, fundamental concepts and the basic results of circuit theory are presented within a framework which is sufficiently general to reveal their scope and their power.

We want to give the student an ability to write the differential equations of any reasonably complicated circuit, including ones with nonlinear and time-varying elements. Our aim is to give him an ability to approach any lumped circuit knowing which facts of circuit theory apply and which do not, so that he can effectively use his knowledge of circuit theory as a predictive tool in design and in the laboratory.

These aims require an overhaul of the traditional teaching of circuit theory. The main reason that such an overhaul became mandatory is that in recent years science and technology have been advancing at unbelievable speeds, and the development of new electronic devices has paced this advance in a dramatic fashion. As a consequence, we must not only teach the basic facts and techniques that are usable today, but we must also give the fundamental concepts required to understand and tackle the engineering problems of tomorrow.

In the past decade, the engineering of large sophisticated systems has made great strides, with concomitant advances in communications and control. Thus a modern curriculum requires that a course in basic circuit theory introduce the student to some basic concepts of system theory, to the idea of stability, to the modeling of devices, and to the analysis of electronic circuits. In such a course the student should be exposed early to nonlinear and time-varying circuit elements, biasing circuits, and small-signal analysis.

v

The computer revolution also makes heavy demands on engineering education. Computers force us to emphasize *systematic* formulation and algorithmic solution of problems rather than special tricks or special graphical methods.    Also, computers have changed the meaning of the word *solution*.    When a set of several ordinary nonlinear, differential equations can be solved on a computer in a few seconds, the absence of closed-form solutions to nonlinear circuit problems is no longer an obstacle.

One other factor cannot be overlooked.    The preparation of our students is steadily improving.    This is partly a consequence of the new high school programs in mathematics (SMSG) and physics (PSSC), and of the new freshman and sophomore courses in mathematics, physics, and engineering.    It is up to us to offer a course in circuit theory to our students whose intellectual caliber is commensurate with these new courses.

**Organization of the book**

These considerations make the development of circuit theory follow a natural course in this book. The book can be thought of as consisting of three parts.    Part I (Chapters 1 to 7) treats simple circuits.    We start with a precise statement of Kirchhoff's laws and their limitations.    We then introduce two-terminal elements and classify them according to whether they are linear or nonlinear, time-invariant or time-varying.    We give examples of simple nonlinear circuits using the tunnel diode and the varactor diode.    In Chapters 3 to 6, we use simple circuits to illustrate many of the principal facts and techniques of linear system theory and some properties of nonlinear circuits.    Linearity and time invariance are stressed and are the bases of our derivation of the convolution integral, thus establishing the fact that the zero-state response is a linear function of the input.    The state-space method is introduced and illustrated with both linear and nonlinear circuits. Finally, in Chapter 7, we give a straightforward and systematic treatment of the phasor method for sinusoidal steady-state analysis.

Part II (Chapters 8 to 12) deals with the analysis of complex networks.    We first introduce the standard coupled elements—transformers, coupled inductors, and controlled sources.    Elements of graph theory are explained concisely, and Tellegen's theorem is introduced and used to derive general properties of impedance functions.    This gives the student a taste of the power of general methods.    General network analysis, comprising the node, mesh, loop, and cut-set methods, is systematically presented, using graph theory as background and including the necessary formalism for computer solution.    The state-variable method is described and then shown to be a powerful tool in formulating equations for nonlinear and time-varying networks.

Part III (Chapters 13 to 17) develops the main results of circuit theory.    A brief introduction to Laplace transforms leads to the fundamental properties of linear time-invariant networks.    Then follow natural frequencies, network functions, the four standard network theorems, and two-ports.    Throughout the discussion, the existence of nonlinear and time-varying networks is constantly in mind.    For example, we show the use of network functions in oscillator design and also demonstrate by telling examples the limits of the applicability of the network theorems. The chapter on two-ports discusses carefully the relation between the small-signal two-port model and the device characteristics.    The brief chapter on resistive networks shows that general properties of nonlinear resistive networks can be formulated even though no closed form solution is known.    The last chapter on energy and passivity considers the energy balance in time-varying elements, parametric amplifiers, and the characterization of passive one-ports.

The book ends with three appendixes devoted to basic mathematical topics: the concept of a function, matrices and determinants, and differential equations. Each appendix is a concise summary of definitions and facts that are used in the text.

**How to teach from the book**
The course which is given at Berkeley lasts for two academic quarters (20 weeks) and is intended for beginning *junior* students. The course consists of three lectures and one two-hour recitation per week. From our own experience and from that of others who have taught the course, we can state that the essential material of the first 17 chapters can be comfortably covered in two quarters. A typical breakdown of the amount of lecture time spent on each chapter is given below:

| Chapters | No. of 1-hr lectures |
|:---:|:---:|
| 1 | 1 |
| 2 | 3 |
| 3 | 2 |
| 4 | 6 |
| 5 | 4 |
| 6 | 4 |
| 7 | 6 |
| 8 | 2 |
| 9 | 2 |
| 10 | 5 |
| 11 | 2 |
| 12 | 2 |
| 13 | 4 |
| 14 | 1 |
| 15 | 4 |
| 16 | 4 |
| 17 | 4 |
| | 56 |

It should be pointed out that not all sections in every chapter are covered in class. In particular, sections whose headings are enclosed in unshaded rectangles are omitted. These sections present more advanced topics and can be omitted with no loss in continuity.

In a one-semester course, we suggest that Chapters 16 and 17 be left out. In addition, other chapters such as 11, 12, and 14 can be touched on lightly. This scheme has been tried in other universities, based on the preliminary volumes.

For a two-semester course, we suggest coverage of the first 17 chapters at a more leisurely pace, proceeding lightly through Chapter 18 and covering thoroughly Chapter 19 because it gives an excellent opportunity to interrelate the ideas of time-varying element, nonlinear element, time-domain analysis, frequency-domain analysis, power, energy, and stability.

**Acknowledg-ments**   Even though this book is a systematic introduction to circuit theory, it uses many concepts and techniques which were developed by people doing research in circuit and system theory.   In fact, without our own deep involvement in research, this book could not have been written.   It is a pleasure to publicly acknowledge the research support of the University of California, the National Science Foundation, and the Department of Defense, in particular the Office of Naval Research, the Office of Scientific Research, and the Army Research Office.

We are also indebted to many people who have taught the course from our class notes and the preliminary edition [volume I (1966) and volume II (1967)] and have given us valuable suggestions.   It is a pleasure to mention in particular C. T. Chen, L. Forys, I. T. Frisch, L. A. Gerhardt, J. Katznelson, R. W. Liu, R. N. Newcomb, R. A. Rohrer, L. M. Silverman, R. W. Snelsire, E. Wong, and B. A. Wooley. Special thanks are due to R. W. Liu and R. A. Rohrer who went through the preliminary edition chapter by chapter and gave us detailed comments, many of which are incorporated in the present volume.   We also benefited from discussions with Leon Chua, J. B. Cruz, B. J. Leon, and M. E. Van Valkenburg concerning symbols and notations.   Finally, we feel specially privileged to have had Michael Elia of the McGraw-Hill Book Company as our editor.   His enthusiasm and sense of humor were invaluable to us, and his many novel suggestions on writing style and artwork contributed a great deal to the readability and appearance of the book.

*Charles A. Desoer*
*Ernest S. Kuh*

# To the Student

In this volume we attempt to bring you quickly in contact with the principal facts of circuit theory.  As a means to this end, we have given careful attention to terminology, sign conventions, and notations.  Of course, these are not ends in themselves but merely tools to make ideas precise and to help your comprehension. Often, when your understanding of a sentence is fuzzy, some aspect of a previous definition or result has escaped you in your study.  When this happens it is imperative that you review previous pertinent passages.  To help you in studying, every chapter ends with a *summary* which restates the principal results of the chapter.  You will really understand a chapter when you are able to explain and illustrate by examples each and every statement in the summary.

To help you achieve thorough understanding, each technical term is carefully defined, and in the defining sentence it is printed in **boldface type.**  In the subject index, the number of the page which carries the definition is also printed in boldface.  *Italic type* is used to emphasize certain words.  Also, all conventions concerning reference directions and signs are completely stated.

Our notations are standard except in one respect: we must often differentiate between two closely related ideas.  For example, the current through a resistor as a *waveform* defined over a specified interval of time is distinct from the value of the current at some particular time.  When we wish to emphasize that we mean the current $i$ as a *waveform*, we write $i(\cdot)$, and when we mean the value of the current at some time, say $t_0$, we write $i(t_0)$.  We use conventional notations for units of physical entities.  We use the symbol $\triangleq$ to mean *is defined to be equal;* it is important to distinguish between an equation which asserts the equality of two already defined objects and an equation which defines a new symbol.  Finally, to distinguish between scalars and vectors or matrices, we use boldface for vectors and matrices.

For your convenience, three mathematical appendixes briefly summarize basic definitions and theorems used in the text.  In particular, you should thoroughly familiarize yourself with Appendix A, which covers the concepts of function and of linear function, before you get deeply into Chapter 2.  Appendix C, on differential equations, should be reviewed before studying Chapter 4.  Appendix B, on matrices and determinants, should be studied before Chapter 9.  In other words, you should know most of the material in the appendixes to thoroughly understand the text.

*Charles A. Desoer*
*Ernest S. Kuh*

# Contents

| Chapter 3 | Simple Circuits |
|---|---|

| Chapter 4 | First-order Circuits |
|---|---|

| Chapter 5 | Second-order Circuits |
|---|---|

## Chapter 6   Introduction to Linear Time-invariant Circuits

## Chapter 7   Sinusoidal Steady-state Analysis

**Chapter 8**   **Coupling Elements and Coupled Circuits**

**Chapter 9**   **Network Graphs and Tellegen's Theorem**

**Chapter 10**   **Node and Mesh Analyses**

## Chapter 11  Loop and Cut-set Analysis

## Chapter 12  State Equations

**Appendix C** **Differential Equations**

**List of Tables**

# 1

# Lumped Circuits and Kirchhoff's Laws

Electric circuits are not new to you. Many of us have already encountered them in high school and college physics courses and possibly in some engineering courses. However, the treatment of circuits may have been casual and may have consisted by and large of special cases. In this book the basic theory of electric circuits will be developed *systematically* so that when the reader has finished this book, he should feel confident in his understanding of circuits and in his power to analyze correctly any given circuit. Furthermore, in the process of systematic exposition of circuit theory, you will become acquainted with a number of fundamental ideas important to many engineering fields, for example, communication, control, and mechanical systems. Thus, a systematic course in circuit theory is a keystone in the education of an engineer, especially an *electrical engineer*.

Circuit theory (and any engineering discipline) is based on the concept of modeling. To analyze any complex physical system, we must be able to describe the system in terms of an idealized model that is an interconnection of idealized elements. The idealized elements are simple models that are used to represent or approximate the properties of simple physical elements or physical phenomena. Although physical elements and physical phenomena may be described only approximately, idealized elements are by definition characterized precisely. In circuit theory we study circuits made up of idealized elements, and we also study their general properties. Given a physical circuit, it is possible to obtain a succession of idealized models of this circuit such that the behavior of the model fits more and more closely to the behavior of the physical circuit. By analyzing the circuit model, we can predict the behavior of the physical circuit and design better circuits.

The models of circuit theory are analogous to such familiar models of classical mechanics as the particle and the rigid body. Remember that a particle is a model for a small object. By definition, a particle has zero physical dimensions; but it *does* have a positive mass together with a well-defined position, velocity, and acceleration. Similarly, a rigid body is postulated to have a definite shape, mass, inertia, etc., and it is assumed that, however large the forces acting on the rigid body, the distance between any pair of points of the rigid body does not change. Strictly speaking, there is no such thing in the physical universe as a particle or a rigid body. Yet these idealized models are successfully used in the design of machinery, airplanes, rockets, etc. Circuit elements such as those treated in Chap. 2 are models that have precise characterizations. They are idealizations of the physical properties of practical components that are available commercially. The interconnection of circuit elements forms a circuit. By means of idealized models we analyze and design practical circuits.

There are two kinds of circuits: *lumped circuits* and *distributed circuits.* In this book we shall consider lumped circuits exclusively. We do this for two reasons. First, lumped circuits are simpler to understand and to design; they are analogous to mechanical systems made of a collection of interacting particles. Second, the theory of distributed circuits can be based on that of lumped circuits. Indeed, a distributed circuit may be considered as the limit of a sequence of lumped circuits in the same way that the equations of the string and the membrane can be considered as the limit of systems of interacting particles when the number of particles tends to infinity and the distance between particles goes to zero.

## 1   Lumped Circuits

Lumped circuits are obtained by connecting *lumped elements.* Typical lumped elements are resistors, capacitors, inductors, and transformers. We have encountered them in the laboratory, and we can see them in our radio sets. The key property associated with lumped elements is their small size (compared to the wavelength corresponding to their normal frequency of operation). From the more general electromagnetic field point of view, lumped elements are point singularities; that is, they have negligible physical dimensions. In this way they are similar to a particle. Lumped elements may have two terminals, as in a resistor, or more than two terminals, as in a transformer or a transistor. For *two-terminal* lumped elements, it can be shown that the general laws governing the electromagnetic field, together with the restriction on physical size indicated above, imply that at all times the current entering one terminal is equal to the current leaving the other terminal, and that the voltage difference between the two terminals can be unambiguously defined by physical measurements. Thus, *for two-terminal lumped elements, the current through the element and the voltage across it are well-defined quantities. For lumped elements with more than two terminals, the current entering any terminal and the voltage across any pair of terminals are well defined at all times.*

For the remainder of this book, any interconnection of lumped elements such that the dimensions of the circuit are small compared with the wavelength associated with the highest frequency of interest will be called a **lumped circuit.**

As long as this restriction on the size of the circuit holds, Kirchhoff's current and voltage laws (to be discussed in Secs. 3 and 4) are valid. This restriction is a consequence of the fact that Kirchhoff's laws are approximations of Maxwell's celebrated equations, which are the general laws of the electromagnetic field. The approximation is analogous to the fact that Newton's laws of classical mechanics are approximations to the laws of relativistic mechanics. Even though they are approximations, the laws of Newton and Kirchhoff can be applied to a large number of practical

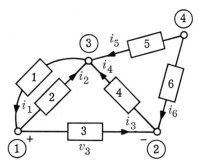

**Fig. 1.1**    A lumped circuit with six branches and four nodes.

problems, which makes them of great theoretical and practical importance.

To exhibit the implications of the restriction on size, consider the following cases: (1) For an audio circuit the highest frequency may be 25 kHz, and the corresponding wavelength is $\lambda = (3 \times 10^8)/(25 \times 10^3)$ m $= 12$ km $\approx 7.5$ miles,† which is much larger than the size of a circuit in a laboratory; (2) for a computer circuit, the frequency may be 500 MHz, in which case $\lambda = (3 \times 10^8)/(5 \times 10^8) = 0.6$ m $\approx 2$ ft, and consequently the lumped approximation may not be good; (3) for a microwave circuit, where $\lambda$ is anything between 10 cm and 1 mm, we shall encounter cavity resonators, and we shall learn that Kirchhoff's laws do not apply in these resonators because they are operated at frequencies whose wavelengths are of the same order of magnitude as the dimensions of the cavity.

As we said before, a **lumped circuit** is by definition an interconnecting lumped element. In a lumped circuit the two-terminal elements are called **branches,** and the terminals of the elements are called **nodes.**‡  Figure 1.1 shows a lumped circuit which has four nodes (labeled ①, ②, ③, and ④) and six branches (labeled 1, 2, 3, 4, 5, and 6).  The voltage across a branch (called **branch voltage**) and the current through a branch (called **branch current**) are the basic variables of interest in circuit theory.  Thus, $i_3$ is the branch current of branch 3, and $v_3$ is the branch voltage of branch 3.  To keep track of polarities, we *arbitrarily* assign a reference direction for the current and a reference direction for the voltage.  We devote the next section to these reference directions.

---

† The symbol $\approx$ means *approximately equals.*
‡ Often we use the terms "node" and "terminal" interchangeably; later on the term node will suggest the additional implication of a terminal at which several elements are connected. Presently, we also use the terms "branch" and "element" interchangeably; the term branch is somewhat more general.

## 2   Reference Directions

Consider any two-terminal lumped element with terminals $A$ and $B$ as shown in Fig. 2.1. It may be a resistor, inductor, or diode; its nature is of no importance at present. To suggest this generally, we refer to the two-terminal element as a *branch*. It is of fundamental importance for an engineer to be very precise concerning the meaning of the reference directions of the branch voltage $v$ and the branch current $i$. The reference direction for the voltage is indicated by the plus and minus symbols located near the terminals $A$ and $B$ in Fig. 2.1. The reference direction for the current is indicated by the arrow.

Given the reference direction for the voltage shown in Fig. 2.1, by convention *the branch voltage $v$ is positive at time $t$* [that is, $v(t) > 0$] whenever the electrical potential of $A$ at time $t$ is larger than the electrical potential of $B$ at time $t$, with both potentials measured with respect to the same reference. If we call these two potentials $v_A$ and $v_B$, respectively, then

$$v(t) = v_A(t) - v_B(t)$$

Given the reference direction for the current shown in Fig. 2.1, by convention *the branch current $i$ is positive at time $t$* [that is, $i(t) > 0$] whenever (at time $t$) a net flow of positive charges enters the branch at node $A$ and leaves it at node $B$.

It is important to realize that reference directions may be assigned arbitrarily, for they do not constitute an assertion of what physically happens in the circuit. For example, it is only when the statement $v(t) > 0$ is coupled with the reference direction for the voltage that we obtain an assertion about the relative voltages of nodes $A$ and $B$.

It is clear from the above that we may assign to a given branch an arbitrary reference direction for the voltage and an arbitrary reference direction for the current. In principle, these reference directions are independent. It is customary to choose directions called *associated reference directions;* the reference direction for the branch voltage and the reference

**Fig. 2.1**

A two-terminal lumped element (or a branch) with nodes $A$ and $B$; the reference directions for the branch voltage $v$ and the branch current $i$ are associated reference directions.

direction for the current are said to be **associated** if a positive current enters the branch by the terminal marked with a plus sign and leaves the branch by the terminal marked with a minus sign. The reference directions shown in Figs. 1.1 and 2.1 are all associated reference directions. Recalling a basic fact from the physics courses, we note that if associated reference directions are used, the product $v(t)i(t)$ is the *power delivered at time t to the branch.*

We turn now to the statement and detailed consideration of the basic laws that apply to lumped circuits.

## 3   Kirchhoff's Current Law (KCL)

We shall first state Kirchhoff's current law for a special case. Later we shall broaden the concept and give a more general statement.

Kirchhoff's
current
law

> For any lumped electric circuit, for any of its nodes, and at any time, the algebraic sum of all branch currents leaving the node is zero.

In applying KCL to a particular node, we first assign a reference direction to each branch current. In the algebraic sum, we assign the plus sign to those branch currents whose reference direction points away from the node; similarly, we assign the minus sign to those branch currents whose reference direction points into the node. For example, in the circuit shown in Fig. 1.1, KCL applied to node ② asserts that

(3.1)   $i_4(t) - i_3(t) - i_6(t) = 0$      for all $t$

since the branch current $i_4$ has a reference direction pointing away from the node, whereas the branch currents $i_3$ and $i_6$ have reference directions pointing into the node. Similarly, for node ① KCL asserts that

(3.2)   $-i_1(t) + i_2(t) + i_3(t) = 0$      for all $t$

where the first term has to be assigned a minus sign because the reference direction of the current $i_1$ points into node ①. In these introductory chapters Eqs. (3.1) and (3.2) will be referred to as *node equations,* i.e., equations obtained from KCL at various nodes.

KCL is of extreme importance. Its simplicity and our familiarity with it may hide some of its most important features. For emphasis we present the following features of KCL.

**Remarks**   1.   KCL imposes a *linear* constraint on the branch currents. In other words, Eqs. (3.1) and (3.2) are *linear homogeneous* algebraic equations (with constant coefficients) in the variables $i_1$, $i_2$, $i_3$, $i_4$, and $i_6$.

2.  KCL applies to any lumped electric circuit; it does not matter whether the circuit elements are linear, nonlinear, active, passive, time-varying, time-invariant, etc. (the precise meaning of these adjectives will be given in later chapters). Another way of stating this idea is to say that *KCL is independent of the nature of the elements.*

3.  If we recall that the current through a branch measures the rate at which electric charges flow through that branch, it is clear that KCL asserts that charges do not accumulate at any node. In other words, *KCL expresses the conservation of charge at every node.*

4.  An example of a case where KCL does not apply is a whip antenna, say, on a policeman's motorcycle. Clearly, when the antenna is transmitting, there is a current at the base of the antenna; however, the current is zero at all times at the tip of the antenna. On the other hand, it is a fact that the length of the antenna is about a quarter of the wavelength corresponding to the frequency of operation; hence, the antenna is not a lumped circuit. Therefore, we should not expect KCL to apply to it.

| 4 | **Kirchhoff's Voltage Law (KVL)** |

In order to state KVL we must know what we mean by a loop. The precise definition of a loop will be given in Chap. 9 when we introduce general networks. Intuitively, a loop means a closed path. Thus, if we consider a circuit as a bunch of branches connected at nodes, a path is formed by starting at one node, traversing one or more branches in succession, and ending at another node. A closed path is a path whose starting node is the same as its ending node.

Kirchhoff's
voltage
law

> For any lumped electric circuit, for any of its loops, and at any time, the algebraic sum of the branch voltages around the loop is zero.

In order to apply KVL we assign a reference direction to the loop. In the algebraic expressing KVL, we assign the plus sign to the branch voltages whose reference directions agree with that of the loop, and we assign the minus sign to the branch voltages whose reference directions do not agree with that of the loop.

**Example**   Consider the circuit of Fig. 4.1.

*a.*   KVL applied to the loop I, which consists of branches 4, 5, and 6, asserts that

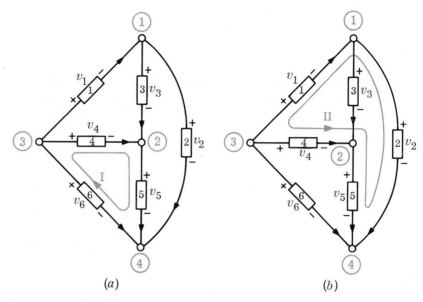

(a)                                         (b)

**Fig. 4.1**   Example illustrating KVL; loops I and II are indicated.

(4.1)   $v_4(t) + v_5(t) - v_6(t) = 0$      for all $t$

The reference direction selected for this loop (marked I) is shown in Fig. 4.1a. The reference directions of branches 4 and 5 agree with the reference direction of loop I, whereas the reference direction of branch 6 does not agree with that of loop I; we therefore assign plus signs to $v_4$ and $v_5$ and a minus sign to $v_6$.

b.   KVL applied to the loop II, which consists of branches 1, 4, 5, and 2, asserts that

(4.2)   $-v_1(t) + v_4(t) + v_5(t) - v_2(t) = 0$      for all $t$

The reference direction for the loop (marked II) is shown in Fig. 4.1b.

In these introductory chapters, equations such as (4.1) and (4.2) will be referred to as loop equations, i.e., equations obtained from KVL for various loops. For emphasis, we present the following features of this important law.

**Remarks**   1.   KVL imposes a *linear* constraint between branch voltages of a loop.

2.   KVL applies to any lumped electric circuit; it does not matter whether the circuit elements are linear, nonlinear, active, passive, time-varying, time-invariant, etc. In other words, *KVL is independent of the nature of the elements.*

| 5† | **Wavelength and Dimension of the Circuit** |
|---|---|

† Sections and subsections with boxed headings may be omitted without loss of continuity.

The purpose of this section is to discuss intuitively what happens when the dimensions of a circuit become comparable to or even larger than the wavelength associated with the highest frequencies of interest. Let us examine this condition. Let $d$ be the largest dimension of the circuit, $c$ the velocity of propagation of electromagnetic waves, $\lambda$ the wavelength of the highest frequency of interest, and $f$ the frequency. The condition states that

(5.1)   $d$ is of the order of or larger than $\lambda$

Now $\tau \triangleq d/c$ is the time required for electromagnetic waves to propagate from one end of the circuit to the other.‡ Since $f\lambda = c$, $\lambda/c = 1/f = T$, where $T$ is the period of the highest frequency of interest. Thus, the condition in terms of the dimension of the circuit and the wavelength can be stated alternately in terms of time as follows:

(5.2)   $\tau$ is of the order of or larger than $T$

Thus, recalling the remarks concerning the applicability of KCL and KVL at high frequencies, we may say that KCL and KVL hold for any lumped circuit as long as the propagation time of electromagnetic waves through the medium surrounding the circuit is negligibly small compared with the period of the highest frequency of interest.

**Example**  To get a feeling for the importance of the conditions stated in (5.1) and (5.2), consider a dipole antenna of an FM broadcast receiver and the 300-ohm transmission line that connects it to the receiver. If we examine the transmission line, we observe that it consists of two parallel copper wires that are held at a constant distance from one another by some insulating plastic in which they are embedded. Suppose for simplicity that the transmission line is infinitely long to the right (see Fig. 5.1). If the electromagnetic field is propagated at infinite velocity, then as soon as a voltage is induced on the antenna, the same voltage appears simultaneously everywhere down the line. To see what happens if the velocity of propagation is not infinite but is $3 \times 10^8$ m/sec, suppose a 100-MHz sinusoidal voltage appears at the antenna. Thus,

$$v_A(t) = V_0 \sin (2\pi \times 10^8 t)$$

where $V_0$ is a constant expressed in volts and $t$ is expressed in seconds. Consider what happens at $B$ on the transmission line, say, 5 ft $\approx 1.5$ m down the line. Since the velocity is $3 \times 10^8$ m/sec, the voltage at $B$ is

‡ The symbol $\triangleq$ means *equals by definition.*

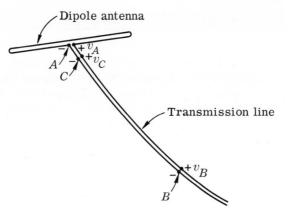

**Fig. 5.1**   A dipole antenna connected to a transmission line.

delayed by $1.5/(3 \times 10^8) = 5 \times 10^{-9}$ sec, with respect to the voltage at $A$. Thus,

$$v_B(t) = V_0 \sin (2\pi \times 10^8)(t - 0.5 \times 10^{-8})$$
$$= V_0 \sin (2\pi \times 10^8 t - \pi)$$
$$= - V_0 \sin (2\pi \times 10^8 t) = -v_A(t)$$

and, at point $B$, the voltage at time $t$ across the line is precisely the opposite of that at point $A$ !   The important fact is that the difference between $v_A(t)$ and $v_B(t)$ is due to the propagation time, which is not negligible in this case. Indeed, the propagation time from $A$ to $B$ is 5 nsec ($10^{-9}$ sec), and a full period of the sinusoidal signal $v_A$ is 10 nsec.

If we think in terms of wavelengths, we find that at 100 MHz

$$\lambda = \frac{c}{f} = \frac{3 \times 10^8}{10^8} = 3 \text{ m}$$

Thus, the distance from $A$ to $B$ is one-half of the wavelength.

Of course, if we were comparing $v_A$ and $v_C$, where the point $C$ is, say, half an inch to the right of $A$, then the propagation time from $A$ to $C$ is about $4 \times 10^{-11}$ sec, and

$$v_C(t) = V_0 \sin (2\pi \times 10^8)(t - 4 \times 10^{-11})$$
$$= V_0 \sin (2\pi \times 10^8 t - 0.025)$$

that is, the phase of $v_C$ lags behind that of $v_A$ by 0.025 rad $= 1.3°$.   Consequently, $v_A(t) \approx v_C(t)$ for all $t$.

## Summary

■ Kirchhoff's laws and the lumped-element model of a circuit are valid provided that the largest physical dimension of the circuit is small compared with the wavelength corresponding to the highest frequency under consideration. Under these conditions the voltage across any branch or pair of nodes is well defined; also, the current entering any two-terminal element through one terminal is well defined and is equal to the current leaving it through the other terminal.

■ KCL states that for any lumped electric circuit, for any of its nodes, and at any time, the algebraic sum of all the branch currents leaving the node is zero.

■ KVL states that for any lumped electric circuit, for any of its loops, and at any time, the algebraic sum of all the branch voltages around the loop is zero.

■ Kirchhoff's laws are linear constraints on the branch voltages and branch currents. Furthermore, they are independent of the nature of the elements.

■ The reference direction of the branch voltage and the reference direction of the branch current of an element are said to be in the associated reference direction if a positive current enters the branch by the terminal marked with a plus sign and leaves the branch by the terminal marked with a minus sign. With associated reference direction, the product of the branch voltage and the branch current is the power delivered to the branch.

## Problems

Wavelength calculation

**1.** An FM receiver is connected to its antenna by a piece of cable 2 m long. Considering that the receiver is tuned to 100 MHz, can you say that the instantaneous currents at the input of the receiver and at the antenna terminals are equal? If not, for what approximate cable lengths would they be equal?

KCL

**2.** Some of the branch currents (in amperes) of the circuit shown in Fig. P1.2 are known, namely, $i_1 = 2$, $i_3 = 1$, $i_7 = 2$, and $i_8 = 3$. Can you determine all the remaining branch currents with this information? Explain. (Find the values of those that can be computed and indicate any

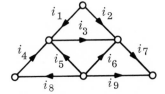

Fig. P1.2

additional data needed for those that you cannot compute.)

KVL    **3.** Suppose that in the circuit of Prob. 2 we use associated reference directions for branch voltages and branch currents. We are given the following branch voltages: $v_1 = v_3 = v_6 = v_9 = 1$ volt. Can you determine the remaining branch voltages with this information? Explain.

KCL and KVL    **4.** In the circuit shown in Fig. P1.4 we use associated reference directions for the reference directions of the branch variables.

a.    Apply KCL to nodes ①, ②, ③, and ④. Show that KCL applied to node ④ is a consequence of the preceding three equations.

b.    If we call any loop which has no internal branch a *mesh*, write KVL for the three meshes of the circuit shown. Write also KVL for loops *afe*, *abdf*, *acde*, and *bcfe*. Show that these equations are consequences of the preceding three mesh equations.

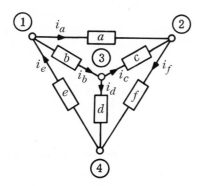

**Fig. P1.4**

Loop    **5.** In the circuit shown in Fig. P1.4 list all possible loops.

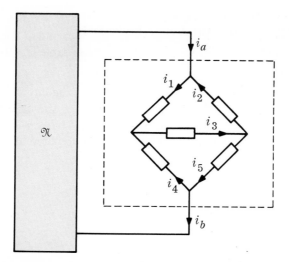

**Fig. P1.6**

KCL   **6.** The portion of the circuit in the dashed line (see Fig. P1.6) can be considered as a two-terminal element connected to the remainder of the circuit $\mathcal{N}$. Is $i_a = i_b$? Prove your answer.

KVL   **7.** In the circuit shown in Fig. P1.7, the following voltages are given in volts: $v_1 = 10$, $v_2 = 5$, $v_4 = -3$, $v_6 = 2$, $v_7 = -3$, and $v_{12} = 8$. Determine as many branch voltages as possible.

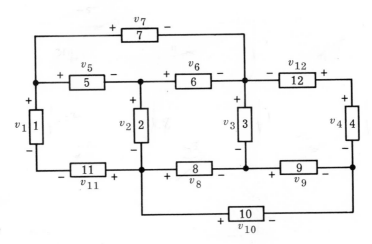

**Fig. P1.7**

KCL   **8.** In the circuit shown in Fig. P1.7, the branch currents are measured in the associated reference directions. The following currents are given in amperes: $i_1 = 2$, $i_7 = -5$, $i_4 = 5$, $i_{10} = -3$, and $i_3 = 1$. Is it possible to determine the remaining branch currents? Determine as many as you can.

KCL   **9.** In the circuit shown in Fig. P1.7, the branch currents are measured in the associated reference directions. Prove that

$$i_1 + i_2 + i_3 + i_4 \equiv 0$$
$$i_7 + i_6 + i_8 + i_{10} \equiv 0$$

# 2 Circuit Elements

The components used to build lumped electric circuits are resistors, diodes, transistors, vacuum tubes, capacitors, inductors, transformers, etc. Each component is designed to make use of a primary physical property. Unfortunately, it is usually impossible to build a physical component that exhibits only a primary property. For example, a resistor, being a two-terminal conducting body, transforms electric energy into heat, and the voltage $v(t)$ across its terminals depends only on the current $i(t)$ through it. This physical picture is approximate because any current creates some magnetic field; consequently, any resistor stores some energy in its magnetic field. The stored energy is usually so small that it can be neglected in analysis and design. Thus, it is only in an approximate sense that we can think of a resistor in terms of a model that satisfies Ohm's law. This approximate modeling illustrates the fundamental fact that in order to analyze and design electric circuits, we shall have to perform *approximations* and select appropriate models, because to study in detail the physics of the majority of circuit components is an almost impossible task. We are in the same position as the physicist who can never describe accurately and completely the experimental setup he is using. For example, he introduces the concept of a particle even though he knows that every physical object has physical dimension. He also introduces the concept of a rigid body even though he knows that every physical object exhibits elastic properties. Similarly, in circuit theory we shall introduce ideal elements (in contrast to physical components) that we shall refer to as **circuit elements** (or **elements** for short). All these circuit elements will be lumped elements in the sense discussed in Chap. 1. They are conceptual models in terms of which we shall interpret our experimental results and design our practical circuits. In this chapter we shall define and discuss the properties of those circuit elements that have two terminals; we call these elements **two-terminal elements.** In Chap. 8 we shall introduce additional circuit elements that have more than two terminals.

## 1 Resistors

In sophomore physics the only resistor that was considered was one that satisfied Ohm's law; i.e., the voltage across such a resistor is proportional to the current flowing through it. In engineering there are many electronic devices that do not satisfy Ohm's law but have similar properties.

**Fig. 1.1**   Symbol for a resistor; note that the voltage reference direction and the current reference direction are associated reference directions.

These devices are used in increasing numbers in computer, control, and communication systems. Therefore, it is important to approach the basic circuit elements from a broader point of view. In this way we shall be much better prepared to analyze and design the large variety of circuits we are likely to encounter now and in the future.

A two-terminal element will be called a **resistor** if at any instant time $t$, its voltage $v(t)$ and its current $i(t)$ satisfy a relation defined by a curve in the $vi$ plane (or $iv$ plane). This curve is called the **characteristic of the resistor at time t.** It specifies the set of all possible values that the pair of variables $v(t)$ and $i(t)$ may take at time $t$. The most commonly used resistor is **time-invariant;** that is, its characteristic does not vary with time. A resistor is called **time-varying** if its characteristic varies with time. In circuit diagrams a resistor is drawn as shown in Fig. 1.1. The key idea of a resistor is that there is a relation between the *instantaneous* value of the voltage and the *instantaneous* value of the current. Typical characteristic curves are shown in Figs. 1.2 to 1.4, Fig. 1.6, and Figs. 1.8 to 1.12.

Any resistor can be classified in four ways depending upon whether it is linear or nonlinear and whether it is time-varying or time-invariant. A resistor is called **linear** if its characteristic is at all times a straight line through the origin. Accordingly, a resistor that is not linear is called **nonlinear.** We turn now to a detailed study of these four classes of resistors.

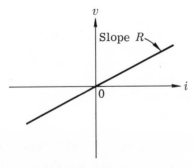

**Fig. 1.2**   The characteristic of a *linear* resistor is at all times a straight line through the origin; the slope $R$ in the $iv$ plane gives the value of the resistance.

## 1.1    The Linear Time-invariant Resistor

A linear time-invariant resistor, by definition, has a characteristic that does not vary with time and is also a straight line through the origin, as shown in Fig. 1.2. Therefore, the relation between its instantaneous voltage $v(t)$ and current $i(t)$ is expressed by Ohm's law as follows:

(1.1)    $v(t) = Ri(t)$      or      $i(t) = Gv(t)$

where

(1.2)    $R = \dfrac{1}{G}$

$R$ and $G$ are constants independent of $i$, $v$, and $t$.   $R$ is called the **resistance** and $G$ is called the **conductance.**   In Eqs. (1.1) and (1.2) the units for voltage, current, resistance, and conductance are volts, amperes, ohms, and mhos, respectively.   We note that in Eq. (1.1) the relation between $i(t)$ and $v(t)$ for the linear time-invariant resistor is expressed by a *linear function.* The first equation in (1.1) expresses $v(t)$ as a linear function of $i(t)$; the second expresses $i(t)$ as a linear function of $v(t)$.   Because the linear time-invariant resistor is of great importance in circuits, we emphasize the following statement: *A linear time-invariant resistor is a resistor that satisfies Ohm's law as given by Eq. (1.1) where R and G are constants.*

A carbon-deposited resistor whose temperature is held constant can be modeled by a linear time-invariant resistor provided that the range of voltage and current is suitably restricted.   Obviously, if the current or the voltage is beyond the specified value, the resistor will overheat and may even burn out.

Two special types of linear time-invariant resistors of particular interest are the *open circuit* and the *short circuit.*   A two-terminal element is called an **open circuit** if it has a branch current identical to zero, whatever the branch voltage may be.   The characteristic of an open circuit is the $v$ axis of the $iv$ plane as shown in Fig. 1.3.   This characteristic has an infinite

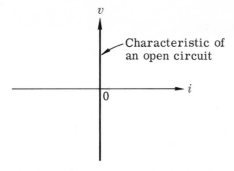

**Fig. 1.3**    The characteristic of an open circuit coincides with the $v$ axis since the current is identically zero.

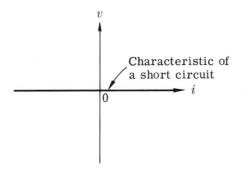

**Fig. 1.4**   The characteristic of a short circuit coincides
with the *i* axis since the voltage is identically
zero.

slope; that is, $R = \infty$, and, equivalently, $G = 0$. A two-terminal element
is called a **short circuit** if it has a branch voltage identical to zero, whatever
the branch current may be. The characteristic of a short circuit is the *i*
axis of the *iv* plane as shown in Fig. 1.4. The slope of the characteristic is
zero; that is, $R = 0$, and, equivalently, $G = \infty$.

**Exercise**   Justify the following statements by Kirchhoff's laws:

*a.*   A branch formed by the series connection of any resistor $\mathfrak{R}$ and an
open circuit has the characteristic of an open circuit.

*b.*   A branch formed by the series connection of any resistor $\mathfrak{R}$ and a short
circuit has the characteristic of the resistor $\mathfrak{R}$.

*c.*   A branch formed by the parallel connection of any resistor $\mathfrak{R}$ and an
open circuit has the characteristic of the resistor $\mathfrak{R}$.

*d.*   A branch formed by the parallel connection of any resistor $\mathfrak{R}$ and a
short circuit has the characteristic of a short circuit.

**1.2**   **The Linear Time-varying Resistor**

The characteristic of a linear time-varying resistor is described by the
following equations:

(1.3)   $v(t) = R(t)i(t)$     or     $i(t) = G(t)v(t)$

where $R(t) = 1/G(t)$. The characteristic obviously satisfies the linear
property, but it changes with time. An illustration of a linear time-
varying resistor is shown in Fig. 1.5. The sliding contact of the potentiometer is moved back and forth by a servomotor so that the characteristic
at time $t$ is given by

(1.4)   $v(t) = (R_a + R_b \cos 2\pi ft)i(t)$

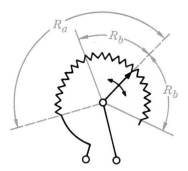

**Fig. 1.5**   Example of a linear time-varying resistor: a potentiometer with a sliding contact, $R(t) = R_a + R_b \cos 2\pi ft$.

where $R_a$, $R_b$, and $f$ are constants and $R_a > R_b > 0$. In the *iv* plane, the characteristic of this linear time-varying resistor is a straight line that passes at all times through the origin; its slope, however, depends on the time $t$. As time changes, the characteristic swings back and forth between two lines with slope $R_a - R_b$ and $R_a + R_b$, as shown in Fig. 1.6.

**Example 1**   Linear time-varying resistors differ from time-invariant resistors in a fundamental way. Let $i(t)$ be a sinusoid with frequency $f_1$; that is

(1.5)   $i(t) = A \cos 2\pi f_1 t$

where $A$ and $f_1$ are constants. Then for a linear time-invariant resistor with resistance $R$, the branch voltage due to this current is given by Ohm's law as follows:

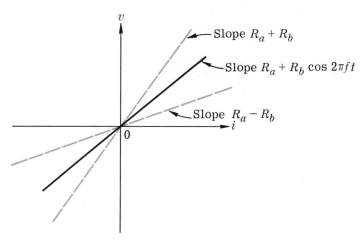

**Fig. 1.6**   Characteristic at time $t$ of the potentiometer of Fig. 1.5.

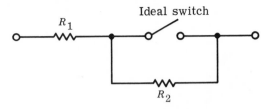

**Fig. 1.7**   Model for a physical switch which has a resistance $R_1 + R_2$ when opened and a resistance $R_1$ when closed; usually $R_1$ is very small, and $R_2$ is very large.

(1.6)   $v(t) = RA \cos 2\pi f_1 t$

Thus, the input current and the output voltage are both sinusoids having the *same* frequency $f_1$. However, for a linear time-varying resistor the result is different. The branch voltage due to the sinusoidal current described by (1.5) for the linear time-varying resistor specified by (1.4) is

(1.7)   $v(t) = (R_a + R_b \cos 2\pi f t)A \cos 2\pi f_1 t$

$$= R_a A \cos 2\pi f_1 t + \frac{R_b A}{2} \cos 2\pi (f + f_1)t + \frac{R_b A}{2} \cos 2\pi (f - f_1)t$$

We can see that this particular linear time-varying resistor can generate signals at two new frequencies which are, respectively, the sum and the difference of the frequencies of the input signal and the time-varying resistor. Thus, linear time-varying resistors can be used to generate or convert sinusoidal signals. This property of linear time-varying resistors is referred to as "modulation" and is of great importance in communication systems.

**Example 2**   A switch can be considered a linear time-varying resistor that changes from one resistance level to another at its opening or closing. An ideal switch is an open circuit when it is opened and a short circuit when it is closed. A practical switch can be modeled in terms of an ideal switch and two resistors, as shown in Fig. 1.7. A periodically operating switch that opens and closes at regular intervals is a key component in digital communication systems.

**1.3   The Nonlinear Resistor**

Recall that a resistor that is not linear is said to be nonlinear. A typical example of a nonlinear resistor is a germanium diode. For the *pn*-junction diode shown in Fig. 1.8, the branch current is a nonlinear function of the branch voltage, according to

(1.8)   $i(t) = I_s(\epsilon^{qv(t)/kT} - 1)$

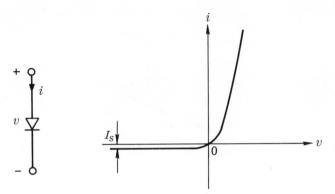

**Fig. 1.8**    Symbol for a *pn*-junction diode and its characteristic plotted in the *vi* plane.

where $I_s$ is a constant that represents the reverse saturation current, i.e., the current in the diode when the diode is reverse-biased (i.e., with $v$ negative) with a large voltage.   The other parameters in (1.8) are $q$ (the charge of an electron), $k$ (Boltzmann's constant), and $T$ (temperature in degrees Kelvin).   At room temperature the value of $kT/q$ is approximately 0.026 volt.   The *vi*-plane characteristic is also shown in Fig. 1.8.

**Exercise**    Plot the characteristic of a typical *pn*-junction diode in the *vi* plane by means of Eq. (1.8).   Given $I_s = 10^{-4}$ amp, $kT/q \approx 0.026$ volt.

By virtue of its nonlinearity, a nonlinear resistor has a characteristic that is not at all times a straight line through the origin of the *vi* plane. Other typical examples of nonlinear two-terminal devices that may be modeled as nonlinear resistors are the tunnel diode and the gas tube. Their characteristics are shown in the *vi* plane of Figs. 1.9 and 1.10.   Note

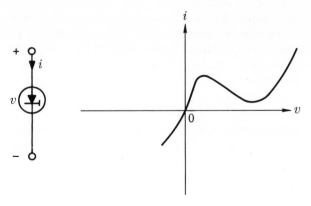

**Fig. 1.9**    Symbol for a tunnel diode and its characteristic plotted in the *vi* plane.

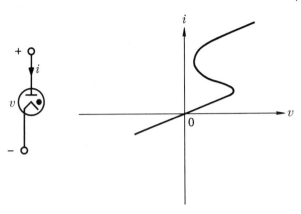

**Fig. 1.10**   Symbol for a gas diode and its characteristic plotted in the $vi$ plane.

that in the first case the current $i$ is a (single-valued) function of the voltage $v$; consequently, we can write $i = f(v)$. Indeed, as shown by the characteristic for each value of the voltage $v$, there is one and only one value possible for the current.† Such a resistor is said to be **voltage-controlled.** On the other hand, in the characteristic of the gas tube the voltage $v$ is a (single-valued) function of the current $i$ because for each $i$ there is one and only one possible value of $v$. Thus, we can write $v = g(i)$. Such a resistor is said to be **current-controlled.** These nonlinear devices have a unique property in that the slope of the characteristic is negative in some range of voltage or current; they are often called negative-resistance devices and are of importance in electronic circuits. They can be used in amplifier circuits, oscillators, and computer circuits. The diode, the tunnel diode, and the gas tube are time-invariant resistors because their characteristics do not vary with time.

A nonlinear resistor can be both voltage-controlled and current-controlled as shown by the characteristic of Fig. 1.11. We can characterize such a resistor by either $i = f(v)$ or $v = g(i) = f^{-1}(i)$, where $g$ is the function inverse to $f$. Note that the slope $df/dv$ in Fig. 1.11 is positive for all $v$; we call such a characteristic *monotonically increasing*. A linear resistor with positive resistance is a special case of such a resistor; it has the monotonically increasing characteristic and is both voltage-controlled and current-controlled.

To analyze circuits with nonlinear resistors, we often depend upon the method of piecewise linear approximation. In this approximation nonlinear characteristics are approximated by piecewise straight-line segments. An often-used model in piecewise linear approximation is the **ideal diode.** A two-terminal nonlinear resistor is called an ideal diode if its

† See Sec. 1.2 of Appendix A.

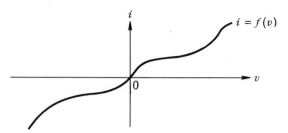

**Fig. 1.11**   A resistor which has a monotonically increasing characteristic is both voltage-controlled and current-controlled.

characteristic in the $vi$ plane consists of two straight-line segments, the negative $v$ axis and the positive $i$ axis. The symbols of the ideal diode and its characteristic are shown in Fig. 1.12. When $v < 0$, $i = 0$; that is, for negative voltages the ideal diode behaves as an open circuit. When $i > 0$, $v = 0$; that is, for positive currents the ideal diode behaves as a short circuit.

At this point it is appropriate to introduce a distinct property of the linear resistor that is not usually present in the nonlinear resistor. A resistor is called **bilateral** if its characteristic is a curve that is symmetric with respect to the origin; in other words, whenever the point $(v,i)$ is on the characteristic, so is the point $(-v, -i)$. Clearly, all linear resistors are bilateral, but most nonlinear resistors are not. It is important to realize the physical consequence of the bilateral property. For a bilateral element it is not important to keep track of the two terminals of the element; the element can be connected to the remainder of the circuit in either way. However, for a nonbilateral element such as a diode, one must know the terminal designation exactly.

**Exercise 1**   Indicate whether the characteristics of Figs. 1.2 to 1.4, Fig. 1.6, and Figs. 1.8 to 1.12 are bilateral.

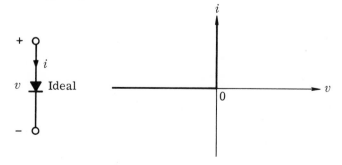

**Fig. 1.12**   Symbol for an ideal diode and its characteristic plotted in the $vi$ plane.

**Exercise 2**   Draw a nonlinear resistor characteristic that is bilateral.

Let us introduce an example to illustrate the behavior of a nonlinear resistor and, in particular, emphasize its difference with that of a linear resistor.

**Example**   Consider a physical resistor whose characteristic can be approximated by the nonlinear resistor defined by

$$v = f(i) = 50i + 0.5i^3$$

where $v$ is in volts and $i$ is in amperes.

a.   Let $v_1$, $v_2$, and $v_3$ be the voltages corresponding to currents $i_1 = 2$ amp, $i_2(t) = 2 \sin 2\pi 60t$ amp, and $i_3 = 10$ amp.   Calculate $v_1$, $v_2$, and $v_3$.   What frequencies are present in $v_2$?   Let $v_{12}$ be the voltage corresponding to the current $i_1 + i_2$.   Is $v_{12} = v_1 + v_2$?   Let $v_2'$ be the voltage corresponding to the current $ki_2$, where $k$ is a constant.   Is $v_2' = kv_2$?

b.   Suppose we were considering only currents of at most 10 mA (milliamperes).   What would be the maximum percentage error in $v$ if we were to calculate $v$ by approximating the nonlinear resistor by a 50-ohm linear resistor?

*Solution*   All voltages below are expressed in volts.

a.   $v_1 = 50 \times 2 + 0.5 \times 8 = 104$

$v_2(t) = 50 \times 2 \sin 2\pi 60t + 0.5 \times 8 \sin^3 2\pi 60t$

$\quad = 100 \sin 2\pi 60t + 4 \sin^3 2\pi 60t$

Recalling that for all $\theta$, $\sin 3\theta = 3 \sin \theta - 4 \sin^3 \theta$, we obtain

$v_2(t) = 100 \sin 2\pi 60t + 3 \sin 2\pi 60t - \sin 2\pi 180t$

$\quad = 103 \sin 2\pi 60t - \sin 2\pi 180t$

$v_3 = 50 \times 10 + 0.5 \times 1{,}000 = 1{,}000$

Frequencies present in $v_2$ are 60 Hz (the fundamental) and 180 Hz (the third harmonic of the frequency of $i_2$).

$v_{12} = 50(i_1 + i_2) + 0.5(i_1 + i_2)^3$

$\quad = 50(i_1 + i_2) + 0.5(i_1{}^3 + i_2{}^3) + 0.5(i_1 + i_2)3i_1i_2$

$\quad = v_1 + v_2 + 1.5i_1i_2(i_1 + i_2)$

Obviously, $v_{12} \neq v_1 + v_2$, and the difference is given by

$v_{12} - (v_1 + v_2) = 1.5i_1i_2(i_1 + i_2)$

Hence

$$v_{12}(t) - [v_1(t) + v_2(t)] = 1.5 \times 2 \times 2 \sin(2\pi60t)(2 + 2\sin 2\pi60t)$$
$$= 12\sin 2\pi60t + 12\sin^2 2\pi60t$$
$$= 6 + 12\sin 2\pi60t - 6\cos 2\pi120t$$

$v_{12}$ thus contains the *third* harmonic as well as the *second* harmonic.

$$v_2' = 50ki_2 + 0.5k^3i_2{}^3 = k(50i_2 + 0.5i_2{}^3) + 0.5k(k^2 - 1)i_2{}^3$$

Therefore,

$$v_2' \neq kv_{21}$$

and

$$v_2' - kv_2 = 0.5k(k^2 - 1)i_2{}^3 = 4k(k^2 - 1)\sin^3 2\pi60t$$

*b.* For $i = 10$ mA, $v = 50 \times 0.01 + 0.5 \times (0.01)^3 = 0.5(1 + 10^{-6})$. The percentage error due to linear approximation equals 0.0001 percent at the maximum current of 10 mA. Therefore, for small currents the nonlinear resistor may be approximated by a linear 50-ohm resistor.

This example illustrates some major properties of nonlinear resistors. First, it is seen that a nonlinear resistor can generate signals at frequencies different from that of the input. In this respect it is similar to the linear time-varying resistor discussed earlier. Second, a nonlinear resistor can often be modeled approximately by using a linear resistor if the range of operation is sufficiently small. Third, the calculations clearly indicate that neither the property of homogeneity nor the property of additivity is satisfied.† In Appendix A we state that a function $f$ is called **homogeneous** if for every $x$ in the domain and for every scalar $\alpha$, $f(\alpha x) = \alpha f(x)$. A function $f$ is said to be **additive** if for every pair of elements $x_1$, $x_2$ of its domain, $f(x_1 + x_2) = f(x_1) + f(x_2)$. A function is said to be **linear** if (1) its domain and its range are linear spaces, (2) it is homogeneous, and (3) it is additive.

Finally, a nonlinear resistor can again be classified according to whether it is time-invariant or time-varying. For example, if a nonlinear germanium diode is submerged in an oil bath whose temperature varies according to a certain schedule, the germanium diode has the characteristic of a nonlinear time-varying resistor.

## 2   Independent Sources

In this section we introduce two new elements, the independent voltage source and the independent current source. We call voltage and current sources *independent* to distinguish them from *dependent* sources, which we shall encounter later. For convenience, however, we shall often use the

† See Sec. 2.3 of Appendix A.

terms "voltage source" and "current source" without the adjective "independent." This should not cause confusion because whenever we encounter dependent sources we shall say specifically that they are dependent.

## 2.1  Voltage Source

A two-terminal element is called an **independent voltage source** if it maintains a prescribed voltage $v_s(t)$ across the terminals of the arbitrary circuit to which it is connected; that is, whatever the current $i(t)$ flowing through the source, the voltage across its terminals is $v_s(t)$. The complete description of the voltage source requires the specification of the function $v_s$. The symbols of the voltage source and the arbitrary circuit connected to it are shown in Fig. 2.1a. If the prescribed voltage $v_s$ is constant (that is, if it does not depend on time), the voltage source is called a *constant voltage source*† and is represented as shown in Fig. 2.1b.

It is customary and convenient to use reference directions for the branch voltage and the branch current of an independent source that are *opposite from the associated reference directions.* Under these conditions the product $v_s(t)i(t)$ is the power *delivered* by the source to the arbitrary circuit to which it is connected (see Fig. 2.1a).

From its definition, a voltage source has a characteristic at time $t$ which is a straight line parallel to the $i$ axis with $v_s(t)$ as the ordinate in the $iv$ plane, as shown in Fig. 2.2. A voltage source can be considered as a nonlinear resistor because whenever $v_s(t) \neq 0$, the straight line does *not* go through the origin. It is a current-controlled nonlinear resistor because to each value of the current there corresponds a unique voltage. It is time varying if $v_s$ is not a constant, and it is time invariant if $v_s$ is a constant.

*If the voltage $v_s$ of a voltage source is identically zero, the voltage source is effectively a short circuit.* Indeed, its characteristic coincides with the $i$ axis; the voltage across the source is zero whatever the current through it may be.

† A constant voltage source is frequently referred to as a dc source or simply a battery.

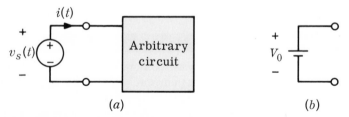

(a)  (b)

**Fig. 2.1** (a) Independent voltage source connected to an arbitrary circuit; (b) symbol for a constant voltage source of voltage $V_0$.

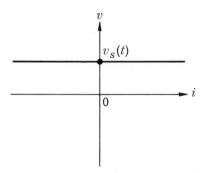

**Fig. 2.2**   Characteristic at time $t$ of a voltage source.   A voltage source may be considered as a current-controlled nonlinear resistor.

In the physical world there is no such thing as an independent voltage source.† However, certain devices over certain ranges of current approximate quite effectively a voltage source.

**Example**   An automobile battery has a voltage and a current which depend on the load to which it is connected, according to the equation

(2.1)   $v = V_0 - R_s i$

where $v$ and $-i$ are the branch voltage and the branch current, respectively, as shown in Fig. 2.3$a$.   The characteristic of Eq. (2.1), plotted in the $iv$ plane, is shown in Fig. 2.3$b$.   The intersection of the characteristic with

---

† The definition of the independent voltage source given above might be more precisely described as an *ideal* independent voltage source.   Some authors call our independent voltage source an "ideal voltage source."   The adjective "ideal" is clearly redundant since all models are "ideal."

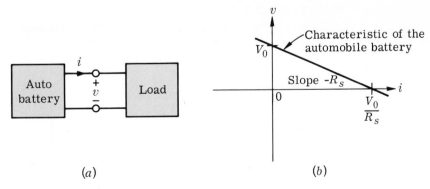

(a)                                           (b)

**Fig. 2.3**   Automobile battery connected to an arbitrary load and its characteristic plotted in the $iv$ plane.

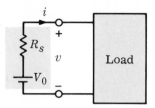

**Fig. 2.4**   Equivalent circuit of the automobile battery.

the $v$ axis is $V_0$. $V_0$ can be interpreted as the open-circuit voltage of the
battery, that is, the voltage across its terminals when $i$ is zero. The constant $R_s$ can be considered as the internal resistance of the battery. Thus,
the automobile battery can be represented by an equivalent circuit that
consists of the series connection of a constant voltage source $V_0$ and a
linear time-invariant resistor with resistance $R_s$, as shown in Fig. 2.4. One
can justify the equivalent circuit by writing the KVL equation for the loop
in Fig. 2.4 and obtaining Eq. (2.1). If the resistance $R_s$ is very small, the
slope in Fig. 2.4$b$ is approximately zero, and the intersection of the characteristic with the $i$ axis will occur far off this sheet of paper. If $R_s = 0$,
the characteristic is a horizontal line in the $iv$ plane, and the battery is a
constant voltage source as defined above.

## 2.2   Current Source

A two-terminal element is called an **independent current source** if it maintains a prescribed current $i_s(t)$ into the arbitrary circuit to which it is connected; that is, whatever the voltage $v(t)$ across the terminals of the circuit
may be, the current into the circuit is $i_s(t)$. Again we note the reference
directions used. The complete description of the current source requires
the specification of the function $i_s$. The symbol of a current source is
shown in Fig. 2.5.

At time $t$ the characteristic of a current source is a vertical line of
abscissa $i_s(t)$ shown in Fig. 2.6. Thus, a current source may be considered
as a nonlinear time-varying resistor that is voltage-controlled.

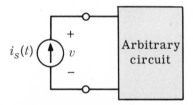

**Fig. 2.5**   Independent current source connected to an arbitrary circuit.

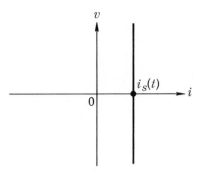

**Fig. 2.6** Characteristic of a current source. A current source may be considered as a voltage-controlled nonlinear resistor.

*If the current $i_s$ is identically zero, the current source is effectively an open circuit.* Indeed, $i_s = 0$ implies that the characteristic coincides with the $v$ axis, and that the current through the device is zero whatever the voltage across it may be.

**2.3    Thévenin and Norton Equivalent Circuits**

We have learned about the independent voltage source and the independent current source. They are ideal circuit models. Most practical sources are like the automobile battery illustrated in the previous example; i.e., they can be represented in the form of a series connection of an ideal voltage source and an ideal linear time-invariant resistor $R_s$.† At this juncture we find it convenient to introduce an alternative but equivalent representation of the automobile battery in terms of a current source.

If we consider the characteristic of the automobile battery plotted in Fig. 2.3*b*, we may think of it either as representing a constant voltage source $V_0$ *in series* with a linear time-invariant resistor $R_s$ or as representing a constant current source $I_0 \triangleq V_0/R_s$ *in parallel* with a linear time-invariant resistor $R_s$, as shown in Fig. 2.7.

We say that the two circuits shown are equivalent because they have the same characteristic. Indeed, writing the Kirchhoff voltage law for the circuit in Fig. 2.7*a*, we have

(2.2a)    $v = V_0 - R_s i$

Similarly, writing the Kirchhoff current law for the circuit in Fig. 2.7*b*, we have

---

† Strictly speaking we should say "a linear time-invariant resistor with resistance $R_s$." In circuit diagrams such as that of Fig. 2.7*a* we usually designate a linear resistor by its resistance $R_s$, and for simplicity, we refer to it by saying "the resistor $R_s$."

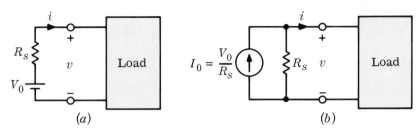

**Fig. 2.7**   (*a*) Thévenin equivalent circuit; (*b*) Norton equivalent circuit of the automobile battery.

(2.2b)   $i = I_0 - \dfrac{1}{R_s} v$

Since $I_0 = V_0/R_s$, the two equations are the same; hence they represent the same straight line in the $iv$ plane.

The series connection of a voltage source and the linear time-invariant resistor $R_s$ in Fig. 2.7a is called the Thévenin equivalent circuit, and the parallel connection of a current source and the linear time-invariant resistor $R_s$ in Fig. 2.7b is called the Norton equivalent circuit. In some cases we find it more convenient to use voltage sources than current sources. In other situations we find it more convenient to use current sources. The Thévenin and Norton equivalent circuits thus give us flexibility.

The equivalence of these two circuits is a special case of the Thévenin and Norton equivalent circuit theorem, which we shall discuss in great detail in Chap. 16.

## 2.4   Waveforms and Their Notation

As mentioned previously, the complete description of a voltage source $v_s$ or a current source $i_s$ requires the specification of the complete time function, that is, $v_s(t)$ for all $t$ or $i_s(t)$ for all $t$. Thus, the specification of the voltage source $v_s$ must include either a complete tabulation of the function $v_s$ or a rule that allows us to calculate the voltage $v_s(t)$ for any $t$ we might consider later on. We encounter here a difficulty of notation which we shall face throughout the course; that is, sometimes we think of the "whole function $v_s$," say, as a waveform traced on a scope, and sometimes we think only of a particular ordinate, say $v_s(t)$ for some given $t$. The difference between the two concepts is illustrated in Fig. 2.8. Whenever we want to emphasize the fact that we are talking about the whole function we shall use the locution "**the waveform** $v_s(\,\cdot\,)$." We leave a dot instead of a letter such as $t$ because we do not consider any particular $t$, but we consider the "whole function."

Unfortunately, to rigorously follow the scheme would lead us to very involved expressions. So for convenience, we shall often say "the wave-

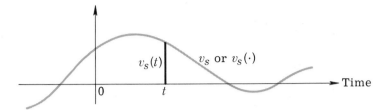

**Fig. 2.8**   This figure illustrates the difference between the waveform $v_s(\,\cdot\,)$ and the number $v_s(t)$ which is the value assumed by the function $v_s$ at the instant $t$.

form cos $\omega t$" when we should have said "the waveform $f(\,\cdot\,)$, where $f(t) =$ cos $\omega t$ for all $t$."

A typical use of the distinction between the concept of "whole function" and the value assumed by a function at some time $t$ is the following. Consider a complicated circuit made of resistors, inductors, and capacitors and driven by a single current source $i_s$.   Call $v_c$ the voltage across one of the capacitors.   We might state that the response $v_c(t)$ (meaning "the value of the response at time $t$") depends on the waveform $i_s(\,\cdot\,)$ (meaning "the whole function $i_s$").   We use this language to emphasize that $v_c(t)$ depends not only on $i_s(t)$ (the value of $i_s$ at the same instant $t$) but also on all past values of $i_s$.

## 2.5   Some Typical Waveforms

Let us now define some of the more useful waveforms that we shall use repeatedly later.

*The constant*   This is the simplest waveform; it is described by

$$f(t) = K \qquad \text{for all } t$$

where $K$ is a constant.

*The sinusoid*   To represent a sinusoidal waveform or **sinusoid** for short, we use the traditional notation

$$f(t) = A \cos (\omega t + \phi)$$

where the constant $A$ is called the **amplitude** of the sinusoid, the constant $\omega$ is called the (angular) **frequency** (measured in radians per second), and the constant $\phi$ is called the **phase.**   The sinusoid is illustrated in Fig. 2.9.

*The unit step*   The **unit step function** as shown in Fig. 2.10 is denoted by $u(\,\cdot\,)$ and is defined by

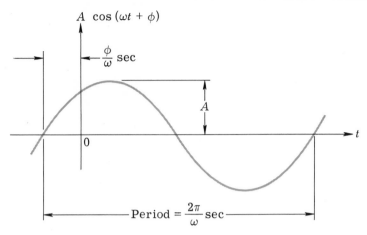

**Fig. 2.9**   A sinusoidal waveform of amplitude $A$ and phase $\phi$.

$$(2.3) \quad u(t) = \begin{cases} 0 & \text{for } t < 0 \\ 1 & \text{for } t > 0 \end{cases}$$

and its value at $t = 0$ may be taken to be 0, ½, or 1.  For the purposes of this course it does not matter.  However, when using Fourier or Laplace transforms, $u(0) = ½$ is preferable.  Throughout this book we shall use the letter $u$ exclusively for the unit step.

Suppose we delayed a unit step by $t_0$ sec.  The resulting waveform has $u(t - t_0)$ as an ordinate at time $t$.  Indeed, for $t < t_0$, the argument is negative, and hence the ordinate is zero; for $t > t_0$, the argument is positive, and the ordinate is equal to 1.  This is shown in Fig. 2.11.

*The pulse*   We shall frequently have to use a rectangular pulse.  For this purpose we define the **pulse function** $p_\Delta(\cdot)$ by

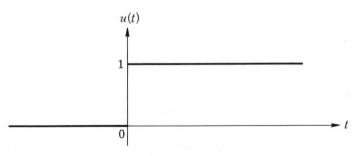

**Fig. 2.10**   The unit step function $u(\cdot)$.

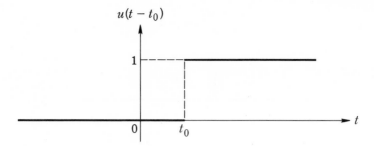

**Fig. 2.11**   The delayed unit step function.

$$(2.4) \quad p_\Delta(t) = \begin{cases} 0 & t < 0 \\ \dfrac{1}{\Delta} & 0 < t < \Delta \\ 0 & \Delta < t \end{cases}$$

In other words, $p_\Delta$ is a pulse of height $1/\Delta$, of width $\Delta$, and starting at $t = 0$. Note that whatever the value of the positive parameter $\Delta$, the area under $p_\Delta(\cdot)$ is 1 (see Fig. 2.12). Note that

$$(2.5) \quad p_\Delta(t) = \frac{u(t) - u(t - \Delta)}{\Delta} \qquad \text{for all } t$$

*The unit impulse*   The **unit impulse** $\delta(\cdot)$ (also called the Dirac delta function) is not a function in the strict mathematical sense of the term (see Appendix A). For our purposes we state that

$$(2.6) \quad \delta(t) = \begin{cases} 0 & \text{for } t \neq 0 \\ \text{singular} & \text{at } t = 0 \end{cases}$$

and the singularity at the origin is such that for any $\xi > 0$

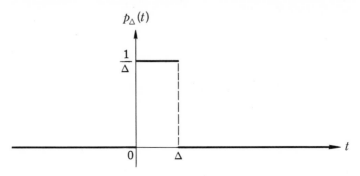

**Fig. 2.12**   A pulse function $p_\Delta(\cdot)$.

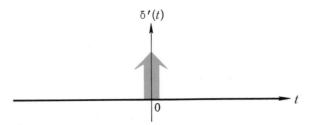

**Fig. 2.13**   A unit impulse function $\delta(\,\cdot\,)$.

(2.7)   $\displaystyle\int_{-\xi}^{\xi} \delta(t)\, dt = 1$

Intuitively, we may think of the impulse function $\delta$ as the limit, as $\Delta \to 0$, of the pulse $p_\Delta$. This fact will be used frequently. Physically we may think of $\delta$ as representing the charge density of a *unit* point charge located at $t = 0$ on the $t$ axis.

From the definition of $\delta$ and $u$ we get formally

(2.8)   $\boxed{\; u(t) = \displaystyle\int_{-\infty}^{t} \delta(t')\, dt' \;}$

and

(2.9)   $\boxed{\; \dfrac{du(t)}{dt} = \delta(t) \;}$

These two equations are very important and will be used repeatedly in later chapters. The impulse function is shown graphically in Fig. 2.13.

Another frequently useful property is the *sifting property* of the unit impulse. Let $f$ be a continuous function. Then

(2.10)   $\displaystyle\int_{-\xi}^{\xi} f(t)\delta(t)\, dt = f(0)$

for any positive $\xi$.

This is easily made reasonable by approximating $\delta$ by $p_\Delta$ as follows:

$$\int_{-\xi}^{\xi} f(t)\delta(t)\, dt = \lim_{\Delta \to 0} \int_{-\xi}^{+\xi} f(t)p_\Delta(t)\, dt$$

$$= \lim_{\Delta \to 0} \int_{0}^{\Delta} f(t)\frac{1}{\Delta}\, dt$$

$$= f(0)$$

**Remarks**   1.   Related to the unit step function is the **unit ramp** $r(\,\cdot\,)$, defined by

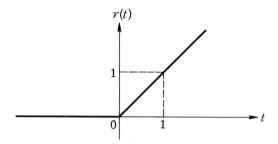

**Fig. 2.14** A unit ramp function $r(\cdot)$.

(2.11)  $r(t) = tu(t)$    for all $t$

The waveform $r(\cdot)$ is shown in Fig. 2.14. From (2.3) and (2.11) we can show that

(2.12)  $r(t) = \int_{-\infty}^{t} u(t')\,dt'$

and

(2.13)  $\dfrac{dr(t)}{dt} = u(t)$

2.  Closely related to the unit impulse function is the **unit doublet** $\delta'(\cdot)$, which is defined by

(2.14)  $\delta'(t) = \begin{cases} 0 & \text{for } t \neq 0 \\ \text{singular} & \text{at } t = 0 \end{cases}$

and the singularity at $t = 0$ is such that

(2.15)  $\delta(t) = \int_{-\infty}^{t} \delta'(t')\,dt'$

and

(2.16)  $\dfrac{d\delta(t)}{dt} = \delta'(t)$

The symbol for a unit doublet is shown in Fig. 2.15.

**Exercise 1**  Sketch the waveforms specified by

  *a.*  $3u(t) - 3u(t - 2)$
  *b.*  $5p_{0.1}(t) - 3p_{0.1}(t - 0.1) + 2p_{0.2}(t - 3)$
  *c.*  $r(t) - u(t - 1) - r(t - 1)$

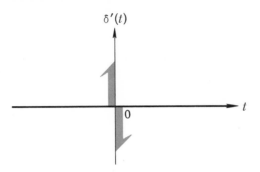

$\delta'(t)$

0

$t$

**Fig. 2.15**   A doublet $\delta'(\cdot)$.

**Exercise 2**   Express $\sin t$ and $3 \sin (2t + 1)$ in the standard form for the sinusoid (here the phase is given in radians).

---

### 3   Capacitors

Capacitors are used in electric circuits because they store electric charges. The element called a capacitor is an idealized model of the physical capacitor, for example, a parallel-plate condenser. A physical capacitor is a component which, in addition to its dominant property of storing electric charge, has some leakage (usually a very small amount). A two-terminal element is called a **capacitor** if at any time $t$ its stored charge $q(t)$ and its voltage $v(t)$ satisfy a relation defined by a curve in the $vq$ plane. This curve is called the **characteristic of the capacitor at time** $t$. The basic idea is that there is a relation between the *instantaneous* value of the charge $q(t)$ and the *instantaneous* value of the voltage $v(t)$. As in the case of the resistor, the characteristic of the capacitor may vary with time. Typically, the characteristic would have the shape shown in Fig. 3.1. The characteristic of nearly all physical capacitors is monotonically increasing; that is, as $v$ increases, $q$ increases.

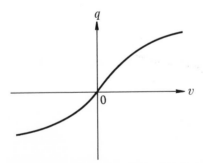

$q$

0

$v$

**Fig. 3.1**   Characteristic of a (nonlinear) capacitor plotted on the $vq$ plane.

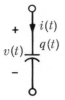

**Fig. 3.2**    Symbol for a capacitor.

In circuit diagrams a capacitor is represented symbolically as shown in Fig. 3.2; note that we shall always call $q(t)$ the charge at time $t$ on the plate to which the reference arrow of the current $i(t)$ points. When $i(t)$ is positive, positive charges are brought (at time $t$) to the top plate whose charge is labeled $q(t)$; hence the rate of change of $q$ [that is, the current $i(t)$] is also positive. Thus, we have

(3.1)    $$i(t) = \frac{dq}{dt}$$

In this equation currents are given in amperes and charges in coulombs. From Eq. (3.1) we obtain the branch-voltage and branch-current characterization of a capacitor using the specified charge-voltage relation.

A capacitor whose characteristic is at all times a straight line through the origin of the $vq$ plane is called a **linear capacitor.** Conversely, if at any time the characteristic is not a straight line through the origin of the $vq$ plane, the capacitor is called **nonlinear.** A capacitor whose characteristic does not change with time is called a **time-invariant capacitor.** If the characteristic changes with time, the capacitor is called a **time-varying capacitor.**

Just as in the case of resistors, we have a four-way classification of capacitors, depending on whether they are linear or nonlinear and time-invariant or time-varying.

**3.1    The Linear Time-invariant Capacitor**

From the definition of linearity and time invariance, the characteristic of a linear time-invariant capacitor can be written as

(3.2)    $$q(t) = Cv(t)$$

where $C$ is a constant (independent of $t$ and $v$) which measures the slope of the characteristic and which is called **capacitance.** The units in Eq. (3.2) are coulombs, farads, and volts, respectively. The equation relating the terminal voltage and the current is

(3.3)   $i(t) = \dfrac{dq}{dt} = C\dfrac{dv}{dt} = \dfrac{1}{S}\dfrac{dv}{dt}$

where $S = C^{-1}$, and is called the **elastance.** Integrating (3.3) between 0 and $t$, we get

(3.4)   $v(t) = v(0) + \dfrac{1}{C}\displaystyle\int_0^t i(t')\, dt'$

or, in terms of the elastance $S$,

(3.5)   $v(t) = v(0) + S\displaystyle\int_0^t i(t')\, dt'$

Thus, a linear time-invariant capacitor is completely specified as a circuit element only if the capacitance $C$ (the slope of its characteristic) and the initial voltage $v(0)$ are given.

It should be stressed that Eq. (3.3) defines a function expressing $i(t)$ in terms of $dv/dt$; that is, $i(t) = f(dv/dt)$. It is fundamental to observe that this function $f(\,\cdot\,)$ is linear. On the other hand, Eq. (3.4) defines a function expressing $v(t)$ in terms of $v(0)$ and the current waveform $i(\,\cdot\,)$ over the interval $[0,t]$. It is important to note that only if $v(0) = 0$, the function defined by (3.4) is a *linear function* that gives the value of $v(t)$, the voltage at time $t$, in terms of the current *waveform* over the interval $[0,t]$. The integral in (3.4) represents the net area under the current curve between time 0 and $t$; we say "net area" to remind ourselves that sections of the curve $i(\,\cdot\,)$ above the time axis contribute positive areas, and those below contribute negative areas. It is interesting to note that the value of $v$ at time $t$, $v(t)$, depends on its initial value $v(0)$ and all the values of the current between time 0 and time $t$; this fact is often alluded to by saying that "capacitors have memory."

**Exercise 1**   Let a current source $i_s(t)$ be connected to a linear time-invariant capacitor with capacitance $C$ and $v(0) = 0$. Determine the voltage waveform $v(\,\cdot\,)$ across the capacitor for

    *a.*   $i_s(t) = u(t)$

    *b.*   $i_s(t) = \delta(t)$

    *c.*   $i_s(t) = A\cos(\omega t + \phi)$

**Exercise 2**   Let a voltage source $v_s(t)$ be connected to a linear time-invariant capacitor with capacitance $C$ and $v(0) = 0$. Determine the current waveform $i(\,\cdot\,)$ in the capacitor for

    *a.*   $v_s(t) = u(t)$

    *b.*   $v_s(t) = \delta(t)$

    *c.*   $v_s(t) = A\cos(\omega t + \phi)$

**Example**   A current source is connected to the terminals of a linear time-invariant capacitor with a capacitance of 2 farads and an initial voltage of $v(0) = -\frac{1}{2}$ volt (see Fig. 3.3a).   Let the current source be given by the simple waveform $i(\,\cdot\,)$ shown in Fig. 3.3b.   The branch voltage across the capacitor can be computed immediately from Eq. (3.4) as

$$v(t) = -\frac{1}{2} + \frac{1}{2} \int_0^t i(t')\, dt'$$

and the voltage waveform $v(\,\cdot\,)$ is plotted in Fig. 3.3c.   The voltage is $-\frac{1}{2}$ volt for $t$ negative.   At $t = 0$ it starts to increase and reaches $\frac{1}{2}$ volt at $t = 1$ sec as a result of the contribution of the positive portion of the current waveform.   The voltage then decreases linearly to $-\frac{1}{2}$ volt because of the constant negative current for $1 < t < 2$, and stays constant for $t \geq 2$ sec.   This simple example clearly points out that $v(t)$ for $t \geq 0$ depends on the initial value $v(0)$ and on all the values of the waveform $i(\,\cdot\,)$ between time 0 and time $t$.   Furthermore it is easy to see that $v(t)$ is not a linear function of $i(\,\cdot\,)$ when $v(0)$ is not zero.   On the other hand if the initial value $v(0)$ is zero, the branch voltage at time $t$, $v(t)$, is a linear function of the current waveform $i(\,\cdot\,)$.

**Exercise**   Assume that the current waveform in Fig. 3.3b is doubled in value for all $t$. Calculate the voltage $v(t)$ for $t \geq 0$.   Show that linearity is violated unless $v(0) = 0$.

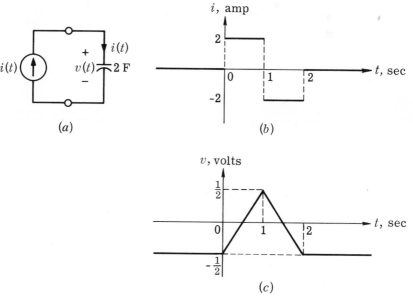

**Fig. 3.3**   Voltage and current waveform across a linear time-invariant capacitor.

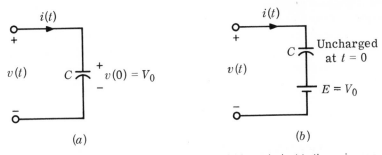

$i(t)$

$v(t)$     $C$  $v(0) = V_0$

$(a)$

$i(t)$

$v(t)$     $C$  Uncharged at $t = 0$

$E = V_0$

$(b)$

**Fig. 3.4**    The initially charged capacitor with $v(0) = V_0$ in $(a)$ is equivalent to the series connection of the same capacitor (which is initially uncharged) and a constant voltage source $E = V_0$ in $(b)$.

**Remarks**

1. Equation (3.4) states that at time $t$ the branch voltage $v(t)$, where $t \geq 0$, across a linear time-invariant capacitor is the sum of two terms. The first term is the voltage $v(0)$ at $t = 0$, that is, the initial voltage across the capacitor. The second term is the voltage at time $t$ across a capacitor of $C$ farads if at $t = 0$ this capacitor is initially uncharged. Thus, any linear time-invariant capacitor with an initial voltage $v(0)$ can be considered as the series connection of a dc voltage source $E = v(0)$ and the same capacitor with zero initial voltage, as shown in Fig. 3.4. This result is very useful and will be repeatedly used in later chapters.

2. Consider a linear time-invariant capacitor with zero initial voltage; that is, $v(0) = 0$. It is connected in series with an arbitrary independent voltage source $v_s(t)$ as shown in Fig. 3.5a. The series connection is equivalent to the circuit (as shown in Fig. 3.5b) in which the same capacitor is connected in parallel with a current source $i_s(t)$, and

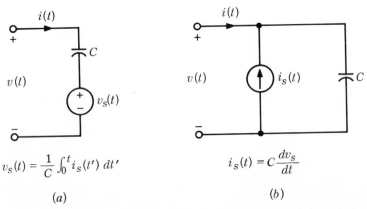

$i(t)$

$v(t)$     $C$     $v_s(t)$

$$v_s(t) = \frac{1}{C} \int_0^t i_s(t') \, dt'$$

$(a)$

$i(t)$

$v(t)$     $i_s(t)$     $C$

$$i_s(t) = C \frac{dv_s}{dt}$$

$(b)$

**Fig. 3.5**    Thévenin and Norton equivalent circuits for a capacitor with an independent source.

(3.6)   $i_s(t) = C\dfrac{dv_s}{dt}$

The voltage source $v_s(t)$ in Fig. 3.5a is given in terms of the current source $i_s(t)$ in Fig. 3.5b by

(3.7)   $v_s(t) = \dfrac{1}{C} \displaystyle\int_0^t i_s(t')\,dt'$

The results in Fig. 3.5a and b are referred to as the Thévenin and the Norton equivalent circuits, respectively. The proof is similar to that of the resistor case in Sec. 2.3. In particular, if the voltage source $v_s$ in Fig. 3.4a is a unit step function, by Eq. (3.6) the current source $i_s$ in Fig. 3.5b is an impulse function $C\delta(t)$.

3.   Consider Eq. (3.4) again at instant $t$ and at instant $t + dt$; by subtraction we get

(3.8)   $v(t + dt) - v(t) = \dfrac{1}{C} \displaystyle\int_t^{t+dt} i(t')\,dt'$

Let us assume that $i(t)$ is bounded for all $t$; that is, there is a *finite* constant $M$ such that $|i(t)| \le M$ for all $t$ under consideration. The area under the waveform $i(\,\cdot\,)$ over the interval $[t, t + dt]$ will go to zero as $dt \to 0$. Also from (3.8), as $dt \to 0$, $v(t + dt) \to v(t)$, or stated in another way, the voltage waveform $v(\,\cdot\,)$ is continuous.

We can thus state an important property of the linear time-invariant capacitor: *If the current $i(\,\cdot\,)$ in a linear time-invariant capacitor remains bounded for all time in the closed interval $[0,T]$, the voltage $v$ across the capacitor is a continuous function in the open interval $(0,T)$; that is, the branch voltage for such a capacitor cannot jump instantaneously from one value to a different value (as in a step function) as long as the current remains bounded.* This property is very useful in solving problems in which pulses or step functions of voltage or current are applied to a circuit. Applications of this property will be illustrated in later chapters.

**Exercise**   Prove the statement in Remark 2.

### 3.2   The Linear Time-varying Capacitor

If the capacitor is linear but time-varying, its characteristic is at all times a straight line through the origin, but its slope depends on time. Therefore, the charge at time $t$ can be expressed in terms of the voltage at time $t$ by an equation of the form

(3.9)   $q(t) = C(t)v(t)$

where $C(\,\cdot\,)$ is a prescribed function of time that specifies for each $t$ the

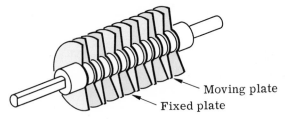

Moving plate

Fixed plate

**Fig. 3.6**   With the moving plate driven mechanically, this capacitor becomes a time-varying capacitor.

slope of the capacitor characteristic. This function $C(\cdot)$ is part of the specification of the linear time-varying capacitor. Equation (3.1) then becomes

$$(3.10) \quad i(t) = \frac{dq}{dt} = C(t)\frac{dv}{dt} + \frac{dC}{dt}v(t)$$

A simple example of a linear time-varying capacitor is shown in Fig. 3.6, where a parallel-plate capacitor contains a fixed and a moving plate. The moving plate is driven mechanically in a periodic fashion. The capacitance of this periodically varying capacitor may be expressed in a Fourier series as

$$(3.11) \quad C(t) = C_0 + \sum_{k=1}^{\infty} C_k \cos(2\pi f k t + \phi_k)$$

where $f$ represents the frequency of rotation of the moving plate.

Periodically varying capacitors are of great importance in the study of parametric amplifiers. A different type of periodically varying capacitor will be mentioned in the next section.

**Exercise**   Consider the circuit shown in Fig. 3.7. Let the voltage be a sinusoid, $v(t) = A \cos \omega_1 t$, where the constant $\omega_1 = 2\pi f_1$ is the angular frequency. Let the linear time-varying capacitor be specified by

$$C(t) = C_0 + C_1 \cos 3\omega_1 t$$

where $C_0$ and $C_1$ are constants. Determine the current $i(t)$ for all $t$.

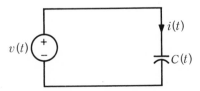

$i(t)$

$v(t)$

$C(t)$

**Fig. 3.7**   A linear time-varying capacitor is driven by a sinusoidal voltage source.

## 3.3 The Nonlinear Capacitor

A varactor diode is a device used in many modern communication systems as a key circuit component of parametric amplifiers, oscillators, and signal converters; the varactor diode can be modeled essentially by a nonlinear capacitor. An accurate model of transistors also includes a nonlinear capacitor. In high-speed switching applications the effect of the nonlinear capacitor is often of great significance. The analysis of circuits, which include *nonlinear* elements is, in general, much more difficult than that of linear circuits. In nonlinear analysis various techniques exist, each suitable for a special situation. Among them and probably the most useful is the so-called *small-signal analysis*. We shall introduce this main concept in the following example.

**Example** Consider a nonlinear capacitor specified by its characteristic $q = f(v)$ (see Fig. 3.8). Let us assume that the voltage $v$ is the sum of two terms, as shown in Fig. 3.9. The first term $v_1$ is a constant voltage applied to the capacitor by the biasing battery (it is often called "dc bias"), and the second term $v_2$ is a small varying voltage. For example, $v_2$ might be a small voltage in an input stage of a receiver. Using a Taylor series expansion, we have

$$q = f(v) = f(v_1 + v_2)$$

(3.12) $$\approx f(v_1) + \left.\frac{df}{dv}\right|_{v_1} v_2$$

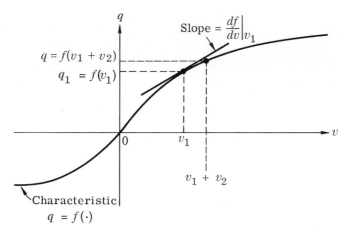

**Fig. 3.8** Characteristic of the nonlinear capacitor and the small-signal approximation about the operating point $(v_1, f(v_1))$.

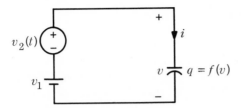

**Fig. 3.9**   A nonlinear capacitor is driven by a voltage $v$ which is the sum of a dc voltage $v_1$ and a small-varying voltage $v_2(t)$.

In Eq. (3.12) we neglected second-order terms; this introduces negligible errors provided $v_2$ is sufficiently small.   More precisely, $v_2$ must be sufficiently small so that the part of the characteristic corresponding to the abscissa $v_1 + v_2$ is well approximated by a straight-line segment passing through the point $(v_1, f(v_1))$ and having slope $\dfrac{df}{dv}\Big|_{v_1}$.   The current $i(t)$ from Eq. (3.1) is

$$(3.13) \qquad i(t) = \frac{dq}{dt} = \frac{df}{dv}\Big|_{v_1} \frac{dv_2}{dt}$$

The equation is of the form

$$(3.14) \qquad i(t) = C(v_1)\frac{dv_2}{dt}$$

Note that $v_1$ is a constant.   Thus, as far as the small-signal $v_2$ is concerned, the capacitance $C(v_1) = \dfrac{df}{dv}\Big|_{v_1}$ is a linear time-invariant capacitance and is equal to the slope of the capacitor characteristic in the $vq$ plane at the operating point, as shown in Fig. 3.8.   Hence, the capacitance depends on the dc voltage $v_1$.

If the nonlinear capacitor is used in a parametric amplifier, the voltage $v_1$ is not a constant; however $v_2$, which represents the time-varying signal, is still assumed to be small so that the approximations used in writing (3.12) are still valid. We must then modify the above analysis slightly.

The voltage across the capacitor is $v_1(t) + v_2(t)$.   Consequently, the charge is

$$q(t) = f(v_1(t) + v_2(t))$$

Since $v_2(t)$ is small for all $t$, we have

$$q(t) \approx f(v_1(t)) + \frac{df}{dv}\Big|_{v_1(t)} v_2(t)$$

Let

(3.15)    $q_1(t) \triangleq f(v_1(t))$

The charge $q_1(t)$ can be considered to be the charge due to $v_1(t)$. The remaining charge $q_2(t) \triangleq q(t) - q_1(t)$ is given approximately by

(3.16)    $q_2(t) \approx \left. \dfrac{df}{dv} \right|_{v_1(t)} v_2(t)$

This charge $q_2$ is proportional to $v_2$ and can be considered as the small-signal charge variation due to $v_2$. Since $v_1$ is now a given function of time, $\left. \dfrac{df}{dv} \right|_{v_1(t)}$ can be identified as a linear time-varying capacitor $C(t)$, where

$C(t) = \left. \dfrac{df}{dv} \right|_{v_1(t)}$. Therefore, we have demonstrated that a nonlinear capacitor can be modeled as a linear time-varying capacitor in the small-signal analysis. This type of analysis is basic to the understanding of parametric amplifiers.

---

**Exercise**    Given a nonlinear capacitor characterized by the equation

$q = 1 - \epsilon^{-v}$

determine the small-signal capacitance $C$ that is defined by $\left. \dfrac{df}{dv} \right|_{v_1}$ of (3.16)

for

a.    $v_1 = 10$ volts

b.    $v_1(t) = 10 + 5 \cos \omega_1 t$

Let $v_2(t) = 0.1 \cos 10 \, \omega_1 t$, and determine the approximate capacitor current due to $v_2$ for both cases.

---

## 4    Inductors

Inductors are used in electric circuits because they store energy in their magnetic fields. The element called an inductor is an idealization of the physical inductor. More precisely a two-terminal element will be called an **inductor** if at any time $t$ its flux $\phi(t)$ and its current $i(t)$ satisfy a relation defined by a curve in the $i\phi$ plane. This curve is called the **characteristic of the inductor at time** $t$. The basic idea is that there is a relation between the *instantaneous* value of the flux $\phi(t)$ and the *instantaneous* value of the current $i(t)$. In some cases the characteristic may vary with time. In circuit diagrams an inductor is represented symbolically, as shown in Fig. 4.1. Since in circuit theory the fundamental characterization of a two-terminal element is in terms of voltage and current, we need to relate the

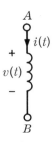

*A*

*i(t)*

+

*v(t)*

−

*B*

**Fig. 4.1**   Symbol for an inductor.

flux and the branch voltage. The voltage across the inductor (measured with the reference direction indicated in Fig. 4.1) is given by Faraday's induction law as

$$(4.1) \quad v(t) = \frac{d\phi}{dt}$$

where $v$ is in volts and $\phi$ is in webers.

Let us verify that (4.1) agrees qualitatively with Lenz' law, which states that the electromotive force induced by a rate of change of flux will have a polarity such that it will oppose the cause of that rate of change of flux. Consider the following case: The current $i$ increases; that is, $di/dt > 0$. The increasing current creates an increasing magnetic field; hence the flux $\phi$ increases; that is, $d\phi/dt > 0$. According to (4.1), $v(t) > 0$, which means that the potential of node $A$ is larger than the potential of node $B$; this is precisely the polarity required to oppose any further increase in current.

The four-way classification of inductors, according to whether they are linear or nonlinear, time-invariant or time-varying, is similar to that of resistors and capacitors. An inductor is called **time-invariant** if its characteristic does not change with time; an inductor is called **linear** if at all times its characteristic is a straight line through the origin of the $i\phi$ plane.

---

**4.1**   **The Linear Time-invariant Inductor**

By definition the characteristic of the linear time-invariant inductor has an equation of the form

$$(4.2) \quad \phi(t) = Li(t)$$

where $L$ is a constant (independent of $t$ and $i$) and is called the **inductance.** The characteristic is a fixed straight line through the origin whose slope is $L$. The units in this equation are webers, henrys, and amperes, respectively. The equation relating the terminal voltage and current is easily obtained from (4.1) and (4.2). Thus

(4.3)   $v(t) = L\dfrac{di}{dt}$

Integrating Eq. (4.3) between 0 and $t$, we get

(4.4)   $i(t) = i(0) + \dfrac{1}{L}\displaystyle\int_0^t v(t')\,dt'$

Let $\Gamma \triangleq L^{-1}$, and let $\Gamma$ be called the **reciprocal inductance.**  Then

(4.5)   $i(t) = i(0) + \Gamma \displaystyle\int_0^t v(t')\,dt'$

In Eqs. (4.4) and (4.5) the integral is the net area under the voltage curve between time 0 and time $t$.  Clearly, the value of $i$ at time $t$, $i(t)$, depends on its initial value $i(0)$ and on all the values of the voltage waveform $v(\,\cdot\,)$ in the interval $[0,t]$.  This fact, as in the case of capacitors, is often alluded to by saying that "inductors have memory."

In considering Eq. (4.4) it is important to note that a linear time-invariant inductor is completely specified as a circuit element only if the inductance $L$ (the slope of its characteristic) and the initial current $i(0)$ are given.  We shall encounter this important fact throughout our study of circuit theory.

It should be stressed that Eq. (4.3) defines a *linear* function expressing the instantaneous voltage $v(t)$ in terms of the derivative of the current evaluated at time $t$.  Equation (4.4) defines a function expressing the instantaneous current $i(t)$ in terms of $i(0)$ and the waveform $v(\,\cdot\,)$ over $[0,t]$.  It is important to note that only if $i(0) = 0$ is the function defined by Eq. (4.4) a *linear function* which gives the value of the current $i$ at time $t$, $i(t)$, in terms of the voltage waveform $v(\,\cdot\,)$ over the interval $[0,t]$.

**Exercise 1**  Let a current source $i_s(t)$ be connected to a linear time-invariant inductor with inductance $L$ and $i(0) = 0$.  Determine the voltage waveform $v(\,\cdot\,)$ across the inductor for

    *a.*   $i_s(t) = u(t)$

    *b.*   $i_s(t) = \delta(t)$

**Exercise 2**  Let a voltage source $v_s(t)$ be connected to a linear time-invariant inductor with inductance $L$ and $i(0) = 0$.  Determine the current waveform $i(\,\cdot\,)$ in the inductor for

    *a.*   $v_s(t) = u(t)$

    *b.*   $v_s(t) = \delta(t)$

    *c.*   $v_s(t) = A \cos \omega t$, where $A$ and $\omega$ are constants

**Remarks**  1.  Equation (4.4) states that at time $t$ the branch current $i(t)$ (where $t \ge 0$) in a linear time-invariant inductor is the sum of two terms.

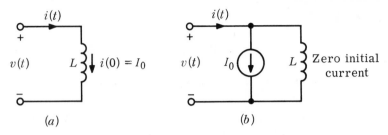

**Fig. 4.2**   The inductor with an initial current $i(0) = I_0$ in (a) is equivalent to the nonparallel connection of the same inductor with zero initial current and a constant current source $I_0$ in (b).

The first term is the current $i(0)$ at $t = 0$, that is, the initial current in the inductor. The second term is the current at time $t$ in an inductor $L$ if, at $t = 0$, this inductor has zero initial current. Thus, given any linear time-invariant inductor with an initial current $i(0)$, we can consider the inductor as the parallel connection of a dc current source $I_0 = i(0)$ and the same inductor with zero initial current. See Fig. 4.2. This useful result will be encountered often in later chapters.

2.   Consider a linear time-invariant inductor with zero initial current; that is, $i(0) = 0$. It is connected in parallel with an arbitrary current source $i_s(t)$ as shown in Fig. 4.3a. The parallel connection is equivalent to the circuit shown in Fig. 4.3b where the same inductor is connected in series with a voltage source $v_s(t)$ and

(4.6)   $v_s(t) = L \dfrac{di_s}{dt}$

The current source $i_s(t)$ in Fig. 4.3a (in terms of the voltage source in Fig. 4.3b) is

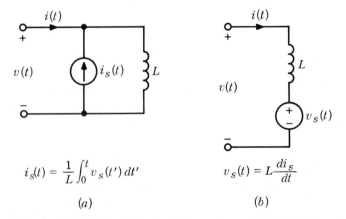

$$i_s(t) = \frac{1}{L} \int_0^t v_s(t')\, dt'$$

(a)

$$v_s(t) = L \frac{di_s}{dt}$$

(b)

**Fig. 4.3**   The Norton (a) and Thévenin (b) equivalent circuits for an inductor with source.

(4.7)   $i_s(t) = \dfrac{1}{L} \displaystyle\int_0^t v_s(t')\,dt'$

The results in Fig. 4.3a and b are referred to as the Norton and Thévenin equivalent circuits, respectively. In particular, if $i_s$ in Fig. 4.3a is a unit step function, the voltage source $v_s$ in Fig. 4.3b is an impulse function $L\delta(t)$.

3.   Following reasoning similar to that used in the case of capacitors, we may conclude with the following important property of inductors: *If the voltage v across a linear time-invariant inductor remains bounded for all times in the closed interval* [0,t], *the current i is a continuous function in the open interval* (0,t); that is, the current in such an inductor cannot jump instantaneously from one value to a different value as long as the voltage across it remains bounded.

### 4.2    The Linear Time-varying Inductor

If the inductor is linear but time-varying, its characteristic is at all times a straight line through the origin, but its slope is a function of time. The flux is expressed in terms of the current by

(4.8)   $\phi(t) = L(t)i(t)$

where $L(\,\cdot\,)$ is a prescribed function of time. Indeed, this function $L(\,\cdot\,)$ is a part of the specification of the time-varying inductance. Equation (4.1) becomes

(4.9)   $v(t) = L(t)\dfrac{di}{dt} + \dfrac{dL}{dt}\,i(t)$

### 4.3    The Nonlinear Inductor

Most physical inductors have nonlinear characteristics. Only for certain specified ranges of currents can inductors be modeled by linear time-invariant inductors. A typical characteristic of a physical inductor is shown in Fig. 4.4. For large currents the flux saturates; that is, it increases extremely slowly as the current becomes very large.

**Example**   Suppose the characteristic of a nonlinear time-invariant inductor can be represented by

$\phi = \tanh i$

Let us calculate the voltage across the inductor, where the current is sinusoidal and is given by

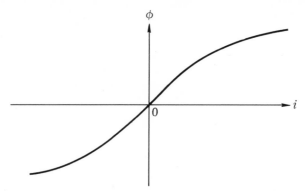

**Fig. 4.4**   Characteristic of a nonlinear inductor.

$i(t) = A \cos \omega t$

The flux is thus

$\phi(t) = \tanh (A \cos \omega t)$

By (4.1) we have

$$v(t) = \frac{d}{dt} \phi(i(t)) = \frac{d\phi}{di}\bigg|_{i(t)} \frac{di}{dt}$$

$$= \frac{d \tanh i}{di}\bigg|_{i(t)} \frac{dA \cos \omega t}{dt}$$

$$= \frac{1}{\cosh^2 (A \cos \omega t)} (-A\omega \sin \omega t)$$

We conclude that

$$v(t) = -A\omega \frac{\sin \omega t}{\cosh^2 (A \cos \omega t)}$$

Thus, given the amplitude $A$ and the angular frequency $\omega$ of the current, the voltage across the inductor is completely specified as a function of time.

## 4.4   Hysteresis

A special type of nonlinear inductor, such as a ferromagnetic-core inductor, has a characteristic that exhibits the *hysteresis phenomenon.* In terms of the current-flux plot, a hysteresis characteristic is shown in Fig. 4.5. Assume that we start at the origin in the $i\phi$ plane; as current is increased, the flux builds up according to curve 1. If the current is decreased at the

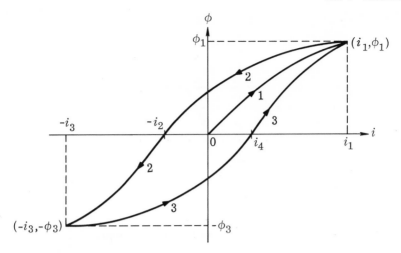

**Fig. 4.5**   Hysteresis phenomenon.

point $(i_1, \phi_1)$, the flux follows curve 2 rather than retracing curve 1. As the current reaches the point $-i_2$, the flux finally becomes zero. If the current is increased at the point $(-i_3, -\phi_3)$, the flux follows curve 3 and becomes zero as the current reaches a positive value at $i_4$.

Our formal definition of the inductor as a circuit element does not include the case of physical inductors that exhibit the hysteresis phenomenon since a characteristic shown in Fig. 4.5 is not, strictly speaking, a curve. To our best knowledge there is no effective way of describing the general hysteresis phenomenon. Nevertheless, we show in the example below how, by suitable idealization and for certain types of current waveforms, it is easy to determine the terminal voltage across an inductor that exhibits hysteresis.

**Example**   Let us assume that a nonlinear inductor has an idealized hysteresis characteristic such as is shown in Fig. 4.6. Assume that the operating point at time 0 is at $A$ on the characteristic, where $i = 0$ and $\phi = -1$, and that the current waveform is that shown in Fig. 4.7a. The voltage across the inductor is to be determined. We should emphasize that when we use the idealized characteristic, we assume that when $|i| > 3$, the flux is constant. When $|i| < 3$, the flux may take two values for each $i$, depending upon whether the current is increasing or decreasing. Using the given current waveform, we may easily plot the curve giving $\phi$ as a function of time (see Fig. 4.7b). The differentiation of the function $\phi(t)$ thus obtained gives the voltage across the inductor; the result is shown in Fig. 4.7c. This kind of idealization and calculation is in common use in the analysis of magnetic amplifiers and some computer circuits.

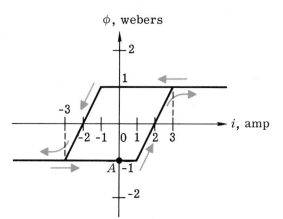

**Fig. 4.6** Characteristic of an inductor which exhibits hysteresis.

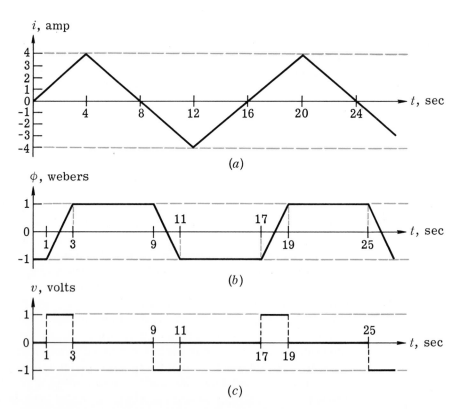

**Fig. 4.7** Waveform of $i$, $\phi$, and $v$ for the nonlinear inductor whose hysteresis characteristic is shown in Fig. 4.6.

**5**    ## Summary of Two-terminal Elements

In this brief section we would like to bring together some of the concepts that are common to all the circuit elements we have considered thus far. These elements are resistors, independent sources, capacitors, and inductors.

All are *two-terminal elements.*  Resistors and independent sources are characterized by a curve on the *iv* plane, capacitors by a curve on the *vq* plane, and inductors by a curve on the *iϕ* plane.  In each case the curve is called *the characteristic of the two-terminal element at time t.*  The characteristic specifies the set of all possible values that the pair of variables (appropriate for that two-terminal element) may take at time *t*.

If we consider the four-way classification, we can see that we have used the following concepts:

1.    A two-terminal element will be called *linear* if its characteristic is at all times a straight line through the origin; equivalently, the instantaneous value of one of the variables is a linear function of the instantaneous value of the other.

2.    A two-terminal element will be said to be *time-invariant* if its characteristic does not change with time.  Consequently, a two-terminal element will be called *linear time-invariant* if it is both linear and time-invariant; by definition, this means that its characteristic is a fixed straight line through the origin, and the characteristic is completely specified by a single number, its slope.

In Table 2.1 the analytic expressions specifying the characteristic and the equations relating the voltage to the current are given for each of the elements.  As we mentioned earlier, usual physical capacitors have a *vq* characteristic that is monotonically increasing; therefore, the instantaneous value of the charge $q(t)$ can always be expressed as a (single-valued) function of the instantaneous value of the voltage $v(t)$.  Thus, the capacitor characteristic can always be put in the form $q = f(v)$ if the capacitor is time-invariant, and in the form $q(t) = f(v(t),t)$ if the capacitor is time-varying.  If we exclude the hysteresis phenomenon, we can make similar comments for inductors; their characteristic can always be put in the form $\phi = f(i)$ for the time-invariant case, and in the form $\phi(t) = f(i(t),t)$ for the time-varying case.

For resistors the situation is more complicated.  Refer to Fig. 1.9, and observe that the characteristic of a tunnel diode can be represented by an equation of the form $i = f(v)$, where *f* is a (single-valued) function; indeed, for each value of the voltage *v* the characteristic allows one and only one value for the instantaneous current *i*.  Such a resistor is called *voltage-controlled.*  On the other hand, if we refer to Fig. 1.10, we observe that the characteristic of a gas tube has the property that for each value of *i* there is one and only one value of *v* allowed by the characteristic; we then have

**Table 2.1  Summary of Four-way Classification of Two-terminal Elements**

| | Linear | | Nonlinear | |
|---|---|---|---|---|
| | Time-invariant | Time-varying | Time-invariant | Time-varying |
| **Resistors** | $v(t) = Ri(t)$ $i(t) = Gv(t)$ $R = 1/G$ | $v(t) = R(t)i(t)$ $i(t) = G(t)v(t)$ $R(t) = 1/G(t)$ | $v(t) = f(i(t))$ Current-controlled $i(t) = g(v(t))$ Voltage-controlled | $v(t) = f(i(t),t)$ Current-controlled $i(t) = g(v(t),t)$ Voltage-controlled |
| **Capacitors** $i = \dfrac{dq}{dt}$ | $q(t) = Cv(t)$ $i(t) = C\dfrac{dv}{dt}$ $v(t) = v(0) + \dfrac{1}{C}\displaystyle\int_0^t i(t')\,dt'$ | $q(t) = C(t)v(t)$ $i(t) = \dfrac{dC}{dt}v(t) + C(t)\dfrac{dv}{dt}$ | $q(t) = f(v(t))$ $i(t) = \left.\dfrac{df}{dv}\right|_{v(t)}\dfrac{dv}{dt}$ | $q(t) = f(v(t),t)$ $i(t) = \dfrac{\partial f}{\partial t} + \left.\dfrac{\partial f}{\partial v}\right|_{v(t)}\dfrac{dv}{dt}$ |
| **Inductors** $v = \dfrac{d\phi}{dt}$ | $\phi(t) = Li(t)$ $v(t) = L\dfrac{di}{dt}$ $i(t) = i(0) + \dfrac{1}{L}\displaystyle\int_0^t v(t')\,dt'$ | $\phi(t) = L(t)i(t)$ $v(t) = \dfrac{dL}{dt}i(t) + L(t)\dfrac{di}{dt}$ | $\phi(t) = f(i(t))$ $v(t) = \left.\dfrac{df}{di}\right|_{i(t)}\dfrac{di}{dt}$ | $\phi(t) = f(i(t),t)$ $v(t) = \dfrac{\partial f}{\partial t} + \left.\dfrac{\partial f}{\partial i}\right|_{i(t)}\dfrac{di}{dt}$ |

$v = f(i)$, where $f$ is a (single-valued) function. Such a resistor is called *current-controlled.* Some resistors are neither current-controlled nor voltage-controlled, for example, the ideal diode. If $v = 0$, the current may have any value, provided it is nonnegative (hence, it cannot be a voltage-controlled resistor); if $i = 0$, the voltage may have any value provided it is nonpositive (hence, it cannot be a current-controlled resistor). A linear resistor is both voltage-controlled and current-controlled provided $0 < R < \infty$.

## 6    Power and Energy

In physics we learned that a resistor does not store energy but absorbs electrical energy, a capacitor stores energy in its electric field, and an inductor stores energy in its magnetic field. In this section we shall discuss power and energy from a viewpoint that is most convenient for lumped circuits.

In our study of lumped circuits we have thus far concentrated our attention on two-terminal elements. We now want to take a broader view. Suppose that we have a circuit, and from this circuit we draw two wires which we connect to another circuit which we call a generator (see Fig. 6.1). For example, the circuit we start with may be a loudspeaker that we connect to the two terminals of the cable coming from the power amplifier; the power amplifier is then considered to be a generator. We shall call the circuit we are considering a **two-terminal circuit** since, from our point of view, we are only interested in the voltage and the current at the two terminals and the power transfer that occurs at these terminals.

In modern terminology a two-terminal circuit is called a **one-port.** The term *one-port* is appropriate since by **port** we mean a pair of terminals of a circuit in which, at all times, the instantaneous current flowing into one terminal is equal to the instantaneous current flowing out of the other. This fact is illustrated in Fig. 6.1. Note that the current $i(t)$ entering the top terminal of the one-port $\mathfrak{N}$ is equal to the current $i(t)$ leaving the bottom terminal of the one-port $\mathfrak{N}$. The current $i(t)$ entering the port is called the **port current,** and the voltage $v(t)$ across the port is called the **port voltage.** The port concept is very important in circuit theory. When we use the term one-port, we want to indicate that we are only interested in the port voltage and the port current. Other network variables pertaining to elements inside the one-port are not accessible. When we consider a net-

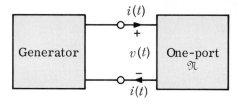

**Fig. 6.1**    Instantaneous power entering the one-port $\mathfrak{N}$ at time $t$ is $p(t) = v(t)i(t)$.

work $\mathfrak{N}$ as a one-port, then as far as we are concerned, the port is a pair of leads coming out of a black box. The box is black because we are not allowed to see what is inside! With this concept in mind it is clear that resistors, independent voltage sources, capacitors, and inductors are simple and special examples of *one-ports* consisting of a single element.

It is a fundamental fact of physics that the **instantaneous power** *entering the one-port is equal to the product of the port voltage and the port current* provided the reference directions of the port voltage and the port current are associated reference directions, as indicated in Fig. 6.1. Let $p(t)$ denote the instantaneous power (in watts) delivered by the generator to the one-port at time $t$. Then

(6.1)
$$p(t) = v(t)i(t)$$

where $v$ is in volts and $i$ is in amperes. Since the energy (in joules) is the integral of power (in watts), it follows that *the* **energy delivered** *by the generator to the one-port from time $t_0$ to time $t$ is*

(6.2)
$$W(t_0,t) \triangleq \int_{t_0}^{t} p(t')\,dt' = \int_{t_0}^{t} v(t')i(t')\,dt'$$

### 6.1　Power Entering a Resistor, Passivity

Since a resistor is characterized by a curve in the $vi$ plane (or $iv$ plane), the instantaneous power entering a resistor at time $t$ is uniquely determined once the *operating point* $(i(t),v(t))$ on the characteristic is specified; the instantaneous power is equal to the area of the rectangle formed by the operating point and the axes of the $iv$ plane, as shown in Fig. 6.2. If the

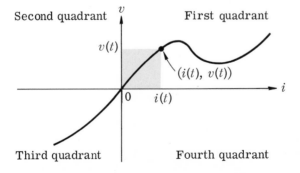

**Fig. 6.2**　The power entering the resistor at time $t$ is $v(t)i(t)$.

operating point is in the first or third quadrant (hence, $iv > 0$), the power entering the resistor is positive; that is, the resistor receives power from the outside world. If the operating point is in the second or fourth quadrant (hence $iv < 0$), the power entering the resistor is negative; that is, the resistor delivers power to the outside world. For these reasons we say that a resistor is **passive** if for all time $t$ the characteristic lies in the first and third quadrants. Here the first and third quadrants include the $i$ axis and the $v$ axis. The geometrical constraint on the characteristic of a passive resistor is equivalent to $p(t) \geq 0$ at all times irrespective of the current waveform through the resistor. This is the fundamental property of passive resistors: *a passive resistor never delivers power to the outside world.* It is easy to see that a germanium diode, a tunnel diode,† an open circuit, a short circuit, and a linear time-invariant resistor with $R \geq 0$ are passive resistors.

A resistor is said to be **active** if it is not passive. For example, any voltage source (for which $v_s$ is not identically zero) and any current source (for which $i_s$ is not identically zero) is an active resistor since its characteristic at all time is parallel to either the $i$ axis or the $v$ axis, and thus it is not restricted to the first and third quadrants. It is interesting to note that a *linear resistor* (either time-invariant or time-varying) *is active if and only if $R(t)$ is negative for some time $t$.* This is so because the characteristic of a linear resistor is a straight line passing through the origin, and the slope is equal to the resistance $R$; thus, if $R < 0$, the characteristic lies in the second and fourth quadrant. It follows that if a current is driven through the resistor (say, by a current source) when $R(t) < 0$, the resistor will deliver power to the outside world at the rate of $|R(t)|i^2(t)$ watts. It is true that one seldom finds a physical component that behaves as a linear active resistor as defined above. However, the model of a linear active resistor is important because a nonlinear resistor such as a tunnel diode behaves as a linear active resistor in the small-signal analysis. This will be explained in the next chapter.

## 6.2   Energy Stored in Time-invariant Capacitors

Let us apply Eq. (6.2) to calculate the energy stored in a capacitor. For simplicity we assume that it is time-invariant, but it can be nonlinear.‡

Suppose that the one-port of Fig. 6.1, which is connected to the generator, is a capacitor. The current through the capacitor is

---

† A tunnel diode has its characteristic in the first and third quadrants hence is a passive element. We shall see in Chap. 3 that it can be used as an amplifier only if an external active element is connected to it. In practice, this is done by using a bias circuit containing a battery.

‡ The energy stored in time-varying inductors and capacitors requires more subtle considerations; its calculation will be treated in Chap. 19.

(6.3)   $i(t) = \dfrac{dq}{dt}$

Let the capacitor characteristic be described by the function $\hat{v}(\,\cdot\,)$, namely

(6.4)   $v = \hat{v}(q)$

The energy delivered by the generator to the capacitor from time $t_0$ to $t$ is then

(6.5)   $W(t_0,t) = \displaystyle\int_{t_0}^{t} v(t')i(t')\,dt' = \int_{q(t_0)}^{q(t)} \hat{v}(q_1)\,dq_1$

To obtain Eq. (6.5), we first used Eq. (6.3) and wrote $i(t')\,dt' = dq_1$ according to (6.3), where $q_1$ is a dummy integration variable representing the charge. We used (6.4) to express the voltage $v(t')$ by the characteristic of the capacitor, i.e., the function $\hat{v}(\,\cdot\,)$ in terms of the integration variable $q_1$. The lower and upper integration constants were changed accordingly from $t_0$ to $q(t_0)$ and from $t$ to $q(t)$. Let us assume that the capacitor is initially uncharged; that is, $q(t_0) = 0$. It is natural to use the uncharged state of the capacitor as the state corresponding to zero energy stored in the capacitor. Since a capacitor stores energy but does not dissipate it, we conclude that the energy stored at time $t$, $\mathcal{E}_E(t)$, is equal to the energy delivered to the capacitor by the generator from $t_0$ to $t$, $W(t_0,t)$. Thus, the **energy stored in the capacitor** is, from (6.5),

(6.6)   $\boxed{\;\mathcal{E}_E(t) = \displaystyle\int_{0}^{q(t)} \hat{v}(q_1)\,dq_1\;}$

In terms of the capacitor characteristic on the $vq$ plane, the shaded area in Fig. 6.3 represents the energy stored [note that on this graph $q$ is the ordinate and $v$ is the abscissa; thus the integral in (6.6) represents the shaded area *above* the curve]. Obviously, if the characteristic passes through the

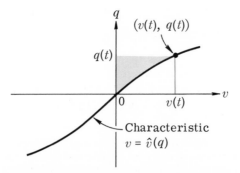

**Fig. 6.3**   The shaded area gives the energy stored at time $t$ in the capacitor.

origin of the $vq$ plane and lies in the first and third quadrant, the stored energy is always nonnegative.   A capacitor is said to be **passive** if its stored energy is always nonnegative.   For a linear time-invariant capacitor, the equation of the characteristic is

(6.7)   $q = Cv$

where $C$ is a constant independent of $t$ and $v$.   Equation (6.6) reduces to the familiar expression

(6.8)
$$\mathcal{E}_E(t) = \int_0^{q(t)} \frac{q_1}{C} \, dq_1 = \frac{1}{2} \frac{q^2(t)}{C} = \frac{1}{2} Cv^2(t)$$

Accordingly, a linear time-invariant capacitor is passive if its capacitance is nonnegative and active if its capacitance is negative.   An active capacitor stores negative energy; that is, it can deliver energy to the outside.   Of course, it is not physically realizable.   However, it is possible to obtain a negative capacitance within a small operating range and a narrow frequency band by means of a suitably designed electronic circuit.

We shall see in Chap. 19 that a linear time-varying capacitor may be active even if its $C(t)$ is positive for all $t$.

## 6.3   Energy Stored in Time-invariant Inductors

The calculation of the energy stored in an inductor is very similar to the same calculation for the capacitor.   As a matter of fact, if in the preceding derivation we simply change variables (i.e., change $i$ into $v$, $q$ into $\phi$, and $v$ into $i$), we obtain the results for an inductor.   This procedure, an aspect of the method of duality, is of great importance in circuit theory.   Duality will be studied in great detail later.

For an inductor Faraday's law states that

(6.9)   $v(t) = \dfrac{d\phi}{dt}$

Let the inductor characteristic be described by the function $\hat{\imath}(\,\cdot\,)$; namely,

(6.10)   $i = \hat{\imath}(\phi)$

Let the inductor be the one-port that is connected to the generator in Fig. 6.1.   Then the energy delivered by the generator to the inductor from $t_0$ to $t$ is

(6.11)   $W(t_0,t) = \displaystyle\int_{t_0}^{t} v(t')i(t') \, dt' = \int_{\phi(t_0)}^{\phi(t)} \hat{\imath}(\phi_1) \, d\phi_1$

To obtain (6.11), we used Eq. (6.9) and wrote $v(t') \, dt' = d\phi_1$, where the dummy integration variable $\phi_1$ represents flux.   Equation (6.10) was used to express current in terms of flux.   The procedure is similar to the derivation of (6.5).   Suppose that initially the flux is zero; that is, $\phi(t_0) = 0$. Again choosing this state of the inductor to be the state corresponding to zero energy stored, and observing that an inductor stores but does not dissipate energy, we conclude that the magnetic energy stored at time $t$, $\mathscr{E}_M(t)$, is equal to the energy delivered to the inductor by the generator from $t_0$ to $t$, $W(t_0,t)$.   Thus the **energy stored in the inductor is**

(6.12)
$$\mathscr{E}_M(t) = \int_0^{\phi(t)} \hat{\imath}(\phi_1)\, d\phi_1$$

In terms of the inductor characteristic on the $i\phi$ plane, the shaded area in Fig. 6.4 represents the energy stored.   Similarly, if the characteristic in the $i\phi$ plane passes through the origin and lies in the first and third quadrant, the stored energy is always nonnegative.   An inductor is said to be **passive** if its stored energy is always nonnegative.   A linear time-invariant inductor has a characteristic of the form

$$\phi = Li$$

(6.13)   where $L$ is a constant independent of $t$ and $i$.   Hence Eq. (6.12) leads to the familiar form

(6.14)
$$\mathscr{E}_M(t) = \int_0^{\phi(t)} \frac{\phi_1}{L}\, d\phi_1 = \frac{1}{2}\frac{\phi^2(t)}{L} = \frac{1}{2}Li^2(t)$$

Accordingly, a linear time-invariant inductor is passive if its inductance is nonnegative and is active if its inductance is negative.

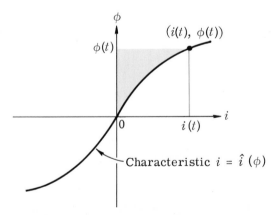

**Fig. 6.4**   The shaded area gives the energy stored at time $t$ in the inductor.

## 7    Physical Components versus Circuit Elements

As mentioned at the beginning of this chapter, the circuit elements which we have defined are circuit models which have simple but precise characterizations. These circuit models are the analogs to the particle and the rigid body of the physicist. These circuit models are indispensable in both analysis and synthesis of physical circuits and systems. However, we must realize that *physical components* such as physical resistors (to be distinguished from model resistors), diodes, coils, and condensers, which we have in the laboratory or use in practical circuits, can only be approximated with our circuit models. Engineering is not a precise subject like mathematics; therefore, it is important and necessary to use approximation in almost all problems. The crucial thing is to know the right modeling and to use valid approximations in solving problems.

In this section we shall give a brief discussion of the modeling problem for some commonly used physical components. Many physical components can be modeled more or less exactly by their primary physical functions. For example, a parallel-plate condenser under normal operation (to be explained) can be modeled as a linear time-invariant capacitor. At low frequencies a junction diode can be considered as a nonlinear resistor and can then be approximated by the combination of an ideal diode and linear resistors. However, in using these components we must understand under what conditions the model is valid and, more importantly, under what situation the model needs to be modified. In the following we shall discuss three principal considerations that are of importance in modeling physical components.

*Range of operation*   Any physical component is specified in terms of its normal range of operation. Typically the maximum voltage, the maximum current, and the maximum power are almost always specified for any device. If in a circuit the voltage, current, or power exceed the specified value, the component cannot be modeled in its usual way and may actually be destroyed when so used.

Another commonly specified range of operation is the range of frequencies. For example, at very high frequencies a physical resistor cannot be modeled only as a resistor. Strictly speaking, whenever there is a voltage difference, there is an electric field; hence, some electrostatic energy is stored. Similarly, the presence of current implies that some magnetic energy is stored. At low frequencies these effects are negligible, and hence a physical resistor can be modeled as a single circuit element, a resistor. However, at high frequencies a more accurate model will include some capacitive and inductive elements in addition to the resistor. Thus, to model one physical component, we use two or more circuit elements. By specifying the frequency range we know that within the range the physical resistor can be modeled only by, say, a 100-ohm resistor.

*Temperature effect*   Resistors, diodes, and almost all circuit components are temperature-sensitive. If they are used in an environment where temperature changes, their characteristics are time-varying. Semiconductor devices are also very sensitive to temperature changes. Circuits made up of semiconductor devices often contain additional circuit schemes, such as feedback, which counteract the changes due to temperature variation.

*Parasitic effect*   Probably the most noticeable phenomenon in a physical inductor, in addition to its magnetic field when current passes through, is its dissipation. The wiring of a physical inductor has a resistance that may have substantial effects in some circuits. Thus, in modeling a physical inductor we often use a series connection of an inductor and resistor. Similarly, a junction diode at high frequency must be modeled by the parallel connection of a nonlinear resistor and capacitor. The need for the capacitor is caused primarily by the charge storage at the junction. We have already mentioned that a practical battery is not an (ideal) voltage source. However, a model that includes the parasitic resistive effect can be used to approximate the external behavior.

Engineers must use their experience and common sense in choosing physical components. For example, high-quality coils having negligible dissipation are available, but they may be economically unfeasible in a particular design. Instead, one must use a more complicated circuit with cheaper components to achieve the same purpose.

In summary, it is important to appreciate the difference between a circuit element that is an idealized model, and a physical component that is a real-life object. We must know the assumptions under which models are derived to represent physical components. However, in this book our principal aim is to develop the theory of circuits made up with models. It is even more important to know that only through modeling are we able to develop precise methods of analysis, concrete theorems, and deep understanding of physical circuits and systems.

*Typical element size*   At this juncture we shall mention briefly the size of element values that are encountered in practice. For resistors the commonly used range of values varies from a few ohms to megohms, the accuracy of specification depending on the particular application. For a precise physical experiment we may want to measure resistances to a few tenths or hundredths of an ohm. On the other hand, in designing a biasing circuit for an audio amplifier a 10 percent precision requirement on the resistors is usually sufficient.

The useful range of values for capacitors varies from a few picofarads ($10^{-12}$ farad) in the case of parasitic capacitances of an electronic device to microfarads ($10^{-6}$ farad). A practical inductor ranges from microhenrys in the case of the lead inductance of a short wire to a few henrys in power transformers.

In the examples treated in this book we consistently use round and simple figures, such as a resistor of 10 ohms, a capacitor of 1 farad, and an inductor of ½ henry. It is important to know that these numbers do not correspond to practical numbers of physical elements. The purpose for using these numbers is, of course, to help us focus our attention on the ideas and the methods rather than on complicated numerical evaluations. In Chap. 7 we shall give a brief discussion of element normalization, which is useful in analysis and design of circuits. By means of element normalization we can design a practical circuit by carrying out all our calculations in "normalized" values such as 1 farad or 0.7 henry. This procedure has the further advantage of reducing the effect of round-off errors in numerical calculations.

## Summary

- Circuit elements are ideal models that are used to analyze and design circuits. Physical components can be approximately modeled by circuit elements.

- Each two-terminal element is defined by a characteristic, that is, by a curve drawn in an appropriate plane. Each element can be subjected to a four-way classification according to its linearity and its time invariance. An element is said to be *time-invariant* if its characteristic does not change with time; otherwise it is called *time-varying*. A two-terminal element is said to be *linear* if, for each $t$, its characteristic at time $t$ is a straight line through the origin; otherwise, it is called *nonlinear*.

- A resistor is characterized, for each $t$, by a curve in the $iv$ (or $vi$) plane. An independent voltage source is characterized by a line parallel to the $i$ axis. An independent current source is characterized by a line parallel to the $v$ axis.

- A capacitor is characterized, for each $t$, by a curve in the $vq$ plane. An inductor is characterized, for each $t$, by a curve in the $i\phi$ plane.

- A *one-port* (or two-terminal circuit) is formed by two terminals of a circuit if the current entering one terminal is equal to the current leaving the other terminal at all times. When using the term "one-port," we are interested in only the port voltage and the port current. The *instantaneous power* entering the one-port is given by $p(t) = v(t)i(t)$. The *energy delivered* to the one-port from time $t_0$ to time $t$ is given by $W(t_0,t] = \int_{t_0}^{t} v(t')i(t')\, dt'$.

- Circuit elements may also be classified according to their passivity. An element is said to be *passive* if it never delivers a net amount of energy to the outside world. An element which is not passive is said to be *active*.

- In the linear time-invariant case, resistors, capacitors, and inductors are passive if and only if $R \geq 0$, $C \geq 0$, and $L \geq 0$, respectively.

- The magnetic energy stored in a linear time-invariant inductor is

$$\mathcal{E}_M = \frac{1}{2} L i^2 = \frac{1}{2} \frac{\phi^2}{L}$$

- The electric energy stored in a linear time-invariant capacitor is

$$\mathcal{E}_E = \frac{1}{2} C v^2 = \frac{1}{2} \frac{q^2}{C}$$

## Problems

Property of nonlinear resistor

**1.** Suppose that the nonlinear resistor $\mathcal{R}$ has a characteristic specified by the equation

$$v = 20i + i^2 + \tfrac{1}{2} i^3$$

*a.*   Express $v$ as a sum of sinusoids for

$$i(t) = \cos \omega_1 t + 2 \cos \omega_2 t$$

*b.*   If $\omega_2 = 2\omega_1$, what frequencies are present in $v$?

Characterization of resistors

**2.** The equations below specify the characteristics of some resistors.  Indicate whether they are linear or nonlinear, time-varying or time-invariant, bilateral, voltage-controlled, current-controlled, passive or active.

*a.*   $v + 10i = 0$           *f.*   $i + 3v = 10$

*b.*   $v = (\cos 2t)i + 3$       *g.*   $i = 2 + \cos \omega t$

*c.*   $i = \epsilon^{-v}$             *h.*   $i = \ln (v + 2)$

*d.*   $v = i^2$                *i.*   $i = v + (\cos 2t)\dfrac{v}{|v|}$

*e.*   $i = \tanh v$

Waveforms

**3.** Plot the waveforms specified below:

*a.*   $3\delta(t - 2)$                       *h.*   $3p_2(t)$

*b.*   $\delta(t) - \delta(t - 1) + \delta(t - 2)$     *i.*   $p_{1/2}(t - 2)$

*c.*   $u(2t)$                            *j.*   $\epsilon^{2t} \cos t$

*d.*   $u(t) \cos (2t + 60°)$              *k.*   $u(t) - 2u(t - 1)$

*e.*   $u(-t)$                           *l.*   $r(t) \sin t$

*f.*   $u(3 - 2t)$                        *m.*   $u(t)\epsilon^{-2t} \sin (t - 90°)$

*g.*   $u(t)\epsilon^{-t}$

Waveforms

**4.** Give functional representations for the waveforms given in Fig. P2.4. (See pages 64 and 65.)

Linear time-
invariant
inductor and
capacitor

**5.** Assuming that the waveforms given in Fig. P2.4 are branch currents, sketch on graph paper the branch voltages when

*a.*  the element is a linear time-invariant inductor of 1 henry.

*b.*  the element is a linear time-invariant capacitor of 1 farad $[v(0) = 0]$.

Linear time-
invariant
inductor and
capacitor

**6.** Assuming that the waveforms given in Fig. P2.4 are branch voltages, sketch on graph paper the branch currents when

*a.*  the element is a linear time-invariant inductor of 2 henrys $[i(0) = 0]$.

*b.*  the element is a linear time-invariant capacitor of 2 farads.

Linear resis-
tors and sources

**7.** Determine the voltages across the linear resistors for each of the three circuits shown in Fig. P2.7.

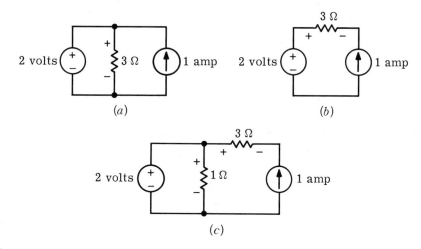

**Fig. P2.7**

Power

**8.** For the three circuits in Fig. P2.7 calculate the power dissipated in each resistor. Determine where this power comes from by calculating the contribution due to the voltage source and that due to the current source.

Power and
energy

**9.** The branch voltage and the branch current of an element are measured with respect to associated reference directions, and they turn out to be (for all $t$)

$$v = \cos 2t \qquad i = \cos (2t + 45°)$$

Determine and plot the power delivered to this element. Determine the energy delivered to the branch from $t = 0$ to $t = 10$ sec.

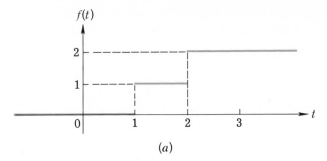

(a)

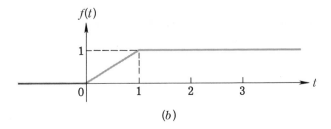

(b)

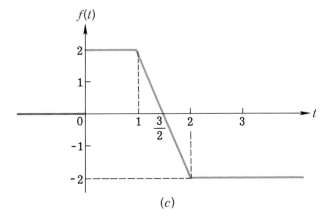

(c)

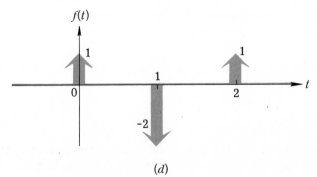

**Fig. P2.4**

(d)

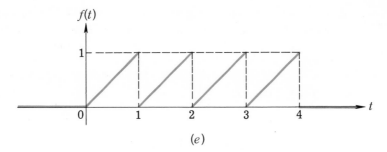

(e)

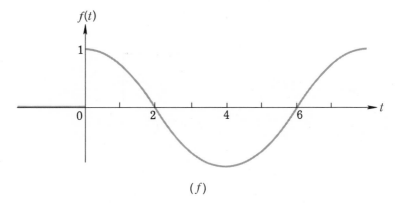

(f)

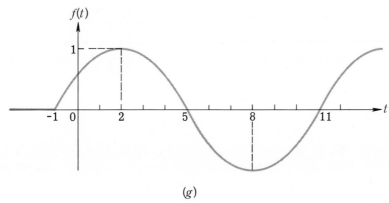

(g)

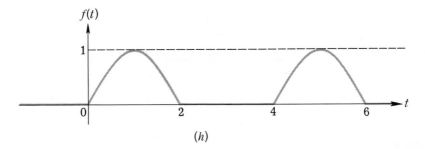

(h)

Linear time-
invariant RLC
elements
**10.** The circuit shown in Fig. P2.10a is made up of linear time-invariant elements. Calculate $v_R$, $v_L$, and $v_C$ for $t > 0$, and then sketch the corresponding waveforms for each of the following input currents (given in amperes):

a.  $i(t) = 0.2 \cos\left(2t + \dfrac{\pi}{4}\right)$

b.  $i(t) = \epsilon^{-1/2t}$

c.  $i(\cdot)$ is given as shown in Fig. P2.10b.

d.  $i(\cdot)$ is given as shown in Fig. P2.10c.

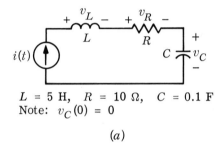

$$L = 5 \text{ H}, \quad R = 10 \text{ }\Omega, \quad C = 0.1 \text{ F}$$
Note: $v_C(0) = 0$

(a)

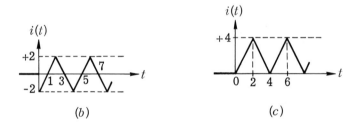

(b)                    (c)

**Fig. P2.10**

Linear time-invari-
ant RLC circuit
with sources
**11.** In the linear time-invariant circuit shown in Fig. P2.11 the voltage $v_s(t)$ and the current $i_s(t)$ are given by $v_s(t) = A \cos \omega t$ and $i_s = B\epsilon^{-\alpha t}$ (where $A$, $B$, $\alpha$, and $\omega$ are constants); calculate $v_L(t)$ and $i_C(t)$.

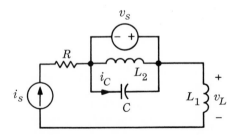

**Fig. P2.11**

Nonlinear inductor, linear approximation

**12.** Suppose that an inductor has the characteristic

$$\phi = 10^{-2}(1 - i^2)i$$

*a.* Calculate the voltage across the inductor if the current (in amperes) is

$$i(t) = 2 \times 10^{-2} \cos 2\pi 60t$$

*b.* Suppose that the application under consideration requires 1 percent components; i.e., the tolerance on element values is 1 percent. Within this tolerance can you consider the inductor above as linear if the typical current is

$$i(t) = 2 \times 10^{-2} \cos 2\pi 60t$$

Small-signal inductance and capacitance

**13.** *a.* A nonlinear time-invariant inductor has a characteristic given by

$$\phi = 10^{-2} \tanh i + 10^{-4}i$$

Plot the value of the small-signal (linear) inductance versus the bias current.

*b.* A nonlinear time-invariant capacitor has a characteristic given by

$$q = 1 - \epsilon^{-|v|}$$

This equation is only valid for $v$ larger than a few tenths of a volt. Plot the value of the small-signal (linear) capacitance vs. the bias voltage.

Nonlinear inductor

**14.** The $i\phi$ characteristic of a given inductor fits very closely the function

$$\phi = \beta \tanh \alpha i$$

where $\alpha = 10^2$ amp and $\beta = 10^{-7}$ weber. By using a suitable approximation, calculate the voltage resulting from the simultaneous flow of sinusoidal and constant currents ($i_{ac}$ and $I_{dc}$, respectively) as specified by each pair below:

*a.*  $I_{dc} = 16 \times 10^{-3}$ amp, $i_{ac}(t) = 10^{-4} \sin 10^7t$ amp
*b.*  $I_{dc} = -4 \times 10^{-3}$ amp, $i_{ac}(t) = 10^{-4} \sin 10^7t$ amp

Linear time-varying inductor

**15.** Through a linear time-varying inductor, whose time dependence is specified by the curve shown in Fig. P2.15, flows a constant current $i(t) = I_0$ amp ($I_0$ a constant and $-\infty < t < +\infty$). Calculate and sketch $v(t)$.

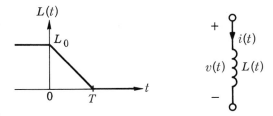

**Fig. P2.15**

Energy stored
in linear
capacitor

**16.** The current $i(t)$ specified by the curve shown in Fig. P2.16 flows through a linear time-invariant capacitor with capacitance $C = 2\ \mu F$. Given $v_C(0) = 0$, calculate and sketch for $t \geq 0$ the voltage $v_C(t)$, the instantaneous power delivered by the source $p(t)$, and the energy stored in the capacitor $\mathcal{E}_E(t)$.

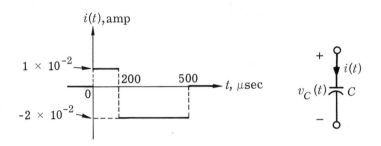

**Fig. P2.16**

Power and
energy stored
in linear
inductor

**17.** A linear time-invariant inductor with inductance $L = 10$ mH operates in a circuit that constrains $i_L(t)$ to the time dependence shown in the graph of Fig. P2.17. Calculate and sketch for $t \geq 0$ the voltage $v_L(t)$, the instantaneous power delivered by the source $p(t)$, and the energy stored in the inductor $\mathcal{E}_M(t)$.

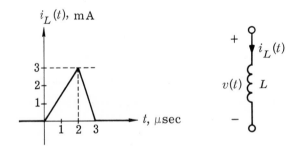

**Fig. P2.17**

Nonlinear
time-invariant
RLC elements

**18.** The voltage $v(t)$ defined by the curve shown in Fig. P2.18 is applied to a time-invariant parallel $RLC$ circuit which has each component specified by a characteristic curve. Given $i_L(0) = 0$, calculate and sketch the currents $i_L(t)$, $i_C(t)$, and $i_R(t)$.

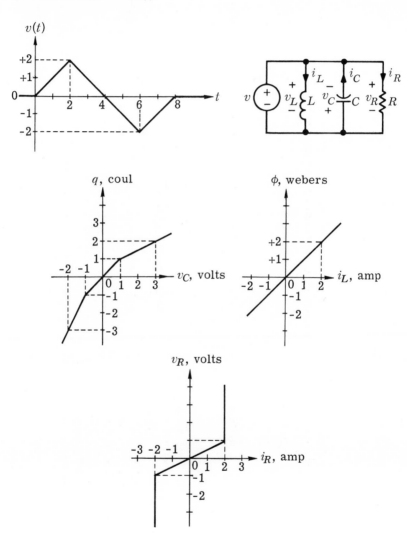

**Fig. P2.18**

**Table P2.19**

| | Measurements (current in amperes, voltage in volts) | | | |
|---|---|---|---|---|
| Test description | Component 1 | Component 2 | Component 3 | Component 4 |
| 1. dc tests<br>Voltage source $v(t) = V_0$ where $V_0$ has been tried for many constant values | $i(t) = 0$ | $i(t) = 10^{-2} V_0 (2 + \sin \Omega t)$ | $i(t) = 10^{-3} V_0^3$ | $i(t) = 10^{-3}$ |
| 2. ac tests<br>Voltage source $v(t) = A \sin \omega t$ where the amplitude $A$ and the angular frequency $\omega$ have been tried for many values | $i(t) =$ $\begin{cases} 5 \times 10^{-6} A\omega \cos \omega t \\ \quad \text{for} \\ (2n-1)\pi \leq \omega t \leq 2n\pi \\ \quad \text{and} \\ 2 \times 10^{-6} A\omega \cos \omega t \\ \quad \text{for} \\ 2n\pi \leq \omega t \leq (2n+1)\pi \end{cases}$ | $i(t) = 2 \times 10^{-2} A \sin \omega t$ $+ 5 \times 10^{-3} A[\cos (\Omega - \omega)t$ $- \cos (\Omega + \omega)t]$ | $i(t) =$ $10^{-3} A^3 \sin^3 \omega t$ | $i(t) = 10^{-3}$ |

Modeling     **19.** A set of unknown two-terminal circuit elements (resistors, capacitors, inductors, and sources) is being tested for specification. A sample of the test sheets corresponding to four components is presented in Table P2.19. Determine the characteristic of each component.

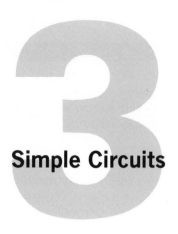

# Simple Circuits

In Chap. 1 we introduced the two Kirchhoff laws for lumped circuits and emphasized the fact that these laws do not depend on the nature of the circuit elements. Rather, they constitute linear constraints on the instantaneous values that the branch voltages and branch currents can take. Since these constraints depend only on the way the circuit elements are connected, they are referred to as *topological constraints.*

In Chap. 2 we studied the properties of two-terminal circuit elements in detail. Each branch in a given circuit is characterized by its branch relation, that is, the relation between the branch voltage and branch current. The topological constraints and the branch relations for all the branches in a circuit describe the circuit completely. The circuit analysis problem is to determine all branch currents of the circuit.† These voltages and currents will be called **network variables.** The many fundamental concepts and basic methods that are useful in solving the circuit analysis problem are the principal subjects of this text. In this chapter we shall present some introductory ideas and techniques for analysis of simple circuits. These circuits are made of only *one kind* of circuit element; that is, they contain only resistors, inductors, or capacitors.

It is convenient to introduce the concept of equivalence in the following discussion. We say that *one-ports are* **equivalent** *if their characterizations in terms of the port voltage and port current are identical.* In the preceding chapter we have already discussed the simple forms of the Thévenin and Norton equivalent circuits in order to convert voltage source into current source and vice versa. Those equivalent circuits are special cases of equivalent one-ports. We shall derive more general equivalent one-ports in this chapter. The term *equivalent* is used frequently to signify the fact that different circuits have the same electric characterization in terms of pertinent voltage and current variables. Often we use the term *equivalent branches;* then the pertinent variables are the branch voltage and the branch current.

† Of course, in many cases we want to know only certain branch voltages and currents, or some linear combinations of branch voltages and currents.

| **1** | **Series Connection of Resistors** |

The meaning of a series connection of circuit elements is intuitively clear. We have already discussed in the previous chapter the series connection of a resistor and a voltage source. In this section we shall give a more general treatment of a series connection of resistors.

**Example 1**   Consider the circuit in Fig. 1.1, where two nonlinear resistors $\mathscr{R}_1$ and $\mathscr{R}_2$ are connected at node $B$. Nodes $A$ and $C$ are connected to the rest of the circuit, which is designated by $\mathscr{N}$. The one-port, consisting of resistors $\mathscr{R}_1$ and $\mathscr{R}_2$, whose terminals are nodes $A$ and $C$, is called the *series connection of resistors $\mathscr{R}_1$ and $\mathscr{R}_2$*. For our present purposes the nature of $\mathscr{N}$ is of no importance. The two resistors $\mathscr{R}_1$ and $\mathscr{R}_2$ are specified by their characteristics, as shown in the *iv* plane of Fig. 1.2. We wish to determine the characteristic of the series connection of $\mathscr{R}_1$ and $\mathscr{R}_2$, that is, the characteristic of a resistor equivalent to the series connection.

First, KVL for the mesh $ABCA$ requires that

(1.1)   $v = v_1 + v_2$

Next, KCL for the nodes $A$, $B$, and $C$ requires that

$$i = i_1 \qquad i_1 = i_2 \qquad i_2 = i$$

Clearly, one of the above three equations is redundant; they may be summarized by

(1.2)   $i_1 = i_2 = i$

Thus, Kirchhoff's laws state that $\mathscr{R}_1$ and $\mathscr{R}_2$ are traversed by the *same* current, and the voltage across the series connection is the sum of the voltages across $\mathscr{R}_1$ and $\mathscr{R}_2$. Thus, the characteristic of the series connection is easily obtained graphically; for each fixed $i$ we add the values of the voltages allowed by the characteristics of $\mathscr{R}_1$ and $\mathscr{R}_2$. The process is illustrated in Fig. 1.2. The characteristic thus obtained is called the characteristic of the resistor equivalent to the series connection of $\mathscr{R}_1$ and $\mathscr{R}_2$. Observe that

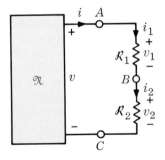

**Fig. 1.1**   The series connection of $\mathscr{R}_1$ and $\mathscr{R}_2$.

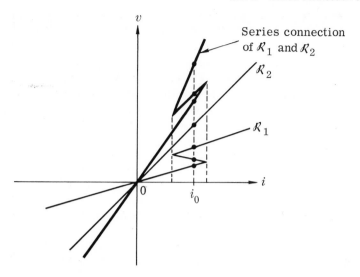

**Fig. 1.2**   Series connection of two resistors $\Re_1$ and $\Re_2$ of Example 1.

in this example $\Re_2$ is a linear resistor, and $\Re_1$ is a voltage-controlled nonlinear resistor; i.e., the current in $\Re_1$ is specified by a (single-valued) function of the voltage. In Fig. 1.2 it is seen that if the current is $i_0$, the characteristic of $\Re_1$ allows three possible values for the voltage; hence, $\Re_1$ is not current-controlled. It is interesting to note that the series connection has a characteristic that is neither voltage-controlled nor current-controlled.

In the above example we obtained the characteristic of the series connection of two resistors graphically by adding the corresponding voltages across the resistors for the same current. Analytically, we can determine the characteristic of the resistor that is equivalent to the series connection of two resistors $\Re_1$ and $\Re_2$ only if both are current-controlled. Current-controlled resistors $\Re_1$ and $\Re_2$ have characteristics that may be described by equations of the form

(1.3)   $v_1 = f_1(i_1) \qquad v_2 = f_2(i_2)$

where the reference directions are shown in Fig. 1.1. In view of Eqs. (1.1) and (1.2) the series connection has a characteristic given by

(1.4)   $v = f_1(i_1) + f_2(i_2)$
$= f_1(i) + f_2(i)$

Therefore, we conclude that the two-terminal circuit as characterized by the voltage-current relation of Eq. (1.4) is another resistor specified by

(1.5a)   $v = f(i)$

where

(1.5b)   $f(i) = f_1(i) + f_2(i)$      for all $i$

Equations (1.5a) and (1.5b) show that the series connection of the two current-controlled resistors is equivalent to a current-controlled resistor $\mathcal{R}$, and its characteristic is described by the function $f(\cdot)$ defined in (1.5b); this characteristic is shown in Fig. 1.3.

Using analogous reasoning, we can state that the *series connection of m current-controlled resistors with characteristics described by* $v_k = f_k(i_k)$, $k = 1, 2, \ldots, m$, *is equivalent to a single current-controlled resistor whose characteristic is described by* $v = f(i)$, *where* $f(i) = \sum_{k=1}^{m} f_k(i)$ *for all i.*   If, in particular, all resistors are linear; that is, $v_k = R_k i_k$, $k = 1, 2, \ldots, m$, the equivalent resistor is also linear, and $v = Ri$, where

(1.6)   $R = \sum_{k=1}^{m} R_k$

---

**Example 2**   Consider the circuit in Fig. 1.4 where $m$ voltage sources are connected in series.   Clearly, this is only a special case of the series connection of $m$ current-controlled resistors.   Extending Eq. (1.1), we see that the series combination of the $m$ voltage sources is equivalent to a single voltage source whose terminal voltage is $v$, where

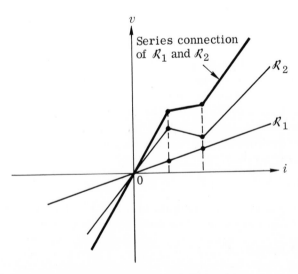

**Fig. 1.3**   Series connection of two current-controlled resistors.

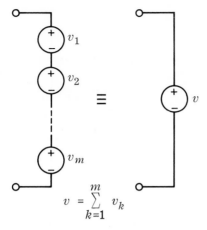

**Fig. 1.4**    Series connection of voltage sources.

$$(1.7) \quad v = \sum_{k=1}^{m} v_k$$

**Example 3**    Consider the series connection of $m$ current sources as shown in Fig. 1.5. It is seen immediately that such a connection usually violates KCL; indeed, KCL applied to nodes $B$ and $C$ requires that $i_1 = i_2 = i_3 = \ldots$. Therefore, it does not make sense physically to consider the series connection of current sources unless this condition is satisfied. Then the series connection of $m$ identical current sources is equivalent to one such current source.

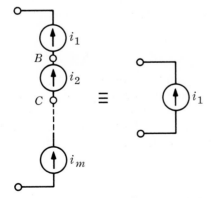

**Fig. 1.5**    The series connection of current sources can be made only if $i_1 = i_2 = \cdots = i_m$.

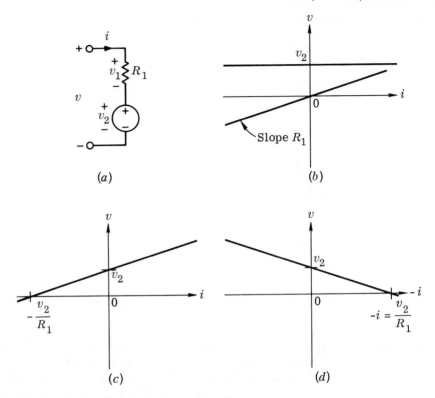

**Fig. 1.6** Series connection of a linear resistor and a voltage source.

**Example 4** Consider the series connection of a linear resistor $R_1$ and a voltage source $v_2$, as shown in Fig. 1.6a. Their characteristics are plotted on the same $iv$ plane and are shown in Fig. 1.6b. The series connection has a characteristic as shown in Fig. 1.6c. In terms of functional characterization we have

$$(1.8) \qquad v = v_1 + v_2 = R_1 i + v_2$$

Since $R_1$ is a known constant and $v_2$ is known, Eq. (1.8) relates all possible values of $v$ and $i$. It is the equation of a straight line as shown in Fig. 1.6c. In Fig. 1.6d we plot the characteristic in the $(-i)v$ plane; we recognize the characteristic of the automobile battery discussed in Chap. 2, Sec. 2, where the opposite to the associated reference direction was used for the battery.

**Example 5** Consider the circuit of Fig. 1.7a where a linear resistor is connected to an ideal diode. Their characteristics are plotted on the same graph and are shown in Fig. 1.7b. The series connection has a characteristic as shown in Fig. 1.7c; it is obtained by reasoning as follows.

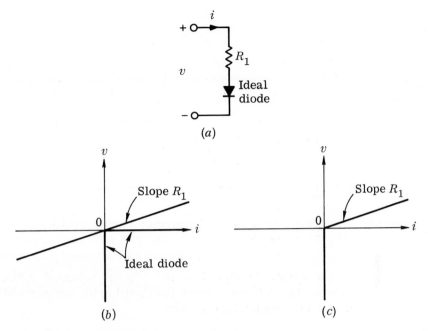

**Fig. 1.7**   Series connection of an ideal diode and a linear resistor.   (*a*) The circuit; (*b*) the characteristic of each element; (*c*) the characteristic of the series connection.

First, for positive current we can simply add the ordinates of the two curves.   Next, for negative voltage across the diode the ideal diode is an open circuit; hence, the series connection is again an open circuit.   The current *i* cannot be negative.

To illustrate that an ideal diode is not a bilateral device, suppose we turn it around as shown in Fig. 1.8*a*.   Following the same reasoning, we obtain the characteristic shown in Fig. 1.8*b*.

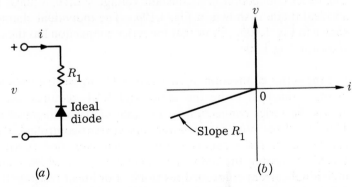

**Fig. 1.8**   The series connection is analogous to that of Fig. 1.7 except that the diode is reversed.   (*a*) The circuit; (*b*) the characteristic of the series connection.

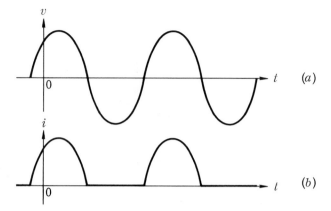

**Fig. 1.9**   For an applied voltage shown in (a) the resulting current is shown in (b) for the circuit of Fig. 1.7a.

The circuits in Figs. 1.7a and 1.8a are idealized rectifiers. Let us assume that a voltage source is connected to the one-port of Fig. 1.7a and that it has a sinusoidal waveform

(1.9)   $v_s(t) = A \cos (\omega_0 t + \phi)$

as shown in Fig. 1.9a. The current $i$ passing through the series connection is a periodic function of time and is shown in Fig. 1.9b. Observe that the applied voltage $v(\cdot)$ is a periodic function of time with zero average value. The current $i(\cdot)$ is also a periodic function of time with the same period, but it is always nonnegative. By the use of filters it is possible to make this current almost constant; hence, a sinusoidal signal can be converted into a dc signal.

**Exercise**   The series connection of a constant voltage source, a linear resistor, and an ideal diode is shown in Fig. 1.10a. The individual characteristics are shown in Fig. 1.10b. Show that the series connection has the characteristic shown in Fig 1.10c.

**Summary**   For the series connection of elements, KCL forces the currents in all elements (branches) to be the same, and KVL requires that the voltage across the series connection be the sum of the voltages of all branches. Thus, if all the nonlinear resistors are current-controlled, the equivalent resistor of the series connection has a characteristic $v = f(i)$ which is obtained by adding the individual functions $f_k(\cdot)$ which characterize the individual current-controlled resistors. For linear resistors the sum of individual resistances gives the resistance of the equivalent resistor; i.e., for $m$ linear resistors in series

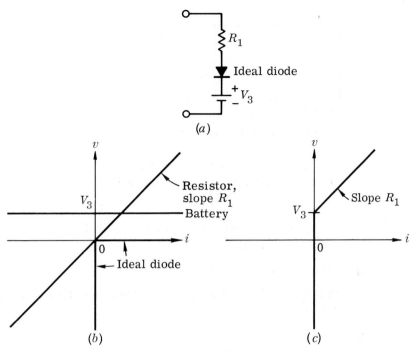

(a)

(b)                           (c)

**Fig. 1.10**  Series connection of a linear resistor, an ideal diode, and a battery. (a) The connection; (b) the individual characteristics; (c) the overall characteristic.

$$R = \sum_{k=1}^{m} R_k$$

where $R_k$ is the individual resistance and $R$ is the equivalent resistance.

## 2  Parallel Connection of Resistors

Consider the circuit in Fig. 2.1 where two resistors $\mathcal{R}_1$ and $\mathcal{R}_2$ are connected in parallel at nodes $A$ and $B$. Nodes $A$ and $B$ are also connected to the rest of the circuit designated by $\mathcal{R}$. The exact description of $\mathcal{R}$ is of no importance for our present purposes. Let the two resistors be specified by their characteristics, which are shown in Fig. 2.2 where they are plotted on the $vi$ plane. We want to find the characteristic of the parallel connection of $\mathcal{R}_1$ and $\mathcal{R}_2$. Thus, Kirchhoff's laws imply that $\mathcal{R}_1$ and $\mathcal{R}_2$ have the same branch voltage, and the current through the parallel connection is the sum of the currents through each resistor. The characteristic of the parallel connection is thus obtained by adding, for each fixed $v$, the values of the current allowed by the characteristics of $\mathcal{R}_1$ and $\mathcal{R}_2$. The process is illustrated in Fig. 2.2. The characteristic thus obtained is that of the resistor *equivalent* to the parallel connection.

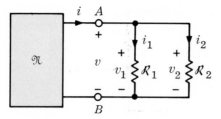

**Fig. 2.1**   Parallel connection of two resistors.

Analytically, if $\mathfrak{R}_1$ and $\mathfrak{R}_2$ are voltage-controlled, their characteristics may be described by equations of the form

(2.1)   $i_1 = g_1(v_1) \qquad i_2 = g_2(v_2)$

and in view of Kirchhoff's laws, the parallel connection has a characteristic described by

(2.2)   $i = i_1 + i_2 = g_1(v) + g_2(v)$

In other words the parallel connection is described by the function $g(\,\cdot\,)$, defined by

(2.3a)   $i = g(v)$

where

(2.3b)   $g(v) = g_1(v) + g_2(v) \qquad$ for all $v$

Extending this result to the general case, we can state that *the parallel connection of m voltage-controlled resistors with characteristics* $i_k = g_k(v_k)$,

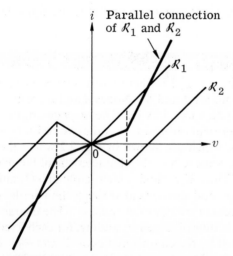

**Fig. 2.2**   Characteristics of $\mathfrak{R}_1$ and $\mathfrak{R}_2$ and of their parallel connection.

$k = 1, 2, \ldots, m$, *is equivalent to a single voltage-controlled resistor with characteristic $i = g(v)$, where $g(v) = \sum\limits_{k=1}^{m} g_k(v)$ for all $v$.* If, in particular, all resistors are linear, that is, $i_k = G_k v_k$, $k = 1, 2, \ldots, m$, the equivalent resistor is also linear, and $i = Gv$, where

(2.4)   $$G = \sum_{k=1}^{m} G_k$$

$G$ is the conductance of the equivalent resistor.  In terms of resistance value

$$R = \frac{1}{G} = \frac{1}{\sum\limits_{k=1}^{m} G_k}$$

or

(2.5)   $$\frac{1}{R} = \sum_{k=1}^{m} \frac{1}{R_k}$$

**Example 1**   As shown in Fig. 2.3, the parallel connection of $m$ current sources is equivalent to a single current source whose source current is

(2.6)   $$i = \sum_{k=1}^{m} i_k$$

**Example 2**   The parallel connection of voltage sources violates KVL with the exception of the trivial case where all voltage sources are equal.

**Example 3**   The parallel connection of a current source $i_1$ and a linear resistor with resistance $R_2$ as shown in Fig. 2.4a can be represented by an equivalent resistor that is characterized by

(2.7)   $$i = -i_1 + \frac{1}{R_2}v$$

Eq. (2.7) can be written as

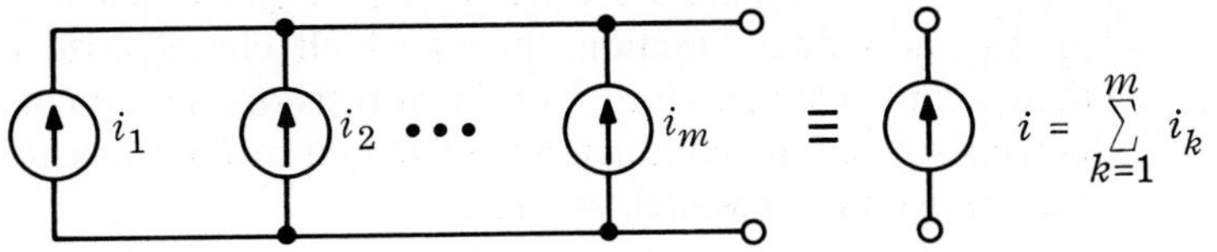

**Fig. 2.3**   Parallel connection of current sources $i = \sum\limits_{k=1}^{m} i_k$.

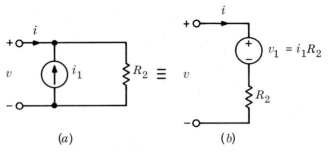

**Fig. 2.4**   Equivalent one-ports illustrating a simple case of the Thévenin and Norton equivalent circuit theorems.

(2.8)   $v = i_1 R_2 + i R_2$

The alternative equivalent circuit can be drawn by interpreting the voltage $v$ as the sum of two terms, a voltage source $v_1 = i_1 R_2$ and a linear resistor with resistance $R_2$, as shown in Fig. 2.4b. This equivalence, which was also discussed in Chap. 2, Sec. 2, represents a special case of the Thévenin and Norton equivalent circuit theorem and is extremely useful in circuit analysis.

**Example 4**   The parallel connection of a current source, a linear resistor, and an ideal diode is shown in Fig. 2.5a. Their characteristics are shown in Fig. 2.5b. The equivalent resistor has the characteristic shown in Fig. 2.5c. Again it should be pointed out that for an ideal diode the current is not a function of the voltage. However, we can use physical reasoning to obtain the resulting characteristic; that is, for $v$ negative the characteristic of the equivalent resistor is obtained by the addition of the three curves. For $i_3$ positive the ideal diode is a short circuit; thus, the voltage $v$ across it is always zero. Consequently the parallel connection has the characteristic shown in Fig. 2.5c.

**Summary**   For the parallel connection of elements, KVL requires that all the voltages across the elements be the same, and KCL requires that the current through the parallel connection be the sum of the currents in all branches. For nonlinear voltage-controlled resistors the equivalent resistor of the parallel connection has a characteristic $i = g(v)$ which is obtained by adding the individual functions $g_k(\cdot)$ which characterize each individual voltage-controlled resistor. For linear resistors the sum of individual conductances gives the conductance of the equivalent resistor. Thus, for $m$ linear resistors in parallel, we have

$$G = \sum_{k=1}^{m} G_k$$

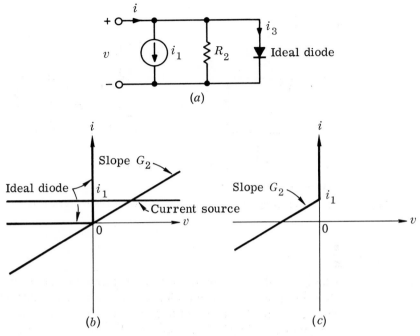

**Fig. 2.5**   Parallel connection of a current source, a linear resistor, and an ideal diode.   (a) The circuit; (b) the characteristic of each element; (c) the characteristic of the parallel connection.

where $G_k$ is the individual conductance and $G$ is the equivalent conductance.

---

## 3   Series and Parallel Connection of Resistors

**Example 1**   Consider the circuit in Fig. 3.1, where a resistor $\Re_1$ is connected in series with the parallel connection of $\Re_2$ and $\Re_3$.   The problem is to determine the characteristic of the equivalent resistor.

  If the characteristics of $\Re_1$, $\Re_2$, and $\Re_3$ are specified graphically, we need first to determine graphically the characteristic of $\Re^*$, the resistor equivalent to the parallel connection of $\Re_2$ and $\Re_3$, and second, to determine graphically the characteristic of $\Re$, the resistor equivalent to the series connection of $\Re_1$ and $\Re^*$.   The necessary steps are shown in Fig. 3.1.

  Let us assume that the characteristics of $\Re_2$ and $\Re_3$ are voltage-controlled and specified by

$$(3.1) \quad i_2 = g_2(v_2) \quad \text{and} \quad i_3 = g_3(v_3)$$

where $g_2(\,\cdot\,)$ and $g_3(\,\cdot\,)$ are single-valued functions.   The parallel connection has an equivalent resistor $\Re^*$, which is characterized by

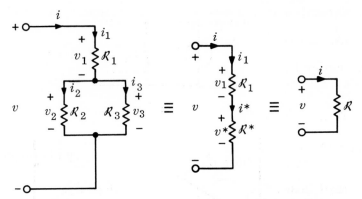

**Fig. 3.1**   Series-parallel connection of resistors and its successive reduction.

(3.2)   $i^* = g(v^*)$

where $i^*$ and $v^*$ are the branch current and voltage of the resistor $\mathcal{R}^*$ as shown in Fig. 3.1. The parallel connection requires the voltages $v_2$ and $v_3$ to be equal to $v^*$. The resulting current $i^*$ is the sum of $i_2$ and $i_3$. Thus, the characteristic of $\mathcal{R}^*$ is related to those of $\mathcal{R}_2$ and $\mathcal{R}_3$ by

(3.3)   $g(v^*) = g_2(v^*) + g_3(v^*)$      for all $v^*$

Let $g_2(\cdot)$ and $g_3(\cdot)$ be specified as shown in Fig. 3.2a. Then $g(\cdot)$ is obtained by adding the two functions.

The next step is to obtain the series connection of $\mathcal{R}_1$ and $\mathcal{R}^*$. Let us assume that the characteristic of $\mathcal{R}_1$ is current-controlled and specified by

(3.4)   $v_1 = f_1(i_1)$

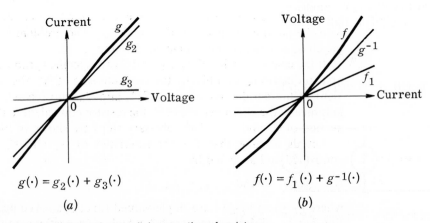

$g(\cdot) = g_2(\cdot) + g_3(\cdot)$        $f(\cdot) = f_1(\cdot) + g^{-1}(\cdot)$

(a)                                        (b)

**Fig. 3.2**   Example 1: the series-parallel connection of resistors.

where $f_1(\cdot)$ is a single-valued function as shown in Fig. 3.2b. The series connection of $\mathcal{R}_1$ and $\mathcal{R}^*$ has an equivalent resistor $\mathcal{R}$, as shown in Fig. 3.1. The characteristic of $\mathcal{R}$ as specified by

(3.5)   $v = f(i)$

is to be determined. Obviously the series connection forces the currents $i_1$ and $i^*$ to be the same and equal to $i$. The voltage $v$ is simply the sum of $v_1$ and $v^*$. However, in order to add the two voltages we must first be able to express $v^*$ in terms of $i^*$. From (3.2) we can write

(3.6)   $v^* = g^{-1}(i^*)$

where $g^{-1}(\cdot)$ is the inverse of the function $g(\cdot)$. For our present example, the inverse function $g^{-1}(\cdot)$ is plotted on the current-voltage plane of Fig. 3.2b directly from the function $g(\cdot)$ on the voltage-current plane of Fig. 3.2a. This is easily done by inverting the curve $g(\cdot)$ to form the mirror image with respect to the straight line that passes through the origin and is at an angle 45° from the axis. Thus the series connection of $\mathcal{R}_1$ and $\mathcal{R}^*$ is characterized by $f(\cdot)$ of (3.5), where

$$f(i) = f_1(i) + g^{-1}(i) \qquad \text{for all } i$$

This is also plotted in Fig. 3.2b. Thus, the crucial step in the derivation is the question of whether $g^{-1}(\cdot)$ exists as a single-valued function. If the inverse does not exist, the reduction procedure fails; indeed, no equivalent representation exists in terms of single-valued functions. One simple criterion that guarantees the existence of such a representation is that all resistors have strictly monotonically increasing characteristics. For example, linear resistors with positive resistances are monotonically increasing. The equivalent resistance of $\mathcal{R}$ for the circuit in Fig. 3.1, assuming that all resistors are linear, is

(3.7)   $R = R_1 + \dfrac{1}{1/R_2 + 1/R_3}$

where $R_1$, $R_2$, and $R_3$ are, respectively, the resistances of $\mathcal{R}_1$, $\mathcal{R}_2$, and $\mathcal{R}_3$.

---

**Exercise**    The circuit shown in Fig. 3.3 is called an infinite-ladder network. All resistors are linear; the series resistors have resistances $R_s$, and the shunt resistors have resistances $R_p$. Determine the input resistance $R$, that is, the resistance of the equivalent one-port. Hint: Since the ladder consists of an infinite chain of identical sections (a series $R_s$ and a shunt $R_p$), we can consider the first section as being terminated by an infinite chain of the very same sections. Thus, the input resistance $R$ will not be changed if the first section is terminated by a resistor with resistance $R$.

Up to now we have dealt with the problem of determining the charac-

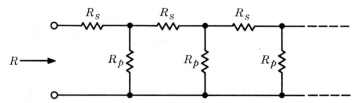

**Fig. 3.3**    An infinite ladder consisting of linear resistors.    $R_s$ is called the series resistance, and $R_p$ is called the shunt resistance.    $R$ is the input resistance, i.e., the resistance of the equivalent one-port.

teristics of a resistor equivalent to that obtained by a series, parallel, or series-parallel connection of resistors.   In circuit analysis we are often interested in finding out the voltages and currents at various locations in a circuit when sources are applied.   The following examples will illustrate how to solve these problems.

**Example 2**    Consider the simple circuit, shown in Fig. 3.4, where $\mathcal{R}_1$ and $\mathcal{R}_2$ are voltage-controlled resistors characterized by

(3.8)    $i_1 = 6 + v_1 + v_1{}^2$      and      $i_2 = 3v_2$

$i_0$ is a constant current source of 2 amp.   We wish to determine the currents $i_1$ and $i_2$ and the voltage $v$.   Since $v = v_1 = v_2$, the characteristic of the equivalent resistor of the parallel combination is simply

$i = i_1 + i_2$

(3.9)    $= 6 + v + v^2 + 3v = 6 + 4v + v^2$

To obtain the voltage $v$ for $i = i_0 = 2$ amp, we need to solve Eq. (3.9). Thus,

$v^2 + 4v + 6 = 2$

or

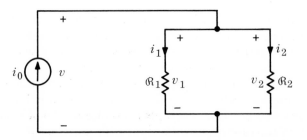

**Fig. 3.4**    Example 2: parallel connection of resistors and a current source.

(3.10)   $v = -2$ volts

Since $v = v_1 = v_2$, substituting (3.10) in (3.8), we obtain

$i_1 = 8$ amp

and

$i_2 = -6$ amp

---

**Exercise**   Determine the power dissipated in each resistor and show that the sum of their power dissipations is equal to the power delivered by the current source.

---

**Example 3**   In the ladder of Fig. 3.5, where all resistors are linear and time-invariant, there are four resistors shown. A voltage source of $V_0 = 10$ volts is applied. Let $R_s = 2$ ohms and $R_p = 1$ ohm. Determine the voltages $v_a$ and $v_b$.

We first compute the input resistance $R$ of the equivalent one-port that is faced by the voltage source $V_0$. Based on the method of series-parallel connection of resistors we obtain immediately a formula similar to Eq. (7.7); thus

$$R = R_s + \frac{1}{1/R_p + 1/(R_s + R_p)}$$

$$= 2 + \frac{1}{1 + \frac{1}{3}}$$

$$= 2\frac{3}{4} \text{ ohms}$$

Thus, the current $i_1$ is given by

$$i_1 = \frac{V_0}{R} = \frac{10}{2\frac{3}{4}} = \frac{40}{11} \text{ amp}$$

The branch voltage $v_1$ is given by

$$v_1 = R_s i_1 = \frac{80}{11} \text{ volts}$$

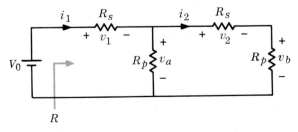

**Fig. 3.5**   Example 3: a ladder with linear resistors.

Using KVL for the first mesh, we obtain

$$v_a = V_0 - v_1 = \frac{30}{11} \text{ volts}$$

Knowing $v_a$, we determine immediately

$$i_2 = \frac{v_a}{R_s + R_p} = \frac{30/11}{3} = \frac{10}{11} \text{ amp}$$

From Ohm's law, we have

$$v_b = R_p i_2 = \frac{10}{11} \text{ volts}$$

---

Thus, by successive use of Kirchhoff's laws and Ohm's law we can determine the voltages and currents of any series-parallel connection of linear resistors. For nonlinear resistors the problem is more complex, as evidenced by possible difficulties such as the necessity of finding the inverse of a function (as in Example 1) and obtaining solution of a polynomial equation (as in Example 2).

**Exercise 1**   For the infinite ladder of Fig. 3.3, determine the ratio $R_s/R_p$ such that each node voltage is half of the preceding node voltage.

**Exercise 2**   Suppose we want to design a finite ladder, say, a chain consisting of 10 sections with the ratio of $R_s$ and $R_p$ found in Exercise 1; how do we terminate the chain so that the property described in Exercise 1 holds?

For resistive circuits that are not in the form of series-parallel connection, the analysis is again more complex. We shall give general methods of analysis for circuits with linear resistors in Chaps. 10 and 11. However, it is useful to introduce an example of the non-series-parallel type that we can solve at present by simple physical reasoning.

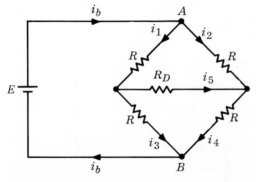

**Fig. 3.6**   Example 4: a symmetric bridge circuit.

**Example 4**   Consider the bridge circuit of Fig. 3.6.   Note that it is not of the form of a series-parallel connection.   Assume that the four resistances are the same. Obviously, because of symmetry the battery current $i_b$ must divide equally at node $A$ and also at node $B$; that is, $i_1 = i_2 = i_b/2$, and $i_3 = i_4 = i_b/2$. Consequently, the current $i_5$ must be zero.

**Exercise**   Twelve linear resistors, each with resistance $R$, are laid along the edges of a cube.   At each corner of the cube the resistors are soldered together.   Call ① and ② two nodes that are at diagonally opposite corners of the cube. What is the equivalent resistance between node ① and node ②?   (Hint: Draw the cube in perspective, and use symmetry arguments to decide how the current splits at each node.)

## 4 | Small-signal Analysis

As mentioned in the previous section, the analysis of circuits with non-linear resistors is difficult.   We obtained the equivalent characteristic of the series-parallel connection of nonlinear resistors, and we introduced a simple example to show how to calculate the current distribution of two nonlinear resistors in parallel.   General analysis of circuits made of non-linear resistors is beyond the scope of this volume; however, in Chap. 18 some basic facts concerning these circuits will be developed.

One particular technique of great importance in engineering is the small-signal analysis of a nonlinear system.   We shall illustrate the basic idea with a simple resistive circuit containing a tunnel diode.   The concept will be discussed further in Chap. 17, when we discuss nonlinear two-ports.

**Example**   Consider the circuit in Fig. 4.1, where a tunnel diode (a voltage-controlled nonlinear resistor) is connected to a linear resistor with resistance $R_s$ and an input consisting of a constant voltage source $E$ and a time-varying voltage source $v_s(t)$.   For our present discussion we shall assume that

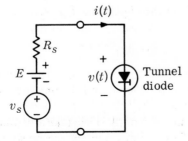

**Fig. 4.1**   Tunnel-diode amplifier circuit.

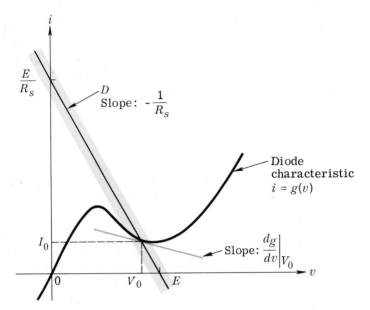

**Fig. 4.2**   Tunnel-diode characteristic and characteristic of the remaining circuit.

$|v_s(t)| \ll E$ for all $t$, which means that the time-varying voltage is at all times much smaller (in absolute value) than the dc source. In practical application the time-varying source corresponds to the signal, and the dc source is called the bias. The problem is to determine the voltage $v(t)$ and the current $i(t)$ for the tunnel diode shown in Fig. 4.1.

Let us use Kirchhoff's laws and the branch equations of all the elements in the circuit to obtain the necessary equations. First, KCL implies that each element of the circuit of Fig. 4.1 is traversed by the same current $i(t)$. Next, KVL applied for the mesh gives

(4.1)   $E + v_s(t) = R_s i(t) + v(t)$      for all $t$

Let the characteristic of the tunnel diode be described by

(4.2)   $i = g(v)$

The characteristic is plotted on the $vi$ plane in Fig. 4.2. Combining (4.1) and (4.2), we have

(4.3)   $E + v_s(t) = R_s g[v(t)] + v(t)$      for all $t$

This is an equation in which $v(t)$ is the only unknown. Since it holds for all $t$, Eq. (4.3) should be solved for each $t$, and the unknown function $v(\cdot)$ would be obtained point by point.

Before proceeding to the solution of (4.3), we take advantage of the fact that the input is the sum of two terms, namely the dc source $E$ and the time-varying source $v_s(t)$. The solution of the problem can be more easily

obtained by considering first the dc source only.    After the solution to the
dc problem is found, we then include the time-varying source and analyze
the whole problem by means of small-signal analysis.

*Step 1*    $v_s(t) = 0$ for all $t$.    The independent voltage source $v_s$ in Fig. 4.1 becomes a
short circuit.    KVL gives

(4.4)    $E - R_s i = v$

The tunnel diode is described by its characteristic according to (4.2).    The
two equations, (4.2) and (4.4), have two unknowns, $v$ and $i$.    We solve the
problem graphically.    In Fig. 4.2, the straight line labeled $D$ is the locus of
all points $(v,i)$ that satisfy Eq. (4.4).    Similarly, the tunnel-diode characteristic is the locus of all points $(v,i)$ that satisfy Eq. (4.2).    Hence any point
$(v,i)$ that lies both on the line of $D$ and on the tunnel-diode characteristic
has coordinates $(v,i)$ that satisfy Eqs. (4.2) and (4.4).    Thus, every intersection of $D$ and the characteristic gives a solution of the system described
by Eqs. (4.2) and (4.4).    In the present situation, there is only one solution
$(V_0, I_0)$, as indicated in Fig. 4.2.    Thus, $(V_0, I_0)$ satisfies Eqs. (4.2) and (4.4);
that is,

(4.5a)    $E - R_s I_0 = V_0$

and

(4.5b)    $I_0 = g(V_0)$

$(V_0, I_0)$ is called the *operating point*.    We now proceed to the whole problem.

*Step 2*    The voltage $v_s$ is not identically zero.    The equations describing this
situation are

(4.6a)    $E + v_s(t) - R_s i(t) = v(t)$      for all $t$

and

(4.6b)    $i(t) = g[v(t)]$      for all $t$

For each $t$, the locus of all points $(v(t), i(t))$ that satisfy (4.6a) is a straight
line parallel to the line $D$ in the $vi$ plane of Fig. 4.2.    This line is above $D$ if
$v_s(t) > 0$ and below $D$ if $v_s(t) < 0$.    The locus of all points $(v(t), i(t))$ that
satisfy (4.6b) is the tunnel-diode characteristic that remains fixed in time.
Therefore, any point $(v(t), i(t))$ that lies on both the straight line and the
characteristic satisfies (4.6a) and (4.6b).    In short, the intersection determines the solution.    Thus, Eqs. (4.6) can always be solved graphically.

We have assumed that $|v_s(t)| \ll E$ for all $t$.    Small-signal analysis is an
approximate method of solution which is valid as long as $|v_s(t)|$ is small.
The first step is to write the solution $(v(t), i(t))$ as the sum of two terms.
Thus,

(4.7a)  $v(t) = V_0 + v_1(t)$

(4.7b)  $i(t) = I_0 + i_1(t)$

Note that $(V_0, I_0)$ is the operating point, i.e., the solution when $v_s(t) = 0$. Since $v_s(t)$ is assumed small, the solution $(v(t), i(t))$ lies in the neighborhood of $(V_0, I_0)$; thus, $v_1(t)$ and $i_1(t)$ can be considered as a perturbation from the dc solution $(V_0, I_0)$. This perturbation is caused by the small-signal source $v_s(t)$. Let us now determine $v_1(t)$ and $i_1(t)$ for all $t$.

First, consider the tunnel-diode characteristic $i = g(v)$. Using Eq. (4.7a) and (b), we have

(4.8)  $I_0 + i_1(t) = g[V_0 + v_1(t)]$

Since $v_1(t)$ is small by assumption, we can expand the right-hand side of (4.8) by Taylor's series and take only the first two terms as an approximation. Thus,

(4.9)  $I_0 + i_1(t) \approx g(V_0) + \dfrac{dg}{dv}\bigg|_{V_0} v_1(t)$

Substituting (4.5b) in (4.9), we obtain a simple equation for $i_1(t)$ and $v_1(t)$. Thus,

(4.10)  $i_1(t) \approx \dfrac{dg}{dv}\bigg|_{V_0} v_1(t)$

The term $\dfrac{dg}{dv}\bigg|_{V_0}$ is the slope of the tunnel-diode characteristic curve at the operating point $(V_0, I_0)$, as shown in Fig. 4.2. Let us designate

(4.11)  $\dfrac{dg}{dv}\bigg|_{V_0} \triangleq G = \dfrac{1}{R}$

and call $G$ *the small-signal conductance of the tunnel diode at the operating point* $(V_0, I_0)$. Note that $G$ is *negative*. Thus, to the small-signal source $v_s$ the tunnel diode is a linear active resistor since as far as $v_s$ is concerned the resistor characteristic of the tunnel diode has an "origin" at $(V_0, I_0)$, and the characteristic in the neighborhood of $(V_0, I_0)$ is that of a linear resistor with negative resistance. Thus,

(4.12)  $i_1(t) = Gv_1(t)$      or      $v_1(t) = Ri_1(t)$

In order to calculate $v_1(t)$ and $i_1(t)$ we must first go back to our original KVL equation, i.e., Eq. (4.6a), and combine it with Eqs. (4.7a) and (4.7b). We obtain

(4.13)  $E + v_s(t) - R_s[I_0 + i_1(t)] = V_0 + v_1(t)$

Using the information of (4.5a), which relates $I_0$ and $V_0$, we obtain the following equation relating $i_1(t)$ and $v_1(t)$:

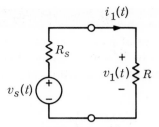

$i_1(t)$

$R_s$

$+$

$v_1(t) \gtrless R$

$-$

$v_s(t)$

**Fig. 4.3**    Small-signal equivalent circuit.

(4.14)    $v_s(t) - R_s i_1(t) = v_1(t)$

Equations (4.12) and (4.14) constitute a system of two *linear* algebraic equations in two unknowns $v_1(t)$ and $i_1(t)$ and can be solved easily. Since $G$ in (4.12) is a constant, (4.12) describes the branch equation of a linear time-invariant (active) resistor. Equation (4.14) represents simply the KVL equation for a circuit shown in Fig. 4.3. This circuit is called *the small-signal equivalent circuit* [about the operating point $(V_0, I_0)$] of the tunnel-diode circuit of Fig. 4.1. From (4.12) and (4.14) we calculate easily the solution

(4.15)    $i_1(t) = \dfrac{v_s(t)}{R_s + R}$

and

(4.16)    $v_1(t) = R i_1(t) = \dfrac{R v_s(t)}{R_s + R}$

In the present situation $R = 1/G$, and $G$ is negative; hence $v_1(t)$ can be made much larger than $v_s(t)$ by an appropriate choice of $R_s$. The varying voltage $v_1(t)$ across the diode is then much larger than the applied voltage $v_s(t)$. Since the time-varying currents in the voltage source $v_s$ and the resistor $R$ are the same, the signal power delivered to the resistor has been amplified. Actually the circuit in Fig. 4.1 is a simple tunnel-diode amplifier. The dc source and the resistance $R_s$, which constitute the "biasing circuit," determine the operating point $(V_0, I_0)$ according to Eqs. (4.5a) and (4.5b). The slope of the tunnel diode at the operating point, that is, the small-signal equivalent conductance $G$, and the value of $R_s$ determine $R/(R + R_s)$, the amplification factor of the amplifier.

Of course, the analysis is highly simplified since it neglects all parasitic elements (such as the parasitic capacitor) of the tunnel diode. However, it does illustrate how the basic laws can be applied to solve some interesting problems.

## 5   Circuits with Capacitors or Inductors

The series and parallel connections of pure capacitors or pure inductors can be treated in a similar fashion to those of resistors. For simplicity we shall illustrate this fact with linear time-invariant cases.

### 5.1   Series Connection of Capacitors

Consider the series connection of capacitors as illustrated by Fig. 5.1. The branch characterization of linear time-invariant capacitors is

$$(5.1) \quad v_k(t) = v_k(0) + \frac{1}{C_k} \int_0^t i_k(t') \, dt'$$

Using KCL at all nodes, we obtain

$$(5.2) \quad i_k(t) = i(t) \qquad k = 1, 2, \ldots, m$$

Using KVL, we have

$$(5.3) \quad v(t) = \sum_{k=1}^m v_k(t)$$

At $t = 0$

$$(5.4) \quad v(0) = \sum_{k=1}^m v_k(0)$$

Combining Eqs. (5.1) to (5.4), we obtain

$$(5.5) \quad v(t) = v(0) + \sum_{k=1}^m \frac{1}{C_k} \int_0^t i(t') \, dt'$$

Therefore the equivalent capacitor is given by

$$(5.6) \quad \frac{1}{C} = \sum_{k=1}^m \frac{1}{C_k}$$

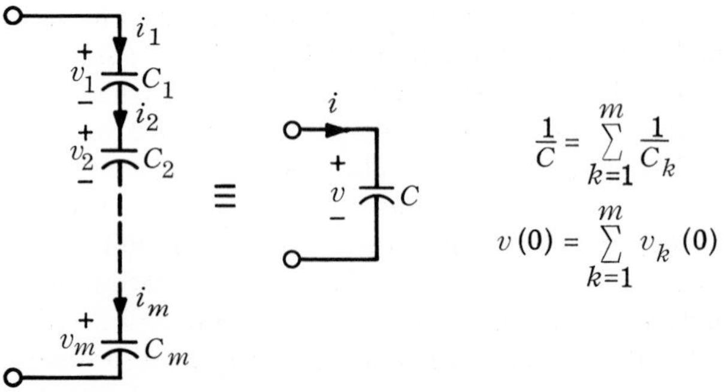

**Fig. 5.1**   Series connection of linear capacitors.

We state, therefore, that *the series connection of m linear time-invariant capacitors, each with value $C_k$ and initial voltage $v_k(0)$, is equivalent to a single linear time-invariant capacitor with value C, which is given by Eq. (5.6), and initial voltage*

(5.7)    $v(0) = \sum\limits_{k=1}^{m} v_k(0)$

If we use elastance instead of capacitance, that is, $S_k = 1/C_k$, then Eq. (5.6) becomes

(5.8)    $S = \sum\limits_{k=1}^{m} S_k$

which means that the elastance of the linear time-invariant capacitor, which is equivalent to the series connection of $m$ linear time-invariant capacitors with elastances $S_k$, $k = 1, 2, \ldots, m$, is equal to the sum of the $m$ elastances. Thus, elastance serves the same role for the capacitor as resistance does for the resistor.

**Exercise**    Compute the total energy stored in the capacitors for the series connection and compare it with the energy stored in the equivalent capacitor.

### 5.2    Parallel Connection of Capacitors

For the parallel connection of $m$ capacitors we must assume that all capacitors have the same initial voltages, for otherwise KVL is violated at $t = 0$. It is easy to show that for the parallel connection of $m$ linear time-invariant capacitors with the same initial voltage $v_k(0)$, the equivalent capacitor is equal to

(5.9)    $C = \sum\limits_{k=1}^{m} C_k$

and

(5.10)    $v(0) = v_k(0)$

This is shown in Fig. 5.2.

$$v(0) = v_k(0) \quad k = 1, 2, 3, \ldots, m$$

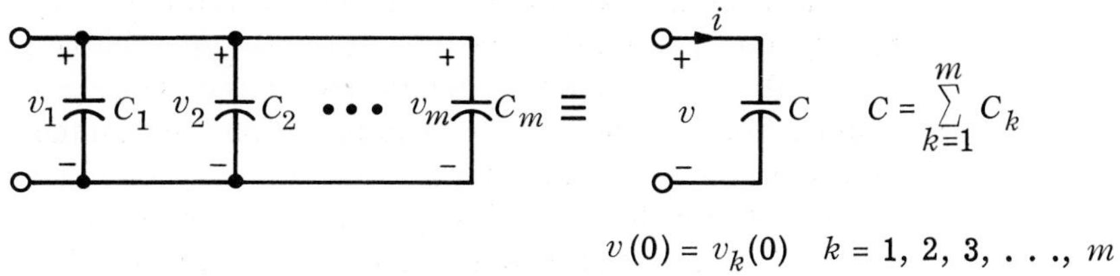

**Fig. 5.2**    Parallel connection of linear capacitors.

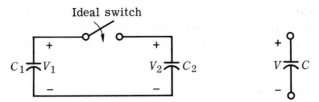

**Fig. 5.3**   The parallel connection of two capacitors with different voltages.

**Example**   Let us consider the parallel connection of two linear time-invariant capacitors with different voltages. In Fig. 5.3, capacitor 1 has capacitance $C_1$ and voltage $V_1$, and capacitor 2 has capacitance $C_2$ and voltage $V_2$. At $t = 0$, the switch is closed so that the two capacitors are connected in parallel. What can we say about the voltage across the parallel connection right after the closing of the switch?

First, from (5.9) we know that the parallel connection has an equivalent capacitance

(5.11)   $C = C_1 + C_2$

At $t = 0-$ (immediately before the closure of the switch) the charge stored in the two capacitors is

(5.12)   $Q(0-) = Q_1(0-) + Q_2(0-)$
$\qquad\qquad = C_1V_1 + C_2V_2$

Since it is a fundamental principle of physics that electric charge is conserved, at $t = 0+$ (immediately after the closure of the switch)

(5.13)   $Q(0+) = Q(0-)$

From (5.11) through (5.13) we can derive the new voltage across the parallel connection of the capacitors. Let the new voltage be $V$; then

$CV = C_1V_1 + C_2V_2$

or

(5.14)   $V = \dfrac{C_1V_1 + C_2V_2}{C_1 + C_2}$

Physically, we can explain the phenomenon as follows: Assume $V_1$ is larger than $V_2$ and $C_1$ is equal to $C_2$; thus, at $t = 0-$, the charge $Q_1(0-)$ is bigger than $Q_2(0-)$. At the time when the switch is closed, $t = 0$, some charge is dumped from the first capacitor to the second instantaneously. This implies that an impulse of current flows from capacitor 1 to capacitor 2 at $t = 0$. As a result, at $t = 0+$, the voltages across the two capacitors are equalized to the intermediate value $V$ required by the conservation of charge.

This phenomenon is analogous to the collision of two particles of different mass $m_1$ and $m_2$ with velocities $v_1$ and $v_2$, respectively. Before the collision, the momentum is $m_1 v_1 + m_2 v_2$; after the collision the momentum is $(m_1 + m_2)v$. Since momentum is conserved, the velocity $v$ after collision is given by

$$v = \frac{m_1 v_1 + m_2 v_2}{m_1 + m_2}$$

This equation is analogous to Eq. (5.14).

**Exercise** Compute the total energy stored in the capacitors before and after the close of the switch. If the two energy values are not the same, where did the energy difference go? This question will become clear after we study Chap. 4.

### 5.3   Series Connection of Inductors

The series connection of $m$ linear time-invariant inductors is shown in Fig. 5.4. Let the inductors be specified by

(5.15)   $\quad v_k = L_k \dfrac{d}{dt} i_k \qquad k = 1, 2, \ldots, m$

and let the initial currents be $i_k(0)$. Using KCL at all nodes, we have

(5.16)   $\quad i = i_k \qquad k = 1, 2, \ldots, m$

Thus, at $t = 0$, $i(0) = i_k(0)$, $k = 1, 2, \ldots, m$. KCL requires that in the series connection of $m$ inductors all the initial value of the currents through the inductors must be the same. Using KVL, we obtain

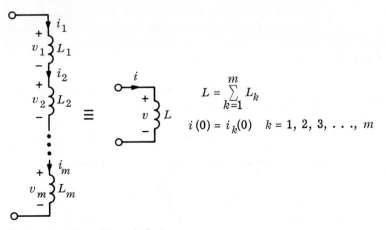

Fig. 5.4   Series connection of linear inductors.

(5.17)   $v = \sum_{k=1}^{m} v_k$

Combining Eqs. (5.15) to (5.17), we have

(5.18)   $v = \sum_{k=1}^{m} L_k \dfrac{di}{dt}$

Therefore, the equivalent inductance is given by

(5.19)   $L = \sum_{k=1}^{m} L_k$

We conclude then that the *series connection of m linear time-invariant inductors, each with inductance $L_k$ and initial current $i(0)$, is equivalent to a single inductor of inductance $L = \sum_{k=1}^{m} L_k$ with the same initial current $i(0)$.*

## 5.4   Parallel Connection of Inductors

We can similarly derive the parallel connection of linear time-invariant inductors shown in Fig. 5.5. The result is simply expressed by the following equations:

(5.20)   $\dfrac{1}{L} = \sum_{k=1}^{m} \dfrac{1}{L_k}$

and

(5.21)   $i(0) = \sum_{k=1}^{m} i_k(0)$

**Remarks**   1.   If we define reciprocal inductance $\Gamma_k \triangleq 1/L_k, k = 1, 2, \ldots, m$, (5.20) states that the equivalent reciprocal inductance $\Gamma$ of the parallel connection of $m$ linear time-invariant inductors each with reciprocal inductance $\Gamma_k$ is the sum of the $m$ reciprocal inductances. Thus,

(5.22)   $\Gamma = \sum_{k=1}^{m} \Gamma_k$

The reciprocal inductance thus plays the same role for an inductor as conductance does for a resistor.

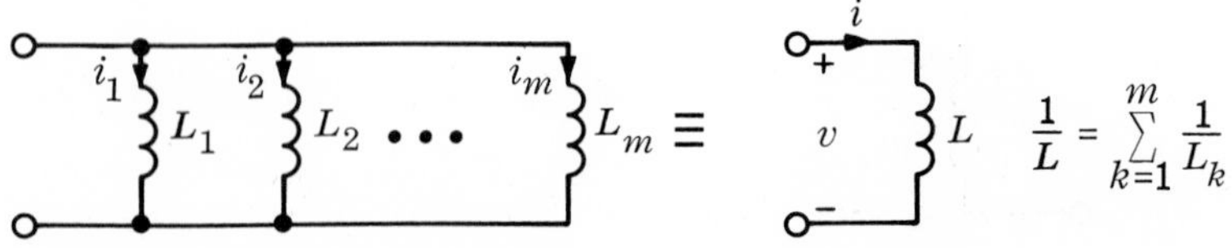

**Fig. 5.5**   Parallel connection of linear inductors.

2.  For the connection of inductors, the analog of conservation of charge is the conservation of flux. For linear time-invariant inductors the total flux in $m$ inductors is

(5.23)     $$\phi = \sum_{k=1}^{m} L_k I_k$$

where $L_k$ and $I_k$ are, respectively, the inductance and the instantaneous current of the $k$th inductor. *Conservation of flux* states that irrespective of the manner in which the connection is made for the $m$ inductors, the total flux $\phi$ remains fixed.

## Summary

- In a series connection of elements, the current in all elements is the same. The voltage across the series connection is the sum of the voltages across each individual element.

- In a parallel connection of elements, the voltage across all elements is the same. The current through the parallel connection is the sum of the currents through each individual element.

- Table 3.1 summarizes the formulas for series and parallel connections of linear resistors, capacitors, and inductors.

**Table 3.1**   **Series and Parallel Connection of Linear Elements**

| Type of elements | Series connection of $m$ elements | Parallel connection of $m$ elements |
|---|---|---|
| Resistors $\quad R$ = resistance $\quad G$ = conductance | $R = \sum_{k=1}^{m} R_k$ | $G = \sum_{k=1}^{m} G_k$ |
| Capacitors $\quad C$ = capacitance $\quad S$ = elastance | $S = \sum_{k=1}^{m} S_k$ | $C = \sum_{k=1}^{m} C_k$ |
| Inductors $\quad L$ = inductance $\quad \Gamma$ = reciprocal $\qquad$ inductance | $L = \sum_{k=1}^{m} L_k$ | $\Gamma = \sum_{k=1}^{m} \Gamma_k$ |

## Problems

Series-parallel connection of linear resistors

**1.** The ladder circuit shown in Fig. P3.1 contains linear resistors with resistances as specified on the figure. What is the resistance of the one-port seen at terminals ① and ①'?

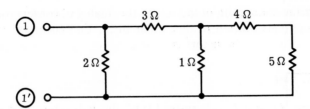

**Fig. P3.1**

**2.** A constant voltage source of 10 volts is applied to the one-port of Fig. P3.1.   Determine all the branch currents.

**3.** For the circuit shown in Fig. P3.3

*a.*   Determine the characteristic of the one-port ① ①′, that is, the equation describing the one-port in terms of the port voltage and the port current.

*b.*   Plot the characteristic in the *vi* plane;

*c.*   Draw the Thévenin equivalent circuit.

*d.*   Draw the Norton equivalent circuit.

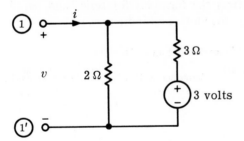

**Fig. P3.3**

**4.** For the circuit shown in Fig. P3.4, repeat (*a*), (*b*), (*c*), and (*d*) of Prob. 3.

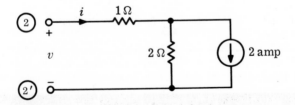

**Fig. P3.4**

**5.** If the two one-ports in Fig. P3.3 and P3.4 are connected back to back as shown in Fig. 3.5, what is the resulting voltage *v*?   If the terminal ②′ is connected to terminal ① and the terminal ①′ is connected to terminal ②, what is the voltage $v$?

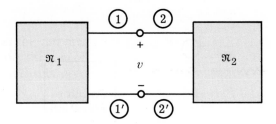

**Fig. P3.5**

Resistor-
source-diode
circuit

**6.** Describe analytically and graphically the $vi$ characteristic of the circuit shown in Fig. P3.6, where $D$ is an ideal diode.

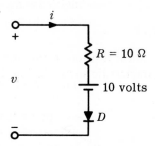

**Fig. P3.6**

Diode circuit

**7.** Suppose that the connection of the diode $D$ in Fig. P3.6 is reversed. Describe analytically and graphically the characteristic of the new circuit.

Synthesis of
resistive
circuit

**8.** Find a circuit which is formed by the parallel connection of a resistor, an ideal diode, and a current source, given that it must have the $vi$ character-istic shown in Fig. P3.8.

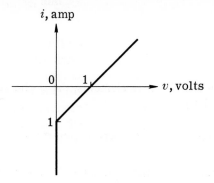

**Fig. P3.8**

Solution of resistive circuit

**9.** Fig. P3.9 shows the circuit of Prob. 8 connected to the series connection of a constant voltage source of 2 volts and a resistor of 2 ohms. Determine the current through the voltage source and the power delivered to the circuit.

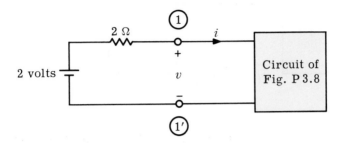

**Fig. P3.9**

Synthesis of resistive circuit

**10.** Design a resistive one-port with linear resistors, ideal diodes, and independent sources that has the *vi* characteristic shown in Fig. P3.10.

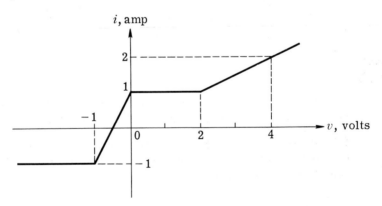

**Fig. P3.10**

Series and parallel connection of nonlinear resistors

**11.** Suppose we are given two resistive elements, the *vi* characteristics for which are shown in Fig. P3.11.

*a.* Find the *vi* characteristic of the series connection of these two elements.

*b.* Find the *vi* characteristic of the parallel connection of these two elements.

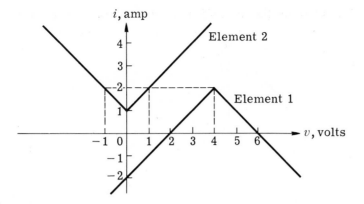

**Fig. P3.11**

**12.** In the circuit shown in Fig. P3.12, the germanium diode has a $vi$ characteristic as follows:

$$i = I_s(\epsilon^{qv/kT} - 1) \qquad I_s = 0.05 \text{ mA}; \; kT/q \approx 0.026 \text{ volt}$$

The signal source $v_1$ is a sinusoid

$$v_1 = 10^{-3} \sin 2\pi 60t \text{ volts}$$

Determine the small-signal equivalent circuits for biasing voltages $V_0 = 0.1, 0,$ and $-0.1$ volts, respectively.

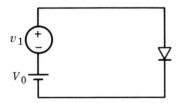

**Fig. P3.12**

**13.** For the circuit shown in Fig. P3.13, determine the currents in all the resistors. (Hint: Can you find the solution of this circuit by using symmetry?)

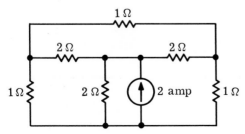

**Fig. P3.13**

Symmetrical
circuit

**14.** Determine the current $i$ in the circuit shown in Fig. P3.14.

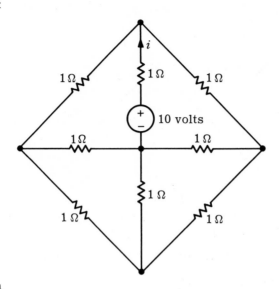

**Fig. P3.14**

Solution of
nonlinear
resistive circuit

**15.** The circuit shown in Fig. P3.15 contains two nonlinear resistors and a current source. The characteristics of the two resistors are given in the figure. Determine the voltage $v$ for

*a.*  $i_s = 1$ amp

*b.*  $i_s = 10$ amp

*c.*  $i_s = 2 \cos t$ amp

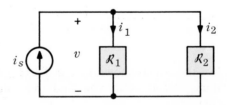

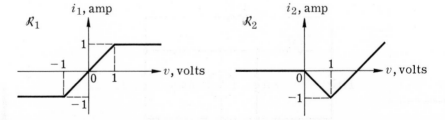

**Fig. P3.15**

Solution of
nonlinear
resistive
circuit

**16.** Replace the current source $i_s$ in the circuit shown in Fig. P3.15 by the series connection of a voltage source $v_s$ and a linear resistor with a resistance of 2 ohms. Determine the voltage $v$ for

*a.* $v_s = 1$ volt

*b.* $v_s = 10$ volts

*c.* $v_s = 2 \cos t$ volts

Switching in
of capacitors

**17.** We are given three isolated linear time-invariant capacitors with capacitances 1, 2, and 3 farads and initial voltages 1, 2, and 3 volts, respectively. The three capacitors are simultaneously connected in parallel by means of instantaneous switching. What is the resulting voltage across the parallel connection? Calculate the electric energy stored in the capacitors before and after the connection.

Switching in of
inductors

**18.** Two linear inductors with inductances 1 and 2 henrys and currents 2 and 1 amp, respectively, are switched to a series connection. What is the resulting current? Calculate the magnetic energy stored in the inductors before and after the connection.

Connection of
nonlinear
inductors

**19.** The characteristics of two nonlinear inductors are specified by the corresponding $\phi i$ curve as shown in Fig. P3.19. At $t = 0$, no current flows through the inductors. Plot the characteristic of the connections shown in the figure.

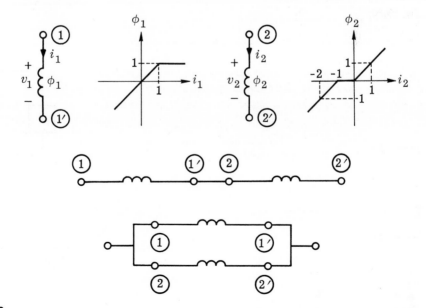

**Fig. P3.19**

**20.** The characteristics of two nonlinear capacitors are specified by the corresponding $qv$ curve as shown in Fig. P3.20. At $t = 0$, the charge on each capacitor is zero. Plot the characteristics of the connections shown in the figure. Determine the energy stored in each connection when the voltage across the connections is $-1$, 1, and 2 volts.

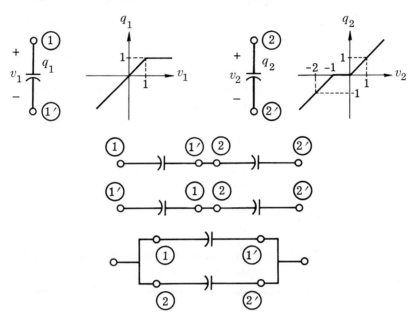

**Fig. P3.20**

# 4 First-order Circuits

In the previous two chapters we studied in detail the three basic types of circuit element and analyzed some simple circuits. We considered series and parallel connections of circuit elements of the same kind. We showed in examples how to obtain equivalent one-ports and how to find their solutions. In these examples we used both graphical and analytical approaches. In either approach we needed only algebraic operations; differential equations are not involved in the solution of circuits with one kind of element, no matter how complicated these circuits may be.

In this chapter we shall analyze circuits with more than one kind of element; as a consequence, we shall have to use differentiation and/or integration. We shall restrict ourselves to circuits that can be described by first-order differential equations; hence, we give them the name *first-order circuits*. As a starting point we shall analyze a circuit that contains a linear time-invariant resistor and capacitor. We shall use this simple example throughout the chapter to introduce some basic facts concerning linear time-invariant circuits and systems. First, we shall present the concepts of the zero-input response, the zero-state response, and the complete response, along with a review of solutions of linear nonhomogeneous differential equations. We shall then review the step and impulse functions and show how to find the step and impulse responses. The treatment of higher-order circuits, that is, circuits described by a higher-order differential equation, will be covered in later chapters. Finally, simple nonlinear and time-varying first-order circuits will be treated briefly at the end of this chapter. Our purpose is mainly to introduce simple but useful techniques that are effective for solving circuits with time-varying and nonlinear elements, and thereby to point out the differences between these circuits and circuits containing linear and time-invariant elements.

To simplify a number of descriptions in the remainder of this book, we adopt the following terminology. A lumped circuit is said to be **linear** if each of its elements is either a linear element or an independent source. Similarly, a lumped circuit is said to be **time-invariant** if each of its elements is either a time-invariant element or an independent source. Thus, the elements of a *linear time-invariant circuit* are either linear time-invariant elements or independent sources. Similarly, a circuit containing one or more nonlinear elements that are not independent sources is called **nonlinear**. A circuit containing one or more time-varying elements that are not independent sources is called **time-varying.** The reason why independent sources are considered separately will become clear later on.

# 1    Linear Time-invariant First-order Circuit, Zero-input Response

## 1.1    The $RC$ (Resistor-Capacitor) Circuit

In the circuit of Fig. 1.1, the linear time-invariant capacitor with capacitance $C$ is charged to a potential $V_0$ by a constant voltage source. At $t = 0$ the switch $k_1$ is opened, and switch $k_2$ is closed simultaneously. Thus, the charged capacitor is disconnected from the source and connected to the linear time-invariant resistor with resistance $R$ at $t = 0$. Let us describe physically what is going to happen. Because of the charge stored in the capacitor ($Q_0 = CV_0$) a current will flow in the direction specified by the reference direction assigned to $i(t)$, as shown in Fig. 1.1. The charge across the capacitor will decrease gradually and eventually will become zero; the current $i$ will do the same. During the process the electric energy stored in the capacitor is dissipated as heat in the resistor.

Let us use our knowledge of circuit theory to analyze this problem. Restricting our attention to $t \geq 0$, we redraw the $RC$ circuit as shown in Fig. 1.2. Note that the reference directions for branch voltages and branch currents are clearly indicated. $V_0$, along with the positive and negative signs next to the capacitor, specifies the magnitude and polarity of the initial voltage. Kirchhoff's laws and topology (the parallel connection of $R$ and $C$) dictate the following equations:

(1.1)    KVL:    $\qquad v_C(t) = v_R(t) \qquad t \geq 0$

(1.2)    KCL:    $\quad i_C(t) + i_R(t) = 0 \qquad t \geq 0$

The two branch equations for the two circuit elements are

(1.3)    Resistor:    $\quad v_R = Ri_R$

(1.4a)    Capacitor:    $\quad i_C = C\dfrac{dv_C}{dt} \qquad$ and $\qquad v_C(0) = V_0$

or, equivalently,

(1.4b)    $v_C(t) = V_0 + \dfrac{1}{C}\displaystyle\int_0^t i_C(t')\, dt'$

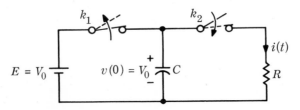

**Fig. 1.1**    A charged capacitor is connected to a resistor ($k_1$ opens and $k_2$ closes at $t = 0$).

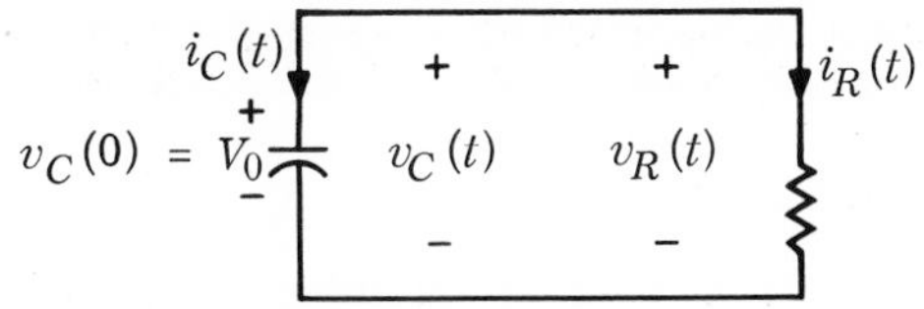

**Fig. 1.2**   An $RC$ circuit, $v_C(0) = V_0$.

In Eq. (1.4a) we want to emphasize that the initial condition of the capacitor voltage must be written together with $i_C = dv_C/dt$; otherwise, the state of the capacitor is not completely specified. This is made obvious by the alternate branch equation (1.4b).

We have four equations for the four unknowns in the circuit, namely, the two branch voltages $v_C$ and $v_R$ and the two branch currents $i_C$ and $i_R$. A complete mathematical description of the circuit has been given, and we can go on to solve for any or all of the unknown variables. Suppose we wish to find the voltage across the capacitor. Combining Eqs. (1.1) to (1.4a), we obtain, for $t \geq 0$,

$$C\frac{dv_C}{dt} = i_C = -i_R = -\frac{v_R}{R} = -\frac{v_C}{R} \quad \text{and} \quad v_C(0) = V_0$$

or

(1.5)   $$C\frac{dv_C}{dt} + \frac{v_C}{R} = 0 \quad t \geq 0 \quad \text{and} \quad v_C(0) = V_0$$

This is a first-order linear homogeneous differential equation with constant coefficients. Its solution is of the exponential form

(1.6)   $$v_C(t) = K\epsilon^{s_0 t}$$

where

(1.7)   $$s_0 = -\frac{1}{RC}$$

This is easily verified by direct substitution of Eqs. (1.6) and (1.7) in the differential equation (1.5). In (1.6) $K$ is a constant to be determined from the initial condition. Setting $t = 0$ in Eq. (1.6), we obtain $v_C(0) = K = V_0$. Therefore, the solution to the problem is given by

(1.8)   $$v_C(t) = V_0\epsilon^{-(1/RC)t} \quad t \geq 0$$

It is important to note that in Eq. (1.8), $v_C(t)$ is specified for $t \geq 0$ since for negative $t$ the voltage across the capacitor is a constant, according to our original physical specification. Yet Eq. (1.8), without the qualification of $t \geq 0$, gives an exponential expression even for negative $t$. The voltage $v_C$ is plotted in Fig. 1.3 as a function of time. Of course, we can immedi-

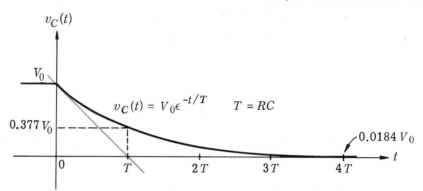

**Fig. 1.3**   The discharge of the capacitor of Fig. 1.2 is given by an exponential curve.

ately obtain the other three branch variables once $v_C$ is known.  From Eq. (1.4a) we have

(1.9)   $i_C(t) = C\dfrac{dv_C}{dt} = -\dfrac{V_0}{R}\epsilon^{-(1/RC)t} \qquad t \geq 0$

From Eq. (1.2) we have

(1.10)   $i_R(t) = -i_C(t) = \dfrac{V_0}{R}\epsilon^{-(1/RC)t} \qquad t \geq 0$

From Eq. (1.3) we have

(1.11)   $v_R(t) = v_C(t) = V_0\epsilon^{-(1/RC)t} \qquad t \geq 0$

These curves are plotted in Fig. 1.4.

**Exercise**   Show that the light line in Fig. 1.3, which is tangent to the curve $v_C$ at $t = 0+$, intersects the time axis at the abscissa $T$.

Let us study the waveform $v_C(\cdot)$ more carefully.  We say that the voltage across the capacitor decreases exponentially with time, as shown in Fig. 1.3.  Since exponential curves and simple $RC$ circuits occur very often in the everyday life of electrical engineers, it is important to know their properties precisely.  An exponential curve can be characterized by two numbers, namely, the ordinate of the curve at a reference time, say $t = 0$, and the **time constant** $T$, which is defined by $f(t) = f(0)\epsilon^{-t/T}$.  In the curve of Fig. 1.3 we have $f(0) = V_0$ and $T = RC$.  It is convenient to remember some simple facts about the exponential curve.  Assume that $V_0 = 1$; that is, $v_C(0) = 1$; at $t = T$, $v_C(T) = \epsilon^{-1} \approx 0.377$, and at $t = 4T$, $v_C(4T) = \epsilon^{-4} \approx 0.0184$.  Thus, at a time equal to the time constant, the exponential curve reaches approximately 38 percent of the starting value, and at four times the time constant, the exponential curve reaches approximately 2 percent of the starting value.

**Remark**   The term $s_0 = -1/T = -1/RC$ in Eqs. (1.6) and (1.7) has a dimension of reciprocal time or frequency and is measured in radians per second. It is called the *natural frequency* of the circuit. The concept of "natural frequency" is of great significance in linear time-invariant circuits, as will be illustrated in later chapters.

**Exercise**   Recall that the unit of capacitance is the farad and the unit of resistance is the ohm. Show that the unit of $T = RC$ is the second.

In circuit analysis we are almost always interested in the behavior of a particular network variable called the **response** (it is sometimes called the output). Recall that a network variable is either a branch voltage, a branch current, or a linear combination of branch voltages and branch currents. A network variable can also be a charge on a capacitor or a flux in an inductor. In the present example, the response could be any of the curves in Figs. 1.3 and 1.4. Usually the responses are due to either independent

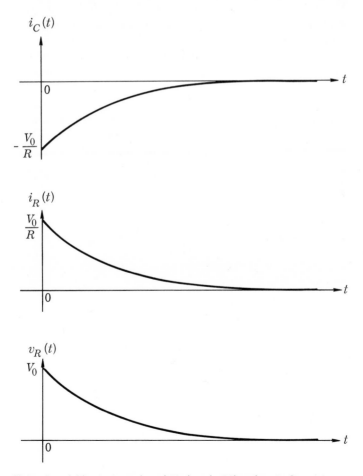

**Fig. 1.4**   Network variables $i_C$, $i_R$, and $v_R$ plotted against time for $t \geq 0$.

sources that we consider as inputs or to the initial condition, or to both. In the present example, there is no input, and the response is due to the initial voltage of the capacitor; therefore, we call this response the zero-input response. In general we give the name **zero-input response** to the response of a circuit with no applied input. This zero-input response depends on the initial condition and the characteristics of the circuit. The zero-input response of the simple $RC$ circuit is an exponential curve; it is completely specified by the natural frequency $s_0 = -1/RC$ and the initial voltage $V_0$.

## 1.2 The *RL* (Resistor-Inductor) Circuit

The other typical first-order circuit is the $RL$ circuit. We shall study its zero-input response. As shown in Fig. 1.5 for $t < 0$, switch $k_1$ is on terminal $B$, $k_2$ is open, and the linear time-invariant inductor with inductance $L$ is supplied with a constant current $I_0$. At $t = 0$ switch $k_1$ is flipped to terminal $C$, and $k_2$ is closed. Thus, for $t \geq 0$ the inductor with initial current $I_0$ is connected to a linear time-invariant resistor with resistance $R$. The energy stored in the magnetic field as a result of $I_0$ in the inductance decreases gradually and dissipates in the resistor in the form of heat. The current in the $RL$ loop decreases monotonically and eventually tends to zero.

We can similarly analyze this circuit by writing Kirchhoff's laws and the branch equations. For this purpose we redraw the circuit for $t \geq 0$, as shown in Fig. 1.6. Note that the reference directions of all branch voltages and branch currents are clearly indicated. KCL says $i_R = -i_L$, and KVL states $v_L - v_R = 0$. Using the branch equations for both elements, that is, $v_L = L(di_L/dt)$, $i_L(0) = I_0$, and $v_R = Ri_R$, we obtain the following differential equation in terms of the current $i_L$:

(1.12) $$L\frac{di_L}{dt} + Ri_L = 0 \qquad t \geq 0 \qquad i_L(0) = I_0$$

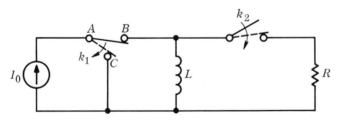

**Fig. 1.5** For $t < 0$, switch $k_1$ connects terminal $A$ to terminal $B$, and $k_2$ is open; therefore, for $t < 0$, the current $I_0$ goes through the inductor $L$. At $t = 0$, switch $k_1$ is flipped to $C$, and switch $k_2$ closes; the current source is then shorted on itself and the inductor current must now go through the resistor $R$.

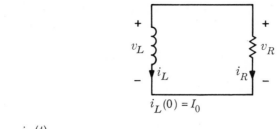

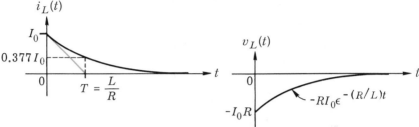

**Fig. 1.6**    An $RL$ circuit with $i_L(0) = I_0$ and the waveforms for $t \geq 0$.

This is a first-order linear homogeneous differential equation with constant coefficients; it has precisely the same form as the previous Eq. (1.5). Therefore the solution is the same except for notation:

(1.13)    $i_L(t) = I_0\epsilon^{-(R/L)t} \qquad t \geq 0$

where $L/R = T$ is the time constant and $s_0 \triangleq -R/L$ is the natural frequency. The current $i_L$ and the voltage $v_L$ are plotted in Fig. 1.6.

**1.3    The Zero-input Response as a Function of the Initial State**

For the $RC$ circuit and the $RL$ circuit considered above, the zero-input responses are, respectively,

(1.14)    $v(t) = V_0\epsilon^{-t/RC} \qquad i(t) = I_0\epsilon^{-(R/L)t} \qquad t \geq 0$

The initial conditions are specified by $V_0$ and $I_0$, respectively. The numbers $V_0$ and $I_0$ are also called the *initial state* of the $RC$ circuit and of the $RL$ circuit, respectively. Now if we consider the way in which the waveform of the zero-input response depends on the initial state, we reach the following conclusion:

*For first-order linear time-invariant circuits, the zero-input response considered as a waveform defined for $0 \leq t < \infty$ is a linear function of the initial state.*

Let us prove this statement by considering the $RC$ circuit. We wish to show that the waveform $v(\,\cdot\,)$ in Eq. (1.14) is a linear function of the initial state $V_0$. It is necessary to check the requirements of homogeneity and additivity for the function (see Appendix A, Sec. 2.3). Homogeneity is obvious; if the initial state is multiplied by a constant $k$, Eq. (1.14) shows that the whole waveform is multiplied by $k$. Additivity is just as simple. The zero-input response corresponding to the initial state $V_0'$ is

$$v'(t) = V_0' \epsilon^{-t/RC} \qquad t \geq 0$$

and the zero-input response corresponding to some other initial state $V_0''$ is

$$v''(t) = V_0'' \epsilon^{-t/RC} \qquad t \geq 0$$

Then the zero-input response corresponding to the initial state $V_0' + V_0''$ is

$$(V_0' + V_0'')\epsilon^{-t/RC} \qquad t \geq 0$$

This waveform is the sum of the two preceding waveforms. Hence, additivity holds. Since the dependence of the zero-input response on the initial state satisfies the requirements of homogeneity and additivity, the dependence is a linear function.

**Remark**   This property does not hold in the case of nonlinear circuits. Consider the $RC$ circuit shown in Fig. 1.7a. The capacitor is linear and time-invariant and has a capacitance of 1 farad, and the resistor is nonlinear with a characteristic

$$i_R = v_R{}^3$$

The two elements have the same branch voltage $v$, and expressing the branch currents in terms of $v$, we obtain from KCL

$$C\frac{dv}{dt} + i_R = \frac{dv}{dt} + v^3 = 0 \qquad v(0) = V_0$$

Hence

$$\frac{dv}{v^3} = -dt$$

If we integrate between 0 and $t$, the voltage takes the initial value $V_0$ and the final value $v(t)$; hence

$$-\frac{1}{2[v(t)]^2} + \frac{1}{2V_0{}^2} = -t$$

or

$$(1.15) \qquad v(t) = \frac{V_0}{\sqrt{1 + 2V_0{}^2 t}} \qquad t \geq 0$$

This is the zero-input response of this nonlinear $RC$ circuit starting from the initial state $V_0$ at time 0. The waveforms corresponding to $V_0 = 0.5$ and $V_0 = 2$ are plotted in Fig. 1.7b. It is obvious that the top curve (for $V_0 = 2$) cannot be obtained from the lower one (for $V_0 = 0.5$) by multi-

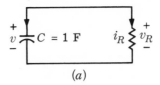

(a)

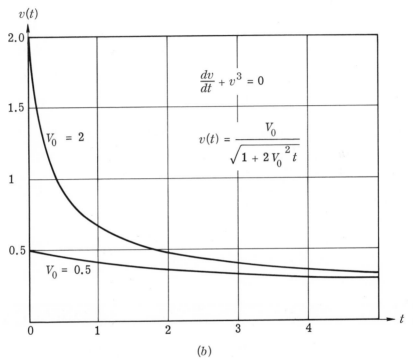

(b)

**Fig. 1.7**    Nonlinear $RC$ circuit and two of its zero-input responses.  The capacitor is linear with capacitance $C = 1$ farad, and the resistor characteristic is $i_R = v_R^3$.

plying its ordinates by 4.   Clearly the zero-input response is not a linear function of the initial state.   From an experimental point of view, this is very important.   Suppose we have in our laboratory notebook the scope picture of the zero-input response of a first-order circuit, say, for $V_0 = 1$. If the circuit is linear, the ordinates of the zero-input response for any initial state, say, $V_0 = k$, are simply $k$ times the ordinates of the recorded curve.   In the nonlinear case we have to go back to the laboratory or solve again the differential equation for the initial condition $V_0 = k$.

<table>
<tr><td>**1.4**</td><td>**Mechanical Example**</td></tr>
</table>

Let us consider a familiar mechanical system that has a behavior similar to that of the linear time-invariant $RC$ and $RL$ circuits above.   Figure 1.8

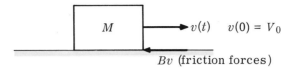

**Fig. 1.8**   A mechanical system which is described by a first-order differential equation.

shows a block of mass $M$ moving at an initial velocity $V_0$ at $t = 0$. As time proceeds, the block will slow down gradually because friction tends to oppose the motion. Friction is represented by friction forces that are always in the direction opposite to the velocity $v$, as shown in the figure. Let us assume that these forces are proportional to the magnitude of the velocity; thus, $f = Bv$, where the constant $B$ is called the damping coefficient. From Newton's second law of motion we have, for $t \geq 0$,

$$(1.16) \quad M\frac{dv}{dt} = -Bv \qquad v(0) = V_0$$

Therefore,

$$(1.17) \quad v(t) = V_0 \epsilon^{-(B/M)t} \qquad t \geq 0$$

where $M/B$ represents the time constant for the mechanical system and $-B/M$ is the natural frequency.

## 2   Zero-state Response

### 2.1   Constant Current Input

In the circuit of Fig. 2.1 a current source $i_s$ is switched to a parallel linear time-invariant $RC$ circuit. For simplicity we consider first the case where the current $i_s$ is constant and equal to $I$. Prior to the opening of the switch, the current source produces a circulating current in the short circuit. At $t = 0$, the switch is opened and thus the current source is connected to the $RC$ circuit. From KVL we see that the voltage across all three elements is the same. Let us designate this voltage by $v$ and assume that $v$ is the response of interest. Writing the KCL equation in terms of $v$, we obtain the following network equation:

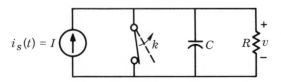

**Fig. 2.1**   $RC$ circuit with a current-source input. At $t = 0$, switch $k$ is opened.

(2.1)   $C\dfrac{dv}{dt} + \dfrac{1}{R}v = i_s(t) = I \qquad t \geq 0$

where $I$ is a constant. We assume that the capacitor is initially uncharged. Thus, the initial condition is

(2.2)   $v(0) = 0$

Before we solve Eqs. (2.1) and (2.2), let us figure out what will happen after we open the switch. At $t = 0+$, that is, immediately after the opening of the switch, the voltage across the capacitor remains zero because, as we learned in Chap. 2, the voltage across a capacitor cannot jump abruptly unless there is an infinitely large current. At $t = 0+$, since the voltage is still zero, the current in the resistor must be zero by Ohm's law. Therefore all the current from the source enters the capacitor at $t = 0+$. This implies a rate of increase of the voltage specified by Eq. (2.1); thus

(2.3)   $\left.\dfrac{dv}{dt}\right|_{0+} = \dfrac{I}{C}$

As time proceeds, $v$ increases, and $v/R$, the current through the resistor, increases also. Long after the switch is opened, the capacitor is completely charged, and the voltage is practically constant. Then and thereafter, $dv/dt \approx 0$. All the current from the source goes through the resistor, and the capacitor behaves as an open circuit; that is,

(2.4)   $v \approx RI$

This fact is clear from Eq. (2.1), and it is also shown in Fig. 2.2. The circuit is said to have reached a *steady state*. It only remains to show how the whole change of voltage takes place. For that we rely on the following analytical treatment.

The solution of a linear nonhomogeneous differential equation can be written in the following form:

(2.5)   $v = v_h + v_p$

where $v_h$ is a solution of the homogeneous differential equation and $v_p$ is

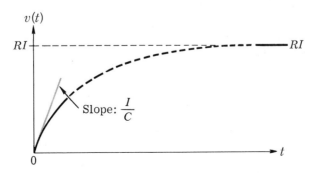

**Fig. 2.2**   Initial and final behavior of the voltage across the capacitor.

any particular solution of the nonhomogeneous differential equation. Of course, $v_p$ depends on the input. For our problem the general solution of the homogeneous equation is of the form

(2.6)   $v_h = K_1 \epsilon^{s_0 t}$     $s_0 = -\dfrac{1}{RC}$

where $K_1$ is any constant. The most convenient particular solution for a constant current input is a constant

(2.7)   $v_p = RI$

since the constant $RI$ satisfies the differential equation (2.1). Substituting (2.6) and (2.7) in (2.5), we obtain the general solution of (2.1):

(2.8)   $v(t) = K_1 \epsilon^{-(1/RC)t} + RI$     $t \geq 0$

where $K_1$ is to be evaluated from the initial condition specified by Eq. (2.2). Setting $t = 0$ in (2.8), we have

$v(0) = K_1 + RI = 0$

Thus,

(2.9)   $K_1 = -RI$

The voltage as a function of time is then

(2.10)   $v(t) = RI(1 - \epsilon^{-(1/RC)t})$     $t \geq 0$

The graph in Fig. 2.3 shows the voltage approaching its steady-state value exponentially. At about four times the time constant, the voltage is within 2 percent of its final value $RI$.

**Exercise 1**   Sketch with appropriate scales the zero-state response of the circuit of Fig. 2.1 with

a.   $I = 200$ mA, $R = 1$ k$\Omega$ ($10^3$ ohms), and $C = 1$ $\mu$F ($10^{-6}$ farad).

b.   $I = 2$ mA, $R = 50$ ohms, and $C = 5$ pF ($10^{-9}$ farad).

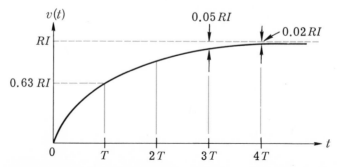

**Fig. 2.3**   Voltage response for the $RC$ circuit due to a constant source $I$ as shown in Fig. 2.1, where $v(0) = 0$.

**Exercise 2**   Discuss the capacitor charging in the circuit of Fig. 2.1 from an energy point of view.   More precisely,

*a.*   Calculate and sketch the waveforms $p_s(\cdot)$ (the power delivered by the source), $p_R(\cdot)$ (the power dissipated by the resistor), and $\mathcal{E}_C(\cdot)$, (the energy stored in the capacitor).

*b.*   Calculate the efficiency of the process, i.e., the ratio of the energy eventually stored in the capacitor to the energy delivered by the source $\left[\text{that is, } \int_0^\infty p_s(t)\,dt\right]$.

| 2.2 | Sinusoidal Input |
|-----|------------------|

We consider now the same circuit but with a different input; the source is now given by a sinusoid

(2.11)   $i_s(t) = A_1 \cos (\omega t + \phi_1) \qquad t \geq 0$

where the constant $A_1$ is called the *amplitude* of the sinusoid and the constant $\omega$ is called the (angular) *frequency*.   The frequency is measured in radians per second.   The constant $\phi_1$ is called the *phase*.   Let us proceed immediately to the solution; the physical interpretation will be given in the next section.   First, the solution of the homogeneous differential equation is of the same form as before [see Eq. (2.6)], since the circuit is the same except for the input.   Thus, we need only find a particular solution for the sinusoidal input.   The most convenient particular solution of a linear differential equation with a constant coefficient for a sinusoidal input is a sinusoid of the same frequency.   Thus, $v_p$ is taken to be of the form

(2.12)   $v_p(t) = A_2 \cos (\omega t + \phi_2)$

where $A_2$ and $\phi_2$ are constants to be determined.   To evaluate them, we substitute (2.12) in the given differential equation, namely,

(2.13)   $C\dfrac{dv_p}{dt} + \dfrac{1}{R}v_p = A_1 \cos (\omega t + \phi_1)$

We obtain

$$-CA_2\omega \sin (\omega t + \phi_2) + \frac{1}{R}A_2 \cos (\omega t + \phi_2)$$

$$= A_1 \cos (\omega t + \phi_1) \qquad \text{for all } t \geq 0$$

Using standard trigonometric identities to express $\sin (\omega t + \phi_2)$, $\cos (\omega t + \phi_2)$, and $\cos (\omega t + \phi_1)$ as a linear combination of $\cos \omega t$ and $\sin \omega t$, and equating separately the coefficients of $\cos \omega t$ and $\sin \omega t$, we obtain the following results:

(2.14)   $A_2 = \dfrac{A_1}{\sqrt{(1/R)^2 + (\omega C)^2}}$

and

(2.15)   $\phi_2 = \phi_1 - \tan^{-1} \omega RC$

Here $\tan^{-1} \omega RC$ denotes the angle between 0 and 90° whose tangent is equal to $\omega RC$. This particular solution and the input current are plotted in Fig. 2.4. A more general and more elegant derivation of this particular solution will be given in Chap. 7.

**Exercise**   Derive Eqs. (2.14) and (2.15) in detail.

The general solution of (2.13) is therefore of the form

(2.16)   $v(t) = K_1 \epsilon^{-(1/RC)t} + A_2 \cos(\omega t + \phi_2)$     $t \geq 0$

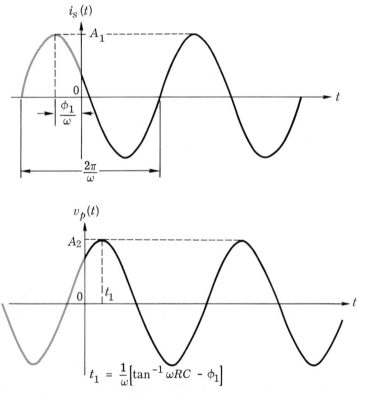

**Fig. 2.4**   Input current and a particular solution for the output voltage of the $RC$ circuit in Fig. 2.1.

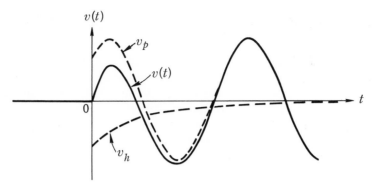

**Fig. 2.5**   Voltage response of the circuit in Fig. 2.1 with $v(0) = 0$ and $i_s(t) = A_1 \cos (\omega t + \phi_1)$.

Setting $t = 0$, we have

(2.17)   $v(0) = K_1 + A_2 \cos \phi_2 = 0$

that is,

(2.18)   $K_1 = -A_2 \cos \phi_2$

Therefore, the response is given by

(2.19)   $v(t) = -A_2 \cos \phi_2 \epsilon^{-(1/RC)t} + A_2 \cos (\omega t + \phi_2) \qquad t \geq 0$

where $A_2$ and $\phi_2$ are defined in Eqs. (2.14) and (2.15). The graph of $v$, that is, the zero-state response to the input $A_1 \cos (\omega t + \phi_1)$, is plotted in Fig. 2.5.

In the two cases treated in this section we considered the voltage $v$ as the response and the current source $i_s$ as the input. The initial condition in the circuit is zero; that is, the voltage across the capacitor is zero before the application of the input. In general, we say that a circuit is in the **zero state** if all the initial conditions in the circuit are zero.† The response of a circuit, which starts from the zero state, is due exclusively to the input. By definition, **the zero-state response** is the response of a circuit to an input applied at some arbitrary time, say, $t_0$, subject to the condition that the circuit be in the zero state just prior to the application of the input (that is, at time $t_0-$). In calculating zero-state responses, our primary interest is the behavior of the response for $t \geq t_0$. For this reason we adopt the following *convention*: The input and the zero-state response are taken to be identically zero for $t < t_0$.

---

† We shall prove in Chap. 13 that a linear time-invariant circuit is in the zero state if the initial voltages across all capacitors and the initial currents through all inductors are zero.

| **3** | **Complete Response: Transient and Steady-state** |

| **3.1** | **Complete Response** |

The response of the circuit to both an input and the initial conditions is called the **complete response** of the circuit. Thus, the zero-input response and the zero-state response are special cases of the complete response. In this section we demonstrate that

*for the simple linear time-invariant RC circuit considered, the complete response is the sum of the zero-input response and the zero-state response.†*

Consider the circuit in Fig. 3.1 where the capacitor is initially charged; that is, $v(0) = V_0 \neq 0$, and a current input is switched into the circuit at $t = 0$. By definition, the complete response is the waveform $v(\cdot)$ caused by both the input $i_s(\cdot)$ and the initial state $V_0$. Mathematically, it is the solution of the equation

(3.1)   $C\dfrac{dv}{dt} + Gv = i_s(t) \qquad t \geq 0$

with

(3.2)   $v(0) = V_0$

where $V_0$ is the initial voltage on the capacitor. Let $v_i$ be the zero-input response; by definition, it is the solution of

$C\dfrac{dv_i}{dt} + Gv_i = 0 \qquad t \geq 0$

with

$v_i(0) = V_0$

Let $v_0$ be the zero-state response; by definition, it is the solution of

† This statement is, in fact, true for any linear circuit (time-varying or time-invariant).

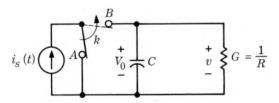

**Fig. 3.1**   *RC circuit with $v(0) = V_0$ is excited by a current source $i_s(t)$. The switch $k$ is flipped from $A$ to $B$ at $t = 0$.*

$$C \frac{dv_0}{dt} + Gv_0 = i_s(t) \qquad t \geq 0$$

with

$$v_0(0) = 0$$

From these four equations we obtain, by addition,

$$C \frac{d}{dt}(v_i + v_0) + G(v_i + v_0) = i_s(t) \qquad t \geq 0$$

and

$$v_i(0) + v_0(0) = V_0$$

However these two equations show that the waveform $v_i(\cdot) + v_0(\cdot)$ satisfies both the required differential equation (3.1) and the initial condition (3.2). Since the solution of a differential equation such as (3.1), subject to initial conditions such as (3.2), is unique, it follows that the complete response $v$ is given by

$$v(t) = v_i(t) + v_0(t) \qquad t \geq 0$$

that is, the complete response $v$ is the sum of the zero-input response $v_i$ and the zero-state response $v_0$.

**Example**   If we assume that the input is a constant current source applied at $t = 0$, that is, $i_s = I$, the complete response of the circuit can be written immediately since we have already calculated the zero-input response and the zero-state response. Thus,

$$v(t) = v_i(t) + v_0(t) \qquad t \geq 0$$

From Eq. (1.8) we have

$$v_i(t) = V_0 \epsilon^{-(1/RC)t} \qquad t \geq 0$$

and from Eq. (2.10) we have

$$v_0(t) = RI(1 - \epsilon^{-(1/RC)t}) \qquad t \geq 0$$

Thus the complete response is

(3.3)   $$\underbrace{v(t)}_{\substack{\text{Complete} \\ \text{response}}} = \underbrace{V_0 \epsilon^{-(1/RC)t}}_{\substack{\text{Zero-input} \\ \text{response } v_i}} + \underbrace{RI(1 - \epsilon^{-(1/RC)t})}_{\substack{\text{Zero-state} \\ \text{response } v_0}} \qquad t \geq 0$$

The responses are shown in Fig. 3.2.

Of course, from a purely computational point of view, the calculation

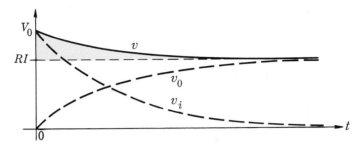

**Fig. 3.2**    Zero-input, zero-state, and complete response of the simple $RC$ circuit. The input is a constant current source $I$ applied at $t = 0$.

of the complete response requires only the solution of a nonhomogeneous differential equation with specified initial conditions; the decomposition into the zero-input response and the zero-state response may not be necessary. On the other hand, taking a physical point of view, it is extremely interesting to note that the complete response is the sum of the zero-state response (due to the input only) and of the zero-input response (due to the initial conditions only). This decomposition is a fundamental result of linear circuit theory and, in fact, of linear system theory.

**Remark**    We shall prove in Chap. 6 that for the linear time-invariant parallel $RC$ circuit the complete response can be explicitly written in the following form for any arbitrary input $i_s$:

$$\underbrace{v(t)}_{\substack{\text{Complete}\\\text{response}}} = \underbrace{V_0 \epsilon^{-t/RC}}_{\substack{\text{Zero-input}\\\text{response}}} + \underbrace{\int_0^t \frac{1}{C} \epsilon^{-(t-t')/RC} i_s(t')\, dt'}_{\substack{\text{Zero-state}\\\text{response}}}$$

**Exercise**    By direct substitution show that the expression for the complete response given in the above remark satisfies Eqs. (3.1) and (3.2).

**3.2    Transient and Steady State**

In the previous example we can also partition the complete response in a different way. The complete response due to the initial state $V_0$ and the constant current input $I$ in Eq. (3.3) is rewritten as follows:

(3.4)    $$\underbrace{v(t)}_{\substack{\text{Complete}\\\text{response}}} = \underbrace{(V_0 - RI)\epsilon^{-(1/RC)t}}_{\text{Transient}} + \underbrace{RI}_{\substack{\text{Steady}\\\text{state}}} \qquad t \geq 0$$

The first term is a decaying exponential as represented by the shaded area, i.e., the difference of the waveform $v(\cdot)$ and the constant $RI$ in Fig. 3.2.

For very large $t$, the first term is negligible, and the second term dominates. For this reason we call the first term the *transient* and the second term the *steady state*. In this example it is evident that transient is contributed by both the zero-input response and the zero-state response, whereas the steady state is contributed only by the zero-state response. Physically, the transient is a result of two causes, namely, the initial conditions in the circuit and the sudden application of the input. If the circuit is well behaved as time goes on, the transient eventually dies out. The steady state is a result of only the input and has a waveform closely related to that of the input. For example, if the input is a constant, the steady-state response is also a constant; if the input is a sinusoid of angular frequency $\omega$, the steady-state response is also a sinusoid of the same frequency. In the example of Sec. 2.2 the input is $i_s = A_1 \cos(\omega t + \phi_1)$; the response [as seen in Eq. (2.19)] has a steady-state portion $A_2 \cos(\omega t + \phi_2)$ and a transient portion $-A_2 \cos \phi_2 \epsilon^{-(1/RC)t}$. A more thorough discussion of the transient and the steady state will be given in Chap. 7.

**Exercise**   The circuit shown in Fig. 3.3 contains a 1-farad linear capacitor and a linear resistor with a negative resistance. When the current source is applied, it is in the zero state at time $t = 0$, so that for $t \geq 0, i_s = I_m \cos \omega t$ (where $I_m$ and $\omega$ are constants). Calculate and sketch the response $v$. Is there a sinusoidal steady state? Explain.

**Remark**   It is interesting to note that with sinusoidal input it is sometimes possible to eliminate the transient completely by choosing a particular time for the application of the input. We shall illustrate this fact with the same example as appeared in Sec. 2.2. Recall the problem was to find the zero-state response of an $RC$ circuit to the input current $A_1 \cos(\omega t + \phi_1)$. The solution of the problem was expressed in Eq. (2.16) in terms of a constant $K_1$, which was to be evaluated from the given initial condition. Clearly, if $K_1$ were zero, there would be no transient, and $v$ in Eq. (2.16) would be a pure sinusoid. In Eq. (2.17) it is seen that $K_1$ depends on the initial voltage across the capacitor as well as on the value of the input waveform at $t = 0$; in fact, $K_1 = 0$ if and only if $\phi_2 = \pm 90°$. Physically this means that the zero-state response contains no transient if at $t = 0$ the steady-state voltage across the capacitor $A_2 \cos \phi_2$ is equal to the initial voltage across the

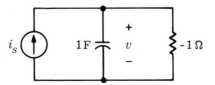

**Fig. 3.3**   Exercise on steady state.   Note the circuit contains a resistor with *negative* resistance.

capacitor $v(0)$. Equation (2.15) implies that for $\phi_2$ to be $\pm 90°$, the input phase $\phi_1$ must be picked to be $\pm 90° + \tan^{-1} \omega CR$. We may conclude this discussion by saying that if at $t = 0$ the capacitor voltage is specified, the sudden application of a sinusoidal current source will introduce a transient unless the sinusoidal input is adjusted to a suitable amplitude and phase angle so that the steady-state portion of $v$ at time $t = 0$ is equal to the specified initial voltage.

## 3.3  Circuits with Two Time Constants

Problems involving the calculation of transients occur frequently in circuits with switches. Let us illustrate such a problem with the circuit shown in Fig. 3.4. Assume that the capacitor and the resistors are linear and time-invariant, and that the capacitor is initially uncharged. For $t < 0$ switch $k_1$ is closed and switch $k_2$ is open. Switch $k_1$ is opened at $t = 0$ and thus connects the constant current source to the parallel $RC$ circuit. The capacitor is gradually charged with the time constant $T_1 \triangleq R_1 C$. Suppose that at $t = T_1$, switch $k_2$ is closed. The problem is to determine the voltage waveform across the capacitor for $t \geq 0$. We can divide the problem into two parts, the interval $[0, T_1]$ and the interval $[T_1, \infty)$. First, we determine the voltage in $[0, T_1]$ before switch $k_2$ closes. Since $v(0) = 0$ by assumption, the zero-state response can be found immediately. Thus,

$$(3.5) \quad v(t) = \begin{cases} 0 & t \leq 0 \\ R_1 I(1 - \epsilon^{-t/T_1}) & 0 \leq t \leq T_1 \end{cases}$$

At $t = T_1$

$$(3.6) \quad v(T_1) = R_1 I \left( 1 - \frac{1}{\epsilon} \right)$$

which represents the initial condition for the second part of our problem. For $t > T_1$, since switch $k_2$ is closed, we have a parallel combination of $C$, $R_1$, and $R_2$; the time constant is

$$(3.7) \quad T_2 = C \left( \frac{R_1 R_2}{R_1 + R_2} \right)$$

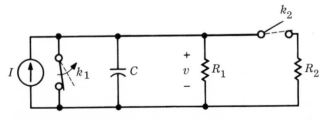

**Fig. 3.4**  A simple transient problem. The switch $k_1$ is opened at $t = 0$; the switch $k_2$ is closed at $t = T_1 \triangleq R_1 C$.

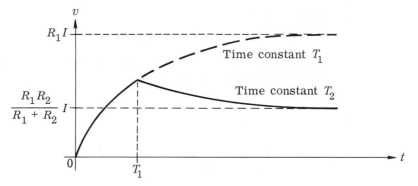

**Fig. 3.5**    Waveform of voltage for the circuit in Fig. 3.4.

and the input is $I$.    The complete response for this second part is, for $t \geq T_1$,

$$(3.8) \quad v(t) = R_1 I \left( 1 - \frac{1}{\epsilon} \right) \epsilon^{-(t - T_1)/T_2} + \frac{R_1 R_2}{R_1 + R_2} I(1 - \epsilon^{-(t - T_1)/T_2}) \qquad t \geq T_1$$

The waveform $v(\,\cdot\,)$ is plotted in Fig. 3.5.

<br>

**4**    ## The Linearity of the Zero-state Response

It is a fact that the zero-state response of *any* linear circuit is a linear function of the input; that is, the dependence of the waveform of the zero-state response on the input waveform is expressed by a linear function.    Note that any independent source in a linear circuit is considered as an input.    Let us illustrate this fact with the linear time-invariant $RC$ circuit that we studied (see Fig. 4.1).    Let the input be the current waveform $i_s(\,\cdot\,)$, and let the response be the voltage waveform $v(\,\cdot\,)$.    We are going to show the following in detail:

*The zero-state response of the linear time-invariant parallel RC circuit (shown in Fig. 4.1) is a linear function of the input; that is, the dependence of the zero-state response waveform on the input waveform has the property of additivity and homogeneity.*

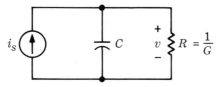

**Fig. 4.1**    Linear time-invariant $RC$ circuit with input $i_s$ and response $v$.

1.  Let us check additivity.  Consider two input currents $i_1$ and $i_2$ that are both applied at $t_0$.  Note that by $i_1$ (and also $i_2$) we mean a current waveform that starts at $t_0$ and goes on forever.  Call $v_1$ and $v_2$ the corresponding zero-state responses.  By definition, $v_1$ is the unique solution of the differential equation

(4.1)   $C\dfrac{dv_1}{dt} + Gv_1 = i_1(t) \qquad t \geq t_0$

with

(4.2)   $v_1(t_0) = 0$

Similarly, $v_2$ is the unique solution of

(4.3)   $C\dfrac{dv_2}{dt} + Gv_2 = i_2(t) \qquad t \geq t_0$

with

(4.4)   $v_2(t_0) = 0$

Adding (4.1) and (4.3), and taking (4.2) and (4.4) into account, we see that the function $v_1 + v_2$ satisfies

(4.5)   $C\dfrac{d}{dt}(v_1 + v_2) + G(v_1 + v_2) = i_1(t) + i_2(t) \qquad t \geq t_0$

with

(4.6)   $v_1(t_0) + v_2(t_0) = 0$

Now, by definition, the zero-state response to the input $i_1 + i_2$ applied at $t = t_0$ is the unique solution of the differential equation

(4.7)   $C\dfrac{dy}{dt} + Gy = i_1(t) + i_2(t) \qquad t \geq t_0$

with

(4.8)   $y(t_0) = 0$

By the uniqueness theorem for the solution of such differential equations and by comparing (4.5) and (4.6) with (4.7) and (4.8), we arrive at the conclusion that the waveform $v_1(\cdot) + v_2(\cdot)$ is the zero-state response to the input waveform $i_1(\cdot) + i_2(\cdot)$.  Since this reasoning applies to *any* input $i_1$ and *any* $i_2$ applied at *any* time $t_0$, we have shown that *the zero-state response of the RC circuit is a function of the input, which obeys the additivity property.*

2.  Let us check homogeneity.  We consider the input $i_1$ (applied at $t_0$) and the input $ki_1$, where $k$ is an arbitrary real constant.  By definition,

the zero-state response due to $i_1$ satisfies (4.1) and (4.2).   Similarly, the zero-state response due to $ki_1$ satisfies the differential equation

(4.9)   $C\dfrac{dy}{dt} + Gy = ki_1(t) \qquad t \geq t_0$

with

(4.10)   $y(t_0) = 0$

By multiplying (4.1) and (4.2) by the *constant k*, we obtain

(4.11)   $C\dfrac{d}{dt}(kv_1) + G(kv_1) = ki_1(t) \qquad t \geq 0$

with

(4.12)   $kv_1(t_0) = 0$

Again, the comparison of the four equations above, together with the uniqueness theorem of ordinary differential equations, leads to the conclusion that the zero-state response due to $ki_1$ is $kv_1$.   Since this reasoning applies to *any* input waveform $i_1(\,\cdot\,)$, *any* initial time $t_0$, and *any* constant $k$, we have shown that *the zero-state response of the RC circuit is a function of the input, which obeys the homogeneity property.*

The zero-state response, being both an additive and homogeneous function of the input is, therefore, by the very definition of linear function, a *linear function* of the input.   Thus our assertion is proved.

*The $\mathfrak{T}_{t_0}$ operator*   The linearity of the zero-state response can be expressed symbolically by introducing the operator $\mathfrak{T}_{t_0}$.   For the *RC* circuit shown in Fig. 4.1, let $\mathfrak{T}_{t_0}(i_s)$ denote the *waveform* of the zero-state response of the *RC* circuit to the input waveform $i_s(\,\cdot\,)$.   The subscript $t_0$ in $\mathfrak{T}_{t_0}$ is used to indicate that the *RC* circuit is in the zero state at time $t_0$ and that the input is applied at $t_0$.   Therefore, the linearity of the zero-state response means precisely the following:

1.   For all input waveforms $i_1(\,\cdot\,)$ and $i_2(\,\cdot\,)$ (defined for $t \geq t_0$ and taken to be identically zero for $t < t_0$), the zero-state response due to the input $i_1(\,\cdot\,) + i_2(\,\cdot\,)$ is the sum of the zero-state response due to $i_1(\,\cdot\,)$ alone and the zero-state response due to $i_2(\,\cdot\,)$ alone; that is,

(4.13)   $\mathfrak{T}_{t_0}(i_1 + i_2) = \mathfrak{T}_{t_0}(i_1) + \mathfrak{T}_{t_0}(i_2)$

2.   For all real numbers $\alpha$ and all input waveforms $i(\,\cdot\,)$, the zero-state response due to the input $\alpha i(\,\cdot\,)$ is equal to $\alpha$ times the zero-state response due to the input $i(\,\cdot\,)$; that is,

(4.14)   $\mathfrak{T}_{t_0}(\alpha i) = \alpha \mathfrak{T}_{t_0}(i)$

**Remarks**  1. If the capacitor and resistor in Fig. 4.1 are linear and *time-varying*, the differential equation is, for $t \geq t_0$,

(4.15)    $\dfrac{d}{dt}[C(t)v(t)] + G(t)v(t) = i_s(t)$

The zero-state response is still a linear function of the input; indeed, the proof of additivity and homogeneity would require only slight modifications. This proof still works because

$$\frac{d}{dt}[C(t)v_1(t)] + \frac{d}{dt}[C(t)v_2(t)] = \frac{d}{dt}\{C(t)[v_1(t) + v_2(t)]\}$$

2. The following fact is true although we have only proven it for a special case. Consider any circuit that contains linear (time-invariant or time-varying) elements. Let the circuit be driven by a single independent source, and let the response be any branch voltage or branch current. Then the zero-state response is a linear function of the input. The proof depends on the general analysis of networks (and will be presented in Chap. 6). For example, the linear $RL$ circuit shown in Fig. 4.2 with the source voltage $e_s$ as input and the current $i_R$ as response has the property that its zero-state response $i_R(\cdot)$ is a linear function of the input $e_s(\cdot)$.

3. It is easy to see from the proof of the simple linear $RC$ circuit above that the *complete response* is *not* a linear function of the input (unless, of course, the circuit starts from the zero state). Let us go back to the proof and note that if the circuit is in an initial state $V_0 \neq 0$, that is, $v_1(t_0) = V_0$ in Eq. (4.2) and $v_2(t_0) = V_0$ in Eq. (4.4), then in Eq. (4.6) $[v_1(t_0) + v_2(t_0)] = 2V_0$, which is not the specified initial state. This emphasizes again the important fact that initial conditions, together with the differential equation, characterize the input-response relation of a circuit. In Chap. 6 we shall show that the complete response of any linear circuit can be written explicitly in terms of the input waveform and the zero-input response; the latter depends only on the initial conditions.

**Exercise**  The purpose of this exercise is to show that if a circuit includes nonlinear elements, the zero-state response is not necessarily a linear function of the

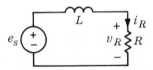

**Fig. 4.2**   Linear $RL$ circuit with input $e_s$ and response $i_R$.

input.   Consider the circuit shown in Fig. 4.2, but let the resistor be non-linear with the characteristic

$$v_R = a_1 i_R + a_3 i_R^3$$

where $a_1$ and $a_3$ are positive constants.   Show that the operator $\mathcal{T}_{t_0}$ does not possess the additivity property.

## 5   Linearity and Time Invariance

In Chap. 2 we classified circuit elements according to whether they were linear or nonlinear, time-varying or time-invariant.   In the previous section we demonstrated in a simple case that for linear circuits, the zero-state response is a linear function of the input; we noted also that this holds for both time-varying and time-invariant circuits.   In this section we shall bring out the differences between responses of circuits with time-invariant elements and circuits with time-varying elements.   These will help us understand the significance of *time invariance*.

### 5.1   Step Response

Up to this point, whenever we connected an independent source to a circuit, we used a switch to indicate that at a certain time $t = 0$ the switch closes or opens, and the input starts acting on the circuit.   An alternate description of the operation of applying an input starting at a specified time, say $t = 0$, can be supplied by using a step function.   For example, a constant current source that is applied to a circuit at $t = 0$ can be represented by a current source permanently connected to the circuit (without the switch) but with a step-function waveform plotted in Fig. 5.1.   Thus, for $t < 0$, $i(t) = 0$, and for $t > 0$, $i(t) = I$.   At $t = 0$ the current jumps from 0 to $I$.

We call the **step response** of a circuit its zero-state response to the unit step input $u(\,\cdot\,)$; we denote the step response by $\Delta$.   More precisely, $\Delta(t)$ is the response at time $t$ of the circuit provided that (1) its input is the step

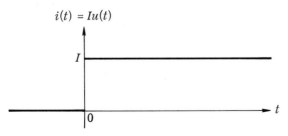

**Fig. 5.1**   Step function of magnitude $I$.

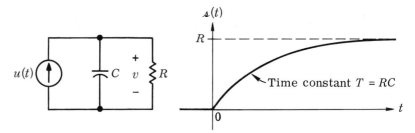

**Fig. 5.2**    Step response of simple *RC* circuit.

function $u(\cdot)$ and (2) the circuit is in the zero state just prior to the application of the unit step. As mentioned before, we adopt the convention that $\lambda(t) = 0$ for $t < 0$. For the linear time-invariant *RC* circuit in Fig. 5.2 the step response is, for all $t$,

(5.1)    $\lambda(t) = u(t)R(1 - \epsilon^{-(1/RC)t})$

Note that the presence of $u(t)$ in Eq. (5.1) makes it unnecessary to indicate as before that the result is true only for $t \geq 0$.

## 5.2    The Time-invariance Property

Our purpose is to focus now on a fundamental property of linear time-invariant circuits. We start with an intuitive discussion and continue with a formal description of the time-invariance property.

Consider any linear time-invariant circuit driven by a single independent source, and pick a network variable as a response. For example, we might use the parallel *RC* circuit previously considered. Let the voltage $v_0$ be the zero-state response of the circuit due to the current source input $i_0$ starting at $t = 0$. In terms of the operator $\mathcal{Z}_0$ we have

(5.2a)    $v_0 \triangleq \mathcal{Z}_0(i_0)$

The subscript 0 of the operator $\mathcal{Z}_0$ denotes specifically the starting time $t = 0$. Thus, $v_0$ is the unique solution of the differential equation

(5.2b)    $C\dfrac{dv_0}{dt} + Gv_0 = i_0(t) \qquad t \geq 0$

with

(5.2c)    $v_0(0) = 0$

In solving (5.2b) and (5.2c), we are only interested in $t \geq 0$. By a previous convention, we assume $i_0(t) = 0$ and $v_0(t) = 0$ for $t < 0$. Suppose that without changing the shape of the waveform $i_0(\cdot)$, we shift it horizontally so that it starts now at time $\tau$, with $\tau \geq 0$ (see Fig. 5.3). The new graph defines a new function $i_\tau(\cdot)$; the subscript $\tau$ represents the new starting time. Obviously from the graph, the ordinate of $i_\tau$ at time $\tau + t_1$ is equal to the ordinate of $i_0$ at time $t_1$; thus, since $t_1$ is arbitrary,

$$i_\tau(\tau + t_1) = i_0(t_1) \qquad \text{for all } t_1$$

If we set $t = \tau + t_1$, we obtain

(5.3) $\quad i_\tau(t) = \begin{cases} i_0(t - \tau) & t \geq \tau \\ 0 & t < \tau \end{cases}$

Consider now $v_\tau$, the response of the $RC$ circuit to $i_\tau$, given that the circuit is in the zero state at time 0; that is,

(5.4a) $\quad v_\tau \triangleq \mathcal{Z}_0(i_\tau)$

More precisely, $v_\tau$ is the unique solution of

(5.4b) $\quad C\dfrac{d}{dt}v_\tau(t) + Gv_\tau(t) = i_\tau(t) \qquad t \geq 0$

with

(5.4c) $\quad v_\tau(0) = 0$

Intuitively, we expect that the waveform $v_\tau$ will be the waveform $v_0$ shifted by $\tau$. Indeed, the circuit is time-invariant; therefore, its response to $i_\tau$ applied at time $\tau$ is, except for a shift in time, the same as its response to $i_0$ applied at time $t = 0$. This fact is illustrated in Fig. 5.4.

For students who desire a more detailed reasoning we proceed with the following proof in two steps.

1.   On the interval $(0,\tau)$, $v_\tau$ is identical to zero; indeed, $v_\tau \equiv 0$ satisfies Eq. (5.4b) for $0 \leq t \leq \tau$ (because $i_\tau \equiv 0$ on that interval) and the initial condition (5.4c). Since $v_\tau \equiv 0$ on $0 \leq t \leq \tau$, it follows that

(5.5) $\quad v_\tau(\tau) = 0$

2.   Now we must determine $v_\tau$ for $t \geq \tau$. In this task we use Eq. (5.5) as our initial condition. We assert that the waveform obtained by shift-

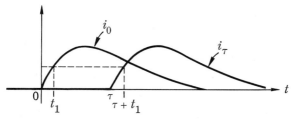

**Fig. 5.3**    The waveform $i_\tau$ is the result of shifting the waveform $i_0$ by $\tau$ sec.

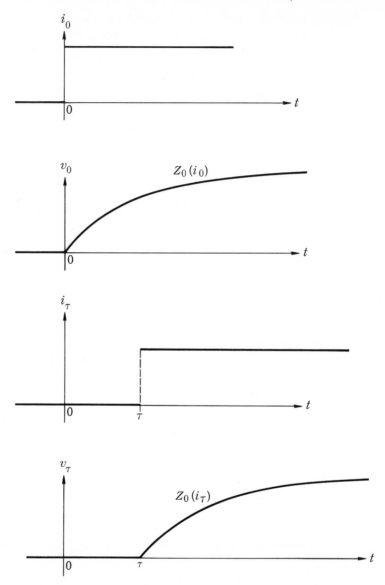

**Fig. 5.4**    Illustration of the time-invariance property.

ing $v_0$ by $\tau$ satisfies Eq. (5.4b) for $t \geq \tau$ and Eq. (5.5).    To prove this statement, let us verify that the function $y$, defined by $y(t) \overset{\Delta}{=} v_0(t - \tau)$, satisfies the differential equation (5.4b) for $t \geq \tau$ and the initial condition (5.5).

Replacing $t$ by $t - \tau$ in Eq. (5.2b), we obtain

(5.6a) $\quad C\dfrac{d}{dt}[v_0(t-\tau)] + Gv_0(t-\tau) = i_0(t-\tau) = i_\tau(t) \qquad t \geq \tau$

or, by definition,

(5.6b) $\quad C\dfrac{d}{dt}[y(t)] + Gy(t) = i_\tau(t) \qquad t \geq \tau$

which is precisely Eq. (5.4b) for $t \geq \tau$. The initial condition is obviously satisfied since

$$y(\tau) \overset{\Delta}{=} v_0(t-\tau)\Big|_{t=\tau} = v_0(0) = 0$$

In other words, the function $y(t) \overset{\Delta}{=} v_0(t-\tau)$ satisfies the differential equation (5.4b) for $t \geq \tau$ and the initial condition (5.5). This fact, together with $v_\tau \equiv 0$ on $(0,\tau)$, implies that *the waveform $v_0$ shifted by $\tau$ is $\mathfrak{Z}_0(i_\tau)$, the zero-state response to $i_\tau$.*

---

**Example**   If $i_0(t) = Iu(t)$, then

$$v_0(t) = u(t)RI(1 - \epsilon^{-t/RC}) \qquad \text{for all } t$$

and the zero-state response to $i_\tau(t) = i_0(t-\tau) = Iu(t-\tau)$ is

$$v_\tau(t) = u(t-\tau)RI(1 - \epsilon^{-(t-\tau)/RC}) \qquad \text{for all } t$$

---

**Remarks**  1.  The reasoning outlined above does not depend upon the particular value of $\tau \geq 0$, nor does it depend upon the shape of the input waveform $i_0$. In other words, for all $\tau \geq 0$ and all $i_0$, $\mathfrak{Z}_0(i_\tau)$ is identical with the waveform $\mathfrak{Z}_0(i_0)$ shifted by $\tau$. This fact is called the *time-invariance property* of the linear time-invariant $RC$ circuit.

2.  It is crucial to observe that the constancy of $C$ and $G$ was used in arguing that Eq. (5.6) was simply Eq. (5.2b) in which $t - \tau$ was substituted for $t$.

## 5.3   The Shift Operator

The idea of time invariance can be expressed precisely by the use of a *shift operator*. Let $f(\cdot)$ be any waveform defined for all $t$. Let $\mathfrak{T}_\tau$ be an operator which when applied to $f$ yields an identical waveform, except that it has been delayed by $\tau$; the shifted waveform is called $f_\tau(\cdot)$ and its ordinates are given by

$$f_\tau(t) = f(t-\tau) \qquad \text{for all } t$$

In other words, the result of applying the operator $\mathfrak{T}_\tau$ to the waveform $f$ is a new waveform denoted by $\mathcal{T}_\tau f$, such that the value at any time $t$ of the new waveform, denoted by $(\mathcal{T}_\tau f)(t)$, is related to the values of $f$ by

$$(\mathcal{T}_\tau f)(t) = f(t - \tau) \qquad \text{for all } t$$

In the notation of our previous discussion we have $\mathcal{T}_\tau f = f_\tau$. The operator $\mathcal{T}_\tau$ is called a **shift operator;** some authors call it a translation operator. It is a very important fact that a shift operator is a linear operator. Indeed, it is additive. Thus,

$$\mathcal{T}_\tau(f + g) = \mathcal{T}_\tau f + \mathcal{T}_\tau g$$

that is, the result of shifting $f + g$ is equal to the sum of the shifted $f$ and the shifted $g$. It is also homogeneous. If $\alpha$ is any real number and $f$ is any waveform,

$$\mathcal{T}_\tau[\alpha f] = \alpha \mathcal{T}_\tau f$$

that is, if we multiply the waveform $f$ by the number $\alpha$ and shift the result, we have the very same waveform that we would have had if we first shifted $f$ and then multiplied it by $\alpha$.

Let us use the shift operator to express the time-invariance property. As before let $\mathcal{Z}_0(i_0)$ be the response of the circuit to the input $i_0$ provided that the circuit is in the zero state at time 0. Previously, we used $v_0(t)$ to denote the value of the zero-state response at time $t$ [see Eq. (5.2a)]. The reason that $\mathcal{Z}_0(i_0)$ is used now is to emphasize the dependence of the zero-state response on the whole input waveform $i_0(\cdot)$ and to emphasize the time at which the circuit is in the zero state. It is very important to keep in mind that $\mathcal{Z}_0(i_0)$ is the whole waveform, not its value at time $t$. With these notations the time-invariance property demonstrated above can be written as

(5.7) $\qquad \mathcal{T}_\tau[\mathcal{Z}_0(i_0)] = \mathcal{Z}_0[\mathcal{T}_\tau i_0] \qquad$ for all inputs $i_0$ and for all $\tau \geq 0$

Although we have only proved (5.7) for a circuit with a linear time-invariant resistor and capacitor in parallel, it is, in fact, valid for *any* linear time-invariant circuit, for *any* input $i$, and for *any* $\tau \geq 0$. Equation (5.7) states the time-invariance property of linear time-invariant circuits. It will play a key role in the derivation of the convolution representation of the zero-state response in Chap. 6.

**Remark** The time-invariance property as expressed by (5.7) may be interpreted as that the operators $\mathcal{T}_\tau$ and $\mathcal{Z}_0$ *commute;* i.e., the order of applying the two operations is immaterial. Although you have seen many operations that commute (addition of real numbers, addition of matrices, etc.), there are many that do not (the multiplication of $n \times n$ matrices, for example). It is a remarkable fact that the operators $\mathcal{T}_\tau$ and $\mathcal{Z}_0$ commute for linear time-invariant circuits, because in the large majority of cases if the order of two operations is interchanged, the results are drastically different. For ex-

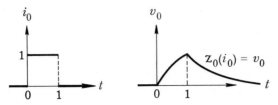

**Fig. 5.5**   Current $i_0$ and corresponding zero-state response $v_0$.

ample, if (1) you load a pistol and (2) you bring it to your temple and pull the trigger, the result is vastly different from that obtained by performing step 2 before step 1!

**Example**   We give an example to illustrate the consequence of linearity and time invariance.   Let us consider an arbitrary linear time-invariant circuit. Suppose that we have measured the zero-state response $v_0$ to the pulse $i_0$ shown in Fig. 5.5 and have a record of the waveform $v_0$.   Using our previous notation, this means that $v_0 = \mathcal{Z}_0(i_0)$.   The problem is to find the zero-state response $v$ to the input $i$ shown in Fig. 5.6, where

$$i(t) = \begin{cases} 1 & \text{for } 0 < t \le 1 \\ 3 & \text{for } 1 < t \le 2 \\ 0 & \text{for } 2 < t \le 3 \\ -2 & \text{for } 3 < t \le 4 \\ 0 & \text{for } 4 < t \end{cases}$$

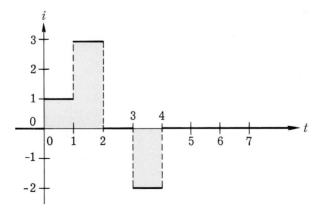

**Fig. 5.6**   Input $i(t)$.

The key observation is that the given input can be represented as a linear combination of $i_0$ and multiples of $i_0$ shifted in time. The process is illustrated in Fig. 5.7; the sum of the three functions shown is $i$. It is obvious from the graphs of $i$ and $i_0$ that

$$i = i_0 + 3\mathcal{T}_1(i_0) - 2\mathcal{T}_3(i_0)$$

Now call $v$ the zero-state response due to $i$; that is,

$$v = \mathcal{Z}_0(i)$$
$$= \mathcal{Z}_0[i_0 + 3\mathcal{T}_1(i_0) - 2\mathcal{T}_3(i_0)]$$

By the linearity of the zero-state response we get

$$v = \mathcal{Z}_0(i_0) + 3\mathcal{Z}_0[\mathcal{T}_1(i_0)] - 2\mathcal{Z}_0[\mathcal{T}_3(i_0)]$$

and by the time-invariance property

$$v = \mathcal{Z}_0(i_0) + 3\mathcal{T}_1[\mathcal{Z}_0(i_0)] - 2\mathcal{T}_3[\mathcal{Z}_0(i_0)]$$

Since

$$v_0 = \mathcal{Z}_0(i_0)$$
$$v = v_0 + 3\mathcal{T}_1(v_0) - 2\mathcal{T}_3(v_0)$$

or

$$v(t) = v_0(t) + 3v_0(t - 1) - 2v_0(t - 3) \qquad \text{for } t \geq 0$$

**Remark**   The method used to calculate $v$ in terms of $v_0$ is usually referred to as the *superposition* method. It is fundamental to realize that we have to invoke the time-invariance property and the fact that the zero-state response is a *linear function* of the input.

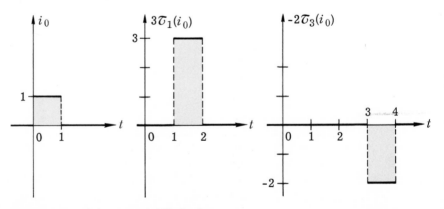

**Fig. 5.7**   Decomposition of $i$ in terms of shifted pulses.

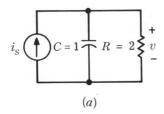

(a)

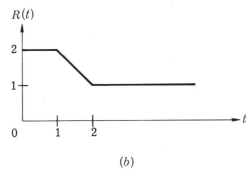

(b)

**Fig. 5.8**   (a) A simple linear $RC$ circuit; (b) time-varying resistor characteristic.

**Exercise**   Consider the familiar linear time-invariant $RC$ circuit shown in Fig. 5.8a; $i_s$ is its input, and $v$ is its response.

a.   Calculate and sketch the zero-state response to the following inputs:

$$i_1(t) = \begin{cases} 1 & 0 < t \le 0.5 \\ \\ 0 & 0.5 < t \end{cases} \qquad i_2(t) = \begin{cases} 3 & 0 < t \le 0.5 \\ 0 & 0.5 < t \le 2 \\ -0.5 & 2 < t \le 2.5 \\ 0 & 2.5 < t \end{cases}$$

b.   Suppose now that the resistor is time-varying but still linear. Let its resistance be a function of time as shown in Fig. 5.8b. Suppose we were to calculate the response of this circuit to the input $i_1$; may we still use the method discussed previously? If not, state briefly why.

---

**6**   **Impulse Response**

The zero-state response of a time-invariant circuit to a *unit* impulse applied at $t = 0$ is called the **impulse response** of a circuit and is denoted by $h$. More precisely, $h(t)$ is the response at time $t$ of the circuit provided that (1) its input is the unit impulse $\delta$ and (2) it is in the *zero state* just prior to the application of the impulse. For convenience in later formulations we shall define $h$ to be zero for $t < 0$. Since the calculation of impulse re-sponse is of great importance to electrical engineers, we shall present three methods.

*First method*   We approximate the impulse by the pulse function $p_\Delta$. In order to obtain a first acquaintance with the impulse response, let us calculate the impulse response of the parallel $RC$ circuit shown in Fig. 6.1. The input to the cir-cuit is the current source $i_s$, and the response is the output voltage $v$. Since the impulse response is defined to be the zero-state response to $\delta$, the im-pulse response is the solution of the differential equation

(6.1)   $$C \frac{dv}{dt} + Gv = \delta(t)$$

with

(6.2)   $$v(0-) = 0$$

where the symbol $0-$ designates the time immediately before $t = 0$.

   We have to distinguish between $0-$ and $0+$ because of the presence of the impulse on the right-hand side of (6.1). At time $t = 0$ an infinitely large current goes through the circuit for an infinitesimal interval of time. The situation is analogous to the golf ball sitting on the tee and being hit by the club at $t = 0$; it is obviously of great importance to distinguish be-tween the velocity of the ball at $0-$ just prior to being hit, and its velocity at $0+$ just after being hit.

   Equation (6.2) states that the circuit is in the zero state just prior to the application of the input. In order to solve (6.2) we run into some diffi-culties since, strictly speaking, $\delta$ is *not* a function. Therefore, the solution will be obtained by approximating unit impulse $\delta$ by the pulse function $p_\Delta$, computing the resulting solution, and then letting $\Delta \to 0$. Recall that $p_\Delta$ is defined by

$$p_\Delta(t) = \begin{cases} 0 & \text{for } t < 0 \\ \dfrac{1}{\Delta} & \text{for } 0 < t < \Delta \\ 0 & \text{for } \Delta < t \end{cases}$$

and it is plotted in Fig. 6.2. The first step is to solve for $h_\Delta$, the zero-state

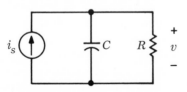

**Fig. 6.1**   Linear time-invariant $RC$ circuit.

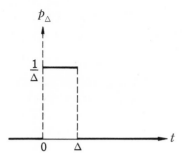

**Fig. 6.2**    Pulse function $p_\Delta(\,\cdot\,)$.

response of the $RC$ circuit to $p_\Delta$, where $\Delta$ is chosen to be much smaller than the time constant $RC$. The waveform $h_\Delta$ is the solution of

$$(6.3a) \quad C\frac{dh_\Delta}{dt} + \frac{1}{R}h_\Delta = \frac{1}{\Delta} \qquad 0 < t < \Delta$$

$$(6.3b) \quad C\frac{dh_\Delta}{dt} + \frac{1}{R}h_\Delta = 0 \qquad t > \Delta$$

with $h_\Delta(0) = 0$.   Clearly, $1/\Delta$ is a constant; hence from (6.3a)

$$(6.4a) \quad h_\Delta(t) = \frac{R}{\Delta}(1 - \epsilon^{-t/RC}) \qquad 0 < t < \Delta$$

and it is the zero-state response due to a step $(1/\Delta)u(t)$. From (6.3b), $h_\Delta$ for $t > \Delta$ is the zero-input response that starts from $h_\Delta(\Delta)$ at $t = \Delta$; thus

$$(6.4b) \quad h_\Delta(t) = h_\Delta(\Delta)\epsilon^{-(t-\Delta)/RC} \qquad t > \Delta$$

The total response $h_\Delta$ from (6.4a) and (6.4b) is shown on Fig. 6.3a.   From (6.4a)

$$h_\Delta(\Delta) = \frac{R}{\Delta}(1 - \epsilon^{-\Delta/RC})$$

Since $\Delta$ is much smaller than $RC$, using

$$\epsilon^{-x} = 1 - x + \frac{x^2}{2!} - \frac{x^3}{3!} + \cdots$$

we obtain

$$h_\Delta(\Delta) = \frac{R}{\Delta}\left[\frac{\Delta}{RC} - \frac{1}{2!}\left(\frac{\Delta}{RC}\right)^2 + \cdots\right]$$

$$= \frac{1}{C}\left[1 - \frac{1}{2!}\left(\frac{\Delta}{RC}\right) + \cdots\right]$$

Similarly, from (6.4a) for $\Delta$ very small and $0 < t < \Delta$, expanding the exponential function, we obtain

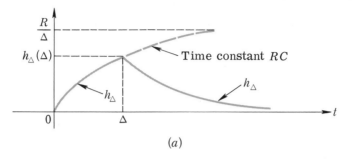

(a)

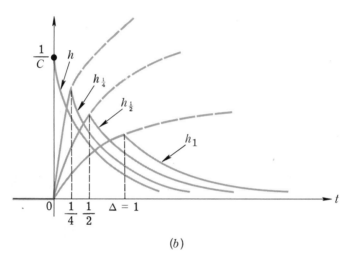

(b)

**Fig. 6.3**   (a) Zero-state response of $p_\Delta$; (b) the responses as $\Delta \to 0$.

$$h_\Delta(t) = \frac{1}{C}\frac{t}{\Delta} + \cdots \qquad 0 < t < \Delta$$

Note that the slope of the curve $h_\Delta$ over $(0,\Delta)$ is $1/C\Delta$. This slope is very large since $\Delta$ is small. As $\Delta \to 0$, the curve $h_\Delta$ over $(0,\Delta)$ becomes steeper and steeper, and $h_\Delta(\Delta) \to 1/C$. In the limit, $h_\Delta$ jumps from 0 to $1/C$ at the instant $t = 0$. For $t > 0$, we obtain, from (6.4b),

$$h_\Delta(t) \to \frac{1}{C}\epsilon^{-t/RC}$$

As $\Delta$ approaches zero, $h_\Delta$ approaches the impulse response $h$ as shown in Fig. 6.3b. Recalling that by convention we set $h(t) = 0$ for $t < 0$, we can therefore write

(6.5)  $$h(t) = u(t)\frac{1}{C}\epsilon^{-t/RC} \qquad \text{for all } t$$

The impulse response $h$ is shown in Fig. 6.4.

The above calculation of $h$ calls for two remarks.

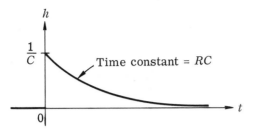

**Fig. 6.4**   Impulse response of the $RC$ circuit of Fig. 6.1.

**Remarks**   1.   Our purpose in calculating the impulse response in this manner has been to exhibit the fact that it is a straightforward procedure; it requires only the approximation of $\delta$ by a suitable pulse, here $p_\Delta$. The only requirements that $p_\Delta$ must satisfy are that it be zero outside the interval $(0,\Delta)$ and the area under $p_\Delta$ be equal to 1; that is,

$$\int_0^\Delta p_\Delta(t)\,dt = 1$$

It is a fact that the shape of $p_\Delta$ is irrelevant; therefore, we choose a shape that requires the least amount of work. We might very well have chosen a triangular pulse as shown in Fig. 6.5. Observe that the maximum amplitude of the triangular pulse is now $2/\Delta$; this is required in order that the area under the pulse be unity for all $\Delta > 0$.

2.   Since $\delta(t) = 0$ for $t > 0$ (that is, the input is identically zero for $t > 0$), it follows that the impulse response $h(t)$ is, for $t > 0$, identical to a particular zero-input response. We shall use this fact later.

*Relation between impulse response and step response*   We wish now to establish a very important relation between the step response and the impulse response of a linear time-invariant circuit. More precisely we wish to show that the following is true:

*The impulse response of a linear time-invariant circuit is the time derivative of its step response.*

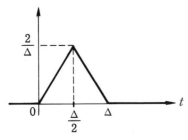

**Fig. 6.5**   A triangular pulse can also be used to approximate the impulse.

Symbolically,

(6.6)   $h = \dfrac{d\delta}{dt}$   or equivalently   $\delta(t) = \displaystyle\int_{-\infty}^{t} h(t')\, dt'$

We prove this important statement by approximating the impulse by the pulse function $p_\Delta$. Let $h_\Delta$ be the zero-state response to the input $p_\Delta$; that is,

$$h_\Delta \triangleq \mathcal{Z}_0(p_\Delta)$$

As $\Delta \to 0$, the pulse function $p_\Delta$ approaches $\delta$, the unit impulse, and $h_\Delta$, the zero-state response to the pulse input, approaches the impulse response $h$. Now consider $p_\Delta$ as a superposition of a step and a delayed step as shown in Fig. 6.6. Thus,

$$p_\Delta = \frac{1}{\Delta}[u(t) - u(t - \Delta)] = \frac{1}{\Delta}u + \frac{-1}{\Delta}\mathcal{T}_\Delta u$$

By the linearity of the zero-state response, we have

$$\mathcal{Z}_0(p_\Delta) = \mathcal{Z}_0\!\left(\frac{1}{\Delta}u + \frac{-1}{\Delta}\mathcal{T}_\Delta u\right)$$

(6.7)   $$= \frac{1}{\Delta}\mathcal{Z}_0(u) + \frac{-1}{\Delta}\mathcal{Z}_0(\mathcal{T}_\Delta u)$$

Since the circuit is linear and time-invariant, the $\mathcal{Z}_0$ operator and the shift operator commute; thus,

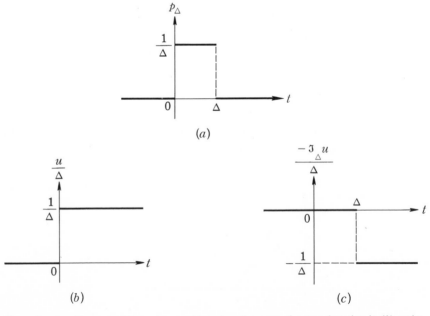

(a)

(b)                              (c)

**Fig. 6.6**   The pulse function $p_\Delta$ in (a) can be considered as the sum of a step function in (b) and a delayed step function in (c).

(6.8)    $\mathfrak{Z}_0(\mathfrak{T}_\Delta u) = \mathfrak{T}_\Delta \mathfrak{Z}_0(u)$

Let us denote the step response by

$\mathfrak{s} \triangleq \mathfrak{Z}_0(u)$

Equations (6.7) and (6.8) can be combined to yield

$h_\Delta \triangleq \mathfrak{Z}_0(p_\Delta) = \dfrac{1}{\Delta}\mathfrak{s} - \dfrac{1}{\Delta}\mathfrak{T}_\Delta\,\mathfrak{s}$

or

$h_\Delta(t) = \dfrac{1}{\Delta}\mathfrak{s}(t) - \dfrac{1}{\Delta}\mathfrak{s}(t - \Delta)$

$= \dfrac{\mathfrak{s}(t) - \mathfrak{s}(t - \Delta)}{\Delta}$ for all $t$

Now with $\Delta \to 0$, the right-hand side becomes the derivative; hence

$\lim_{\Delta \to 0} h_\Delta(t) = h(t) = \dfrac{d\mathfrak{s}}{dt}$

**Remark**    The two equations in (6.6) do not hold for linear *time-varying* circuits; this should be expected since time invariance is used in a key step of the derivation. Thus, for linear *time-varying* circuits the time derivative of the step response is *not* the impulse response.

*Second method*    We use $h = d\mathfrak{s}/dt$. Again considering the parallel $RC$ circuit of Fig. 6.1, we recall that its step response $\mathfrak{s}$ is given by

$\mathfrak{s}(t) = u(t)R(1 - \epsilon^{-(1/RC)t})$

If we consider the right-hand side as a product of two functions and use the rule of differentiation $(uv)' = u'v + uv'$, we obtain the impulse response

$h(t) = \delta(t)R(1 - \epsilon^{-(1/RC)t}) + \dfrac{1}{C}u(t)\epsilon^{-(1/RC)t}$

The first term is identically zero because for $t \neq 0$, $\delta(t) = 0$, and for $t = 0$, $1 - \epsilon^{-(1/RC)t} = 0$. Therefore,

$h(t) = \dfrac{1}{C}u(t)\epsilon^{-(1/RC)t}$

This result, of course, checks with the previously obtained result in (6.5).

*Third method*    We use the differential equation directly. We propose to show that $h$ defined by

$h(t) = \dfrac{1}{C}u(t)\epsilon^{-t/RC}$ for all $t$

is the solution to the differential equation

(6.9)   $C\dfrac{d}{dt}(v) + Gv = \delta$    with $v(0-) = 0$

In order not to prejudice the case, let us call $y$ the solution to (6.9). Thus, we propose to show that $y = h$. Since $\delta(t) = 0$ for $t > 0$ and $y$ is the solution of (6.9), we must have

(6.10)   $y(t) = y(0+)\epsilon^{-t/RC}$     for $t > 0$

This is shown in Fig. 6.7a. Since $\delta(t) = 0$ for $t < 0$ and the circuit is in the zero state at time $0-$, we must also have

(6.11)   $y(t) = 0$     for $t < 0$

This is shown in Fig. 6.7b. Combining (6.10) and (6.11), we conclude that

(6.12)   $y(t) = u(t)y(0+)\epsilon^{-t/RC}$     for all $t$

It remains to calculate $y(0+)$, that is, the magnitude of the jump in the curve $y$ at $t = 0$. In order to do this we use the known fact that

$$\delta(t) = \frac{du(t)}{dt}$$

From (6.12) and by considering the right-hand side as a product of functions, we obtain

$$\frac{dy}{dt}(t) = \delta(t)y(0+)\epsilon^{-t/RC} + u(t)y(0+)\frac{-1}{RC}\epsilon^{-t/RC}$$

In the first term, since $\delta(t)$ is zero everywhere except at $t = 0$, we may set $t$ to zero in the factor of $\delta(t)$; thus

$$\frac{dy}{dt}(t) = \delta(t)y(0+) + u(t)y(0+)\frac{-1}{RC}\epsilon^{-t/RC}$$

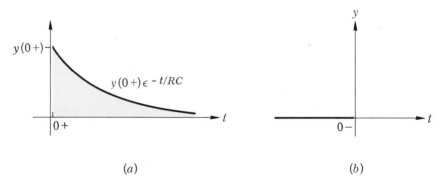

(a)                                   (b)

**Fig. 6.7**   Impulse response for the parallel $RC$ circuit.   (a) $y(t)$ for $t > 0$; (b) $y(t)$ for $t < 0$.

Substituting in (6.9), we obtain

$$\delta(t)Cy(0+) - u(t)y(0+)G\epsilon^{-t/RC} + Gu(t)y(0+)\epsilon^{-t/RC} = \delta(t)$$

After cancellation the only term that remains on the left-hand side is $Cy(0+)\delta(t)$; since it must balance the term $\delta(t)$ in the right-hand side, we obtain $y(0+)C = 1$; equivalently,

$$y(0+) = \frac{1}{C}$$

Inserting this value of $y(0+)$ into (6.12), we conclude that the solution of (6.9) is actually $h$, the impulse response calculated previously.

**Remark**   We have just shown that the solution of the differential equation

$$C\frac{d}{dt}(v) + Gv = \delta \qquad \text{with } v(0-) = 0$$

for $t > 0$ is identical with the solution of

(6.13)   $$C\frac{d}{dt}(v) + Gv = 0 \qquad \text{with } v(0+) = \frac{1}{C}$$

for $t > 0$. This can be seen by integrating both sides of (6.9) from $t = 0-$ to $t = 0+$ to obtain

$$Cv(0+) - Cv(0-) + G\int_{0-}^{0+} v(t')\, dt' = 1$$

Since $v$ is finite, $G\int_{0-}^{0+} v(t')\, dt' = 0$, and since $v(0-) = 0$, we obtain

$$v(0+) = \frac{1}{C}$$

In Eq. (6.13) the effect of the impulse at $t = 0$ has been taken care of by the initial condition at $t = 0+$.

---

**7**   **Step and Impulse Responses for Simple Circuits**

**Example 1**   Let us calculate the impulse response and the step response of the $RL$ circuit shown in Fig. 7.1. The series connection of the linear time-invariant resistor and inductor is driven by a voltage source. As far as the impulse response is concerned, the differential equation for the current $i$ is

(7.1)   $$L\frac{di}{dt} + Ri = \delta \qquad i(0-) = 0$$

If we confine our attention to the values of $t > 0$, this problem is equiva-

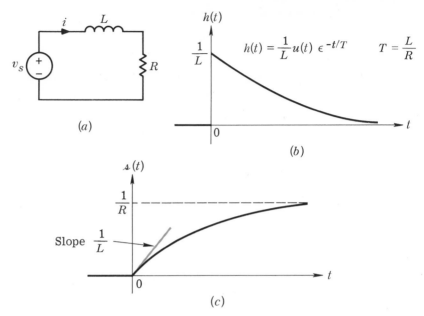

**Fig. 7.1**    (*a*) Linear time-invariant *RL* circuit; $v_s$ is the input and *i* is the response; (*b*) impulse response; (*c*) step response.

lent to that of the same circuit with no voltage source but with the initial condition $i(0+) = 1/L$; that is, for $t > 0$,

(7.2)    $L\dfrac{di}{dt} + Ri = 0 \qquad i(0+) = \dfrac{1}{L}$

The solution is

(7.3)    $i(t) = h(t) = \dfrac{1}{L} u(t)\epsilon^{-(R/L)t}$

The step response can be obtained either from integration of (7.3) or directly from the differential equation

(7.4)    $\jmath(t) = \dfrac{1}{R} u(t)(1 - \epsilon^{-(R/L)t})$

The physical explanation of the step response of the series *RL* circuit is now given. As the step of voltage is applied to the circuit, that is, at $0+$, the current in the circuit remains zero because, as we noted earlier, the current through an inductor cannot change instantaneously unless there is an infinitely large voltage across it. Since the current is zero, the voltage across the resistor must be zero. Therefore, at $0+$ all the voltage of the voltage source appears across the inductor; in fact $\dfrac{di}{dt}\bigg|_{0+} = 1/L$. As time increases, the current increases monotonically, and after a very long time, the current becomes practically constant. Thus, for large $t$, $di/dt \approx 0$; that is, the voltage across the inductor is zero, and all the voltage of the source is across the resistor. Therefore, the current is approximately $1/R$. In the limit we reach what is called the *steady state* and $i = 1/R$. We conclude that the inductor behaves as a short circuit in the steady state for a step-voltage input.

**Example 2**   Consider the circuit in Fig. 7.2, where the series connection of a linear time-invariant resistor $R$ and a capacitor $C$ is driven by a voltage source. The current through the resistor is the response of interest, and the problem is to find the impulse and step responses. The equation for the current $i$ is given by writing KVL for the loop; thus

(7.5)   $\dfrac{1}{C} \displaystyle\int_0^t i(t')\, dt' + Ri(t) = v_s(t)$

Let us use the charge on the capacitor as the variable; then (7.5) becomes

(7.6)   $\dfrac{q}{C} + R\dfrac{dq}{dt} = v_s(t)$

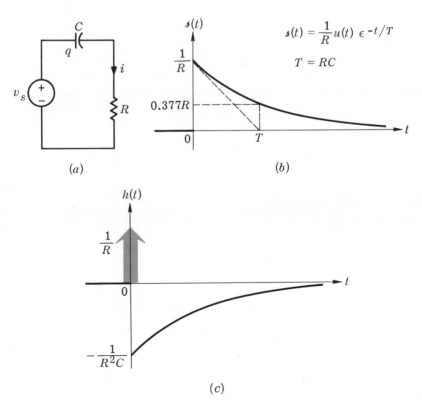

$(a)$                        $(b)$

$(c)$

**Fig. 7.2**   $(a)$ Linear time-invariant $RC$ circuit; $v_s$ is the input and $i$ is the response; $(b)$ step response; $(c)$ impulse response.

**Table 4.1  Step and Impulse Responses for Simple Linear Time-invariant Circuits**

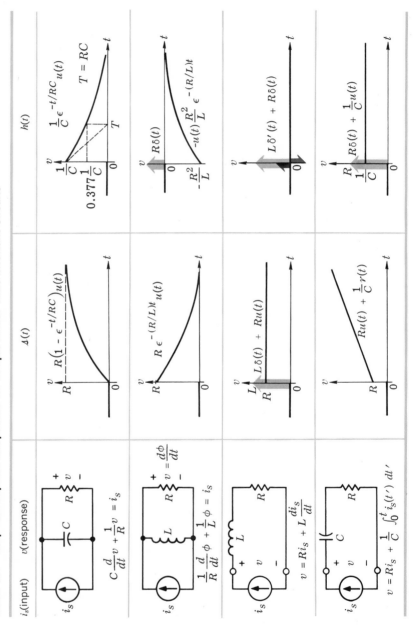

152

**Table 4.1  Step and Impulse Responses for Simple Linear Time-invariant Circuits** (*Continued*)

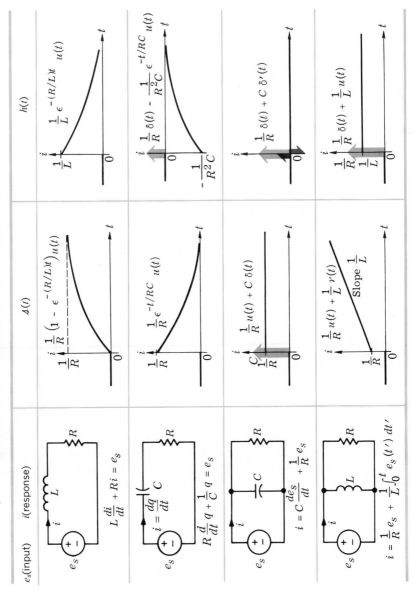

Since we have to find the step and impulse responses, the initial condition is $q(0-) = 0$. If $v_s$ is a unit step, (7.6) gives

$$q_{\Delta}(t) = u(t)C(1 - \epsilon^{-t/RC})$$

and by differentiation, the step response for the current is

$$i_{\Delta}(t) = \Delta(t) = \frac{1}{R}u(t)\epsilon^{-t/RC}$$

If $v_s$ is a unit impulse, (7.6) gives

$$q_{\delta}(t) = \frac{1}{R}u(t)\epsilon^{-t/RC}$$

and, by differentiation, the impulse response for the current is

$$i_{\delta}(t) = h(t) = \frac{1}{R}\delta(t) - \frac{1}{R^2C}u(t)\epsilon^{-t/RC}$$

We observe that in response to a step, the current is discontinuous at $t = 0$; $i_{\Delta}(0+) = 1/R$ as we expect, since at $t = 0$ there is no charge (hence no voltage) on the capacitor. In response to an impulse, the current includes an impulse of value $1/R$, and, for $t > 0$, the capacitor discharges through the resistor.

The step and impulse responses for simple first-order linear time-invariant circuits are tabulated in Table 4.1.

## 8  Time-varying Circuits and Nonlinear Circuits

Up to this point we have analyzed almost exclusively linear time-invariant circuits. We have studied the implications of the linearity and of the time invariance of element characteristics as far as the relation between input and output is concerned. In this section we shall first summarize the main implications of linearity and of time invariance of element characteristics. Next we shall consider examples of circuits with nonlinear and of time-varying elements to demonstrate that without linearity and time invariance these main implications are no longer true.

In our study of first-order circuits we have seen that if the circuits are *linear* (time-invariant or time-varying), then

1. The zero-input response is a linear function of the initial state.

2. The zero-state response is a linear function of the input.

3. The complete response is the sum of the zero-input response and of the zero-state response.

We have also seen that if the circuit is *linear* and *time-invariant,* then

1. $\mathcal{Z}_0[\mathcal{T}_\tau(i)] = \mathcal{T}_\tau[\mathcal{Z}_0(i)] \qquad \tau \geq 0$

which means that the zero-state response (starting in the zero state at time zero) to the shifted input is equal to the shift of the zero-state response (starting also in the zero state at time zero) to the original input.

2.   The impulse response is the derivative of the step response.

For *time-varying* circuits and *nonlinear* circuits the analysis problem is in general difficult. Furthermore there exists no general method of analysis except numerical integration of the differential equations. Consequently, we shall give only simple examples to point out techniques that may be useful in simple cases. Our main emphasis is, however, to demonstrate certain properties of the solutions.

**Example 1**   Consider the parallel $RC$ circuit of Fig. 8.1, where the capacitor is linear and time-invariant with $C = 1$ farad and the initial voltage at $t = 0$ is 1 volt. The zero-input responses are to be determined for the following types of resistor:

*a.*   A linear time-invariant resistor with $R = 1$ ohm

*b.*   A linear time-varying resistor with $R(t) = 1/(1 + 0.5 \cos t)$ ohm

*c.*   A nonlinear time-invariant resistor having a characteristic $i_R = v_R^2$

*Solution*   *a.*   The solution has been discussed before and is of the form

$$v(t) = u(t)\epsilon^{-t}$$

*b.*   The differential equation is given by

$$\frac{dv}{dt} + (1 + 0.5 \cos t)v = 0 \qquad t \geq 0$$

and

$$v(0) = 1$$

The equation can be put in the following form

$$\frac{dv}{v} = -(1 + 0.5 \cos t)\, dt$$

Integrating the right-hand side from zero to $t$ and the left-hand side from $v(0) = 1$ to $v(t)$, we obtain

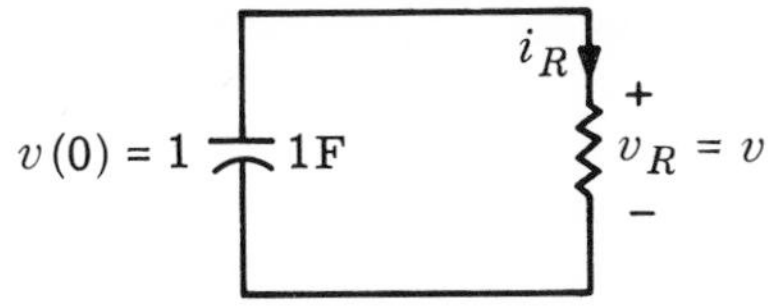

**Fig. 8.1**   Illustration of the zero-input responses of a simple $RC$ circuit.

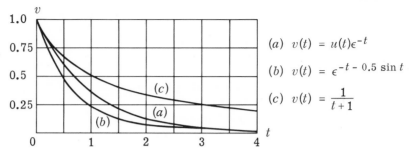

**Fig. 8.2**   Zero-input responses for three different types of resistors.

$$\ln v(t) = -(t + 0.5 \sin t)$$

or

$$v(t) = u(t)\epsilon^{-t-0.5 \sin t}$$

*c.*   The differential equation is nonlinear and is given by

$$\frac{dv}{dt} + v^2 = 0 \qquad t \geq 0$$

and

$$v(0) = 1$$

The equation can be solved again by separating the variables and integrating both sides as

$$\int_1^{v(t)} \frac{d\hat{v}}{\hat{v}^2} = -\left(\frac{1}{v(t)} - 1\right) = \int_0^t - dt' = -t$$

or

$$v(t) = u(t)\frac{1}{t + 1}$$

The three results are plotted in Fig. 8.2 for comparison.

---

**Example 2**   Consider the same circuit as given in the previous example with $v(0) = 0$. For the same three cases as above, we want to calculate the unit step response due to a current source connected in parallel.

*a.*   The step response for the linear time-invariant $RC$ circuit is

$$v(t) = u(t)(1 - \epsilon^{-t})$$

*b.*   The differential equation is given by

$$\frac{dv}{dt} + (1 + 0.5 \cos t)v = u(t) \qquad v(0) = 0$$

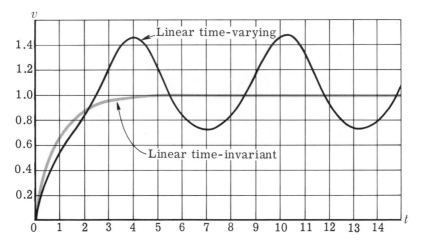

**Fig. 8.3**   Step responses of two $RC$ circuits. In both cases the capacitor is linear and time-invariant; the resistor is linear and time-varying (dark curve) or linear and time-invariant (light curve).

We cannot integrate the impulse response to obtain the step response because of the time-varying resistor. The solution can be obtained by means of a step-by-step calculation using a computer, and is plotted in Fig. 8.3 along with the step response obtained in ($a$) for the time-invariant case. Note that for $t > 2$, the response is asymptotic to a periodic function, and we can say that the circuit has reached a steady-state behavior. For $t < 2$ the circuit is undergoing a transient. The steady-state part contains a constant term and a periodic term with angular frequency $\omega = 1$. The periodic term is caused by the constant current input and the sinusoidal variation of the time-varying resistor. This example indicates the main difference between a linear circuit with time-invariant elements and a linear circuit with a time-varying element; that is, for a step input the steady-state solution for the time-invariant case contains only a constant term but for the time-varying case it contains a constant term and, in addition, a periodic term.

Let us use this example to illustrate the fact that for time-varying circuits the shifting property no longer holds. Consider two other inputs as given by

$$i_s(t) = \begin{cases} u\!\left(t - \dfrac{\pi}{2}\right) \\ u(t - \pi) \end{cases}$$

In the first situation the input waveform jumps from zero to unity at $t = \pi/2$, whereas in the second situation it jumps at $t = \pi$. The zero-state responses are plotted in Fig. 8.4. Let $v_{\pi/2}$ and $v_{\pi}$ be the zero-state responses for the two cases, respectively, and let $v_0$ be the zero-state response for the original example due to the unit step input $u(t)$. The figure shows

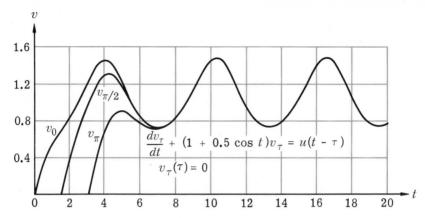

**Fig. 8.4**    Plots of three responses to step inputs; the waveforms $v_\pi(\cdot)$ and $v_{\pi/2}(\cdot)$ cannot be obtained from the waveform $v_0(\cdot)$ by a time shift.

clearly that the two new responses cannot be obtained by shifting the response $v_0$ in time; in symbols

$$v_{\pi/2} \neq \Im_{\pi/2} v_0$$

and

$$v_\pi \neq \Im_\pi v_0$$

where $\Im$ is the shift operator.

*c.*    The differential equation is

$$\frac{dv}{dt} + v^2 = u(t) \qquad v(0) = 0$$

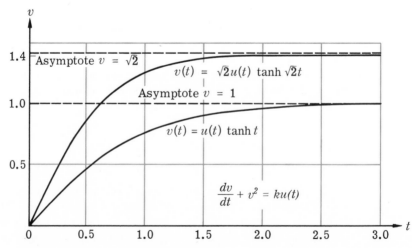

**Fig. 8.5**    Zero-state responses of an *RC* circuit with nonlinear resistor to $i_s(t) = u(t)$ and $i_s(t) = 2u(t)$.

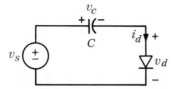

**Fig. 8.6**   A simple circuit with a *pn*-junction diode and a linear time-invariant capacitor.

The solution can be obtained by separating the variables and integrating both sides.  Thus,

$$\int_0^v \frac{d\hat{v}}{1 - \hat{v}^2} = \int_0^t dt'$$

or

$$v = u(t) \tanh t$$

Similarly, if the input were $ku(t)$ where $k$ is a constant, the zero-state response would be

$$v = \sqrt{k}\, u(t) \tanh \sqrt{k}t$$

The plots for $k = 1$ and $2$ are shown in Fig. 8.5.  Clearly, the homogeneity property is not satisfied; indeed, the ordinates of the second curve are not always twice as large as the corresponding ordinates of the first curve.

**Example 3**   The purpose of this example is to demonstrate the usefulness and simplicity of the piecewise linear analysis method in electronic circuits.  Figure 8.6 shows a voltage source $v_s$ connected to a linear time-invariant capacitor $C$ in series with a *pn*-junction diode.  The voltage source has a waveform that is a rectangular wave, as shown on Fig. 8.8.  We wish to

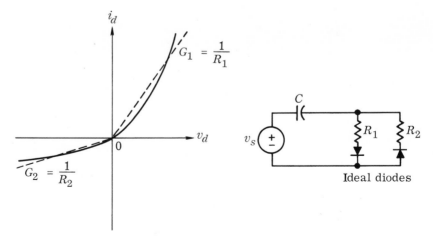

**Fig. 8.7**   Piecewise linear approximation of the *pn*-junction diode characteristic.

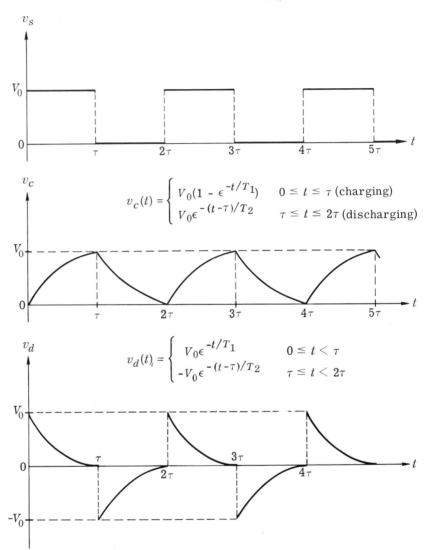

**Fig. 8.8**   Waveforms of voltages $v_s$, $v_c$, and $v_d$.   Note: $v_s = v_c + v_d$.   The capacitor is completely charged or discharged at the end of each half cycle.

determine the waveforms $v_c(\,\cdot\,)$ and $v_d(\,\cdot\,)$.   We settle for an approximate solution because the closed-form solutions are too complicated.   We use piecewise linear analysis; we approximate the diode characteristic by two straight-line segments starting at the origin, as shown on Fig. 8.7.   For positive voltages the approximating line segment has slope $G_1$, and for negative voltages the approximating line segment has slope $G_2$.   Suppose that the value of $C$ is such that the two time constants $T_1 = R_1C$ and $T_2 = R_2C$ have the property that $T_1 \ll \tau$ and $T_2 \ll \tau$, where $2\tau$ is the period of the rectangular waveform $v_s$, which is shown on Fig. 8.8.   As a consequence, at the end of each half cycle the capacitor is practically completely charged or completely discharged. The waveforms of the voltages across the capacitor and the diode are shown in Fig. 8.8.

Next, assume that $T_1$ and $T_2$ are of a magnitude comparable to $\tau$. Then in alternate half cycles, the capacitor is neither completely charged nor discharged. However, as time proceeds, the response, say the voltage across the capacitor, will reach a steady state as shown in Fig. 8.9. Assume that at $t = 0$, the response has reached a steady state. It is straightforward to calculate the voltages $V_1$ and $V_2$ that characterize the steady-state behavior and are defined in the figure. The waveforms for the response in the charging half cycle and the discharging half cycle are, respectively,

$$v_1(t) = V_1 + (V_0 - V_1)(1 - \epsilon^{-t/T_1}) \qquad 0 \le t \le \tau$$

and

$$v_2(t) = V_2 \epsilon^{-(t-\tau)/T_2} \qquad \tau \le t \le 2\tau$$

Clearly, at $t = \tau$

$$v_1(\tau) = V_2 = V_1 + (V_0 - V_1)(1 - \epsilon^{-\tau/T_1})$$

and at $t = 2\tau$

$$v_2(2\tau) = V_1 = V_2 \epsilon^{-\tau/T_2}$$

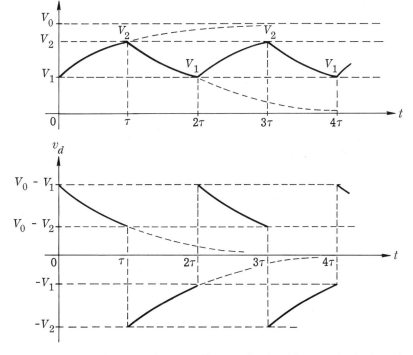

**Fig. 8.9**   Voltage waveforms in the steady state. The capacitor is neither completely charged nor discharged at the end of each half cycle.

Solving for $V_1$ and $V_2$ from the two equations, we have

$$V_2 = V_0 \frac{1 - \epsilon^{-\tau/T_1}}{1 - \epsilon^{-\tau/T_1}\epsilon^{-\tau/T_2}}$$

$$V_1 = V_0 \frac{(1 - \epsilon^{-\tau/T_1})\epsilon^{-\tau/T_2}}{1 - \epsilon^{-\tau/T_1}\epsilon^{-\tau/T_2}}$$

Of course, the degree to which the results of the piecewise linear analysis approximate the actual response depends upon the degree to which the piecewise linear approximation of the diode characteristic is a good model for its nonlinear characteristic.

**Example 4**  Let us consider the zero-input response of the parallel $RC$ circuit shown in Fig. 8.10$a$. The capacitor is linear and time-invariant and has a capacitance $C$ and an initial voltage $V_0$. The resistor is time-invariant but nonlinear; its characteristic is shown in Fig. 8.10$b$. Note that the nonlinear resistor is current-controlled; that is, the branch voltage $v_R$ is a single-valued function of the branch current $i_R$. Thus

(8.1)  $v_R = f(i_R)$

However, since for some voltages several different values for the current are allowed by the characteristic, the inverse function does not exist. We wish to determine the zero-input response due to the initial voltage $V_0$.

As usual we use Kirchhoff's laws and the branch equations to obtain a differential equation in terms of some network variable in the circuit. We proceed by writing

KCL:  $\qquad\qquad\qquad i_C + i_R = 0$

KVL:  $\qquad\qquad\qquad v_R = v_C$

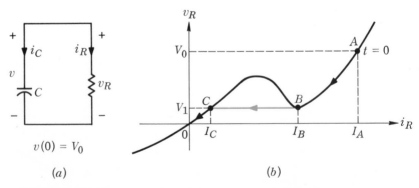

(a)  $\qquad\qquad\qquad\qquad\qquad$ (b)

**Fig. 8.10**  ($a$) An $RC$ circuit where the capacitor is linear and time-invariant and the resistor is nonlinear and time-invariant with the characteristic shown in ($b$); $V_0$ is the initial voltage on the capacitor.

Capacitor branch:     $i_C = C \dfrac{dv_C}{dt}$

Combining the above three equations, we have

(8.2)   $C \dfrac{dv_R}{dt} + i_R = 0$

Had we been able to express $i_R$ in terms of $v_R$, we would have obtained a differential equation in $v_R$. However, since the resistor, as characterized by Fig. 8.10b, is not voltage-controlled, this is impossible to achieve. In the following we shall attempt to solve the circuit by reasoning physically. Suppose that at $t = 0$, the initial voltage is $V_0$. This corresponds to point $A$ on the characteristic of the nonlinear resistor in Fig. 8.10b. Thus, the initial current flowing in the resistor is $I_A$. As time increases, the capacitor discharges through the nonlinear resistor. Both the voltage $v_R$ and the current $i_R$ decrease according to the characteristic curve of the resistor and at a rate specified by Eq. (8.2). When point $B$ is reached, a dilemma appears. Had the voltage and current followed the characteristic curve, Eq. (8.2) would have been violated. This is clear since the voltage $v_R$ would increase after point $B$ has been reached, which implies that $dv_R/dt$ is positive; but $i_R$ is also positive. Consequently, (8.2) cannot be satisfied. The only possible conclusion is as follows: When point $B$ is reached, $v_R = V_1$, and $i_R = I_B$; the voltage will remain at $V_1$, but the current jumps instantaneously to $I_C$, which corresponds to point $C$ on the characteristic.† This instantaneous change of current is referred to as a *jump phenomenon*. After point $C$ is reached, the voltage and current will follow the characteristic curve once again until the origin is reached. The waveforms for $v_R$ and $i_R$ are shown in Fig. 8.11.

We should comment that the jump phenomenon frequently occurs in electronic circuits. Actually, it is indispensable in computer circuits. In a later chapter we shall see that if we modify the circuit slightly by introducing a small inductance in series with the resistor, we shall be able to write the differential equation. The solution to the modified circuit will not have an abrupt jump, but it will be a smooth curve with a very rapidly varying portion. Physically, the small inductance may represent the lead inductance of the connecting wire.

**Exercise**   Let the nonlinear resistor in Fig. 8.10a be voltage-controlled; that is, $i_R = g(v_R)$, where $g$ is a (single-valued) function. Determine the zero-input response of the voltage due to an initial voltage $V_0$. Write the appropriate differential equation, and discuss physically what happens.

---

† The voltage across the capacitor cannot jump instantaneously. For the capacitor voltage to jump, an infinitely large current is required; clearly, the resistor is not capable of supplying such a current.

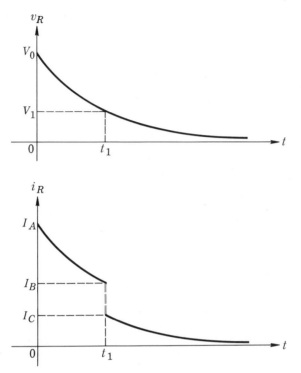

**Fig. 8.11**    Voltage and current waveforms for the parallel $RC$ circuit of Fig. 8.10.

## Summary

- A lumped circuit is said to be linear if each of its elements is either a linear element or an independent source. A lumped circuit is said to be time-invariant if each of its elements is either time-invariant or an independent source.

- The zero-input response of a circuit is defined to be the response of the circuit when no input is applied to it; thus, the zero-input response is due to the initial state only.

- The zero-state response of a circuit is defined to be the response of the circuit due to an input applied at some time, say $t_0$, subject to the condition that the circuit be in the zero state just prior to the application of the input (that is, at time $t_{0-}$); thus the zero-state response is due to the input only.

- The step response is defined to be the zero-state response due to a unit step input.

- The impulse response is defined to be the zero-state response due to a unit impulse.

■   For linear first-order circuits (time-invariant or time-varying) we have shown that

1.   The zero-input response is a linear function of the initial state.

2.   The zero-state response is a linear function of the input waveform.

3.   The complete response is the sum of the zero-input response and the zero-state response.

■   For linear time-invariant first-order circuits we have shown that

1.   The zero-state response to the shifted input is equal to the shift of the zero-state response to the original input:

$$\mathcal{T}_\tau[\mathcal{Z}_0(i_0)] = \mathcal{Z}_0[\mathcal{T}_\tau i_0]$$

2.   The impulse response is the derivative of the step response:

$$h = \frac{d\mathfrak{s}}{dt} \quad \text{or equivalently} \quad \mathfrak{s}(t) = \int_{-\infty}^t h(t')\, dt'$$

■   The step and impulse responses for first-order circuits are tabulated in Table 4.1 (see pages 152 and 153).

## Problems

Zero-input response, energy calculation

**1.** The circuit shown in Fig. P4.1 is made of linear time-invariant elements. Prior to time 0, the left capacitor is charged to $V_0$ volts, and the right capacitor is uncharged. The switch is closed at time 0. Calculate the following:

*a.*   The current $i$ for $t \geq 0$.

*b.*   The energy dissipated during the interval $(0,T)$.

*c.*   The limiting values for $t \to \infty$ of (1) the capacitor voltages $v_1$ and $v_2$, (2) the current, (3) the energy stored in the capacitor and the energy dissipated in the resistor.

*d.*   Is there any relation between these energies? If so, state what it is.

*e.*   What happens as $R \to 0$?

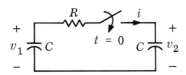

**Fig. P4.1**

Steady-state response and zero-input response

**2.** The switch $S$ in Fig. P4.2 is closed until the steady state prevails, and then it is opened. Assuming that the opening occurs at $t = 0$, find $i(t)$ and $v(t)$ for $t > 0$.

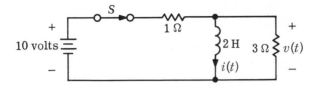

**Fig. P4.2**

Initial condition, transient, and steady state

**3.** In the circuit shown in Fig. P4.3, the switch is caused to snap back and forth between the two positions $A$ and $B$ at regular intervals equal to $L/R$ sec. After a large number of cycles the current becomes *periodic,* as shown in the accompanying plot. Determine the current levels $I_1$ and $I_2$ characterizing this periodic waveform.

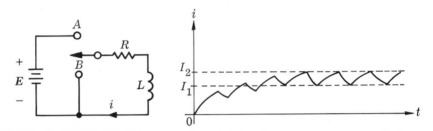

**Fig. P4.3**

Impulse response

**4.** Let the circuit given in Fig. P4.4 be linear and time-invariant. Let $e_s$ be a unit impulse. Find the zero-state response $i$.

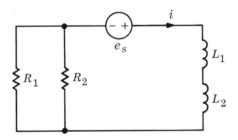

**Fig. P4.4**

Linearity of the zero-state response

**5.** The purpose of this problem is to illustrate the important fact that although the *zero-state response* of a linear circuit is a linear function of its input, the *complete response* is *not.* Consider the linear time-invariant circuit shown in Fig. P4.5.

**Fig. P4.5**

*a.* Let $i(0) = 2\,\text{mA}$.  Let $i_1$ and $i_2$ be the responses corresponding to voltages $e_1$ and $e_2$ applied by the voltage source $e_s$.  The voltages $e_1$ and $e_2$ are given as

$$e_1 = 10 \text{ volts} \qquad \text{for } t \geq 0$$

$$e_2 = 30 \text{ volts} \qquad \text{for } t \geq 0$$

Plot $i_1$ and $i_2$ as functions of $t$.  Is it true that $i_2(t) = 3i_1(t)$ for all $t \geq 0$?

*b.* Consider now the zero-state responses due to $e_1$ and $e_2$; call them $i_1'$ and $i_2'$.  Plot $i_1'$ and $i_2'$ as functions of $t$.  Is it true that $i_2'(t) = 3i_1'(t)$ for all $t \geq 0$?

Pulse response   **6.** A pulse of voltage of 10 volts magnitude and 5 $\mu$sec duration is applied to the linear time-invariant $RL$ circuit shown in Fig. P4.6.  Find the current waveform $i(\,\cdot\,)$, and plot it for the following sets of values of $R$ and $L$:

*a.*  $R = 2$ ohms, $L = 10\ \mu\text{H}$

*b.*  $R = 2$ ohms, $L = 5\ \mu\text{H}$

*c.*  $R = 2$ ohms, $L = 2\ \mu\text{H}$

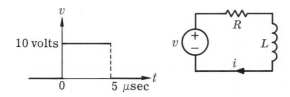

**Fig. P4.6**

Complete response and transient due to a sinusoidal input   **7.** Consider the linear time-invariant circuit shown on Fig. P4.7.  Let $v_C(0) = 1$ volt.

*a.*  If the sinusoidal voltage $e_s = 30 \cos 2\pi \times 10^3 t$ volts is applied at $t = 0$, calculate and sketch $i(t)$ for $t \geq 0$.

*b.*  Suppose we have control over the phase $\phi$ of the generator voltage; is there a value of $\phi$ such that the input $30 \cos (2\pi \times 10^3 t + \phi)$ does not cause any transient?  If so, what is an appropriate value of $\phi$?

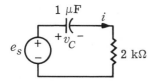

**Fig. P4.7**

Transient and
steady-state
response due
to a constant
input

**8.** Consider the linear time-invariant circuit shown in Fig. P4.8. At $t = 0$ a constant voltage source of 10 volts is applied to the circuit. Find all branch voltages and all branch currents at $t = 0$ and at $t = \infty$, given $i_1(0) = 2$ amp and $v_4(0) = 4$ volts.

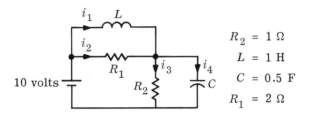

$R_2 = 1\ \Omega$
$L = 1\ H$
$C = 0.5\ F$
$R_1 = 2\ \Omega$

**Fig. P4.8**

Pulse
responses and
an elementary
design prob-
lem

**9.** The zero-state responses $v_1$, $v_2$, and $v_3$ of three linear time-invariant one-ports $\mathfrak{N}_1$, $\mathfrak{N}_2$, and $\mathfrak{N}_3$ to the same input current $i$ are shown in the Fig. P4.9. Knowing that these circuits can be described by first-order differential equations, propose a circuit topology and appropriate element values.

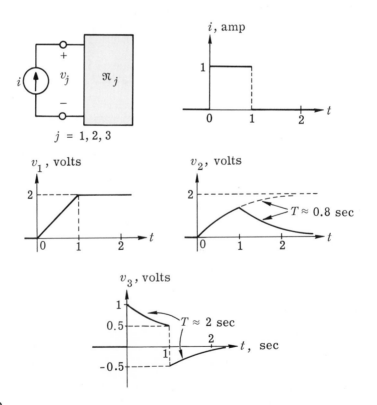

**Fig. P4.9**

Complete re-
sponse, power,
and energy
**10.** The source shown in the linear time-invariant circuit of Fig. P4.10 delivers current according to the equation $i(t) = 5r(t)$ mA. Knowing that $v(0) = -5$ volts, calculate and sketch the voltage $v(t)$, the power delivered by the source, $p(t)$, and the energy stored in the circuit, $\mathcal{E}(t)$, for $t \geq 0$.

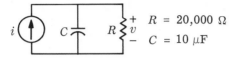

$R = 20,000\ \Omega$

$C = 10\ \mu\text{F}$

**Fig. P4.10**

Nonlinear RL
circuit
**11.** *a.* Given the circuit shown in Fig. P4.11, where $L = 10$ mH, $R = 100$ ohms, and $D$ is a nonlinear time-invariant resistor defined by $i = I_s(\epsilon^{qv/kT} - 1)$ amp, where $I_s = 10^{-6}$ amp and $kT/q = 0.026$ volt, write the differential equation for the current $i$.

*b.* Is the voltage across the resistor $R$ at time $t = 0$ sufficient to establish the initial condition?

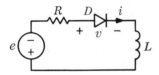

**Fig. P4.11**

Complete
response
**12.** The circuit shown in Fig. P4.12 is linear and time-invariant. Knowing that $i(0) = -10$ mA,

*a.* Calculate and sketch the current $i$ for $t \geq 0$.

*b.* Identify the zero-state response and the zero-input response from the solution obtained in (*a*).

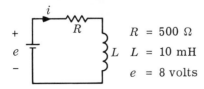

$R = 500\ \Omega$

$L = 10$ mH

$e = 8$ volts

**Fig. P4.12**

Complete
response,
sinusoidal
steady state
*c.* Does a steady state exist for the solution obtained in (*a*)? If there is one, indicate what it is.

**13.** Solve the preceding problem for a value of $R = -500$ ohms.

Waveforms<br>and shifted<br>waveforms

**14.** The function $f$ is specified by the graph of Fig. P4.14.

 *a.* Define the function $f$ by appropriate mathematical expressions for the intervals $(-\infty,0)$, $(0,1)$, $(1,2)$, $(2,3)$, and $(3,\infty)$.

 *b.* Graph the functions $g_1, g_2, g_3$, and $g_4$, which are defined from $f$ by

$$g_1(t) = f(-t) \qquad\qquad g_3(t) = f(-3 - t)$$
$$g_2(t) = f(2 - t) \qquad\qquad g_4(t) = -4f(t - 7)$$

 for all $t$.

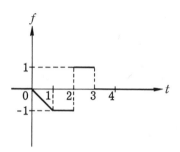

**Fig. P4.14**

Transient and<br>steady state

**15.** Consider the linear time-invariant circuit shown in Fig. P4.15.  The voltage source is an oscillator delivering a voltage

$$e_s(t) = \begin{cases} 2 \sin 10^3 t \text{ volts} & \text{for } t \geq 0 \\ 0 & \text{for } t < 0 \end{cases}$$

Knowing that $i(0) = 2$ mA, calculate and sketch the current $i$ for $t \geq 0$. Assume now that the voltage delivered by the oscillator is

$$e_s(t) = \begin{cases} 2 \sin (10^3 t + \phi) \text{ volts} & \text{for } t \geq 0 \\ 0 & \text{for } t < 0 \end{cases}$$

where the phase angle $\phi$ can be adjusted to any desired value.  Given $i(0) = 2$ mA, is there any value of $\phi$ for which a transient does not occur (i.e., steady state takes place at $t = 0$)?  If so, give an appropriate value for $\phi$.

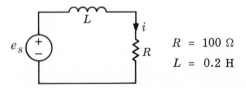

$$R = 100 \ \Omega$$
$$L = 0.2 \ H$$

**Fig. P4.15**

**16.** The zero-state response $v$ of a linear time-invariant $RC$ circuit to a unit step of current is $v(t) = u(t)2(1 - \epsilon^{-t})$. Sketch with reasonable accuracy the zero-state response to a current having the waveform $i(\cdot)$ (see Fig. P4.16).

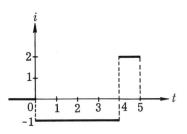

**Fig. P4.16**

**17.** Find the exact solution for the voltage $v(t)$ resulting from the discharge of a linear capacitor $C$ through a nonlinear resistor $\Re$. The capacitor and the resistor are time-invariant, the latter being characterized by the $vi$ curve given in Fig. P4.17. (Hint: For $|v| > 2$ volts, the resistor $\Re$ is equivalent to a combination of linear resistor and voltage source, and for $|v| < 2$, $\Re$ is equivalent to a 2-ohm resistor. Compute the part of the response when $v > 2$ volts and when $v < 2$ volts.)

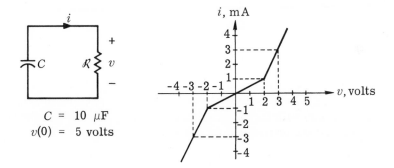

$$C = 10 \ \mu F$$
$$v(0) = 5 \text{ volts}$$

**Fig. P4.17**

**18.** Consider the linear time-invariant series $RL$ circuit driven by a voltage source as shown in Fig. P4.18.

a. Calculate the zero-state response $i$ to a unit step of voltage.

b. Calculate and sketch the zero-state response $h_\Delta(t)$ to an input voltage $p_\Delta$ for values of $\Delta = 0.5$, 0.2, and 0.05 sec.

c. Determine the unit impulse response $h$.

d. With regard to the solution of (c), explain the discontinuity of the current through the inductor at time $t = 0$.

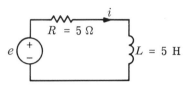

**Fig. P4.18**

Nonlinear circuit

**19.** Consider the nonlinear circuit shown in Fig. P4.19. The voltage source $e_s$ is the input, and the capacitor voltage $v_C$ is the response. The resistor and the capacitor are linear and time-invariant.

a. Calculate the impulse response and the step response.

b. Show that the derivative of the step response is *not* equal to the impulse response.

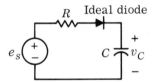

**Fig. P4.19**

Linear time-varying circuit

**20.** Consider the linear circuit shown in Fig. P4.20; the resistor is time-varying. Its resistance at time $t$ is $R(t) = t + 1$, and the capacitor is time-invariant with the capacitance of 1 farad. The input is the voltage source $e_s$, and the response is the capacitor charge $q$.

a. Calculate the impulse response and the step response.

b. Show that the derivative of the step response is *not* equal to the impulse response.

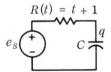

**Fig. P4.20**

Relation be-
tween the
impulse re-
sponse and
the step
response

**21.** A linear time-invariant circuit has an impulse response $h$ as shown in Fig. P4.21.  Find the step response of the circuit at times

*a.*   $t = 1$

*b.*   $t = \infty$

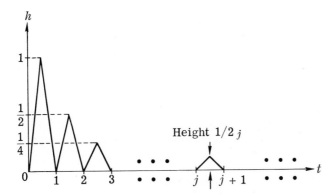

**Fig. P4.21**

Linear time-
varying circuit

**22.** The circuit in the Fig. P4.22 is linear and time-varying.  Show that the zero-state response is a linear function of the input and that the zero-input response is a linear function of the initial state.

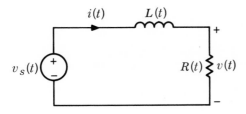

**Fig. P4.22**

Complete
response

**23.** For the linear time-invariant circuit shown in Fig. P4.23, the input is a unit step applied at time $t = 0$.   The initial conditions are $v_C(0-) = 1$ volt and $i_L(0-) = 2$ amp.   Find the output voltage response $v(\cdot)$, $t \geq 0$.

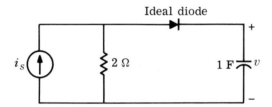

**Fig. P4.23**

Nonlinear circuit

**24.** For the circuit shown in Fig. P4.24, calculate the ramp response $v(\,\cdot\,)$, that is, the response $v$ due to an input $i_s(t) = r(t)$ with $v(0) = 0$. Is the step response for the circuit the derivative of the ramp response?

**Fig. P4.24**

Piecewise linear analysis of nonlinear network

**25.** The circuit shown in the figure contains a linear capacitor that is charged at $t = 0$ with $v(0) = 3$ volts. The nonlinear resistor has a $vi$ characteristic as shown in Fig. P4.25. Determine the waveform $v(\,\cdot\,)$ and the time it takes for the voltage to reach 1 volt.

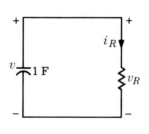
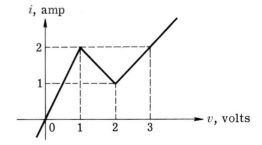

**Fig. P4.25**

Piecewise linear analysis of nonlinear network

**26.** For the circuit given in Fig. P4.25, the resistor characteristic is changed to that of Fig. P4.26. Determine the waveform $v(\,\cdot\,)$ if $v(0) = 2$ volts.

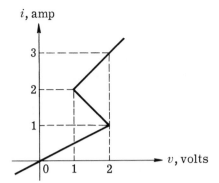

$i$, amp

**Fig. P4.26**

**27.** Consider the linear time-varying circuit shown in Fig. P4.27. The sliding contact $S$ is moved up and down in a sinusoidal fashion, so that as long as the terminals ② and ②′ are left open-circuited we have the relation

$$\frac{v_2(t)}{v_1(t)} = \frac{1 + \cos \pi t}{2}$$

*a.* A current source is connected to terminals ① and ①′, and the open-circuit voltage is observed across terminals ② and ②′. If the current source applies at a unit impulse of current at time $\tau$, calculate the voltage $h_{21}(t,\tau)$ observed across ② ②′.

*b.* Now connect the current source to terminals ② and ②′, and observe the open-circuit voltage across terminals ① and ①′. If the current source applies a unit impulse of current at time $\tau$, calculate the voltage $h_{12}(t,\tau)$ across the terminals ① and ①′.

*c.* Sketch for $t \geq \tau$, the waveforms $h_{21}(t,0)$, $h_{21}(t,0.5)$, $h_{21}(t,1)$, $h_{12}(t,0)$, $h_{12}(t,0.5)$, and $h_{12}(t,1)$.

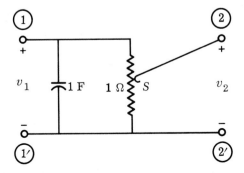

**Fig. P4.27**

# 5

# Second-order Circuits

In Chap. 4 we studied first-order electric circuits in detail and encountered both linear and nonlinear circuits. We studied linear circuits and calculated their complete response, zero-input response, and zero-state response. We established that for linear circuits the zero-input response is a linear function of the initial state and that the zero-state response is a linear function of the input. These facts are valid for general linear networks and will be proven in Chap. 13. In this chapter we shall study second-order circuits. We shall use a simple parallel $RLC$ (resistor-inductor-capacitor) circuit to illustrate the calculation of the zero-input response and the zero-state response. We shall also encounter a new way of describing a circuit, the state-space method. We shall apply this method not only to linear circuits, but also to nonlinear circuits.

## 1   Linear Time-invariant $RLC$ Circuit, Zero-input Response

In Fig. 1.1 we have a parallel connection of three linear time-invariant and passive elements: a resistor, an inductor, and a capacitor. Their branch equations are

$(1.1a)$    $v_R = Ri_R$    or    $i_R = Gv_R$

$(1.1b)$    $v_L = L\dfrac{di_L}{dt}$    $i_L(0) = I_0$    or    $i_L(t) = I_0 + \dfrac{1}{L}\displaystyle\int_0^t v_L(t')\,dt'$

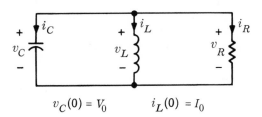

$$v_C(0) = V_0 \qquad i_L(0) = I_0$$

**Fig. 1.1**   Parallel $RLC$ circuit; the three elements are linear time-invariant and passive.

(1.1c)   $v_C(t) = V_0 + \dfrac{1}{C} \displaystyle\int_0^t i_C(t') \, dt'$   or   $i_C = C\dfrac{dv_C}{dt}$   $v_C(0) = V_0$

where $R$, $G$, $L$, and $C$ are *positive* numbers representing, respectively, the resistance, conductance, inductance, and capacitance. $I_0$ represents the initial current in the inductor, and $V_0$ represents the initial voltage across the capacitor. $v_R$, $v_L$, $v_C$, $i_R$, $i_L$, and $i_C$ are the six network variables. From KVL we have

(1.2)   $v_C = v_R = v_L$

and from KCL we have

(1.3)   $i_C + i_R + i_L = 0$

Altogether we have six equations, three in (1.1), two in (1.2), and one in (1.3). This leads us to expect that the six unknown network variables can be uniquely determined. In fact, our development will show that they are indeed uniquely determined.

The problem is to pick the most convenient variable, write the most convenient equation in terms of that variable, solve for that variable, and then calculate the remaining five variables. One way is to use the capacitor voltage $v_C$ as the most convenient variable. Using Eqs. (1.1) to (1.3), we obtain the following integrodifferential equation in terms of the voltage $v_C$:

(1.4)   $C\dfrac{dv_C}{dt} + Gv_C + I_0 + \dfrac{1}{L}\displaystyle\int_0^t v_C(t') \, dt' = 0$

and

(1.5)   $v_C(0) = V_0$

Once the voltage $v_C$ is obtained, the five other network variables can be found from Eqs. (1.1) and (1.2). An alternate approach is to choose the inductor current $i_L$ as the variable. If we use the branch equations for the capacitor and resistor, we obtain from Eq. (1.3)

$C\dfrac{dv_C}{dt} + Gv_R + i_L = 0$

Since in (1.2) $v_C = v_R = v_L$, the above equation becomes

(1.6)   $C\dfrac{dv_L}{dt} + Gv_L + i_L = 0$

Now we use the branch equation for the inductor to obtain the following second-order differential equation with $i_L$ as the dependent variable:

(1.7)   $LC\dfrac{d^2 i_L}{dt^2} + GL\dfrac{di_L}{dt} + i_L = 0$

The necessary initial conditions are

(1.8)    $i_L(0) = I_0$

and

(1.9)    $\dfrac{di_L}{dt}(0) = \dfrac{v_L(0)}{L} = \dfrac{v_C(0)}{L} = \dfrac{V_0}{L}$

The differential equation (1.7) with initial conditions (1.8) and (1.9) has a unique solution $i_L$. Once the current $i_L$ is obtained, we can find the five other network variables from Eqs. (1.1) and (1.2). Let us proceed to solve for $i_L$ from Eqs. (1.7) to (1.9). Since no source is driving the circuit, the response $i_L$ is the *zero-input response*.

For convenience in manipulation let us define two parameters $\alpha$ and $\omega_0$ as

(1.10)    $\alpha \triangleq \dfrac{G}{2C} \qquad \omega_0 \triangleq \dfrac{1}{\sqrt{LC}}$

The parameter $\alpha$ is called the **damping constant,** and the parameter $\omega_0$ (in radians per second) is called the (angular) **resonant frequency.** $\omega_0 = 2\pi f_0$, where $f_0$ (in hertz) is the resonant frequency of the inductor and the capacitor. These two parameters $\alpha$ and $\omega_0$ characterize the behavior of the *RLC* circuit. Dividing Eq. (1.7) by *LC*, we obtain

(1.11)    $$\boxed{\dfrac{d^2 i_L}{dt^2} + 2\alpha \dfrac{di_L}{dt} + \omega_0{}^2 i_L = 0}$$

This is a second-order homogeneous differential equation with constant coefficients. The *characteristic polynomial* for this differential equation is

(1.12)    $s^2 + 2\alpha s + \omega_0{}^2$

The zeros of the characteristic polynomial are called the characteristic roots or, better, the *natural frequencies of the circuit;* they are

(1.13)    $\left.\begin{array}{c} s_1 \\ s_2 \end{array}\right\} = -\alpha \pm \sqrt{\alpha^2 - \omega_0{}^2} = \begin{cases} -\alpha + \alpha_d \\ -\alpha - \alpha_d \end{cases}$

where

$\alpha_d \triangleq \sqrt{\alpha^2 - \omega_0{}^2}$

The form of the zero-input response of the circuit depends upon the relative values of $\alpha$ and $\omega_0$. According to the relative values of $\alpha$ and $\omega_0$, we can classify the zero-input response into four cases: overdamped, critically damped, underdamped, and lossless. The first three cases give waveforms $i_L(\cdot)$ that are some forms of damped exponentials, whereas the last case corresponds to a sinusoidal waveform.

1. **Overdamped** $(\alpha > \omega_0)$. The two natural frequencies $s_1$ and $s_2$ are *real* and *negative*. The response is the sum of two damped exponentials

(1.14)   $$i_L(t) = k_1 \epsilon^{s_1 t} + k_2 \epsilon^{s_2 t}$$

where the constants $k_1$ and $k_2$ depend on the initial conditions.

2. **Critically damped** $(\alpha = \omega_0)$. The two natural frequencies are equal and real; that is, $s_1 = s_2 = -\alpha$. The response is

(1.15)   $$i_L(t) = (k + k't)\epsilon^{-\alpha t}$$

where $k$ and $k'$ are constants that depend upon the initial conditions.

3. **Underdamped** $(\alpha < \omega_0)$. The two natural frequencies are complex conjugate $(s_1 = -\alpha + j\omega_d$, and $s_2 = -\alpha - j\omega_d$, where $\omega_d^2 \triangleq \omega_0^2 - \alpha^2)$. The response is of the form

(1.16)   $$i_L(t) = k\epsilon^{-\alpha t} \cos(\omega_d t + \theta)$$

where $k$ and $\theta$ are real constants that depend upon the initial conditions. A typical plot of the waveform $i_L(\cdot)$ is shown in Fig. 1.2, where

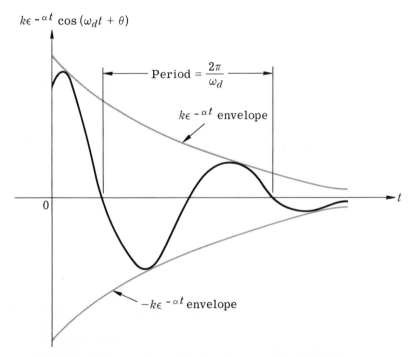

**Fig. 1.2**   Waveform $i_L(\cdot)$ for the underdamped case $(\alpha < \omega_0)$ of the parallel *RLC* circuit.

the light exponential curves are called the envelopes.  Note that the peaks of the waveform decrease in amplitude according to the damped exponential envelopes.

4.  Lossless ($\alpha = 0$, hence $G = 0$).  The two natural frequencies are imaginary ($s_1 = j\omega_0$, and $s_2 = -j\omega_0$).  The response is

(1.17)    $$i_L(t) = k \cos (\omega_0 t + \theta)$$

where $k$ and $\theta$ are real constants that depend upon the initial conditions.

It can be easily shown by direct substitution that Eqs. (1.14) to (1.17) are the general solutions of the homogeneous differential equation in (1.11).  In each case the two arbitrary constants are to be determined from the given initial conditions in Eqs. (1.8) and (1.9).  The evaluation of the arbitrary constants from the given initial conditions is straightforward.

The four cases can also be classified in terms of the natural frequencies, i.e., the two roots $s_1$ and $s_2$ of the characteristic polynomial of the differential equation.  Since natural frequencies can be real, complex, or imaginary, it is instructive to locate them in the complex plane, called the *complex frequency plane*.  In the complex frequency plane ($s$ plane), the horizontal axis represents the real part, and the vertical axis represents the imaginary part.  The four cases are illustrated in Fig. 1.3, where the location of the natural frequencies is plotted in the $s$ plane on the left, and the corresponding waveform $i_L(\cdot)$ is plotted on the right.  The significance of the complex frequency plane will become clearer in Chap. 13 when the Laplace transform is introduced.  However, it should be recognized now that the locations of the natural frequencies in the complex frequency plane dictate the form of the response.

**Exercise**    The solution of the homogeneous differential equation (1.11) for the underdamped case can also be written as

$$i_L(t) = k_1 \epsilon^{s_1 t} + k_2 \epsilon^{s_2 t}$$

where $s_1$, $s_2$, $k_1$, and $k_2$ are complex numbers, and

$$s_2 = \bar{s}_1 = -\alpha - j\omega_d \qquad k_2 = \bar{k}_1$$

The bars denote the complex conjugate.  Derive Eq. (1.16) from the above, and show that

$$k = |k_1| \qquad \text{and} \qquad \theta = \measuredangle \bar{k}_1$$

*Evaluation of arbitrary constants*    Let us consider the overdamped case.  The current $i_L$ is given by (1.14) as

$$i_L(t) = k_1 \epsilon^{s_1 t} + k_2 \epsilon^{s_2 t}$$

We wish to determine the constants $k_1$ and $k_2$ from the initial conditions

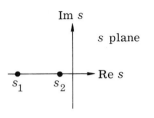 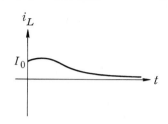

$$(a)$$

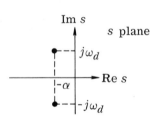 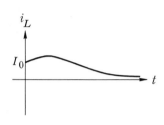

$$(b)$$

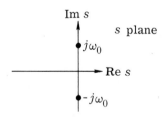 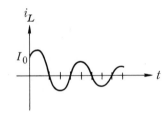

$$(c)$$

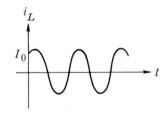

$$(d)$$

**Fig. 1.3**    Zero-input responses of the parallel $RLC$ circuit classified according to the locations of natural frequencies on the left and the waveforms on the right. $(a)$ Overdamped $(\alpha > \omega_0)$; $(b)$ critically damped $(\alpha = \omega_0)$; $(c)$ underdamped $(\alpha < \omega_0)$; $(d)$ lossless $(\alpha = 0)$.

specified in Eqs. (1.8) and (1.9).   Evaluating $i_L(t)$ in (1.14) at $t = 0$, we obtain

$$(1.18) \quad i_L(0) = k_1 + k_2 = I_0$$

Differentiating (1.14) and evaluating the derivative at $t = 0$, we obtain

(1.19)   $\dfrac{di_L}{dt}(0) = k_1 s_1 + k_2 s_2 = \dfrac{V_0}{L}$

Solving for $k_1$ and $k_2$ in Eqs. (1.18) and (1.19), we obtain

(1.20)   $k_1 = \dfrac{1}{s_1 - s_2}\left(\dfrac{V_0}{L} - s_2 I_0\right)$

and

(1.21)   $k_2 = \dfrac{1}{s_2 - s_1}\left(\dfrac{V_0}{L} - s_1 I_0\right)$

Substituting $k_1$ and $k_2$ in (1.14), we obtain a general expression of the current waveform $i_L(\,\cdot\,)$ in terms of the *initial state* of the circuit, i.e., the initial current $I_0$ in the inductor and the initial voltage $V_0$ across the capacitor. Thus,

(1.22)   $i_L(t) = \dfrac{V_0}{(s_1 - s_2)L}(\epsilon^{s_1 t} - \epsilon^{s_2 t}) + \dfrac{I_0}{s_1 - s_2}(s_1 \epsilon^{s_2 t} - s_2 \epsilon^{s_1 t})$

The voltage $v_C$ across the capacitor can be calculated from $i_L$ since $v_C = v_L$ and $v_L = L\, di_L/dt$. Thus,

(1.23)   $v_C(t) = \dfrac{V_0}{s_1 - s_2}(s_1 \epsilon^{s_1 t} - s_2 \epsilon^{s_2 t}) + \dfrac{LI_0 s_1 s_2}{s_1 - s_2}(\epsilon^{s_2 t} - \epsilon^{s_1 t})$

Similarly, we can derive, for the underdamped case, the inductor current and the capacitor voltage as

(1.24)   $i_L(t) = \dfrac{V_0}{\omega_d L}\epsilon^{-\alpha t}\sin \omega_d t + I_0 \epsilon^{-\alpha t}\left(\cos \omega_d t + \dfrac{\alpha}{\omega_d}\sin \omega_d t\right)$

(1.25)   $v_C(t) = V_0 \epsilon^{-\alpha t}(\cos \omega_d t - \dfrac{\alpha}{\omega_d}\sin \omega_d t) - \dfrac{L\omega_0^2}{\omega_d}I_0 \epsilon^{-\alpha t}\sin \omega_d t$

**Exercise 1**   Prove the formulas in Eqs. (1.24) and (1.25).

**Exercise 2**   Show that for the lossless case the inductor current and the capacitor voltage are given by

(1.26)   $i_L(t) = \dfrac{V_0}{\omega_0 L}\sin \omega_0 t + I_0 \cos \omega_0 t$

(1.27)   $v_C(t) = V_0 \cos \omega_0 t - \omega_0 L I_0 \sin \omega_0 t$

**Exercise 3**   Given $I_0 = 1$ amp and $V_0 = 1$ volt, determine the zero-input responses and plot the waveforms $i_L(\,\cdot\,)$ and $v_C(\,\cdot\,)$ vs. $t$ for each of the following parallel *RLC* circuits:

    *a.*  $R = 1$ ohm, $L = 1$ henry, and $C = 1$ farad

    *b.*  $R = 1$ ohm, $L = 4$ henrys, and $C = \frac{1}{4}$ farad

    *c.*  $R = \infty$, $L = 4$ henrys, and $C = 1$ farad

It should be noted that the lossless case is actually a limiting case of the underdamped case. If we let $R$ approach infinity ($\alpha = 0$), the damped oscillation becomes a sinusoidal oscillation with angular frequency $\omega_0$.

*Energy and the Q factor*   Recall that the initial state is given by the initial current $I_0$ in the inductor and the initial voltage $V_0$ across the capacitor at $t = 0$. Thus, the initial stored energy is the sum of $\frac{1}{2}LI_0^2$ (in the magnetic field) and $\frac{1}{2}CV_0^2$ (in the electric field). Let us consider the underdamped case. As time proceeds, the energy is being transferred back and forth from the capacitor to the inductor. Meanwhile the resistor dissipates part of the energy into heat as the oscillation goes on. Thus, the total energy left in the electric and magnetic fields gradually diminishes. For $R = \infty$, the current in the resistor is always zero, and there is no energy loss; hence we have a sustained oscillation.

Note that the parameter $\omega_0$ is related to the frequency of the damped oscillation, $\omega_d = \sqrt{\omega_0^2 - \alpha^2}$, whereas the parameter $\alpha$ determines the rate of exponential decaying. The relative damping in a damped oscillation is often characterized by a number $Q$, defined by

$$(1.28) \qquad Q \triangleq \frac{\omega_0}{2\alpha} = \frac{\omega_0 C}{G} = \frac{R}{\omega_0 L} = \frac{R}{\sqrt{L/C}}$$

$Q$ can be considered as a *quality factor* of a physical resonant circuit. The less damping, the larger $Q$. Note that for the parallel $RLC$ circuit, to *decrease* the damping, we must *increase* the resistance. A lossless resonant circuit has zero damping or infinite $Q$. In Chap. 7 we shall show that $Q$ can be related to the ratio of energy stored and average power dissipated per cycle.

The four cases we have studied can also be classified according to the value of $Q$. The overdamped case has a $Q < \frac{1}{2}$, the critically damped case has a $Q = \frac{1}{2}$, the underdamped case has a $Q > \frac{1}{2}$, and the lossless case has a $Q = \infty$. In Fig. 1.4 the values of $Q$ are related to the locations of the natural frequencies in the four cases.

The lossless case ($\alpha = 0$, $R = \infty$, and $Q = \infty$) is an ideal case, for a physical inductor always has some dissipation. Thus, in practical *passive* circuits $Q = \infty$ cannot be found, which means that a sinusoidal oscillation due only to the initial state is, in fact, impossible to achieve. In Sec. 4 we shall show that if some active circuit element is employed in addition to a lossy $L$ and $C$, we can obtain sustained oscillation.

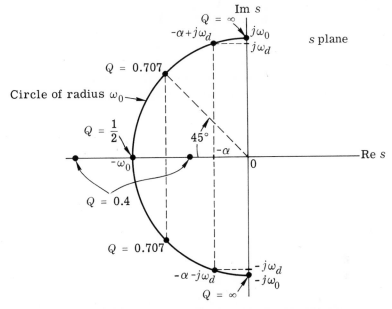

**Fig. 1.4**    Locus of the natural frequencies of the four cases; in the characteristic equation $s^2 + 2\alpha s + \omega_0^2 = s^2 + (\omega_0/Q)s + \omega_0^2 = 0$.   The resonant frequency $\omega_0 = 1/\sqrt{LC}$ is kept constant, and $Q$ varies.   This corresponds to a circuit with $L$ and $C$ fixed and $R$ varying.

**Exercise**    In practical lumped circuits, a $Q$ of the order of few hundred is available. To get a feeling of the meaning of $Q$, assume that $Q \gg 1$, and show that the amplitude of the damped oscillation decreases to 4.3 percent of its initial value after $Q$ periods.

## 2    Linear Time-invariant *RLC* Circuit, Zero-state Response

We continue with the same linear time-invariant parallel *RLC* circuit to illustrate the computation and properties of the *zero-state* response.   Thus, we are now in the case in which the initial conditions are zero and the input is not identically zero; in the previous section, the input was identically zero and the initial conditions were not all zero.

Indeed, by zero-state response we mean the response of a circuit due to an input applied at some arbitrary time $t_0$, subject to the condition that the circuit is in the zero state at $t_0-$.   We say $t_0-$ rather than $t_0$ to emphasize the fact that immediately prior to the application of the input, the initial conditions (current in the inductor and voltage across the capacitor) are zero.

KCL for the circuit of Fig. 2.1 gives

(2.1)    $i_C + i_R + i_L = i_s$

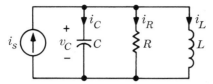

**Fig. 2.1** Parallel $RLC$ circuit with current source as input.

Following the same procedure as in Sec. 1, we obtain the network equation in terms of the inductor current $i_L$. Thus

(2.2) $\quad LC\dfrac{d^2 i_L}{dt^2} + LG\dfrac{di_L}{dt} + i_L = i_s(t) \qquad t \geq 0$

and

(2.3) $\quad i_L(0-) = 0$

(2.4) $\quad \dfrac{di_L}{dt}(0-) = \dfrac{v_C(0-)}{L} = 0$

The three equations above correspond to Eqs. (1.7), (1.8), and (1.9) of the previous section. The differences are that previously the input was zero and the initial conditions were nonzero and presently the forcing function is $i_s(t)$ as in (2.2) and the initial conditions are zero, as given by (2.3) and (2.4). We remember that the solution of a linear nonhomogeneous differential equation with constant coefficients is the sum of two terms; that is,

(2.5) $\quad i_L = i_h + i_p$

where $i_h$ is a solution of the homogeneous differential equation, that is, Eq. (2.2) with $i_s = 0$ and $i_p$ as a particular solution of the nonhomogeneous differential equation. For our problem, $i_h$ has been calculated in the previous section since it is the zero-input response; recall that it contains two arbitrary coefficients. Except for the critically damped case, $i_h$ can be written in the form

(2.6) $\quad i_h = k_1 \epsilon^{s_1 t} + k_2 \epsilon^{s_2 t}$

Of course if the natural frequencies are complex, then

(2.7) $\quad s_2 = \bar{s}_1 = -\alpha - j\omega_d \qquad \text{and} \qquad k_2 = \bar{k}_1$

and $i_h$ can also be written as

(2.8) $\quad i_h = |k_1| \epsilon^{-\alpha t} \cos(\omega_d t + \measuredangle k_1)$

On the other hand, $i_p$ depends upon the input. It is convenient to pick $i_p$ to be a constant if the input is a step function and to be a sinusoid if the input is a sinusoid.

In the remainder of this section we shall calculate only the step response and the impulse response. The calculation of the zero-state response for a sinusoidal input will be given in Chap. 7 and that for arbitrary inputs in Chap. 6.

An important property of the zero-state response for a linear circuit is that it is a linear function of the input. We shall not go through the proof now since it is similar to that of the first-order circuits and was done in Chap. 4.

| **2.1** | **Step Response** |

Let us calculate the step response of the parallel *RLC* circuit shown in Fig. 2.1. By definition the input is a unit step, and the initial conditions are zero; hence, from Eqs. (2.2) to (2.4) we have

(2.9) $\quad LC\dfrac{d^2 i_L}{dt^2} + LG\dfrac{di_L}{dt} + i_L = u(t)$

(2.10) $\quad i_L(0) = 0$

(2.11) $\quad \dfrac{di_L}{dt}(0) = 0$†

The most convenient particular solution of (2.9) is

(2.12) $\quad i_p(t) = 1 \qquad$ for $t \geq 0$

Therefore, the general solution is of the form

(2.13) $\quad i_L(t) = k_1 \epsilon^{s_1 t} + k_2 \epsilon^{s_2 t} + 1$

if the natural frequencies are distinct, and

(2.14) $\quad i_L(t) = (k + k't)\epsilon^{-\alpha t} + 1$

if they are equal. Let us determine the constants $k_1$ and $k_2$ in (2.13) using initial conditions (2.10) and (2.11). At $t = 0$, Eqs. (2.10) and (2.13) yield

(2.15) $\quad i_L(0) = k_1 + k_2 + 1 = 0$

Differentiating (2.12) and evaluating the derivative at $t = 0$, we obtain

(2.16) $\quad \dfrac{di_L}{dt}(0) = k_1 s_1 + k_2 s_2 = 0$

Solving the two equations above for $k_1$ and $k_2$, we have

(2.17) $\quad k_1 = \dfrac{s_2}{s_1 - s_2} \qquad$ and $\qquad k_2 = \dfrac{-s_1}{s_1 - s_2}$

---

† Since there is no impulse in Eq. (2.9), there is no need to make the distinction between $0-$ and $0+$.

The unit step response is therefore

(2.18)   $i_L(t) = \left[ \dfrac{1}{s_1 - s_2}(s_2 \epsilon^{s_1 t} - s_1 \epsilon^{s_2 t}) + 1 \right] u(t)$

In the underdamped case the natural frequencies are complex; thus,

$$\left. \begin{array}{c} s_1 \\ s_2 \end{array} \right\} = -\alpha \pm j \omega_d$$

or, in polar coordinates (see Fig. 2.2),

$$\left. \begin{array}{c} s_1 \\ s_2 \end{array} \right\} = \omega_0 \epsilon^{\pm j(\pi/2 + \phi)}$$

where

(2.19)   $|s_1| = |s_2| = \sqrt{\alpha^2 + \omega_d^2} = \omega_0$    and    $\phi = \tan^{-1} \dfrac{\alpha}{\omega_d}$

The first term of (2.18) can be expressed as follows:

$$\dfrac{1}{s_1 - s_2}(s_2 \epsilon^{s_1 t} - s_1 \epsilon^{s_2 t}) = \dfrac{1}{2j\omega_d} \omega_0 \epsilon^{-\alpha t}\left( \epsilon^{j(\omega_d t - \pi/2 - \phi)} - \epsilon^{-j(\omega_d t - \pi/2 - \phi)} \right)$$

$$= \dfrac{\omega_0}{2j\omega_d} \epsilon^{-\alpha t} 2j \sin\left( \omega_d t - \dfrac{\pi}{2} - \phi \right)$$

(2.20)   $\qquad\qquad\qquad = -\dfrac{\omega_0}{\omega_d} \epsilon^{-\alpha t} \cos(\omega_d t - \phi)$

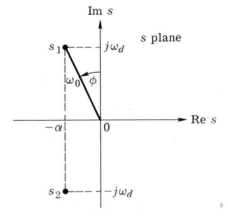

**Fig. 2.2**   Representing natural frequencies $s_1$ and $s_2$ in terms of rectangular and polar coordinates, we write

$$\left. \begin{array}{c} s_1 \\ s_2 \end{array} \right\} = -\alpha \pm j\omega_d = \omega_0 \epsilon^{\pm j(\pi/2 + \phi)}, \ \omega_0 = \sqrt{\omega_d^2 + \alpha^2},$$

$\sin \phi = \alpha/\omega_0$, $\cos \phi = \omega_d/\omega_0$, and $\tan \phi = \alpha/\omega_d$.

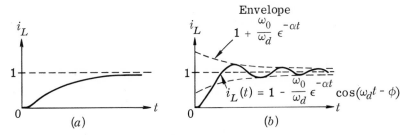

**Fig. 2.3**    Step responses for the inductor current of the parallel *RLC* circuit.  (*a*) Overdamped; (*b*) underdamped.

The unit step response becomes

(2.21)    $i_L(t) = \left[ -\dfrac{\omega_0}{\omega_d} \epsilon^{-\alpha t} \cos(\omega_d t - \phi) + 1 \right] u(t)$

Typical plots of the step response for the overdamped and the underdamped cases are given in Fig. 2.3.

It is instructive to separate the step response into two parts; the term that is either a damped exponential or a damped sinusoid represents the *transient,* and the constant term equal to unity is the *steady state.* In both cases, the current $i_L$ starts at zero and reaches unity as $t = \infty$.

The voltage across the capacitor of the parallel *RLC* circuit can be determined immediately by calculating $L di_L/dt$.  Thus,

(2.22)    $v_C(t) = L \dfrac{s_1 s_2}{s_1 - s_2} (\epsilon^{s_1 t} - \epsilon^{s_2 t}) u(t)$

and for the underdamped case

(2.23)    $v_C(t) = u(t) \sqrt{\dfrac{L}{C}} \dfrac{\omega_0}{\omega_d} \epsilon^{-\alpha t} \sin \omega_d t$

These are plotted in Fig. 2.4.  In this case the *steady state* is identically zero.  Eventually all the current from the source goes through the inductor, and since the current is constant, the voltage across the inductor is identically zero.

*Physical interpretation*    With the parallel *RLC* circuit in the zero state, a constant current source is applied in parallel to the circuit.  Clearly, the voltage across the capacitor and the current through the inductor cannot change instantaneously, so they stay at zero immediately after the input is applied.  This implies that initially the current in the resistor must also be zero, since the voltage $v_R(0) = v_C(0) = 0$.  Thus, at $t = 0$ *all* the current from the source goes through the capacitor, which causes a gradual rise of the voltage.  At $t = 0+$ *the capacitor acts as a short circuit to a suddenly applied finite constant current source.*  As time progresses, the voltage across the capacitor

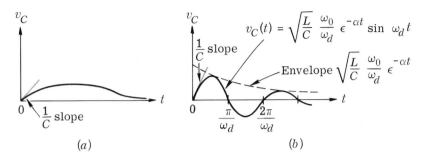

**Fig. 2.4**   Step responses for the capacitor voltage of the parallel $RLC$ circuit.

increases, and the current will flow in both the resistor and inductor. After a long time the circuit reaches a steady state; that is,

$$\frac{di_L}{dt} = 0 \qquad \frac{d^2i_L}{dt^2} = 0$$

Hence, according to Eq. (2.2), all current from the source goes through the inductor. Therefore, the voltage across the parallel circuit is zero because the current in the resistor is zero. At $t = \infty$ the inductor acts as a short circuit to a constant current source. The currents through the capacitor, resistor, and inductor are plotted in Fig. 2.5 for the overdamped case ($Q < \frac{1}{2}$).

**Exercise**   For the parallel $RLC$ circuit with $R = 1$ ohm, $C = 1$ farad, and $L = 1$ henry, determine the currents in the inductor, the capacitor, and the resistor as a result of an input step of current of 1 amp. The circuit is in the zero state at $t = 0-$. Plot the waveforms.

| 2.2 | **Impulse Response** |
|---|---|

We now calculate the impulse response for the parallel $RLC$ circuit. By definition, the input is a unit impulse, and the circuit is in the zero state at $0-$; hence, the impulse response $i_L$ is the solution of

$$(2.24) \qquad LC\frac{d^2i_L}{dt^2} + LG\frac{di_L}{dt} + i_L = \delta(t)$$

$$(2.25) \qquad i_L(0-) = 0$$

$$(2.26) \qquad \frac{di_L}{dt}(0-) = 0$$

Since the computation and physical understanding of the impulse response are of great importance in circuit theory, we shall again present several methods and interpretations, treating only the underdamped case, that is, the circuit with complex natural frequencies.

*First method*   We use the differential equation directly.   Since the impulse function $\delta(t)$ is identically zero for $t > 0$, we can consider the impulse response as a zero-input response starting at $t = 0+$.   The impulse at $t = 0$ creates an initial condition at $t = 0+$, and the impulse response for $t > 0$ is essentially the zero-input response due to that initial condition.   The problem then is to determine this initial condition.   Let us integrate both sides of Eq. (2.24) from $t = 0-$ to $t = 0+$.   We obtain

$$LC \frac{di_L}{dt}(0+) - LC \frac{di_L}{dt}(0-) + LGi_L(0+) - LGi_L(0-)$$

(2.27)
$$+\int_{0-}^{0+} i_L(t')\, dt' = 1$$

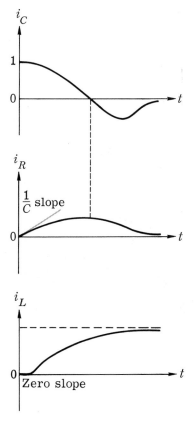

**Fig. 2.5**   Plots of $i_C$, $i_R$, and $i_L$ due to a step-current input for the parallel *RLC* circuit (overdamped case, $Q < \frac{1}{2}$).

where the right-hand side is obtained by using the fact that

$$\int_{0-}^{0+} \delta(t') \, dt' = 1$$

We know that $i_L$ cannot jump at $t = 0$, or equivalently, that $i_L$ is a continuous function; that is,

$$\int_{0-}^{0+} i_L(t') \, dt' = 0 \quad \text{and} \quad i_L(0+) = i_L(0-)$$

If it were not continuous, $di_L/dt$ would contain an impulse, $d^2 i_L/dt^2$ would contain a doublet, and (2.24) could never be satisfied since there is no doublet on the right-hand side. From (2.27) we obtain

(2.28)   $$\frac{di_L}{dt}(0+) = \frac{di_L}{dt}(0-) + \frac{1}{LC} = \frac{1}{LC}$$

As far as $t > 0$ is concerned, the nonhomogeneous differential equation (2.24), with the initial condition given in (2.25) and (2.26), is equivalent to

(2.29)   $$LC \frac{d^2 i_L}{dt^2} + LG \frac{di_L}{dt} + i_L = 0$$

with

(2.30)   $$i_L(0+) = 0$$

and

(2.31)   $$\frac{di_L}{dt}(0+) = \frac{1}{LC}$$

For $t \leq 0$, clearly, $i_L(t)$ is zero. The solution of the above is therefore

(2.32)   $$i_L(t) = u(t) \frac{\omega_0^2}{\omega_d} \epsilon^{-\alpha t} \sin \omega_d t$$

The waveform is shown in Fig. 2.6a. Note that (2.32) can also be obtained from the zero-input response of (1.24) for a given initial state $I_0 = 0$ and $V_0 = 1/C$.

**Remark**   Consider the parallel connection of the capacitor and the current source $i_s$. In Chap. 2 we showed that the parallel connection is equivalent to the series connection of the same capacitor and a voltage source $v_s$, where

$$v_s(t) = \frac{1}{C} \int_{0-}^{t} i_s(t') \, dt' \qquad t \geq 0$$

Thus, for an impulse current source, the equivalent voltage source is $(1/C)u(t)$. For $t < 0$, the voltage source is identically zero, and for $t > 0$, the voltage source is a constant $1/C$. The series connection of an uncharged capacitor and a constant voltage source is equivalent to a charged capacitor with initial voltage $1/C$. Therefore, the impulse response of a

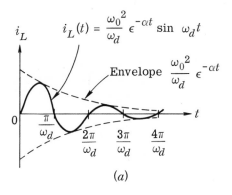

$$i_L(t) = \frac{\omega_0^2}{\omega_d} \epsilon^{-\alpha t} \sin \omega_d t$$

(a)

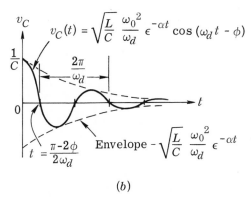

$$v_C(t) = \sqrt{\frac{L}{C}} \frac{\omega_0^2}{\omega_d} \epsilon^{-\alpha t} \cos (\omega_d t - \phi)$$

(b)

**Fig. 2.6**    Impulse response of the parallel *RLC* circuit for the underdamped case ($Q < \frac{1}{2}$).

parallel *RLC* circuit due to a current impulse in parallel is the same as a zero-input response with $v_C(0+) = 1/C$. These equivalences are illustrated in Fig. 2.7.

Let us verify by direct substitution into Eqs. (2.24) to (2.26) that (2.32) is the solution. This is a worthwhile exercise for getting familiar with manipulations involving impulses. First, $i_L$ as given by (2.32) clearly satisfies the initial conditions of (2.25) and (2.26); that is, $i_L(0-) = 0$ and $(di_L/dt)(0-) = 0$. It remains for us to show that (2.32) satisfies the differential equation (2.24). Differentiating (2.32), we obtain

(2.33)    $$\frac{di_L}{dt} = \delta(t)\left(\frac{\omega_0^2}{\omega_d} \epsilon^{-\alpha t} \sin \omega_d t\right) + \frac{u(t)\omega_0^3}{\omega_d} \epsilon^{-\alpha t} \cos (\omega_d t + \phi)$$

Now the first term is of the form $\delta(t)f(t)$. Since $\delta(t)$ is zero whenever $t \neq 0$, we may set $t = 0$ in the factor and obtain $\delta(t)f(0)$; however $f(0) = 0$. Hence the first term of (2.33) disappears, and

(2.34)    $$\frac{di_L}{dt} = \frac{u(t)\omega_0^3}{\omega_d} \epsilon^{-\alpha t} \cos (\omega_d t + \phi)$$

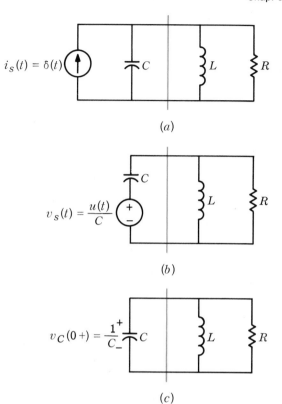

$i_s(t) = \delta(t)$

$C$   $L$   $R$

(a)

$v_s(t) = \dfrac{u(t)}{C}$   $C$   $L$   $R$

(b)

$v_C(0+) = \dfrac{1^+}{C_-}$   $C$   $L$   $R$

(c)

**Fig. 2.7**   The problem of determining the impulse response of a parallel $RLC$ circuit is reduced to that of determining the zero-input response of an $RLC$ circuit.   Note that the parallel connection of the capacitor and the impulse current source in (a) is reduced to the series connection of the capacitor and a step voltage source in (b) and is further reduced to a charged capacitor in (c).

Differentiating again, we obtain

$$\frac{d^2 i_L}{dt^2} = \delta(t)\frac{\omega_0{}^3}{\omega_d}\cos\phi - u(t)\frac{\omega_0{}^4}{\omega_d}\epsilon^{-\alpha t}\sin(\omega_d t + 2\phi)$$

(2.35)

$$= \omega_0{}^2\delta(t) - u(t)\frac{\omega_0{}^4}{\omega_d}\epsilon^{-\alpha t}[\sin(\omega_d t + \phi)\cos\phi + \cos(\omega_d t + \phi)\sin\phi]$$

Substituting Eqs. (2.32), (2.34), and (2.35) in (2.24), which is rewritten below in terms of $\omega_0$ and $\alpha$,

$$\frac{1}{\omega_0{}^2}\frac{d i_L{}^2}{dt^2} + \frac{2\alpha}{\omega_0{}^2}\frac{d i_L}{dt} + i_L = \delta(t)$$

we shall see that the left-hand side is equal to $\delta(t)$ as it should be.   Thus, we have verified by direct substitution that (2.32) is the impulse response of the parallel *RLC* circuit.

**Exercise**   Show that the impulse response for the capacitor voltage of the parallel *RLC* circuit is

(2.36)   $v_C(t) = u(t) \sqrt{\dfrac{L}{C}} \dfrac{\omega_0^2}{\omega_d} \epsilon^{-\alpha t} \cos(\omega_d t + \phi)$

The waveform is shown in Fig. 2.6*b*.

*Second method*   We use the relation between the impulse response and the step response. This method is applicable only to circuits with linear time-invariant elements for it is only for such circuits that the impulse response is the derivative of the step response.

**Exercise**   Show that the impulse responses for $i_L$ in Eq. (2.32) and $v_C$ in Eq. (2.36) are obtainable by differentiating the step response for $i_L$ in Eq. (2.21) and $v_C$ in Eq. (2.23).

*Physical interpretation*   Let us use the pulse input $i_s(t) = p_\Delta(t)$ as shown in Fig. 2.8*a* to explain the behavior of all the branch currents and voltages in the parallel *RLC* circuit.   Remember that as $\Delta \to 0$, pulse $p_\Delta$ approaches an impulse, and the response approaches the impulse response.   To start with, we assume $\Delta$ is finite and positive, but very small.   From the discussion of the step

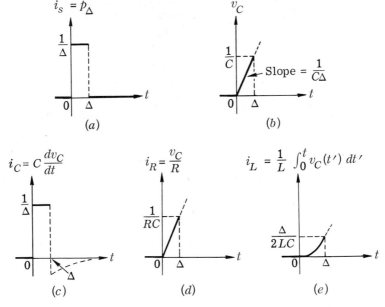

**Fig. 2.8**   Physical explanation of impulse response of a parallel *RLC* circuit; $p_\Delta$ is the input pulse; the resulting $v_C$, $i_C$, $i_R$, and $i_L$ are shown.

response, we learned that at $t = 0+$ all current from the source goes into the capacitor; that is, $i_C(0+) = i_s(0+) = 1/\Delta$, and $i_R(0+) = i_L(0+) = 0$. The current in the capacitor forces a gradual rise of the voltage across it at an initial rate of $(dv_C/dt)(0+) = i_C(0+)/C = 1/C\Delta$. Since our primary interest is in a small $\Delta$, let us assume that during the short interval $(0,\Delta)$ the slope of the voltage curve remains constant; then the voltage reaches $1/C$ at time $\Delta$, as shown in Fig. 2.8$b$. The current through the resistor is proportional to the voltage $v_C$, and hence it is linear in $t$ (see Fig. 2.8$d$). The inductor current, being proportional to the integral of $v_L$, is parabolic in $t$ (see Fig. 2.8$e$). The current through the capacitor remains constant during the interval, as shown in Fig. 2.8$c$. Of course, the assumption that during the whole interval $(0,\Delta)$ all the current from the source goes through the capacitor is false; however, the error consists of higher-order terms in $\Delta$. Therefore, as $\Delta \to 0$ the error becomes zero. Going back to Fig. 2.8$a$, we see that as $\Delta \to 0$, $i_s$ becomes an impulse $\delta$, $v_C$ undergoes a jump from 0 to $1/C$, $i_C$ becomes an impulse $\delta$, $i_R$ undergoes a jump from 0 to $1/RC$, and $i_L$ is such that $i_L(0-) = i_L(0+) = (di_L/dt)(0-) = 0$ and $(di_L/dt)(0+) = 1/LC$. Finally, as $\Delta \to 0$, from KCL we see that

$$i_C(0+) = -i_R(0+) - i_L(0+) = \frac{-1}{RC}$$

Note that these conditions check with those found earlier by other methods, as in (2.31).

## 3 | The State-space Approach

The analysis carried out in the previous sections was a straightforward extension of the method used for first-order circuits; that is, pick one appropriate variable ($i_L$ in the case above), and write one differential equation in this variable. Once this equation is solved, the remaining variables are easily calculated. However, there is another way of looking at the problem. It is clear that the zero-input response is completely determined once the initial conditions of the inductor current $I_0$ and of the capacitor voltage $V_0$ are known. Thus, we are led to think of $I_0$ and $V_0$ as specifying the *initial state* of the circuit; and the present state $(i_L(t),v_C(t))$ can be expressed in terms of the initial state $(I_0,V_0)$. In other words, we may think of the behavior of the circuit as a trajectory in a two-dimensional space starting from the initial state $(I_0,V_0)$, and for every $t$ the corresponding point of the trajectory specifies $i_L(t)$ and $v_C(t)$.

We may legitimately ask why we need to learn this new point of view. The reason is fairly simple. First, it gives a clear pictorial description of the complete behavior of the circuit, and second, it is the only effective way to analyze nonlinear and time-varying circuits. In these more general cases, to try to select one appropriate variable and write one higher-order differential equation in terms of that variable leads to many unnecessary complications. Thus, we have a strong incentive to learn the state-space approach in the simple context of second-order linear time-invariant circuits. A further advantage is that computationally the system of equations obtained from the state-space approach is readily programmed for numerical solution on a digital computer and readily set up for solution on an analog computer. A more detailed treatment of the state-space approach will be given in Chap. 12.

**3.1    State Equations and Trajectory**

Consider the same parallel $RLC$ circuit as was illustrated in Sec. 1. Let there be no current source input. We wish to compute the zero-input response. Let us use $i_L$ and $v_C$ as variables and rewrite Eqs. (1.1$b$) and (1.6) as follows:

(3.1)    $$\frac{di_L}{dt} = \frac{1}{L} v_C \qquad t \geq 0$$

(3.2)    $$\frac{dv_C}{dt} = -\frac{1}{C} i_L - \frac{G}{C} v_C \qquad t \geq 0$$

The reason that we write the equations in the above form (two simultaneous first-order differential equations) will be clear later. The variables $v_C$ and $i_L$ have great physical significance since they are closely related to the energy stored in the circuit. Equations (3.1) and 3.2) are first-order simultaneous differential equations and are called the *state equations* of the circuit. The pair of numbers $(i_L(t),v_C(t))$ is called the *state of the circuit at time t.* The pair $(i_L(0),v_C(0))$ is naturally called the *initial state;* it is given by the initial conditions

(3.3)    $i_L(0) = I_0$

$v_C(0) = V_0$

From the theory of differential equations we know that the initial state specified by (3.3) defines uniquely, by Eqs. (3.1) and (3.2), the value of $(i_L(t),v_C(t))$ for all $t \geq 0$. Thus, if we consider $(i_L(t),v_C(t))$ as the coordinates of a point on the $i_L$-$v_C$ plane, then, as $t$ increases from 0 to $\infty$, the point $(i_L(t),v_C(t))$ traces a curve that starts at $(I_0,V_0)$. The curve is called the *state-space trajectory,* and the plane $(i_L,v_C)$ is called the *state space* for the circuit. We can think of the pair of numbers $(i_L(t),v_C(t))$ as the components of a vector $\mathbf{x}(t)$ whose origin is at the origin of the coordinate axes; thus, we write

$$\mathbf{x}(t) = \begin{bmatrix} i_L(t) \\ v_C(t) \end{bmatrix}$$

The vector $\mathbf{x}(t)$ is called the *state vector* or, briefly, the *state*. Thus, $\mathbf{x}(t)$ is a vector defined for all $t \geq 0$ in the state space. Its components, the current $i_L$ through the inductor and the voltage $v_C$ across the capacitor, are called the *state variables*. Knowing the state at time $t$, that is, the pair of numbers $(i_L(t), v_C(t))$, we can obtain the velocity of the trajectory $((di_L/dt)(t), (dv_C/dt)(t))$ from the state equations (3.1) and (3.2).

**Example 1**  Consider the overdamped, underdamped, and lossless cases of the parallel *RLC* circuit. Let the initial state be $I_0 = 1$ amp and $V_0 = 1$ volt.

*a.*  Overdamped. $R = 1$ ohm, $L = 4$ henrys, and $C = \frac{1}{12}$ farad ($\alpha = 2$, and $\omega_0 = \sqrt{3}$). Thus, the natural frequencies are $s_1 = -1$ and $s_2 = -3$. From Eqs. (1.22) and (1.23), we obtain

$$i_L(t) = \frac{1}{8}(\epsilon^{-t} - \epsilon^{-3t}) + \frac{1}{2}(-\epsilon^{-3t} + 3\epsilon^{-t})$$

$$= \frac{13}{8}\epsilon^{-t} - \frac{5}{8}\epsilon^{-3t}$$

and

$$v_C(t) = \frac{1}{2}(-\epsilon^{-t} + 3\epsilon^{-3t}) + 6(\epsilon^{-3t} - \epsilon^{-t})$$

$$= -\frac{13}{2}\epsilon^{-t} + \frac{15}{2}\epsilon^{-3t}$$

The waveforms are plotted in Fig. 3.1*a*. Next we use $t$ as a parameter, and plot for each value of $t$ the state $(i_L(t), v_C(t))$ in the state space, i.e., the plane with $i_L$ as abscissa and $v_C$ as ordinate. The result is shown in Fig. 3.1*b*. Note that the trajectory starts at $(1,1)$ when $t = 0$ and ends at the origin when $t = \infty$.

*b.*  Underdamped. $R = 1$ ohm, $L = 1$ henry, and $C = 1$ farad ($\alpha = \frac{1}{2}$, $\omega_0 = 1$, and $\omega_d = \sqrt{3}/2$). From Eqs. (1.24) and (1.25) we have

$$i_L(t) = \epsilon^{-t/2}\left(\cos\frac{\sqrt{3}}{2}t + \sqrt{3}\sin\frac{\sqrt{3}}{2}t\right)$$

$$= 2\epsilon^{-t/2}\cos\left(\frac{\sqrt{3}}{2}t - 60°\right)$$

and

$$v_C(t) = \epsilon^{-t/2}\left(\cos\frac{\sqrt{3}}{2}t - \sqrt{3}\sin\frac{\sqrt{3}}{2}t\right)$$

$$= 2\epsilon^{-t/2}\cos\left(\frac{\sqrt{3}}{2}t + 60°\right)$$

The waveforms are plotted in Fig. 3.2*a*, and the trajectory is plotted in Fig. 3.2*b*. Note that the trajectory is a spiral starting at $(1,1)$ and terminating at the origin.

*c.*  Lossless. $L = 4$ henrys and $C = 1$ farad ($\alpha = 0$, and $\omega_0 = 2$). From Eqs. (1.26) and (1.27) we have

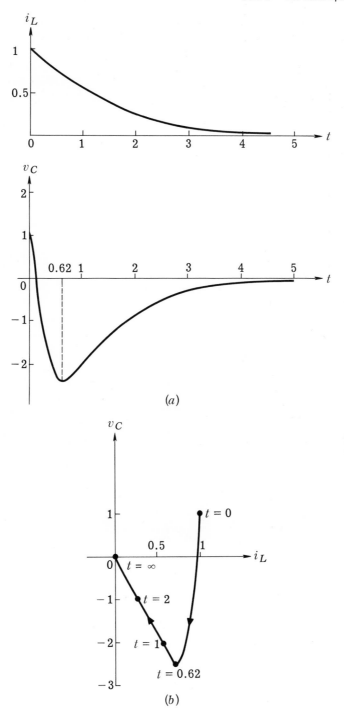

**Fig. 3.1**    Overdamped parallel $RLC$ circuit.    (*a*) Waveforms for $i_L$ and $v_C$;
(*b*) state trajectory.

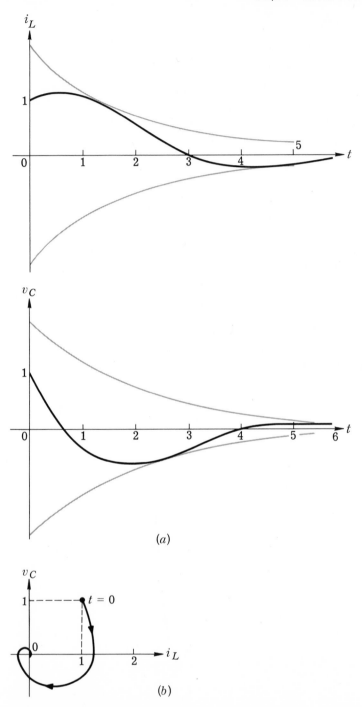

**Fig. 3.2**   Underdamped parallel $RLC$ circuit. ($a$) Waveforms for $i_L$ and $v_C$; ($b$) state trajectory.

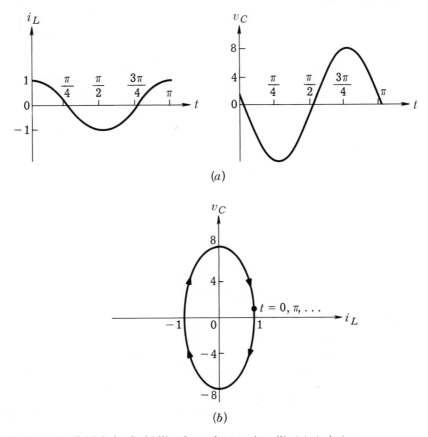

**Fig. 3.3**   Lossless parallel $LC$ circuit. ($a$) Waveforms for $i_L$ and $v_C$; ($b$) state trajectory.

$$i_L(t) = \cos 2t + \tfrac{1}{8} \sin 2t = 1.01 \cos (2t - 7°)$$

and

$$v_C(t) = \cos 2t - 8 \sin 2t = 8.06 \cos (2t + 83°)$$

The waveforms and the trajectory are plotted in Fig. 3.3$a$ and $b$. Note in this case, the trajectory is an ellipse centered at the origin, which indicates that the zero-input response is oscillatory.

**3.2**   **Matrix Representation**

In terms of state variables, Eqs. (3.1) and (3.2) may be written in the matrix form as follows:

(3.4)   $$\frac{d\mathbf{x}(t)}{dt} = \mathbf{A}\mathbf{x}(t) \qquad t \geq 0$$

and

(3.5)   $\mathbf{x}(0) = \mathbf{x}_0$

where

(3.6)   $\mathbf{A} \triangleq \begin{bmatrix} 0 & \dfrac{1}{L} \\ -\dfrac{1}{C} & -\dfrac{G}{C} \end{bmatrix}$

and

(3.7)   $\mathbf{x}_0 = \begin{bmatrix} I_0 \\ V_0 \end{bmatrix}$

The matrix equations (3.4) and (3.5) are very similar to the scalar equation

(3.8)   $\dfrac{dx}{dt} = ax \qquad x(0) = x_0$

The scalar equation has the well-known solution $x(t) = \epsilon^{at}x_0$. Similarly, the matrix equation has a solution

(3.9)   $\mathbf{x}(t) = \epsilon^{\mathbf{A}t}\mathbf{x}_0 \qquad t \geq 0$

where $\epsilon^{\mathbf{A}t}$ is a *matrix* that depends upon $t$ and $\mathbf{A}$. Geometrically speaking, it maps the initial-state vector $\mathbf{x}_0$ into the state vector $\mathbf{x}(t)$ at time $t$. In fact, just as the ordinary exponential $\epsilon^{at}$ is given by the power series (valid for all $t$)

$$\epsilon^{at} = 1 + at + \frac{a^2 t^2}{2!} + \frac{a^3 t^3}{3!} + \cdots$$

the matrix $\epsilon^{\mathbf{A}t}$ is given by the power series (valid for all $t$)

$$\epsilon^{\mathbf{A}t} = \mathbf{I} + \mathbf{A}t + \mathbf{A}^2 \frac{t^2}{2!} + \mathbf{A}^3 \frac{t^3}{3!} + \cdots$$

where $\mathbf{I}$ is the unit matrix. In this last series, each term is a *matrix;* hence $\epsilon^{\mathbf{A}t}$ is also a matrix. Each element of the matrix $\epsilon^{\mathbf{A}t}$ is a function of $t$. It is important to observe that (3.9) represents a *linear function* that maps the vector $\mathbf{x}_0$ (the initial-state vector) into the vector $\mathbf{x}(t)$(the state vector at time $t$). Though we shall not go further into the representation and calculation of $\epsilon^{\mathbf{A}t}$, the idea that the vector equation (3.9) generates the whole state-space trajectory is important.

**3.3   Approximate Method for the Calculation of the Trajectory**

With reference to Eqs. (3.4) and (3.5), we may view Eq. (3.4) as defining, for each $t$, the velocity $(d\mathbf{x}/dt)(t)$ along the trajectory at the point $\mathbf{x}(t)$ of the state space. In particular, given the initial state $\mathbf{x}(0)$, Eq. (3.4) gives the initial velocity of the state vector $(d\mathbf{x}/dt)(0)$. We may use a simple step-by-step method to compute an approximation to the trajectory. This method is based on the assumption that if a sufficiently small interval of time $\Delta t$ is considered, then during that interval the velocity $dx/dt$ is approximately constant; equivalently, the trajectory is approximately a straight-line segment. Thus, starting with the initial state $\mathbf{x}_0$ at time 0, we have

(3.10)    $\dfrac{d\mathbf{x}}{dt}(0) = \mathbf{A}\mathbf{x}_0$

and, since we assume the velocity to be constant during the small interval $(0, \Delta t)$,

(3.11)    $\mathbf{x}(\Delta t) \approx \mathbf{x}_0 + \dfrac{d\mathbf{x}}{dt}(0)\,\Delta t = \mathbf{x}_0 + \mathbf{A}\mathbf{x}_0\,\Delta t$

For the next interval, $(\Delta t, 2\,\Delta t)$, we again assume the velocity to be constant and calculate it on the basis of the approximate value of $\mathbf{x}(\Delta t)$ given by (3.11). Thus,

(3.12)    $\dfrac{d\mathbf{x}}{dt}(\Delta t) = \mathbf{A}\mathbf{x}(\Delta t)$

hence

(3.13)    $\mathbf{x}(2\,\Delta t) \approx \mathbf{x}(\Delta t) + \mathbf{A}\mathbf{x}(\Delta t)\,\Delta t$

We continue to calculate successive approximate values of the state

(3.14)    $\mathbf{x}[(k+1)\,\Delta t] \approx \mathbf{x}(k\,\Delta t) + \mathbf{A}\mathbf{x}(k\,\Delta t)\,\Delta t \qquad k = 0, 1, 2, \ldots, N$

$\qquad\qquad = (\mathbf{1} + \Delta t\,\mathbf{A})\mathbf{x}(k\,\Delta t)$

This process can be easily programmed on a digital computer. In fact, it can be shown that the successive approximate values $\mathbf{x}(\Delta t)$, $\mathbf{x}(2\,\Delta t)$, ..., $\mathbf{x}(N\,\Delta t)$ calculated in this manner tend to the exact points of the exact trajectory if $\Delta t \to 0$. In practice, the value of $\Delta t$ that should be selected depends (1) on the number of significant figures carried in the computation, (2) on the accuracy required, (3) on the constants of the problem, and (4) on the length of the time interval over which the trajectory is desired. Once the trajectory is computed, the response of the circuit is easily obtained since it is either one component of the state or a linear combination of them.

**Example 2**    Let us employ the method to calculate the state trajectory of the underdamped parallel *RLC* circuit in Example 1. The state equation is

$$\begin{bmatrix} \dfrac{dx_1}{dt} \\[2mm] \dfrac{dx_2}{dt} \end{bmatrix} = \begin{bmatrix} 0 & 1 \\ -1 & -1 \end{bmatrix} \begin{bmatrix} x_1 \\ x_2 \end{bmatrix}$$

and the initial state is

$$\begin{bmatrix} x_1(0) \\ x_2(0) \end{bmatrix} = \begin{bmatrix} 1 \\ 1 \end{bmatrix}$$

Let us pick $\Delta t = 0.2$ sec. We can use (3.11) to obtain the state at $\Delta t$; thus

$$\begin{bmatrix} x_1(0.2) \\ x_2(0.2) \end{bmatrix} = \begin{bmatrix} 1 \\ 1 \end{bmatrix} + 0.2 \begin{bmatrix} 0 & 1 \\ -1 & -1 \end{bmatrix} \begin{bmatrix} 1 \\ 1 \end{bmatrix} = \begin{bmatrix} 1.2 \\ 0.6 \end{bmatrix}$$

Next the state at $2\,\Delta t$ is obtained from (3.13), and we obtain

$$\begin{bmatrix} x_1(0.4) \\ x_2(0.4) \end{bmatrix} = \begin{bmatrix} 1.2 \\ 0.6 \end{bmatrix} + 0.2 \begin{bmatrix} 0 & 1 \\ -1 & -1 \end{bmatrix} \begin{bmatrix} 1.2 \\ 0.6 \end{bmatrix} = \begin{bmatrix} 1.32 \\ 0.24 \end{bmatrix}$$

From (3.14) we can actually write the state at $(k + 1)\,\Delta t$ in terms of the state at $k\,\Delta t$ as

$$\mathbf{x}[(k + 1)\,\Delta t] = \begin{bmatrix} 1 & 0.2 \\ -0.2 & 0.8 \end{bmatrix} \mathbf{x}(k\,\Delta t)$$

Figure 3.4 shows the trajectory as a continuous curve and the points computed with $\Delta t = 0.2$ sec. If we had used $\Delta t = 0.002$ sec, the points computed by repeated application of Eq. (3.14) would all be on the trajectory.

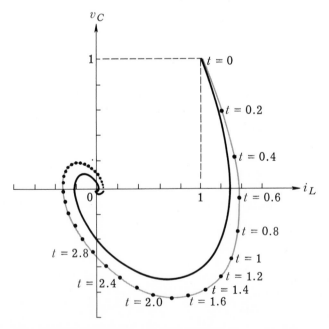

**Fig. 3.4**  State trajectory calculation using the step-by-step method for Example 2 with $\Delta t = 0.2$ sec.

**Exercise**   Compute the state trajectory by using

   *a.*   $\Delta t = 0.1$ sec

   *b.*   $\Delta t = 0.5$ sec

   Comment on the results.

**Remark**   If we consider a parallel *RLC* circuit in which the resistor, inductor, and capacitor are *nonlinear* but time-invariant, then, under fairly general assumptions concerning their characteristics, we have equations of the form

(3.15)   $$\frac{di_L}{dt} = f_1(i_L, v_C) \qquad \frac{dv_C}{dt} = f_2(i_L, v_C)$$

where the functions $f_1$ and $f_2$ are obtained in terms of the branch characteristics.

   It is fundamental to note that the general method of obtaining the approximate calculation of the trajectory still holds; the equations are

(3.16)   $$\frac{d\mathbf{x}(t)}{dt} = \mathbf{f}(\mathbf{x}(t))$$

and the equations corresponding to (3.11) and (3.13) are now

(3.17)   $\mathbf{x}(\Delta t) \approx \mathbf{x}_0 + \mathbf{f}(\mathbf{x}_0)\,\Delta t \qquad \mathbf{x}(2\Delta t) \approx \mathbf{x}(\Delta t) + \mathbf{f}(\mathbf{x}(\Delta t))\,\Delta t$

Examples will be given in Sec. 5.

---

**3.4**   **State Equations and Complete Response**

If the parallel *RLC* circuit is driven by a current source, as in Fig. 2.1, the state equations can be similarly written. First, the voltage across the parallel circuit is the same as if there were no source. We obtain, as in Eq. (3.1),

$$\frac{di_L}{dt} = \frac{1}{L}v_C$$

Next, for the *KCL* equation we must include the effect of the current source. Thus, an additional term is needed in comparison with Eq. (3.2), and we have

$$\frac{dv_C}{dt} = -\frac{1}{C}i_L - \frac{G}{C}v_C + \frac{i_s}{C}$$

The initial state, the same as given by Eq. (3.3), is

$$i_L(0) = I_0$$
$$v_C(0) = V_0$$

If we use the vector **x** to denote the state vector, that is, $\mathbf{x} = \begin{bmatrix} i_L \\ v_C \end{bmatrix}$, the state equation in matrix form is

(3.18)   $\dfrac{d\mathbf{x}}{dt} = \mathbf{A}\mathbf{x} + \mathbf{b}w$

and the initial state is

(3.19)   $\mathbf{x}(0) = \begin{bmatrix} I_0 \\ V_0 \end{bmatrix}$

In (3.18)

(3.20)   $\mathbf{A} = \begin{bmatrix} 0 & \dfrac{1}{L} \\ -\dfrac{1}{C} & -\dfrac{G}{C} \end{bmatrix}$

and

(3.21)   $\mathbf{b}w = \begin{bmatrix} 0 \\ \dfrac{1}{C} \end{bmatrix} i_s$

The matrices **A** and **b** depend upon the circuit elements, whereas the input is denoted by $w$. Equation (3.18) is a first-order nonhomogeneous matrix differential equation and is similar to the first-order scalar nonhomogenous linear differential equation

(3.22)   $\dfrac{dx}{dt} = ax + bw$

The solution of this scalar equation, satisfying the specified initial condition $x(0) = x_0$, is

(3.23)   $x = \epsilon^{at}x_0 + \displaystyle\int_0^t \epsilon^{a(t-t')}bw(t')\,dt'$

Note that the complete response is written as the sum of two terms. The first term, $\epsilon^{at} x_0$, is the zero-input response, and the second term, which is represented by the integral, is the zero-state response. Similarly the matrix equation (3.18) has the solution

(3.24)   $\mathbf{x} = \epsilon^{\mathbf{A}t}\mathbf{x}_0 + \displaystyle\int_0^t \epsilon^{\mathbf{A}(t-t')}\mathbf{b}w(t')\,dt'$

The first term, $\epsilon^{\mathbf{A}t}\mathbf{x}_0$ is the zero-input response, and the second term, which is represented by the integral, is the zero-state response. The proof of (3.24) will not be given here; however, the form of (3.24) is worth noting.

Again, the expression depends upon the evaluation of $\epsilon^{At}$. The approximate method of calculating **x**, as given in Sec. 3.3, is applicable.

## 4   Oscillation, Negative Resistance, and Stability

In previous sections we studied the linear time-invariant parallel $RLC$ circuit in great detail. The solutions for the underdamped case were derived explicitly. Of particular concern in this section is the lossless case, which has an oscillatory zero-input response. We shall study the properties of such a circuit and, moreover, give some specific physical considerations.

The lossless parallel $LC$ circuit can be considered as a special case of the underdamped circuit with $R = \infty$ (or $G = 0$, $\alpha = 0$, $Q = \infty$). The formulation of differential equations, in terms of the capacitor voltage or the inductor current, and state equations is no different from that of the underdamped circuit. Moreover, the zero-input response and the zero-state response can be obtained directly from the responses of the underdamped case by setting $\alpha = 0$ and $\omega_d = \omega_0 = 1/\sqrt{LC}$. Let us review some of these results. The natural frequencies of the lossless circuit are $s = \pm j\omega_0$, $\omega_0 = 1/\sqrt{LC}$. The zero-input response is a sinusoid with the same angular frequency $\omega_0$. This fact was illustrated in Example 1 of Sec. 3. The state trajectory is an ellipse, as shown in Fig. 3.3$b$, which implies that the zero-input response of a lossless circuit is in sustained oscillation. The initial stored energy in the capacitor and/or in the inductor is being transferred back and forth endlessly.

Let us consider next the zero-state response. Referring back to Sec. 2, we recall that the impulse response of a lossless $LC$ circuit is a sinusoid with angular frequency $\omega_0$. The step response also contains a sinusoidal part with the same frequency. As a matter of fact, if the circuit is in the zero state at time 0 and any input is applied during an interval $[0,T]$ (where $T$ is any time later than zero) and set to zero after $T$, then the response beyond time $T$ is of the form $K \sin (\omega_0 t + \theta)$, where the amplitude $K$ and the phase $\theta$ depend on the input.

The lossless $LC$ circuit is called a resonant circuit or a tuned circuit. The word "tuned" implies that the frequency of oscillation is tuned to a given number $f_0 = \omega_0/2\pi$ by adjusting the element value of either the capacitor or the inductor. If the physical circuit were such that its physical inductor and its physical capacitor were identical with our models of linear time-invariant inductor and capacitor, we would have a "linear oscillator" that oscillates at the angular frequency $\omega_0$. Clearly, physical components are not identical with our circuit models. As mentioned in Chap. 2, a *physical* inductor always has a certain amount of dissipation and should be modeled by a series connection of an inductor and resistor. Thus, in practice, a physical tuned circuit (by itself) is not an oscillator and behaves as an underdamped circuit, provided that the dissipation is small enough. For tuned circuits, $Q$'s of several hundred are attainable in practice. With superconductors, an infinite $Q$ is attainable, in principle.

To obtain an oscillator, we need to compensate for the dissipation present in any physical tuned circuit. The obvious means is to introduce some negative resistance to the circuit so the net effect will be again loss-less. A typical oscillator can often be thought of as consisting of a physical tuned circuit connected to a resistor with negative resistance. This is illustrated in Fig. 4.1. We have discussed in Chap. 2 the small-signal negative-resistance property of a tunnel diode. We shall see that it is also possible to obtain negative resistance by means of feedback in a transistor circuit. These negative resistances are all approximations; that is, over certain ranges of voltages or currents and perhaps over certain frequency bands, the devices behave like linear time-invariant resistors with negative resistances. Nevertheless, the model of a linear time-invariant active resistor is useful, and we shall use it to analyze the be-havior of some simple second-order circuits. A clear understanding of these circuits will be useful in the study of nonlinear circuits.

Next, let us consider the linear time-invariant parallel $RLC$ circuit, as shown in Fig. 4.2, where the resistor has a *negative* resistance ($R < 0$ and $G < 0$). The characteristic polynomial for the circuit is $s^2 + 2\alpha s + \omega_0{}^2$, where $\alpha = G/2C$ is negative. $\omega_0$ is, as before, equal to $1/\sqrt{LC}$. The natural frequencies of the circuit are the characteristic roots, which can be written in the form

$$\left.\begin{matrix} s_1 \\ s_2 \end{matrix}\right\} = |\alpha| \pm \sqrt{\alpha^2 - \omega_0{}^2}$$

since $\alpha < 0$. $\sqrt{\alpha^2 - \omega_0{}^2}$ is either purely imaginary or real, in which case it is smaller than $|\alpha|$. Thus, the natural frequencies are located in the right half of the complex frequency plane. We shall study the zero-input re-sponse and make the following classifications:

Physical
tuned circuit
with dissipation

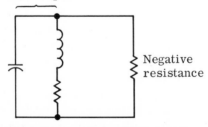

Negative
resistance

**Fig. 4.1**   A simple linear oscillator which comprises a physical tuned circuit and a negative resist-ance.

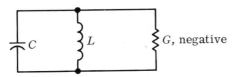

**Fig. 4.2**   Parallel *RLC* circuit.

1. $|\alpha| < \omega_0$; the two natural frequencies are complex conjugate ($s_1 = |\alpha| + j\omega_d$, and $s_2 = |\alpha| - j\omega_d$, where $\omega_d^2 = \omega_0^2 - \alpha^2$). The response is then

$$k\epsilon^{|\alpha|t} \cos(\omega_d t + \theta)$$

where $k$ and $\theta$ are constants that depend on the initial conditions.

2. $|\alpha| > \omega_0$; the two natural frequencies $s_1$ and $s_2$ are real and positive. The response is a sum of two *increasing* exponentials

$$k_1 \epsilon^{s_1 t} + k_2 \epsilon^{s_2 t}$$

where $k_1$ and $k_2$ depend on initial conditions.

In both cases the responses contain growing exponential factors; hence, as time passes, the responses become arbitrarily large. The procedure for determining the waveforms for $i_L$ and $v_C$ is exactly the same as for the case in which the resistance is positive. In Figs. 4.3 and 4.4, plots of $v_C$ versus $t$, $i_L$ versus $t$, and the state trajectory of a case ($|\alpha| < \omega_0$) are given for the initial state $v_C(0) = 1$ and $i_L(0) = 1$.

It is important to understand these responses. The linear resistor with *negative* resistance is an *active* element that delivers energy to the inductor and capacitor instead of dissipating energy as in the passive resistor. Thus, without any input, the responses can grow after they are started by the initial energy in the inductor and/or the capacitor. As we have mentioned, the linear active resistor is only a model that approximates the behavior of some device over specified ranges of voltages and currents. As the voltages and currents grow beyond these ranges, the calculations no longer represent the actual physical behavior of the circuit. In most instances we must consider the nonlinear characterization of the device and modify the mathematical result obtained under linear approximation. The actual physical behavior could end up with a nonlinear oscillation, as will be shown in the next section, or in other cases some of the components in the circuit cannot tolerate the excessive current and eventually burn out.

Let us come back to our linear analysis. Consider both the cases of passive and active linear resistors together. We can divide the zero-input responses of the parallel *RLC* circuits into three categories.

*Case 1*   The natural frequencies are in the *left-half plane;* that is both natural frequencies $s_1$ and $s_2$ have *negative real parts*.  This includes the overdamped, critically damped, and underdamped cases in Sec. 1.  Because of the damped exponential factor, the zero-input response approaches zero as $t \to \infty$.  In the state space, for any initial state the trajectory reaches the origin as $t \to \infty$.  We say that such a circuit is *asymptotically stable*.  The state trajectories of Figs. 3.1*b* and 3.2*b* are typical examples.  Since the concept of asymptotic stability is extremely important, we repeat it: a circuit is said to be **asymptotically stable** if for any initial state and for zero input the trajectory in state space remains bounded and tends to the origin as $t \to \infty$.  The boundedness requirement is important only for certain special nonlinear circuits.

*Case 2*    The natural frequencies are on the *imaginary axis;* that is, $s_1$ and $s_2$ have zero real parts.   $s_1 = j2\pi f_0$, and $s_2 = -j2\pi f_0$.  This is the lossless case.

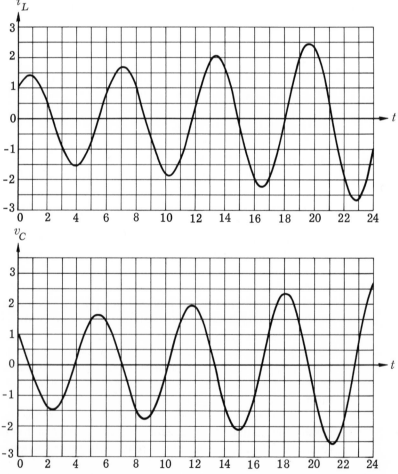

**Fig. 4.3**    Plots of $i_L$ and $v_C$ for the parallel $RLC$ circuit of Fig. 4.2.   Note the active resistor. We assume that $|\alpha| < \omega_0$.

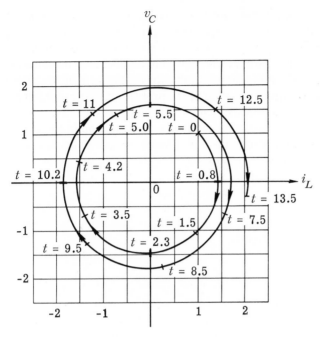

**Fig. 4.4**    State trajectory of the *RLC* circuit of Fig. 4.2.

The zero-input response is sinusoidal with frequency $f_0$. In the state space the trajectory is an ellipse centered on the origin. We call the circuit *oscillatory*.

*Case 3*    The natural frequencies are in the *right-half plane;* that is, $s_1$ and $s_2$ have positive real parts. This corresponds to the negative-resistance situation. The zero-input response becomes unbounded as $t \to \infty$. In the state space the trajectory reaches infinity as $t \to \infty$. We call the circuit *unstable*. A typical example is the trajectory of Fig. 4.4. The associated waveforms of $i_L$ and $v_C$ are shown in Fig. 4.3.

<br>

**5** | **Nonlinear and Time-varying Circuits**

In Chap. 4 when we treated first-order nonlinear and time-varying circuits, we discovered that we could sometimes solve these problems analytically. In addition to showing simple analytical solutions, our main emphasis in Chap. 4 was to demonstrate that the linearity property did not hold for nonlinear circuits and the time-invariance property did not hold for time-varying circuits. In second-order nonlinear and time-varying circuits there are analytical methods for some very special types of circuits; there are also various graphical methods that can be used advantageously for wider classes of circuits. In the literature there are many special equations and methods such as the van der Pol equation, the Mathieu equation, the Duffin equation, the isocline method, and the Liénard method. We shall not present these conventional methods here. First of all, they are specialized subjects and are clearly beyond the scope of this book. Secondly, in the age of digital computers these special equations and methods have somewhat lost their significance; it is cheaper and it makes more engineering sense to solve the best model known rather than to make rough approximations to fit the problem into the mold of another problem whose solution is known.

Our purpose in this section is, first, to explain the physical behavior of some nonlinear circuits and, second, to demonstrate carefully the writing of differential equations of nonlinear circuits. The equations that are most convenient for numerical computation are the systems of two first-order differential equations (rather than a single second-order ordinary differential equation). In the linear case these equations are called state equations, whereas in the nonlinear case, they are called *equations in the normal form;* that is,

(5.1) $$\frac{d\mathbf{x}}{dt} = \mathbf{f}(\mathbf{x},t,w)$$

where $\mathbf{x}$ represents a vector whose components are selected network variables (voltages, currents, charges, and fluxes), $w$ represents the input, and $\mathbf{f}$ is a vector-valued function. Equation (5.1) is a generalization of the linear state equation

(5.2) $$\frac{d\mathbf{x}}{dt} = \mathbf{A}\mathbf{x} + \mathbf{b}w$$

which we discussed in Sec. 3. As mentioned previously, the step-by-step integration method can be used for numerical work. The following two examples illustrate these points.

---

**Example 1**   Consider the parallel $RLC$ circuit as shown in Fig. 5.1, where the inductor and capacitor are linear and time-invariant; the resistor is a nonlinear device and has a characteristic as given in the figure. The nonlinear characteristic might in some cases be approximated by a polynomial such as

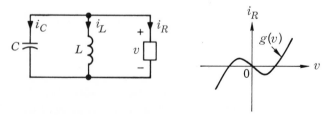

**Fig. 5.1**   Nonlinear oscillator with a nonlinear resistor whose characteristic is the curve shown in $vi_R$ plane.

(5.3)   $g(v) \approx -\alpha v + \beta v^3$

where $\alpha$ and $\beta$ are two constants to fit the curve in Fig. 5.1. First, the voltage $v$ can be related to the inductor current by

(5.4)   $\dfrac{di_L}{dt} = \dfrac{v}{L} \qquad i_L(0) = I_0$

Next, writing the KCL equation for the circuit, we have

$i_c = -i_L - i_R$

or

(5.5)   $\dfrac{dv}{dt} = -\dfrac{i_L}{C} - \dfrac{g(v)}{C} \qquad v(0) = V_0$

Combining Eqs. (5.4) and (5.5), we have the equation in normal form as follows:

(5.6)   $\dfrac{d\mathbf{x}}{dt} = \begin{bmatrix} \dfrac{di_L}{dt} \\[2mm] \dfrac{dv}{dt} \end{bmatrix} = \begin{bmatrix} \dfrac{v}{L} \\[2mm] -\dfrac{i_L}{C} - \dfrac{g(v)}{C} \end{bmatrix} = \mathbf{f}(\mathbf{x})$

with the initial state

(5.7)   $\mathbf{x}(0) = \begin{bmatrix} i_L(0) \\ v(0) \end{bmatrix} = \begin{bmatrix} I_0 \\ V_0 \end{bmatrix} = \mathbf{x}_0$

Given the initial state $\mathbf{x}_0$, the numbers $L$ and $C$, and the characteristic $g(\cdot)$, we can find the solution by means of the step-by-step method described in Sec. 3. Starting with the given initial state $\mathbf{x}(0) = \mathbf{x}_0$ in (5.7), we compute the state at time $\Delta t$, $\mathbf{x}(\Delta t)$, by Eq. (3.17). Thus,

$\mathbf{x}(\Delta t) \approx \mathbf{x}(0) + \mathbf{f}(\mathbf{x}_0)\,\Delta t$

We continue with

$\mathbf{x}(2\Delta t) \approx \mathbf{x}(\Delta t) + \mathbf{f}[\mathbf{x}(\Delta t)]\,\Delta t$

Thus, the trajectory can be drawn in the state space, i.e., the $i_L v$ plane. In Fig. 5.2 we present two such trajectories. The first one, which is shown in Fig. 5.2a, has the initial state

$\mathbf{x}_0 = \begin{bmatrix} I_0 \\ V_0 \end{bmatrix} = \begin{bmatrix} 0 \\ 1 \end{bmatrix}$

Note that as $t$ increases, the trajectory approaches a closed curve called the *limit cycle*. This implies that after some time the zero-input response of the nonlinear circuit is arbitrarily close to a periodic motion; that is,

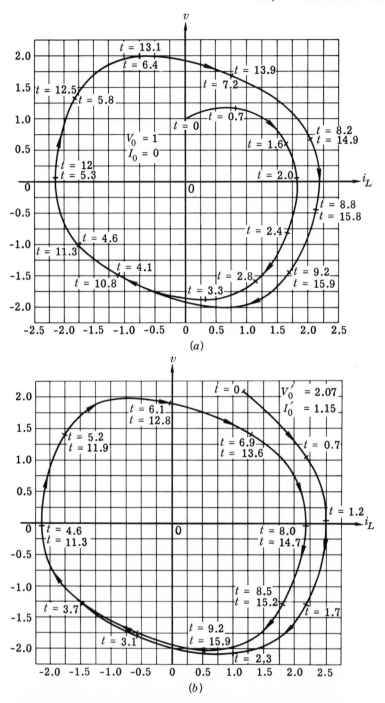

**Fig. 5.2**    Trajectories of the nonlinear oscillator of Fig. 5.1; for both initial conditions the same limit cycle is reached.

both waveforms $i_L(\cdot)$ and $v(\cdot)$ eventually become periodic functions of time. In Fig. 5.2b we start at a different initial state

$$\mathbf{x}_0 = \begin{bmatrix} I_0 \\ V_0 \end{bmatrix} = \begin{bmatrix} 1.15 \\ 2.07 \end{bmatrix}$$

It is interesting to note that in this case the trajectory approaches the *same* limit cycle from the outside as $t$ increases.

**Remark**   It should be pointed out that there are distinct differences between the zero-input response of a linear circuit and the zero-input response of a nonlinear circuit. In the linear parallel *LC* circuit (lossless case), starting with an arbitrary initial state, we reach sinusoidal oscillation immediately. Moreover, the amplitudes of the oscillation for $i_L$ and $v$ depend on the initial state. In the nonlinear case oscillation is reached only after a transient elapses, and the oscillation in this example does not seem to depend on the initial state.

*Piecewise*   Let us explain the physical behavior of the circuit, based on the piece-
*linear*   wise linear approximation of the characteristic of the nonlinear resistor.
*approximation*   In Fig. 5.3a we divide the range of voltage across the resistor into three regions. In region 1, that is, where $-\infty < v \le -E_1$, the characteristic of the nonlinear resistor is approximated by the straight line of positive slope $1/R_1$, which intersects the $i_R$ axis at ordinate $I_1$. Thus, in region 1, the nonlinear resistor can be replaced by the parallel connection of a linear resistor of positive resistance $R_1$ and a constant current source $I_1$. This replacement is shown in the equivalent circuit of Fig. 5.3b. In region 2, that is, where $-E_1 < v < E_2$, the characteristic of the nonlinear resistor is approximated by the straight line through the origin with negative slope $-1/R_0$, as shown in Fig. 5.3a (note that $R_0 > 0$). Thus, in region 2, the nonlinear *active* resistor can be replaced by a linear resistor of *negative* resistance $R_0$. This replacement is shown in the equivalent circuit of Fig. 5.3c. In region 3, that is, where $E_2 < v < \infty$, the characteristic of the nonlinear resistor is approximated by a straight line of positive slope $1/R_2$, which intersects the $i_R$ axis at ordinate $-I_2$ (note that $I_2 > 0$). Thus, in region 3, the nonlinear resistor can be replaced by the parallel connection of a linear resistor of positive resistance $R_2$ and a constant current source $I_2$. This replacement is shown in the equivalent circuit of Fig. 5.3d. Depending on the voltage across the nonlinear resistor, one of the three approximate equivalent circuits in Fig. 5.3 must be used.

Since we are familiar with the analysis of second-order parallel *linear RLC* circuits, we can easily determine the characteristics of the circuit in the three separate regions of the nonlinear resistor. The problem is then to determine what happens to the circuit at the boundary of two regions. Suppose that the initial state is $i_L = 0$ and $v = 2$, which is assumed to be

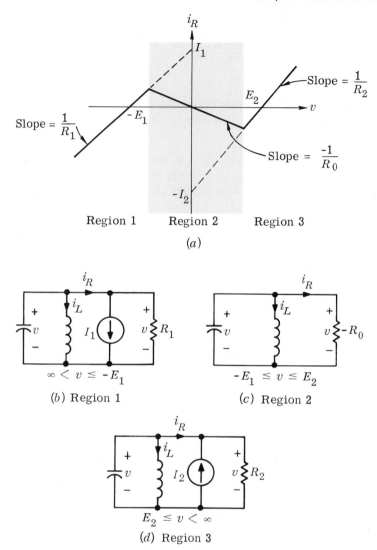

Fig. 5.3   Piecewise linear approximation of the nonlinear oscillator.

in region 2.   The parallel linear $RLC$ circuit can be analyzed (as in the previous section), where the resistor is linear and active.   The trajectory for the *linear* circuit starts at $(0,2)$ moves away from the origin, and is supposed to reach infinity as $t \to \infty$ since the circuit is unstable.   However, at some time $t_1$ the trajectory reaches a point where $v(t_1) = -E_1$ or $E_2$, and the negative-resistance approximation is no longer valid.   As the trajectory passes the point $\big(v(t_1),i_L(t_1)\big)$, we are either in region 1 or region 3,  and we need to use the combination of the linear passive resistor and the constant current source to represent the device.   The circuit is then switched from the piecewise linear approximation of Fig. 5.3c to Fig. 5.3b or d, depending upon whether $v(t_1)$ is equal to $-E_1$ or $+E_2$.

Let us assume that the actual voltage waveform is shown in Fig. 5.4. At $t = 0$, the device is in region 2, and at $t = t_1$, the voltage reaches $-E_1$. Thus, for $t > t_1$ the device is in region 1; we have to use the circuit

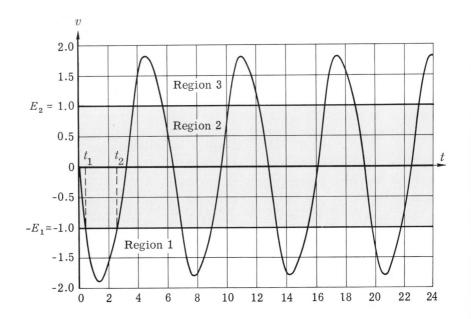

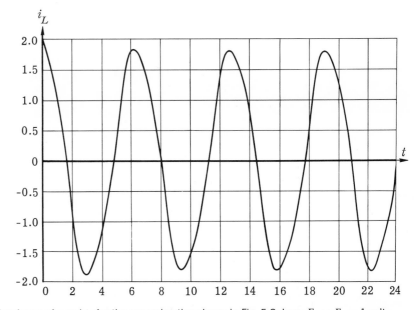

**Fig. 5.4**    Waveforms of $v$ and $i_L$ for the approximation shown in Fig. 5.3; here $E_1 = E_2 = 1$ volt.

in Fig. 5.3*b* and compute the complete response with initial state given by $(v(t_1), i_L(t_1))$, where $v(t_1) = -E_1$. The response can be computed easily with the linear equivalent circuit in Fig. 5.3*b*; it is shown in Fig. 5.4 in terms of $v$ and $i_L$ plotted vs. time. At $t = t_2$ the voltage is $v(t_2) = -E_1$ once again. For $t > t_2$ the device is back to region 2 operation. The equivalent circuit of Fig. 5.3*c* must then be used. Therefore, we compute the response of the active circuit in Fig. 5.3*c* with the initial state given by $(v(t_2), i_L(t_2))$, where $v(t_2) = -E_1$. The device will next enter region 3 operation and then come back to region 2. Continuing this process, the waveforms of voltage and current eventually reach a steady state, that is, a periodic behavior as shown in the figure. In the state space we call the portion of the trajectory that is a closed curve the limit cycle.

**Example 2**   Let us consider the linear parallel $LC$ circuit in Fig. 5.5, where the capacitor is time-invariant but the inductor is time-varying. The KCL equation is

(5.8)   $i_L + i_C = i_s$

Let us use the flux as the network variable; then

(5.9)   $i_L(t) = \dfrac{\phi(t)}{L(t)}$

(5.10)   $v = \dfrac{d\phi}{dt}$

For the linear time-invariant capacitor we have

(5.11)   $i_C = C\dfrac{dv}{dt}$

Combining the four equations, we obtain a second-order linear differential equation with $\phi$ as the dependent variable; thus

$$C\frac{d^2\phi}{dt^2} + \frac{\phi}{L(t)} = i_s(t)$$

(5.12)   If $L(t)$ is a periodic function of the form

(5.13)   $L(t) = \dfrac{1}{a + b\cos\omega_1 t}$

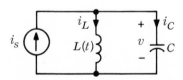

**Fig. 5.5**   Linear time-varying circuit; the capacitor $C$ is time-invariant but the inductor $L(t)$ is time-varying.

where $a$ and $b$ are constants with $b < a$, Eq. (5.12) has the form of the celebrated equation of Mathieu. It can be shown that if $\omega_1$ is chosen appropriately, an oscillation with exponentially increasing amplitude appears in the circuit. This phenomenon is called *parametric oscillation*. The energy of the increasing oscillation is supplied by the agent that varies the inductance. In the early days of radio, alternators were used to supply the varying inductance of the oscillator. A more detailed treatment of this is given in Chap. 19.

Let us now consider the same circuit from the state-space point of view. Let us use the capacitor charge $q$ and the inductor flux $\phi$ as the dependent variables. Combining Eqs. (5.8) and (5.9), we have

(5.14)
$$\frac{dq}{dt} = -\frac{\phi}{L(t)} + i_s(t)$$

Combining Eqs. (5.10) and (5.11), we have

(5.15)
$$\frac{d\phi}{dt} = \frac{q}{C}$$

In matrix form, we have

(5.16)
$$\frac{d}{dt}\begin{bmatrix} q \\ \phi \end{bmatrix} = \begin{bmatrix} 0 & -\dfrac{1}{L(t)} \\ \dfrac{1}{C} & 0 \end{bmatrix}\begin{bmatrix} q \\ \phi \end{bmatrix} + \begin{bmatrix} 1 \\ 0 \end{bmatrix} i_s(t)$$

with the initial state

(5.17)
$$\begin{bmatrix} q(0) \\ \phi(0) \end{bmatrix} = \begin{bmatrix} Q \\ \Phi \end{bmatrix}$$

The equations can again be solved numerically by using the step-by-step integration method.

| 6 | **Dual and Analog Circuits** |

| 6.1 | **Duality** |

Up to now we have considered linear, nonlinear, time-invariant, and time-varying second-order circuits, but we have restricted ourself to the parallel-$RLC$-circuit configuration. Let us consider another simple example, the series $RLC$ circuit. Its behavior is closely related to that of the parallel $RLC$ circuit.

Consider the circuit in Fig. 6.1, where a series connection of a linear time-invariant resistor, inductor, and capacitor is driven by a voltage source. The analysis is similar to that of the parallel $RLC$ circuit. We

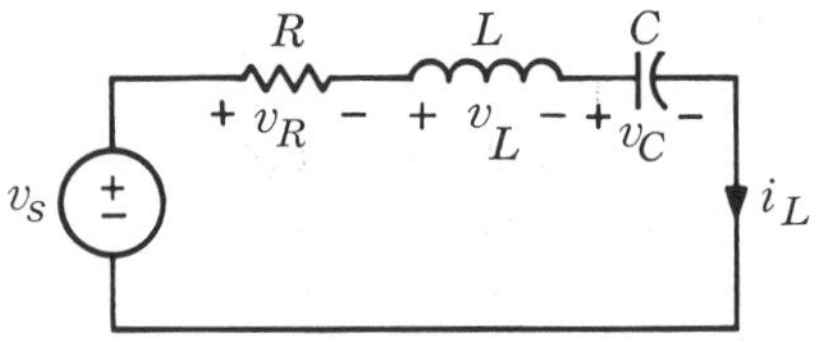

**Fig. 6.1**   Series $RLC$ circuit with a voltage source input.

wish to determine the complete response, that is, the response to both the input and the initial state.   First we need to derive a differential equation in terms of a convenient network variable.   For each of the three branches, the branch voltage and current are related by their branch equations.   The current variables must satisfy the KCL constraints; that is,

(6.1)   $i_L = i_R = i_C$

while the voltage variables must satisfy the KVL constraint

(6.2)   $v_L + v_R + v_C = v_s$

Therefore, in terms of the mesh current (denoted by $i_L$), we have the following integrodifferential equation:

(6.3)   $L \dfrac{di_L}{dt} + Ri_L + V_0 + \dfrac{1}{C} \displaystyle\int_0^t i_L(t') \, dt' = v_s$

with

(6.4)   $i_L(0) = I_0$

It is then possible to solve for $i_L$ from (6.3) and (6.4).   However, if the voltage $v_C$ is considered as the variable of interest, the equation becomes a second-order differential equation.   It is only necessary to substitute the branch equations

(6.5)   $v_C = V_0 + \dfrac{1}{C} \displaystyle\int_0^t i_L(t') \, dt' \qquad i_L = C \dfrac{dv_C}{dt}$

in (6.3).   The second-order differential equation is

(6.6)   $LC \dfrac{d^2 v_C}{dt^2} + RC \dfrac{dv_C}{dt} + v_C = v_s$

with initial conditions

(6.7)   $v_C(0) = V_0$

and

(6.8)   $\dfrac{dv_C}{dt}(0) = \dfrac{i_L(0)}{C} = \dfrac{I_0}{C}$

Equations (6.6) to (6.8) specify completely the capacitor voltage for all $t \geq 0$.

We can easily recognize the similarity between the analysis of the *series RLC* circuit and that of the *parallel RLC* circuit. As a matter of fact, if we introduce some consistent changes in notation, we may end up with identical equations. We may have noticed already, from Eqs. (6.6) to (6.8), that the capacitor voltage in the series *RLC* circuit plays the same role as the inductor current in the parallel *RLC* circuit [see Eqs. (1.7) to (1.9)]. Thus, the solution of the series *RLC* circuit may be obtainable from the solution of the parallel *RLC* circuit if appropriate transcriptions are used. In the following examples we illustrate this concept, which is usually referred to as *duality*. A more general treatment will be given in Chap. 10.

**Example 1** Consider the parallel linear time-invariant *RLC* circuit of Fig. 6.2. We wish to compare it with the series *RLC* circuit of Fig. 6.1. To distinguish between the notations and symbols for the series and for the parallel circuits, we use "hats" ($\hat{i}$) to designate all parameters and variables for the parallel circuit.

For instance, writing the KVL equation for the series circuit, we obtain

$$v_s = v_L + v_R + v_C$$

$$v_s = L\frac{di}{dt} + Ri + \frac{1}{C}\int_0^t i(t')\,dt' + v_C(0)$$

Similarly, writing the KCL equation, we obtain

$$\hat{i}_s = \hat{i}_c + \hat{i}_G + \hat{i}_L$$

$$\hat{i}_s = \hat{C}\frac{d\hat{v}}{dt} + \hat{G}\hat{v} + \frac{1}{\hat{L}}\int_0^t \hat{v}(t')\,dt' + \hat{i}_L(0)$$

Suppose now that $L = \hat{C}$, $R = \hat{G}$, $C = \hat{L}$, and $v_C(0) = \hat{i}_L(0)$. Then the two equations have the same coefficients; they differ only in notation. Consequently if, in addition, $i(0) = \hat{v}(0)$ and $v_s(t) = \hat{i}_s(t)$ for all $t \geq 0$, the responses are identical; that is, $i(t) = \hat{v}(t)$ for all $t \geq 0$. The two circuits are said to be *dual*. In particular, both circuits have identical impulse responses and step responses. A tabulation of the zero-input responses of each circuit is given in Table 5.1.

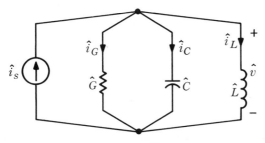

**Fig. 6.2**    Parallel *RLC* circuit with a current source input.

**Table 5.1  Zero-input Response of a Second-order Circuit**

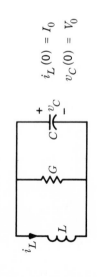

$i_L(0) = I_0$
$v_C(0) = V_0$

$$\frac{d^2 i_L}{dt^2} + \frac{G}{C}\frac{di_L}{dt} + \frac{1}{LC}i_L = 0$$

$Q \triangleq \omega_0/2\alpha$.

$$\omega_0 \triangleq \frac{1}{\sqrt{LC}}$$
$$\alpha \triangleq \frac{G}{2C}$$
$$Q \triangleq \frac{\omega_0}{2\alpha} = \frac{\sqrt{C/L}}{G} = \omega_0 CR$$

$i_L(0) = I_0$
$v_C(0) = V_0$

$$\frac{d^2 v_C}{dt^2} + \frac{R_s}{L}\frac{dv_C}{dt} + \frac{1}{LC}v_C = 0$$

Both equations have the same form: $\dfrac{d^2x}{dt^2} + 2\alpha\dfrac{dx}{dt} + \omega_0^2 x = 0.$  The relation between $\alpha$ and $\omega_0$ is often given by $Q$ which is defined as

$$\omega_0 \triangleq \frac{1}{\sqrt{LC}}$$
$$\alpha \triangleq \frac{R_s}{2L}$$
$$Q \triangleq \frac{\omega_0}{2\alpha} = \frac{\sqrt{L/C}}{R_s} = \frac{\omega_0 L}{R_s}$$

*Case* 1  $\alpha > \omega_0$ or $Q < \frac{1}{2}$.  Overdamped case.  $(s_1 = -\alpha + \alpha_d,\; s_2 = -\alpha - \alpha_d)$ where $\alpha_d \triangleq \sqrt{\alpha^2 - \omega_0^2}$

$$i_L(t) = \frac{I_0}{s_1 - s_2}(s_1\epsilon^{s_2 t} - s_2\epsilon^{s_1 t}) + \frac{V_0}{(s_1 - s_2)L}(\epsilon^{s_1 t} - \epsilon^{s_2 t})$$

$$v_C(t) = I_0\frac{s_1 s_2 L}{s_1 - s_2}(\epsilon^{s_2 t} - \epsilon^{s_1 t}) + \frac{V_0}{s_1 - s_2}(s_1\epsilon^{s_1 t} - s_2\epsilon^{s_2 t})$$

$$v_C(t) = \frac{V_0}{s_1 - s_2}(s_1\epsilon^{s_2 t} - s_2\epsilon^{s_1 t}) + \frac{I_0}{(s_1 - s_2)L}(\epsilon^{s_1 t} - \epsilon^{s_2 t})$$

$$i_L(t) = V_0\frac{s_1 s_2 C}{s_1 - s_2}(\epsilon^{s_2 t} - \epsilon^{s_1 t}) + \frac{I_0}{s_1 - s_2}(s_1\epsilon^{s_1 t} - s_2\epsilon^{s_2 t})$$

*Case 2*   $\alpha = \omega_0$ or $Q = \frac{1}{2}$.   Critically damped case.   $(s_1 = s_2 = -\alpha)$

$$i_L(t) = I_0(1 + \omega_0 t)\epsilon^{-\omega_0 t} + \frac{V_0}{\omega_0 L}\omega_0 t\epsilon^{-\omega_0 t}$$

$$v_C(t) = -I_0\omega_0^2 Lt\epsilon^{-\omega_0 t} + V_0(1 - \omega_0 t)\epsilon^{-\omega_0 t}$$

$$v_C(t) = V_0(1 + \omega_0 t)\epsilon^{-\omega_0 t} + \frac{I_0}{\omega_0 C}\omega_0 t\epsilon^{-\omega_0 t}$$

$$i_L(t) = -V_0\omega_0^2 Ct\epsilon^{-\omega_0 t} + I_0(1 - \omega_0 t)\epsilon^{-\omega_0 t}$$

*Case 3*   $\alpha < \omega_0$ or $Q > \frac{1}{2}$.   Underdamped case.   $(s_1 = -\alpha + j\omega_d,\ s_2 = -\alpha - j\omega_d)$ where $\omega_d \triangleq \sqrt{\omega_0^2 - \alpha^2}$ and $\sin\phi = \dfrac{\alpha}{\omega_0}$

$$i_L(t) = I_0\frac{\omega_0}{\omega_d}\epsilon^{-\alpha t}\cos(\omega_d t - \phi) + \frac{V_0}{\omega_0 L}\frac{\omega_0}{\omega_d}\epsilon^{-\alpha t}\sin\omega_d t$$

$$v_C(t) = -I_0\frac{\omega_0^2 L}{\omega_d}\epsilon^{-\alpha t}\sin\omega_d t + V_0\frac{\omega_0}{\omega_d}\epsilon^{-\alpha t}\cos(\omega_d t + \phi)$$

$$v_C(t) = V_0\frac{\omega_0}{\omega_d}\epsilon^{-\alpha t}\cos(\omega_d t - \phi) + \frac{I_0}{\omega_0 C}\frac{\omega_0}{\omega_d}\epsilon^{-\alpha t}\sin\omega_d t$$

$$i_L(t) = -V_0\frac{\omega_0^2 C}{\omega_d}\epsilon^{-\alpha t}\sin\omega_d t + I_0\frac{\omega_0}{\omega_d}\epsilon^{-\alpha t}\cos(\omega_d t + \phi)$$

*Case 4*   $\alpha = 0$ or $Q = \infty$.   Lossless case.   $(s_1 = j\omega_0,\ s_2 = -j\omega_0)$

$$i_L(t) = I_0\cos\omega_0 t + \frac{V_0}{\omega_0 L}\sin\omega_0 t$$

$$v_C(t) = -I_0\omega_0 L\sin\omega_0 t + V_0\cos\omega_0 t$$

$$v_C(t) = V_0\cos\omega_0 t + \frac{I_0}{\omega_0 C}\sin\omega_0 t$$

$$i_L(t) = -V_0\omega_0 C\sin\omega_0 t + I_0\cos\omega_0 t$$

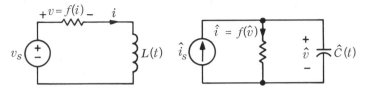

**Fig. 6.3**   Two dual circuits; note that the resistors are nonlinear.

**Example 2**   Two circuits need not be *linear* and *time-invariant* in order to be dual. Consider the two circuits of Fig. 6.3. The linear time-varying inductor of the first circuit is characterized for each $t$ by the slope of its characteristic $L(t)$; similarly, the linear time-varying capacitor of the second circuit is characterized by $\widehat{C}(t)$. The nonlinear resistor of the first circuit is characterized by the function $f(\cdot)$ whose graph is the resistor characteristic plotted as $v$ versus $i$; the nonlinear resistor of the second circuit is characterized by the same curve provided the characteristic is plotted as $\widehat{i}$ versus $\widehat{v}$ (note the interchange of current and voltage). In other words, the two resistors have characteristics described by the same curve, provided that the first characteristic is plotted on the $iv$ plane and the second is plotted on the $\widehat{v}\widehat{i}$ plane. If the current through the first resistor is $i$, the voltage across it is $f(i)$; if the voltage across the second resistor is $\widehat{v}$, the current through it is $f(\widehat{v})$. For the series circuit we have, from KVL,

$$v_s(t) = \frac{d}{dt}[L(t)i(t)] + f(i(t))$$

For the parallel circuit we have, from KCL,

$$\widehat{i}_s(t) = \frac{d}{dt}[\widehat{C}(t)\widehat{v}(t)] + f(\widehat{v}(t))$$

Suppose that $L(t) = \widehat{C}(t)$ for every $t \geq 0$. The two equations have the same form, and the two circuits are called *dual*. Consequently if the initial states are the same $[i(0) = \widehat{v}(0)]$ and if the inputs have the same waveform $[v_s(t) = \widehat{i}_s(t)$, for all $t \geq 0]$, the responses are identical; that is, the waveform of $i(\cdot)$ defined for $t \geq 0$ is identical to the waveform of $\widehat{v}(\cdot)$ defined for $t \geq 0$.

Let us examine these two examples and observe that there are many one-to-one correspondences between them. A KVL equation of one circuit corresponds to a KCL equation of the other; a mesh of one of them corresponds to a node of the other. The following gives typical dual terms:

| KVL | KCL |
| --- | --- |
| Current | Voltage |
| Mesh | Node together with the set of branches connected to it |
| Elements in series | Elements in parallel |
| Inductor | Capacitor |
| Resistor | Resistor |
| Voltage source | Current source |

It is important to observe that some of these correspondences relate to *properties of the graph,* whereas others pertain to *the nature of the branches.* Therefore, in the formulation of duality we must introduce the notion of dual graphs. A thorough treatment of this subject will be given in Chap. 10. At the present we only wish to stress the fact that the concept of duality is of great importance in circuit theory. Many circuits can be understood without a detailed analysis if the properties of a dual circuit are known. We shall take advantage of the duality concept from time to time throughout the text.

## 6.2   Mechanical and Electrical Analog

In classical mechanics we have encountered simple harmonic motion, damped oscillation, and exponential damping similar to what we have studied so far in this course. Let us review the basic mechanical elements and the formulation of equations in mechanical systems, and observe the analogy with electric circuits.

**Example 3**   Consider the mechanical system in Fig. 6.4 in which a block with mass $M$ is tied to the wall by a spring with spring constant $K$. The block is driven by a force designated $f_s$. The contact surface between the block and ground has a friction force that retards the motion of the block; at every instant of time it is opposite to the velocity. The equation of motion can be written by means of a free-body diagram, as shown in Fig. 6.4. Let $f_K$ be the force applied by the spring to the block, and let $f_B$ be the friction

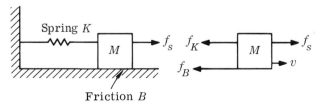

**Fig. 6.4**   Mechanical system and its free-body diagram.

force; then the net force to accelerate the block is $f_s - f_K - f_B$, equal to the derivative of the momentum according to Newton's law.   Thus,

(6.9)   $f_s - f_K - f_B = \dfrac{d}{dt} Mv$

where $v$ is the velocity in the direction of the force $f_s$.   It is well known that the friction force is a function of the velocity denoted by $f_B(\cdot)$, whereas the elastic force is a function of the displacement $x$ denoted by $f_K(\cdot)$.   Let us rewrite Eq. (6.9) in the following form:

(6.10)   $f_s = f_K(x) + f_B(v) + \dfrac{d}{dt} Mv$

We now consider a parallel connection of a resistor, an inductor, a capacitor, and a current source $i_s$.   We can write the KCL equation

(6.11)   $i_s = i_L(\phi) + i_R(\hat{v}) + \dfrac{d}{dt} C\hat{v}$

where $C\hat{v}$ is the charge in the linear capacitor and $\phi(t) = \phi(0) + \int_0^t \hat{v}(t')\, dt'$ is the flux for a nonlinear inductor.   $i_L(\cdot)$ and $i_R(\cdot)$ represent, respectively, the current in the inductor as a function of flux and the current in the resistor as a function of voltage.   In the mechanical system, $v = dx/dt$ is the velocity, and $x = x(0) + \int_0^t v(t')\, dt'$ is the displacement.   If, in addition, $f_s = i_s, f_K = i_L, f_B = i_R$, and $M = C$, the two equations are identical, and we call the parallel $RLC$ circuit the electric analog of the mechanical system.   In the analog circuit, the circuit variable $\hat{v}$ (the voltage) behaves the same way as the mechanical variable $v$ (the velocity).   The concept of analogy is similar to that of duality, except that ordinarily analogy implies only a dynamic equivalence between two systems whereas duality implies, in addition, a topological relation.   The notion of analogy is often useful for explaining and understanding many physical phenomena since, depending on their experience and education, people are invariably more familiar with one type of system than another.   As pointed out, the concept of analogy is not restricted to linear time-invariant systems.   The following summarizes the analogous variables and elements:

| Mechanical systems | Electric circuits |
| --- | --- |
| Force $f_s$ | Current $i_s$ |
| Velocity $v$ | Voltage $\hat{v}$ |
| Displacement $x$ | Flux $\phi$ |
| Spring | Inductor |
| Friction | Resistor |
| Mass | Capacitor |

In particular, if the three basic elements are linear and time-invariant, we have the following familiar analogous relations:

| Mechanical systems | | Electric circuits | |
|---|---|---|---|
| Mass | $f = M \dfrac{dv}{dt}$ | Capacitor | $i = C \dfrac{d\hat{v}}{dt}$ |
| Friction | $f = Bv$ | Conductor | $i = G\hat{v}$ |
| Spring | $f(t) = f(0) + K \displaystyle\int_0^t v(t')\, dt'$ | Inductor | $i(t) = i(0) + \dfrac{1}{L} \displaystyle\int_0^t \hat{v}(t')\, dt'$ |

This set of analogous quantities is not the only one possible. In particular, if instead of relating the mechanical system to the *parallel RLC* circuit we had related it to the *series RLC* circuit, we would have obtained a different set of analogous quantities. For instance, the voltage $v_s$ would correspond to the force $f_s$, and the current $i$ would correspond to the velocity $v$.

## Summary

- The zero-input responses of second-order linear time-invariant passive *RLC* circuits fall into four cases, as shown in Table 5.1 (see pp. 222–223).

- The zero-input responses of a linear time-invariant parallel *RLC* circuit can be expressed in terms of state trajectories plotted on the $i_L v_C$ plane with $t$ as a parameter. The state of the circuit at time $t$ is the vector $\mathbf{x}(t) \triangleq \begin{bmatrix} i_L(t) \\ v_C(t) \end{bmatrix}$ and the initial state is the vector $\mathbf{x}(0) = \begin{bmatrix} i_L(0) \\ v_C(0) \end{bmatrix}$.

- The second-order time-invariant parallel *RLC* circuit (passive and active) can be further categorized according to the location of natural frequencies and the nature of the state trajectories, as shown in Table 5.2.

**Table 5.2** **Classification of Linear Time-invariant Parallel *RLC* Circuits**

| Parallel *RLC* circuit | Passive: $G = \dfrac{1}{R} > 0$ | Lossless: $G = \dfrac{1}{R} = 0$ | Active: $G = \dfrac{1}{R} < 0$ |
|---|---|---|---|
| Location of natural frequencies | Left-half $s$ plane | $j\omega$ axis | Right-half $s$ plane |
| State trajectories | Asymptotically stable | Oscillatory | Unstable |

- The state-space approach is most useful for the analysis of nonlinear and time-varying circuits. The equations are of the form $\dot{\mathbf{x}} = \mathbf{f}(\mathbf{x}, w, t)$ where x is the state and w is the input.   The solution can be obtained by means of the step-by-step calculation.

■   The concept of dual circuits is based on the fact that the equations describing dual circuits have the same form.

■   If a mechanical system is the analog of an electric circuit, then they are both described by equations which have the same form.

## Problems

Calculation with exponentials

**1.** Using the notation of Sec. 1, show that

$$\frac{d}{dt} \epsilon^{-\alpha t} \cos \omega_d t = -\omega_0 \epsilon^{-\alpha t} \sin (\omega_d t + \phi)$$

$$\frac{d}{dt} \epsilon^{-\alpha t} \sin \omega_d t = \omega_0 \epsilon^{-\alpha t} \cos (\omega_d t + \phi)$$

Natural frequencies

**2.** Suppose that the natural frequencies of a linear time-invariant circuit consist of one of the following sets:

*a.*  $s_1 = -2, s_2 = 3$

*b.*  $s_1 = s_2 = -2$

*c.*  $s_1 = j2, s_2 = -j2$

*d.*  $s_1 = 2 + j3, s_2 = 2 - j3$

Give the general expression for the zero-input responses in terms of real-valued time functions.

Q factor

**3.** Given an $RLC$ circuit with a $Q$ of 500, how many periods need one wait to have the envelope of the zero-input response reduced to 10 percent, 1 percent, and 0.1 percent of its peak value during the first period? (Give, in each case, an answer to the nearest ½ period.)

Q factor

**4.** Consider two linear time-invariant $RLC$ circuits.   The first one is a parallel circuit with element values $R'$, $L$, and $C$, and the second one is a series circuit with element values $R$, $L$, and $C$.   If the two circuits must have the same $Q$, what is the relation between $R$ and $R'$?   What happens when $Q \to \infty$?

Determination of arbitrary constants from the initial conditions

**5.** Given a linear time-invariant parallel $RLC$ circuit with $\omega_0 = 10$ rad/sec, $Q = \frac{1}{2}$, and $C = 1$ farad, write the differential equation and determine the zero-input response for the voltage $v_C$ across the capacitor.   The initial conditions are $v_C(0) = 2$ volts and $i_L(0) = 5$ amp.

Step and impulse responses

**6.** For the parallel $RLC$ circuit in Prob. 5, let the input be a current source $i_s$ connected in parallel.   Determine the step response and the impulse response for the voltage $v_C$.

Complete
response

**7.** Connect a current source $i_s$ in parallel with the $RLC$ circuit in Prob. 5. Let the current be $i_s(t) = u(t) \cos 2t$. Determine the zero-state response and the transient response.

Sinusoidal
steady-state
response,
transient, and
complete
response

**8.** For the parallel $RLC$ circuit in Prob. 5, let the input be a current source $i_s(t) = u(t) \cos 2t$ connected in parallel. Determine the complete response for the initial conditions $v_C(0) = 2$ volts and $i_L(0) = 5$ amp. Indicate clearly the transient part and the steady-state part. Demonstrate that the complete response is the sum of the zero-input response of Prob. 5 and the zero-state response of Prob. 7.

Transient
elimination

**9.** Connect a current source $i_s$ in parallel to the $RLC$ circuit in Prob. 5. Let $i_s(t) = u(t) \cos 2t$. Is it possible to choose the initial conditions so that there is no transient? If so, determine the required initial conditions; if not, justify your answer.

Solution of
differential
equations

**10.** Solve the following differential equations:

a. $\dfrac{d^2x}{dt^2} + 2\dfrac{dx}{dt} + x = \epsilon^{-2t}$      $x(0) = 1; \dfrac{dx}{dt}(0) = -1$

b. $\dfrac{d^2x}{dt^2} + 3\dfrac{dx}{dt} + 2x = 5$      $x(0) = 1; \dfrac{dx}{dt}(0) = 0$

c. $\dfrac{d^2x}{dt^2} + x = \cos t$      $x(0) = 0; \dfrac{dx}{dt}(0) = 1$

d. $\dfrac{d^2x}{dt^2} + \dfrac{dx}{dt} + x = tu(t)$      $x(0) = 0; \dfrac{dx}{dt}(0) = 0$

Solution of
matrix
differential
equations,
state
trajectories

**11.** Solve for the following matrix differential equations by the method of successive approximation and plot the state trajectories:

a. $\dfrac{d\mathbf{x}}{dt} = \begin{bmatrix} 1 & 0 \\ -1 & 2 \end{bmatrix} \mathbf{x}$      $\mathbf{x}(0) = \begin{bmatrix} 1 \\ 1 \end{bmatrix}$

b. $\dfrac{d\mathbf{x}}{dt} = \begin{bmatrix} 1 & -2 \\ 1 & 0 \end{bmatrix} \mathbf{x}$      $\mathbf{x}(0) = \begin{bmatrix} 1 \\ -1 \end{bmatrix}$

Impulse re-
sponse and
change of
source

**12.** Given a linear time-invariant parallel $RLC$ circuit with $\omega_0 = 1$ rad/sec and $Q = 10$: the input is a voltage source connected in series with the inductor. Determine the impulse response for the voltage $v_C$ across the capacitor. (Hint: Use Norton's equivalent circuit.)

Step response
and ramp
response

**13.** For the circuit in Prob. 12 determine the step response and the ramp response.

Linearity of
zero-state
response

**14.** Given a linear time-invariant parallel $RLC$ circuit. The zero-state response to the sinusoidal input $i_1(t) = u(t) \cos 2t$ is given by

$v_1(t) = \epsilon^{-t} + 2\epsilon^{-2t} + \cos(2t + 60°)$      $t \geq 0$

The complete response to the sinusoidal input $i_2(t) = 3u(t) \cos 2t$ when the circuit starts from a certain initial state is

$$v_2(t) = -\epsilon^{-t} + 3\epsilon^{-2t} + 3 \cos (2t + 60°) \qquad t \geq 0$$

Determine the complete response to the sinusoidal input $i_3(t) = 5u(t) \cos 2t$ if the circuit starts at the same initial state.

**15.** The circuit shown in Fig. P5.15 is linear and time-invariant.

*Zero-input response, differential equation, and state trajectory*

a.  Write the differential equation with $v_C$ as the dependent variable, and indicate the proper initial conditions as functions of $i_L(0)$ and $v_C(0)$. (Hint: Write a node equation for node ① and a mesh equation for $i_L$, using $v_C$ and $i_L$ as variables.)

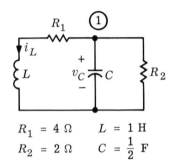

$$R_1 = 4\ \Omega \qquad L = 1\ \text{H}$$
$$R_2 = 2\ \Omega \qquad C = \frac{1}{2}\ \text{F}$$

**Fig. P5.15**

b.  Calculate the zero-input response $i_L(\cdot)$ and $v_C(\cdot)$.

c.  Write the zero-input response as a state vector $\mathbf{x}(\cdot) = \begin{bmatrix} i_L(\cdot) \\ v_C(\cdot) \end{bmatrix}$ and plot the state trajectories $x_1(\cdot)$ and $x_2(\cdot)$ corresponding to $x_1(0) = \begin{bmatrix} 1 \\ 2 \end{bmatrix}$

and $x_2(0) = \begin{bmatrix} -2 \\ 2 \end{bmatrix}$, plotting $i_L$ along the abscissa and $v_C$ along the ordinate.

d.  Does the trajectory $x_1(\cdot)$ have any special property? Which other trajectories, if any, have a similar property?

*State trajectory of nonlinear circuit, approximate integration*

**16.** The nonlinear time-invariant $RLC$ circuit in Fig. P5.16 has components specified by $i_R = \alpha v_R$, $q = \beta v_C + \gamma v_C^3$, and $\phi = \delta i_L$, where $\alpha = 2$ mhos, $\beta = 1$ farad, $\gamma = \frac{1}{3}$ farad/volt$^2$, and $\delta = \frac{1}{2}$ henry. The source drives the circuit with voltage $e(t) = \sin (0.5t)$ volt, and at time $t = 0$ the voltage across the capacitor $v_C(0) = 2$ volts and the current through the inductor $i_L(0) = -2$ amp. Write the state equation for the circuit using

$$\mathbf{x} = \begin{bmatrix} i_L \\ v_C \end{bmatrix}$$ as state vector, and plot the state-space trajectory using successive straight-line approximations; that is, $\mathbf{x}[(n + 1)\,\Delta t] \approx \mathbf{x}(n\,\Delta t) + \dot{\mathbf{x}}(n\,\Delta t)\,\Delta t$ with $\Delta t = 0.2$ sec and $n = 0, 1, \ldots, 10$. Plot $v_C$ and $i_L$ as functions of time.

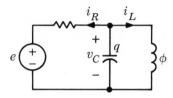

**Fig. P5.16**

Formulation of differential equations, energy

**17. a.** For the circuit shown in Fig. P5.17 set up the differential equations for $v_1(t)$ and $v_2(t)$.

  **b.** Let $\mathscr{E}(t) = \frac{1}{2}[v_1{}^2(t) + v_2{}^2(t)]$; show that $d\mathscr{E}(t)/dt \leq 0$ for all $t$.

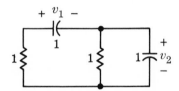

**Fig. P5.17**

Nonlinear circuit and equations in normal form, energy

**18. a.** Write the differential equation for the nonlinear time-invariant circuit given in Fig. P5.18, using $q$ and $\phi$ as variables, where the inductor characteristic is given by $i_L = \phi + \phi^3$ ($\phi$ is the flux) and the capacitor characteristic is given by $v_C = 2q$ ($q$ is the charge).

  **b.** At time $t_0$ let the flux and the charge have the values $\phi_0$ and $q_0$, respectively. What is the energy stored in the circuit?

  **c.** At time $t_0$ let $q = 0$ and $\phi = 2$. What is the maximum value of $q(t)$ for $t \geq t_0$? (Hint: Is there any energy dissipation in the circuit?)

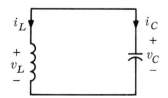

**Fig. P5.18**

**19.** Use $\mathbf{x} = \begin{bmatrix} i_L \\ v_C \end{bmatrix}$ as the state vector for a linear time-invariant series $RLC$
circuit as shown in Fig. P5.19.   The only data available from measure-
ments taken of this circuit are the time derivatives of the state vector at two
different states, namely,

$$\dot{\mathbf{x}} = \begin{bmatrix} -15 \\ 10 \end{bmatrix} \text{ at } \mathbf{x} = \begin{bmatrix} 2 \\ 1 \end{bmatrix} \qquad \text{and} \qquad \dot{\mathbf{x}} = \begin{bmatrix} 3 \\ -5 \end{bmatrix} \text{ at } \mathbf{x} = \begin{bmatrix} -1 \\ 1 \end{bmatrix}$$

a.   Determine the element values $R$, $L$, and $C$.

b.   Calculate the derivative of the state vector at $\mathbf{x} = \begin{bmatrix} 3 \\ 0 \end{bmatrix}$ by two
methods.   First, use the equation $\dot{\mathbf{x}} = \mathbf{Ax}$, and second, equate the unknown derivative to an appropriate linear combination of the given
derivatives.

c.   Calculate the slope $dv_C/di_L$ of the state-space trajectory at $\mathbf{x} = \begin{bmatrix} 3 \\ 0 \end{bmatrix}$.

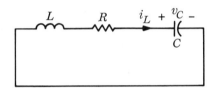

**Fig. P5.19**

**20.** The circuit shown in Fig. P5.20 is made of linear time-invariant elements.   The voltage $e_s$ is the input, and $v_C$ is the response.   The element
values are $R_1 = 2$ ohms, $R_2 = 3$ ohms, $L = 1$ henry, $C = 0.25$ farad.

a.   Calculate the impulse response $h$.

b.   Calculate the complete response to the input $e_s(t) = u(t)$ and the initial
state $i_L(0) = 2$ amp, $v_C(0) = 1$ volt.

c.   Calculate and sketch the sinusoidal steady state for $v_C$, $i_L$, and $v_{R_2}$ to the
input $e_s = 5 \cos 2t$.   Give the results as real-valued functions of time.

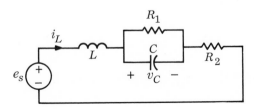

**Fig. P5.20**

**21.** Consider the linear time-invariant $LC$ circuit shown in Fig. P5.21. Before time $t = 0$ the switch is open, and the voltages across the capacitors are $v_1 = 1$ and $v_2 = 4$ volts. The switch is closed at time $t = 0$ and remains in this condition for a time interval of $2\pi$ sec. The switch is opened at $t = 2\pi$ sec and remains open thereafter. What are the values of $v_1$ and $v_2$ for $t > 2\pi$ sec? Sketch the state trajectory in the $i_L v_C$ plane ($v_C = v_2 + v_1$). What can be said about the energy stored in the circuit before time $t = 0$ and after time $t = 2\pi$ sec? (Hint: Analyze the consequences of the particular choice of time interval.)

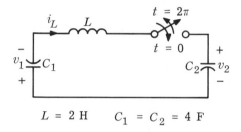

$$L = 2\ \text{H} \qquad C_1 = C_2 = 4\ \text{F}$$

**Fig. P5.21**

**22.** The resistance $R_2$ in the circuit in Fig. P5.15 is changed to $-2$ ohms.

*a.* What are the natural frequencies of the circuit?

*b.* If $i_L(0) = 1$ amp and $v_C(0) = 1$ volt, determine the zero-input responses $i_L(\cdot)$ and $v_C(\cdot)$.

*c.* Plot the state trajectory.

**23.** Draw a circuit that is the dual of the circuit in P5.15, and specify all element values.

**24.** For the mechanical system in Fig. 6.4 draw two electric circuits that are electric analogs to the mechanical system.

**25.** The circuit in Fig. P5.25 is a typical tunnel-diode oscillator circuit. The resistor has a characteristic specified by $i_D = g(v_D)$. Write the differential equation in normal form with $i_L$ and $v_C$ as variables.

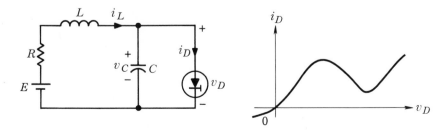

**Fig. P5.25**

# 6

# Introduction to Linear Time-invariant Circuits

In the previous two chapters we studied first-order and second-order circuits and developed many basic concepts and techniques. In this chapter we shall first summarize some of the important results and then make a few generalizations. We shall give a preliminary treatment of node analysis and mesh analysis for linear time-invariant circuits. It will be shown that the result of these analyses leads to an input-output description in terms of an $n$th-order linear differential equation with constant coefficients. A method of calculating the impulse response from the $n$th-order differential equation will be presented. We shall then consider responses to arbitrary inputs. The convolution-integral representation will be carefully derived, and the computation of the convolution integral will be illustrated with examples.

## 1    Some General Definitions and Properties

In Chap. 2 we introduced three basic circuit elements, namely, the resistor, the capacitor, and the inductor. For each element we gave a four-way classification: linear or nonlinear, time-invariant or time-varying.

To facilitate our later formulations, we shall recall our four-way classification of circuits. Any circuit with the property that each of its elements is either a *linear element* or an independent source is called a **linear circuit.** Similarly, any circuit with the property that each of its elements is either a *time-invariant element* or an independent source is called a **time-invariant circuit.** Consequently, a **linear time-invariant circuit** is a circuit with the property that each of its elements is either a *linear time-invariant element* or an independent source. Obviously, if a circuit is not linear, it will be called a **nonlinear circuit;** if a circuit is not time-invariant, it will be called a **time-varying circuit.**

In these definitions the independent sources must be treated separately because (1) The voltage across a voltage source and the current of a current source play a role in the analysis that is different from that of the network variables of other elements, and (2) all the independent sources are nonlinear and time-varying elements (for example, a sinusoidal voltage source may be considered as a nonlinear time-varying resistor since its characteristic is, for each $t$, a horizontal line in the $iv$ plane whose ordinate is a sinusoidal function of time; that is, its characteristic is a straight line that does *not* go through the origin all the time).

Furthermore, it should be emphasized that the set of *all* voltages across independent voltage sources and *all* currents through independent current sources is referred to as the set of *inputs* of the circuit. Thus, a circuit containing only one independent source is called a *single-input circuit.*

In this chapter we shall consider only single-input–single-output circuits, that is, circuits containing only one independent source and having only one variable (the output) that needs to be calculated. The input is the waveform of either an independent voltage source or an independent current source; this waveform may be a constant, a step function, an impulse, a sinusoid, or any arbitrary function of time. The output, which may also be called the *response,* is either a particular branch voltage, a particular branch current, a linear combination of some branch voltages and branch currents, or the charge on a capacitor or the flux of an inductor.

For all the lumped circuits that we shall consider in this text, we shall be able to write a differential equation or a system of differential equations that will allow us to calculate all the branch voltages and all the branch currents of the circuit.

In order to calculate the unique solution of the system of differential equations, we must know the inputs as well as the initial conditions. The particular way in which these initial conditions will be expressed depends upon the way in which the differential equations are written. In particular, we shall show in Chap. 13 that if, at the initial time, all the capacitor voltages and all the inductor currents are known, the required initial conditions are uniquely specified.

*We shall give the name of* **state of a circuit at time** $t_0$ *to any set of initial conditions that, together with the inputs, uniquely determines all the network variables of the circuit for all* $t \geq t_0$.

From the previous statements, we see that the state of a circuit at some time $t_0$ can always be taken to be the set of all capacitor voltages and all inductor currents at time $t_0$. The state that corresponds to having all initial conditions equal to zero is called the **zero state.** For linear circuits, if all the inputs are zero and if the circuit is in the zero state, all network variables remain equal to zero forever after. When the input is applied at some time $t_0$, the set of initial conditions at time $t_0$ required to uniquely determine all network variables is called, according to the previous definition, the state at time $t_0$; for brevity it is also called *initial state.* The word "initial" refers to the fact that it is the state of the circuit when the input is applied.

We give the name **zero-state response** to the response (output) of a circuit to an input applied at some arbitrary time $t_0$, subject to the condition that the circuit be in the zero-state just prior to the application of the input (that is, at $t_0-$). **Zero-input response** is defined as the response of a circuit when its input is identically zero. Clearly, the zero-state response is due only to the input; similarly, the zero-input response is due to the initial state only. It is the response of the circuit due to the energy initially stored in it. **Complete response** is defined as the response of a circuit to both an input and the initial state.

In previous chapters we established the properties of linear time-invariant circuits of the first and second order; we shall see later that these properties also hold for any linear time-invariant or time-varying circuit. For linear circuits (time-invariant or time-varying)

1. The *complete response* is the sum of the zero-input response and the zero-state response.

2. The *zero-state response* is a linear function of the input.

3. The *zero-input response* is a linear function of the initial state.

2 **Node and Mesh Analyses**

In Chap. 3 we analyzed simple resistive circuits that were of the form of series and parallel connections of elements, and we obtained equivalent circuits. In Chaps. 4 and 5 we dealt with circuits containing resistors, inductors, and capacitors. The circuits had simple topology; they either contained only a single loop so that a single-loop equation (KVL) characterized the behavior of the circuit, or they had only two nodes so that a single-node equation (KCL) characterized the behavior of the circuit. For circuits with complex topology we need to develop general and systematic methods of network analysis, which we shall do in Chaps. 9 through 12. In this section we shall employ a circuit slightly more complicated than those in Chap. 5 to illustrate two basic methods of network analysis: node analysis and mesh analysis.

We start with the simple circuit shown in Fig. 2.1. The input is the current source $i_s$, and the output is the voltage $v_2$ across the resistor with

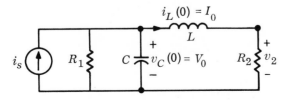

**Fig. 2.1**   A simple example illustrating the node analysis and the mesh analysis. The current source $i_s$ is the input, and the voltage $v_2$ is the output.

resistance $R_2$. The initial state is given by $v_C(0) = V_0$ and $i_L(0) = I_0$; their reference directions are indicated in the figure.

### 2.1   Node Analysis

The first step of node analysis is to count the number of nodes of the circuit. In the present case there are three nodes labeled ①, ②, and ③ (see Fig. 2.2, which is Fig. 2.1 redrawn to emphasize the nodes).

Clearly, we can define three *node-pair* voltages between nodes, namely, $v_{13}$, $v_{23}$, and $v_{12}$. They turn out to be the branch voltages for the branches connecting the nodes ① and ③, ② and ③, and ① and ②, respectively. From KVL the sum of voltages in any loop must be zero. Thus, KVL implies a linear constraint among the three node-pair voltages. For example, if we set $v_{13} = v_1$ and $v_{23} = v_2$, we have $v_{12} = v_1 - v_2$. In general, we pick a particular node as a *reference node* (sometimes called datum node or ground and indicated by the symbol $\perp$). We then call the voltages from all the other nodes to the reference node *node voltages* (or node-to-datum voltages). In the present case, with ③ as the reference node, the node voltages are $v_1$ and $v_2$. Clearly, all other node-pair voltages can be expressed in terms of the node voltages $v_1$ and $v_2$ by means of KVL. Thus, in general, if a circuit has $n + 1$ nodes, there are $n$ node voltages to be determined because once they are known, any node-pair voltages and, in particular, branch voltages can be found immediately. In the above we have used only Kirchhoff's voltage law. We must, in addition, use Kirchhoff's current law and the branch equations in order to be able to calculate all branch voltages and branch currents and, in particular, to calculate the desired response.

Let us consider the implication of Kirchhoff's current law in this circuit. We can, of course, write three node equations for nodes ①, ②,

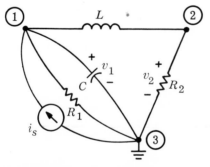

**Fig. 2.2**   The circuit in Fig. 2.1 is redrawn to emphasize the node labeling, which is the first step of node analysis.

and ③.   However, it is obvious that one of the three equations will be redundant; indeed, adding any two of these equations yields the third, possibly except for a factor of $-1$.   Thus, for this three-node circuit, KCL yields only two independent node equations.   In an $(n + 1)$-node circuit, it is not difficult to show that there are $n$ independent node equations (this will, in fact, be shown in Chap. 10).   To save time, instead of explicitly writing the node equations in terms of branch currents, we shall use branch equations and express the currents directly in terms of branch voltages.   Also we shall express all branch voltages in terms of node voltages.   The net result is that we obtain two equations for the two un-known node voltages.   Therefore, we shall be able to solve for all or any of the node voltages.

Let us proceed to write the two node equations for our example, keep-ing in mind the relation $v_{12} = v_1 - v_2$ and all the branch equations of the elements.   Using KCL, for node ① we have,

$$(2.1) \quad C \frac{dv_1}{dt} + \frac{v_1}{R_1} + I_0 + \frac{1}{L} \int_0^t (v_1 - v_2) \, dt' = i_s(t)$$

and for node ②,

$$(2.2) \quad -I_0 + \frac{1}{L} \int_0^t (v_2 - v_1) \, dt' + \frac{v_2}{R_2} = 0$$

The additional given initial condition is

$$(2.3) \quad v_1(0) = V_0$$

Equations (2.1) and (2.2) are the two node equations in which only the two node voltages $v_1$ and $v_2$ appear as variables.   Equation (2.3) is an initial condition required to specify a unique solution to (2.1) and (2.2).

The aim of our problem is to obtain a differential equation with $v_2$ as the dependent variable.   A systematic method will be given in Chap. 13 for obtaining a single differential equation from a set of simultaneous in-tegrodifferential equations.   In the present case, the simplest way is to add (2.1) and (2.2) first to yield

$$(2.4) \quad C \frac{dv_1}{dt} + \frac{v_1}{R_1} + \frac{v_2}{R_2} = i_s$$

Differentiating (2.2), we have

$$\frac{1}{L} v_2 - \frac{1}{L} v_1 + \frac{1}{R_2} \frac{dv_2}{dt} = 0$$

or

$$(2.5) \quad v_1 = v_2 + \frac{L}{R_2} \frac{dv_2}{dt}$$

Differentiating (2.5), we have

(2.6)   $$\frac{dv_1}{dt} = \frac{dv_2}{dt} + \frac{L}{R_2}\frac{d^2v_2}{dt^2}$$

The differential equation for $v_2$ is then obtained by substituting (2.5) and (2.6) in (2.4); thus,

(2.7)   $$LC\frac{d^2v_2}{dt^2} + \left(R_2C + \frac{L}{R_1}\right)\frac{dv_2}{dt} + \left(1 + \frac{R_2}{R_1}\right)v_2 = R_2i_s$$

The initial conditions required to uniquely determine the solution of (2.7) can be obtained from Eq. (2.3) and the original equations (or their equivalents) by substituting $t = 0$.† From (2.2) we obtain

(2.8)   $$v_2(0) = R_2I_0$$

and from (2.5) we have

(2.9)   $$\frac{dv_2}{dt}(0) = \frac{R_2}{L}[v_1(0) - v_2(0)] = \frac{R_2}{L}(V_0 - R_2I_0)$$

**Exercise**   For the circuit of Fig. 2.2, obtain the differential equation relating $v_1$ and $i_s$. Be sure to specify also the initial condition necessary to specify the solution uniquely.

## 2.2   Mesh Analysis

An alternative procedure for analyzing a general network depends on the writing of mesh equations. Let us redraw the circuit in Fig. 2.1 by using the Thévenin equivalent circuit of the parallel combination of $i_s$ and the resistor. The new circuit is shown in Fig. 2.3 and

(2.10)   $$v_s = R_1i_s$$

---

† Assume that $i_s$ is not an impulse or other singular function. If $i_s$ includes an impulse at $t = 0$, care is required; we may integrate the equations from $t = 0-$ to $t = 0+$ to obtain the new initial condition at $0+$.

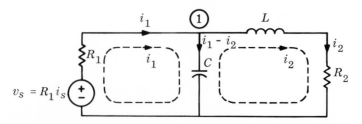

**Fig. 2.3**   The circuit in Fig. 2.1 is redrawn for mesh analysis. Note that the current source in Fig. 2.1 has been replaced with the voltage source by means of Thévenin's equivalent circuit.

Let us designate the current in mesh 1 (which contains $v_s$, $R_1$, and $C$) by $i_1$ and the current in mesh 2 (which contains $C$, $L$, and $R_2$) by $i_2$. The actual branch currents of the source $v_s$ and the resistor $R_1$ are $i_1$, whereas the branch currents of the inductor $L$ and the resistor $R_2$ are $i_2$. The branch current of the capacitor is the algebraic sum of the two mesh currents, $i_1 - i_2$. This can also be seen by using KCL at node ①.

Next we apply KVL to the meshes. In the expressions for KVL we express the branch voltages explicitly in terms of $i_1$ and $i_2$ by means of branch equations. Thus, for mesh 1 we have

$$(2.11) \qquad R_1 i_1 + V_0 + \frac{1}{C} \int_0^t (i_1 - i_2)\, dt' = v_s(t)$$

and for mesh 2

$$(2.12) \qquad L \frac{di_2}{dt} + R_2 i_2 - V_0 + \frac{1}{C} \int_0^t (i_2 - i_1)\, dt' = 0$$

The additional given initial condition is

$$(2.13) \qquad i_2(0) = I_0$$

Equations (2.11) and (2.12) are the mesh equations of the circuit; only the two mesh currents $i_1$ and $i_2$ appear as variables. Equation (2.13) is an initial condition required to uniquely specify the solution. To obtain the differential equation with the output variable $v_2$, we need only to find the differential equation for $i_2$. The simplest way is to add (2.11) and (2.12) to yield

$$R_1 i_1 + L \frac{di_2}{dt} + R_2 i_2 = v_s$$

or

$$(2.14) \qquad i_1 = -\frac{L}{R_1} \frac{di_2}{dt} - \frac{R_2}{R_1} i_2 + \frac{v_s}{R_1}$$

Differentiating (2.12), we have

$$(2.15) \qquad L \frac{d^2 i_2}{dt^2} + R_2 \frac{di_2}{dt} + \frac{i_2}{C} - \frac{i_1}{C} = 0$$

Substituting (2.14) in (2.15) we obtain

$$(2.16) \qquad LC \frac{d^2 i_2}{dt^2} + \left( R_2 C + \frac{L}{R_1} \right) \frac{di_2}{dt} + \left( 1 + \frac{R_2}{R_1} \right) i_2 = \frac{v_s}{R_1}$$

The initial conditions are obtained from (2.13); that is,

$$(2.17) \qquad i_2(0) = I_0$$

and from (2.12) by setting $t = 0$ and obtaining

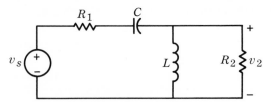

**Fig. 2.4**    Circuit for exercise on node analysis and mesh analysis; $v_s$ is the input, and $v_2$ is the response.

(2.18)    $\dfrac{di_2}{dt}(0) = \dfrac{1}{L}(V_0 - R_2 I_0)$

Since $v_2 = R_2 i_2$ and $v_s = R_1 i_s$, the equations in terms of $v_2$ and $i_s$ are

(2.19)    $LC\dfrac{d^2 v_2}{dt^2} + \left(R_2 C + \dfrac{L}{R_1}\right)\dfrac{dv_2}{dt} + \left(1 + \dfrac{R_2}{R_1}\right)v_2 = R_2 i_s$

and the initial conditions are

(2.20)    $v_2(0) = R_2 I_0$

(2.21)    $\dfrac{dv_2}{dt}(0) = \dfrac{R_2}{L}(V_0 - R_2 I_0)$

This simple example has illustrated the general fact that, given any single-input single-output linear time-invariant circuit, it is always possible to write a single differential equation relating the output to the input. Of course, the more complicated the circuit, the more work is involved. However, as we shall see in Chaps. 10 and 13, for such circuits there are systematic methods of obtaining the simplest differential equation relating the output to the input.

**Exercise**    Use node analysis and mesh analysis to write the differential equation for the voltage $v_2$ in the circuit shown in Fig. 2.4.

## 3    Input-Output Representation ($n$th-order Differential Equation)

For single-input single-output linear time-invariant circuits in general, the relation between input and output can be expressed by an $n$th-order linear differential equation with constant coefficients; thus,

(3.1)    $\dfrac{d^n y}{dt^n} + a_1 \dfrac{d^{n-1} y}{dt^{n-1}} + \cdots + a_n y = b_0 \dfrac{d^m w}{dt^m} + b_1 \dfrac{d^{m-1} w}{dt^{m-1}} + \cdots + b_m w$

where $y$ represents the output and $w$ the input. The constants $a_1, a_2, \ldots,$ $a_n$ and $b_0, b_1, \ldots, b_m$ depend on the element values and the topology of the circuit. The initial conditions are†

---

† As we shall see later, this is strictly true only if the term $d^m w/dt^m$ does not include the impulse function $\delta(t)$ or any of its derivatives.

$$y(0), \frac{dy}{dt}(0), \ldots, \frac{d^{n-1}y}{dt^{n-1}}(0)$$

The differential equation is obtained from Kirchhoff equations and from branch characterizations with node or mesh analysis as illustrated in the previous section. The initial conditions are obtained from the given initial state of the circuit and the network equations. The general method of writing nth-order differential equations and the determination of the initial conditions will be discussed in Chaps. 10, 11, and 13. At the present we wish to assume that the input-output relation has been expressed in (3.1), and we now proceed with the discussion of various types of responses.

## 3.1   Zero-input Response

The zero-input response is the response of the circuit when the input is identically zero. Accordingly, the right-hand side of (3.1) is identically zero; in other words, the differential equation is homogeneous. The characteristic polynomial of this differential equation is the nth-degree polynomial in $s$

$$s^n + a_1 s^{n-1} + \cdots + a_{n-1}s + a_n$$

and the zeros of this polynomial, $s_i$, $i = 1, 2, \ldots, n$, are called *the natural frequencies of the network variable y*. It is well known that if all the natural frequencies are distinct, the solution of the homogeneous equation is given by

$$(3.2) \quad y(t) = \sum_{i=1}^{n} k_i \epsilon^{s_i t}$$

where the constants $k_i$ are determined from the specified initial conditions. If some of the natural frequencies coincide, Eq. (3.2) must be modified to include powers of $t$ in the representation, as discussed in Appendix C. For example, if $s_1$ is a zero of the characteristic polynomial of order 3, Eq. (3.2) would contain $k_1 \epsilon^{s_1 t} + k_2 t \epsilon^{s_1 t} + k_3 t^2 \epsilon^{s_1 t}$.

## 3.2   Zero-state Response

The zero-state response for the variable $y$ in Eq. (3.1) is, in general, of the form (again assuming that all natural frequencies are distinct)

$$(3.3) \quad y(t) = \sum_{i=1}^{n} k_i \epsilon^{s_i t} + y_p(t)$$

where $y_p$ is *any* particular solution of (3.1) and depends upon the input $w$. The $n$ constants $k_i$ are specified by the requirement that all initial conditions $y(0-), dy/dt\,(0-), \ldots, d^{n-1}y/dt^{n-1}\,(0-)$ be zero; that is, the circuit must be in the zero state just prior to being hit by the input.

**Example**  Consider the $RC$ circuit shown in Fig. 3.1; $v_s$ is the input, and the resistor current $i$ is the output. Just prior to the application of the input, the capacitor is uncharged. From $t = 0$ on, the input $v_s(t) = V_m \cos t$ is applied; equivalently, we may use the step function $u(\cdot)$ and put $v_s(t) = u(t)V_m \cos t$, for all $t$. From KVL we obtain

$$(3.4) \quad Ri(t) + \frac{1}{C}\int_0^t i(t')\,dt' = v_s(t) = u(t)V_m \cos t$$

or

$$(3.5) \quad R\frac{di}{dt} + \frac{1}{C}i = \frac{dv_s}{dt}$$

Note that (3.5) is in the form of (3.1). Let us calculate the right-hand side of (3.5). Thus

$$\frac{dv_s}{dt} = V_m \frac{du}{dt}\cos t + V_m u(t)\frac{d}{dt}\cos t = V_m\,\delta(t) - V_m u(t)\sin t$$

The presence of $V_m\,\delta(t)$ in the right-hand side of (3.5) will cause the current $i$ to be discontinuous at $t = 0$. Indeed, for the left-hand side of (3.5) to balance the impulse $V_m\,\delta(t)$ of the right-hand side, $R(di/dt)$ must include the impulse $V_m\,\delta(t)$; hence $i$ will include the term $(V_m/R)u(t)$, which is a step function. Physically, this is easily explained. Since the voltage waveform $v_s(\cdot)$ is bounded, the voltages across the capacitor $C$ and the resistor $R$ are bounded; consequently, the current is also bounded, and, finally, the charge and voltage across $C$ are continuous. Thus, $v_C(0-) = v_C(0+)$, and, by KVL, $v_R(0+) = v_s(0+) - v_C(0+) = V_m$. In other words,

$$i(0+) = \frac{v_R(0+)}{R} = \frac{V_m}{R}$$

Thus, we see that even though just prior to turning on the voltage source $v_s$ (that is, at $0-$) the initial condition is zero, $i(0-) = 0$; it turns out that at $t = 0+$ the initial condition is not zero but depends on the input!

It is also important to point out that the term $y_p$ in (3.3) is *any* particular solution, that is, by definition, any solution which satisfies the nonhomogeneous differential equation (3.1). Some particular solutions are more

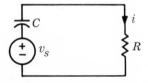

**Fig. 3.1**    A simple $RC$ circuit.

convenient than others. For step input, $y_p$ is chosen to be a constant;† for a sinusoidal input, $y_p$ is chosen to be a sinusoid of the same frequency; and for an input which is a polynomial in $t$, $y_p$ is chosen to be a polynomial in $t$ of the same degree (see Appendix C). In the next chapter we shall give a detailed discussion of the case in which the input is sinusoidal.

## 3.3   Impulse Response

The calculation of the impulse response is somewhat delicate because the right-hand side of (3.1) includes impulses and derivatives of impulses. In this subsection we shall use an example to illustrate the calculation of the impulse response directly from the differential equation. In Sec. 4 we shall show that the determination of the zero-state response for an arbitrary input depends only on knowledge of the impulse response. It is therefore of paramount importance to feel comfortable with the impulse response and to understand the method used to calculate it.

We propose now to show how one can obtain the impulse response directly from the differential equation (3.1):

$$y^{(n)} + a_1 y^{(n-1)} + \cdots + a_n y = b_0 w^{(m)} + b_1 w^{(m-1)} + \cdots + b_m w$$

with initial conditions

$$(3.6) \quad y(0-) = y^{(1)}(0-) = y^{(2)}(0-) = \cdots = y^{(n-1)}(0-) = 0$$

For simplicity in notation we have used the superscript $^{(n)}$ to represent $d^n/dt^n$, etc. Clearly, if the input $w$ is a unit impulse, the right-hand side includes the impulse function and its successive derivatives. The successive derivatives of the impulse function are sometimes called *singular functions*. Formally, we have

$$\frac{du}{dt} = \delta \qquad \text{and} \qquad \int_{-\infty}^{t} \delta(t')\,dt' = u(t)$$

$$\frac{d\delta}{dt} = \delta^{(1)} \qquad \text{and} \qquad \int_{-\infty}^{t} \delta^{(1)}(t')\,dt' = \delta(t)$$

$$\frac{d\delta^{(1)}}{dt} = \delta^{(2)} \qquad \text{and} \qquad \int_{-\infty}^{t} \delta^{(2)}(t')\,dt' = \delta^{(1)}(t)$$

. . . . . . . . . . . . . . . . . . . . . . . . . . . . . . . . . . . . . .

$$\frac{d\delta^{(n)}}{dt} = \delta^{(n+1)} \qquad \text{and} \qquad \int_{-\infty}^{t} \delta^{(n+1)}(t')\,dt' = \delta^{(n)}(t)$$

The direct determination of the impulse response $h$ is based on balancing the singular functions of the right-hand side by singular functions of

---

† If the degree $m$ is larger than the degree $n$ in the differential equation, then for a step input, $y_p$ must include, in addition to a constant, an impulse and some of its derivatives. This will be explained later.

the left-hand side of (3.1). Since in (3.1) $w(t) = \delta(t)$, the highest-order singular function of the right-hand side is $\delta^{(m)}$, and the behavior of the impulse response $h$ will depend on the relative magnitudes of $m$ and $n$.

1.  $n > m$ (the proper case). The impulse response $h$ does not include any singular functions, but $d^n h/dt^n$ includes $\delta^{(m)}$, as required by (3.1).

2.  $n = m$. The impulse response $h$ will include an impulse, $b_0 \delta$ [here, $b_0$ is the coefficient of $w^{(m)}$ in (3.1)].

3.  $n < m$. The impulse response $h$ will include more than one singular function, and the weighting coefficient assigned to each singular function is easily obtained by balancing both sides of the equation.

In the following discussion we shall restrict ourself to the case of $n > m$ (the proper case). Recall that the impulse function $\delta(t)$ is, by definition, identically zero for $t > 0$. Clearly, the successive derivatives of the impulse, that is, the singular functions, have the same property. Thus, for a unit-impulse input, the right-hand side of (3.1) is identically zero for $t > 0$; consequently, as far as $t > 0$ is concerned, the impulse response is identical to a zero-input response. The singular functions on the right-hand side of (3.1) essentially specify the initial conditions at $t = 0+$, that is, the conditions immediately after the impulse is applied. These conditions are

$$h(0+), h^{(1)}(0+), \ldots, h^{(n-1)}(0+)$$

Therefore, as far as $t > 0$ is concerned, we may express the impulse response $h$ in the same form as the solution of the homogeneous equation in terms of $n$ arbitrary constants $k_i$. Assuming that all characteristic roots of (3.1) are distinct, we have

$$(3.7) \quad h(t) = \sum_{i=1}^{n} k_i \epsilon^{s_i t} \qquad t > 0$$

Since, by convention, $h(t) = 0$ for $t < 0$, and since $h$ does not include any singular function, we may write (for all $t$)

$$(3.8) \quad h(t) = \left( \sum_{i=1}^{n} k_i \epsilon^{s_i t} \right) u(t)$$

The remaining task is to substitute (3.8) in the differential equation (3.1) and evaluate $n$ constants $k_i$. However, care must be taken in differentiating singular functions.

**Example**   Suppose that the differential equation relating the response $y$ to the input $w$ of a given circuit is

$$(3.9) \quad \frac{d^2 y}{dt^2} + 4 \frac{dy}{dt} + 3y = \frac{dw}{dt} + 2w$$

Let us find the impulse response $h$ of this circuit. Note that in Eq. (3.9)

$n = 2$ and $m = 1$; hence, it is a proper case. Consequently, the impulse response does not contain any singular function. The characteristic roots of the differential equation (3.9) are $s_1 = -1$ and $s_2 = -3$. Therefore, we may express the impulse response as

(3.10)    $h(t) = (k_1\epsilon^{-t} + k_2\epsilon^{-3t})u(t)$

Differentiating $h$ once, we obtain

$$h^{(1)}(t) = (k_1\epsilon^{-t} + k_2\epsilon^{-3t})\,\delta(t) + (-k_1\epsilon^{-t} - 3k_2\epsilon^{-3t})u(t)$$
$$= (k_1 + k_2)\,\delta(t) + (-k_1\epsilon^{-t} - 3k_2\epsilon^{-3t})u(t)$$

Differentiating it once again, we obtain

$$h^{(2)}(t) = (k_1 + k_2)\,\delta^{(1)}(t) + (-k_1 - 3k_2)\,\delta(t) + (k_1\epsilon^{-t} + 9k_2\epsilon^{-3t})u(t)$$

Substituting $w = \delta(t)$ and $y = h(t)$ in Eq. (3.9), we have

$$h^{(2)}(t) + 4h^{(1)}(t) + 3h(t) = (k_1 + k_2)\,\delta^{(1)}(t) + (3k_1 + k_2)\delta(t)$$
$$= \delta^{(1)}(t) + 2\delta(t)$$

We now equate the coefficients of $\delta^{(1)}(t)$ and those of $\delta(t)$; thus,

$k_1 + k_2 = 1$

and

$3k_1 + k_2 = 1$

Therefore, the constants $k_1$ and $k_2$ turn out to be

$k_1 = \frac{1}{2}$     and     $k_2 = \frac{1}{2}$

Therefore, the impulse response is, from (3.10),

$h(t) = \frac{1}{2}(\epsilon^{-t} + \epsilon^{-3t})u(t)$

---

**Exercise**    Determine the impulse responses for the variable $y$, which is characterized by the following differential equations,

$$\frac{dy}{dt} + 2y = w$$

$$\frac{d^2y}{dt^2} + \frac{dy}{dt} + y = \frac{dw}{dt} + w$$

$$\frac{d^3y}{dt^3} + 6\frac{d^2y}{dt^2} + 11\frac{dy}{dt} + 6y = \frac{d^2w}{dt^2} + w$$

---

**4    Response to an Arbitrary Input**

We know now how to calculate the impulse response of a linear time-invariant circuit. We shall show in this section that the impulse response of such a circuit can be used to calculate the zero-state response to any arbitrary input. *Linearity* and *time invariance* are the two crucial properties used in the derivation.

## 4.1    Derivation of the Convolution Integral

We propose to calculate the zero-state response $v(\cdot)$ of a linear time-invariant circuit to an input $i_s(\cdot)$. We assume that the input is applied at time $t_0$ and that the circuit is in the zero state at $t_0$; thus we may consider $i_s(t) = 0$ for $t < t_0$.

The problem is to calculate $v(t)$, the response $v$ at time $t$, for any $t > t_0$, assuming that we know the impulse response $h$ of the circuit. As a first step let us approximate the input $i_s$ in the following manner. As shown in Fig. 4.1, divide the interval $(t_0, t)$ in a large number, say $n$, of small equal intervals of duration $\Delta$. Call the subdivision points $t_1, t_2, \ldots, t_k$, $t_{k+1}, \ldots, t_{n-1}$. Thus, $t_1 - t_0 = t_2 - t_1 = \cdots = t_{k+1} - t_k = \cdots = \Delta$. Consider the step approximation $i_{sa}$ to the curve $i_s$ such that the ordinate of the approximating curve $i_{sa}(\cdot)$ at the abscissa $t'$ is given by

$$(4.1) \quad i_{sa}(t') = \begin{cases} i_s(t_0) & t_0 \le t' < t_1 \\ i_s(t_1) & t_1 \le t' < t_2 \\ \cdots\cdots\cdots\cdots\cdots\cdots\cdots\cdots \\ i_s(t_k) & t_k \le t' < t_{k+1} \\ \cdots\cdots\cdots\cdots\cdots\cdots\cdots\cdots \\ i_s(t_{n-1}) & t_{n-1} \le t' < t_n = t \end{cases}$$

Note specifically that $t'$ denotes any time in the interval $[t_0, t]$. The relation of the waveform $i_{sa}(\cdot)$ to the given waveform $i_s(\cdot)$ is shown in Fig. 4.1. It is clear that (for a wide class of inputs) as $n \to \infty$ (hence as $\Delta \to 0$), the difference between the response of the circuit to $i_s(\cdot)$ and to $i_{sa}(\cdot)$ also goes to zero (this can easily be shown to be the case for any piecewise continuous $i_s$).

Observe that the step approximation $i_{sa}$ may be considered as a sum of rectangular pulses, as shown in Fig. 4.2; all pulses have the same width $\Delta$

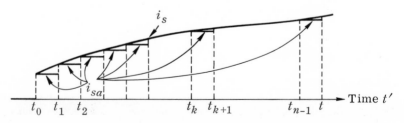

**Fig. 4.1**    Approximating $i_s$ by $i_{sa}$, which is a succession of pulses of equal duration.

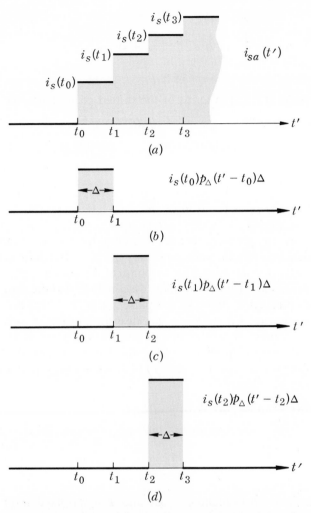

**Fig. 4.2**  The approximating function $i_{sa}$ in (a) can be interpreted as the sum of rectangular pulses in (b), (c), (d), etc.

but differ in height and position along the time axis.  Recall that in Chap. 2 we defined the pulse function $p_\Delta$ as

$$p_\Delta(t') = \begin{cases} 0 & t' \leq 0 \\ \dfrac{1}{\Delta} & 0 < t' < \Delta \\ 0 & \Delta \leq t' \end{cases}$$

If we shift this pulse function $p_\Delta$ to the *right* by $t_k$, we obtain the shifted pulse function represented by

$$p_\Delta(t' - t_k) = \begin{cases} 0 & t' \leq t_k \\ \dfrac{1}{\Delta} & t_k < t' < t_k + \Delta \\ 0 & t_k + \Delta \leq t' \end{cases}$$

Representing $i_{sa}(t')$ in terms of the shifted pulse functions, we obtain

$$
\begin{aligned}
i_{sa}(t') = \; & i_s(t_0)p_\Delta(t' - t_0)\Delta + i_s(t_1)p_\Delta(t' - t_1)\Delta \\
& + i_s(t_2)p_\Delta(t' - t_2)\Delta + \cdots + i_s(t_k)p_\Delta(t' - t_k)\Delta \\
& \qquad\qquad\qquad + \cdots + i_s(t_{n-1})p_\Delta(t' - t_{n-1})\Delta
\end{aligned}
$$

(4.2)

Since the circuit is linear, the zero-state response (at time $t$, the observation time) to $i_{sa}$ is the sum of the zero-state responses (at time $t$) to the pulses $i_s(t_0)p_\Delta(t' - t_0)\Delta$, $i_s(t_1)p_\Delta(t' - t_1)\Delta, \ldots,$ $i_s(t_{n-1})p_\Delta(t' - t_{n-1})\Delta$. Therefore, the problem reduces to finding the zero-state response (at time $t$) of the circuit to one such pulse, say the $(k + 1)$st pulse $i_s(t_k)p_\Delta(t' - t_k)\Delta$. If we call $h_\Delta(\cdot)$ the zero-state response of the circuit to $p_\Delta(\cdot)$, then, invoking linearity and time invariance, we conclude that the zero-state response at the observation time $t$ to the pulse $i_s(t_k)p_\Delta(t' - t_k)\Delta$ is $i_s(t_k)h_\Delta(t - t_k)\Delta$. The argument of $h_\Delta$ is $t - t_k$ because the pulse $p_\Delta(t' - t_k)$ is applied at time $t_k$. Therefore, at the time of observation, which is called $t$ here, only $t - t_k$ sec have elapsed since the pulse has been applied. Repeating this reasoning for each of the pulses of (4.2), we get the zero-state response due to $i_{sa}(\cdot)$ as follows:

$$
\begin{aligned}
i_s(t_0)h_\Delta(t - t_0)\Delta & + i_s(t_1)h_\Delta(t - t_1)\Delta \\
& + \cdots + i_s(t_k)h_\Delta(t - t_k)\Delta + \cdots + i_s(t_{n-1})h_\Delta(t - t_{n-1})\Delta \\
& \qquad\qquad\qquad\qquad = \sum_{k=0}^{n-1} i_s(t_k)h_\Delta(t - t_k)\Delta
\end{aligned}
$$

(4.3)

The next step is to make $n \to \infty$; since $t - t_0$ is fixed and $t - t_0 = n\Delta$, as $n \to \infty$, $\Delta \to 0$. Now as $\Delta \to 0$, the following events occur:

1. The step approximation $i_{sa}(\cdot)$ becomes the actual input $i_s(\cdot)$.

2. The zero-state response to $i_{sa}(\cdot)$ becomes the zero-state response to $i_s(\cdot)$, namely, $v(\cdot)$.

3. The zero-state response $h_\Delta(\cdot)$ to $p_\Delta(\cdot)$ becomes the impulse response $h$.

4. The sum in (4.3) becomes an integral; in other words,

$$v(t) = \int_{t_0}^{t} i_s(t')h(t - t')\,dt' \qquad \text{for } t \geq t_0$$

This equation gives, for any $t \geq t_0$, the *zero-state output voltage at* time $t$ caused by the input current $i_s$ applied at time $t_0$.

*Conclusion*    The calculation of the *zero-state* response of any linear time-invariant circuit to an "arbitrary" input reduces to

1.    The determination of the *impulse response h*

2.    The calculation of the integral

(4.4)

$$\int_{t_0}^{t} h(t - t') i_s(t') \, dt' = v(t) \qquad \text{for } t \geq t_0$$

where $t_0$ is the instant at which the input $i_s$ is applied.  An integral of this type is called a **convolution integral.**

**COROLLARY**    As a direct consequence of (4.4) it follows that the waveform $v(\cdot)$, *the zero-state response of a linear time-invariant circuit to an "arbitrary" input, is a linear function of the input waveform $i_s(\cdot)$.*  (See the definition of a linear function in Appendix A and Example 4 of Sec. 2.3 in that appendix.)

**Remarks**    1.    Each new value of $t$ at which we wish to find the output voltage $v(t)$ requires a new integration because the integrand also depends upon $t$.

2.    Note that we start integrating at $t_0$, the instant at which the circuit is in the zero state; also note that the upper limit of the integral is $t$, the instant at which we want to calculate $v$.  We must not integrate beyond $t$ because the values taken by the input current after $t$ do not affect the response at time $t$.

3.    Let us consider again the reason why we stated that the zero-state response at time $t$ due to an impulse applied at $t_k$ is a function of $t - t_k$.  In general, we might write $h(t, t_k)$; that is, the response is a function of two variables, $t$ the instant of observation and $t_k$ the instant at which the impulse is applied.  Now let us recall that the circuit is *time-invariant*, which means that for any $T$ the results of an experiment performed now are identical to those obtained from the same experiment performed $T$ sec later.  In particular, the zero-state response at time $t$ due to an impulse applied at time $t_k$ is equal to the zero-state response at time $t + T$ due to an impulse applied at $t_k + T$.  Thus,

$$h(t, t_k) = h(t + T, t_k + T) \qquad \text{for all } T$$

Since this equation holds for all $T$, the number $h(t, t_k)$ is uniquely defined by the difference $t - t_k$.  Hence, we are justified in writing it in the form $h(t - t_k)$.

4.    It is interesting to note that since the calculation of the zero-state response by (4.4) does not use the differential-equation representation of *lumped* circuits, it follows that if by some method one knows the

impulse response of a distributed linear time-invariant circuit, then (4.4) can be used to calculate its *zero-state response* to *any* input.

## 4.2   Example of a Convolution Integral in Physics

We have probably already encountered the convolution integral in physics. Suppose we have a taut nylon thread on which we have sprayed some electric charges, say, the belt of a Van de Graaff generator. Suppose we wish to calculate the electrostatic potential at the point $x$ of the thread due to a continuous charge distribution with charge density $\rho$ as shown in Fig. 4.3. The charge contained in the small interval $(x', x' + \Delta x')$ is $\rho(x')\,\Delta x'$, where $\rho(x')$ is the charge density at $x'$ in coulombs per meter and $\Delta x'$ is the length of the interval in meters. If this charge were 1 coul, the potential at $x$ would be $1/(4\pi\varepsilon_0|x - x'|)$. (Note the use of the absolute value of $x - x'$, which is necessary since the distance between two points is a positive number.) Now we use the fact that the potential at a point is a linear function of the charge. By the homogeneity property, the contribution to the potential at $x$ of the charge $\rho(x')\,\Delta x'$ is

$$\frac{\rho(x')\,\Delta x'}{4\pi\varepsilon_0|x - x'|}$$

Using the additivity property and going to the limit, we obtain the potential

$$(4.5) \qquad \phi(x) = \int_{-\infty}^{+\infty} \frac{\rho(x')\,dx'}{4\pi\varepsilon_0|x - x'|}$$

For convenience let $h(r) = 1/(4\pi\varepsilon_0|r|)$, where $r$ represents the distance. Then (4.5) may be written in the form

$$\phi(x) = \int_{-\infty}^{+\infty} h(x - x')\rho(x')\,dx'$$

The interpretation of the function $h$ is as follows: $h(r)$ is the potential created by a *unit* charge at a distance $r$ from the charge. The convolution integral must here be carried from $-\infty$ to $+\infty$ because any charge on the

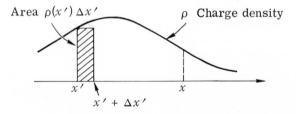

**Fig. 4.3**   Electrostatic-potential illustration of the convolution integral.

taut thread, whether it is to the left or to the right of the point $x$, contributes to the potential at $x$.

## 4.3   Comments on Linear Time-varying Circuits

So far we have calculated the impulse response of linear time-invariant circuits. The concept of the impulse response applies also to linear time-varying circuits. By definition, the zero-state response to a unit impulse is called the impulse response. Consider, as an example of a linear time-varying circuit, a linear amplifier whose gain is varied slowly in time. Thus, the zero-state response to a unit impulse applied at $\tau_1$, that is, $\delta(t - \tau_1)$, will not be the same as the zero-state response to a unit impulse applied at some later time $\tau_2$, that is, $\delta(t - \tau_2)$; this is because the gain of the amplifier at time $\tau_1$ and shortly thereafter is different from the gain at time $\tau_2$ and shortly thereafter. Consequently, in denoting the impulse response, we must keep track of the instant at which the impulse is applied. In the case under discussion, the impulse response might be similar to that shown in Fig. 4.4. In general, $h(t,\tau)$ denotes the *zero-state response at time t* due to a *unit impulse applied at time $\tau$*.

For *linear time-varying* circuits, it can be shown that the zero-state response is a linear function of the input. In fact, using the properties of homogeneity and additivity, we can show that *the zero-state response to an "arbitrary" input $i_s$ applied at time $t_0$ is given by*

$$(4.6) \quad v(t) = \int_{t_0}^{t} h(t,t') i_s(t') \, dt' \qquad t \geq t_0$$

If we compare the above formula with (4.4), we see that the only difference is that the impulse response is now a function of two variables $t$ and $t'$ rather than a function of the difference $t - t'$.

Similarly, in the electrostatic problem, if say, the dielectric constant $\varepsilon$ is a function of $x'$, the potential at $x$ would be given by the formula

$$\phi(x) = \int_{-\infty}^{+\infty} g(x,x') \rho(x') \, dx'$$

where $g(x,x')$ is the potential at $x$ contributed by a unit point charge at $x'$.

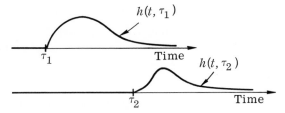

**Fig. 4.4**   Impulse responses of a time-varying circuit. In the first case a unit impulse is applied at time $\tau_1$; in the second case it is applied at time $\tau_2$.

**4.4**    **The Complete Response**

In Chap. 4 we proved that for a linear time-invariant first-order $RC$ circuit, the complete response is the sum of the zero-state response and the zero-input response. As a matter of fact, for any linear circuit, time-invariant or time-varying, the statement is still true. A general and complete proof of the statement will be given in Chap. 13. Let us for the present simply state the fact in terms of the following equation:

$$y(t) = z(t) + v(t)$$

or

(4.7)    $$y(t) = z(t) + \int_{t_0}^t h(t,t')w(t')\,dt' \qquad \text{for } t \geq t_0$$

where $z$ is the zero-input response, $v$ is the zero-state response, $w$ is the input, and $y$ is the complete response. From Eq. (4.7) it is clear that *the complete response is a linear function of the input only if the zero-input response is identically zero.*

**Exercise 1**    Suppose that in the circuit of Fig. 2.1, the inductor $L$ and the resistor $R_2$ were removed. Let $i_s$ be the input, and let $v_C$ be the response. Give the expression for $v_C(t)$, the complete response of the circuit to $i_s$, and the initial capacitor charge $V_0$.

**Exercise 2**    Remove the capacitor $C$ from the circuit of Fig. 2.1. Consider $i_L$ as the response, and give the expression for $i_L(t)$, the complete response of the circuit to $i_s$, and the initial inductor current $I_0$.

**5**    **Computation of Convolution Integrals**

In the previous section we have shown that the zero-state response $v$ of a linear time-invariant circuit to an arbitrary input $i_s(\,\cdot\,)$ applied at time $t_0$ is given by the convolution integral

(5.1)    $$v(t) = \int_{t_0}^t h(t - t')i_s(t')\,dt' \qquad \text{for all } t \geq t_0$$

where $h$ is the unit-impulse response. Thus, given the impulse response $h$, we can determine $v(t)$ for $t \geq t_0$ due to $i_s(\,\cdot\,)$ applied at $t_0$ by performing the integration in (5.1). In this section we shall use several examples to illustrate the computation of the convolution integral. However, first let us derive two simple but useful results.

1.    Suppose that the input $i_s$ is a unit impulse applied at $t_1 > t_0$; that is, $i_s(t) = \delta(t - t_1)$. We wish to show from (5.1) that the response $v$ is given by $h(t - t_1)$. From (5.1) we have

(5.2)  $v(t) = \int_{t_0}^{t} h(t - t')\delta(t' - t_1)\,dt'$     for $t > t_0$

From the definition of the impulse function, we know that $\delta(t' - t_1)$ is identically zero except at $t' = t_1$; at the point $t' = t_1$, $\delta$ is singular with the property $\int_{t_1-}^{t_1+} \delta(t' - t)\,dt' = 1$. We can therefore replace (5.2) by

$v(t) = \int_{t_1-}^{t_1+} h(t - t')\delta(t' - t_1)\,dt'$

where $t_1-$ and $t_1+$ denote the time immediately before and immediately after $t_1$, respectively. Now for lumped linear time-invariant circuits, $h$ is a continuous function on $(0,\infty)$; therefore we can write

(5.3)  $v(t) = h(t - t_1) \int_{t_1-}^{t_1+} \delta(t' - t_1)\,dt' = h(t - t_1)$     for $t > 0$

Thus, the convolution integral has the important property that (for $t > t_1 > t_0$)

(5.4)  $\int_{t_0}^{t} h(t - t')\delta(t' - t_1)\,dt' = h(t - t_1)$

Eq. (5.4) may also be considered to be a direct consequence of the linearity and the time invariance of the circuit. Since, by definition, $h(t)$ is the zero-state response at time $t$ to an impulse *applied at* 0, time invariance implies that if the impulse is applied at $t_1$, the zero-state response will be the same waveform but shifted by $t_1$ sec; in other words, it is $h(t - t_1)$ [as predicted by (5.4).]

2.  The convolution integral of (5.1) can be written in another form by introducing a change of variable. Let $t - t' = \tau$, a new dummy variable; then $t' = t - \tau$, and $-dt' = d\tau$. The lower integration limit, in terms of the new variable, becomes $\tau = t - t_0$, and the upper integration limit becomes $\tau = 0$. Therefore,

$v(t) = \int_{t-t_0}^{0} h(\tau)i_s(t - \tau)(-d\tau)$

(5.5)  $= \int_{0}^{t-t_0} h(\tau)i_s(t - \tau)\,d\tau$

Since $\tau$ and $t'$ are dummy integration variables, we can write (5.5) in terms of $t'$ again in order to compare it with (5.1). Thus,

(5.6)  $v(t) = \int_{0}^{t-t_0} h(t')i_s(t - t')\,dt'$     for $t \geq t_0$

Therefore, if $t_0 = 0$, (5.1) and (5.6) both have the same integration limits; that is, from 0 to $t$,

(5.7)  $\int_{0}^{t} h(t - t')i_s(t')\,dt' = \int_{0}^{t} h(t')i_s(t - t')\,dt'$     for $t \geq 0$

It is interesting to note in (5.7) the symmetric role that the input and impulse response have in the convolution integral. In computations one can often take advantage of this symmetry, as illustrated by the following examples.

**Example 1**  Let the input be a unit step function and the impulse response be a triangular waveform, as shown in Figs. 5.1*a* and 5.2*a*. We wish to determine the step response using the integrals in (5.7). We begin by calculating the first integral in (5.7), namely,

$$(5.8) \quad v(t) = \int_0^t h(t - t')i_s(t')\, dt' \qquad \text{for } t \geq 0$$

Figure 5.1*a* gives the graph of the impulse response. It is repeated on Fig. 5.1*b*, where the variable has been relabeled $t'$ instead of $t$. Figure 5.1*c* shows $h(-t')$ versus $t'$; note that this graph is the mirror image of the preceding one with respect to the ordinate axis. Figure 5.1*d* shows $h(t - t')$ plotted versus $t'$; note that $t$ is a constant (in the figure $t = 1$). Note also that the waveform of Fig. 5.1*d* is obtained by shifting that of Fig. 5.1*c* by $t$ sec to the right. Figure 5.1*d* gives the graph of the integrand of (5.8), the product of the step function $i_s(t')$, and $h(t - t')$. The area under this graph gives $v(t)$ for $t = 1$.

Let us run through the calculation of the second integral in (5.7), namely,

$$(5.9) \quad v(t) = \int_0^t i_s(t - t')h(t')\, dt' \qquad \text{for } t \geq 0$$

We start by plotting $i_s$ versus $t'$ (Fig. 5.2*b*). Then we plot its mirror image with respect to the ordinate axis, namely, $i_s(-t')$ versus $t'$. We shift the whole graph by $t$ sec, thus obtaining $i_s(t - t')$ versus $t'$ (see Fig. 5.2*d*). We then plot the product of the impulse response $h(t')$ and $i_s(t - t')$ (see Fig. 5.2*e*). The area under this graph gives us $v(t)$ for $t = 1$. Clearly, the results obtained in both cases are the same.

In Fig. 5.3, we have sketched the plots used to evaluate the integral in (5.9) for $t = 0, 1, 2,$ and 3.

**Example 2**  Determine and sketch the zero-state response for the input and impulse response given in Fig. 5.4*a*. We have

$$i_s(t) = u(t) - u(t - 1)$$

$$h(t) = \epsilon^{-t}u(t)$$

Clearly, the response $v(t)$ is zero for $t$ negative. For $t \geq 0$ we use

$$v(t) = \int_0^t h(t - t')i_s(t')\, dt'$$

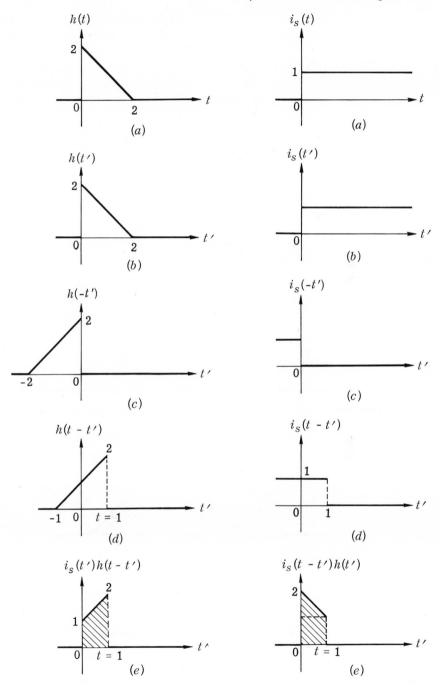

**Fig. 5.1**  Example to illustrate the evaluation of a convolution integral using Eq. (5.8). The calculation is performed for $t = 1$.

**Fig. 5.2**  Example to illustrate the evaluation of a convolution integral using Eq. (5.9).

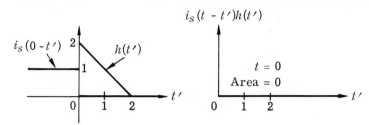

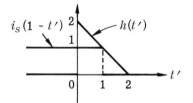

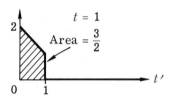

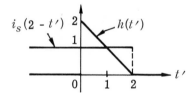

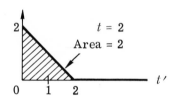

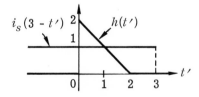

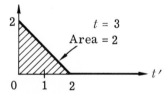

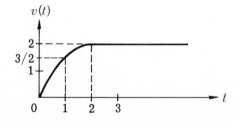

**Fig. 5.3**    Example 1: Illustration of convolution calculation.

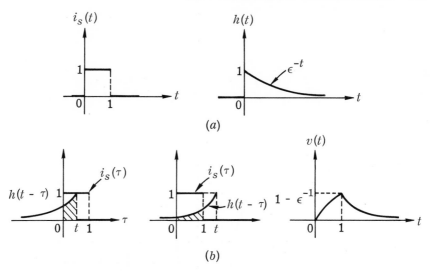

**Fig. 5.4**　Example 2 of a convolution integral.

For $0 \leq t < 1$, since $i_s(t) = 1$, we have

$$v(t) = \int_0^t \epsilon^{-(t-t')} \, dt' = 1 - \epsilon^{-t}$$

For $t \geq 1$, since $i_s(t) = 0$, we only need to integrate up to $t = 1$. Thus,

$$v(t) = \int_0^1 \epsilon^{-(t-t')} \, dt' = (\varepsilon - 1)\epsilon^{-t}$$

The graphical interpretation of these two steps and the response are shown in Fig. 5.4b.

---

**Example 3**　Determine the zero-state response for the input and impulse response given in Fig. 5.5a. We have

$$i_s(t) = u(t) \sin \pi t$$
$$h(t) = u(t) - u(t - 1)$$

For $t \leq 0$

$$v(t) = 0$$

We use the graphical method to evaluate the convolution integral. For $0 \leq t \leq 1$

$$v(t) = \int_0^t \sin \pi t' \, dt' = \frac{1}{\pi}(1 - \cos \pi t)$$

For $1 \leq t$

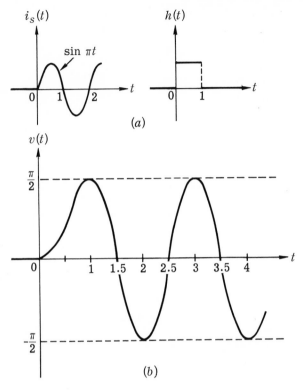

**Fig. 5.5**   Example 3 of a convolution integral.

$$v(t) = \int_{t-1}^{t} \sin \pi t' \, dt' = \frac{1}{\pi} [\cos \pi(t-1) - \cos \pi t]$$

$$= -\frac{2}{\pi} \cos \pi t$$

The result is shown in Fig. 5.5b.   Note that the response is sinusoidal after the *transient* which extends over the interval $0 \leq t \leq 1$ has disappeared.

---

**Example 4**   Determine the zero-state response for the same input as that in Example 3. The impulse response is a rectangular pulse which lasts 2 sec, as shown in Fig. 5.6a.   Thus,

$$i(t) = (\sin \pi t)u(t) \qquad h(t) = u(t) - u(t-2)$$

For $t < 0$,

$$v(t) = 0$$

For $0 \leq t \leq 2$ (see Fig. 5.6b),

$$v(t) = \int_{0}^{t} \sin \pi t' \, dt' = \frac{1}{\pi}(1 - \cos \pi t)$$

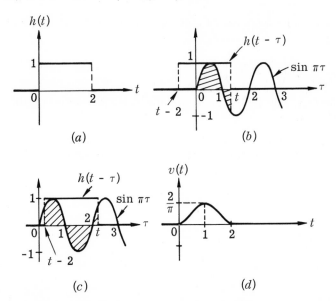

**Fig. 5.6**    Example of the calculation of a convolution integral. (*a*) The impulse response; (*b*) the calculation for $0 \leq t \leq 2$; (*c*) the calculation for $2 \leq t$; (*d*) the output.

For $2 \leq t$ (see Fig. 5.6*c*),

$$v(t) = \int_{t-2}^{t} \sin \pi t' \, dt' = \frac{1}{\pi} [\cos \pi(t-2) - \cos \pi t] = 0$$

The response $v(\cdot)$ is shown in Fig. 5.6*d*. Note that the response is identically zero for $t \geq 0$. In fact, for $t \geq 2$, the zero-state response can be interpreted as a sinusoid of amplitude zero.

## Summary

- This chapter, in contrast to the previous ones, deals mainly with techniques useful in analyzing linear time-invariant circuits. The three principal ones are (1) the node and mesh analysis, (2) determination of the impulse response for $n$th-order differential equations, and (3) the calculation of the convolution integrals.

- Node analysis of a circuit with $n + 1$ nodes is based on writing $n$ KCL equations at $n$ nodes in terms of a set of $n$ node-pair voltages.

- Mesh analysis of a network with $m$ meshes is based on writing $m$ KVL equations for the $m$ meshes in terms of $m$ mesh currents.

- For a single-input, single-output, linear time-invariant circuit, we obtain an $n$th-order linear differential equation with constant coefficients for the output variable by manipulating the equation of the mesh or node analysis.

- The impulse response of a proper $n$th-order differential equation is of the form

$$h(t) = u(t)\left(\sum_{i=1}^{n} k_i \epsilon^{s_i t}\right)$$

  where $s_i$, $i = 1, 2, \ldots, n$, are the $n$ *distinct* zeros of the characteristic polynomial.

- For linear time-invariant circuits, the zero-state response $v$ to any input $i_s$ applied at $t_0$ is equal to the *convolution* of the input $i_s$ applied at $t_0$ with the impulse response $h$. Thus,

$$v(t) = \int_{t_0}^{t} h(t - t')i_s(t') \, dt' \qquad \text{for } t \geq t_0$$

- For all $t \geq 0$

$$\int_{0}^{t} h(t - t')i_s(t') \, dt' = \int_{0}^{t} i_s(t - t')h(t') \, dt'$$

## Problems

Node analysis   **1.** For the circuit shown in Fig. P6.1, use the node analysis to obtain the differential equation in terms of the voltage $v$, given the initial conditions $i_L(0) = I_0$ and $v_C(0) = V_0$.

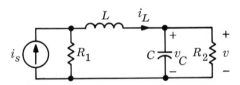

**Fig. P6.1**

Mesh analysis   **2.** For the circuit in the preceding problem, transform the current source to an equivalent voltage source, and then use the mesh analysis to obtain the differential equation in terms of the voltage $v$.

State equa-   **3.** Give the state equations for the circuit in the preceding problem. Use
tions   $v$ and $i_L$ as state variables.

Node analysis   **4.** Write the node equations for the linear time-invariant circuit shown in Fig. P6.4. Determine the differential equations for the voltages $v_1$ and $v_2$. Indicate the needed initial conditions for each case in terms of $v_C(0)$ and $i_L(0)$.

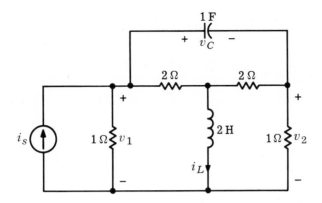

**Fig. P6.4**

**5.** Repeat the preceding problem using mesh analysis.

**6.** Write the state equations for the circuit of Fig. P6.4.

**7.** Consider the linear time-invariant $RLC$ circuit shown in Fig. P6.7. Let $h$ be the current in the inductor corresponding to $i_s = \delta$ and $h(0-) = dh/dt\,(0-) = 0$.

*a.* Evaluate (in terms of $R$, $L$, and $C$) $h(0+)$ and $dh/dt\,(0+)$.

*b.* Show directly (by checking the initial conditions and substituting in the differential equation) that

$$i_L(t) = \int_0^t h(t - t')i_s(t')\,dt' \qquad \text{for } t \geq 0$$

is the zero-state response to $i_s$.

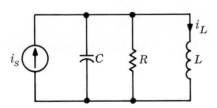

**Fig. P6.7**

**8.** Given the differential equation of a linear time-invariant circuit,

$$\frac{d^3y}{dt^3} + 4\frac{d^2y}{dt^2} + 5\frac{dy}{dt} + 2y = \frac{d^2w}{dt^2} + 3w$$

The initial conditions are

$$y(0-) = 1 \qquad \frac{dy}{dt}(0-) = 2 \qquad \frac{d^2y}{dt^2}(0-) = -1$$

Determine the zero-input response, the impulse response, and the step response.

Zero-state response and complete response
**9.** For Prob. 8, if the input $w$ is a sinusoid, $w(t) = \cos t$, determine the zero-state response using two different methods. Find the complete response for the initial conditions indicated.

Impulse response
**10.** Find the impulse response of the following differential equations:

a. $\dfrac{d^2y}{dt^2} + \dfrac{dy}{dt} + y = w$

b. $\dfrac{d^2y}{dt^2} + \dfrac{dy}{dt} + y = \dfrac{dw}{dt} + w$

c. $\dfrac{d^2y}{dt^2} + 2\dfrac{dy}{dt} + y = \dfrac{dw}{dt} + 2w$

d. $\dfrac{d^3y}{dt^3} + 3\dfrac{d^2y}{dt^2} + 3\dfrac{dy}{dt} + y = \dfrac{d^2w}{dt^2} + 2w$

Impulse responses of improper systems
**11.** Determine the impulse response of the following differential equations. The responses will contain singular functions. Determine the required singular functions by balancing the leading terms on both sides of the equations.

a. $\dfrac{d^2y}{dt^2} + 3\dfrac{dy}{dt} + 2y = 2\dfrac{d^2w}{dt^2} + 5\dfrac{dw}{dt} + w$

b. $\dfrac{dy}{dt} + 2y = \dfrac{d^2w}{dt^2} + 3\dfrac{dw}{dt} + 3w$

c. $\dfrac{d^2y}{dt^2} + 5\dfrac{dy}{dt} + 6y = 3\dfrac{d^3w}{dt^3} + 2\dfrac{d^2w}{dt^2} + w$

Zero-state response
**12.** A linear time-invariant circuit has impulse response $h(\cdot)$ as shown in Fig. P6.12a.

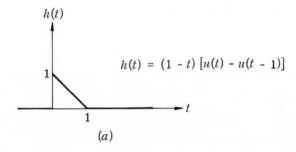

$h(t)$

$h(t) = (1 - t)\,[u(t) - u(t - 1)]$

1

1

$t$

(a)

**Fig. P6.12**

a.   Find the step response $s(\cdot)$.

b.   If the system is in zero state at $0-$, find the response due to the wave-form $f$ shown in Fig. P6.12$b$.

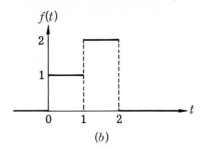

**Fig. P6.12**
(*continued*)

Zero-state
response
**13.** Sketch accurately the zero-state response for the following cases without using convolution integrals (see Fig. P6.13); $h$ denotes the impulse response of the linear time-invariant circuit under consideration; $i$ denotes the input.

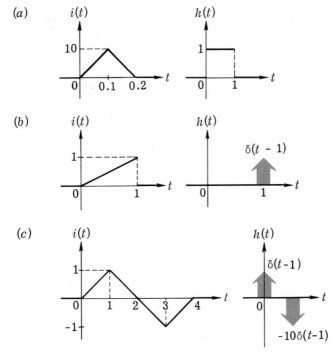

**Fig. P6.13**

($d$)        $i(t)$ as in ($c$)

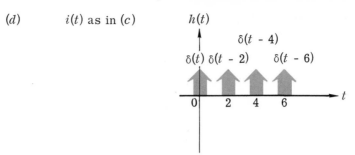

**Fig. P6.13**
(*continued*)

Convolution **14.** Repeat Prob. 13 by the use of the convolution integral.
integral
Step response **15.** The zero-state response $h$ of a network to a unit impulse of current in-
and pulse put is graphed in Fig. P6.15.
response

*a.*   Calculate and sketch the zero-state response to a unit step input,
       $i(t) = u(t)$.

*b.*   Calculate and sketch the zero-state response to pulses $i(t) = 5\Delta p_\Delta(t)$
       for values of $\Delta = 0.2$, 1, and 5 sec.

*c.*   Suppose that by redesign of the circuit with available components we
       can modify $h$ to have any shape provided that

       (1)   $h(t) = 0$       for all $t < 0$
       (2)   $0 \le |h(t)| \le 5$       for all $t \ge 0$

       (3)   $\int_0^\infty h(t)\,dt = 1$

Given this constraint, what shape would you choose for $h$ if you wished to
have the step response of the modified circuit reach its steady state in the
shortest possible time?

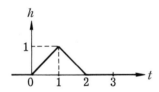

**Fig. P6.15**

Ramp **16.** The impulse response of a linear time-invariant circuit is specified by
response

$$h(t) = \begin{cases} \dfrac{1}{1+t} & \text{for all } t \ge 0 \\ 0 & \text{for all } t < 0 \end{cases}$$

Calculate and sketch the zero-state response $v$ of the circuit to a unit ramp $r$ applied at time $t_0 = 1$.

Convolution integral **17.** If the impulse response of a linear time-invariant circuit is given as

$$h(t) = \begin{cases} 2\epsilon^{-t} & 0 \le t < 3 \\ 0 & t \ge 3 \end{cases}$$

find the zero-state response to the circuit due to an input

$$i_s(t) = \begin{cases} 4u(t) & 0 \le t < 2 \\ 0 & t \ge 2 \end{cases}$$

Time-varying circuit **18.** For a linear time-varying circuit, if the response at time $t$ for a unit impulse applied at time $\tau$ is

$$h(t,\tau) = t - \tau^2$$

Calculate, by using convolution, the response for an input

$$i_s(t) = tu(t) + 2u(t) - \delta(t)$$

Complete response **19.** For the circuit in Fig. P6.1, let $R_1 = 1$ ohm, $L = 1$ henry, $C = 2$ farads, $R_2 = 1$ ohm, $I_0 = 1$ amp, and $V_0 = 1$ volt. Determine the impulse response and the complete response due to a pulse $i_{s1}(t) = u(t) - u(t - 1)$ for the output voltage $v$. If the input is changed to $i_{s2}(t) = 3i_{s1}(t)$, what is the complete response?

Convolution integral **20.** Determine the zero-input response of the linear time-invariant circuit from the impulse response $h$ and the input $i_s$, as shown in Fig. P6.20.

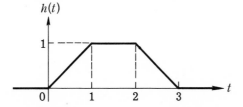
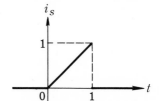

**Fig. P6.20**

Convolution integral **21.** Repeat Prob. 20 for the same impulse response, but for a different input $i_s$, as shown in Fig. P6.21.

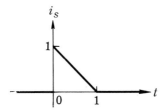

**Fig. P6.21**

Impulse
response,
complete
response, and
convolution
**22.** Consider the linear time-invariant series $RLC$ circuit with input $e_s$ and response $i$, as shown in Fig. P6.22a.

*a.*   Calculate and sketch the impulse response.

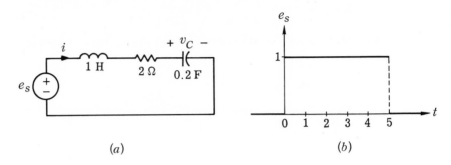

$(a)$                                        $(b)$

**Fig. P6.22**

*b.*   Write down an expression that would permit you to calculate the complete response for any given input voltage $e_s$ applied at $t = 0$, and for any initial state $i_L(0) = I_0$ and $v_C(0) = V_0$.

*c.*   Calculate and sketch the complete response for $I_0 = 1$ amp, $V_0 = -1$ volt, and $e_s$ as shown in Fig. P6.22b.

# 7

# Sinusoidal Steady-state Analysis

Sinusoidal waveforms play an important role in science and engineering. In electric circuits the frequencies of sinusoids of interest may vary from a few hertz (cycles per second) to kilohertz, megahertz, and gigahertz. We are all familiar with the 60-Hz sinusoidal current used in power transmission and in the home. In the laboratory we have used sinusoidal signal generators and detectors covering several frequency ranges. As electrical engineers, we know that sinusoidal waveforms are bread and butter in our professional life because, as we shall see later, if we know the response of a linear time-invariant circuit to *any sinusoid*, we know, in principle, its response to *any signal*. It is therefore important to learn the most efficient means of dealing with sinusoids.

In Chap. 4 we introduced examples in which we calculated the response of simple circuits to sinusoidal inputs. The method used to determine a particular solution was straightforward yet extremely clumsy. In this chapter we shall develop a much simpler and more elegant method, which is based on the idea of representing a sinusoid of a given frequency by a complex number.

## 1 Review of Complex Numbers

### 1.1 Description of Complex Numbers

Let us first summarize some principal facts about complex numbers. Let $z$ be a complex number, and let $x$ and $y$ be its real part and its imaginary part, respectively. We then have

(1.1) $\quad z = x + jy$

where $j = \sqrt{-1}$. We may also write

(1.2) $\quad \text{Re}(z) = x \qquad \text{Im}(z) = y$

where $\text{Re}(\cdots)$ means "real part of ..." and $\text{Im}(\cdots)$ means "imaginary part of ...." The right-hand side of Eq. (1.1) is called the **rectangular coordinate representation** of the complex number $z$. The **polar representation** of the complex number $z$ is

(1.3)   $z = |z|\epsilon^{j\theta}$

where $|z|$ is called the **magnitude** or **amplitude** of $z$, with a value

(1.4)   $|z| = (x^2 + y^2)^{1/2}$

and $\theta$ is called the **angle** or **phase** of $z$, with a value

(1.5)   $\theta = \tan^{-1}\dfrac{y}{x}$

Sometimes we write $\sphericalangle z$ for the angle $\theta$.  In terms of $|z|$ and $\theta$, we have

(1.6)   $x = |z| \cos \theta \qquad y = |z| \sin \theta$

These facts are illustrated in Fig. 1.1, where the complex number $z$ is associated with the point which has Re $(z)$ and Im $(z)$ as coordinates. Note that the phase $\theta$ is measured from the $x$ axis to the vector which starts from the origin and ends at the point $z$.

**Remark**   The angle $\theta$ restricted to the interval $[0,2\pi)$ or $(-\pi,\pi].\dagger$   As a consequence, $\theta$ is uniquely defined by $x$ and $y$.   In calculating $\theta$ by Eq. (1.5), we should keep in mind that if we know $\tan \theta$, the angle $\theta$ is not uniquely defined in $[0,2\pi)$.   For example, $\tan 26.6° = 0.5$; but $\tan 206.6°$ also equals 0.5.   In

$\dagger$ $[0,2\pi)$ denotes the interval $0 \leq \theta < 2\pi$; $(-\pi,\pi]$ denotes the interval $-\pi < \theta \leq \pi$.

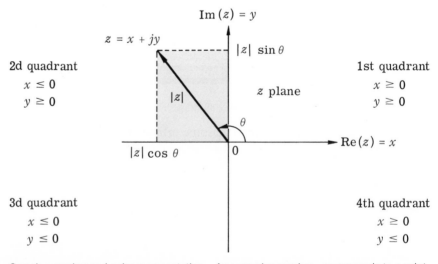

**Fig. 1.1**   Complex number and polar representation.   Any complex number $z$ corresponds to a point in the $z$ plane; it can be characterized by either its real and imaginary parts or its magnitude and phase.

order to uniquely determine $\theta$ we must consider the signs of Re $(z)$ and Im $(z)$, which specify the quadrant of the complex plane in which $z$ lies.

**Exercise 1**   Sketch $\tan\theta$ versus $\theta$ for $0 \le \theta < 2\pi$.

**Exercise 2**   Express in polar form the following complex numbers: $1 + j0.5$, $1 + j10$, $1 - j2$, $-1 + j2$, and $-1 - j3$.

**Exercise 3**   Express the following complex numbers in rectangular form (that is, $z = x + jy$): $5\epsilon^{j30°}$, $5\epsilon^{j150°}$, $10\epsilon^{-j45°}$, $10\epsilon^{j240°}$, and $2\epsilon^{j180°}$.

### 1.2   Operations with Complex Numbers

The rules of operation of complex numbers are identical with those of real numbers, provided one uses the fact that $j^2 = -1$. The rules are identical because the real numbers and the complex numbers obey the axioms of a field (see Appendix A, Sec. 2.1). Let

$$z_1 = x_1 + jy_1 = |z_1|\epsilon^{j\theta_1}$$
$$z_2 = x_2 + jy_2 = |z_2|\epsilon^{j\theta_2}$$

be two complex numbers. The operations of complex numbers are defined as follows:

*Addition*
$$z_1 + z_2 = (x_1 + jy_1) + (x_2 + jy_2)$$
$$= (x_1 + x_2) + j(y_1 + y_2)$$

*Multiplication*
$$z_1 z_2 = (x_1 + jy_1)(x_2 + jy_2)$$
$$= (x_1 x_2 - y_1 y_2) + j(x_1 y_2 + x_2 y_1)$$

and in terms of their polar representations

$$z_1 z_2 = |z_1|\epsilon^{j\theta_1}|z_2|\epsilon^{j\theta_2}$$
$$= |z_1||z_2|\epsilon^{j(\theta_1 + \theta_2)}$$

**Exercise**   Show that

$$\frac{z_1}{z_2} = \frac{(x_1 x_2 + y_1 y_2) + j(-x_1 y_2 + x_2 y_1)}{x_2^2 + y_2^2}$$

and

$$\frac{z_1}{z_2} = \frac{|z_1|}{|z_2|}\epsilon^{j(\theta_1 - \theta_2)}$$

*Complex conjugate*    Given the complex number $z = x + jy$, we say that the complex number $x - jy$, denoted by $\bar{z}$, is the **complex conjugate** of $z$. It is easy to see that if $z = |z|\epsilon^{j\theta}$, then

$$\bar{z} = |z|\epsilon^{-j\theta}$$

and

$$z + \bar{z} = 2x \qquad z - \bar{z} = 2jy$$

More importantly,

$$z\bar{z} = |z|^2 = x^2 + y^2$$

and

$$x = \frac{1}{2}(z + \bar{z}) \qquad y = \frac{1}{j2}(z - \bar{z})$$

**Exercise**    Evaluate

$$\frac{(1 + j1)(1 + j2)}{j5(1 - j1)} \qquad \text{and} \qquad 2\epsilon^{j30°} - \epsilon^{-j45°}$$

and express the answers in terms of both the polar and the rectangular coordinates.

---

## 2    Phasors and Ordinary Differential Equations

### 2.1    The Representation of a Sinusoid by a Phasor

We have defined a *sinusoid of angular frequency* $\omega$ to be any function of time $t$ defined on $(-\infty,\infty)$ and of the form

(2.1)    $A_m \cos(\omega t + \phi)$

where the real constants $A_m$, $\omega$, and $\phi$ are called the **amplitude,** the **angular frequency,** and the **phase** of the sinusoid, respectively.

The purpose of the development to follow is to establish the following important theorem.

**MAIN THEOREM**    The algebraic sum of any number of sinusoids of the *same angular frequency,* say $\omega$, and of any number of their derivatives of any order is also a sinusoid of the *same* angular frequency $\omega$.

---

**Example 1**    Consider the function $f(\cdot)$ defined for all $t$ by

$$f(t) = 2\cos(2t + 60°) - 4\sin 2t + \frac{d}{dt}2\sin 2t$$

Note that $f$ is the sum of two sinusoids and the derivative of a third one; each of these sinusoids has the same angular frequency $\omega = 2$ rad/sec. The main theorem asserts that the function $f$ can be represented by a single sinusoid with the *same* angular frequency. Checking this fact by direct expansion of the cosine term, we obtain

$$f(t) = 2 \cos 2t \cos 60° - 2 \sin 2t \sin 60° - 4 \sin 2t + 4 \cos 2t$$

$$= \cos 2t - \sqrt{3} \sin 2t - 4 \sin 2t + 4 \cos 2t$$

$$= 5 \cos 2t - (4 + \sqrt{3}) \sin 2t$$

$$= \sqrt{5^2 + (4 + \sqrt{3})^2} \cos \left( 2t + \tan^{-1} \frac{4 + \sqrt{3}}{5} \right)$$

$$= 7.6 \cos (2t + 48.8°)$$

which is of the form given in Eq. (2.1).

The proof of the main theorem will be given at the end of this subsection. First, we wish to discuss the implication of the main theorem. It suggests that we could treat sinusoids by algebraic methods. First, observe that a sinusoid with angular frequency $\omega$ is completely specified by its amplitude $A_m$ and its phase $\phi$. Thus, we are led to the idea of *representing* the sinusoid by the complex number $A \triangleq A_m \epsilon^{j\theta}$. Note that $A_m = |A|$ is the magnitude of the complex number $A$, and $\phi = \angle A$ is the phase. More precisely, the sinusoid $x(t) \triangleq A_m \cos (\omega t + \phi)$ is *represented* by the complex number $A \triangleq A_m \epsilon^{j\theta}$ and, conversely, given the complex number $A = A_m \epsilon^{j\theta}$ *and the angular frequency* $\omega$, we may recover the sinusoid as follows:

(2.2)    $x(t) = \text{Re} (A \epsilon^{j\omega t})$

Indeed,

$$\text{Re} (A \epsilon^{j\omega t}) = \text{Re} (A_m \epsilon^{j(\omega t + \phi)})$$

$$= \text{Re} [A_m \cos (\omega t + \phi) + j A_m \sin (\omega t + \phi)]$$

(2.3)    $$= A_m \cos (\omega t + \phi) = x(t)$$

Note that in the last step we used the fact that $A_m$, $\omega$, $t$, and $\phi$ are *real* numbers. The complex number $A$, which represents the sinusoid $A_m \cos (\omega t + \phi)$, is called, for convenience, the **phasor** representing the sinusoid. By definition, the phasor $A$ is given by $A = A_m \epsilon^{j\phi}$.

**Example 2**    Let $v(t) = \sqrt{2} \, 110 \cos (2\pi 60 t + \pi/3)$ volts; then the phasor representing the sinusoid is

$$A = \sqrt{2} \, 110 \epsilon^{j(\pi/3)}$$

that is,

$$v(t) = \mathrm{Re}\,(A\epsilon^{j2\pi 60t})$$

**Remarks**   1.   It should be stressed that the knowledge of the phasor representing a sinusoid determines the amplitude and the phase, but not the frequency. Thus, it is important, when performing calculations with phasors, to keep in mind the frequency of the phasors.

2.   Alternately, if we specify a sinusoid by a sine function rather than by a cosine function, we have

$$y(t) = A_m \sin\,(\omega t + \phi)$$

Then a phasor representation $A \triangleq A_m\epsilon^{j\phi}$ works as well. However, we must recover the sinusoid from the following equation:

$$y(t) = \mathrm{Im}\,(A\epsilon^{j\omega t})$$

In this book we shall use exclusively the real-part representation.

3.   On the complex plane let us graph the function $A\epsilon^{j\omega t}$. The coordinates of the complex number $A\epsilon^{j\omega t}$ are

$$x(t) = \mathrm{Re}\,(A\epsilon^{j\omega t}) \qquad y(t) = \mathrm{Im}\,(A\epsilon^{j\omega t})$$

We may think of $x(t)$ as the projection on the $x$ axis of the point $A\epsilon^{j\omega t}$, which rotates on the circle of radius $A_m$ at an angular velocity of $\omega$ rad/sec in the counterclockwise direction, as shown in Fig. 2.1. Thus, $A\epsilon^{j\omega t}$ may be called a rotating phasor. Similarly, the projection on the $y$ axis of the point $A\epsilon^{j\omega t}$ gives $y(t)$.

The phasor representation of sinusoids is used mainly in the computation of a *particular solution* of *ordinary linear differential equations* with *real constant coefficients* when the forcing function is a sinusoid. In other words, the differential equation is of the form

$$\alpha_0 \frac{d^n x}{dt^n} + \alpha_1 \frac{d^{n-1}x}{dt^{n-1}} + \cdots + \alpha_{n-1}\frac{dx}{dt} + \alpha_n x = A_m \cos\,(\omega t + \phi)$$

where $\alpha_0, \alpha_1, \ldots, \alpha_n, A_m, \omega,$ and $\phi$ are *real constants*. Indeed, according to the theorem stated above, if we substitute a sinusoid of angular frequency $\omega$ for $x$ in the left-hand side, then the left-hand side will also be equal to a sinusoid of frequency $\omega$. This is precisely what the right-hand side requires. Therefore, the only real problem is to calculate the amplitude and phase of the sinusoid that is the particular solution. To do so we shall use phasors, and the method is called the **phasor method.**

Instead of plunging directly into calculations, we shall carefully exhibit three lemmas that explain why the phasor method works.

**LEMMA 1**   Re $[\cdots]$ is *additive* and *homogeneous*. In other words, let $z_1$ and $z_2$ be any complex-valued functions of the real variable $t$, and let $\alpha$ be a *real* number; then *additivity* means

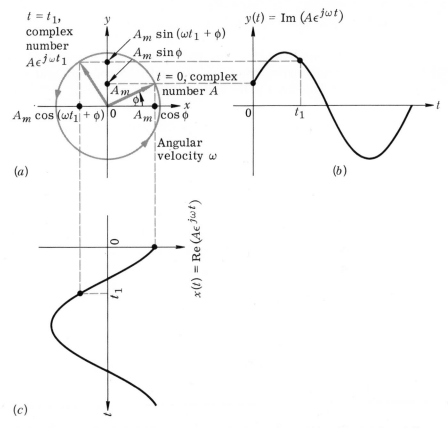

**Fig. 2.1**   Representation of the rotating phasor $A\epsilon^{j\omega t}$.   (a) $A\epsilon^{j\omega t}$ may be considered as a vector rotating clockwise with angular frequency $\omega$; (b) its projection on the $y$ axis; (c) its projection on the $x$ axis.

(2.4a)   $\mathrm{Re}\,[z_1(t) + z_2(t)] = \mathrm{Re}\,[z_1(t)] + \mathrm{Re}\,[z_2(t)]$

and *homogeneity* means that

(2.4b)   $\mathrm{Re}\,[\alpha z_1(t)] = \alpha\,\mathrm{Re}\,[z_1(t)]$

for all such functions $z_1$ and $z_2$, all *real* numbers $\alpha$, and all values of $t$.   The two conditions (2.4a) and (2.4b) are equivalent to

$\mathrm{Re}\,[\alpha_1 z_1(t) + \alpha_2 z_2(t)] = \alpha_1\,\mathrm{Re}\,[z_1(t)] + \alpha_2\,\mathrm{Re}\,[z_2(t)]$

for all *real* numbers $\alpha_1$ and $\alpha_2$ and all complex-valued functions $z_1$ and $z_2$.

    The proof is simple.   It is omitted because it follows directly from the use of the rectangular representation of $z_1(t)$ and $z_2(t)$.

**LEMMA 2**   Let $A$ be a complex number whose polar representation is $A_m\epsilon^{j\phi}$; that is, $A_m \triangleq |A|$ and $\phi \triangleq \measuredangle A$.   Then

(2.5)    $\dfrac{d}{dt}\, \text{Re}\,(A\epsilon^{j\omega t}) = \text{Re}\!\left(\dfrac{d}{dt}\, A\epsilon^{j\omega t}\right) = \text{Re}\,(j\omega A\epsilon^{j\omega t})$

Lemma 2 teaches us two facts: The operations of taking the real part and differentiating commute (Re and $d/dt$ commute), and $d/dt$ applied to $A\epsilon^{j\omega t}$ amounts to the multiplication of $A\epsilon^{j\omega t}$ by $j\omega$.

Let us calculate the left-hand side of Eq. (2.5). We obtain

$$\frac{d}{dt}\, \text{Re}\,(A\epsilon^{j\omega t}) = \frac{d}{dt}\, \text{Re}\,(A_m\epsilon^{j(\omega t+\phi)})$$

$$= \frac{d}{dt}\,[A_m \cos{(\omega t + \phi)}]$$

$$= -\omega A_m \sin{(\omega t + \phi)}$$

$$= \text{Re}\,(j\omega A_m\epsilon^{j(\omega t+\phi)})$$

$$= \text{Re}\,(j\omega A\epsilon^{j\omega t})$$

$$= \text{Re}\!\left(\frac{d}{dt}\, A\epsilon^{j\omega t}\right)$$

**LEMMA 3**    Let $A$ and $B$ be complex numbers, and let $\omega$ be an angular frequency. Under these conditions, the statement

(2.6)    $\text{Re}\,(A\epsilon^{j\omega t}) = \text{Re}\,(B\epsilon^{j\omega t})$      for all $t$

implies that $A = B$, and conversely $A = B$ implies

$\text{Re}\,(A\epsilon^{j\omega t}) = \text{Re}\,(B\epsilon^{j\omega t})$      for all $t$

*Proof*    The proof is divided into two parts. For the first part let us assume that

(2.7)    $\text{Re}\,(A\epsilon^{j\omega t}) = \text{Re}\,(B\epsilon^{j\omega t})$      for all $t$

We have to show that the complex numbers $A$ and $B$ are equal. Let us exhibit the real and imaginary parts of $A$ and $B$ as follows:

$A \triangleq A_r + jA_i$      $B \triangleq B_r + jB_i$

Consider first the case of $t = 0$. Since $\epsilon^{j\omega t}|_{t=0} = 1$, Eq. (2.7) implies

$\text{Re}\,(A) = \text{Re}\,(B)$

which means that

(2.8)    $A_r = B_r$

Next let $t = \pi/2\omega$. Thus, $\epsilon^{j\omega t}|_{t=\pi/2\omega} = j$, and Eq. (2.7) becomes

$\text{Re}\,(jA) = \text{Re}\,(jB)$      or      $\text{Re}\,(jA_r - A_i) = \text{Re}\,(jB_r - B_i)$

hence

(2.9)  $A_i = B_i$

Finally, Eqs. (2.8) and (2.9) mean, by the definition of equality of complex numbers, that $A = B$.

Let us next prove the converse. The assumption is now that $A = B$, and we must show that

$$\text{Re}\,(A\epsilon^{j\omega t}) = \text{Re}\,(B\epsilon^{j\omega t}) \quad \text{for all } t$$

This is immediate since $A = B$ implies that

$$A\epsilon^{j\omega t} = B\epsilon^{j\omega t} \quad \text{for all } t$$

and hence

$$\text{Re}\,(A\epsilon^{j\omega t}) = \text{Re}\,(B\epsilon^{j\omega t}) \quad \text{for all } t$$

*Proof of the*
*main theorem*

For simplicity let us consider a particular case of three sinusoids

$$x(t) \overset{\Delta}{=} A_m \cos\,(\omega t + \phi_1) = \text{Re}\,(A\epsilon^{j\omega t})$$

$$y(t) \overset{\Delta}{=} B_m \cos\,(\omega t + \phi_2) = \text{Re}\,(B\epsilon^{j\omega t})$$

$$z(t) \overset{\Delta}{=} C_m \cos\,(\omega t + \phi_3) = \text{Re}\,(C\epsilon^{j\omega t})$$

Thus,

$$A \overset{\Delta}{=} A_m\epsilon^{j\phi_1} = A_r + jA_i$$

$$B \overset{\Delta}{=} B_m\epsilon^{j\phi_2} = B_r + jB_i$$

$$C \overset{\Delta}{=} C_m\epsilon^{j\phi_3} = C_r + jC_i$$

where $A$, $B$, and $C$ are the three phasors which represent the sinusoids $x$, $y$, and $z$, respectively. Let us calculate $x(t) + y(t) + (d/dt)z(t)$. Call this sum $\Sigma(t)$; then

$$\Sigma(t) = \text{Re}\,(A\epsilon^{j\omega t}) + \text{Re}\,(B\epsilon^{j\omega t}) + \frac{d}{dt}\,\text{Re}\,(C\epsilon^{j\omega t})$$

By Lemma 2, the third term may be written as

$$\text{Re}\,(j\omega C\epsilon^{j\omega t})$$

Using Lemma 1, we get

$$\Sigma(t) = \text{Re}\,[(A + B + j\omega C)\epsilon^{j\omega t}]$$

Therefore $\Sigma(\,\cdot\,)$ is a sinusoid of angular frequency $\omega$. Also, $\Sigma(\,\cdot\,)$ is of the form $\text{Re}\,(S\epsilon^{j\omega t})$, where

$$S = S_m\epsilon^{j\measuredangle S} = S_r + jS_i$$

is the phasor which represents the sinusoid $\Sigma(\,\cdot\,)$. The complex number $S$ is determined by $S = A + B + j\omega C$, according to Lemma 3. The last equation implies the following: If we consider real and imaginary parts, we get

$$S_r = A_r + B_r - \omega C_i$$

$$S_i = A_i + B_i + \omega C_r$$

$$S_m = \sqrt{(A_r + B_r - \omega C_i)^2 + (A_i + B_i + \omega C_r)^2}$$

$$\angle S = \tan^{-1} \frac{A_i + B_i + \omega C_r}{A_r + B_r - \omega C_i}$$

where the $\angle S$ lies in the quadrant chosen according to the rule stated previously.

Clearly, we can extend the proof to the sum of any number of sinusoids of the same frequency and to the sum of any number of their derivatives of any order.

**Exercise 1**    Using standard trigonometric formulas, show that

$$A_m \cos \omega t + B_m \sin \omega t = \sqrt{A_m^2 + B_m^2} \cos (\omega t - \phi)$$

where $\phi$ is determined by $\tan \phi = B_m/A_m$, and the quadrant in which $\phi$ lies is specified by

$$\cos \phi = \frac{A_m}{\sqrt{A_m^2 + B_m^2}} \qquad \sin \phi = \frac{B_m}{\sqrt{A_m^2 + B_m^2}}$$

**Exercise 2**    Derive the same result using phasors.

## 2.2    Application of the Phasor Method to Differential Equations

As mentioned in the beginning of this section, the phasor method is the most convenient method for obtaining a particular solution of a linear differential equation with real constant coefficients when the forcing function is a sinusoid.  Consider the equation

(2.10)    $$\alpha_0 \frac{d^n x}{dt^n} + \alpha_1 \frac{d^{n-1} x}{dt^{n-1}} + \cdots + \alpha_{n-1} \frac{dx}{dt} + \alpha_n x = A_m \cos (\omega t + \phi)$$

where $\alpha_0, \alpha_1, \ldots, \alpha_n, A_m, \omega$, and $\phi$ are real constants.  Introducing phasors, we set

(2.11)    $$A \triangleq A_m \epsilon^{j\phi} \qquad \text{and} \qquad X \triangleq X_m \epsilon^{j\psi}$$

Substituting in the differential equation $x(t)$ by $\mathrm{Re}\,(X\epsilon^{j\omega t})$, we get, successively,

$$\alpha_0 \frac{d^n}{dt^n} \mathrm{Re}\,(X\epsilon^{j\omega t}) + \cdots + \alpha_n \mathrm{Re}\,(X\epsilon^{j\omega t}) = \mathrm{Re}\,(A\epsilon^{j\omega t})$$

By Lemma 1, we may write

$$\frac{d^n}{dt^n} \mathrm{Re}\,(\alpha_0 X\epsilon^{j\omega t}) + \cdots + \mathrm{Re}\,(\alpha_n X\epsilon^{j\omega t}) = \mathrm{Re}\,(A\epsilon^{j\omega t})$$

and, by Lemma 2, applied repeatedly, we obtain

$$\text{Re}\left[\alpha_0(j\omega)^n X \epsilon^{j\omega t}\right] + \cdots + \text{Re}\left(\alpha_n X \epsilon^{j\omega t}\right) = \text{Re}\left(A \epsilon^{j\omega t}\right)$$

Using Lemma 1 again, we have

$$\text{Re}\left\{[\alpha_0(j\omega)^n + \alpha_1(j\omega)^{n-1} + \cdots + \alpha_{n-1}(j\omega) + \alpha_n] X \epsilon^{j\omega t}\right\} = \text{Re}\left(A \epsilon^{j\omega t}\right)$$

Lemma 3 gives the algebraic equation for $X$ as follows:

(2.12a)    $[\alpha_0(j\omega)^n + \alpha_1(j\omega)^{n-1} + \cdots + \alpha_{n-1}(j\omega) + \alpha_n] X = A$

or

(2.12b)    $$X = \frac{A}{\alpha_0(j\omega)^n + \alpha_1(j\omega)^{n-1} + \cdots + \alpha_{n-1}(j\omega) + \alpha_n}$$

Hence, the magnitude is

(2.13a)    $$X_m = \frac{A_m}{\left[\underbrace{(\alpha_n - \alpha_{n-2}\omega^2 + \cdots)^2}_{\text{Even powers of } \omega} + \underbrace{(\alpha_{n-1}\omega - \alpha_{n-3}\omega^3 + \cdots)^2}_{\text{Odd powers of } \omega}\right]^{1/2}}$$

and the phase is

(2.13b)    $$\psi = \phi - \tan^{-1}\frac{\alpha_{n-1}\omega - \alpha_{n-3}\omega^3 + \cdots}{\alpha_n - \alpha_{n-2}\omega^2 + \cdots}$$

where the angle represented by $\tan^{-1}(\cdot)$ lies in the quadrant chosen according to the rule stated previously.

**Remark**    Equation (2.12a) can be solved for $A$ and gives the answer given by Eq. (2.12b) provided that $\omega$ is such that

$$\alpha_0(j\omega)^n + \alpha_1(j\omega)^{n-1} + \cdots + \alpha_{n-1}j\omega + \alpha_n \neq 0$$

If this polynomial is equal to zero for the value of $\omega$ under consideration, $j\omega$ is a natural frequency, and, consequently, a particular solution of the form $tA \cos(\omega t + \phi)$ should be considered (see Appendix C, Sec. 3.2).

The preceding development can clearly be generalized to a linear time-invariant circuit with a single input $w$ and a single output $y$, as described by the following differential equation:

(2.14)    $$\frac{d^n y}{dt^n} + a_1\frac{d^{n-1}y}{dt^{n-1}} + \cdots + a_n y = b_0\frac{d^m w}{dt^m} + b_1\frac{d^{m-1}w}{dt^{m-1}} + \cdots + b_m w$$

where $a_1, a_2, \ldots, a_n$ and $b_0, b_1, \ldots, b_m$ are real numbers.    If the input is a sinusoid given by

(2.15a)    $w(t) = \text{Re}(A\epsilon^{j\omega t}) = |A| \cos(\omega t + \phi)$

where

(2.15b)   $A \triangleq |A|\epsilon^{j\phi}$

then a particular solution of Eq. (2.14) is of the form

(2.16a)   $y(t) = \text{Re}\,(B\epsilon^{j\omega t}) = |B|\cos(\omega t + \psi)$

where

(2.16b)   $B \triangleq |B|\epsilon^{j\psi}$

The relation between the input expressed in terms of phasor $A$ and the portion of the output (particular solution only) expressed in terms of phasor $B$ can be obtained from the following equation:

(2.17)   $[(j\omega)^n + a_1(j\omega)^{n-1} + \cdots + a_n]B = [b_0(j\omega)^m + b_1(j\omega)^{m-1} + \cdots + b_m]A$

Equation (2.17) is obtained directly from Eq. (2.14) *by replacing the kth derivatives of $w(t)$ with $(j\omega)^k A$, for $k = 0$ to $m$, and by replacing the kth derivatives of $y(t)$ with $(j\omega)^k B$, for $k = 0$ to $n$.* Thus, in essence the determination of a particular solution as expressed by Eq. (2.16) is immediate from Eq. (2.17). We only need to manipulate complex numbers to put the solution in the form of Eq. (2.16a).

**Example 3**   Consider the linear time-invariant series $RLC$ circuit shown in Fig. 2.2. Let the input be the sinusoidal voltage source

$$e_s(t) = \text{Re}\,(E\epsilon^{j\omega t}) = |E|\cos(\omega t + \phi)$$

Let the output variable be the voltage across the capacitor. Then the differential equation is, for all $t$,

(2.18)   $LC\dfrac{d^2v_C(t)}{dt^2} + RC\dfrac{dv_C(t)}{dt} + v_C(t) = e_s(t)$

and a particular solution is of the form

(2.19)   $v_C(t) = \text{Re}\,(V_C\epsilon^{j\omega t}) = |V_C|\cos(\omega t + \psi)$

The relation between the output phasor $V_C$, which is to be determined, and the input phasor $E$, which is known, is as follows:

(2.20)   $[LC(j\omega)^2 + RC(j\omega) + 1]\,V_C = E$

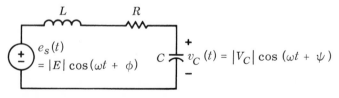

**Fig. 2.2**   Series $RLC$ circuit in sinusoidal steady state.

Note that Eq. (2.20) is obtained from Eq. (2.18) by replacing $e_s(t)$ with $E$ and the $k$th derivatives of $v_C(t)$ with $(j\omega)^k V_C$. Thus,

(2.21) $\quad V_C = \dfrac{E}{1 - \omega^2 LC + j\omega RC}$

Hence the magnitude and phase of $V_C$ are

$$|V_C| = \frac{|E|}{[(1 - \omega^2 LC)^2 + (\omega RC)^2]^{1/2}}$$

$$\psi = \phi - \tan^{-1}\frac{\omega RC}{1 - \omega^2 LC}$$

The solution $v_C(t)$, expressed as a real function of time, is then readily obtained from Eq. (2.19).

## 3 Complete Response and Sinusoidal Steady-state Response

### 3.1 Complete Response

A linear time-invariant circuit with sinusoidal input has a complete response of the form

(3.1) $\quad y(t) = y_h(t) + y_p(t) \qquad$ for all $t$

where the particular solution chosen, $y_p(\cdot)$, is a sinusoid of the same frequency as the input, and $y_h(\cdot)$ is a solution of the homogeneous differential equation. Assuming that all natural frequencies of the circuit are distinct (i.e., the characteristic polynomial has no multiple zeros), we have

(3.2) $\quad y_h(t) = \displaystyle\sum_{i=1}^{n} k_i \epsilon^{s_i t}$

where $s_i$ are the natural frequencies and $k_i$ are the arbitrary constants to be determined by the initial conditions. The particular solution $y_p(\cdot)$ is easily obtained using the phasor representation of a sinusoid, according to the method shown in the previous section. This decomposition of the complete response is illustrated by the following example.

**Example 1** Consider the series $RLC$ circuit of Fig. 3.1. The input is the sinusoidal voltage source $e_s(\cdot)$ which is applied at $t = 0$. The output is the capacitor voltage waveform $v_C(\cdot)$. Let us illustrate the calculation of the complete response with the following specifications:

$e_s(t) = u(t) \cos 2t$

$C = 1$ farad $\qquad R = \frac{3}{2}$ ohms $\qquad L = \frac{1}{2}$ henry

$i_L(0-) = I_0 = 1$ amp $\qquad v_C(0-) = V_0 = 1$ volt

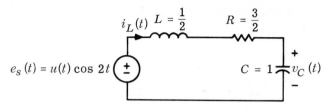

**Fig. 3.1**    Series *RLC* circuit illustrating the calculation of a complete response. The initial state is specified by $i_L(0-) = 1$ and $v_C(0-) = 1$.

Note the presence of the unit step function $u(\cdot)$ as a factor in the expression for $e_s$. The factor is necessary to describe the fact that the input $e_s$ is applied at $t = 0$; that is, $e_s(t) = 0$ for $t < 0$.

First we review the writing of the differential equation and the determination of the necessary initial conditions. From KVL we have

(3.3)    $$L \frac{di_L(t)}{dt} + Ri_L(t) + v_C(t) = e_s(t)$$

Since the current $i_L$ is also the current through the capacitor, we have

(3.4)    $$i_L(t) = C \frac{dv_C(t)}{dt}$$

Hence, Eq. (3.3) becomes

$$LC \frac{d^2v_C}{dt} + RC \frac{dv_C}{dt} + v_C = e_s(t)$$

or, upon insertion of the numerical values,

(3.5)    $$\frac{1}{2} \frac{d^2v_C}{dt^2} + \frac{3}{2} \frac{dv_C}{dt} + v_C = u(t) \cos 2t$$

The initial conditions are

(3.6a)    $$v_C(0-) = 1 \text{ volt}$$

and

(3.6b)    $$\frac{dv_C(0-)}{dt} = \frac{i_L(0-)}{C} = 2 \text{ volts/sec}$$

Equations (3.5) and (3.6) describe completely the output $v_C$. The complete solution can be obtained readily. The characteristic polynomial is $\frac{1}{2}s^2 + \frac{3}{2}s + 1$, and the natural frequencies are $s_1 = -1$ and $s_2 = -2$. Therefore, the solution to the homogeneous equation is of the form

(3.7)    $$v_h(t) = k_1\epsilon^{-t} + k_2\epsilon^{-2t}$$

The most convenient particular solution is of the form

(3.8)   $v_p(t) = \text{Re}\,(V\epsilon^{j\,2t}) = |V|\cos{(2t + \psi)}$

where $V$ represents the phasor of the output variable and is called the output voltage phasor. Let us also represent the input, in terms of the voltage phasor $E$, as

$e_s(t) = \text{Re}\,(E\epsilon^{j\,2t}) = \cos 2t$

with $E = 1\epsilon^{j0}$. The voltage phasor $V$ is found immediately from Eq. (3.5) according to the rule stated in the previous section [substituting the $k$th derivative of $v_C$ by $(j\omega)^k V$]. Thus,

$[\tfrac{1}{2}(j\omega)^2 + \tfrac{3}{2}(j\omega) + 1]V = E$

or

(3.9)   $V = \dfrac{1}{1 - \tfrac{1}{2}\omega^2 + j\tfrac{3}{2}\omega}$

With $\omega = 2$

$V = \dfrac{1}{-1 + j3} = 0.316\epsilon^{-j108.4°}$

From Eq. (3.8) we obtain the particular solution

(3.10)   $v_p(t) = 0.316 \cos{(2t - 108.4°)}$

The complete solution is

$v_C(t) = v_h(t) + v_p(t)$

(3.11)   $\qquad = k_1\epsilon^{-t} + k_2\epsilon^{-2t} + 0.316 \cos{(2t - 108.4°)}$

The constants $k_1$ and $k_2$ are next determined from Eqs. (3.6$a$) and (3.6$b$). From (3.6$a$) and (3.11)

$v_C(0) = 1 = k_1 + k_2 + 0.316 \cos{(-108.4°)}$

or

$k_1 + k_2 = 1.1$

From Eqs. (3.6$b$) and (3.11)

$\dfrac{dv_C}{dt}(0) = 2 = -k_1 - 2k_2 - 0.316 \times 2 \sin{(-108.4°)}$

or

$k_1 + 2k_2 = -1.4$

Therefore,

$k_1 = 3.6 \qquad \text{and} \qquad k_2 = -2.5$

The complete solution is

(3.12)   $v_C(t) = 3.6\epsilon^{-t} - 2.5\epsilon^{-2t} + 0.316 \cos (2t - 108.4°)$

The sketch of $v_C(t)$ is shown in Fig. 3.2. Note that the complete response can be meaningfully separated into two components, namely the transient and the steady state. The transient is identical to $v_h$ of Eq. (3.7), whereas the steady state is the $v_p$ of Eq. (3.10). Note that for $t > 4$ sec, the complete response is essentially the sinusoidal steady-state response.

**Remark**   In some simple circuits it is possible to choose the initial state so that the sinusoidal steady-state response is reached immediately after the input is applied; in other words, the transient term is identically zero. The way to choose the initial state for this particular purpose depends on two facts (see Chap. 2, Secs. 3 and 4): (1) the voltage across a capacitor can-

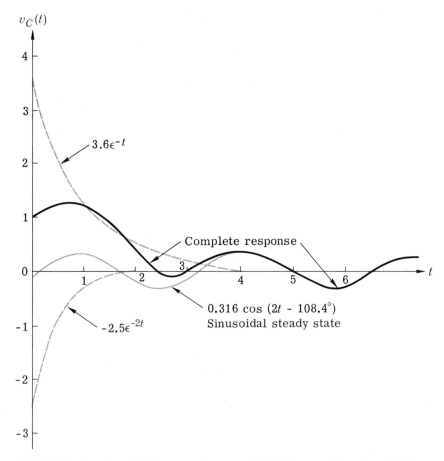

**Fig. 3.2**   The complete response $v_C(\cdot)$ (shown by a heavy solid line) is the sum of the sinusoidal steady state (light solid line) and the transient terms (dotted lines).

not change instantaneously if the current is bounded, and (2) the current in an inductor cannot change instantaneously if the voltage is bounded.

**Exercise 1**    Let the inductance $L$ in Fig. 3.1 be zero.   Thus, we have a series $RC$ circuit.   Determine the initial voltage $v_C(0-)$ across the capacitor for which no transient exists after the input $e_s$ is applied.

**Exercise 2**    Consider the circuit in Fig. 3.1 with $L = \frac{1}{2}$ henry, as in Example 1.   Is it possible to choose the initial state $i_L(0-)$ and $v_C(0-)$ such that no transient exists after the input $e_s$ is applied?   If so, determine the initial state.

## 3.2    Sinusoidal Steady-state Response

Let us consider an arbitrary linear time-invariant circuit driven by a single sinusoidal source.   Suppose we are interested in a particular network variable, say, $y$.   The response $y$ to the sinusoidal input and to the specified initial state is of the form

$$(3.13) \quad y(t) = k_1 \epsilon^{s_1 t} + k_2 \epsilon^{s_2 t} + \cdots + k_n \epsilon^{s_n t} + A_m \cos(\omega t + \psi)$$

where, for simplicity, we have assumed that the natural frequencies are simple, $k_1, k_2, \ldots, k_n$ are constants which depend on the initial state, and the amplitude $A_m$ and the angle $\psi$ of the particular solution can easily be determined by the phasor method.

The following observation is extremely important.   Suppose that all the natural frequencies lie within the open left-half complex frequency plane.†   Then, in Eq. (3.13), the terms $k_1 \epsilon^{s_1 t}, k_2 \epsilon^{s_2 t}, \ldots, k_n \epsilon^{s_n t}$ tend to zero as $t \to \infty$; equivalently, as $t \to \infty$, $y(t)$ becomes arbitrarily close to the sinusoid $A_m \cos(\omega t + \psi)$.   This allows us to state the following very important fact:

*Irrespective of the initial state and provided that all the natural frequencies are in the open left-half plane, the response will become sinusoidal as $t \to \infty$. This sinusoidal response is called the* **sinusoidal steady-state response**. *The sinusoidal steady-state response can easily be calculated by the phasor method.*

In Example 1, as we have seen, the sinusoidal steady state is described by the phasor $V$ in Eq. (3.9), and the sinusoidal steady-state response is given by the particular solution obtained by the phasor method; that is, $v_p(t) = \text{Re}(V\epsilon^{j2t})$.

On the basis of the above considerations, we can adopt the following language, as we did in Chap. 5.   When a linear time-invariant circuit has all its natural frequencies in the *open left-half* plane we say that the circuit

---

† The *open* left-half plane consists of the left half of the complex plane with the imaginary axis *excluded*.   In other words, the open left-half plane includes all points with negative real parts.

is **asymptotically stable.** If one or more of its natural frequencies are in the *open right-half* plane, we say that the circuit is **unstable.** Thus, any zero-input response of an asymptotically stable circuit approaches zero as $t \to \infty$. For unstable circuits, we can only state that for most initial states, the zero-input response $\to \infty$ as $t \to \infty$.

Thus, the important conclusion is that for *asymptotically stable* circuits driven by a single sinusoidal input, whatever the initial state may be, any network variable tends to the corresponding sinusoidal steady state as $t \to \infty$. We refer to this fact by the statement "asymptotically stable circuits have a sinusoidal steady-state response."

**Remark**   If the circuit has natural frequencies which are purely imaginary besides those which are in the open left-half plane, the steady-state response can sometimes still be defined. To understand this remark, we need to review the solution of differential equations that contain purely imaginary characteristic roots and multiple characteristic roots. Let us use the following two examples to illustrate two different situations.

**Example 2**   Let the characteristic polynomial of a differential equation be of the form

$$(s^2 + \omega_0^2)^2 = s^4 + 2\omega_0^2 s^2 + \omega_0^4$$

The characteristic roots are $s_1 = s_2 = j\omega_0$ and $s_3 = s_4 = -j\omega_0$. The solution of the homogeneous differential equation is of the form

$$y_h(t) = (k_1 + k_2 t)\epsilon^{j\omega_0 t} + (k_3 + k_4 t)\epsilon^{-j\omega_0 t}$$

which can also be expressed in terms of cosines as

$$y_h(t) = K_1 \cos(\omega_0 t + \phi_1) + K_2 t \cos(\omega_0 t + \phi_2)$$

where $K_1$, $K_2$, $\phi_1$, and $\phi_2$ are real constants. Clearly, as $t$ increases, $y_h(t)$ will take on arbitrarily large values, showing that the circuit is unstable. In the complete solution $y = y_h + y_p$, $y_p$ is negligible in comparison with $y_h$ for large $t$. From this example we conclude that if a circuit has multiple natural frequencies that lie on the imaginary axis, it is unstable, and it does not have a sinusoidal steady-state response.

**Example 3**   Let the characteristic polynomial of a differential equation be of the form $s^2 + \omega_0^2$. The characteristic roots are $s_1 = j\omega_0$ and $s_2 = -j\omega_0$. Let the forcing function be a sinusoid of angular frequency $\omega$ with $\omega \neq \omega_0$. Then the complete solution is of the following form:

$$y(t) = y_h(t) + y_p(t)$$

where

$$y_h(t) = k_1 \epsilon^{j\omega_0 t} + k_2 \epsilon^{-j\omega_0 t}$$
$$= K \cos(\omega_0 t + \phi)$$

with $K$ and $\phi$ real constants, and where the particular solution obtained by the phasor method is of the form

$$y_p(t) = B \cos(\omega t + \psi)$$

with $B$ and $\psi$ real constants.   Note that $y_h$ is oscillatory and hence cannot be considered as the transient of the complete response.   Yet the term $y_p$ is a sinusoid with the same frequency as the input and can thus be defined as the sinusoidal steady-state response, even though the complete response contains another sinusoid at a different frequency.   This type of sinusoidal steady-state response can be detected by a suitably tuned receiver.

On the other hand, if the angular frequency $\omega$ of the input coincides with $\omega_0$, the complete response will contain a term $Kt \cos(\omega t + \phi)$ and will become arbitrarily large as $t$ increases; hence, the steady-state response does not exist.   From this example we conclude that if the circuit has an imaginary natural frequency, say at $j\omega_0$, which is a simple zero of the characteristic polynomial, and if the angular frequency $\omega$ of the input sinusoid is not equal to $\omega_0$, then the sinusoidal steady-state response is well defined.

To summarize, *a linear time-invariant circuit whose natural frequencies are all within the open left half of the complex frequency plane has a sinusoidal steady-state response when driven by a sinusoidal input.   If, in addition, the circuit has imaginary natural frequencies that are simple and if these are different from the angular frequency of the input sinusoid, the steady-state response also exists.*

*The sinusoidal steady-state response always has the same frequency as the input and can be obtained most efficiently by the phasor method.*

**Remark**    For linear *time-varying* circuits or *nonlinear* circuits the steady-state response (if it exists) to a sinusoidal input is usually not sinusoidal.   It may include many sinusoids, even sinusoids with frequencies equal to a fraction of that of the input (see, for example, Probs. 3, 4, and 5 of this chapter).

## 3.3    Superposition in the Steady State

Consider a situation in which a linear time-invariant circuit with all its natural frequencies in the open left-half plane is driven by two sources with *different* frequencies.   For example, this would be the case when an audio amplifier boosts a single note played by, say, a flute; the sinusoids are the fundamentals and the harmonics of the flute.

In order to perform the analysis easily, let us consider the series $RLC$ circuit of Fig. 3.3.   The differential equation is

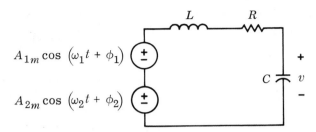

**Fig. 3.3**   Series $RLC$ circuit driven by two sinusoidal voltage sources.

(3.14)   $$LC\frac{d^2v}{dt^2} + RC\frac{dv}{dt} + v = A_{1m}\cos(\omega_1 t + \phi_1) + A_{2m}\cos(\omega_2 t + \phi_2)$$

where the input voltages have amplitudes $A_{1m}$ and $A_{2m}$, frequencies $\omega_1$ and $\omega_2$, and phases $\phi_1$ and $\phi_2$, respectively. The solution is of the form $v_p + v_h$, where $v_h$ is the solution of the homogeneous equation. To obtain a convenient particular solution $v_p$, observe that if $v_{p1}$ is the particular solution computed by the phasor method when the sinusoid $A_{1m}\cos(\omega_1 t + \phi_1)$ alone drives the circuit, and if $v_{p2}$ is the corresponding solution when $A_{2m}\cos(\omega_2 t + \phi_2)$ alone drives the circuit, then $v_p = v_{p1} + v_{p2}$. Indeed, by the definition of $v_{p1}$,

$$LC\frac{d^2v_{p1}}{dt^2} + RC\frac{dv_{p1}}{dt} + v_{p1} = A_{1m}\cos(\omega_1 t + \phi_1)$$

and, by the definition of $v_{p2}$,

$$LC\frac{d^2v_{p2}}{dt^2} + RC\frac{dv_{p2}}{dt} + v_{p2} = A_{2m}\cos(\omega_2 t + \phi_2)$$

Hence, by addition

$$LC\frac{d^2}{dt^2}(v_{p1} + v_{p2}) + RC\frac{d}{dt}(v_{p1} + v_{p2}) + (v_{p1} + v_{p2})$$

$$= A_{1m}\cos(\omega_1 t + \phi_1) + A_{2m}\cos(\omega_2 t + \phi_2)$$

from which we conclude that $v_{p1} + v_{p2}$ is a particular solution of Eq. (3.14). Using the results of phasor analysis, we see that

(3.15)   $$v_p(t) = V_{1m}\cos(\omega_1 t + \phi_1 + \theta_1) + V_{2m}\cos(\omega_2 t + \phi_2 + \theta_2)$$

where

$$V_{1m}\epsilon^{j(\theta_1 + \phi_1)} \triangleq \frac{A_{1m}\epsilon^{j\phi_1}}{1 - \omega_1^2 LC + j\omega_1 RC}$$

and

$$V_{2m}\epsilon^{j(\theta_2 + \phi_2)} \triangleq \frac{A_{2m}\epsilon^{j\phi_2}}{1 - \omega_2^2 LC + j\omega_2 RC}$$

Observe that in the denominators of these two expressions, we used the frequencies $\omega_1$ and $\omega_2$, respectively. We must use the frequency of the appropriate sinusoidal input. It is important to observe that whatever the initial conditions may be, as $t \to \infty$, the voltage $v$ becomes arbitrarily close to the value of $v_p$ given by Eq. (3.15). The waveform $v_p(\,\cdot\,)$ is called the *steady state* and not the *sinusoidal* steady state, since the sum of two sinusoids of different frequencies is not a sinusoid.

Note the important fact that the *steady state* resulting from the two input sinusoids is the sum of the sinusoidal steady states that would exist if each input sinusoid were acting alone on the circuit. Even though this result was proved only for the *RLC* circuit of Fig. 3.3, it is not difficult to see that the same method of proof could be applied to any linear time-invariant circuit.

## 4    Concepts of Impedance and Admittance

In the previous two sections we demonstrated that the sinusoidal steady-state response can be simply obtained by using the phasor representation of a sinusoid. We also learned that in determining the sinusoidal steady-state response we need only to solve an algebraic equation rather than a differential equation. Instead of adding, subtracting, or differentiating sinusoids, we can add or subtract complex numbers which represent them. In this section we shall explore further properties of the phasor representation of sinusoids and develop the important concepts of *impedance* and *admittance*. We shall see that when we only need to know the sinusoidal steady-state response of a linear time-invariant circuit, we can bypass the formulation of differential equations. Instead we can obtain the necessary linear algebraic equations directly from a network in terms of phasors, which represent the input, the output, and other network variables.

### 4.1    Phasor Relations for Circuit Elements

The voltage-current characterizations of simple circuit elements were studied in detail in Chap. 2. For *linear time-invariant* circuit elements, if we are only interested in the sinusoidal-state response, we can give a characterization in terms of the phasor representation of voltages and currents. In this subsection we shall derive the characterizations for the three basic elements: resistors, capacitors, and inductors. In each case let us assume that the element under consideration is connected to a linear time-invariant circuit, as shown in Fig. 4.1, and that the circuit is in sinusoidal steady state with angular frequency $\omega$. Let the sinusoidal steady-state branch voltage and branch current for the element be

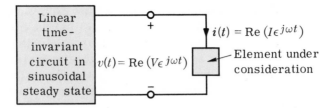

**Fig. 4.1**   A linear time-invariant circuit in sinusoidal steady state drives the element under consideration.

(4.1)   $v(t) = \text{Re}\,(Ve^{j\omega t}) = |V|\cos\,(\omega t + \angle V)$

and

(4.2)   $i(t) = \text{Re}\,(Ie^{j\omega t}) = |I|\cos\,(\omega t + \angle I)$

We wish to obtain the relation between the voltage phasor $V$ and the current phasor $I$ for each of the three elements.

*Resistor*   A linear time-invariant resistor with resistance $R$ or conductance $G = 1/R$ is characterized by

(4.3)   $v(t) = Ri(t) \qquad i(t) = Gv(t)$

To obtain the relation between the voltage phasor and the current phasor, we substitute Eqs. (4.1) and (4.2) in Eq. (4.3). Using Lemma 3 of Sec. 2, we obtain

(4.4)   $V = RI \qquad I = GV$

Although the resistance and the conductance of a resistor are always real numbers, the voltage phasor $V$ and the current phasor $I$ are usually complex numbers   It is instructive to plot the voltage phasor and the current phasor in a complex plane as shown in Fig. 4.2*b*. Since $R$ is a real number, the complex numbers $V$ and $I$ are collinear and must have the same angle; that is, $\angle I = \angle V$. The voltage and current waveforms are shown in Fig. 4.2*c*. They are said to be in phase to each other; that is, they cross the time axis at the same time and reach their maxima and minima simultaneously.

*Capacitor*   A linear time-invariant capacitor with capacitance $C$ is characterized by

(4.5)   $i = C\dfrac{dv}{dt}$

Using the phasor representations of $i$ and $v$ as in Eqs. (4.1) and (4.2) and substituting them in (4.5), we obtain

(4.6)   $I = j\omega CV \qquad \text{or} \qquad V = \dfrac{1}{j\omega C}I$

In deriving (4.6), we have employed Lemma 2 of Sec. 2 (that is, $d/dt$ applied to $V\epsilon^{j\omega t}$ amounts to the multiplication of $V\epsilon^{j\omega t}$ by $j\omega$). In Eq. (4.6), because of the factor $j\omega$, the current phasor $I$ and the voltage phasor $V$ differ by an angle of 90° when plotted on a complex plane, as shown in Fig. 4.3b. The current phasor *leads* the voltage phasor because $I = j\omega CV$, $\angle I = 90° + \angle V$. In Fig. 4.3c the current and voltage waveforms are plotted, and the current waveform leads the voltage waveform by one-quarter cycle.

It should also be pointed out that, unlike the resistor case, the relation between the current phasor and the voltage phasor depends on the angular frequency $\omega$.

*Inductor*   A linear time-invariant inductor with inductance $L$ is characterized by

$$(4.7) \quad V = L\frac{di}{dt}$$

As in the capacitor case, we obtain the following relations between the current phasor and voltage phasor (for an inductor):

$$(4.8) \quad V = j\omega LI \qquad I = \frac{1}{j\omega L}V$$

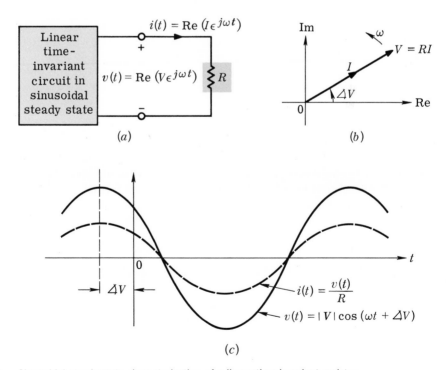

(a)

(b)

(c)

**Fig. 4.2**   Sinusoidal steady-state characterization of a linear time-invariant resistor.

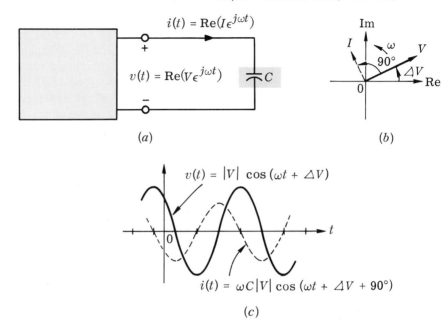

$$i(t) = \text{Re}(I\epsilon^{j\omega t})$$

$$v(t) = \text{Re}(V\epsilon^{j\omega t})$$

*(a)*

*(b)*

$$v(t) = |V| \cos(\omega t + \angle V)$$

$$i(t) = \omega C|V| \cos(\omega t + \angle V + 90°)$$

*(c)*

**Fig. 4.3**    Sinusoidal steady-state characterization of a linear time-invariant capacitor.

In this case the current phasor *lags* the voltage phasor by 90°, which means that the current waveform lags the voltage waveform by one-quarter cycle. These phasors are illustrated in Fig. 4.4*b* and *c*. Also, as in the capacitor case, the relation between the voltage and current phasors is dependent on frequency.

**4.2    Definition of Impedance and Admittance**

The discussion of phasor relations for circuit elements can be extended to general one-ports with linear time-invariant elements. Consider the circuit in Fig. 4.5*a*, where the one-port $\mathfrak{N}$ is formed by an arbitrary interconnection of linear time-invariant elements. The input is a sinusoidal current source at angular frequency $\omega$. Thus,

(4.9)   $i_s(t) = \text{Re}\,(I_s\epsilon^{j\omega t}) = |I_s| \cos(\omega t + \angle I_s)$

Let the sinusoidal steady-state voltage response be

(4.10)   $v(t) = \text{Re}\,(V\epsilon^{j\omega t}) = |V| \cos(\omega t + \angle V)$

We define the **driving-point impedance** *of the one-port $\mathfrak{N}$ at the angular frequency $\omega$* (or simply impedance) to be *the ratio of the output voltage phasor $V$ and the input current phasor $I_s$*; that is,

(4.11)   $Z(j\omega) \triangleq \dfrac{V}{I_s}$

Thus, the magnitude and the phase of the impedance are related to the magnitudes and the phases of the voltage phasor and the current phasor according to the relations

(4.12)    $|Z(j\omega)| = \dfrac{|V|}{|I_s|}$    and    $\measuredangle Z(j\omega) = \measuredangle V - \measuredangle I_s$

In terms of the impedance the output voltage waveform is given by

(4.13)    $v(t) = |Z(j\omega)||I_s| \cos (\omega t + \measuredangle Z(j\omega) + \measuredangle I_s)$

This equation leads to extremely important conclusions which are basic to any interpretation of impedance calculations. Thus,

*If the one-port $\mathfrak{N}$ has $Z(j\omega)$ as a driving-point impedance and if its input current is $|I_s| \cos (\omega t + \measuredangle I_s)$ then, in the sinusoidal steady state, its port voltage is a sinusoid of amplitude $|Z(j\omega)||I_s|$ and of phase $\measuredangle V = \measuredangle Z(j\omega) + \measuredangle I_s$.*

In other words, to obtain the amplitude of the sinusoidal voltage, we *multiply* the amplitude of the current by the magnitude of the impedance (evaluated at the appropriate frequency), and to obtain the phase of the sinusoidal voltage, we *add* to the phase of the current the phase $\measuredangle Z(j\omega)$ of the impedance (also evaluated at the appropriate frequency).

In Fig. 4.5b the input is a sinusoidal voltage source:

(4.14)    $v_s(t) = \mathrm{Re}\,(V_s \epsilon^{j\omega t}) = |V_s| \cos (\omega t + \measuredangle V_s)$

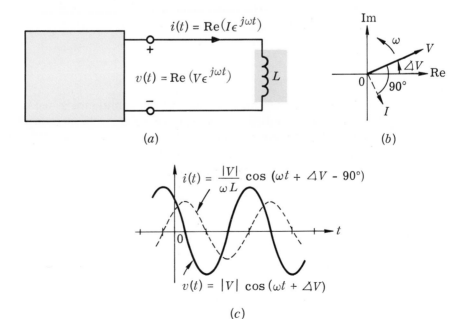

(a)

(b)

(c)

**Fig. 4.4**    Sinusoidal steady-state characterization of a linear time-invariant inductor.

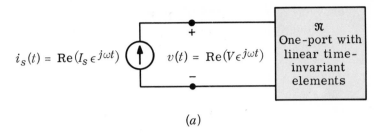

$(a)$

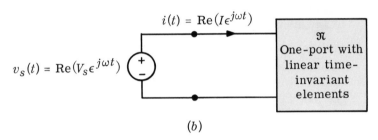

$(b)$

**Fig. 4.5**   The one-port $\mathfrak{N}$, made of linear time-invariant elements, is connected to a sinusoidal current source in $(a)$ and to a sinusoidal voltage source in $(b)$.

and the sinusoidal steady-state response is the current $i$, given by

(4.15)   $i(t) = \mathrm{Re}\,(I\epsilon^{j\omega t}) = |I|\cos(\omega t + \angle I)$

We define the **driving-point admittance** *of the one-port $\mathfrak{N}$ at the frequency $\omega$* (or simply admittance) to be *the ratio of the output current phasor $I$ and the input voltage phasor $V_s$;* that is,

(4.16)   $Y(j\omega) \triangleq \dfrac{I}{V_s}$

Thus, the magnitude and phase of the admittance $Y(j\omega)$ is related to the magnitudes and the phases of the voltage phasor and current phasor according to the relations

(4.17)   $|Y(j\omega)| = \dfrac{|I|}{|V_s|}$   and   $\angle Y(j\omega) = \angle I - \angle V_s$

**Remark**   If the voltage source of Fig. 4.5$b$ is adjusted so that its phasor $V_s$ becomes equal to the output voltage phasor $V$ of Fig. 4.5$a$, it is expected that the current response phasor $I$ in Fig. 4.5$b$ is equal to the current source phasor $I_s$ in Fig. 4.5$a$. Thus, from (4.11) we have

(4.18)   $V = Z(j\omega)I$

From (4.16), we have

(4.19)   $I = Y(j\omega)V$

From Eqs. (4.18) and (4.19), it is clear that for all $\omega$

(4.20) $\quad Z(j\omega) = \dfrac{1}{Y(j\omega)}$

and

(4.21) $\quad |Z(j\omega)| = \dfrac{1}{|Y(j\omega)|} \qquad \sphericalangle Z(j\omega) = -\sphericalangle Y(j\omega)$

A rigorous proof of this reciprocal relation between $Z$ and $Y$ will be given in Chap. 16.

**Exercise**  Give in words the rule which gives the amplitude and phase of the current in terms of that of the voltage and $Y(j\omega)$.

From the above definitions of impedance and admittance, we can obtain immediately the impedances and admittances of the elements $R$, $L$, and $C$:

| Angular frequency $\omega$ | $Z$ (impedance) | $Y$ (admittance) |
|---|---|---|
| Resistor with resistance $R$ | $R$ | $G = \dfrac{1}{R}$ |
| Capacitor with capacitance $C$ | $\dfrac{1}{j\omega C}$ | $j\omega C$ |
| Inductor with inductance $L$ | $j\omega L$ | $\dfrac{1}{j\omega L}$ |

**5**  **Sinusoidal Steady-state Analysis of Simple Circuits**

Kirchhoff's laws state that at any instant of time the algebraic sum of certain branch voltages or the algebraic sum of certain branch currents is zero. If we are only interested in the sinusoidal steady state and if we only need to deal with sinusoidal waveforms of the same frequency, we can write the equations in terms of phasors rather than in terms of the sinusoids themselves. Thus *in the sinusoidal steady state Kirchhoff's equations can be written directly in terms of voltage phasors and current phasors.* For example, let a mesh equation be of the form

$$v_1(t) + v_2(t) + v_3(t) = 0$$

Let each voltage be a sinusoid with the *same* angular frequency $\omega$. Then we have

$V_{1m} \cos(\omega t + \phi_1) + V_{2m} \cos(\omega t + \phi_2) + V_{3m} \cos(\omega t + \phi_3)$

$$= \text{Re}\,(V_1 \epsilon^{j\omega t}) + \text{Re}\,(V_2 \epsilon^{j\omega t}) + \text{Re}\,(V_3 \epsilon^{j\omega t})$$

$$= \text{Re}\,[(V_1 + V_2 + V_3)\epsilon^{j\omega t}] = 0$$

From Lemma 3 of Sec. 2 we can immediately write an equivalent equation in terms of the voltage phasors $V_1$, $V_2$, and $V_3$; thus,

$$V_1 + V_2 + V_3 = 0$$

Of course, knowing the phasor and the frequency $\omega$, we can always obtain the sinusoidal functions of time.   For example, if a voltage phasor is given by $V$ at an angular frequency $\omega$, the sinusoidal function is simply

$$v(t) = \text{Re}\,(V\epsilon^{j\omega t}) = V_m \cos(\omega t + \phi)$$

where

$$V \triangleq V_m \epsilon^{j\phi}$$

Similarly, we can write node equations in terms of current phasors.

## 5.1   Series-Parallel Connections

We first consider the series and parallel connections.   In Fig. 5.1 we have circuit elements connected in series.   In the sinusoidal steady state at a given frequency $\omega$, each element is characterized by an impedance.   Writing a KCL equation at each node, we see immediately that the currents are the same for all elements; in terms of phasors,

$$I_1 = I_2 = \cdots = I_n = I$$

Using KVL with the phasor representation of voltages, we have

$$V = V_1 + V_2 + \cdots + V_n$$

Since

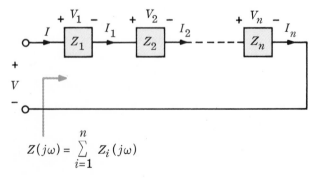

$$Z(j\omega) = \sum_{i=1}^{n} Z_i(j\omega)$$

**Fig. 5.1**   Impedances in series.

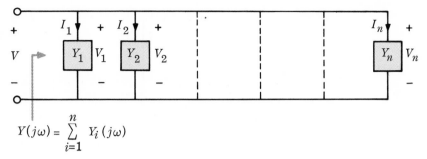

$$Y(j\omega) = \sum_{i=1}^{n} Y_i(j\omega)$$

**Fig. 5.2**   Admittances in parallel.

$$V_i = Z_i I_i \qquad i = 1, 2, \ldots n$$

we have

$$Z(j\omega) = \sum_{i=1}^{n} Z_i(j\omega)$$

where $Z = V/I$ is the impedance of the one-port shown in Fig. 5.1.
    Similarly, in Fig. 5.2 we have simple circuit elements connected in parallel.  Each element is characterized by its impedance or admittance.  Using KVL, we have

$$V_1 = V_2 = \cdots = V_n = V$$

Thus, all branch voltages are the same.  Using KCL, we have

$$I = I_1 + I_2 + \cdots + I_n$$

Since

$$I_i = Y_i V_i \qquad i = 1, 2, \ldots n$$

we have

$$Y(j\omega) = \sum_{i=1}^{n} Y_i(j\omega)$$

where $Y = I/V$ is the admittance of the one-port.

**Exercise 1**   Determine the driving-point impedances as functions of $\omega$ for the one-ports shown in Fig. 5.3.

**Exercise 2**   Plot the magnitude and phase versus $\omega$ for each of the impedances.

**Exercise 3**   Assuming that a current source $i_s$ is connected to each of the one-ports, determine the steady-state voltage responses (across nodes ① and ②) for

    *a.*   $i_s = \cos t$
    *b.*   $i_s = \cos 2t$

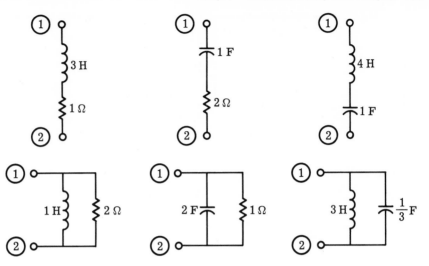

**Fig. 5.3**   The driving-point impedances of the one-ports are to be determined.

We can obviously analyze more complicated circuits by alternatively combining elements in series and in parallel. For example, the circuit in Fig. 5.4 is usually referred to as a *ladder circuit*. The driving-point impedance can be expressed by the form

(5.1)   $$Z = Z_1 + \cfrac{1}{Y_2 + \cfrac{1}{Z_3 + \cfrac{1}{Y_4 + \cfrac{1}{Z_5}}}}$$

It can be rewritten as follows:

$$Z = Z_1 + \cfrac{1}{Y_2 + \cfrac{1}{Z_3 + \cfrac{Z_5}{1 + Y_4 Z_5}}}$$

$$= Z_1 + \cfrac{1}{Y_2 + \cfrac{1 + Y_4 Z_5}{Z_5 + Z_3(1 + Y_4 Z_5)}}$$

$$= Z_1 + \frac{Z_5 + Z_3(1 + Y_4 Z_5)}{1 + Y_4 Z_5 + Y_2[Z_5 + Z_3(1 + Y_4 Z_5)]}$$

$$= \frac{Z_1[1 + Y_4 Z_5 + Y_2 Z_5 + Y_2 Z_3(1 + Y_4 Z_5)] + Z_5 + Z_3(1 + Y_4 Z_5)}{1 + Y_4 Z_5 + Y_2[Z_5 + Z_3(1 + Y_4 Z_5)]}$$

Equation (5.1) is called a *continued-fraction expansion*. This expansion is useful in the synthesis of circuits.

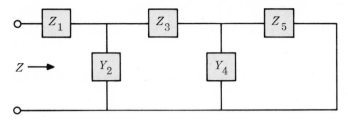

**Fig. 5.4**   A simple ladder network.

**Exercise**   Determine the driving-point impedances for the one-ports shown in Fig. 5.5.

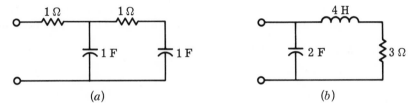

**Fig. 5.5**   The driving-point impedances for the one-ports are to be determined.

From the above examples we see that in analyzing networks formed by series and parallel connections of circuit elements, we only need to combine elements in series by adding the impedances of all branches in series and combine elements in parallel by adding the admittances of all branches in parallel.   Since driving-point impedance is simply the reciprocal of driving-point admittance, we can be flexible in choosing either impedance or admittance, according to which is most convenient in a particular situation.   In the parallel combination of Fig. 5.2 we choose admittance; in the ladder network shown in Fig. 5.3 we deal alternately with impedances and admittances.

| 5.2 | **Node and Mesh Analyses in the Sinusoidal Steady State** |
|---|---|

For linear time-invariant circuits which are not in the form of a series-parallel connection of circuit elements, we can use the two general methods of network analysis, namely, the node analysis and the mesh analysis.   First, we wish to emphasize again that *we are dealing with the sinusoidal steady-state analysis only.*   Thus, we may use voltage phasors, current phasors, impedances, and admittances in writing KCL and KVL equations.   The resulting equations are linear algebraic and can be solved by means of Cramer's rule.   We shall give two examples to illustrate the methods.

**Example 1**    In the circuit in Fig. 5.6 let the input be the current source

(5.2)    $i_s(t) = 10(\cos 2t + 30°)$

We wish to determine the sinusoidal steady-state voltage $v_3$ across the 2-ohm resistor.

We shall use node analysis. Let us pick the datum node as shown in the figure and denote the node-to-datum voltages $v_1$, $v_2$, and $v_3$. Since there are altogether four nodes in the circuit, we can write three KCL equations. Thus, the three node-to-datum voltages constitute the three unknowns that can be determined from the three KCL equations. Before proceeding to the writing of these equations, we wish to define the current source phasor $I_s$, which represents the source waveform $i_s(\,\cdot\,)$, and the three voltage phasors $V_1$, $V_2$, and $V_3$, which represent the three sinusoidal steady-state voltages $v_1$, $v_2$, and $v_3$, respectively. From (5.2) we have

$$i_s(t) = \text{Re}\,(I_s\epsilon^{j2t}) = 10\cos(2t + 30°)$$

or

(5.3)    $I_s \triangleq 10\epsilon^{j30°}$

Note that the angular frequency is $\omega = 2$ rad/sec. Let the voltage phasors be defined by the following equations:

$$v_1(t) = \text{Re}\,(V_1\epsilon^{j2t})$$

(5.4)    $v_2(t) = \text{Re}\,(V_2\epsilon^{j2t})$

$$v_3(t) = \text{Re}\,(V_3\epsilon^{j2t})$$

Recall that to obtain a current phasor, we multiply the voltage phasor by the admittance of the element. For example, let the current in the

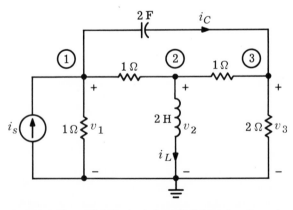

**Fig. 5.6**    Example 1: Sinusoidal steady-state analysis based on the node method.

inductor be $i_L$, which is represented by the current phasor $I_L$. If $V_2$ is given, $I_L$ can be obtained by

$$I_L = Y_L V_2 = \frac{1}{j\omega L} V_2 = \frac{1}{j4} V_2$$

Similarly, let the current in the capacitor be $i_C$, which is represented by the current phasor $I_C$. Noting that the voltage across the capacitor is $v_1 - v_3$, we obtain, in terms of phasors,

$$I_C = Y_C(V_1 - V_3) = j\omega C(V_1 - V_3) = j4(V_1 - V_3)$$

Following this scheme, we can express all branch-current phasors in terms of the phasors of node-to-datum voltages.

We then write the KCL node equations at the three nodes in terms of the three node-to-datum-voltage phasors. Thus, at node one

$$V_1 + j4(V_1 - V_3) + (V_1 - V_2) = I_s$$

at node two

$$\frac{1}{j4} V_2 + (V_2 - V_1) + (V_2 - V_3) = 0$$

and at node three

$$\tfrac{1}{2} V_3 + j4(V_3 - V_1) + (V_3 - V_2) = 0$$

Rearranging the equations, we obtain

$$(2 + j4)V_1 - V_2 - j4V_3 = I_s$$

$$-V_1 + \left(2 + \frac{1}{j4}\right)V_2 - V_3 = 0$$

$$-j4V_1 - V_2 + (\tfrac{3}{2} + j4)V_3 = 0$$

These results form a set of three linear algebraic equations with complex coefficients. The desired voltage phasor $V_3$ can be obtained by means of Cramer's rule. Thus,

$$V_3 = \frac{\begin{vmatrix} 2 + j4 & -1 & I_s \\ -1 & 2 + \dfrac{1}{j4} & 0 \\ -j4 & -1 & 0 \end{vmatrix}}{\begin{vmatrix} 2 + j4 & -1 & -j4 \\ -1 & 2 + \dfrac{1}{j4} & -1 \\ -j4 & -1 & \tfrac{3}{2} + j4 \end{vmatrix}} = \frac{2 + j8}{6 + j11.25} I_s$$

Since $I_s = 10\epsilon^{j30°}$

$V_3 = 6.45\epsilon^{j44°}$

Thus, the sinusoidal steady-state voltage output is

$v_3(t) = 6.45\cos(2t + 44°)$

---

**Example 2**   We wish to use the mesh analysis to solve the same problem.  First, we convert the current source into a voltage source by means of the Norton equivalent circuit.  The resulting circuit is shown in Fig. 5.7, and the voltage source is

$v_s(t) = 10(\cos 2t + 30°)$

Thus, the phasor representing $v_s$ is

$V_s = 10\epsilon^{j30°}$

   In mesh analysis we use mesh currents as network variables.  These currents are $i_1$, $i_2$, and $i_3$, as shown in Fig. 5.7.  The phasor representations for $i_1$, $i_2$, and $i_3$ are defined by

$i_1(t) = \text{Re}\,(I_1\epsilon^{j2t})$

$i_2(t) = \text{Re}\,(I_2\epsilon^{j2t})$

$i_3(t) = \text{Re}\,(I_3\epsilon^{j2t})$

We shall write mesh equations by means of KVL in terms of the phasors $I_1$, $I_2$, $I_3$, and $V_s$.  First, we need to express all branch-voltage phasors in terms of the mesh-current phasors $I_1$, $I_2$, and $I_3$.  To achieve this, we multiply the branch-current phasors by the branch impedances.  For ex-

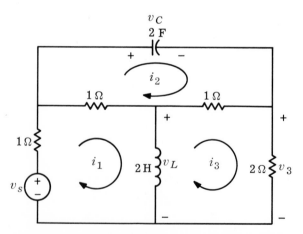

**Fig. 5.7**   Example 2: The same circuit as in Fig. 5.6 except that the current source has been replaced by an equivalent voltage source to facilitate the mesh analysis.

ample, the voltage phasor $V_C$ for the capacitor is equal to $(1/j4)I_2$. Similarly, the voltage phasor $V_L$ for the inductor is equal to $j4(I_1 - I_3)$. The KVL equations in terms of the mesh-current phasors are written next. Thus, for mesh 1

$$I_1 + (I_1 - I_2) + j4(I_1 - I_3) = V_s$$

for mesh 2

$$\frac{1}{j4}I_2 + (I_2 - I_3) + (I_2 - I_1) = 0$$

and for mesh 3

$$2I_3 + (I_3 - I_2) + j4(I_3 - I_1) = 0$$

The three equations are linear algebraic. After rearrangement we obtain

$$(2 + j4)I_1 - I_2 - j4I_3 = V_s$$

$$-I_1 + \left(2 + \frac{1}{j4}\right)I_2 - I_3 = 0$$

$$-j4I_1 - I_2 + (3 + j4)I_3 = 0$$

We solve for $I_3$ by means of Cramer's rule. Thus,

$$I_3 = \frac{\begin{vmatrix} 2 + j4 & -1 & V_s \\ -1 & 2 + \dfrac{1}{j4} & 0 \\ -j4 & -1 & 0 \end{vmatrix}}{\begin{vmatrix} 2 + j4 & -1 & -j4 \\ -1 & 2 + \dfrac{1}{j4} & -1 \\ -j4 & -1 & 3 + j4 \end{vmatrix}} = \frac{2 + j8}{12 + j22.5}V_s$$

Since $V_s = 10\epsilon^{j30°}$ and $V_3 = 2I_3$, we have

$$V_3 = 6.45\epsilon^{j44°}$$

or

$$v_3(t) = 6.45 \cos(2t + 44°)$$

The answer, of course, checks with that obtained by node analysis.

---

**Exercise 1**    Write the mesh equations for the ladder circuit shown in Fig. 5.8. The circuit is assumed to be in the sinusoidal steady state.

**Exercise 2**    Solve for the sinusoidal steady-state voltage $v_2$ across the 1-ohm resistor.

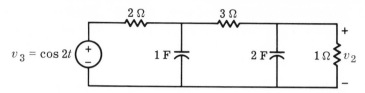

**Fig. 5.8**   A ladder circuit in sinusoidal steady state.

**Exercise 3**   Change the voltage source to a current source, and write the node equations in terms of phasors.

**Exercise 4**   Solve for the sinusoidal steady-state voltage $v_2$ based on the node analysis.

## 6   Resonant Circuits

We shall use a resonant circuit to further illustrate the sinusoidal steady-state analysis, with the concepts of phasor, impedance, admittance, and a new notion, that of *network function*. We shall give various graphical representations to demonstrate many properties of resonant circuits. These graphical methods will be useful for sinusoidal steady-state analysis of more complex circuits.

Two types of resonant circuits, the parallel resonant and the series resonant circuit, are important in applications. We shall analyze the parallel $RLC$ resonant circuit of Fig. 6.1. The series resonant circuit is the dual of the parallel resonant circuit. Since we have discussed briefly the concept of duality, a detailed treatment of series resonant circuits will be omitted. However, for reference, the results for both circuits are summarized in Table 7.1 at the end of this section.

### 6.1   Impedance, Admittance, and Phasors

Consider the resonant circuit of Fig. 6.1, which is driven by a sinusoidal current source

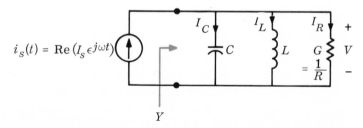

**Fig. 6.1**   Parallel resonant circuit.

(6.1)    $i_s(t) = \text{Re}\,(I_s \epsilon^{j\omega t}) = |I_s|\,\cos\,(\omega t + \angle I_s)$

The admittance of the one-port at the angular frequency $\omega$ is

(6.2)    $Y(j\omega) = G + j\omega C + \dfrac{1}{j\omega L}$

$$= G + j\left(\omega C - \dfrac{1}{\omega L}\right)$$

Thus, the real part of $Y(j\omega)$ is a constant, and the imaginary part is a function of $\omega$.  The imaginary part of an admittance is called the **susceptance** and is denoted by $B$; thus,

(6.3)    $B(\omega) = \omega C - \dfrac{1}{\omega L}$

The susceptance is a function of $\omega$ and is plotted in Fig. 6.2 versus $\omega$.  At the frequency $f_0 = \omega_0/(2\pi) = 1/(2\pi\,\sqrt{LC})$ the susceptance is zero, and the circuit is said to be in **resonance**; the frequency $f_0$ is called the **resonant frequency**.  The significance of the word "resonance" will be discussed later in this section.

*Admittance and impedance planes*    Equation (6.2) shows that the admittance is a function of the angular frequency $\omega$.  Splitting Eq. (6.2) into its real and imaginary components, we obtain

(6.4*a*)    $\text{Re}\,[Y(j\omega)] = G$

(6.4*b*)    $\text{Im}\,[Y(j\omega)] = B(\omega) = \omega C - \dfrac{1}{\omega L}$

The characteristic behavior of the admittance $Y(j\omega)$ may be described

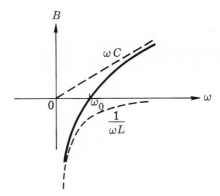

**Fig. 6.2**    Susceptance plot of the parallel resonant circuit; $B(\omega)$ versus $\omega$.  Note that at the angular resonant frequency $\omega_0 = 1/\sqrt{LC}$ rad/sec, and $B(\omega_0) = 0$.

graphically. For each fixed $\omega$, we may plot $Y(j\omega)$ as a point in the complex plane called, in this case, the **admittance plane.** As $\omega$ varies, the point $Y(j\omega)$ varies, and Eqs. (6.4a) and (6.4b) constitute the parametric equations of the curve traced by $Y(j\omega)$ (see Fig. 6.3). This curve is called **the locus of** $Y$. Since in the present case the abscissa $G$ is a constant, the locus is a straight line parallel to the imaginary axis which intersects the real axis at $G$. The distance from $Y(j\omega)$ to the origin is equal to the magnitude $|Y(j\omega)|$. The angle from the real axis to the line joining the origin to $Y(j\omega)$ is the phase $\sphericalangle Y(j\omega)$. Since $\mathrm{Im}\,[Y(j\omega_0)] = 0$, $Y(j\omega_0) = G$. Thus, at resonance ($\omega = \omega_0$), the admittance is *minimum,* and its phase is *zero.* It is interesting to note that the admittance of the parallel resonant circuit *at resonance* is equal to the admittance of the resistor alone; that is, the inductor and capacitor combination behaves as an open circuit.

**Exercise**   Consider a parallel resonant circuit with $L = 1$ henry, $C = 1$ farad, and $R = 100$ ohms. Plot the locus of $Y$. In particular, plot the points corresponding to $\omega = 0, 0.30, 0.995, 1.00, 1.005, 1.30$, and $\infty$ rad/sec.

The impedance of the parallel resonant circuit is

$$Z(j\omega) = \frac{1}{Y(j\omega)} = \frac{1}{G + jB(\omega)} = \frac{1}{G + j(\omega C - 1/\omega L)}$$

(6.5)

$$= \frac{G}{G^2 + B^2(\omega)} + j\,\frac{-B(\omega)}{G^2 + B^2(\omega)}$$

Similarly, the impedance can be plotted in a complex *impedance plane.* From Eq. (6.5) we have

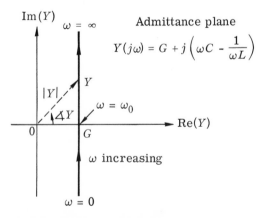

$$Y(j\omega) = G + j\left(\omega C - \frac{1}{\omega L}\right)$$

**Fig. 6.3**   Locus of $Y$ in the admittance plane.

(6.6a)   $\text{Re}\,[Z(j\omega)] = \dfrac{G}{G^2 + B^2(\omega)}$

and

(6.6b)   $\text{Im}\,[Z(j\omega)] \triangleq X(\omega) = \dfrac{-B(\omega)}{G^2 + B^2(\omega)}$

The imaginary part of an impedance is called the **reactance** and is usually denoted by $X(\omega)$. Equations (6.6a) and (6.6b) can be regarded as parametric equations of a curve in the impedance plane; this curve is called **the locus of** $Z$.

**Exercise 1**   Plot the locus of $Z$ for the parallel $RLC$ circuit with $L = 1$ henry, $C = 1$ farad, and $R = 100$ ohms.

**Exercise 2**   Prove that the locus of $Z$ of any parallel $RLC$ circuit is a circle in the complex impedance plane. The center is located at $(1/2G, 0)$, and the radius is $1/2G$, as shown in Fig. 6.4. Hint: The equation of the circle is given by

(6.7)   $\left[\text{Re}\,(Z) - \dfrac{1}{2G}\right]^2 + \left[\text{Im}\,(Z)\right]^2 = \left(\dfrac{1}{2G}\right)^2$

The significance of resonance will be clear if we examine either the locus of $Y$ in Fig. 6.3 or the locus of $Z$ in Fig. 6.4. As a function of $\omega$, the magnitude of the impedance, $|Z(j\omega)|$, starts at zero for $\omega = 0$, increases monotonically, and reaches a *maximum* at resonance ($\omega = \omega_0$). At resonance, the reactance $X(\omega_0)$ is zero, and $Z(j\omega_0)$ is said to be purely resistive. For $\omega > \omega_0$, $|Z(j\omega)|$ decreases monotonically and tends to zero as $\omega \to \infty$. Physically, at resonance all the current from the current source goes through the resistor, and the currents in the inductor and capacitor add to zero. At low frequencies ($\omega \ll \omega_0$) most of the current goes through the inductor, and at high frequencies ($\omega \gg \omega_0$) most of the current goes through the capacitor.

Let us now consider the phasors of the branch voltages and the branch currents. First the voltage phasor $V$ is given by

(6.8)   $V = ZI_s$

*Phasor diagram*   Let the current phasors for the resistor, inductor, and capacitor branches be $I_R$, $I_L$, and $I_C$, respectively. Then

(6.9)   $I_R = GV \qquad I_L = \dfrac{1}{j\omega L}\,V \qquad I_C = j\omega CV$

Clearly,

(6.10)   $I_R + I_L + I_C = I_s$

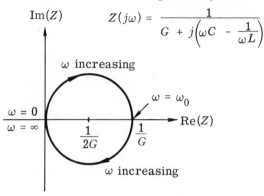

Impedance plane

Fig. 6.4   Locus of $Z$ in the impedance plane.

To illustrate the above relations, let us specify the following:

$$i_s(t) = \cos t = \text{Re}\,(I_s\epsilon^{jt})$$

that is,

$I_s = 1\epsilon^{j0}$ amp      $\omega = 1$ rad/sec

Let the element values be given by

$R = 1$ ohm      $L = ¼$ henry      $C = 1$ farad

The admittance of the resonant circuit at the angular frequency $\omega = 1$ rad/sec is (in mhos)

$$Y(j1) = 1 + j(1 - 4) = 1 - j3 = \sqrt{10}\epsilon^{-j71.6°}$$

Thus, the impedance is (in ohms)

$$Z(j1) = \frac{1}{Y(j1)} = \frac{1}{\sqrt{10}}\,\epsilon^{j71.6°}$$

and the voltage phasor is (in volts)

$$V = Z(j1)I_s = \frac{1}{\sqrt{10}}\,\epsilon^{j71.6°}$$

From Eq. (6.9) with $\omega = 1$, we have (in amperes)

$$I_R = \frac{1}{\sqrt{10}}\,\epsilon^{j71.6°} \qquad I_L = \frac{4}{\sqrt{10}}\,\epsilon^{-j18.4°} \qquad I_C = \frac{1}{\sqrt{10}}\,\epsilon^{j161.6°}$$

The phasors of voltage $V$ and currents are plotted in Fig. 6.5.   It is seen that $I_R + I_L + I_C = I_s$.

Next let us apply a sinusoidal input at the resonant frequency $\omega_0 = 1/\sqrt{LC} = 2$ rad/sec.   Let the input be

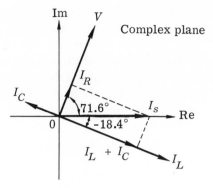

**Fig. 6.5**   Plots of voltage and current phasors in a complex plane ($I_s$ is the source current).

$$i_s(t) = \cos 2t = \text{Re}\,(I_s \epsilon^{j2t})$$

that is,

$$I_s = 1\epsilon^{j0}\ \text{amp} \qquad \omega = 2\ \text{rad/sec}$$

The input has a frequency equal to the resonant frequency of the circuit. The admittance is seen to be

$$Y(j2) = 1\ \text{mho}$$

Thus, the phasor voltage is

$$V = 1\ \text{volt}$$

and

$$I_R = 1 \qquad I_L = 2\epsilon^{-j90°} \qquad I_C = 2\epsilon^{j90°}$$

in amperes.  The phasors are plotted in Fig. 6.6.  It is interesting to note that the branch currents in the inductor and capacitor have magnitudes

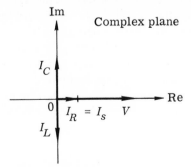

**Fig. 6.6**   Plots of voltage and current phasors at resonance.

twice as large as the input current. This is not surprising, since Eq. (6.10) is an equation with complex numbers; in the present case, $I_L$ and $I_C$ are, respectively, $-90°$ and $+90°$ out of phase with $I_s$.

It is also interesting to observe the effect of the resistance to the overall behavior of the resonant circuit. For example, if the 1-ohm resistor in the case given above is replaced by a resistor of 250 ohms while the inductor and capacitor remain unchanged, the resonant frequency is still 2 rad/sec, and

$$Y(j2) = 4 \times 10^{-3} \text{ mho} \qquad \text{or} \qquad Z(j2) = 250 \text{ ohms}$$

Hence, with the same input current, that is, $I_s = 1$ amp, we get

$$V = 250 \text{ volts}$$
$$I_C = j500 = 500\epsilon^{j90°} \text{ amp}$$
$$I_L = -j500 = 500\epsilon^{-j90°} \text{ amp}$$
$$I_R = 1 \text{ amp}$$

We can think of these currents and voltages as follows: A huge current of 500 amp flows through the $LC$ mesh and the 1-amp current from the source goes through the resistor. In fact, *at resonance* the ratio of the magnitude of the current in the inductor (or capacitor) to the magnitude of the source current is equal to the quality factor $Q$ of the resonant circuit; that is,

$$\frac{|I_L|}{|I_s|} = \frac{|I_C|}{|I_s|} = Q$$

Because of this phenomenon caution is required when measuring currents and voltages of a resonant circuit. For example, in a *series* resonant circuit with an input voltage source of just a few volts in amplitude, the voltage across the inductance or capacitance may have an amplitude of several hundred volts!

**Remark**   In all the discussions of this section, we considered exclusively the sinusoidal steady state, where all the branch voltages and all the branch currents vary sinusoidally with time at the same frequency. For example, when we say that, at resonance $Q \gg 1$, the inductor current is very large compared with the source current, we in fact *mean* that the *amplitude* of the sinusoidal inductor current is very large compared with the *amplitude* of the source current. In fact, at resonance, these two currents are 90° out of phase; when one of them is maximum, the other is zero.

---

**6.2**   **Network Function, Frequency Response**

We are still considering the parallel $RLC$ circuit shown in Fig. 6.1. Let us assume now that the actual output of interest for the resonant circuit is the steady-state current in the resistor, $i_R(t) = \text{Re}\,(I_R\epsilon^{j\omega t})$. The input is still given by the sinusoidal current source $i_s(t) = \text{Re}\,(I_s\epsilon^{j\omega t})$. We define a **network function** to be *the ratio of the output phasor to the input phasor.* Let us denote the network function by $H$; then the network function $H$ evaluated at $j\omega$ is given by

$$H(j\omega) = \frac{I_R}{I_s} = \frac{GV}{I_s} = GZ(j\omega) = \frac{1}{1 + jR(\omega C - 1/\omega L)}$$

(6.11)
$$= \frac{1}{1 + jQ(\omega/\omega_0 - \omega_0/\omega)}$$

where

(6.12)    $$Q \triangleq \frac{\omega_0}{2\alpha} = \omega_0 CR \qquad \omega_0 = \frac{1}{\sqrt{LC}}$$

Observe that network functions usually depend on the angular frequency $\omega$, as is the case for $H$ in (6.11). The magnitude of the network function $H$ is

(6.13)    $$|H(j\omega)| = \frac{1}{[1 + Q^2(\omega/\omega_0 - \omega_0/\omega)^2]^{1/2}}$$

and its phase is

$$\angle H(j\omega) = -\tan^{-1} Q\!\left(\frac{\omega}{\omega_0} - \frac{\omega_0}{\omega}\right)$$

with

(6.14)    $$-\frac{\pi}{2} \leq \angle H(j\omega) \leq \frac{\pi}{2}$$

The two parameters $Q$ and $\omega_0$ characterize completely the network function $H$. In Fig. 6.7 we plot the magnitude and phase of $H$ versus $\omega/\omega_0$ with $Q$ as a parameter. The two sets of curves, i.e., the magnitude and the phase versus $\omega$, are extremely useful since they give all the needed information for any resonant circuit at all frequencies. To find the sinusoidal steady-state response $i_R$ due to the input $i_s = \text{Re}\,(I_s\epsilon^{j\omega t})$, we need only find the magnitude and phase of $H(j\omega)$ from the sets of curves. Since $I_R = H(j\omega)I_s$,

(6.15)    $$i_R(t) = \text{Re}\,[H(j\omega)I_s\epsilon^{j\omega t}]$$
$$= |H(j\omega)||I_s|\cos\,[\omega t + \angle I_s + \angle H(j\omega)]$$

**Remark**    1.    The driving-point impedance and admittance are special cases of the general concept of network functions. If we compare Eq. (6.15) with Eq. (4.13), we notice that the same rule applies in determining the sinusoidal steady-state output waveform from the sinusoidal input waveform and the network function.

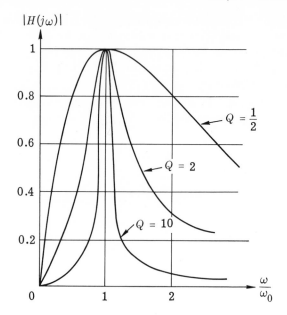

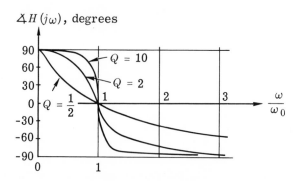

**Fig. 6.7**   Frequency response of resonant circuits.

2. The sets of curves in Fig. 6.7 are also valid for the series resonant circuit shown in Fig. 6.8. It is only necessary to use the corresponding definition for $Q$; that is, $Q \triangleq \omega_0 L/R_s$ (see Table 5.1). The network function for a series circuit is defined by $H = V_R/V_s$.

**Exercise**   Let the current source input be specified by $i_s(t) = \cos 2t$. Determine the current phasors $I_R$ in the parallel $RLC$ circuits specified by $\omega_0 = 1$ rad/sec and $Q = \frac{1}{2}$, 2, and 10, respectively. (Hint: Use Fig. 6.7.)

*Frequency*   Since $H(j\omega)$ contains all the needed information concerning the sinusoi-
*response*   dal steady-state response, we call the curves of the magnitude and phase

of $H(j\omega)$ (versus $\omega$ or log $\omega$) the **frequency response** of the circuit for that specified input and output ($I_s$ and $I_R$, respectively, in the case of parallel resonant circuits). In order to obtain a physical interpretation of the frequency response, we shall consider the sinusoidal steady-state response of the circuit for several values of the frequency. As above, let $I_s(j\omega)$ be the phasor representation of the input current at the angular frequency $\omega$. Then the output phasor, which represents the sinusoidal steady-state response at the angular frequency $\omega$, is $I_R(j\omega)$, and from the definition of the network function,

(6.16a)   $I_R(j\omega) = H(j\omega)I_s(j\omega)$

Thus, the magnitude of the output phasor is related to the magnitude of the input phasor by

(6.16b)   $|I_R(j\omega)| = |H(j\omega)||I_s(j\omega)|$

Similarly, the phase of the output phasor is related to the phase of the input phasor by

(6.16c)   $\sphericalangle I_R(j\omega) = \sphericalangle H(j\omega) + \sphericalangle I_s(j\omega)$

In particular, if $H(j\omega) = 1$, the output phasor is identical to the input phasor; if $H(j\omega) = 0$, the output phasor is zero. For the resonant circuit, Eq. (6.11) shows that the network function $H$ is equal to 1 at the resonant frequency and is zero at $\omega = 0$ and $\infty$. Therefore, we say that a resonant circuit passes signals at the resonant frequency but stops signals at zero and infinite frequencies. At other frequencies, the magnitude and phase are modified according to the curves of Fig. 6.7. Thus, in the immediate neighborhood of the resonant frequency, input signals get through with only a slight reduction in magnitude and a slight change in phase. At low frequencies ($\omega \ll \omega_0$) and at high frequencies ($\omega \gg \omega_0$), the output is considerably reduced in magnitude. Because of this fact, we call a resonant circuit a *bandpass filter*. Only signals with frequencies in the neighborhood of the resonant frequency are not "filtered out" by the resonant circuit. The shapes of the magnitude and phase curves of a resonant circuit depend on the quality factor $Q$. A higher $Q$ produces a narrower passband. An ideal bandpass filter has a magnitude curve as

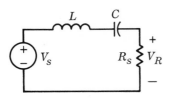

**Fig. 6.8**   Series resonant circuit with $\omega_0 = 1/\sqrt{LC}$, and $Q = \omega_0 L/R_s$.

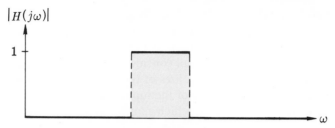

**Fig. 6.9**   Magnitude curve for an ideal bandpass filter.

shown in Fig. 6.9.   Ideally, all signals inside the passband pass with no change in magnitude and phase; outside the passband, the output is uniformly zero.   However, the magnitude curve of Fig. 6.9 is not physically realizable.   For a practical filter circuit (such as a resonant circuit) the passband can be defined in various ways.   The most commonly used definition is the **3-db passband,†** which means that at the edges of the passband, $|H(j\omega)|$ is $1/\sqrt{2}$ of the passband's maximum value.   From Eq. (6.13) the maximum magnitude of $|H(j\omega)|$ occurs at $\omega = \omega_0$, and is equal to 1. Setting $|H(j\omega)| = 1/\sqrt{2}$, we have

$$|H(j\omega)| = \frac{1}{[1 + Q^2(\omega/\omega_0 - \omega_0/\omega)^2]^{1/2}} = \frac{1}{\sqrt{2}}$$

or

$$Q^2\left(\frac{\omega}{\omega_0} - \frac{\omega_0}{\omega}\right)^2 = 1$$

Solving for positive values of $\omega$ in terms of $Q$, we obtain

(6.17)   $$\frac{\omega}{\omega_0} = \sqrt{1 + \frac{1}{4Q^2}} \pm \frac{1}{2Q}$$

---

† The abbreviation db stands for decibel.   Voltages and currents can be expressed in decibels according to the formula

$$\text{Voltage}\Big|_{\text{in decibels}} = 20 \log \text{voltage}\Big|_{\text{in volts}}$$

(and similarly for currents).   The transfer function $H$, being a ratio of currents is also expressible in decibels as follows:

$$|H(j\omega)|\Big|_{\text{in decibels}} = 20 \log|H(j\omega)|$$

Since in the present case, $H(j\omega_0) = 1$, at $\omega_0$ the transfer function is 0 db and 0°.   Since 20 log $(1/\sqrt{2}) \approx -3$, if for some frequency $\omega_1$, $|H(j\omega_1)|$ is $-3$ db, it means that

$$\frac{|H(j\omega_1)|}{|H(j\omega_0)|} = \frac{1}{\sqrt{2}} = 0.707$$

In the case of large $Q$ ($Q \gg 1$), using

$$\sqrt{1 + x} = 1 + \frac{x}{2} - \frac{x^2}{8} + \cdots$$

we obtain

(6.18)   $\dfrac{\omega}{\omega_0} \approx 1 \pm \dfrac{1}{2Q}$

We can thus define the passband as the band from $\omega_1$ to $\omega_2$, where

(6.19)   $\omega_1 \approx \omega_0\left(1 - \dfrac{1}{2Q}\right)$     $\omega_2 \approx \omega_0\left(1 + \dfrac{1}{2Q}\right)$     for $Q \gg 1$

The frequencies

$$f_1 = \frac{\omega_1}{2\pi} \quad \text{and} \quad f_2 = \frac{\omega_2}{2\pi}$$

are called the **3-db cutoff frequencies,** whereas $f_2 - f_1 = \Delta f$ is called the **3-db bandwidth** and is given in hertz by

(6.20)   $\Delta f = f_2 - f_1 = \dfrac{\omega_2 - \omega_1}{2\pi} \approx \dfrac{\omega_0}{2\pi Q} = \dfrac{f_0}{Q} = \dfrac{\alpha}{\pi}$

In Chap. 5 we classified second-order circuits either according to the location of their natural frequencies in the complex frequency plane or according to the value of the quality factor $Q$. For $\infty > Q > 1/2$, we called the circuit underdamped, and we labeled the natural frequencies

$$\begin{Bmatrix} s_1 \\ s_2 \end{Bmatrix} = -\alpha \pm j\omega_d$$

where

$$\alpha = \frac{\omega_0}{2Q}$$

and

(6.21)   $\omega_d = \sqrt{\omega_0{}^2 - \alpha^2} = \omega_0\sqrt{1 - \dfrac{1}{4Q^2}}$

In Fig. 6.10, the complex frequency plane is shown along with the magnitude curve to indicate many interesting relations among the locations of natural frequencies at $\alpha \pm j\omega_d$, the resonant frequency $\omega_0$, the bandwidth $\omega_2 - \omega_1$, and the cutoff frequencies $\omega_1$ and $\omega_2$. Figure 6.10 is drawn for a case in which $Q$ is large. With $Q \gg 1$, by dropping terms in $1/Q^2$, we obtain from Eqs. (6.19) and (6.21) the relations

(6.22)   $\omega_d \approx \omega_0$      $\omega_1 \approx \omega_0 - \alpha$      $\omega_2 \approx \omega_0 + \alpha$

For convenience we summarize the principal results of parallel and series resonant circuits in Table 7.1.

**Table 7.1**   **Sinusoidal Steady-state Properties of Resonant Circuits**

| Parallel resonant circuit | Series resonant circuit |
|---|---|

$$Q \triangleq \frac{\omega_0}{2\alpha} = \omega_0 CR = \frac{R}{\omega_0 L} = \frac{R}{\sqrt{L/C}}$$

$$\alpha = \frac{1}{2RC}$$

$$H(j\omega) \triangleq \frac{I_R}{I_s} \qquad Y(j\omega) = \frac{1}{RH(j\omega)}$$

$$Q \triangleq \frac{\omega_0}{2\alpha} = \frac{\omega_0 L}{R_s} = \frac{\sqrt{L/C}}{R_s}$$

$$\alpha = \frac{R_s}{2L}$$

$$H(j\omega) \triangleq \frac{V_R}{V_s} \qquad Z(j\omega) = \frac{R_s}{H(j\omega)}$$

$$\omega_0 \triangleq \frac{1}{\sqrt{LC}} \qquad Q \triangleq \frac{\omega_0}{2\alpha}$$

$$H(j\omega) = \frac{1}{1 + jQ\left(\dfrac{\omega}{\omega_0} - \dfrac{\omega_0}{\omega}\right)}$$

If $Q > \frac{1}{2}$ (underdamped case), the natural frequencies are $-\alpha \pm j\omega_d$, where $\omega_d \triangleq \sqrt{\omega_0^2 - \alpha^2} = \omega_0 \sqrt{1 - 1/4Q^2}$. If $Q \gg 1$, $\omega_d \approx \omega_0$.

$$\text{3-db angular cutoff frequencies} \begin{cases} \omega_1 \approx \omega_0 - \alpha = \omega_0\left(1 - \dfrac{1}{2Q}\right) \\ \omega_2 \approx \omega_0 + \alpha = \omega_0\left(1 + \dfrac{1}{2Q}\right) \end{cases}$$

$$\text{3-db angular bandwidth } \Delta\omega = \omega_2 - \omega_1 \approx 2\alpha = \frac{\omega_0}{Q} \qquad \text{rad/sec}$$

$$\Delta f = f_2 - f_1 \approx \frac{f_0}{Q} \qquad \text{Hz}$$

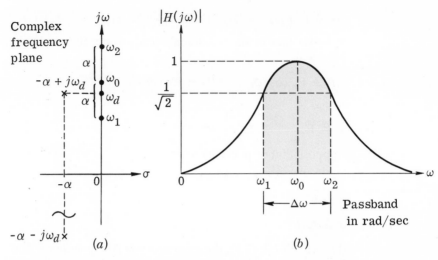

**Fig. 6.10** (a) Natural frequencies in the complex frequency plane and the corresponding passband for a high-$Q$ resonant circuit; (b) magnitude curve (for $Q \gg 1$, $\Delta\omega \simeq \omega_0/Q$).

## 7 Power in Sinusoidal Steady State

In Chap. 2 we calculated the instantaneous power entering a one-port at time $t$ and the energy delivered to a one-port from $t_0$ to $t$. With reference to Fig. 7.1, the instantaneous *power* entering the one-port $\mathfrak{N}$ is

(7.1) $\quad p(t) = v(t)i(t)$

and then *energy* delivered to $\mathfrak{N}$ in the interval $(t_0,t)$ is

(7.2) $\quad W(t_0,t) = \int_{t_0}^{t} p(t') \, dt'$

In this section we shall use the above equations to calculate the power and the energy in the sinusoidal steady state.

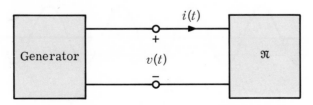

**Fig. 7.1** The one-port $\mathfrak{N}$ is made of linear time-invariant elements. The port voltage is $v(t)$, and its port current is $i(t)$.

**7.1     Instantaneous, Average, and Complex Power**

Suppose that in the sinusoidal steady state, the port voltage of the one-port $\mathfrak{N}$ is

(7.3a)   $v(t) = V_m \cos (\omega t + \sphericalangle V) = \mathrm{Re}\,(Ve^{j\omega t})$

where

(7.3b)   $V \triangleq V_m \epsilon^{j \sphericalangle V} \qquad V_m = |V|$

Suppose that its port current is

(7.4a)   $i(t) = I_m \cos (\omega t + \sphericalangle I) = \mathrm{Re}\,(Ie^{j\omega t})$

where

(7.4b)   $I \triangleq I_m \epsilon^{j \sphericalangle I} \qquad I_m = |I|$

Then from Eq. (7.1), the *instantaneous power* entering $\mathfrak{N}$ is

(7.5)   $p(t) = v(t)i(t)$

$\qquad = V_m I_m \cos (\omega t + \sphericalangle V) \cos (\omega t + \sphericalangle I)$

$\qquad = \tfrac{1}{2} V_m I_m \cos (\sphericalangle V - \sphericalangle I) + \tfrac{1}{2} V_m I_m \cos (2\omega t + \sphericalangle V + \sphericalangle I)$

The current $i$, the voltage $v$, and the instantaneous power $p$ are plotted on Fig. 7.2. The first term in the power expression of Eq. (7.5) is a constant, whereas the second is a sinusoid with angular frequency $2\omega$. If we calculate the average power over one period $T = 2\pi/\omega_0$, the second term will always equal zero (since the average of any sinusoid over any integral number of its periods is zero). Hence, labeling the *average power* with $P_{av}$, we get

(7.6a)   $P_{av} \triangleq \dfrac{1}{T} \displaystyle\int_0^T p(t')\, dt'$

hence

(7.6b)   $P_{av} = \tfrac{1}{2} V_m I_m \cos (\sphericalangle V - \sphericalangle I)$

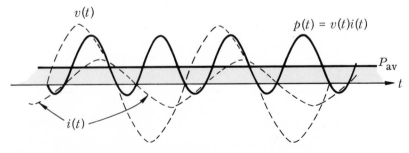

**Fig. 7.2**   Sinusoidal steady-state current and voltage waveforms and instantaneous and average power.

**Remarks**    1.   The angle $\angle V - \angle I$, which is the argument of the cosine in Eq. (7.6b), is the phase difference between the voltage sinusoid and the current sinusoid.   Since $V = ZI$, $\angle V - \angle I = \angle Z$.   $\angle V - \angle I$ is also the angle of the impedance of the one-port under consideration.   Therefore, it is possible to change the average power received by a one-port by changing the angle of the impedance while keeping the magnitude of the impedance constant.

2.   $P_{av}$ is the average *over one period* of the instantaneous power $p(\cdot)$ entering the one-port.   A typical graph of $p$ versus time is shown in Fig. 7.2.   In most instances the one-port $\mathfrak{N}$ contains only passive elements; i.e., all resistances, capacitances, and inductances are positive. Consequently, the inductors and the capacitors store energy, and the resistors dissipate energy.   By the principle of conservation of energy, in the sinusoidal steady state the average power entering the one-port $\mathfrak{N}$ must be nonnegative ($\geq 0$).   The fact that the *average* power is greater than or equal to zero does not imply that $p(t) \geq 0$ for all $t$.   As shown in Fig. 7.2 the instantaneous power $p(t)$ may be negative over some intervals of time in each period.

3.   The easiest way to calculate the average power delivered to the one-port $\mathfrak{N}$ is to proceed as follows.   In the sinusoidal steady state, we define

$$P \triangleq \tfrac{1}{2} V \bar{I}$$

as the **complex power** delivered to the one-port $\mathfrak{N}$.   Here we used the overbar to denote the complex conjugate.   Then

$$P = \tfrac{1}{2}|V||I|\epsilon^{j(\angle V - \angle I)}$$
$$= \tfrac{1}{2}|V||I| \cos (\angle V - \angle I) + j\tfrac{1}{2}|V||I| \sin (\angle V - \angle I)$$

By Eq. (7.6), the real part of the complex power $P$ is the average power

(7.7a)    $P_{av} = \mathrm{Re}\,(P) = \mathrm{Re}\,(\tfrac{1}{2}V\bar{I})$

4.   Let $Z(j\omega)$ and $Y(j\omega)$ be, respectively, the driving-point impedance and the driving-point admittance of the one-port at frequency $\omega$. Since $V = ZI$ and $I = YV$, Eq. (6.7a) becomes

(7.7b)    $P_{av} = \tfrac{1}{2}|I|^2 \, \mathrm{Re}\,[Z(j\omega)] = \tfrac{1}{2}|V|^2 \, \mathrm{Re}\,[Y(j\omega)]$

Eq. (7.7b) leads to an important conclusion.   Suppose the one-port is made of *passive* elements.   Then it is intuitively obvious that $P_{av}$ must be nonnegative.†   Thus, the *driving-point* impedance $Z$ and the *driving-point* admittance $Y$ of *any* one-port made of *passive* elements satisfy the inequalities

---

† This will be proved in Chap. 9.

(7.8a)  $\text{Re}\,[Z(j\omega)] \geq 0 \qquad \text{Re}\,[Y(j\omega)] \geq 0 \qquad$ for all $\omega$

Or from Eq. (7.6), $\cos(\angle V - \angle I) \geq 0$, which is equivalent to

(7.8b)  $|\angle Z(j\omega)| \leq 90° \qquad |\angle Y(j\omega)| \leq 90° \qquad$ for all $\omega$

Equations (7.8a) and (7.8b) are so important that they will be derived by a different method in Chap. 9.

## 7.2 Additive Property of Average Power

Suppose the one-port $\mathfrak{N}$ is driven by an input which is the sum of several sinusoids at *different* frequencies, and suppose that the one-port is in the steady state; then each sinusoidal input produces a sinusoidal output at the same frequency, and the output consists of a sum of sinusoids. Suppose that the input current is

$$i(t) = I_{1m} \cos(\omega_1 t + \psi_1) + I_{2m} \cos(\omega_2 t + \psi_2)$$

and that the input impedance is a known function $Z(j\omega)$; then, in the steady state

$$v(t) = I_{1m}|Z(j\omega_1)| \cos[\omega_1 t + \psi_1 + \angle Z(j\omega_1)]$$
$$+ I_{2m}|Z(j\omega_2)| \cos[\omega_2 t + \psi_2 + \angle Z(j\omega_2)]$$

For simplicity we shall write $v(t)$ in the form

$$v(t) = V_{1m} \cos(\omega_1 t + \phi_1) + V_{2m} \cos(\omega_2 t + \phi_2)$$

where

$$\phi_1 \overset{\Delta}{=} \psi_1 + \angle Z(j\omega_1)$$
$$\phi_2 \overset{\Delta}{=} \psi_2 + \angle Z(j\omega_2)$$

The instantaneous power entering the one-port $\mathfrak{N}$ is

$$p(t) = v(t)i(t) = \tfrac{1}{2}V_{1m}I_{1m}\cos(\phi_1 - \psi_1) + \tfrac{1}{2}V_{2m}I_{2m}\cos(\phi_2 - \psi_2)$$
$$+ \tfrac{1}{2}V_{1m}I_{1m}\cos(2\omega_1 t + \phi_1 + \psi_1)$$
$$+ \tfrac{1}{2}V_{2m}I_{2m}\cos(2\omega_2 t + \phi_2 + \psi_2)$$
$$+ \tfrac{1}{2}V_{1m}I_{2m}\cos[(\omega_1 + \omega_2)t + \phi_1 + \psi_2]$$
$$+ \tfrac{1}{2}V_{1m}I_{2m}\cos[(\omega_1 - \omega_2)t + \phi_1 - \psi_2]$$
$$+ \tfrac{1}{2}V_{2m}I_{1m}\cos[(\omega_1 + \omega_2)t + \psi_1 + \phi_2]$$

(7.9) $$+ \tfrac{1}{2}V_{2m}I_{1m}\cos[(\omega_1 - \omega_2)t + \psi_1 - \phi_2]$$

Equation (7.9) shows that the *instantaneous power* is *not* the sum of the instantaneous power due to the currents at $\omega_1$ and $\omega_2$ acting alone. Indeed, the sum consists only of the first four terms of the right-hand side of Eq. (7.9). On the other hand, the *average power*† is the sum of the average power at $\omega_1$ and the average power at $\omega_2$. In fact, once averaging has been accomplished, only the first two terms of the right-hand side remain. In other words, in the steady state, superposition holds for the *average* power provided the frequencies are different.

**Exercise**    Show, by an example, that if two sinusoidal sources have the *same* frequency and deliver power to the same linear time-invariant circuit, the average power delivered by both sources acting together is not necessarily equal to the sum of the average power delivered by each source acting alone. Call $Z$ the driving-point impedance of the circuit at the frequency of interest.

**7.3    Effective or Root-Mean-Square Values**

Consider the sinusoidal steady-state response of a linear time-invariant resistor with resistance $R$. From Eq. (7.1),

$$p(t) = v(t)i(t) = R^2 i(t) = R^2 I_m{}^2 \cos^2{(\omega t + \psi)}$$

The average power is, from Eq. (7.6) or (7.7),

$$P_{av} = \tfrac{1}{2} I_m{}^2 R = \tfrac{1}{2} I_m V_m$$

Let us define the **effective value of a sinusoidal waveform** to be its amplitude or peak value divided by $\sqrt{2}$. Thus,

(7.10)    $$I_{\text{eff}} \triangleq \frac{I_m}{\sqrt{2}} \qquad V_{\text{eff}} \triangleq \frac{V_m}{\sqrt{2}}$$

---

† The calculation of the average of the right-hand side of Eq. (7.9) is not always a trivial matter. Consider the case in which a single sinusoid is present [Eq. (7.5)]; in that case, the right-hand side is a periodic function, and $T \triangleq 2\pi/\omega$ is its period. So the average power $P_{av}$ is given by (7.6a). The situation of Eq. (7.9) is simple if the frequencies $\omega_1$ and $\omega_2$ are harmonically related; i.e., if there are *integers* $n_1$ and $n_2$ such that $n_1\omega_1 = n_2\omega_2$. Consider the least common multiple of $n_1$ and $n_2$, and call it $n$. Let $p_1 \triangleq n/n_1$ and $p_2 \triangleq n/n_2$; $p_1$ and $p_2$ are integers. Then sinusoids of frequencies $\omega_1$, $\omega_2$, $\omega_1 + \omega_2$, and $\omega_1 - \omega_2$ have a *common* period $T_c = p_1 \, (2\pi/\omega_1) = p_2(2\pi/\omega_2)$. Therefore, the right-hand side of (7.9) is periodic over the period $T_c$. It is therefore calculated by Eq. (7.6a), in which $T$ is replaced by $T_c$, and the results stated in the text follow immediately. If the frequencies $\omega_1$ and $\omega_2$ are not harmonically related (for example, if $\omega_1 = 1$ rad/sec and $\omega_2 = \sqrt{2}$ rad/sec), then the right-hand side of Eq. (7.9) is not a periodic function, and (7.6a) cannot be applied. The concept of average power can still be defined by a limiting process as

$$P_{av} \triangleq \lim_{T \to \infty} \frac{1}{2T} \int_{-T}^{+T} p(t) \, dt$$

The results stated in the text follow directly from this modified definition. It requires, however, some lengthy calculation.

then

(7.11)    $P_{av} = I_{eff}^2 R = I_{eff} V_{eff}$

For example, the common domestic line voltage is 110 volts effective, which corresponds to an amplitude of $110\sqrt{2}$ volts. Similarly, many voltmeters and ammeters give readings in terms of effective values. To obtain the amplitude, or peak value, we have to multiply the effective value by $\sqrt{2}$.

For a nonsinusoidal but *periodic waveform*, the effective value can be defined in terms of the following integrals:

(7.12a)    $I_{eff} \triangleq \left[ \dfrac{1}{T} \displaystyle\int_0^T i^2(t)\, dt \right]^{1/2}$

(7.12b)    $V_{eff} \triangleq \left[ \dfrac{1}{T} \displaystyle\int_0^T v^2(t)\, dt \right]^{1/2}$

where $i(\,\cdot\,)$ and $v(\,\cdot\,)$ are periodic functions with period $T$. The significance of the definitions (7.12a) and (7.12b) is that the average power delivered by a periodic function to a resistor with resistance $R$ is equal to

$$P_{av} = I_{eff}^2 R = \frac{V_{eff}^2}{R} = I_{eff} V_{eff}$$

This is clear since $P_{av}$, by the definition given in Eq. (7.6), is

(7.13)    $P_{av} = \dfrac{1}{T} \displaystyle\int_0^T p(t)\, dt = \dfrac{1}{T} \displaystyle\int_0^T v(t)i(t)\, dt$

(7.14)    $= \dfrac{1}{T} \displaystyle\int_0^T Ri^2(t)\, dt = \dfrac{1}{T} \displaystyle\int_0^T \dfrac{v^2(t)}{R}\, dt$

Comparing Eq. (7.12) with (7.14), we obtain immediately Eq. (7.13). In Eq. (7.12) the effective values are defined in terms of the square root of the mean of the squares of the voltage and current values; hence the name "root-mean-square" value.

## 7.4    Theorem on the Maximum Power Transfer

A problem of great practical importance is illustrated by Fig. 7.3. In this circuit, $Z_s$ represents a *given* passive impedance, and $V_s$ is the phasor representing the *given* sinusoidal voltage source at angular frequency $\omega$. Thus,

$v_s(t) = \text{Re}\, (V_s \epsilon^{j\omega t})$

The impedance $Z_L$ represents a passive load impedance whose value must be selected so that the average power entering the load impedance $Z_L$ (in the sinusoidal steady state) is maximum. For example, we might wish to

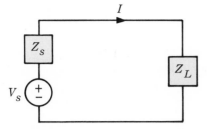

**Fig. 7.3**  A circuit illustrating power transfer from a source to a load.

design the first stage of a radar or radio telescope.  The voltage source $v_s$ represents the incoming electromagnetic wave, and the impedance $Z_s$ is the impedance of free space, the cables, waveguides, etc., leading to the first stage.  The problem is to select the best input impedance $Z_L$ for the first stage so that as much power as possible is fed to that stage.

The maximum-power-transfer theorem states that the *optimum load impedance $Z_{Lo}$ is equal to the complex conjugate of $Z_s$, that is, $Z_{Lo} = \overline{Z}_s$.*

*Proof*  All the calculations that follow involve impedances at the angular frequency $\omega$ of the source.  To simplify the notation, we shall write $Z_L$ for $Z_L(j\omega)$.  In terms of the current phasor $I$, the average power delivered to $Z_L$ is

$$P_{av} = \tfrac{1}{2}|I|^2 \operatorname{Re}(Z_L)$$

Since

$$I = \frac{V_s}{Z_s + Z_L}$$

it follows that

$$P_{av} = \frac{1}{2}|V_s|^2 \frac{\operatorname{Re}(Z_L)}{|Z_s + Z_L|^2}$$

Letting the real and imaginary parts of $Z_s$ and $Z_L$ be $R_s$, $R_L$, $X_s$ and $X_L$, respectively, we have

$$P_{av} = \frac{1}{2}|V_s|^2 \frac{R_L}{(R_L + R_s)^2 + (X_L + X_s)^2}$$

Here $V_s$, $R_s$, $X_s$ are given and $R_L$ and $X_L$ have to be chosen to maximize $P_{av}$.  Since the reactance $X_L$ can either be positive or negative, we can choose $X_L = -X_s$ so that the term $(X_L + X_s)^2$ in the denominator becomes zero.  For example, let $Z_s$ be the series connection of a resistor and an inductor with an inductance of 1 henry, and let $\omega = 2$ rad/sec.  Then $X_s = \omega L = 2$ ohms.  The required $X_L$ is $-2$ ohms, which can be realized by a capacitor of $\frac{1}{4}$ farad.  With this choice of $X_L$, $P_{av}$ becomes

(7.15)   $P_{\text{av}} = \dfrac{1}{2}|V_s|^2 \dfrac{R_L}{(R_L + R_s)^2}$

We must now determine the optimum $R_L$. Taking the partial derivative of $P_{\text{av}}$ with respect to $R_L$, we obtain

(7.16)   $\dfrac{\partial P_{\text{av}}}{\partial R_L} = \dfrac{1}{2}|V_s|^2 \dfrac{(R_L + R_s)^2 - 2(R_L + R_s)R_L}{(R_L + R_s)^4}$

For optimum $P_{\text{av}}$, we set $\partial P_{\text{av}}/\partial R_L$ equal to zero. Thus, we find from (7.16) that $R_L = R_s$. The maximum power is, from (7.15),

(7.17)   $\max P_{\text{av}} = \dfrac{|V_s|^2}{8R_s}$

and is attained when

(7.18)   $Z_{Lo} = R_s - jX_s = \bar{Z}_s$

When this condition is satisfied, we say that the load impedance is **conjugately matched** with the source impedance or, for short, that the load is **matched** to the source.

Equation (7.17) gives the maximum average power delivered to the load. It will be interesting to compare it with the average power delivered by the source. Clearly, the average power delivered by the source is given by

(7.19)   $P_s = \frac{1}{2}|I|^2 \operatorname{Re}(Z_s + Z_L)$

Under the conjugately matched condition of (7.18),

$I = \dfrac{V_s}{Z_{Lo} + Z_s} = \dfrac{V_s}{2R_s}$

Thus, Eq. (7.19) becomes

(7.20)   $P_s = \dfrac{1}{2}\dfrac{|V_s|^2}{4R_s^2}2R_s = \dfrac{|V_s|^2}{4R_s}$

We may define the *efficiency* of the circuit as the ratio of the average power delivered to the load and the average power delivered by the source. Comparing Eq. (7.17) with Eq. (7.20), we find the efficiency of the conjugately matched circuit is 50 percent. For radars and radio telescopes this fact is of no importance because the energy in the incoming electromagnetic wave would be lost if it were not absorbed by the first stage. For power engineers the situation is reversed. The energy delivered by the source costs money, and power companies are extremely interested in efficiency; they want to deliver as much of the average power they produce to the load (i.e., the customer). Consequently, large alternators are never conjugately matched.

**7.5**   **_Q_ of a Resonant Circuit**

We shall give an energy interpretation of the quality factor $Q$ of a resonant circuit. For the parallel resonant circuit shown in Table 7.1 (page 316),

$$Q \triangleq \frac{\omega_0}{2\alpha} = \omega_0 CR$$

If $V$ is the voltage phasor *at resonance,* we can write

(7.21) $\quad Q = \omega_0 \dfrac{\frac{1}{2}C|V|^2}{\frac{1}{2}G|V|^2}$

The denominator $\frac{1}{2}G|V|^2$ is the average power dissipated in the resistor at resonance. To interpret the numerator, recall that in Chap. 2 we showed that the electric energy stored in a linear capacitor is

(7.22) $\quad \mathcal{E}_E(t) = \frac{1}{2}Cv_C^2(t)$

and that the magnetic energy stored in a linear inductor is

(7.23) $\quad \mathcal{E}_M(t) = \frac{1}{2}Li_L^2(t)$

For the resonant circuit at resonant frequency, the voltage across the capacitor is

(7.24) $\quad v_C(t) = \mathrm{Re}\,(Ve^{j\omega_0 t}) = |V|\cos(\omega_0 t + \angle V)$

and the current in the inductor is

$$i_L(t) = \mathrm{Re}\left(\frac{V}{j\omega_0 L}e^{j\omega_0 t}\right) = \frac{|V|}{\omega_0 L}\cos(\omega_0 t + \angle V - 90°)$$

(7.25)
$$= \frac{|V|}{\omega_0 L}\sin(\omega_0 t + \angle V)$$

The total stored energy is, from Eqs. (7.22) to (7.25),

$$\mathcal{E}(t) = \mathcal{E}_E(t) + \mathcal{E}_M(t)$$

$$= \frac{1}{2}C|V|^2\cos^2(\omega_0 t + \angle V) + \frac{1}{2}L\frac{|V|^2}{\omega_0^2 L^2}\sin^2(\omega_0 t + \angle V)$$

Since $\omega_0^2 = 1/(LC)$,

(7.26) $\quad \mathcal{E}(t) = \frac{1}{2}C|V|^2$

Thus, at resonance, the total energy stored is *constant,* meaning that the total stored energy $\mathcal{E}(t)$ does not depend on $t$. From Eq. (7.21) we can interpret $Q$ as follows: *at resonance*

(7.27) $\quad Q = \omega_0 \dfrac{\text{energy stored}}{\text{average power dissipated in the resistor}}$

This formula also holds for the series *RLC* circuit at resonance.

**Exercise** Show that for the parallel $RLC$ circuit, at resonance

(7.28) $\quad Q = 2\pi \dfrac{\text{energy stored}}{\text{energy dissipated in one cycle}}$

Note that at resonance the period of all waveforms is $2\pi/\omega_0$ sec.

---

| 8 | **Impedance and Frequency Normalization** |

The resonant circuits we studied in Sec. 6 have three parameters, namely resistance, inductance, and capacitance. Such resonant circuits are usually used as filters. A typical design problem might read as follows: Design a series resonant circuit with impedance level $Z_0$ (i.e., the impedance at resonance), resonant frequency $\omega_0$, and a 3-db bandwidth $\Delta\omega$, where $Z_0$, $\omega_0$, and $\Delta\omega$ are assigned numerical values. Let us use Table 7.1 to write down the relations between the specified items and the element values $L$, $R$, and $C$ for the series resonant circuit. We find

(8.1a) $\quad$ Impedance level $= Z_0 = R$

(8.1b) $\quad$ Angular resonant frequency $= \omega_0 = \dfrac{1}{\sqrt{LC}}$

(8.1c) $\quad$ 3-db bandwidth $= \Delta\omega = \dfrac{\omega_0}{Q} = \dfrac{R}{L}$

To find $L$, we use (8.1c), and obtain

(8.2a) $\quad L = \dfrac{R}{\Delta\omega}$

To find $C$, we use (8.1b), and obtain

(8.2b) $\quad C = \dfrac{\Delta\omega}{\omega_0^2 R}$

There is an alternate design procedure which is usually preferred by experienced designers. This procedure starts with the design of a *normalized* series resonant circuit, i.e., a series resonant circuit with an impedance level equal to 1 ohm, an angular resonant frequency equal to 1 rad/sec, and a fractional bandwidth

(8.3) $\quad \dfrac{\Delta\omega}{\omega_0} = \dfrac{1}{Q}$

Let $L_0$, $R_0$, and $C_0$ be the element values of the normalized circuit. From Eqs. (8.1a) and (8.2) we have

(8.4) $\quad R_0 = 1 \qquad L_0 = Q \qquad C_0 = \dfrac{1}{Q}$

To obtain the element values of the desired circuit, we have to make two adjustments.   First, change the impedance level to $Z_0$, and then change the resonant frequency to $\omega_0$.   It can be shown that the desired resistance is obtained by multiplying $R_0$ by $Z_0$; the desired inductance is obtained by multiplying $L_0$ by $Z_0/\omega_0$, and the desired capacitance is obtained by multiplying $C_0$ by $1/(Z_0\omega_0)$.   Thus, we end up with

$$(8.5a) \qquad R = Z_0$$

$$(8.5b) \qquad L = \frac{QZ_0}{\omega_0} = \frac{Z_0}{\Delta\omega}$$

$$(8.5c) \qquad C = \frac{1}{Q\omega_0 Z_0} = \frac{\Delta\omega}{Z_0\omega_0{}^2}$$

The final results, of course, agree with Eqs. (8.1) and (8.2).

There are two reasons for the popularity of normalized designs.   First, if an engineer has in his files a normalized design of a bandpass filter (of special desirable characteristics), he has at his finger tips the element values of any such bandpass filter with any impedance level and with any center frequency.   The second reason is the well-known plague of numerical computations.   It is far easier to add, subtract, multiply, and divide numbers whose values are within an order of magnitude of unity.   Furthermore the roundoff errors which always occur in numerical calculations are much less severe.   Circuits encountered in practice often exhibit resistances in the hundreds of ohms, capacitances in the picofarads, inductances in the microhenrys, and frequencies in the megacycles.   It turns out that by impedance and frequency normalization these element values can usually be brought to within an order of magnitude of unity, making otherwise long and tedious computations relatively simple.

Let us state the general rule that has to be applied in order to obtain the desired $R$, $L$, and $C$ element values of an arbitrary network from the normalized element values $R_0$, $L_0$, and $C_0$ of the normalized network .   Let $r_n$ be the impedance normalization factor; more precisely, let

$$r_n \triangleq \frac{\text{desired impedance level}}{\text{impedance level of normalized design}}$$

Let $\Omega_n$ be the frequency normalization factor; more precisely, let

$$\Omega_n \triangleq \frac{\text{desired typical frequency}}{\text{typical frequency of normalized design}}$$

Then, the desired element values are given by

$$(8.6a) \qquad R = r_n R_0$$

$$(8.6b) \qquad L = \frac{r_n}{\Omega_n} L_0$$

$$(8.6c) \qquad C = \frac{C_0}{r_n \Omega_n}$$

A systematic proof of this rule must be based on the general methods of analysis, which will be covered in Chaps. 10 and 11. We may, however, give an heuristic justification of the three relations above. For simplicity, we consider a normalized network $\mathfrak{N}_0$ which contains no sources. We may think of proceeding from the normalized element values of $\mathfrak{N}_0$ to the desired element values in two steps. First, the impedance level is adjusted, and second, the frequency scale is adjusted. Consider the first step. Starting with the normalized network $\mathfrak{N}_0$, we multiply the impedance of each element by $r_n$ and obtain the network $\mathfrak{N}'$. Every resistance and inductance of $\mathfrak{N}'$ is $r_n$ times larger than the corresponding resistance and inductance of $\mathfrak{N}_0$; every capacitance of $\mathfrak{N}'$ is $r_n$ times smaller than the corresponding one of $\mathfrak{N}_0$. Observe that if we drive $\mathfrak{N}_0$ and $\mathfrak{N}'$ by two identical current sources connected at corresponding node pairs, then the node voltages of $\mathfrak{N}'$ are $r_n$ times the corresponding node voltages of $\mathfrak{N}_0$.

The second step is the adjustment of frequency. The network $\mathfrak{N}''$ is obtained from $\mathfrak{N}'$ by dividing all inductances and capacitances by $\Omega_n$. Observe that the impedance at the frequency $\omega''$ of any branch $\mathfrak{N}''$ is still $r_n$ times the impedance at the frequency $\omega'$ of the corresponding branch of $\mathfrak{N}_0$, where $\omega''/\omega' = \Omega_n$. Thus, if the two networks $\mathfrak{N}''$ and $\mathfrak{N}_0$ are driven at corresponding node pairs by two sinusoidal current sources of frequency $\omega''$ and $\omega'$, respectively, and if both networks are in the sinusoidal steady state, then any node-pair voltage of $\mathfrak{N}''$ is represented by a phasor which is $r_n$ times the phasor representing the corresponding node-pair voltage of $\mathfrak{N}_0$.

---

**Example**   Fig. 8.1$a$ shows a low-pass filter whose transfer impedance, defined by $E_2(j\omega)/I_1(j\omega)$, is such that

$$\left|\frac{E_2}{I_1}\right|^2 = \frac{1}{1 + \omega^6}$$

In other words, the gain of the filter $|E_2/I_1|$ is 1 at $\omega = 0$, $1/\sqrt{2}$ at $\omega = 1$, and monotonically decreases to zero as $\omega \to \infty$. For this reason it is called a low-pass filter. It is obvious from the figure that the input impedance of the filter (at $\omega = 0$) is 1 ohm; indeed, at zero frequency the impedance of the capacitors is infinite (open circuit), and the impedance of the inductance is zero (short circuit). Suppose we want an impedance level of 600 ohms with $|E_2/I_1|$ equal to $1/\sqrt{2}$ at 3.5 kHz. Then $r_n = 600$, and $\Omega_n = 2\pi 3.5 \times 10^3 = 2.199 \times 10^4$. The desired element values are easily obtained from Eq. (8.6). The desired filter and its response are shown in Fig. 8.1$b$.

With impedance normalization we have completed our first study of the sinusoidal steady state. In later chapters we shall repeatedly use the methods of this chapter and investigate in detail the properties of network functions.

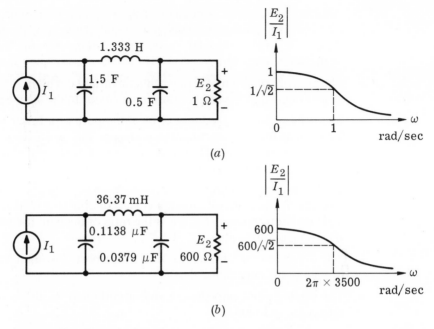

**Fig. 8.1** Low-pass filter illustrating impedance and frequency normalization. (*a*) Normalized design; (*b*) actual design.

## Summary

- A sinusoidal waveform (with angular frequency $\omega$),

$$x(t) = A_m \cos(\omega t + \phi)$$

can be represented by a phasor

$$A \triangleq A_m \epsilon^{j\phi}$$

according to

$$x(t) = \text{Re}(A\epsilon^{j\omega t}) = \text{Re}(A_m \epsilon^{j(\omega t + \phi)})$$

- Conversely, given the phasor $A = A_m \epsilon^{j\phi}$ and the angular frequency $\omega$, we can determine uniquely the sinusoidal waveform $x(\cdot)$; thus,

$$x(t) = \text{Re}(A\epsilon^{j\omega t})$$

$$= A_m \cos(\omega t + \phi)$$

- For linear time-invariant circuits, if the natural frequencies are all in the open left half of the complex frequency plane, the circuit is said to be asymptotically stable.

- For asymptotically stable linear time-invariant circuits, the sinusoidal steady-state response is defined as the response to a sinusoidal input as $t \to \infty$. The sinusoidal steady state is independent of the initial state of the circuit. The sinusoidal steady-state response has the same frequency as the sinusoidal input.

- The network function of a linear time-invariant circuit in sinusoidal steady state is defined as the ratio of the *output phasor* to the *input phasor*.

- The driving-point impedance $Z$ of a one-port with linear time-invariant elements is the network function corresponding to a current source input and voltage response. Thus, it is the ratio of the output-voltage phasor to the current-source phasor.

- The driving-point admittance $Y$ of a one-port with linear time-invariant elements is the network function corresponding to a voltage source input and current response. Thus, it is the ratio of the output-current phasor to the voltage-source phasor.

- The driving-point admittance $Y$ of a one-port $\mathfrak{N}$ is the reciprocal of the driving-point impedance $Z$ of $\mathfrak{N}$.

- The driving-point impedances and admittances for the basic circuit elements are as follows:

|  | $Z(j\omega)$ | $Y(j\omega)$ |
|---|---|---|
| Resistor | $R$ | $G$ |
| Inductor | $j\omega L$ | $\dfrac{1}{j\omega L}$ |
| Capacitor | $\dfrac{1}{j\omega C}$ | $j\omega C$ |

- The *impedance* of a *series* connection of one-ports is the sum of the impedances of the individual one-ports. The *admittance* of a *parallel* connection of one-ports is the sum of the admittance of the individual one-ports.

- Given the network function $H(j\omega) = V_0/I_s$, if the input is the sinusoidal waveform $i_s(t) = |I_s| \cos(\omega t + \phi)$, then the sinusoidal steady-state response is

$$v_0(t) = |H(j\omega)||I_s| \cos[\omega t + \phi + \angle H(j\omega)]$$

that is, the amplitude of the output is obtained by taking the product of the magnitude of the network function and the amplitude of the input. The phase of the output is obtained by adding the phase of the network function and the phase of the input.

- The curves of magnitude and phase versus $\omega$ are called the frequency response of a circuit for a specified input and output.

- In the sinusoidal steady state, if the port voltage and port current of a one-port $\mathfrak{N}$ are

$$v(t) = \mathrm{Re}\,(Ve^{j\omega t})$$
$$i(t) = \mathrm{Re}\,(Ie^{j\omega t})$$

then the *average* power delivered to the one-port is

$$P_{\mathrm{av}} = \tfrac{1}{2}|V||I|\cos(\angle V - \angle I)$$
$$= \tfrac{1}{2}\,\mathrm{Re}\,(V\bar{I})$$
$$= \tfrac{1}{2}|I|^2\,\mathrm{Re}\,[Z(j\omega)]$$
$$= \tfrac{1}{2}|V|^2\,\mathrm{Re}\,[Y(j\omega)]$$

where $Z(j\omega)$ and $Y(j\omega)$ are, respectively, the driving-point impedance and admittance of $\mathfrak{N}$.

- In a linear time-invariant circuit in the steady state, the total *average* power delivered by several sinusoidal sources of *different* frequencies is equal to the sum of the average power delivered by each source if it were driving the circuit alone.

## Problems

Phasor representations

**1.** Determine the phasors which represent the following real-valued time functions:

  a.  $10\cos(2t + 30°) + 5\sin 2t$

  b.  $\sin(3t - 90°) + \cos(3t + 45°)$

  c.  $\cos t + \cos(t + 30°) + \cos(t + 60°)$

Phasor calculation

**2.** The linear time-invariant circuit shown in Fig. P7.2 is in sinusoidal steady state.

  a.  Calculate the phasors representing the following sinusoidal functions of time: $i_s(t)$, $i_L(t)$, $i_R(t)$, $i_C(t)$, and $v(t)$ (that is, $I_s$, $I_L$, $I_R$, $I_C$, and $V$).

  b.  Write down expressions for the real-valued functions of time $i_s(t)$, $i_L(t)$, $i_R(t)$, $i_C(t)$, and $v(t)$, and sketch them to scale.

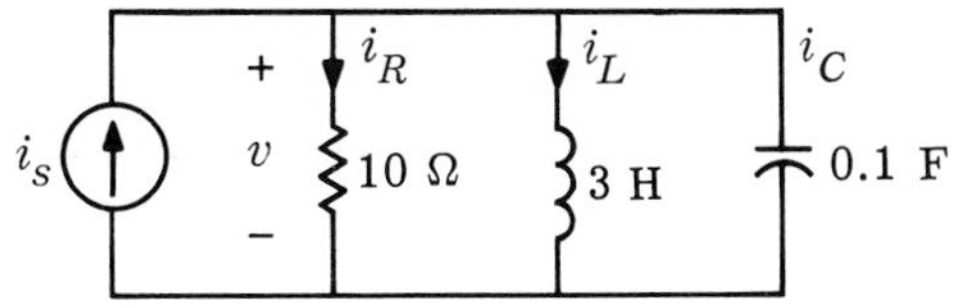

$$i_s(t) = 10\cos\left(2t - \frac{\pi}{3}\right)$$

**Fig. P7.2**

Nonlinear resistor and harmonics

**3.** Let $v$ be the voltage across the nonlinear resistor whose characteristic is

$$v = 50i^3$$

Calculate the voltage $v$ when a current $i = 0.01 \cos 377t$ flows through the nonlinear resistor (express the result as a sum of sinusoids). What frequencies are present in the output?

Nonlinear capacitor and subharmonics

**4.** Consider the nonlinear time-invariant subharmonic generating circuit shown in Fig. P7.4. The inductor is linear and the capacitor has a characteristic

$$v_C = \tfrac{1}{18}q + \tfrac{2}{27}q^3$$

a.   Verify that for an input $e_s = \tfrac{1}{54} \cos t$ volt, a response $q(t) = \cos (t/3)$ coul satisfies the differential equation. (Note that the charge oscillates at *one-third* of the frequency of the source.)

b.   Calculate the current through the source for the charge found in (a).

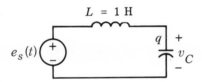

**Fig. P7.4**

Linear time-varying resistor

**5.** Consider the linear time-varying circuit shown in Fig. P7.5. Calculate the voltage $v$ when a current $i_s(t) = 10^{-2} \cos [2\pi 60t + (\pi/6)]$ amp flows through the circuit. (Express the result as a sum of sinusoids.)

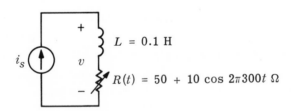

**Fig. P7.5**

Phasor and differential equations

**6.** Find the steady-state solutions of the following differential equations:

a.   $\dfrac{d^2x}{dt^2} + 3\dfrac{dx}{dt} + 10x = \cos (2t + 45°)$

b.   $\dfrac{d^3x}{dt^3} + 6\dfrac{d^2x}{dt^2} + 11\dfrac{dx}{dt} + 6x = \sin 2t$

c.  $\dfrac{d^2x}{dt^2} + 4\dfrac{dx}{dt} + x = \left(\dfrac{d}{dt} + 1\right)\cos 3t$

**7.** Find the complete solution of the following differential equations. Indicate whether the steady-state solution exists for each case.

a.  $\dfrac{d^2x}{dt^2} + \dfrac{dx}{dt} + x = \left(\dfrac{d}{dt} + 1\right)\cos 2t$

$x(0-) = 1 \qquad \dfrac{dx}{dt}(0-) = -1$

b.  $\dfrac{d^2x}{dt^2} + 2\dfrac{dx}{dt} + x = \sin 2t$

$x(0-) = 1 \qquad \dfrac{dx}{dt}(0-) = 2$

c.  $\dfrac{d^2x}{dt^2} + x = \cos 2t$

$x(0-) = 1 \qquad \dfrac{dx}{dt}(0-) = 0$

d.  $\dfrac{d^2x}{dt^2} + x = \cos t$

$x(0-) = 1 \qquad \dfrac{dx}{dt}(0-) = 0$

e.  $\dfrac{d^2x}{dt^2} - 3\dfrac{dx}{dt} + 2x = \cos t$

$x(0-) = 1 \qquad \dfrac{dx}{dt}(0-) = -1$

**8.** The circuit shown in Fig. P7.8 is made of linear time-invariant elements. The input is $e_s$, and the response is $v_C$. Knowing that $e_s(t) = \sin 2t$ volts, and at time $t = 0$ the state is $i_L = 2$ amp and $v_C = 1$ volt, calculate the complete response.

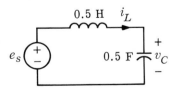

**Fig. P7.8**

**9.** The circuit shown in Fig. P7.9 has linear and time-invariant elements.

a.  Determine the driving-point impedance $Z(j\omega)$.

b.  Calculate the value of the impedance for values of $\omega = 0$ and 1 rad/sec.  (Express the impedance as a magnitude and an angle.)

c.  Explain by physical reasoning the value of the impedance for $\omega = 0$ and $\omega = \infty$.

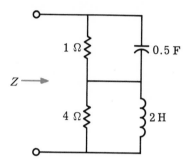

**Fig. P7.9**

Superposition in the steady state

**10.** For the circuit of Fig. P7.10, given that $i_s(t) = 1 + 2 \cos 2t$ for all $t$, determine the steady-state voltage $v$.

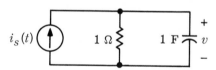

**Fig. P7.10**

Complete response and sinusoidal steady state

**11.** Let a sinusoidal voltage $e_s(t) = 3 \cos 10^6 t$ volts be applied at time $t = 0$ to the linear time-invariant $LC$ circuit shown in Fig. P7.11.

a.  Given $i(0) = 1$ mA and $v(0) = 0$, calculate and sketch $i(t)$ for $t \geq 0$.

b.  Suppose we have control over the phase $\phi$ of the generator voltage $e_s$; that is, suppose $e_s(t) = 3 \cos (10^6 t + \phi)$.  Find the appropriate value of $\phi$, if any, such that the response is of the form

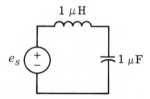

**Fig. P7.11**

$i(t) = 10^{-3} \cos 10^6 t + A \sin 10^6 t$

where $A$ is some constant.

**12.** Consider the linear time-invariant $LC$ series circuit shown in Fig. P7.12, where the input is a sinusoid, $e_s(t) = E_m \cos (t + \phi)$. Set up the differential equation for $v(t)$, and show that the voltage $v$ is not of the form $v(t) = \text{Re}(Ve^{jt})$, where $V$ is the phasor representing $v(t)$. Explain.

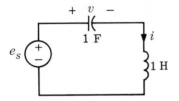

**Fig. P7.12**

**13.** Given, for all $t$,

$e(t) = 50 \sin (10t + \pi/4)$

$i(t) = 400 \cos (10t + \pi/6)$

Find suitable elements of the linear, time-invariant circuit shown in Fig. P7.13, and indicate their values in ohms, henrys, or farads.

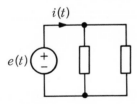

**Fig. P7.13**

**14.** The circuit shown in Fig. P7.14 is linear and time-invariant, and is in the sinusoidal steady state. Find, in terms of $R$ and $C$, the frequency $\omega$ at which $v_2(t)$ lags $45°$ behind $e_1(t)$. Find the amplitude of $v_2(t)$ at that frequency.

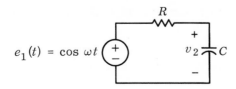

**Fig. P7.14**

Phasor
diagram

**15.** Assuming $e_C(t) = \cos 2t$, construct a phasor diagram showing all voltages and currents indicated (see Fig. P7.15). Find the sinusoidal steady-state voltage $e_1(t)$. (Express it as a real-valued function of time.)

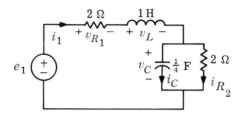

**Fig. P7.15**

Series connection of impedance

**16.** Determine the driving-point impedance $Z(j\omega)$ of the circuit shown in Fig. P7.16. If a sinusoidal voltage source $v_s(t) = 10 \cos 2t$ is applied to the one-port, determine the port current in the sinusoidal steady state.

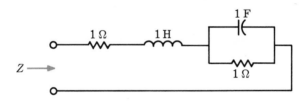

**Fig. P7.16**

Frequency response

**17.** Plot the magnitude and phase of the driving-point impedance $Z(j\omega)$ of the circuit in Fig. P7.16 versus $\omega$. If a current source $i_s(t) = 1 + \cos t + \cos 2t$ is applied to the one-port, determine the steady-state port voltage.

Impedance and admittance loci

**18.** Determine the real and imaginary parts of the impedance $Z(j\omega)$ of the circuit in Fig. P7.16. Determine and plot the susceptance as a function of $\omega$. Plot the loci of impedance and admittance of the one-port.

Ladder circuit and network functions

**19.** For the ladder circuit shown in Fig. P7.19,

*a.* Determine the driving point admittance $Y(j\omega)$.

*b.* Calculate the steady-state current $i_1$ due to the sinusoidal voltage source $e_s(t) = 2 \cos 2t$.

*c.* Determine the transfer impedance $Y_{21}(j\omega) = I_2/E_s$ where $I_2$ and $E_s$ are phasors which represent the sinusoidal current $i_2$ and sinusoidal voltage $e_s$, respectively.

*d.* Calculate the steady-state current $i_2$.

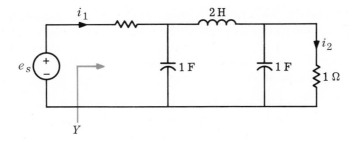

**Fig. P7.19**

Driving-point
admittance
and suscep-
tance plot

**20.** Determine the driving-point admittance $Y(j\omega)$ of the lossless circuit shown in Fig. P7.20.   Plot the susceptance versus $\omega$.   If a sinusoidal voltage source $e_s = \cos \omega t$ is applied to the one-port, what can you say about the current $i_1$ for $\omega = 0, 1, 2,$ and $\infty$?

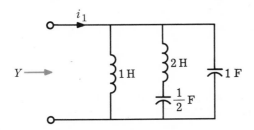

**Fig. P7.20**

Dual circuit

**21.** Determine the dual circuit of the circuit shown in Fig. P7.20.

Mesh analysis

**22.** Determine the sinusoidal steady-state current in the inductor and the sinusoidal steady-state voltage across the 1-farad capacitor for the circuit shown in Fig. P7.22.   The input voltage source is $e_s(t) = \cos 2t$.

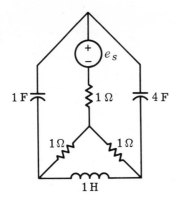

**Fig. P7.22**

Node analysis **23.** Change the series connection of the voltage source and the resistor in the circuit of Fig. P7.22 to the parallel connection of a current source and a resistor. Use node analysis to determine the sinusoidal steady-state current in the inductor and the sinusoidal steady-state voltage across the 1-farad capacitor.

Driving-point impedance and power **24.** *a.* Find the input impedance $Z_{in}(j\omega)$ at the frequency $\omega$.

*b.* If the input voltage is 10 cos $\omega t$ and the circuit is in the sinusoidal steady state, what is the instantaneous power (as a function of time) into the circuit? (See Fig. P7.24.)

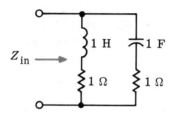

**Fig. P7.24**

Phasor, energy, and power **25.** The series *RLC* circuit shown in Fig. P7.25 is made of linear time-invariant elements.

*a.* Calculate by the phasor method the sinusoidal steady-state response $i$ to $e_s = \sin \omega t$ volts for values of $\omega = 2.0, 2.02,$ and $2.04$ rad/sec. Give each result as a real-valued function of time.

*b.* Calculate the energy stored in the capacitor $\mathscr{E}_E$ and in the inductor $\mathscr{E}_M$ as functions of time for $\omega = 2.00, 2.02,$ and $2.04$ rad/sec.

*c.* Calculate the average power dissipated in the resistor for $\omega = 2.00, 2.02,$ and $2.04$ rad/sec.

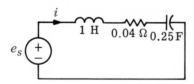

**Fig. P7.25**

Impedance, time response, and superposition **26.** The elements of the circuit shown in Fig. P7.26 are linear and time-invariant. Calculate and sketch the steady-state voltage $v$ as a function of time, given $i_s = 2 \sin t + \cos (3t + \pi/4)$. Explain the basic idea of your method of solution.

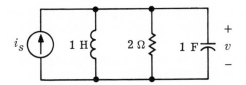

**Fig. P7.26**

Frequency
responses of
resonant
circuits

**27.** Consider the linear time-invariant one-port shown in Fig. P7.27. Calculate the impedances $Z_1(j\omega)$ and $Z_2(j\omega)$. What can be said about the relative shapes of the curves $|Z_1(j\omega)|$ and $|Z_2(j\omega)|$ and the curves $\angle Z_1(j\omega)$ and $\angle Z_2(j\omega)$ if we are interested only in frequencies lying within an octave of the resonant frequency?

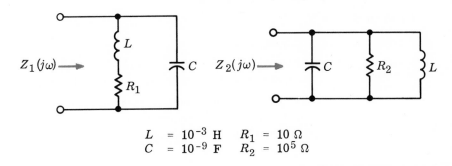

$$L = 10^{-3} \text{ H} \quad R_1 = 10 \ \Omega$$
$$C = 10^{-9} \text{ F} \quad R_2 = 10^5 \ \Omega$$

**Fig. P7.27**

Resonant cir-
cuit, $Q$, and
frequency
response

**28.** The one-port shown in Fig. P7.28 is made of linear time-invariant elements.

*a.* Calculate the resonant frequency $\omega_0$ and the value of $Q$.

*b.* Calculate the driving-point impedance $Z(j\omega)$.

*c.* Calculate graphically the magnitude and the phase angle of the impedance for the following values of $\omega/\omega_0$: $0, 1 - 3/2Q, 1 - 1/2Q$, $1, 1 + 1/2Q, 1 + 3/2Q$, and $2$.

*d.* Plot $|Z(j\omega)|$ and $\angle Z(j\omega)$ against $\omega/\omega_0$ from the results of (*c*).

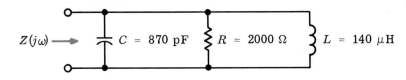

**Fig. P7.28**

Phasor dia-
gram and
power

**29.** The circuit shown in Fig. P7.29 is operating in the sinusoidal steady state. It is determined that $v_a = 10 \cos (1000t + 60°)$ and $v_b = 5 \cos (1000t - 30°)$. The magnitude of the impedance of the capacitance at this frequency is 10. Determine the impedance $Z(j1000)$ of the one-port network $\mathfrak{N}$ and the average power delivered to $\mathfrak{N}$.

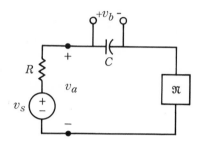

**Fig. P7.29**

Bandwidth of
resonant cir-
cuit, design

**30.** *a.*   Shown in Fig. P7.30 is the resonance curve [$|Z(j\omega)|$ in ohms versus $\omega$ in rad/sec] of a parallel $RLC$ circuit. Find $R$, $L$, and $C$.

   *b.*   The same resonance behavior is desired around a center frequency of 20 kHz. The maximum value of $|Z(j\omega)|$ is to be 0.1 M$\Omega$. Find the new values of $R$, $L$, and $C$.

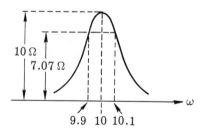

**Fig. P7.30**

# 8

# Coupling Elements and Coupled Circuits

In Chap. 2 we introduced three basic types of circuit elements, namely, resistors, capacitors, and inductors. All of these are two-terminal (or one-port) elements and hence are characterized by the equations relating their branch voltages to their branch currents. In Chaps. 3 through 7 we analyzed special circuits containing these two-terminal elements. Before presenting the general methods of network analysis, we wish to introduce some other useful circuit elements, namely coupled inductors, ideal transformers, and controlled sources (or dependent sources). These elements differ from resistors, inductors, and capacitors in that they have more than one branch and the branch voltages and currents of one branch are related to those of other branches. Because of this they are called **coupling elements.** In this chapter we shall give the characterization and properties of these elements. Furthermore, we shall introduce examples to demonstrate some methods of analysis for circuits containing coupling elements.

## 1 Coupled Inductors

Consider two coils of wire in close physical proximity to one another, as shown in Fig. 1.1. For present purposes it is of no importance whether or not the coils are wrapped around a common core of magnetic material. However, we do assume that the coils do not move with respect to one another or with respect to a core they might be wrapped around.

We adopt the *reference directions* for currents and voltages shown on Fig. 1.1. Note that these reference directions are *associated reference*

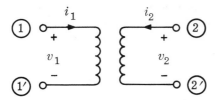

**Fig. 1.1** Coupled coils and their reference directions.

*directions* for each coil. They assert nothing whatsoever about the actual direction of current flow or about the relative voltages of the terminals. The reference directions are only necessary to keep track of the signs of the quantities which represent the actual phenomena.

For the record let us note that if we have some ferromagnetic material in the magnetic circuit of the two coils, then when the currents are sufficiently large, the relations between the fluxes $\phi_1$ and $\phi_2$ and the currents $i_1$ and $i_2$ are no longer linear. In this case the equations have the following form:

$$\phi_1 = f_1(i_1,i_2) \qquad \phi_2 = f_2(i_1,i_2)$$

where $f_1$ and $f_2$ are nonlinear functions of the currents $i_1$ and $i_2$. By Faraday's law

$$v_1 = \frac{d\phi_1}{dt} = \frac{\partial f_1}{\partial i_1}\frac{di_1}{dt} + \frac{\partial f_1}{\partial i_2}\frac{di_2}{dt}$$

$$v_2 = \frac{d\phi_2}{dt} = \frac{\partial f_2}{\partial i_1}\frac{di_1}{dt} + \frac{\partial f_2}{\partial i_2}\frac{di_2}{dt}$$

*Faraday's law*

It should be stressed that the four partial derivatives are functions of $i_1$ and $i_2$. Obviously, when we have such *nonlinear* flux-current equations, the problem is quite complicated. We shall not consider further nonlinear coupled coils until Chap. 17.

---

### 1.1   Characterization of Linear Time-invariant Coupled Inductors

Let us assume that we have a pair of linear time-invariant coupled inductors.† Since the inductors are linear, each flux must be a linear function of the currents; since the inductors are time-invariant, the coefficients of these linear functions must be constants (i.e., they do not depend explicitly on time). Thus, we can write

$$\phi_1(t) = L_{11}i_1(t) + M_{12}i_2(t)$$

$$\phi_2(t) = M_{21}i_1(t) + L_{22}i_2(t)$$

where the constants $L_{11}$, $L_{22}$, $M_{12}$, and $M_{21}$ do not depend on the time $t$ nor on the currents $i_1$ and $i_2$. $L_{11}$ is the **self-inductance** of the inductor ① ①' and $L_{22}$ is the **self-inductance** of the inductor ② ②'. $M_{12}$ and $M_{21}$ are called the **mutual inductances** for the coupled inductors ① ①' and ② ②'. All are measured in henrys when the currents are expressed in amperes and the fluxes are expressed in webers. In physics we learned

---

† We use "inductors" rather than "coils" to indicate that we are dealing with circuit models. Coils designate physical components which usually include a certain amount of energy dissipation and stray capacitance. Coils can be modeled with a combination of inductors, resistors, and capacitors.

from energy considerations that the two mutual inductances are always equal; that is, $M_{12} = M_{21}$. If we call $M$ their common value, we can write

(1.1)  $\phi_1 = L_{11}i_1 + Mi_2$

(1.2)  $\phi_2 = Mi_1 + L_{22}i_2$

It follows immediately from the Faraday law equations above that

(1.3)  $v_1 = L_1 \dfrac{di_1}{dt} + M \dfrac{di_2}{dt}$

(1.4)  $v_2 = M \dfrac{di_1}{dt} + L_2 \dfrac{di_2}{dt}$  $\qquad L_2 \dfrac{di_2}{dt} + M \dfrac{di_1}{dt}.$

In the sinusoidal steady state, if we represent the sinusoidal voltages $v_1$ and $v_2$ and currents $i_1$ and $i_2$ by their corresponding phasors $V_1$ and $V_2$ and $I_1$ and $I_2$, respectively, the equations become

(1.5)  $V_1 = j\omega L_{11}I_1 + j\omega M I_2$

(1.6)  $V_2 = j\omega M I_1 + j\omega L_{22}I_2$

where $\omega$ is the angular frequency.

**Remark**  Although the self-inductances $L_{11}$ and $L_{22}$ are always positive numbers, the mutual inductance $M$ can be either positive or negative depending on the winding of the coils.

*Sign of mutual inductance*  The sign of the mutual inductance $M$ can only be decided by careful consideration of the physical situation and the reference directions. Suppose we leave the second inductor ② ②′ open-circuited; that is, $i_2$ is identically zero. Equations (1.2) and (1.4) become

$\phi_2 = Mi_1$

and

$v_2 = M \dfrac{di_1}{dt}$

If an *increasing* current enters terminal ① of inductor 1, then $di_1/dt > 0$, and the sign of $v_2$ is clearly that of $M$ since $v_2 = M(di_1/dt)$. For example, it may be that terminal ② of inductor 2 is observed to be at a higher potential than terminal ②′. In that case, our reference direction implies that $v_2 > 0$; hence $M > 0$. Thus, the sign of $M$ depends both on the physical situation *and* on the reference directions selected.

Let us consider now the problem of determining the sign of $M$ from an energy viewpoint. We are given a specific pair of coupled inductors together with reference directions for both the currents and the voltages (see Fig. 1.2). We assume that the magnetic permeability of the core is much larger than that of the free space; under this condition almost all the magnetic energy is stored in the core. We wish to decide on the basis of this data whether $M$ in Eqs. (1.3) and (1.4) should be positive or negative.

We shall reach a decision rule by considering the magnetic energy stored when $i_1 > 0$ and $i_2 > 0$. In physics we learned that if $\vec{\mathbf{H}}$ is the magnetic field vector at some point $P$ in the magnetic core, then $(\mu/2)|\vec{\mathbf{H}}|^2\,dv$ is the magnetic energy stored in an element of volume $dv$ which includes the point $P$, where $\mu$ is the permeability of the core. Suppose that by the use of appropriate generators we have constant positive currents $i_1$ and $i_2$ flowing. Let $\vec{\mathbf{H}}_1$ be the magnetic field due to $i_1$ only, and let $\vec{\mathbf{H}}_2$ be the magnetic field due to $i_2$ only. Then the magnetic energy stored in $dv$ is

$$\frac{\mu}{2}|\vec{\mathbf{H}}_1 + \vec{\mathbf{H}}_2|^2\,dv = \left(\frac{\mu}{2}|\vec{\mathbf{H}}_1|^2 + \mu\vec{\mathbf{H}}_1 \cdot \vec{\mathbf{H}}_2 + \frac{\mu}{2}|\vec{\mathbf{H}}_2|^2\right)dv$$

where $\vec{\mathbf{H}}_1 \cdot \vec{\mathbf{H}}_2$ denotes the scalar product of the two vectors $\vec{\mathbf{H}}_1$ and $\vec{\mathbf{H}}_2$. In the equation, $(\mu/2)|\vec{\mathbf{H}}_1|^2\,dv$ is the magnetic energy stored due to the current $i_1$ only, and $(\mu/2)|\vec{\mathbf{H}}_2|^2\,dv$ is the magnetic energy stored due to the current $i_2$ only; then the term $\mu\vec{\mathbf{H}}_1 \cdot \vec{\mathbf{H}}_2\,dv$ is the magnetic energy stored due to the presence of both $i_1$ and $i_2$. Thus, if $\vec{\mathbf{H}}_1 \cdot \vec{\mathbf{H}}_2$ is positive (that is, if the angle between the magnetic fields $\vec{\mathbf{H}}_1$ and $\vec{\mathbf{H}}_2$ is less than 90° in absolute value, giving it a positive cosine) the energy stored in $dv$ when $i_1$ and $i_2$ flow *simultaneously* is larger than the sum of the energy stored in $dv$ when each $i_1$ and $i_2$ flow alone. For example, in Fig. 1.2 the right-hand rule indicates that $\vec{\mathbf{H}}_1$ and $\vec{\mathbf{H}}_2$ have the same direction; hence the energy stored in $dv$ when both $i_1$ and $i_2$ flow is *larger* than the sum of the energies when they flow alone.

Let us next calculate the energy stored not by considering the field, but by considering the circuit. Suppose, for simplicity, that $i_1(0) = 0$ and $i_2(0) = 0$; hence from Eqs. (1.1) and (1.2) the fluxes are zero at $t = 0$, and there is no energy stored at $t = 0$. The energy stored is a function of the instantaneous values of the currents $i_1$ and $i_2$, and we write the energy

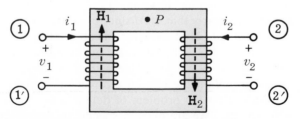

**Fig. 1.2**   Illustration of the determination of sign of $M$.

stored at time $t$ as $\mathcal{E}[i_1(t), i_2(t)]$.   The associated reference directions imply that $v_1(t)i_1(t)$ is the instantaneous power delivered *by* the external world *to* the inductor at terminals ① and ①′, and $v_2(t)i_2(t)$ is the instantaneous power delivered *by* the external world *to* the inductor at terminals ② and ②′.   Therefore,

$$\mathcal{E}[i_1(t), i_2(t)] = \int_0^t [v_1(t')i_1(t') + v_2(t')i_2(t')]\, dt'$$

From Eqs. (1.3) and (1.4), we obtain

$$\mathcal{E}[i_1(t), i_2(t)] = \int_0^t \left[ L_{11}i_1 \frac{di_1}{dt'} + M\left(i_1 \frac{di_2}{dt'} + i_2 \frac{di_1}{dt'}\right) + L_{22}i_2 \frac{di_2}{dt'} \right] dt'$$

Since $i_1(0)$ and $i_2(0)$ are assumed to be zero, we get

(1.7)   $\mathcal{E}[i_1(t), i_2(t)] = \tfrac{1}{2}L_{11}i_1{}^2(t) + Mi_1(t)i_2(t) + \tfrac{1}{2}L_{22}i_2{}^2(t)$

This relation may be rewritten as follows:

(1.8)   $\mathcal{E}(i_1, i_2) = \mathcal{E}(i_1, 0) + Mi_1i_2 + \mathcal{E}(0, i_2)$

where $\mathcal{E}(i_1, 0)$ is the energy stored when $i_2 = 0$ and a current $i_1$ flows in inductor 1, and $\mathcal{E}(0, i_2)$ is the energy stored when $i_1 = 0$ and a current $i_2$ flows in inductor 2.   We conclude from Eq. (1.8) that if $i_1$ and $i_2$ are positive and if $M > 0$, the total energy stored is larger than the sum of the energy stored when the current $i_1$ flows alone and when the current $i_2$ flows alone, respectively.   We have therefore justified the following rule to determine the sign of $M$:

*Let us consider a pair of coupled inductors and assign reference directions for the currents and voltages such that the power delivered to the inductors by the external world is $v_1(t)i_1(t) + v_2(t)i_2(t)$.   (This is guaranteed by the associated reference directions.)   If a current of 1 amp flows in each inductor along the reference direction and if the energy stored under these conditions is larger than the sum of the energies stored when each 1-amp current flows alone, then the mutual inductance $M$ is positive.*

Dots are often used in circuit diagrams as a convention to indicate the sign of $M$.   The convention is as follows:

*First, use associated reference directions for each inductor.   Second, assign one dot to one terminal of each inductor so that $M$ is positive when the reference directions of $i_1$ and $i_2$ both enter (or leave) the windings by the dotted terminals.*

In the situation shown in Fig. 1.2 for the given reference direction of $i_1$ and $i_2$, the mutual inductance $M$ is positive, and a pair of dots should be placed either at terminals ① and ②, or at terminals ①′ and ②′.

## 1.2   Coefficient of Coupling

For linear time-invariant two-winding coupled inductors, we need three parameters $L_{11}$, $L_{22}$, and $M$ to characterize the relations between the currents and fluxes. We know that the self-inductances $L_{11}$ and $L_{22}$ are always positive, whereas the mutual inductance $M$ can be either positive or negative. The ratio of the absolute value of the mutual inductance to the geometric mean of the two self-inductances is a measure of the degree of coupling. Thus, we define the **coefficient of coupling** of two-winding coupled inductors by the equality

$$k \triangleq \frac{|M|}{\sqrt{L_{11}L_{22}}}$$

*self ind. always positive*

The coefficient $k$ is a *nonnegative* number which is independent of the reference directions selected for the inductor currents. If the two inductors are a great distance apart in space, the mutual inductance is very small, and $k$ is close to zero. If the two inductors are tightly coupled, such as is the case when two windings are wound on the same core, most of the magnetic flux is common to both inductors, and $k$ is close to unity. We shall show, by examining the equations for the energy stored in the inductors, that the coefficient of coupling $k$ defined above is always less than or equal to unity. If it is equal to unity, we say that the inductors are **closely coupled.**

Let us examine the expression for the energy stored given by Eq. (1.7). We shall use the algebraic technique of completing the squares; thus,

$$\mathscr{E}(i_1,i_2) = \tfrac{1}{2}L_{11}i_1{}^2 + Mi_1i_2 + \tfrac{1}{2}L_{22}i_2{}^2$$

$$= \frac{1}{2}L_{11}\left(i_1 + \frac{M}{L_{11}}i_2\right)^2 + \frac{1}{2}\left(L_{22} - \frac{M^2}{L_{11}}\right)i_2{}^2$$

Note that the term $[i_1 + (M/L_{11})i_2]^2$ is always nonnegative for any $i_1$ and $i_2$. Recall the energy $\mathscr{E}(i_1,i_2)$ stored in the coupled inductors must be nonnegative for *any* choice of $i_1$ and $i_2$. It follows then that $L_{22} - M^2/L_{11}$ must be nonnegative. The proof is by contradiction. Let us assume that $L_{22} - M^2/L_{11}$ is negative, and let us choose, for example, $i_2 = 1$ and $i_1 = -M/L_{11}$; then $[i_1 + (M/L_{11})i_2]^2$ vanishes, $(L_{22} - M^2/L_{11})i_2{}^2$ is negative, and we arrive at the absurdity that $\mathscr{E}(-M/L_{11}, 1) < 0$. Consequently, we have the condition

$$L_{22} - \frac{M^2}{L_{11}} \geq 0$$

This is equivalent to $L_{11}L_{22} \geq M^2$, or

(1.9)   $$k = \frac{|M|}{\sqrt{L_{11}L_{22}}} \leq 1$$

In summary, energy considerations require that a pair of linear coupled inductors have positive self-inductances and a coefficient of coupling smaller than or equal to unity.

|     | 1.3 | **Multiwinding Inductors and Their Inductance Matrix** |
| --- | --- | --- |

If more than two linear time-invariant inductors are coupled together, as shown in Fig. 1.3, the relation between currents and fluxes is given by a set of linear equations as follows:

$(1.10a)$
$$\phi_1 = L_{11}i_1 + L_{12}i_2 + L_{13}i_3$$
$$\phi_2 = L_{21}i_1 + L_{22}i_2 + L_{23}i_3$$
$$\phi_3 = L_{31}i_1 + L_{32}i_2 + L_{33}i_3$$

In Eq. $(1.10a)$, $L_{11}$, $L_{22}$, and $L_{33}$ are self-inductances of inductors 1, 2, and 3, respectively. $L_{12} = L_{21}$, $L_{23} = L_{32}$, and $L_{13} = L_{31}$ are mutual inductances; more precisely, $L_{12}$ represents the mutual inductance between inductor 1 and inductor 2. It is sometimes more convenient to write Eq. $(1.10a)$ in the matrix form as

$(1.10b)$  $\phi = \mathbf{L}i$

where $\phi$ is called the flux vector, $\mathbf{i}$ the current vector, and $\mathbf{L}$ is a square matrix called the **inductance matrix.** Thus,

$(1.10c)$
$$\phi = \begin{bmatrix} \phi_1 \\ \phi_2 \\ \phi_3 \end{bmatrix} \quad \mathbf{i} = \begin{bmatrix} i_1 \\ i_2 \\ i_3 \end{bmatrix} \quad \mathbf{L} = \begin{bmatrix} L_{11} & L_{12} & L_{13} \\ L_{21} & L_{22} & L_{23} \\ L_{31} & L_{32} & L_{33} \end{bmatrix}$$

The inductance matrix $\mathbf{L}$ has its order equal to the number of inductors. Since the inductors are time-invariant, the elements of $\mathbf{L}$ (the $L_{ij}$'s) are constants. In terms of voltages, we have a voltage vector $\mathbf{v}$ related to $\mathbf{i}$ by

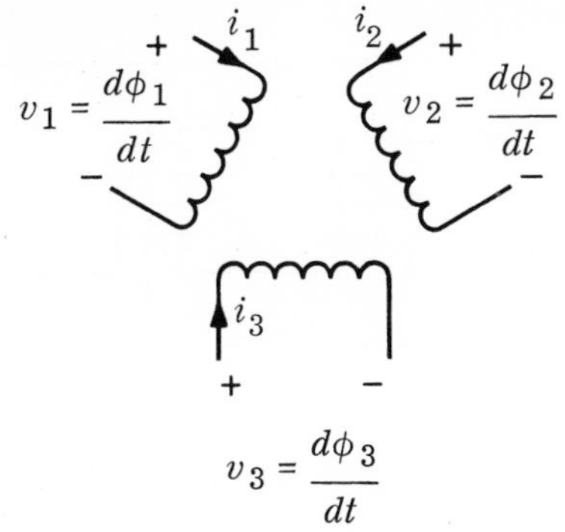

**Fig. 1.3**    Three-winding inductors.

(1.11)   $\mathbf{v} = \mathbf{L}\dfrac{d\mathbf{i}}{dt}$

Since the inductance matrix **L** is always symmetric ($L_{12} = L_{21}$, etc.), a set of three-winding coupled inductors is characterized by six parameters rather than by nine parameters.  The signs of the mutual inductances $L_{12}$, $L_{13}$, and $L_{23}$ can be determined by checking the direction of induced magnetic field.

**Example 1**   Figure 1.4a shows three inductors wound on the same iron core.  Let us choose the reference directions of the voltage and currents for the three pairs of terminals as shown.  We can determine the direction of the magnetic field caused by positive currents flowing in each inductor by the right-hand rule.  For example, the arrow marked $\vec{\mathbf{H}}_1$ indicates the direction of the magnetic field caused by a positive current $i_1$ while $i_2 = i_3 = 0$.  As shown in the figure $\vec{\mathbf{H}}_1$ and $\vec{\mathbf{H}}_2$ have the same direction, but $\vec{\mathbf{H}}_3$ has the opposite direction; thus, $L_{12}$ is positive while $L_{13}$ and $L_{23}$ are negative.

**Exercise**   In the three-winding coupled inductors of Fig. 1.4b determine the signs of the mutual inductances $L_{12}$, $L_{13}$, and $L_{23}$.

It is useful to define a **reciprocal inductance matrix** according to

$\boldsymbol{\Gamma} \triangleq \mathbf{L}^{-1}$

With this definition Eq. (1.10b) becomes

(1.12)   $\mathbf{i} = \boldsymbol{\Gamma}\boldsymbol{\phi}$

For example, the scalar equations for two-winding coupled inductors in terms of the elements of reciprocal inductance matrix are

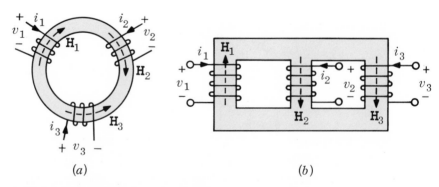

(a)                                    (b)

**Fig. 1.4**    Examples of coupled three-winding inductors.

(1.13)
$$i_1 = \Gamma_{11}\phi_1 + \Gamma_{12}\phi_2$$
$$i_2 = \Gamma_{21}\phi_1 + \Gamma_{22}\phi_2$$

where, from the relation of an inverse matrix or from Cramer's rule,

(1.14a)   $$\Gamma_{11} = \frac{L_{22}}{\det \mathbf{L}} \qquad \Gamma_{22} = \frac{L_{11}}{\det \mathbf{L}}$$

and

(1.14b)   $$\Gamma_{12} = \Gamma_{21} = \frac{-L_{12}}{\det \mathbf{L}}$$

where $\det \mathbf{L}$ denotes the determinant of the inductance matrix $\mathbf{L}$. The $\Gamma_{ij}$ are called **reciprocal inductances.** In terms of voltages, Eq. (1.13) becomes

(1.15a)
$$i_1(t) = \Gamma_{11} \int_0^t v_1(t')\, dt' + \Gamma_{12} \int_0^t v_2(t')\, dt' + i_1(0)$$
$$i_2(t) = \Gamma_{21} \int_0^t v_1(t')\, dt' + \Gamma_{22} \int_0^t v_2(t')\, dt' + i_2(0)$$

These equations express the inductor currents at any time $t$ in terms of the voltages and the initial currents. For this reason, in node analysis the reciprocal inductance matrix is more useful than the inductance matrix.

In the sinusoidal steady state, the current phasors $I_1$ and $I_2$ can be written in terms of the voltage phasors $V_1$ and $V_2$ as follows:

(1.15b)
$$I_1 = \frac{\Gamma_{11}}{j\omega} V_1 + \frac{\Gamma_{12}}{j\omega} V_2$$
$$I_2 = \frac{\Gamma_{21}}{j\omega} V_1 + \frac{\Gamma_{22}}{j\omega} V_2$$

where $\omega$ is the angular frequency.

**Remark**   As a final thought it should be stressed that the inductance matrix by itself does not completely specify the behavior of the branch voltages and currents. In order to write circuit equations correctly we must know both the inductance matrix *and* the reference directions of the inductor currents. The reference directions of the voltages follow from our previous convention that when we use associated reference directions, $v_1(t)i_1(t) + v_2(t)i_2(t)$ is the power delivered *by* the external world *to* the inductors.

## 1.4   Series and Parallel Connections of Coupled Inductors

We now turn our attention to the problem of calculating the equivalent inductance of two coupled linear inductors connected in series or in parallel.

**Example 2**

Figure 1.5a shows two coupled inductors connected in series. To determine the inductance between terminals ① and ②, let us first determine the sign of the mutual inductance. From the chosen reference directions of the two currents $i_1$ and $i_2$, we see that the magnetic fields $\vec{H}_1$ and $\vec{H}_2$ due to $i_1$ and $i_2$, respectively, are in the same direction, which makes $M$ positive. (The dot convention may also be used. Since both $i_1$ and $i_2$ enter their respective inductors at the dotted terminals, $M$ is positive.) From the current and flux equations, we have

(1.16)

$$\phi_1 = L_{11}i_1 + Mi_2 = 5i_1 + 3i_2$$

$$\phi_2 = Mi_1 + L_{22}i_2 = 3i_1 + 2i_2$$

where the inductances are expressed in henrys. Since the two inductors are connected in series, KCL requires that

$$i = i_1 = i_2$$

KVL requires that $v = v_1 + v_2$. Faraday's law states that $v_1 = d\phi_1/dt$ and $v_2 = d\phi_2/dt$. Thus, if $\phi$ is a flux such that $v = d\phi/dt$, we obtain

$$\frac{d\phi}{dt} = \frac{d\phi_1}{dt} + \frac{d\phi_2}{dt}$$

If initially the fluxes are zero, by integration we obtain

$$\phi = \phi_1 + \phi_2$$

and by Eq. (1.16)

$$\phi = 8i_1 + 5i_2 = 13i$$

Therefore, the inductance of the series connection is

$$L = \frac{\phi}{1} = 13 \text{ henrys}$$

Suppose that we connect the two inductors as shown in Fig. 1.5b; terminal ①′ of the first inductor is now connected to terminal ②′ of the second. To determine the inductance of the series connection between terminals ① and ②, we observe by KCL that

$$i = i_1 = -i_2$$

KVL requires that $v = v_1 - v_2$. Since $v_1 = d\phi_1/dt$ and $v_2 = d\phi_2/dt$ and letting $v = d\phi/dt$, we obtain

$$\frac{d\phi}{dt} = \frac{d\phi_1}{dt} - \frac{d\phi_2}{dt}$$

Again we assume that the initial fluxes are zero; on integrating, we obtain

$$\phi = \phi_1 - \phi_2 = 2i_1 + i_2 = i$$

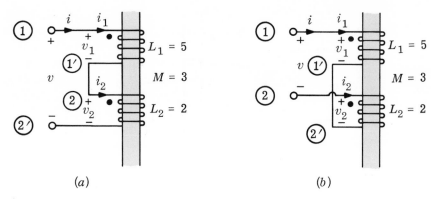

**Fig. 1.5**   Example 2: Series connection of coupled inductors.

Therefore, the inductance of the series connection in Fig. 1.5b is

$$L = \frac{\phi}{i} = 1 \text{ henry}$$

*In conclusion,* the inductance of the series connection of two coupled inductors is easily found by the following rule:

$$L = L_1 + L_2 \pm 2|M|$$

where the plus sign holds if the fluxes produced by the common current in each inductor have the same direction, and the minus sign holds if these fluxes have opposite directions.

**Example 3**   In Fig. 1.6 we connect the two inductors in parallel. To determine the inductance of the parallel connection, it is more convenient to obtain the reciprocal inductances of the coupled inductors in order to calculate the reciprocal inductance of the parallel connection. The reciprocal inductances are, from Eq. (1.14),

$$\Gamma_{11} = \frac{2}{\det\begin{bmatrix} 5 & 3 \\ 3 & 2 \end{bmatrix}} = 2 \qquad \Gamma_{22} = \frac{5}{\det\begin{bmatrix} 5 & 3 \\ 3 & 2 \end{bmatrix}} = 5$$

The mutual reciprocal inductance is denoted by $\Gamma_{12}$, which has the value

$$\Gamma_{12} = \frac{-3}{\det\begin{bmatrix} 5 & 3 \\ 3 & 2 \end{bmatrix}} = -3$$

The current and flux equations are

$$i_1 = \Gamma_{11}\phi_1 + \Gamma_{12}\phi_2 = 2\phi_1 - 3\phi_2$$
$$i_2 = \Gamma_{12}\phi_1 + \Gamma_{22}\phi_2 = -3\phi_1 + 5\phi_2$$

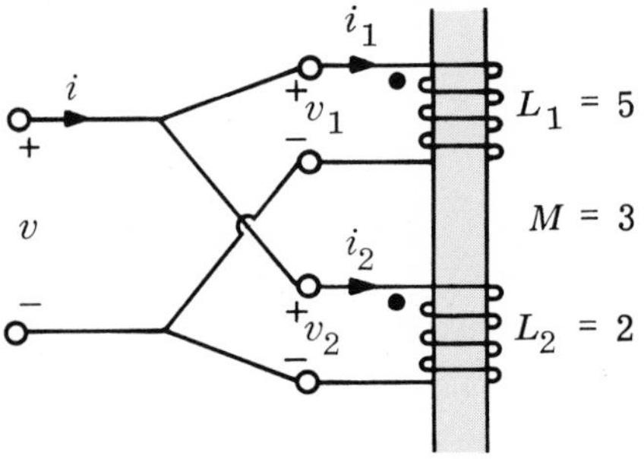

**Fig. 1.6**   Example 3: Parallel connection of coupled inductors.

Referring to Fig. 1.6, we see that KVL requires that

$$v_1(t) = v_2(t) \qquad \text{for all } t$$

If we assume that $i_1(0) = \phi_1(0) = 0$ and $i_2(0) = \phi_2(0) = 0$, by integration of the voltages we obtain

$$\phi_1(t) = \phi_2(t) \qquad \text{for all } t$$

Calling $\phi$ the common value of $\phi_1$ and $\phi_2$, from the flux-current relations we have

$$i = i_1 + i_2 = -\phi_1 + 2\phi_2 = \phi$$

Therefore, the inductance of the parallel connection of Fig. 1.6 is

$$L = \frac{\phi}{i} = 1 \text{ henry}$$

---

*In conclusion,* the reciprocal inductance of the parallel connection of two coupled inductors is given by the following rule:

$$\Gamma = \Gamma_{11} + \Gamma_{22} \pm 2|\Gamma_{22}|$$

where the plus sign holds if the fluxes produced by each inductor current (caused by the common voltage $v$) have opposite directions, and the minus sign holds if these fluxes have the same direction.

**Exercise**   Compute the inductances of the circuits shown in Fig. 1.7a and b.  The inductance matrix for the three-winding coupled inductors is

$$\mathbf{L} = \begin{bmatrix} 10 & -1 & 2 \\ -1 & 3 & 1 \\ 2 & 1 & 5 \end{bmatrix}$$

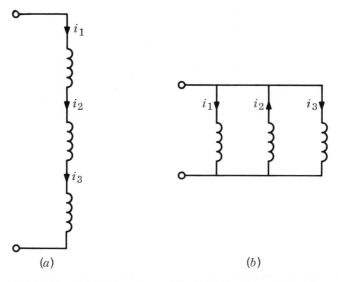

**Fig. 1.7**   Series and parallel connections of three-winding coupled inductors.

| 1.5 | **Double-tuned Circuit** |

A very useful circuit used in communication systems is the double-tuned circuit of Fig. 1.8.  We shall give a simplified analysis of this circuit to demonstrate how node analysis of a circuit with coupled inductors proceeds and to review the steady-state concepts of Chap. 7.

The two parallel resonant circuits are magnetically coupled.  Let us assume for simplicity that the two resonant circuits are identical.  The inductance matrix of the coupled inductors is given by

(1.17)   $\mathbf{L} = \begin{bmatrix} L & M \\ M & L \end{bmatrix} = L \begin{bmatrix} 1 & k \\ k & 1 \end{bmatrix}$

where $k$ is the coefficient of coupling and $M$ is positive (see the data in Fig. 1.8).  In node analysis it is more convenient to use the reciprocal inductance matrix

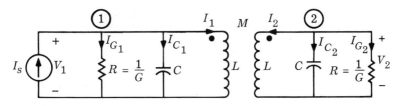

**Fig. 1.8**   Double-tuned circuit.

(1.18)   $\Gamma = L^{-1} = \dfrac{1}{(1 - k^2)L} \begin{bmatrix} 1 & -k \\ -k & 1 \end{bmatrix}$

If we assume that the input is a sinusoid

$i_s(t) = \mathrm{Re}\,(I_s \epsilon^{j\omega t})$

then the sinusoidal steady-state output is a voltage $v_2(t)$ of the form

$v_2(t) = \mathrm{Re}\,(V_2 \epsilon^{j\omega t})$

We are interested in determining the sinusoidal steady-state response for all $\omega$; that is, we wish to determine the network function

(1.19)   $H(j\omega) = \dfrac{V_2}{I_s}$

In steady-state analysis we use phasor representation for all the branch currents and voltages. The relation between current phasors and voltage phasors for two-winding coupled inductors are simply, from Eqs. (1.5), (1.6), and (1.7),

(1.20a)   $V_1 = j\omega L_1 I_1 + j\omega M I_2 = j\omega L I_1 + j\omega L k I_2$

(1.20b)   $V_2 = j\omega M I_1 + j\omega L_2 I_2 = j\omega L k I_1 + j\omega L I_2$

where $V_1$ and $V_2$ are the voltage phasors across the two resonant circuits, and $I_1$ and $I_2$ are the current phasors in the inductors. In terms of reciprocal inductances, using Eqs. (1.15b) and (1.18), we obtain

(1.21a)   $I_1 = \dfrac{1}{j\omega}\Gamma_{11}V_1 + \dfrac{1}{j\omega}\Gamma_{12}V_2 = \dfrac{1}{j\omega L(1 - k^2)}(V_1 - kV_2)$

(1.21b)   $I_2 = \dfrac{1}{j\omega}\Gamma_{12}V_1 + \dfrac{1}{j\omega}\Gamma_{22}V_2 = \dfrac{1}{j\omega L(1 - k^2)}(-kV_1 + V_2)$

In the steady-state node analysis, we choose the two node voltage phasors $V_1$ and $V_2$ as the network variables and write KCL equations in terms of current phasors at the two nodes ① and ②. At node ①, we have

$I_{G_1} + I_{C_1} + I_1 = I_s$

where

$I_{G_1} = GV_1 \qquad I_{C_1} = j\omega C V_1$

and $I_1$ is given by Eq. (1.21a). In terms of the voltage phasors, we have

(1.22)   $\left[ G + j\omega C + \dfrac{1}{j\omega L(1 - k^2)} \right]V_1 - \dfrac{k}{j\omega L(1 - k^2)}V_2 = I_s$

At node ② we have

$I_2 + I_{C_2} + I_{G_2} = 0$

where

$$I_{C_2} = j\omega C V_2 \qquad I_{G_2} = G V_2$$

and $I_2$ is given by Eq. (1.21$b$).   In terms of voltage phasors, we have

(1.23)   $$-\frac{k}{j\omega L(1 - k^2)} V_1 + \left[ G + j\omega C + \frac{1}{j\omega L(1 - k^2)} \right] V_2 = 0$$

Therefore, we have two linear algebraic equations (1.22) and (1.23) with complex coefficients in two unknowns $V_1$ and $V_2$.   The output voltage phasor $V_2$ can be solved immediately by Cramer's rule in terms of the input current phasor $I_2$.   However, because of the symmetry in the circuit and the equations, a simpler method is available.   Let us define two new variables

(1.24)   $$2V_+ = V_1 + V_2 \qquad 2V_- = V_1 - V_2$$

or

(1.25)   $$V_1 = V_+ + V_- \qquad V_2 = V_+ - V_-$$

Adding Eqs. (1.22) and (1.23), we obtain

(1.26)   $$\left[ G + j\omega C + \frac{1}{j\omega L(1 + k)} \right] V_+ = \frac{I_s}{2}$$

Subtracting (1.23) from (1.22), we obtain

(1.27)   $$\left[ G + j\omega C + \frac{1}{j\omega L(1 - k)} \right] V_- = \frac{I_s}{2}$$

Equations (1.26) and (1.27) are exactly in the form of the equations for two single $RLC$ parallel resonant circuits with inductances equal to $L(1 + k)$ and $L(1 - k)$, respectively.   Since the output voltage phasor $V_2 = V_+ - V_-$, it can be considered as the difference of two output phasors of two different single resonant circuits which are uncoupled.

Let us introduce the resonant frequencies and quality factors for the two single resonant circuits to simplify our results.   Let

(1.28)   $$\omega_+^2 = \frac{1}{LC(1 + k)} \qquad \omega_-^2 = \frac{1}{LC(1 - k)}$$

where $\omega_+ < \omega_-$.   Let

(1.29)   $$Q_+ = \omega_+ CR \qquad Q_- = \omega_- CR$$

Then it follows from (1.26) and (1.27) that

(1.30)   $$V_+ = \frac{1}{2} I_s R \frac{1}{1 + jQ_+(\omega/\omega_+ - \omega_+/\omega)}$$

and

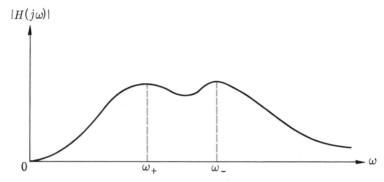

**Fig. 1.9**   A typical magnitude plot of a double-tuned circuit.

(1.31)   $V_- = \dfrac{1}{2} I_s R \dfrac{1}{1 + jQ_-(\omega/\omega_- - \omega_-/\omega)}$

The output voltage phasor is therefore

(1.32)   $V_2 = \dfrac{1}{2} I_s R \left[ \dfrac{1}{1 + jQ_+(\omega/\omega_+ - \omega_+/\omega)} - \dfrac{1}{1 + jQ_-(\omega/\omega_- - \omega_-/\omega)} \right]$

and the network function is

(1.33)   $H(j\omega) = \dfrac{V_2}{I_s} = \dfrac{1}{2} R \left[ \dfrac{1}{1 + jQ_+(\omega/\omega_+ - \omega_+/\omega)} - \dfrac{1}{1 + jQ_-(\omega/\omega_- - \omega_-/\omega)} \right]$

The network function of Eq. (1.33) can further be simplified if the quality factors $Q_+$ and $Q_-$ are much larger than unity and if the coefficient of coupling is small. However, we shall not go into the derivation of simplified formulas. We only wish to point out that the double-tuned circuit can provide a broader bandwidth than that of the resonant circuit which we treated in the previous chapter. A typical magnitude curve $|H(j\omega)|$ is shown in Fig. 1.9. The peaks correspond to approximately $\omega_+$ and $\omega_-$ of the two single *RLC* resonant circuits defined in Eq. (1.28).

## 2   Ideal Transformers

The ideal transformer is an idealization of physical transformers available on the market. With respect to such physical transformers, the ideal transformer is characterized by three idealizations: (1) the ideal transformer does not dissipate energy, (2) it does not have any leakage flux, which means that the coefficient of coupling is equal to unity, and (3) the self-inductance of each winding is infinite. The ideal transformer is a very useful model for circuit calculations because with a few additional elements (*R*, *L*, and *C*) connected to its terminals it can be made to represent fairly accurately the terminal behavior of physical transformers.

### 2.1   Two-winding Ideal Transformer

Let us show heuristically how the ideal transformer would be obtained by winding two coils on a magnetic core as shown in Fig. 2.1 and by having the magnetic permeability $\mu$ of the core become infinite. We assume that the coils have neither losses nor stray capacitance. To simplify the following considerations, let us adopt reference directions for the currents so that the mutual inductance is positive. If the magnetic permeability of the core material were infinite, then all the magnetic field would be constrained to the core and any line of magnetic induction which links a turn of coil 1 would link each and every turn of coil 2. Thus, if $\phi$ is the flux through a one-turn coil located anywhere on the core and if $n_1$ and $n_2$ are the number of turns of coils 1 and 2, respectively, then the total flux $\phi_1$ and $\phi_2$ through coils 1 and 2 respectively, are

$$\phi_1 = n_1\phi \qquad \text{and} \qquad \phi_2 = n_2\phi$$

Since $v_1 = d\phi_1/dt$ and $v_2 = d\phi_2/dt$, we have

(2.1)
$$\boxed{\dfrac{v_1(t)}{v_2(t)} = \dfrac{n_1}{n_2}}$$

for all times $t$ and for all voltages $v_1$ and $v_2$.

Consider now the calculation of $\phi$ in terms of the *magnetomotive force* (mmf) and the *magnetic reluctance* $\mathcal{R}$. Similar to Ohm's law for a linear resistor, the reluctance $\mathcal{R}$ relates the magnetomotive force and the flux $\phi$ according to

$$\text{mmf} = \mathcal{R}\phi$$

In view of our assumption concerning the choice of reference directions for the currents $i_1$ and $i_2$, the mmf is $n_1i_1 + n_2i_2$, and hence

$$n_1i_1 + n_2i_2 = \mathcal{R}\phi$$

Now if the permeability $\mu$ is made infinite, $\mathcal{R}$ becomes zero since the reluctance is inversely proportional to $\mu$. Clearly,

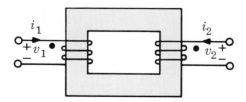

**Fig. 2.1**   A transformer obtained by winding two coils on a common core.

$$n_1 i_1 + n_2 i_2 = 0$$

or

(2.2)
$$\frac{i_1(t)}{i_2(t)} = -\frac{n_2}{n_1}$$

for all $t$ and for all currents $i_1$ and $i_2$.

Equations (2.1) and (2.2) are taken to be the *defining* terminal equations of the ideal transformer. Thus, whenever we use the expression **two-winding ideal transformer,** we shall mean a two-port device whose voltage-current equations are (2.1) and (2.2). Note, in particular, the minus sign in (2.2). On circuit diagrams, ideal transformers are represented by the symbol shown in Fig. 2.2.

**Remarks**   1.   Since Eqs. (2.1) and (2.2) can be interpreted as *linear functions* expressing $v_1$ in terms of $v_2$ and $i_1$ in terms of $i_2$, respectively, and since the coefficients $n_1$ and $n_2$ are independent of time, the ideal transformer is a *linear time-invariant* circuit element.

2.   From (2.1) and (2.2), for all currents and voltages and for all $t$,

(2.3)   $v_1(t)i_1(t) + v_2(t)i_2(t) = 0$

Thus, at all times the sum of the power inputs through each port is zero; no energy is stored, and no energy is dissipated. Whatever power flows into the transformer through one terminal pair flows out through the other terminal pair. These facts are often indicated by saying that an ideal transformer is a *lossless* element *without energy storage,* hence it is memoryless. Note that capacitors, inductors, and pairs of inductors with mutual coupling (even when $k = 1$) are also lossless elements, but ones which *do* store energy.

3.   From (2.1) the voltage $v_1$ across coil 1 does not depend on $i_1$ or on $i_2$; it depends only on $v_2$. Similarly, from (2.2) the current $i_1$ depends

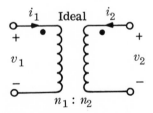

**Fig. 2.2**   Ideal transformer, by definition, $v_1/v_2 = n_1/n_2$, and $i_1/i_2 = -n_2/n_1$.

only on $i_2$ and is independent of $v_1$ and $v_2$.  In particular, if we were to try to measure the self-inductance of inductor 1 (hence with inductor 2 open-circuited and thus $i_2 = 0$), then Eq. (2.2) requires that $i_1(t) = 0$ identically, whatever might be the voltage $v_1$ applied to inductor 1. This fact implies that *the self-inductance of each inductor of an ideal transformer is infinite.*

4. In addition to having infinite self-inductances, a two-winding ideal transformer is a pair of inductors with coefficient of coupling $k = 1$. The energy stored for coupled inductors may be written as (see Eq. 1.7)

$$\mathscr{E}(i_1, i_2) = \tfrac{1}{2}(L_{11}i_1{}^2 + 2\sqrt{L_{11}L_{22}}\, i_1 i_2 + L_{22}i_2{}^2)$$
$$+ \left( \frac{|M|}{\sqrt{L_{11}L_{22}}} - 1 \right) \sqrt{L_{11}L_{22}}\, i_1 i_2$$
$$= \tfrac{1}{2}(\sqrt{L_{11}}\, i_1 + \sqrt{L_{22}}\, i_2)^2 + (k - 1)\sqrt{L_{11}L_{22}}\, i_1 i_2$$

As a consequence of Eq. (2.3), $\mathscr{E}$ is identically zero for an ideal transformer; hence

(2.4)   $k = 1$

and

$$\frac{i_1}{i_2} = -\frac{\sqrt{L_{22}}}{\sqrt{L_{11}}}$$

Note that the last equation agrees with (2.2) since $L_{11}$ and $L_{22}$ are proportional to $n_1{}^2$ and $n_2{}^2$, respectively.

5. As a consequence of our choice of reference directions, Eqs. (2.1) and (2.2) have the signs indicated.  If we choose reference directions as shown in Fig. 2.3 (note that $i_2$ leaves the winding by the dotted terminal), the equations read

$$\frac{v_1}{v_2} = -\frac{n_1}{n_2} \qquad \frac{i_1}{i_2} = \frac{n_2}{n_1}$$

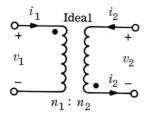

**Fig. 2.3**   Ideal transformer, in view of the dot location, $v_1/v_2 = -n_1/n_2$, and $i_1/i_2 = n_2/n_1$.

**Fig. 2.4**    Mechanical analog of an ideal transformer, with $v_1/v_2 = -d_1/d_2$, and $f_1/f_2 = d_2/d_1$.

6.    The ideal transformer is the electrical analog of a mechanical lever which would be made of a frictionless pivot and a massless, infinitely rigid rod (see Fig. 2.4).   Clearly, under such assumptions, the relations between the forces $f$ and velocities $v$ are

$$\frac{f_1(t)}{f_2(t)} = \frac{d_2}{d_1} \qquad \frac{v_1(t)}{v_2(t)} = -\frac{d_1}{d_2}$$

The absence of friction corresponds to the absence of dissipation of energy in the ideal transformer.   The infinite rigidity of the rod corresponds to the assumed absence of stray capacitance in the ideal transformer; and the masslessness of the rod corresponds to the absence of magnetic energy stored in the ideal transformer.

7.    As a final comment, we should point out that it is possible to consider multiple-winding transformers.   For example, consider the single-core three-winding transformer of Fig. 2.5.   Its equations are

$$\frac{v_1}{n_1} = \frac{v_2}{n_2} = \frac{v_3}{n_3} \qquad n_1 i_1 + n_2 i_2 + n_3 i_3 = 0$$

Again this three-winding single-core ideal transformer is a *linear time-invariant lossless* element *without energy storage.*

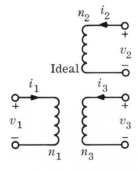

**Fig. 2.5**    A three-winding ideal transformer.

| 2.2 | Impedance-changing Properties |
|---|---|

1.   Consider a resistive load with resistance $R_L$ which is connected on the secondary winding of an ideal transformer as in Fig. 2.6.   The input resistance is

$$R_{in} = \frac{v_1}{i_1} = \frac{(n_1/n_2)v_2}{-(n_2/n_1)i_2} = \left(\frac{n_1}{n_2}\right)^2\left(\frac{v_2}{-i_2}\right)$$

However,

$$v_2 = -R_L i_2$$

·   Therefore,

(2.5)   $$R_{in} = \left(\frac{n_1}{n_2}\right)^2 R_L$$

2.   Let us consider the sinusoidal steady-state behavior of the linear time-invariant circuit shown on Fig. 2.7.   The load is a one-port of impedance $Z_L(j\omega)$,

(2.6)   $$Z_{in}(j\omega) = \frac{V_1}{I_1} = \left(\frac{n_1}{n_2}\right)^2\left(\frac{V_2}{-I_2}\right) = \left(\frac{n_1}{n_2}\right)^2 Z_L(j\omega)$$

Equations (2.5) and (2.6) have an interesting interpretation; that is, ideal transformers change the apparent impedance of a load and can be used to match circuits with different impedances.   For example, the input impedance of a loudspeaker is usually about 8 ohms, too small an impedance for a direct connection to many of the vacuum tube or transistor driving amplifiers with, say, output impedance of 800 ohms.   If a transformer is placed between the output of the power amplifier and the input of the loudspeaker and the turns ratio is selected to compensate for the impedance ratio between the amplifier output and the loudspeaker input, then the amplifier has an appropriate impedance to drive the loudspeaker.   The required turns ratio is $n_1/n_2 = \sqrt{800/8} = 10$.

**Exercise 1**   Show that (2.5) and (2.6) would still be valid if the *dotted* secondary terminal were the bottom one instead of the top one as shown in Fig. 2.7.

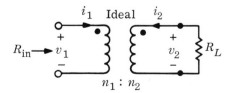

**Fig. 2.6**   Input resistance of a terminated ideal transformer, with $R_{in} = (n_1/n_2)^2 R_L$.

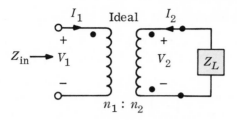

**Fig. 2.7**    Input impedance of a terminated ideal transformer, with $Z_{in} = (n_1/n_2)^2 Z_L$.

**Exercise 2**    Consider the circuit in Fig. 2.8, where an ideal transformer is connected to two linear time-invariant inductors with inductances $L_a$ and $L_b$ as shown.  Prove that the two-port is equivalent to a pair of coupled inductors with the following inductance matrix:

$$
\begin{bmatrix}
L_a + L_b & \dfrac{n_2}{n_1} L_b \\[2ex]
\dfrac{n_2}{n_1} L_b & \left(\dfrac{n_2}{n_1}\right)^2 L_b
\end{bmatrix}
$$

This exercise establishes the important fact that coupled inductors can be replaced by inductors without couplings and an ideal transformer.

# 3    Controlled Sources

## 3.1    Characterization of Four Kinds of Controlled Source

Up to this point we have encountered only independent voltage sources and current sources.  Independent sources constitute the inputs of the circuit.  In this section we shall introduce another type of source called a *controlled source* or *dependent source*.  A controlled source is indispensable in the modeling of electronic devices such as transistors.  By definition, a **controlled source** is an element having two branches, where branch 2 is either a voltage source or a current source, and branch 1 is

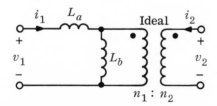

**Fig. 2.8**    A two-port which is equivalent to a pair of coupled inductors.

either an open circuit or a short circuit.   The waveform of the source in branch 2 is a function of the voltage across the open circuit (branch 1) or a function of the current through the short circuit (branch 1).   In other words, the source in branch 2 is *controlled* by a voltage or a current of another branch, namely branch 1.   There are, of course, four possibilities, which are shown on Fig. 3.1, where the diamond-shaped symbols represent the *controlled* sources.   In Fig. 3.1*a* and *b* the sources in branch 2 are current sources; their currents depend, respectively, on the current in branch 1, which is a short circuit, and the voltage in branch 1, which is an open circuit.   These controlled sources are called *current-controlled current source* and *voltage-controlled current source,* respectively.   In Fig. 3.1*c* and *d* the sources in branch 2 are voltage sources; their voltages depend, respectively, on the voltage in branch 1, which is an open circuit, and the current in branch 1, which is a short circuit.   These controlled sources are called *voltage-controlled voltage source* and *current-controlled voltage source,* respectively.

The four kinds of controlled sources are characterized by the equations shown in the figures.   The four proportionality constants $\alpha$, $g_m$, $\mu$, and $r_m$ in Fig. 3.1 represent, respectively, a current ratio, a transfer conductance, a voltage ratio, and a transfer resistance.   Thus, we have

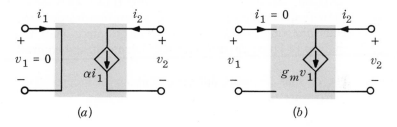

(a)                                   (b)

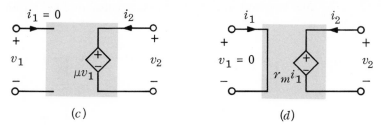

(c)                                   (d)

**Fig. 3.1**   Four types of controlled sources; since the factors $\alpha$, $g_m$, $\mu$, and $r_m$ are constant, these controlled sources are linear time-invariant elements.   (a) $v_1 = 0$, $i_2 = \alpha i_1$, current-controlled current source; (b) $i_1 = 0$, $i_2 = g_m v_1$, voltage-controlled current source; (c) $i_1 = 0$, $v_2 = \mu v_1$, voltage-controlled voltage source; (d) $v_1 = 0$, $v_2 = r_m i_1$, current-controlled voltage source.

Current-controlled current source: $\alpha = \dfrac{i_2}{i_1}$

Voltage-controlled current source: $g_m = \dfrac{i_2}{v_1}$

(3.1)

Voltage-controlled voltage source: $\mu = \dfrac{v_2}{v_1}$

Current-controlled voltage source: $r_m = \dfrac{v_2}{i_1}$

These controlled sources, as specified by Eq. (3.1), where $\alpha$, $g_m$, $\mu$, and $r_m$ are constants, are linear time-invariant elements. A nonlinear controlled source would have a characterization such as $i_2 = f(i_1)$, where $f(\cdot)$ is a nonlinear function. A linear time-varying controlled source would have a characterization such as $i_2 = \alpha(t)i_1$, where $\alpha(\cdot)$ is a given function of time. Linear time-varying controlled sources are useful in representing some modulators. However, for simplicity, only linear time-invariant controlled sources will be considered here. Electronic devices such as transistors and vacuum tubes in small-signal linear operation can be modeled with linear resistors, capacitors, and a linear controlled source, such as those shown in Fig. 3.1. A typical small-signal equivalent circuit of a grounded-emitter transistor is shown in Fig. 3.2, and a low-frequency linear equivalent circuit of a triode is shown in Fig. 3.3. Thus, the small-signal analysis of electronic circuits is reduced to the analysis of linear circuits with $RLC$ elements and controlled sources.

**Remarks**  The reasons for using different symbols for independent and dependent sources are the following:

1. Independent sources play a completely different role from the dependent sources. Independent sources are inputs, and they represent signal generators; i.e., they represent the action of the external world on the circuit. Independent sources are nonlinear elements (usually time-varying) since their characteristics are lines parallel to the $v$ axis

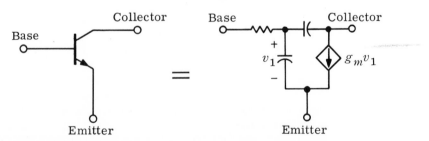

**Fig. 3.2**  Small-signal linear equivalent circuit of a grounded-emitter transistor using a voltage-controlled current source.

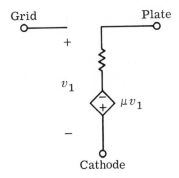

**Fig. 3.3**  Vacuum-tube triode equivalent circuit using a voltage-controlled voltage source.

or the $i$ axis of the $vi$ plane.  Dependent sources are used to model phenomena that occur in electronic devices.  Dependent sources represent the coupling between one network variable in branch 1 to a network variable in branch 2.  Typical dependent sources are those given in Fig. 3.1.  The sources shown in Fig. 3.1 are four-terminal *linear* time-invariant elements.

2.  Linear circuits may contain both independent and dependent sources; however, dependent sources must be linear, whereas independent sources are not linear.

3.  In the Thévenin and Norton equivalent network theorems (Chap. 16), the *dependent* sources play a completely different role from the *independent* sources.

## 3.2  Examples of Circuit Analysis

In circuit analysis, controlled sources are treated like independent sources when the circuit equations are written.  This will be illustrated by the following two examples.

**Example 1**  Consider the simple circuit shown in Fig. 3.4.  The controlled source in this circuit is a voltage-controlled voltage source, where ①① and ②②' represent its two branches, and is specified by

(3.2)  $v_2 = \mu v_1$

Let the input be the independent voltage source $v_s$, and let the output be the voltage $v_L$ across the resistor with resistance $R_L$.  Since there are two meshes, we can write two mesh equations with mesh currents $i_1$ and $i_2$ as variables.  The two equations are

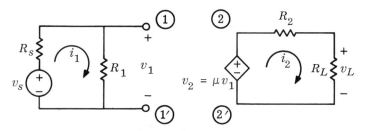

**Fig. 3.4**   Example 1: a simple circuit with a controlled source.

(3.3)   $(R_s + R_1)i_1 = v_s$

(3.4)   $(R_2 + R_L)i_2 = v_2$

Since the controlled source is specified by Eq. (3.2), we can rewrite Eq. (3.4) as follows:

(3.5)   $(R_2 + R_L)i_2 = \mu v_1 = \mu R_1 i_1$

Therefore, Eqs. (3.3) and (3.5) are two linear algebraic equations in two unknown currents $i_1$ and $i_2$. They can be solved immediately; from (3.3)

(3.6)   $i_1 = \dfrac{v_s}{R_s + R_1}$

Substituting (3.6) in (3.5), we obtain

$$i_2 = \frac{\mu v_s R_1}{(R_s + R_1)(R_2 + R_L)}$$

Therefore, the output voltage is

(3.7)   $v_L = R_L i_2 = \dfrac{\mu v_s R_1 R_L}{(R_s + R_1)(R_2 + R_L)}$

**Remarks**
1. If the constant $\mu$ is large and the resistances are suitably chosen, the output voltage $v_L$ can be much larger than the input voltage $v_s$, in which case the circuit represents a simple voltage amplifier.

2. The circuit in Fig. 3.4 contains two meshes which are not connected. The controlled source serves as the coupling element between mesh 1 and mesh 2 or between the input and the output.

**Example 2**   Consider the circuit in Fig. 3.5. The controlled source in the circuit is a voltage-controlled current source, where ①①′ and ②②′ represent its two branches, and is specified by

(3.8)   $i_2 = g_m v_1$

We wish to obtain the differential equation that relates the input current source $i_s$ and the voltage $v_1$.  Let us use node analysis and designate the two node voltages by $v_1$ and $v_2$.  The two node equations are

(3.9)   $$G_1v_1 + C_1\frac{dv_1}{dt} + C_2\frac{d(v_1 - v_2)}{dt} = i_s$$

(3.10)   $$C_2\frac{d(v_2 - v_1)}{dt} + G_2v_2 = -i_2$$

In Eq. (3.10) the current $i_2$ can be substituted with $g_mv_1$ according to Eq. (3.8).  Thus, Eq. (3.10) becomes

(3.11)   $$C_2\frac{d(v_2 - v_1)}{dt} + G_2v_2 + g_mv_1 = 0$$

Equations (3.9) and (3.11) constitute a system of two linear differential equations in $v_1$ and $v_2$ with constant coefficients.  Let us take advantage of the fact that the derivative term is identical (except for sign) in both equations; adding (3.11) to (3.9), we obtain

(3.12)   $$(G_1 + g_m)v_1 + C_1\frac{dv_1}{dt} - i_s = -G_2v_2$$

Differentiating (3.12) and substituting $dv_2/dt$ in (3.9), we obtain the required differential equation in $v_1$.  Thus,

(3.13)   $$\frac{d^2v_1}{dt^2} + \left(\frac{G_1 + g_m + G_2}{C_1} + \frac{G_1}{C_1} + \frac{G_2}{C_2}\right)\frac{dv_1}{dt} + \frac{G_1G_2}{C_1C_2}v_1$$
$$= \frac{1}{C_1}\frac{di_s}{dt} + \frac{G_2}{C_1C_2}i_s$$

The necessary initial condition can be found easily from the given information; that is, $v_1(0) = V_1$, and $v_2(0) = V_2$.  To find $(dv_1/dt)(0)$, we set $t = 0$, in Eq. (3.12) and obtain

$$\frac{dv_1}{dt}(0) = \frac{1}{C_1}[i_s(0) - G_2V_2 - (g_m + G_1)V_1]$$

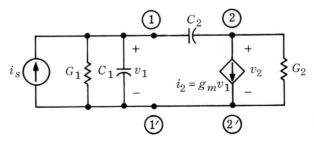

**Fig. 3.5**    Example 2: a simple circuit with a controlled source which is analyzed using node analysis.

With these two initial conditions it is an easy matter to calculate the solution of Eq. (3.13) for any given $i_s$ and then to substitute the result in Eq. (3.12) to obtain $v_2$.

## 3.3    Other Properties of Controlled Sources

As mentioned in Sec. 3.1, the controlled sources shown in Fig. 3.1 are linear time-invariant elements. They are coupling elements because they relate the voltages and currents of two different branches. Since the equations which characterize the controlled sources (see Fig. 3.1) are linear algebraic equations with voltages and currents as variables, controlled sources can be considered as two-port resistive elements. Since we use associated reference directions, the instantaneous power entering the two-port is

$$(3.14) \quad p(t) = v_1(t)i_1(t) + v_2(t)i_2(t)$$

Since branch 1 or the input branch is either a short circuit ($v_1 = 0$) or an open circuit ($i_1 = 0$), the instantaneous power for all four kinds of controlled source is

$$p(t) = v_2(t)i_2(t)$$

Let us connect branch 1 of a voltage-controlled current source to an independent voltage source $v_1$ and connect branch 2, the output branch, to a linear resistor with resistance $R_2$. This is shown in Fig. 3.6. In view of the reference directions for $v_2$ and $i_2$, Ohm's law gives

$$(3.15) \quad v_2 = -i_2 R_2$$

Substituting Eq. (3.15) in (3.14), we obtain

$$p(t) = -i_2{}^2(t)R_2$$

Therefore, the instantaneous power entering the two-port is always negative; in other words, the controlled source delivers power to the re-

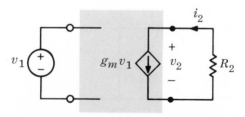

**Fig. 3.6**    A circuit illustrating that controlled sources may deliver energy to the outside world; consequently, they are active elements.

sistor $R_2$ at the rate $R_2 i_2{}^2(t)$ watts. Hence by controlling the input voltage $v_1$ of the circuit in Fig. 3.6, we can have the independent voltage source deliver any amount of energy to the load $R_2$. Recall that in Chap. 2 we defined a *passive* element as an element that cannot deliver energy to the outside world. Since a controlled source can be considered as a two-port resistive element and since it may deliver energy to the outside world, it is an *active element*.

In the previous section we have seen that a circuit consisting of a controlled source and passive resistors can amplify voltages. Let us give one more example to demonstrate another interesting possibility arising from the use of controlled sources.

**Example 3**   Consider the sinusoidal steady-state analysis of the simple circuit in Fig. 3.7. The controlled source is represented by the two branches ① ①' and ② ②'. The impedance $Z_L$ is connected in parallel with the branch ② ②'. The input is the independent current source; its current is represented by the phasor $I_s$. Let us find the input impedance $Z_{in}$ that is faced by the input source.

Using KCL at nodes ① and ②, we have

(3.16)   $I_s = I_1$   and   $I_1 = \alpha I_1 + I_L$

The input impedance is therefore

(3.17)   $Z_{in} = \dfrac{V}{I_s} = \dfrac{Z_L I_L}{I_s}$

Combining Eqs. (3.16) and (3.17), we have

(3.18)   $Z_{in} = (1 - \alpha) Z_L$

It can be seen that if the parameter $\alpha$ is 2, then Eq. (3.18) indicates that $Z_{in}$ is equal to the negative of $Z_L$. Thus if $Z_L$ represents the impedance of a passive one-port, $Z_{in} = -Z_L$ represents the impedance of an active one-port. In Chap. 7 we showed that a necessary property for the driving-

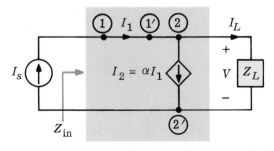

**Fig. 3.7**   Example 3: a two-port formed by a controlled source.

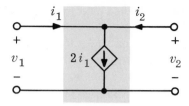

**Fig. 3.8**    A negative-impedance converter which is formed by a controlled source.

point impedance $Z_L$ to be passive is $\text{Re}\,[Z_L(j\omega)] \geq 0$ for all $\omega$. Since $Z_{\text{in}} = -Z_L$, $\text{Re}\,[Z_{\text{in}}(j\omega)] \leq 0$, and hence $Z_{\text{in}}$ is the impedance of an active one-port when $\alpha = 2$. The two-port inside the dotted square in Fig. 3.7 is called a *negative impedance converter*. A **negative impedance converter** is a two-port device which has the property that the input impedance is equal to the negative of whatever impedance is connected at the output port. As a matter of fact, the negative impedance converter is itself a two-port coupling element. If we redraw the circuit of Fig. 3.7 inside the shaded square, as shown in Fig. 3.8 with $\alpha = 2$, and if we redefine the currents and voltages as shown in the figure, then the characterization of a negative impedance converter is

$$(3.19) \qquad v_1 = v_2 \qquad i_1 = i_2$$

As we have seen in Chaps. 2 and 5, a negative resistance is an active element. This point is stressed again in the following exercise. It has been used in the design of amplifiers and in some cable communication systems.

**Exercise**    Consider the circuit shown in Fig. 3.9.

*a.* Calculate the power delivered by the generator, the power received by the negative resistance, and the power received by the load resistance when $e_s$ is a constant equal to 10 volts.

*b.* Repeat the problem when $e_s(t) = 10 \cos \omega t$ (calculate both the instantaneous power and the average power).

*c.* What can you say about the energy flow in the circuit?

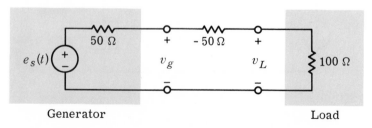

Generator                              Load

**Fig. 3.9**    Exercise showing that the negative resistance delivers power to the load.

## Summary

- Typical coupling elements are coupled inductors, ideal transformers, and dependent sources. Coupling elements consist of more than one branch and have more than two terminals, usually four. They are defined by the equations relating their branch voltages and their branch currents.

- The equations defining a pair of linear time-invariant coupled inductors are

$$v_1 = L_{11} \frac{di_1}{dt} + M \frac{di_2}{dt}$$

$$v_2 = M \frac{di_1}{dt} + L_{22} \frac{di_2}{dt}$$

To complete the specification, the initial currents $i_1(0)$ and $i_2(0)$ are required.

- The energy stored in a pair of linear time-invariant coupled inductors is given by

$$\mathscr{E}(i_1, i_2) = \tfrac{1}{2} L_{11} i_1^2 + M i_1 i_2 + \tfrac{1}{2} L_{22} i_2^2$$

If these inductors are passive, the self-inductances $L_{11}$ and $L_{22}$ are positive, whereas $M$ may be either positive or negative. The magnitude of $M$ is related to the coefficient of coupling $k$, which is defined by

$$k \triangleq \frac{|M|}{\sqrt{L_{11} L_{22}}}$$

Passivity requires that $0 \le k \le 1$.

- Linear time-invariant inductors may be described in terms of the inductance matrix $\mathbf{L}$. Thus,

$$\mathbf{v} = \mathbf{L} \frac{d\mathbf{i}}{dt}$$

They may also be described in terms of the reciprocal inductance matrix $\boldsymbol{\Gamma}$. Thus,

$$\mathbf{i}(t) = \mathbf{i}(0) + \boldsymbol{\Gamma} \int_0^t \mathbf{v}(t') \, dt'$$

It is always the case that

$$\boldsymbol{\Gamma} = \mathbf{L}^{-1}$$

- The defining equations of a two-winding ideal transformer are

$$\frac{v_1(t)}{v_2(t)} = \frac{n_1}{n_2} = \frac{-i_2(t)}{i_1(t)}$$

where $n_1$ and $n_2$ are the number of turns in the first and second windings, respectively. These equations hold when the reference directions of $i_1$ and $i_2$ both enter (or both leave) the dotted terminals; if this is not the case, replace $n_1$ by $-n_1$.

- An ideal transformer is a linear, time-invariant element. It does not dissipate or store energy.

- An ideal transformer may be considered as a pair of linear time-invariant coupled inductors with infinite self-inductances and with a coefficient of coupling equal to 1.

- The four basic linear time-invariant controlled sources are

  Current-controlled current source:   $v_1 = 0$    $i_2 = \alpha i_1$

  Voltage-controlled current source:   $i_1 = 0$    $i_2 = g_m v_1$

  Voltage-controlled voltage source:   $i_1 = 0$    $v_2 = \mu v_1$

  Current-controlled voltage source:   $v_1 = 0$    $v_2 = r_m i_1$

  where $\alpha$, $g_m$, $\mu$, and $r_m$ are constants.

## Problems

Series and parallel connections of coupled inductors

**1.** A pair of coupled inductors has (for the reference directions shown on Fig. P8.1$a$) the inductance matrix

$$L = \begin{bmatrix} 4 & -3 \\ -3 & 6 \end{bmatrix}$$

Find the equivalent inductance of the four connections shown in Fig. P8.1$b$.

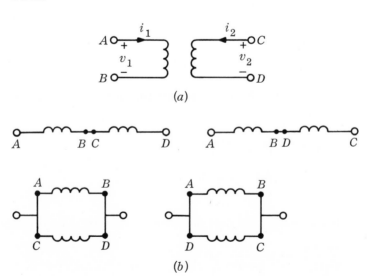

$(a)$

$(b)$

**Fig. P8.1**

Sign of *M*,
series and
parallel
connections

**2.** Sketched in Fig. P8.2 is the physical arrangement of inductor windings on a common core. Each self-inductance is 2 henrys, and each mutual inductance has the absolute value of 1 henry. Compute the net inductance of circuits (*a*) to (*f*). In the sketches (*a*), (*b*), etc., the arrows indicate the direction of the reference arrow of each winding.

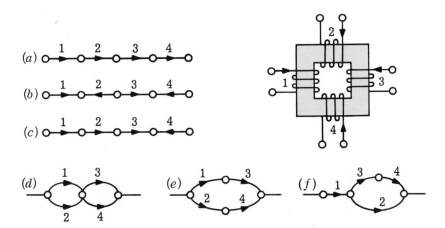

**Fig. P8.2**

Sign of *M*,
equivalent
inductance

**3.** The magnetic coupling between two linear time-invariant inductors is provided by a core as shown in Fig. P8.3. The values of the self-inductances are $L_{11} = 2$ henrys and $L_{22} = 3$ henrys, and the mutual inductance is $M = 1$ henry.

*a.* Calculate the equivalent inductance between terminals ① and ② when ① and ② are tied together.

*b.* Calculate the equivalent inductance between terminals ① and ② when ① and ② are tied together.

*c.* Suggest a procedure for measuring the mutual inductance between windings using only an inductance bridge.

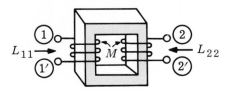

**Fig. P8.3**

Inductance
matrix, equiv-
alent two-ports

**4.** The inductors in the circuits shown in Fig. P8.4 are linear and time-invariant.

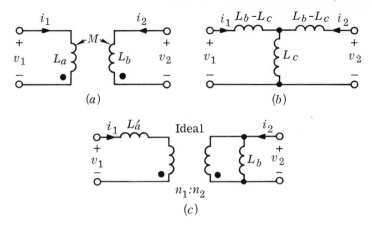

*Fig. P8.4*

a.   Obtain the inductance matrix for each circuit.

b.   Show that if $L_c = M$, circuits (a) and (b) have the same inductance matrix.

c.   How should $L'_a$ and $n_1/n_2$ be related with $L_a$ and $M$ so that circuits (a) and (c) have the same inductance matrix?

Mesh analysis   **5.** The circuit shown in Fig. P8.5 is in sinusoidal steady state, where the input is a voltage source $e_s(t) = \cos(2t + 30°)$.   Determine the steady-state currents $i_1$ and $i_2$.

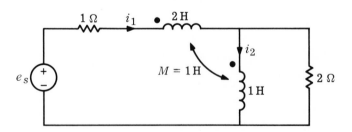

*Fig. P8.5*

Node analysis   **6.** Write the node equations for the circuit shown in Fig. P8.6.   If $i_s(t) = \cos t$, determine the sinusoidal steady-state voltage $v_2(t)$.

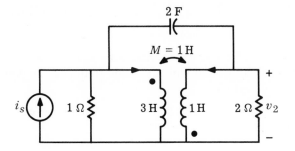

**Fig. P8.6**

**7.** Given the circuit shown in Fig. P8.7, determine the steady-state currents $i_1(t)$, $i_2(t)$, and $i_3(t)$ for the given current source input $i_s(t) = \sin t$. The inductance matrix for the three coupled inductors is

$$L = \begin{bmatrix} 5 & 2 & 1 \\ 2 & 4 & -1 \\ 1 & -1 & 2 \end{bmatrix}$$

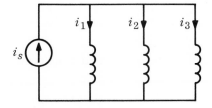

**Fig. P8.7**

Energy stored

**8.** In the circuit of Prob. 7, assume that $i_s$ is a constant current source and $i_1 = 2$, $i_2 = 1$, and $i_3 = -3$ amp. What is the energy stored in the inductors?

Ideal transform-
er and equivalent
two-ports

**9.** Find an expression for $R_2$ such that the two-ports shown in Fig. P8.9 are equivalent.

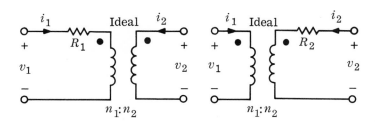

**Fig. P8.9**

**10.** The circuit shown in Fig. P8.10 is linear and time-invariant.

a.   Find $Z(j\omega)$ when $aa'$ and $bb'$ are not connected.

b.   With $aa'$ and $bb'$ connected, assuming that all branch voltages and currents are sinusoidal with the same frequency as that of $e_s$, find $i_1$ when $R_1 = 2$ ohms.

c.   Find the value of $R_1$ that allows the maximum average power to be dissipated in $R$.

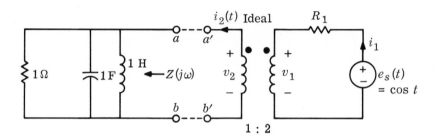

**Fig. P8.10**

**11.** a.   Determine the equivalent resistance of the one-port shown in Fig. P8.11.

b.   Repeat the problem when points $a$ and $a'$ are connected by a short circuit.

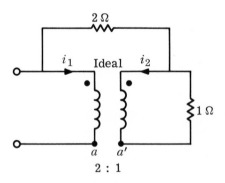

**Fig. P8.11**

**12.** For the circuits shown in Fig. P8.12, calculate [using (a)] the imped-ances

$$Z_{11}(j\omega) = \frac{V_1}{I_1} \qquad \text{and} \qquad Z_{21}(j\omega) = \frac{V_2}{I_1}$$

and [using (b)] the impedances

$$Z_{22}(j\omega) = \frac{V_2}{I_2} \quad \text{and} \quad Z_{12}(j\omega) = \frac{V_1}{I_2}$$

where $V_1$ and $V_2$ are phasors that represent the sinusoidal output voltages $v_1(t)$ and $v_2(t)$, respectively, and $I_1$ and $I_2$ are phasors that represent the sinusoidal input currents $i_1(t)$ and $i_2(t)$, respectively. Note that in (a), terminals ② and ②′ are left open-circuited, and in (b) terminals ① and ①′ are left open-circuited.

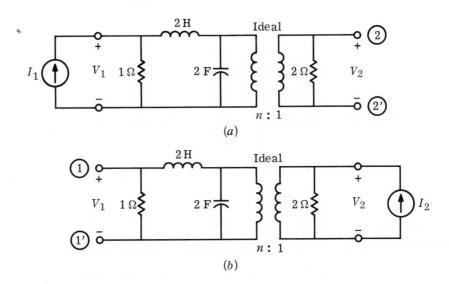

(a)

(b)

**Fig. P8.12**

Driving-point
and transfer
properties of
coupled
inductors

**13.** Consider a pair of linear time-invariant inductors characterized by their inductance matrix (see Fig. P8.13).

a.   Show that the driving-point impedance $Z_{10}(j\omega)$ (seen at terminals ① and ①′ with ② and ②′ open) and the driving-point impedance $Z_{20}(j\omega)$ (seen at ② and ②′ with ① and ①′ open) satisfy

$$\frac{Z_{10}(j\omega)}{L_{11}} = \frac{Z_{20}(j\omega)}{L_{22}}$$

b.   Show that

$$\frac{Z_{1s}(j\omega)}{L_{22}} = \frac{Z_{2s}(j\omega)}{L_{11}}$$

where $Z_{1s}(j\omega)$ is the input impedance seen at terminals ① and ①′ with ② and ②′ short-circuited, and $Z_{2s}(j\omega)$ is the input impedance seen at ② and ②′ with ① and ①′ short-circuited.

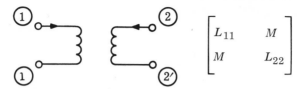

**Fig. P8.13**

**14.** Figure P8.14 shows the small-signal equivalent circuit of a simple transistor amplifier; $V_1$, $I_1$, $V_2$, and $V_0$ are phasor representations of sinusoids of frequency $\omega$.

   *a.*   Calculate the driving-point impedance $Z_{in}(j\omega) = V_1/I_1$.

   *b.*   Calculate the transfer voltage ratio $H(j\omega) = V_2/V_0$.

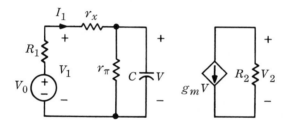

**Fig. P8.14**

**15.** Figure P8.15 shows a vacuum-tube amplifier circuit and its small-signal equivalent circuit. Calculate the voltage ratios $V_k/V_0$ and $V_L/V_0$ in terms of the given resistances and the constant $\mu$.

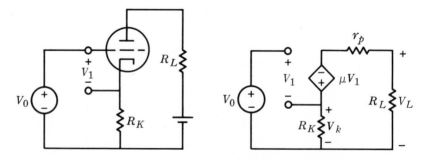

**Fig. P8.15**

**16.** The circuit in Fig. P8.16 represents an alternate model of a transistor amplifier at low frequencies. Determine the voltages $v_1$ and $v_2$.

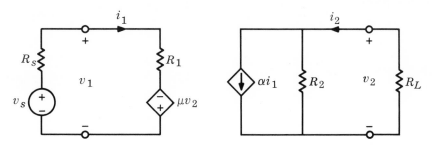

**Fig. P8.16**

**17.** For the circuit shown in Fig. P8.17 determine the network function $H(j\omega) \triangleq V_2/V_0$, where $V_2$ and $V_0$ are phasors that represent the sinusoidal voltages $v_2(t)$ and $v_0(t)$, respectively.

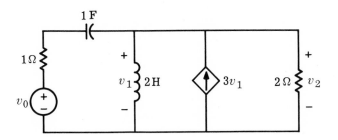

**Fig. P8.17**

**18.** A two-winding ideal transformer with turns ratio $n:1$ can be modeled by two dependent sources. Based on the defining equations of the ideal transformer and the dependent sources, determine an appropriate model for the ideal transformer which uses two properly chosen dependent sources.

# 9

# Network Graphs and Tellegen's Theorem

In the first part of this book we encountered most of the important concepts and properties of circuits. To help us to understand them, we illustrated them only with simple circuits. With the exception of Chap. 7, the typical circuits we considered contained only a few elements and were described by first- or second-order differential equations. In the second part of this book we want to develop systematic procedures to analyze and establish properties of any network of any complexity. Note that we use the word **network,** which has the same meaning as the word *circuit* (i.e., an interconnection of elements); however, the word "network" usually carries the idea of complexity (a network is a circuit with many elements). In practice some networks may be very complicated and may contain several dozens of elements.

An additional reason for our need to develop systematic procedures is that the engineering world has been completely changed by the computer. There are now on the market several computers that multiply two 8-digit numbers in less than a microsecond. For engineers, such capability means that it is now feasible and economical to perform complicated analyses and undertake designs that require a great deal of computation, say, $10^6$ times as many as 15 years ago. It is therefore important to learn systematic procedures so that we can handle any network, however complicated.

As is usual in science and engineering, the first step is a process of abstraction. Since KVL and KCL do not make any assumption whatsoever concerning the nature of the elements of a network, it is natural to overlook the nature of the elements in order to reduce the network to a graph. The first section of this chapter is devoted to developing the concept of a graph. Graph-theoretic ideas are then used to precisely formulate KVL and KCL. Then, as an illustration of the power of the abstract concept of a graph, we derive Tellegen's theorem, which allows us to prove extremely easily several very general properties of networks.

## 1  The Concept of a Graph

Consider any *physical network,* for example, an 80-element lumped delay line or a telemetering repeater. Suppose we consider only those frequencies which permit us to *model* the physical network as a connection of *lumped elements,* namely resistors, capacitors, inductors, couple inductors, transformers, dependent sources, and independent sources. From now on when we say "the network $\mathfrak{N}$ under consideration" we mean this lumped-parameter *model*. In this chapter the network $\mathfrak{N}$ may be linear or nonlinear, active or passive, time-varying or time-invariant. We shall investigate the ways of expressing the constraints imposed on branch voltages and branch currents by KCL and KVL. Since Kirchhoff's laws do not depend on the nature of the elements, it is natural to disregard the nature of the elements. To do so, we replace each element of the network $\mathfrak{N}$ by a **branch** (represented by a line segment), and at the ends of each branch we draw black dots called **nodes** (some authors use the word "edge" for branch and the word "vertex" or "junction" for node). The result of this process is a *graph*. Two examples are shown in Fig. 1.1. In Fig. 1.1*b*, even though the two inductors are mutually coupled, the graph does not indicate the magnetic coupling $M$ since $M$ pertains to the nature of the branches of $\mathfrak{N}$ and is not a property of the graph of $\mathfrak{N}$.

More precisely, by the word **graph** we mean a set of nodes together with a set of branches with the condition that each branch terminates at each end into a node.

Observe that the definition of the graph includes the special case in which a node has no branch connected to it, as shown in the graph of Fig. 1.2*a*. Observe also that since each branch terminates at each end into a node and since these nodes are not required to be distinct according to our definition, a graph may include a *self-loop*, i.e., a loop consisting of a single branch (see Fig. 1.2*b*). In this book we shall not encounter such graphs,

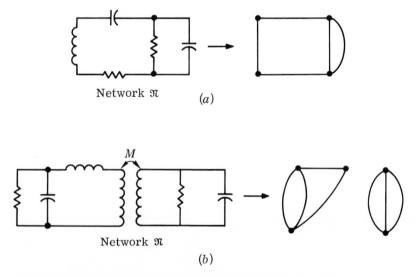

Network $\mathfrak{N}$    (*a*)

Network $\mathfrak{N}$

(*b*)

**Fig. 1.1**    Networks and their graphs. (*a*) Graph with four nodes and five branches; (*b*) graph with two separate parts, five nodes, and seven branches.

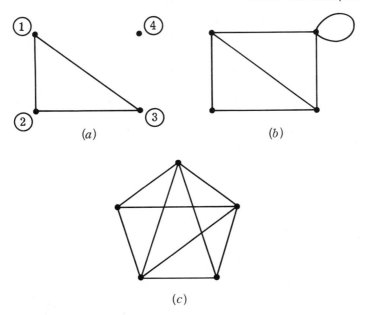

**Fig. 1.2**    (*a*) Graph with an isolated node; (*b*) graph which contains a self-loop; (*c*) a somewhat more complicated graph.

although they appear in engineering work, for example, in flow graphs.

Suppose we have a graph $\mathcal{G}$ in mind.  We then say that $\mathcal{G}_1$ is a **subgraph** of $\mathcal{G}$ if $\mathcal{G}_1$ is itself a graph, if every node of $\mathcal{G}_1$ is a node of $\mathcal{G}$, and if every branch of $\mathcal{G}_1$ is a branch of $\mathcal{G}$.  In other words, given the graph $\mathcal{G}$, we can obtain $\mathcal{G}_1$ by deleting from $\mathcal{G}$ some branches and/or some nodes.  In Fig. 1.3 $\mathcal{G}_1$, $\mathcal{G}_2$, $\mathcal{G}_3$, $\mathcal{G}_4$, and $\mathcal{G}_5$ are subgraphs of $\mathcal{G}$.  Note that $\mathcal{G}_5$ consists of only one node, and is called a **degenerate subgraph.**

Throughout the following discussion we shall adopt reference directions for branch voltages and branch currents that are called **associated;** i.e., the arrowhead that specifies the current reference direction always points toward the terminal labeled with a minus sign for the voltage reference direction.  The branch voltage and current of the $k$th branch will be denoted by $v_k$ and $j_k$ as shown in Fig. 1.4†.  Since in this chapter we always use *associated* reference directions, we need only indicate the arrow specifying the reference direction of the current, and we omit the plus and minus signs for the voltage reference direction.

Given a network $\mathfrak{N}$ with a specified reference direction for each of its branches, the process of abstraction described above leads to a graph whose branches have reference directions.  Such a graph is called an **oriented graph.**  For example, Fig. 1.5 shows a network with reference

† From now on we shall usually use the letter $j$ (as in $j_1, j_2, \ldots, j_k$) to designate *branch currents;* the letter $i$ will usually designate *loop currents* or *mesh currents.*

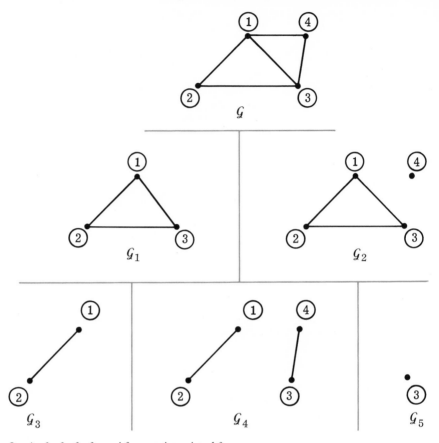

**Fig. 1.3**   Graphs $\mathcal{G}_1$, $\mathcal{G}_2$, $\mathcal{G}_3$, $\mathcal{G}_4$, and $\mathcal{G}_5$ are subgraphs of $\mathcal{G}$.

directions and the corresponding oriented graph.   We think of an *oriented graph* as a set of nodes together with a set of oriented branches, where each branch terminates at each end into a node.   For example, we may number the nodes and the branches of the graph as shown in Fig. 1.5.

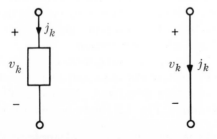

**Fig. 1.4**   Associated reference directions for an element and for a branch.

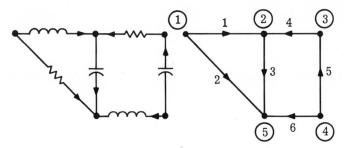

**Fig. 1.5**   Network and its oriented graph.

We say that branch 4 is **incident with** node ② and node ③, branch 4 **leaves** node ③ and **enters** node ②.

From an analytical point of view we may describe the oriented graph of Fig. 1.5 by listing all branches and nodes and indicating which branch is entering and leaving which node. This is conveniently done by writing down a matrix. Suppose that the oriented graph is made up of $b$ branches and $n_t$ nodes. Suppose also that we number arbitrarily all the branches and all the nodes of this graph. We call the **node-to-branch incidence matrix** $\mathbf{A}_a$ a rectangular matrix of $n_t$ rows and $b$ columns whose $(i,k)$th element $a_{ik}$ is defined by

$$a_{ik} = \begin{cases} 1 & \text{if branch } k \text{ } \textit{leaves} \text{ node } ⓘ \\ -1 & \text{if branch } k \text{ } \textit{enters} \text{ node } ⓘ \\ 0 & \text{if branch } k \text{ is } \textit{not incident with} \text{ node } ⓘ \end{cases}$$

Since each branch leaves a single node and enters a single node, each column of the matrix $\mathbf{A}_a$ contains a single $+1$ and a single $-1$, with all other elements equal to zero. The incidence matrix of the graph in Fig. 1.5 is

$$
\overbrace{\hphantom{\qquad\qquad\qquad}}^{b \text{ branches}}
$$

|            | 1 | 2 | 3 | 4 | 5 | 6 |
|------------|---|---|---|---|---|---|
| ①          | 1 | 1 | 0 | 0 | 0 | 0 |
| ②          | -1 | 0 | 1 | -1 | 0 | 0 |
| ③          | 0 | 0 | 0 | 1 | -1 | 0 |
| ④          | 0 | 0 | 0 | 0 | 1 | 1 |
| ⑤          | 0 | -1 | -1 | 0 | 0 | -1 |

$n_t$ nodes

Thus, given any oriented graph, it is a simple matter to number its branches and nodes and to write down the resulting incidence matrix $\mathbf{A}_a$. Conversely to any $n_t \times b$ matrix with the property that each one of its columns contains a single $+1$, a single $-1$, and zeros, we can associate an oriented graph of $b$ branches and $n_t$ nodes. In computer work, the incidence matrix is the standard method used to describe the interconnection of the elements.

**Exercise**    Draw an oriented graph corresponding to the following incidence matrix:

$$\mathbf{A}_a = \begin{bmatrix} -1 & 1 & 1 & 1 & 0 \\ 0 & 0 & 0 & -1 & 0 \\ 1 & -1 & 0 & 0 & -1 \\ 0 & 0 & -1 & 0 & 1 \end{bmatrix}$$

## 2    Cut Sets and Kirchhoff's Current Law

In order to be able to express KCL systematically and without hesitation for any network, we now develop the concept of cut set. Roughly speaking, KCL states that the algebraic sum of all the currents leaving a node is equal to zero. Intuitively then, if we were to partition the nodes of a network into two sets by a closed gaussian surface† so that one set of nodes is inside the surface and the other outside (see Fig. 2.1), then KCL implies that the sum of the currents leaving the gaussian surface is zero. In many cases, the collection of all the branches that cross the gaussian surface will be called a cut set.

In order to make precise this intuitive idea we must proceed step by step and distinguish between connected and unconnected graphs.

We say that a graph is **connected** if there is at least one path (along branches of the graph and disregarding branch orientations) between any two nodes of the graph. By convention, a graph consisting of only one node is connected. A connected graph is also said to be of *one separate*

---

† We use the word "gaussian" to suggest the analogy with the closed surface used in Gauss' law, which states that electric flux out of the closed surface is equal to the charge enclosed.

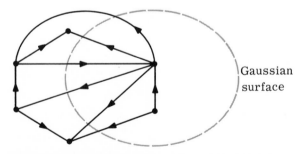

Gaussian surface

**Fig. 2.1**    The gaussian surface which leads intuitively to the concept of cut set.

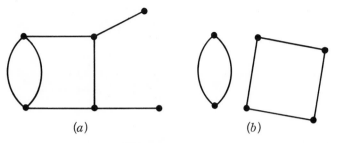

**Fig. 2.2**   (*a*) Connected graph; (*b*) unconnected graph.

*part*. Given an unconnected graph, its maximal connected subgraphs are also called **separate parts.** Thus, an unconnected graph must have at least two separate parts. The graph in Fig. 2.2*a* is connected, whereas the graph in Fig. 2.2*b* is unconnected and has two separate parts.

To explain the concept of a cut set, we must specify what we mean by the expression "to remove a branch." When we say **remove a branch,** we mean that we delete the line segment that joins the nodes but leave the nodes remaining. This is illustrated in Fig. 2.3. The idea of a cut set is related to the idea of cutting a connected graph into two separate parts by removing branches.

A set of branches of a connected graph $\mathcal{G}$ is called a **cut set** if (1) the removal of all the branches of the set causes the remaining graph to have two separate parts and (2) the removal of all but any one of the branches of the set leaves the remaining graph connected. Examples of cut sets are shown in Fig. 2.4. The branches of the cut set are indicated by heavier lines, and the idea of cutting the connected graph into two separate parts is emphasized by the dashed line (which suggests the idea of gaussian surface) which crosses all the branches of the cut set. For the graph of Fig.

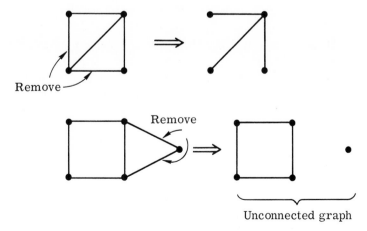

**Fig. 2.3**   Illustrations of the operation of "removing a branch."

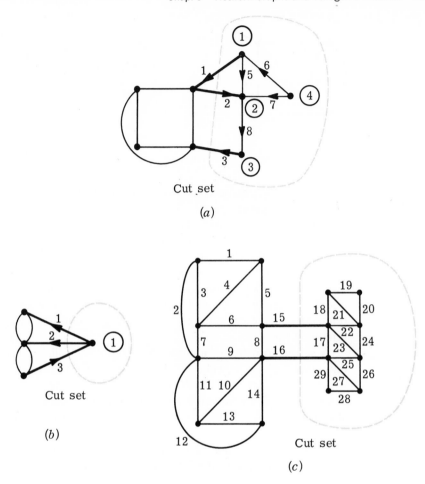

**Fig. 2.4**   Examples of cut sets; the branches of the cut set are indicated by heavier lines.

2.4*b*, we see that the set of branches connected to node ①  is a cut set because an isolated node constitutes a separate part.

In case the graph 𝒢 has *s* separate parts, a **cut set** is defined to be a set of branches that (1) the removal of all the branches of the set causes the remaining graph to have *s* + 1 separate parts, and (2) the removal of all but any one of the branches of the set leaves the remaining graph with *s* separate parts.   With the concept of a cut set well established we can now state KCL with great generality:

| | |
|---|---|
| Kirchhoff's current law | For *any* lumped network, for *any* of its cut sets, and at *any* time, the algebraic sum of all the branch currents traversing the cut-set branches is zero. |

To apply KCL, we proceed as follows: (1) we assign a reference direction to the cut set, namely, from the inside to the outside of the gaussian surface that defines the cut set, and (2) in obtaining the algebraic sum, we assign a plus sign to the branch currents whose reference direction agrees with that of the cut set and a minus sign to the branch currents whose reference direction is opposite to that of the cut set.

**Example 1**   For the cut set shown in Fig. 2.4*a*, KCL gives

$$j_1(t) - j_2(t) + j_3(t) = 0 \qquad \text{for all } t$$

**Example 2**   For the cut set shown in Fig. 2.4*b*, KCL gives

$$j_1(t) + j_2(t) - j_3(t) = 0 \qquad \text{for all } t$$

Kirchhoff's current law, as stated above, is a direct consequence of the node law stated in Chap. 1. Indeed, if we sum all the expressions of KCL applied to the nodes inside the gaussian surface, we obtain the cut-set law. The branch currents of branches joining two internal nodes cancel out! This is easily verified in the following example.

**Example 3**   Let us consider again the cut set shown in Fig. 2.4*a*. The node laws give

$$\begin{aligned}
\text{Node } \textcircled{1}: \quad &+j_1 &&+j_5 \;\; -j_6 &&&= 0 \\
\text{Node } \textcircled{2}: \quad &-j_2 &&-j_5 &&-j_7 \;\; +j_8 &= 0 \\
\text{Node } \textcircled{3}: \quad &+j_3 &&&&-j_8 &= 0 \\
\text{Node } \textcircled{4}: \quad &&&+j_6 \;\; +j_7 &&&= 0
\end{aligned}$$

Sum: $\; j_1 - j_2 + j_3 = 0$

and this equation is the cut-set equation.

**Remarks**   1.   KCL applies to *any lumped network* irrespective of the nature of the elements.

2.   When expressed in terms of branch currents, the equations prescribed by KCL are in the form of *linear* homogeneous algebraic equations with *constant coefficients*.

Let us end this section with an observation. Suppose we apply the Kirchhoff law to the cut sets I, II, and III shown in Fig. 2.5. It is obvious that by adding the equations relative to cut sets I and II we obtain that of cut set III; the three equations are linearly dependent. In other words, the third equation did not supply any information not contained in the preceding ones. Therefore, in our general theory of network analysis we shall

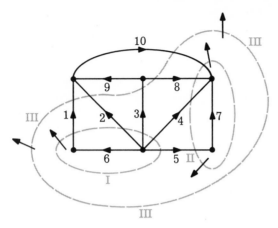

**Fig. 2.5** The cut sets I, II, and III lead to KCL equations that are
linearly dependent.

have to select cut sets in such a way that each equation supplies some new
information. This we shall study in great detail in the succeeding chapter.

**Exercise 1** Refer to Fig. 2.4c. Which of the following sets of branches are cut sets (if
some are not, state carefully why): {5,6,8,17,23,24}, {1,4,6,9,10,14,16},
{1,2,3}, {1,4,5,12,13,14}?

**Exercise 2** Write KCL for cut sets I, II, and III of Fig. 2.5.

**Exercise 3** Consider a connected graph $\mathcal{G}$. Partition the nodes into two sets $A$ and $B$
that are mutually exclusive (that is, $A \cap B$ is the empty set) and exhaus-
tive (that is, $A \cup B$ includes all the nodes of $\mathcal{G}$). What is the name of the
set of branches one of whose terminals is in $A$ and the other is in $B$?
Justify your answer.

## 3 Loops and Kirchhoff's Voltage Law

Thus far in our encounters with simple circuits we have taken the meaning
of a loop to be intuitively clear. For our present systematic approach we
need a precise concept of a loop. Roughly speaking, a loop is a closed
path. However, this vague statement does not tell us whether the sets of
branches emphasized in the graphs of Fig. 3.1 constitute loops. In other
words, do we allow the closed path to go through a node more than once,
as in Fig. 3.1b, or do we allow dangling branches, as in Fig. 3.1c? In order
to simplify many developments in later chapters, we shall formulate the
concept of a loop as follows: a subgraph $\mathcal{L}$ of a graph $\mathcal{G}$ is called a **loop** if
(1) the subgraph $\mathcal{L}$ is connected and (2) precisely two branches of $\mathcal{L}$ are
incident with each node. Figure 3.2 illustrates the definition of a loop.

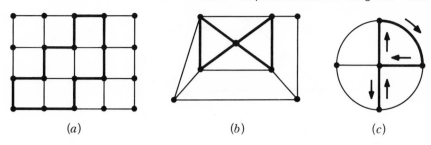

| $(a)$ | $(b)$ | $(c)$ |

**Fig. 3.1**    The emphasized branches in the three figures are closed paths; however, only case $(a)$ qualifies to be called a loop.

We are now in a position to state KVL with as much generality as we shall need in this book:

Kirchhoff's
voltage law

For *any* lumped network, for *any* of its loops, and at *any* time, the algebraic sum of the branch voltages around the loop is zero.

To apply KVL to any loop, we proceed as follows: (1) we assign a reference direction to the loop, and (2) in the algebraic sum of the branch voltages, we assign a plus sign to a branch voltage when its branch reference direction agrees with that of the loop and a minus sign when its branch reference direction disagrees with that of the loop.

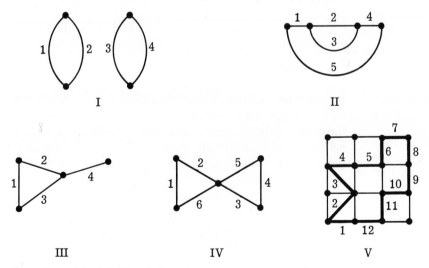

**Fig. 3.2**    Illustrations of the definition of a loop. Case I violates property (1); cases II, III, and IV violate property (2); case V is a loop.

**Example**    Consider the loop indicated on the graph in Fig. 3.3.  For the reference direction specified by the dashed line we have

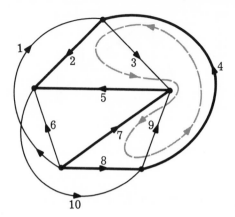

**Fig. 3.3** Illustration for a loop equation; the dashed line indicates the loop reference direction.

$$+ v_4(t) + v_2(t) - v_5(t) - v_7(t) + v_8(t) = 0 \qquad \text{for all } t$$

**Remarks**
1. KVL applies to *any lumped network* irrespective of the nature of the elements.

2. When expressed in terms of the branch voltages, the equations prescribed by KVL are in the form of *linear* homogeneous algebraic equations with *constant coefficients*.

**Exercise** Apply KVL to the loops {5,9,10}, {5,6,7}, {6,8,10}, and {7,8,9}. Are the resulting equations linearly independent?

## 4 Tellegen's Theorem

In this section we introduce our first general network theorem, Tellegen's theorem. This theorem is extremely general; it is valid for any lumped network that contains any elements, *linear or nonlinear, passive or active, time-varying or time-invariant*. This generality follows from the fact that Tellegen's theorem depends only on the two Kirchhoff laws.

Consider an arbitrary lumped network and choose *associated reference directions* for the branch voltages $v_k$ and branch currents $j_k$.† [Hence, $v_k(t)j_k(t)$ is the power delivered at time $t$ by the network to branch $k$.] Next, let us disregard the nature of the branches; in other words, let us think of the network as an oriented graph $\mathcal{G}$, for example, the one shown in Fig. 4.1. Tellegen's theorem asserts that $\sum_{k=1}^{b} v_k j_k = 0$. The only re-

---

† The assumption relative to the associated reference directions is introduced here for *convenience* in later interpretations. It is not required by the proof. The proof requires only that the $v_k$'s and $j_k$'s satisfy the Kirchhoff constraints.

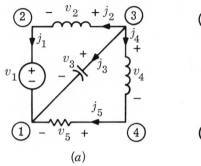
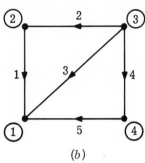

$(a)$                                  $(b)$

**Fig. 4.1**    Network and its oriented graph (arrows indicate reference directions of currents).

quirement on the branch voltages $v_k$ is that they satisfy all the constraints imposed by KVL; similarly, the branch currents $j_k$ must satisfy all the constraints imposed by KCL. The nature of the elements, or, in fact, whether there are any elements that would have these $j_k$'s and $v_k$'s as branch currents and voltages, is *absolutely irrelevant* as far as the truth of Tellegen's theorem is concerned. The power of the theorem lies in the fact that the $v_k$'s and $j_k$'s are arbitrary except for the Kirchhoff constraints.

**THEOREM**    Consider an arbitrary lumped network whose graph $\mathcal{G}$ has $b$ branches and $n_t$ nodes. Suppose that to each branch of the graph we assign arbitrarily a branch voltage $v_k$ and a branch current $j_k$ for $k = 1, 2, \ldots, b$, and suppose that they are measured with respect to arbitrarily picked *associated* reference directions. If the branch voltages $v_1, v_2, \ldots, v_b$ satisfy all the constraints imposed by KVL and if the branch currents $j_1, j_2, \ldots, j_b$ satisfy all the constraints imposed by KCL, then

(4.1)

$$\sum_{k=1}^{b} v_k j_k = 0$$

**Example**    In the network of Fig. 4.1, let us arbitrarily assign branch voltages and branch currents subject only to satisfying Kirchhoff's laws for all the loops and nodes. For example, let us choose

$v_1 = 2 \qquad v_2 = -1 \qquad v_3 = 1 \qquad v_4 = 4 \qquad v_5 = -3$

$j_1 = 1 \qquad j_2 = 1 \qquad j_3 = -3 \qquad j_4 = 2 \qquad j_5 = 2$

Referring to Fig. 4.1, we note that KVL is satisfied since

$v_1 + v_2 = v_3 = v_4 + v_5$

and KCL is satisfied since

$$j_1 = j_2 \qquad j_4 = j_5 \qquad j_1 + j_3 + j_5 = 0$$

To check Tellegen's theorem, we calculate

$$\sum_{k=1}^{5} v_k j_k = 2 - 1 - 3 + 8 - 6 = 0$$

**Remark**  It is of crucial importance to realize that the branch voltages $v_1, v_2, \ldots, v_b$ are picked *arbitrarily* subject only to the KVL constraints. Similarly, the branch currents $j_1, j_2, \ldots, j_b$ are picked *arbitrarily* subject only to the KCL constraints.

For example, suppose $\hat{v}_1, \hat{v}_2, \ldots, \hat{v}_b$ and $\hat{j}_1, \hat{j}_2, \ldots, \hat{j}_b$ are other sets of arbitrarily selected branch voltages and branch currents that obey the *same* KVL constraints and the *same* KCL constraints. Then we may apply Eq. (4.1) to the $\hat{v}_k$'s and the $\hat{j}_k$'s and obtain

(4.2)  $$\sum_{k=1}^{b} \hat{v}_k \hat{j}_k = 0$$

However, we can do the same to the $v_k$'s and $\hat{j}_k$'s and obtain

$$\sum_{k=1}^{b} v_k \hat{j}_k = 0$$

and, also to the $\hat{v}_k$'s and $j_k$'s and obtain

$$\sum_{k=1}^{b} \hat{v}_k j_k = 0$$

*Proof of Tellegen's theorem*  Let us assume, for simplicity, that the graph $\mathcal{G}$ is *connected* and has no branches in parallel; i.e., there exists only one branch between any two nodes. The proof can be easily extended to the general case.† We first pick an arbitrary node as a reference node, and we label it node ①. Thus, $e_1 = 0$. Let $e_\alpha$ and $e_\beta$ be the voltages of the $\alpha$th node and $\beta$th node, respectively, with respect to the reference node. It is important to note that once the branch voltages $(v_1, v_2, \ldots, v_b)$ are chosen, then by KVL the node voltages $e_1, \ldots, e_\alpha, \ldots, e_\beta, \ldots$ are uniquely specified. Let us assume that branch $k$ connects node ⓐ and node ⓑ as shown in Fig. 4.2, and let us denote by $j_{\alpha\beta}$ the current flowing in branch $k$ *from* node ⓐ *to* node ⓑ. Then

$$v_k j_k = (e_\alpha - e_\beta) j_{\alpha\beta}$$

Obviously, $v_k j_k$ can also be written in terms of $j_{\beta\alpha}$, the current *from* node ⓑ *to* node ⓐ, as

(4.3)  $$v_k j_k = (e_\beta - e_\alpha) j_{\beta\alpha}$$

---

† If there are branches in parallel, replace them by a single branch whose current is the sum of the branch currents. If there are several separate parts, the proof shows that Eq. (4.1) holds for each one of them. Hence it holds also when the sum ranges over all branches of the graph.

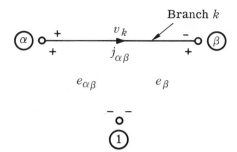

**Fig. 4.2**    An arbitrary branch, branch $k$, connects node
$\alpha$ and node $\beta$; the branch current $j_k$ is also
designated as $j_{\alpha\beta}$ or $-j_{\beta\alpha}$.

Adding the above two equations, we obtain

(4.4)    $v_k j_k = \frac{1}{2}[(e_\alpha - e_\beta)j_{\alpha\beta} + (e_\beta - e_\alpha)j_{\beta\alpha}]$

Now, if we sum the left-hand side of Eq. (4.4) for all the branches in the
graph, we obtain

$$\sum_{k=1}^{b} v_k j_k$$

The corresponding sum of the right-hand side of Eq. (4.4) becomes

$$\frac{1}{2}\sum_{\alpha=1}^{n_t}\sum_{\beta=1}^{n_t}(e_\alpha - e_\beta)j_{\alpha\beta}$$

where the double summation has indices $\alpha$ and $\beta$ carried over all the nodes
in the graph.   The sum leads to the following equation:

(4.5)    $$\sum_{k=1}^{b} v_k j_k = \frac{1}{2}\sum_{\alpha=1}^{n_t}\sum_{\beta=1}^{n_t}(e_\alpha - e_\beta)j_{\alpha\beta}$$

Note that if there is no branch joining node $\alpha$ to node $\beta$, we set
$j_{\alpha\beta} = j_{\beta\alpha} = 0$.   Now that Eq. (4.5) has been established, let us split the
right-hand side of (4.5) as follows:

$$\sum_{k=1}^{b} v_k j_k = \frac{1}{2}\sum_{\alpha=1}^{n_t} e_\alpha\left(\sum_{\beta=1}^{n_t} j_{\alpha\beta}\right) - \frac{1}{2}\sum_{\beta=1}^{n_t} e_\beta\left(\sum_{\alpha=1}^{n_t} j_{\alpha\beta}\right)$$

For each fixed $\alpha$, $\displaystyle\sum_{\beta=1}^{n_t} j_{\alpha\beta}$ is the sum of *all* branch currents leaving node $\alpha$.

For each fixed $\beta$, $\displaystyle\sum_{\alpha=1}^{n_t} j_{\alpha\beta}$ is the sum of *all* branch currents entering node
$\beta$.   By KCL, each one of these sums is zero, hence,

$$\sum_{k=1}^{b} v_k j_k = 0$$

Thus, we have shown that given any set of branch voltages subject to KVL only and any set of branch currents subject to KCL only, the sum of the products $v_k j_k$ is zero.   This concludes the proof.

**Exercise 1**   Suppose that starting from the reference node and following a certain path to node $\textcircled{\small $\alpha$}$, we obtain (by adding appropriate branch voltages) for its node potential the value $e_\alpha$.   Show that if following another path we were to obtain a potential $e'_\alpha \neq e_\alpha$, then the branch voltages of these two paths would violate KVL.

**Exercise 2**   Consider an arbitrary network driven by any number of sources of any kind.   Let $v_1(t), v_2(t), \ldots, v_b(t)$ and $j_1(t), j_2(t), \ldots, j_b(t)$ be its branch voltages and currents at time $t$.   If $t_a$ and $t_b$ are arbitrarily selected instants of time, what can you say about

$$\sum_{k=1}^{b} v_k(t_a) j_k(t_b)$$

## 5 Applications

### 5.1 Conservation of Energy

Considering an arbitrary network, we have, with the notations of Tellegen's theorem,

$$\sum_{k=1}^{b} v_k(t) j_k(t) = 0 \qquad \text{for all } t$$

Since $v_k(t)j_k(t)$ is the power delivered at time $t$ by the network to branch $k$, the theorem may be interpreted as follows: at any time $t$ the sum of the power delivered to each branch of the network is zero.   Suppose the network has several independent sources; separating in the sum the sources from the other branches, we conclude that *the sum of the power delivered by the independent sources to the network is equal to the sum of the power absorbed by all the other branches of the network.*   From a philosophical point of view, this means that as far as lumped circuits are concerned, KVL and KCL imply conservation of energy.

Let us briefly look into the interpretation of this conservation of energy as far as linear time-invariant *RLC* networks are concerned.   The power delivered by the sources is the rate at which energy is absorbed by the network.   The energy is either dissipated in the resistors at the rate $R_k j_k^2(t)$ for the $k$th resistor, or it is stored as magnetic energy in inductors $[\frac{1}{2} L_k j_k^2(t)]$ or as electric energy in capacitors $[\frac{1}{2} C_k v_k^2(t)]$.   When elements are time-varying (as in electric motors and generators or in parametric amplifiers), the discussion is much more complicated and is discussed in Chap. 19.

**Remark**   Tellegen's theorem has some rather astonishing consequences. For example, consider two arbitrary lumped networks whose only constraint is to have the same *graph*. In each one of these networks, let us choose the same reference directions and number the branches in a similar fashion. (The networks may be nonlinear and time-varying and include independent sources as well as dependent sources.) Let $v_k, j_k$ be the branch voltages and currents of the first network and $\hat{v}_k, \hat{j}_k$ be corresponding branch voltages and currents of the second. Since the $v_k$'s and $\hat{v}_k$'s satisfy the same set of KVL constraints and since the $\hat{j}_k$'s and the $j_k$'s satisfy the same set of KCL constraints, Tellegen's theorem guarantees that

$$\sum_{k=1}^{b} v_k j_k = \sum_{k=1}^{b} \hat{v}_k \hat{j}_k = 0$$

and

$$\sum_{k=1}^{b} \hat{v}_k j_k = \sum_{k=1}^{b} v_k \hat{j}_k = 0$$

Note that whereas the first two are expressions of the conservation of energy, the last two expressions do *not* have an energy interpretation because they involve voltages of one network and currents of another.

---

**5.2   Conservation of Complex Power**

Consider a linear time-invariant network. For simplicity let it have only one sinusoidal source in branch 1, as shown in Fig. 5.1. Suppose that the network is in the sinusoidal steady state. For each branch (still using associated reference directions), we represent the branch voltage $v_k$ by the phasor $V_k$ and the branch current $j_k$ by the phasor $J_k$. Clearly, $V_1, V_2, \ldots, V_b$ and $J_1, J_2, \ldots, J_b$ satisfy all the constraints imposed by KVL and KCL. However, the conjugates $\bar{J}_1, \bar{J}_2, \ldots, \bar{J}_b$ also satisfy all the KCL constraints; therefore, by Tellegen's theorem

(5.1)   $$\sum_{k=1}^{b} \tfrac{1}{2} V_k \bar{J}_k = 0$$

Since $V_1$ is the source voltage and $J_1$ is the associated current measured with respect to the associated reference direction, $\tfrac{1}{2} V_1 \bar{J}_1$ is the complex power delivered *to* branch 1 *by* the rest of the network, and hence $-\tfrac{1}{2} V_1 \bar{J}_1$ is the complex power delivered by the source to the rest of the network. We rewrite Eq. (5.1) as follows:

$$-\tfrac{1}{2} V_1 \bar{J}_1 = \sum_{k=2}^{b} \tfrac{1}{2} V_k \bar{J}_k$$

Clearly, the above can be generalized to networks with more than one source. Thus, we state the theorem of *conservation of complex power* as follows.

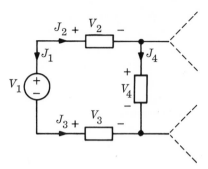

**Fig. 5.1**   Linear time-invariant network in the sinusoidal steady state; the $V_k$'s and the $J_k$'s are
*phasors* which represent the sinusoidal
voltages and currents.

**THEOREM**   Consider a linear time-invariant network which is in the *sinusoidal steady
state* and which is driven by several independent sources that are at the
same frequency (see Fig. 5.2).   Let us pull out all the independent sources
as shown and denote the rest of the network by $\mathfrak{N}$.   Then the sum of the
complex power delivered by each independent source to the network $\mathfrak{N}$ is
equal to the sum of the complex power received by all the other branches
of $\mathfrak{N}$.

**5.3**   **The Real Part and Phase of Driving-point Impedances**

The conservation-of-complex-power theorem can be used to derive many
important properties of driving-point impedances.   With reference to
Fig. 5.3, let us consider the driving-point impedance $Z_{\text{in}}$ of the linear time-
invariant one-port $\mathfrak{N}$ which contains only resistors, inductors, capacitors,
and/or transformers.   Let the network $\mathfrak{N}$ be driven by a sinusoidal current
source at an angular frequency $\omega$.   Let the source current be represented
by the phasor $J_1$.   Let the voltage, measured in the associated reference
direction of the source, be represented by the phasor $V_1$.   Clearly,

$$V_1 = -J_1 Z_{\text{in}}(j\omega)$$

Let the branches inside $\mathfrak{N}$ be numbered from 2 to $b$, and let the branch
phasor currents and branch impedances be denoted by $J_k$ and $Z_k$, $k = 2$,
$3, \ldots, b$.   Let $P$ denote the complex power delivered to the one-port by
the source.   Using Tellegen's theorem, we obtain

$$P = -\tfrac{1}{2} V_1 \bar{J}_1 = \tfrac{1}{2} Z_{\text{in}}(j\omega) |J_1|^2$$

(5.2)
$$= \frac{1}{2} \sum_{k=2}^{b} V_k \bar{J}_k = \frac{1}{2} \sum_{k=2}^{b} Z_k(j\omega) |J_k|^2$$

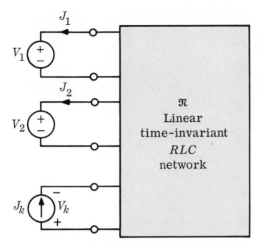

**Fig. 5.2**    Theorem of conservation of complex power.

If we take the real part of Eq. (5.2), we obtain $P_{av}$, the average power delivered by the source to $\mathfrak{N}$, and

$$P_{av} = \tfrac{1}{2}\,\mathrm{Re}\,[Z_{in}(j\omega)]|J_1|^2 = \frac{1}{2}\sum_{k=2}^{b}\mathrm{Re}\,[Z_k(j\omega)]|J_k|^2$$

Note that all these impedances are evaluated at the same angular frequency $\omega$, which is the angular frequency of the source. Let us study the implications of Eq. (5.2) for the following cases.

*Case* 1    Resistive networks made of branches all having positive resistances: Then in Eq. (5.2) all the $Z_k$'s are positive real numbers. Consequently, $Z_{in}$, the input resistance, is a positive real number. In this case, $Z_{in}$ does not depend on the angular frequency $\omega$. Consequently, *the input impedance of a resistive network made of positive resistances is a positive resistance.*

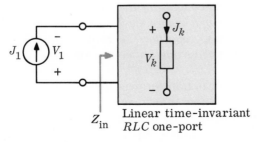

**Fig. 5.3**    Properties of driving-point impedance $Z_{in}(j\omega)$.

*Case 2*   *RL networks made of branches all having positive resistances or positive inductances:*   Then $Z_k$ is either a positive real number or a purely imaginary number of the form $j\omega L_k$, where $\omega > 0$ and $L_k > 0$.   Hence, from Eq. (5.2),

$$\text{Re}\,[Z_{\text{in}}(j\omega)] \geq 0 \qquad \text{and} \qquad \text{Im}\,[Z_{\text{in}}(j\omega)] \geq 0 \qquad \text{for all } \omega \geq 0$$

or equivalently,

$$0 \leq \sphericalangle Z_{\text{in}}(j\omega) \leq 90° \qquad \text{for all } \omega \geq 0$$

Thus, we have shown that *at any positive angular frequency $\omega$, the driving-point impedance of a linear time-invariant RL network made of positive resistances and positive inductances has a phase angle between 0 and 90°.*

*Case 3*   *RC networks made of branches all having either positive resistances or positive capacitances:*   A similar reasoning shows that *at any positive angular frequency $\omega$, the driving-point impedance of a linear time-invariant RC network made of positive resistances and positive capacitances has a phase angle between 0 and $-90°$.*

*Case 4*   *Lossless networks made of capacitors, inductors (including coupled inductors), and/or ideal transformers:*   In Chap. 8, Sec. 2, we stated that coupled inductors can be replaced by inductors without coupling and an ideal transformer.   Thus, inside the one-port $\mathfrak{N}$ we can assume that all branches are either positive inductances, positive capacitances, or ideal transformer windings.   Because ideal transformers neither dissipate nor store energy, the sum $\displaystyle\sum_k V_k \bar{J}_k$ over all the branches with ideal transformers is zero; therefore, ideal transformers contribute nothing to the sum in Eq. (5.2).   The other terms are either of the form $j\omega L_k |J_k|^2$ or $(1/j\omega C_k)|J_k|^2$; in either case, they are purely imaginary.   Therefore,

(5.3)   $$\text{Re}\,[Z_{\text{in}}(j\omega)] = 0 \qquad \text{for all } \omega$$

or equivalently,

$$\sphericalangle Z_{\text{in}}(j\omega) = \pm 90° \qquad \text{for all } \omega$$

Hence we arrive at the conclusion that *at any angular frequency $\omega$, the driving-point impedance of a linear time-invariant network made of inductors (coupled or uncoupled), capacitors, and ideal transformers is purely imaginary; i.e., it has a phase angle of either $-90$ or $+90°$.*

*Case 5*   *RLC networks with ideal transformers having all branches with either positive resistances, positive inductances, positive capacitances, and/or ideal transformer windings:*   As in case 4, the ideal transformers do not contribute anything to the sum in Eq. (5.2).   The other terms are either of the form $R_k |J_k|^2$, $j\omega L_k |J_k|^2$, or $(1/j\omega C_k)|J_k|^2$.   Therefore, each term

$Z_k|J_k|^2$ is either a positive number or an imaginary number. Hence, $Z_{in}$ is a complex number with a real part that is larger than or equal to zero and an imaginary part that may be of either sign. Therefore, we conclude that *at any angular frequency $\omega$, the driving-point impedance of a linear time-invariant RLC network (which may include ideal transformers) has a non-negative real part; equivalently, it has a phase angle between $-90$ and $+90°$:* that is,

(5.4)  $\mathrm{Re}\,[Z_{in}(j\omega)] \geq 0$    for all $\omega$

or equivalently,

$-90° \leq \sphericalangle Z_{in}(j\omega) \leq 90°$    for all $\omega$


**5.4**    **Driving-point Impedance, Power Dissipated, and Energy Stored**

We consider again a linear time-invariant RLC network driven by a single sinusoidal current source (see Fig. 5.3). Again assuming that the network is in the sinusoidal steady state and using the previous notation, we can write the complex power delivered by the source to the network as follows (see Eq. 5.2):

$$P = \tfrac{1}{2}Z_{in}(j\omega)|J_1|^2 = \frac{1}{2}\sum_{m=2}^{b} Z_m(j\omega)|J_m|^2$$

$$= \frac{1}{2}\sum_i R_i|J_i|^2 + \frac{1}{2}\sum_k j\omega L_k|J_k|^2 + \frac{1}{2}\sum_l \frac{1}{j\omega C_l}|J_l|^2$$

where we write out as separate sums the terms corresponding to resistors, inductors, and capacitors. Exhibiting the real and imaginary part of $P$, we obtain

(5.5)  $P = \tfrac{1}{2}\sum_i R_i|J_i|^2 + 2\,j\omega\left(\sum_k \frac{1}{4}L_k|J_k|^2 - \sum_l \frac{1}{4}\frac{1}{\omega^2 C_l}|J_l|^2\right)$

Now we have seen that in the sinusoidal steady state the average (over one period) of $R_i j_i{}^2(t)$ is

$\tfrac{1}{2}R_i|J_i|^2$

Similarly, the average of $\tfrac{1}{2}L_k j_k{}^2(t)$ is

$\tfrac{1}{4}L_k|J_k|^2$

and the average of $\tfrac{1}{2}C_l v_l{}^2(t)$ is

$\tfrac{1}{4}C_l|V_l|^2 = \frac{1}{4}\frac{1}{\omega^2 C_l}|J_l|^2$

Thus, the first term of Eq. (5.5) is the average power dissipated in $\mathfrak{N}$ (denoted by $P_{av}$), and the two terms in the parentheses are, respectively, the average magnetic energy stored $\mathscr{E}_M$ and the average electric energy stored $\mathscr{E}_E$. Therefore, Eq. (5.5) can be rewritten as

(5.6) $\quad Z_{\text{in}}(j\omega) = \dfrac{2P_{\text{av}} + 4j\omega(\mathscr{E}_M - \mathscr{E}_E)}{|J_1|^2}$

It should be stressed (1) that $|J_1|$ is the *peak* amplitude of the sinusoidal input current, (2) $P_{\text{av}}$, $\mathscr{E}_M$, and $\mathscr{E}_E$ are, respectively, the *average* power dissipated, the *average* magnetic energy stored, and the *average* electric energy stored, and (3) these three averages are obtained by averaging over one period of the sinusoidal motion. Thus, we have established the following result.

**THEOREM** Given a linear time-invariant *RLC* network driven by a sinusoidal current source of $|J_1|$ amperes peak amplitude and given that the network is in the sinusoidal steady state, the driving-point impedance seen by the source has a real part that is equal to twice the average power dissipated and an imaginary part that is $4\omega$ times the difference between the average magnetic energy stored and the average electrical energy stored.

**Exercise 1** Write the equation expressing the conservation of complex power for the network of Fig. 4.1

    *a.* in terms of branch-voltage phasors and branch-current phasors

    *b.* in terms of branch impedances and branch currents

**Exercise 2** Prove Case 3 of Sec. 5.3 in detail.

**Exercise 3** Consider the following time-invariant circuits: the series *RL*, the parallel *RC*, the series *LC*, the parallel *LC*, the series *RLC*, the parallel *RLC*. In each case find the frequency (if any) at which $\not{\!\!\!\angle} Z(j\omega) = 0$, $45$, $-45$, and $90°$.

**Exercise 4** An *RLC* network is in the sinusoidal steady state and is driven by a single source delivering the current $j_1(t) = 7 \cos (277t - 30°)$ amp; given that the *average* power dissipated in the resistors is 10 watts, what can you say about the impedance faced by the source at the frequency of 60 Hz?

**Exercise 5** Show that any two-terminal network obtained by interconnecting any number of *passive* two-terminal elements is passive.

## Summary

- The networks under study are connections of *models* of physical elements. These models are resistors, capacitors, inductors, coupled inductors, ideal transformers, and dependent and independent sources. By assigning associated reference directions to the branch voltages and branch currents and by disregarding the nature of the elements, we obtain an *oriented graph* $\mathcal{G}$. Once the branches and nodes of $\mathcal{G}$ are numbered, there is a one-to-one correspondence between the graph $\mathcal{G}$ and its *incidence matrix* $\mathbf{A}_a$. The $n_t \times b$ matrix $\mathbf{A}_a$ is such that $a_{ik} = 1$ or $-1$, depending on whether branch $k$ leaves or enters node $\textcircled{i}$; $a_{ik} = 0$ if branch $k$ is not incident with node $\textcircled{i}$.

- By **cut set** of a connected graph, we mean a set of branches such that (1) the removal of all branches of the set causes the remaining graph to have two separate parts, and (2) the removal of all but any one of the branches of the set leaves the remaining graph connected.

- By KCL, for *any* lumped network, for *any* of its cut sets, and at *any* time, the algebraic sum of all the currents in the cut-set branches is zero.

- Given a graph $\mathcal{G}$, $\mathcal{L}$ is called a loop if (1) $\mathcal{L}$ is a connected subgraph of $\mathcal{G}$ and (2) precisely two branches of $\mathcal{L}$ are incident with each node of $\mathcal{L}$.

- By KVL, for *any* lumped network, for *any* of its loops and at *any* time, the algebraic sum of all the branch voltages around the loop is zero.

- The two Kirchhoff laws hold irrespective of the nature of the elements. When expressed in terms of branch currents (KCL) or branch voltages (KVL), they give *linear* algebraic equations with *constant coefficients*.

- Tellegen's theorem applies to *any lumped* network. Given its oriented graph, if we assign to its branches *arbitrary* branch voltages $v_1, v_2, \ldots, v_b$ subject only to the constraints imposed by KVL, and *arbitrary* branch currents $j_1, j_2, \ldots, j_b$ subject only to the constraints imposed by KCL, then

$$\sum_{k=1}^{b} v_k j_k = 0$$

- From this theorem we proved that

$$Z_{\text{in}}(j\omega) = \frac{2P_{\text{av}} + 4j\omega(\mathcal{E}_M - \mathcal{E}_E)}{|J_1|^2}$$

where $|J_1|$ is the "peak" input current, and all the averages are taken over one period of the sinusoidal steady-state motion at frequency $\omega$.

## Problems

From network to graph

**1.** For the networks shown in Fig. P9.1 find the node-to-branch incidence matrix. The numbers shown refer to the branch number. [For simplicity, associate one branch to the series (or parallel) connection of two or more elements.]

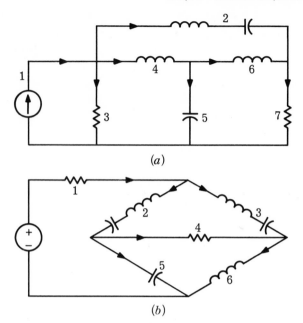

(a)

(b)

**Fig. P9.1**

Incidence
matrix

**2.** Draw an oriented graph whose node-to-branch incidence matrix $\mathbf{A}_a$ is given by

$$\mathbf{A}_a = \begin{bmatrix} 1 & 1 & -1 & 0 & 0 & 0 & 0 & 0 & 0 & 0 & 0 & 0 \\ 0 & 0 & 0 & 0 & -1 & -1 & 1 & 0 & 0 & 0 & 0 & 0 \\ -1 & 0 & 0 & 0 & 0 & 0 & 0 & 0 & -1 & 1 & 0 & 0 \\ 0 & 0 & 1 & 1 & 1 & 0 & 0 & 0 & 0 & 0 & 0 & 0 \\ 0 & 0 & 0 & 0 & 0 & 0 & -1 & 1 & 0 & 0 & 0 & -1 \\ 0 & -1 & 0 & -1 & 0 & 1 & 0 & -1 & 1 & 0 & 1 & 0 \\ 0 & 0 & 0 & 0 & 0 & 0 & 0 & 0 & 0 & -1 & -1 & 1 \end{bmatrix}$$

Incidence
matrix and cut
sets

**3.** Consider the oriented graph shown in Fig. P9.3

*a.*   Write the node-to-branch incidence matrix.

*b.*   Specify which of the following sets of branches are cut sets and justify your answer: {1,9,5,8}, {1,9,4}, {6,8}, {1,9,4,7,6}, or {3,4,5,6}.

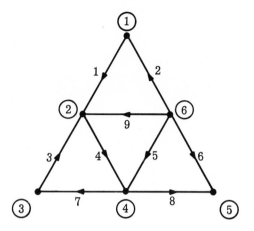

**Fig. P9.3**

Cut sets
and KCL  **4.** Consider the oriented graph shown in Fig. P9.4.

*a.*   Write equations corresponding to the cut sets indicated.

*b.*   Are these cut-set equations linearly independent?

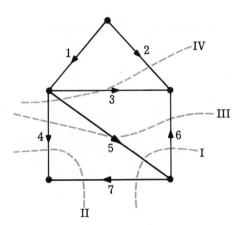

**Fig. P9.4**

Loops and KVL  **5.** Consider the graph shown in Fig. P9.5.   Are the loop equations corresponding to loops *abc, bdf, cde, aef, cehj,* and *bcgi* linearly independent? Justify your answer.

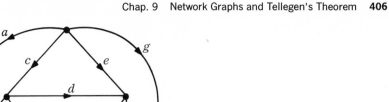

**Fig. P9.5**

Kirchhoff's
laws and
Tellegen's
theorem
**6.** For the networks shown in Fig. P9.6, assign arbitrary branch voltages and branch currents subject only to KVL and KCL constraints.  Verify the conclusion of Tellegen's theorem.

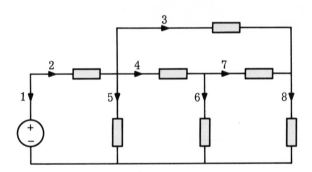

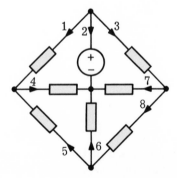

**Fig. P9.6**

Tellegen's
theorem

**7.** The network $\mathfrak{N}$ shown in Fig. P9.7 is made of $n - 2$ linear time-invariant resistors. Voltage and current measurements were taken for two values of $R_2$ and the input. The measurements are tabulated in the figure. Determine the value $\hat{v}_2$.

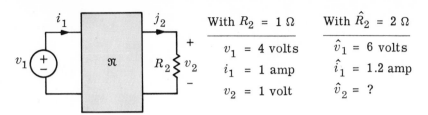

| | With $R_2 = 1\,\Omega$ | With $\hat{R}_2 = 2\,\Omega$ |
|---|---|---|
| | $v_1 = 4$ volts | $\hat{v}_1 = 6$ volts |
| | $i_1 = 1$ amp | $\hat{i}_1 = 1.2$ amp |
| | $v_2 = 1$ volt | $\hat{v}_2 = ?$ |

**Fig. P9.7**

Tellegen's
theorem

**8.** An absent-minded engineer was making sinusoidal steady-state measurements on a linear time-invariant $RLC$ network. He was recording the voltage and the current phasors of the elements shown in Fig. P9.8. At the frequency $\omega_1$ he recorded

$$V_1 = 10\epsilon^{j5°} \qquad J_1 = 5\epsilon^{j40°}$$
$$V_2 = 5\epsilon^{j15°} \qquad J_2 = 7\epsilon^{j45°}$$
$$V_3 = 17\epsilon^{-j10°} \qquad J_3 = 5\epsilon^{-j50°}$$

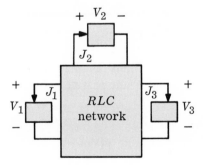

**Fig. P9.8**

He repeated the experiment with a different setting for the oscillator but still at the same frequency $\omega_1$, and he obtained

$$\hat{V}_1 = 7\epsilon^{-j10°} \qquad \hat{J}_1 = 15\epsilon^{-j35°}$$
$$\hat{V}_2 = 3\epsilon^{j5°} \qquad \hat{J}_2 = 5\epsilon^{-j35°}$$
$$\hat{J}_3 = 10\epsilon^{j30°}$$

He forgot to write down $V_3$. Can you find it for him?

Tellegen's
theorem

**9.** Consider the sinusoidal steady-state measurements performed on a linear time-invariant $RLC$ network, shown in Fig. P9.9. In both instances the same voltage source is used (same frequency and same phasor). Show that $\hat{J}_1 = J_2$. (Hint: Show that $V_2\hat{J}_2 + V_1\hat{J}_1 = \hat{V}_2 J_2 + \hat{V}_1 J_1$.)

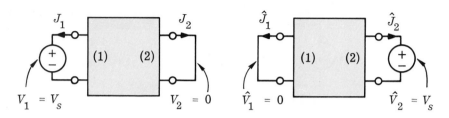

**Fig. P9.9**

Measurements
and properties
of RLC
networks

**10.** Your technician measures the driving-point impedance (or admittance) at a fixed frequency $\omega_0$ of a number of linear time-invariant networks made of passive elements. In each case, state whether or not you have any reasons to believe his results (in ohms or in mhos). Whenever you accept a measurement as plausible, assume $\omega_0 = 1$ rad/sec, and give a linear time-invariant passive network which has the specified network function.

   *a.*   $RC$ network:  $Z = 5 + j2$

   *b.*   $RL$ network:  $Z = 5 - j7$

   *c.*   $RLC$ network: $Y = 2 - j3$

   *d.*   $LC$ network:  $Z = 2 + j3$

   *e.*   $RLC$ network: $Z = -5 - j19$

   *f.*   $RLC$ network: $Z = -j7$

Whenever you accept a measurement as plausible, assume $\omega_0 = 1$ rad/sec, and give a linear time-invariant passive network which has the specified network function.

# 10 Node and Mesh Analyses

This chapter and the succeeding two are devoted to *general* methods of network analysis. The problem of network analysis may be stated as follows: given the network graph, the branch characteristics, the input (i.e., the waveforms of the independent sources), and the initial conditions, calculate all branch voltages and branch currents. In these three chapters we shall consider only the formulation of network equations. The methods of solution and the properties of the solutions will be studied in Chaps. 13 to 16.

In the present chapter we shall systematically develop node and mesh analysis. This systematic treatment is particularly important at present when computers automatically perform the analysis of networks. Also from the results of these systematic analyses we shall obtain the tools that will allow us to develop the properties of these networks.

In Sec. 1, we present source transformations that we shall use in all the following methods of analysis. In Sec. 2, the consequences of Kirchhoff's laws are obtained in the context of node analysis. Section 3 develops the systematic analysis of linear time-invariant networks. Duality is developed in Sec. 4. Finally, mesh analysis is presented in Secs. 5 and 6. Again, only linear time-invariant networks are considered.

## 1 Source Transformations

In the general discussion of the problem of network analysis we assume that the number and the location of independent sources are arbitrary as long as Kirchhoff's laws are not violated (i.e., as long as no independent voltage sources form a loop and no independent current sources form a cut set—for in either case the waveforms of these sources would have to satisfy a linear constraint imposed by KVL and KCL, respectively).

To obviate separating the branches consisting only of sources from the other branches, it is useful to introduce first two network transformations that allow us to relocate sources in the network without affecting the problem. These transformations can be used for both independent and dependent sources. They are illustrated in Figs. 1.1 and 1.2.

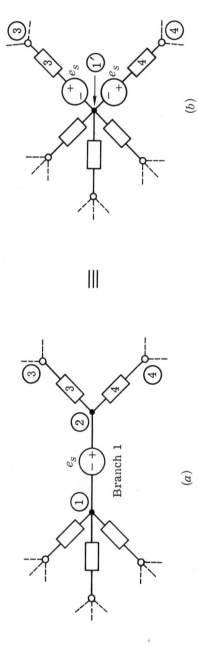

**Fig. 1.1**   Source transformation; a branch consisting of a voltage source alone is eliminated.

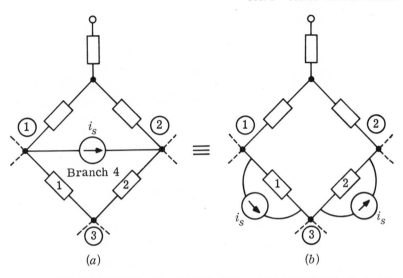

**Fig. 1.2**   Source transformation; a branch consisting of a current source alone is eliminated.

In Fig. 1.1$a$, branch 1 is a voltage source $e_s$ which is connected between nodes ① and ②.  Node ② is connected to node ③ through branch 3 and to node ④ through branch 4.  If the current in branch 1 is of no interest to us, we can replace the circuit in Fig. 1.1$a$ with its equivalent in Fig. 1.1$b$.  In the new circuit, branch 1 has been eliminated, and a new node ①′ is introduced.  This new node ①′ results from the merger of nodes ① and ② of the original circuit.  To be equivalent, two sources $e_s$ must be inserted in branch 3 and branch 4 of the new circuit.

Showing that this transformation does not change the solution of the problem is straightforward.  We need only write the KVL equations for all the loops containing branch 3 and all the loops containing branch 4 in both networks.  It is easily checked that the corresponding equations for the two networks are the same.  Also KCL applied to node ①′ is identical to the sum of the equations obtained by applying KCL to node ① and to node ② of the given network.  Consequently, the KCL equations of both networks impose equivalent constraints.

In Fig. 1.2$a$, branch 4 is a current source $i_s$ connected between nodes ① and ②.  Nodes ① and ② are also connected to node ③ through branches 1 and 2, respectively.  In Fig. 1.2$b$ we show the equivalent circuit, where the current source in branch 4 has been removed; instead, two new current sources $i_s$ are connected in parallel with branches 1 and 2.  That this transformation does not change the solution of the problem can be seen by writing the KCL equations for nodes ①, ②, and ③ in both networks.  Clearly, the corresponding equations are the same.

**Exercise 1**   Show that except for the element affected the following transformations do not affect the branch voltages and the branch currents of a network: (1) if a branch consists of a current source in series with an element, the element may be replaced by a short circuit; (2) if a branch consists of a current source in series with a voltage source, the voltage source may be replaced by a short circuit; (3) if a branch consists of a voltage source in parallel with an element, this element may be replaced by an open circuit; (4) if a branch consists of a voltage source in parallel with a current source, the current source may be replaced by an open circuit. Observe that cases (3) and (4) are the duals of (1) and (2), respectively.

In conclusion, by using these transformations, we can modify any given network in such a way that *each voltage source is connected in series with an element which is not a source and each current source is connected in parallel with an element which is not a source.*

Thus, we find that, without loss of generality, we can assume that for any network a typical branch, say, branch $k$, is of the form shown in Fig. 1.3, where $v_{sk}$ is a voltage source, $j_{sk}$ is a current source, and the rectangular box represents an element which is not a source. As before, the branch voltage is denoted by $v_k$ and the branch current by $j_k$. The characterization of branch $k$ thus includes possible source contributions. In particular, if there is no voltage source in branch $k$, we set $v_{sk} = 0$; similarly, if there is no current source, we set $j_{sk} = 0$.

**Exercise 2**   Suppose that, in Fig. 1.3, the nonsource element is a linear time-invariant resistor of resistance $R_k$. Show that the branch equation is

$$(1.1) \quad v_k = v_{sk} - R_k j_{sk} + R_k j_k$$

Show that this branch can be further simplified to look like Fig. 1.5.

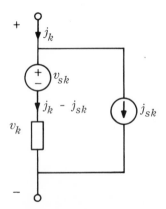

**Fig. 1.3**   Branch $k$, including voltage and current sources.

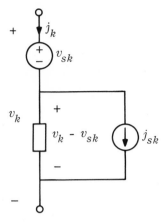

**Fig. 1.4**   Branch $k$, including voltage and
current sources.

**Exercise 3**   Suppose that, in Fig. 1.4, the nonsource element is a linear time-invariant
resistor of conductance $G_k$.  Show that the branch equation is

(1.2)   $j_k = j_{sk} - G_k v_{sk} + G_k v_k$

Show that this branch can be further simplified to look like Fig. 1.6.

These two exercises illustrate useful equivalences.  It is convenient,
though not necessary, in node analysis for all independent sources to be
current sources.  Similarly, it is convenient, though not necessary, in loop
or mesh analysis for all independent sources to be voltage sources.
From now on we shall assume that in dealing with resistive networks
the branches are always of the form of Fig. 1.5 or Fig. 1.6, that is, a resistor
in series with a voltage source or a resistor in parallel with a current source.

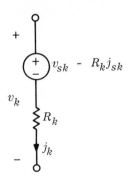

**Fig. 1.5**   A resistive branch with an equivalent
voltage source.

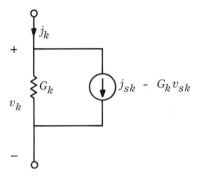

**Fig. 1.6**    A resistive branch with an equivalent current source.

In general networks, the resistor may be replaced by an inductor or a capacitor.

**Exercise 4**    Perform the transformation for the circuits in Fig. 1.7.

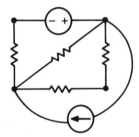 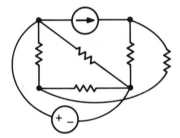

**Fig. 1.7**    Exercises on source transformation.

## 2    Two Basic Facts of Node Analysis

Let us consider any network $\mathfrak{N}$ and let it have $n_t$ nodes and $b$ branches. Altogether there are $b$ branch voltages and $b$ branch currents to be determined. Without loss of generality we may assume that the graph is connected, i.e., that it has one separate part only. (If the network were made of two separate parts, we could connect these two separate parts by tying them together at a common node.)

First, we pick arbitrarily a reference node. This reference node is usually called the **datum node.** We assign to the datum node the label $\textcircled{n_t}$ and to the remaining nodes the labels $\textcircled{1}, \textcircled{2}, \ldots, \textcircled{n}$, where $n \overset{\Delta}{=} n_t - 1$.

### 2.1    Implications of KCL

Let us apply KCL to nodes $\textcircled{1}, \textcircled{2}, \ldots, \textcircled{n}$ (omitting the datum node) and let us examine the form of the equations obtained. Typically, as in the case of node $\textcircled{k}$ shown in Fig. 2.1, we obtain a homogeneous linear algebraic equation in the branch currents; thus,

$$j_4 + j_6 - j_7 = 0$$

Thus, we have a system of $n$ linear algebraic equations in $b$ unknowns $j_1, j_2, \ldots, j_b$. The first basic fact of node analysis is the following statement:

*The $n$ linear homogeneous algebraic equations in $j_1, j_2, \ldots, j_b$, obtained by applying KCL to each node except the datum node, constitute a set of linearly independent equations.*

Let us start by an observation. Consider the $n_t$ equations obtained by writing KCL for each of the $n_t$ nodes of $\mathfrak{N}$. Let $\mathcal{E}_k = 0$ represent the equation pertaining to node $\textcircled{k}$, $k = 1, 2, \ldots, n_t$. For the node $\textcircled{k}$ shown in Fig. 2.1, the equation $\mathcal{E}_k = 0$ gives $j_4 + j_6 - j_7 = 0$. We assert that $\sum_{k=1}^{n_t} \mathcal{E}_k$ reduces identically to zero. In other words, if we add all the $n_t$ KCL equations (written in terms of the branch currents $j_1, j_2, \ldots, j_b$), all terms cancel out. This is obvious. Suppose branch 1 leaves node $\textcircled{2}$ and enters node $\textcircled{3}$. The term $j_1$ appears with a plus sign in $\mathcal{E}_2$ and a minus sign in $\mathcal{E}_3$, and, since $j_1$ appears in no other equation, $j_1$ cancels out in the sum. Since every branch of $\mathfrak{N}$ must leave one node and terminate on another node, all branch currents will cancel out in the sum. We conclude that *the $n_t$ equations obtained by writing KCL for each of the nodes of the network $\mathfrak{N}$ are linearly dependent.*

Let us now prove that the $n$ linear algebraic equations $\mathcal{E}_1 = 0$, $\mathcal{E}_2 = 0, \ldots, \mathcal{E}_n = 0$ are linearly independent. Suppose they were not; i.e., suppose that these $n$ equations are linearly dependent. This would mean that, after some possibly necessary reordering of the equations, there is a linear combination of the first $k$ equation $\mathcal{E}_1, \mathcal{E}_2, \ldots, \mathcal{E}_k$ $(k \leq n)$ with respective *nonzero* weighting factors $\alpha_1, \alpha_2, \ldots, \alpha_k$, which sums *identically* to zero. Thus,

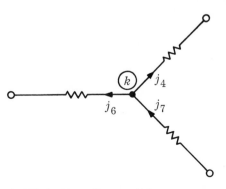

**Fig. 2.1**    A typical node to illustrate KCL.

(2.1)   $\alpha_1 \mathcal{E}_1 + \cdots + \alpha_k \mathcal{E}_k \equiv 0$

Consider the set of all nodes ①, ②, ..., ⓚ corresponding to $\mathcal{E}_1$, $\mathcal{E}_2, \ldots, \mathcal{E}_k$ in Eq. (2.1). Consider the set of remaining nodes as shown in Fig. 2.2. The remainder of $\mathfrak{N}$ contains $n_t - k$ nodes; since $k \leq n$, the remainder includes at least one node. Since $\mathfrak{N}$ is connected, there is at least one branch, say branch $l$, which joins a node of the first set to a node of the second set. Then $j_l$ can appear in one and only one of the equations $\mathcal{E}_1, \mathcal{E}_2, \ldots, \mathcal{E}_k$ since the branch $l$ is connected to only one of the nodes of the first set. Therefore, it cannot cancel out in the sum

$$\sum_{i=1}^{k} \alpha_i \mathcal{E}_i$$

Hence we arrive at a contradiction with Eq. (2.1). This argument holds for any possible linear combination. Therefore, the assumption that $\mathcal{E}_1$, $\mathcal{E}_2, \ldots, \mathcal{E}_n$ are linearly dependent is false; hence the statement that these equations are linearly independent has been proved.

Consider again the system of $n$ linear algebraic equations that express KCL for all the nodes of $\mathfrak{N}$ except the datum node. We assert that this system has the following matrix form:

(2.2)   $\mathbf{Aj} = \mathbf{0}$   (KCL)

where $\mathbf{j}$ represents the branch current vector and is of dimension $b$; that is,

$$\mathbf{j} = \begin{bmatrix} j_1 \\ j_2 \\ \vdots \\ j_b \end{bmatrix}$$

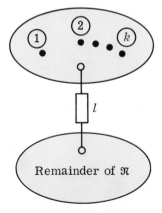

**Fig. 2.2**   Graph used to prove that $n$ KCL equations are linearly independent.

and where $\mathbf{A} = (a_{ik})$ is an $n \times b$ matrix defined by

(2.3)    $a_{ik} = \begin{cases} 1 & \text{if branch } k \text{ \textit{leaves} node } \textcircled{i} \\ -1 & \text{if branch } k \text{ \textit{enters} node } \textcircled{i} \\ 0 & \text{if branch } k \text{ is \textit{not incident with} node } \textcircled{i} \end{cases}$

$\mathbf{A}j$ is therefore a vector of dimension $n$.   This assertion is obvious since when we write that the $i$th component of the vector $\mathbf{A}j$ is equal to zero, we merely assert that the sum of all branch currents leaving node $\textcircled{i}$ is zero.

It is immediately observed that the rule expressed by Eq. (2.3) is identical with the rule specifying the elements of the node-to-branch incidence matrix $\mathbf{A}_a$ defined in the previous chapter.   The only difference is that $\mathbf{A}_a$ has $n_t = n + 1$ rows.   Obviously, $\mathbf{A}$ *is obtainable from $\mathbf{A}_a$ by deleting the row corresponding to the datum node.* $\mathbf{A}$ is therefore called the **reduced incidence matrix.**

**Remark**  The fact that $\mathbf{A}j = \mathbf{0}$ is a set of $n$ linearly independent equations in the variables $j_1, j_2, \ldots, j_b$ implies that the $n \times b$ matrix $\mathbf{A}$ has rank $n$.   Since we always have $b > n$, this conclusion can be restated as follows: the reduced incidence matrix $\mathbf{A}$ is of full rank.

**Example 1**  Consider the graph of Fig. 2.3, which contains four nodes and five branches ($n_t = 4, b = 5$).   Let us number the nodes and branches, as shown on the figure, and indicate that node $\textcircled{4}$ is the datum node by the "ground" symbol used in the figure.   The branch-current vector is

$$\mathbf{j} = \begin{bmatrix} j_1 \\ j_2 \\ j_3 \\ j_4 \\ j_5 \end{bmatrix}$$

The matrix $\mathbf{A}$ is obtained according to Eq. (2.3); thus,

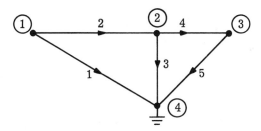

**Fig. 2.3**  Graph for Examples 1 and 2.

Node

$$
\mathbf{A} =
\begin{bmatrix}
1 & 1 & 0 & 0 & 0 \\
0 & -1 & 1 & 1 & 0 \\
0 & 0 & 0 & -1 & 1
\end{bmatrix}
\begin{matrix}
\leftarrow ① \\
\leftarrow ② \\
\leftarrow ③
\end{matrix}
$$

$$
\begin{matrix}
\uparrow & \uparrow & \uparrow & \uparrow & \uparrow \\
\end{matrix}
$$

Branch   1   2   3   4   5

Thus, Eq. (2.2) states that

$$
\mathbf{Aj} =
\begin{bmatrix}
1 & 1 & 0 & 0 & 0 \\
0 & -1 & 1 & 1 & 0 \\
0 & 0 & 0 & -1 & 1
\end{bmatrix}
\begin{bmatrix}
j_1 \\
j_2 \\
j_3 \\
j_4 \\
j_5
\end{bmatrix}
= \mathbf{0}
$$

or

$$
j_1 + j_2 = 0
$$
$$
-j_2 + j_3 + j_4 = 0
$$
$$
-j_4 + j_5 = 0
$$

which are clearly the three node equations obtained by applying KCL to nodes ①, ②, and ③. In the present case it is easy to see that the three equations are linearly independent, since each one contains a variable *not* contained in any of the other equations.

**Exercise 1**   Verify that the $3 \times 5$ matrix **A** above is of full rank. (Hint: You need only exhibit a $3 \times 3$ submatrix whose determinant is nonzero.)

**Exercise 2**   Determine the incidence matrix $\mathbf{A}_a$ of the graph in Fig. 2.3. Observe that **A** is obtained from $\mathbf{A}_a$ by deleting the fourth row.

**2.2**   **Implications of KVL**

Let us call $e_1, e_2, \ldots, e_n$ the node voltages of nodes ①, ②, ..., ⓝ measured with respect to the datum node. The voltages $e_1, e_2, \ldots, e_n$ are called the **node-to-datum voltages.** We are going to use these node-to-datum voltages as variables in node analysis. KVL guarantees that the node-to-datum voltages are defined unambiguously; if we calculate the voltage of any node with respect to that of the datum node by forming an algebraic sum of branch voltages along a path from the datum to the node in question, KVL guarantees that the sum will be independent of the path chosen. Indeed, suppose that a first path from the datum to node ⓚ would give

$e_k$ as the node-to-datum voltage and that a second path would give $e'_k \neq e_k$. This situation contradicts KVL. Consider the loop formed by the first path followed by the second path; KVL requires that the sum of the branch voltages be zero, hence $e'_k = e_k$.

A somewhat roundabout but effective way of expressing the KVL constraints on the branch voltages consists in expressing the $b$ branch voltages in terms of the $n$ node voltages. Since for any networks $b > n$, the $b$ branch voltages cannot be chosen arbitrarily, and they have *only* $n$ degrees of freedom. Indeed, observe that the node-to-datum voltages $e_1, e_2, \ldots, e_n$ are linearly independent as far as the KVL is concerned; this is immediate since the node-to-datum voltages form no loops. Let $\mathbf{e}$ be the vector whose components are $e_1, e_2, \ldots, e_n$. We are going to show that the branch voltages are obtained from the node voltages by the equation

(2.4)   $\mathbf{v} = \mathbf{A}^T \mathbf{e}$

where $\mathbf{A}^T$ is the $b \times n$ matrix which is the transpose of the reduced incidence matrix $\mathbf{A}$ defined in Eq. (2.3).

To show this, it is necessary to consider two kinds of branches, namely, those branches which are incident with the datum node and those which are not. For branches which are incident with the datum node, the branch voltage is equal either to a node-to-datum voltage or its negative. For branches which are not incident with the datum node, the branch voltage must form a loop with two node-to-datum voltages, and hence it can be expressed as a linear combination of the two node-to-datum voltages by KVL. Therefore, in both cases the branch voltages depend linearly on the node-to-datum voltages. To show that the relation is that of Eq. (2.4), let us examine in detail the sign convention. Recall that $v_k$ is the $k$th branch voltage, $k = 1, 2, \ldots, b$, and $e_i$ is the node-to-datum voltage of node $\textcircled{i}$, $i = 1, 2, \ldots, n$. Thus if branch $k$ connects the $i$th node to the datum node, we have

$$v_k = \begin{cases} e_i & \text{if branch } k \text{ \textit{leaves} node } \textcircled{i} \\ -e_i & \text{if branch } k \text{ \textit{enters} node } \textcircled{i} \end{cases}$$

On the other hand, if branch $k$ leaves node $\textcircled{i}$ and enters node $\textcircled{j}$, then we have as is easily seen from Fig. 2.4

$$v_k = e_i - e_j$$

Since in all cases $v_k$ can be expressed as a linear combination of the voltages $e_1, e_2, \ldots, e_n$, we may write

$$
\begin{bmatrix} v_1 \\ v_2 \\ \vdots \\ v_b \end{bmatrix}
=
\begin{bmatrix} c_{11} & c_{12} \cdots c_{1n} \\ c_{21} & c_{22} \cdots c_{2n} \\ \cdots\cdots\cdots\cdots \\ c_{b1} & c_{b2} \cdots c_{bn} \end{bmatrix}
\begin{bmatrix} e_1 \\ e_2 \\ \vdots \\ e_n \end{bmatrix}
$$

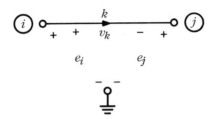

**Fig. 2.4**   Calculation of the branch voltage $v_k$ in terms of the node voltages $e_i$ and $e_j$; $v_k = e_i - e_j$.

where the $c_{ki}$'s are 0, 1, or $-1$ according to the rules above. A little thought will show that

$$(2.5) \quad c_{ki} = \begin{cases} 1 & \text{if branch } k \text{ \textit{leaves} node } \textcircled{i} \\ -1 & \text{if branch } k \text{ \textit{enters} node } \textcircled{i} \\ 0 & \text{if branch } k \text{ is \textit{not incident with} node } \textcircled{i} \end{cases}$$

A comparison of Eq. (2.5) with Eq. (2.3) shows immediately that $c_{ki} = a_{ik}$ for $i = 1, 2, \ldots, n$ and $k = 1, 2, \ldots, b$. Therefore, the matrix **C** (whose elements are the $c_{ki}$'s) is in fact the transpose of the reduced incidence matrix **A**. Hence Eq. (2.4) is established.

**Example 2**   For the circuit in Fig. 2.3,

$$\mathbf{v} = \begin{bmatrix} v_1 \\ v_2 \\ v_3 \\ v_4 \\ v_5 \end{bmatrix} \qquad \mathbf{e} = \begin{bmatrix} e_1 \\ e_2 \\ e_3 \end{bmatrix}$$

According to Eq. (2.5), we have

$$
\begin{array}{cc}
& \text{Branch} \\
\mathbf{A}^T = \begin{bmatrix} 1 & 0 & 0 \\ 1 & -1 & 0 \\ 0 & 1 & 0 \\ 0 & 1 & -1 \\ 0 & 0 & 1 \end{bmatrix} &
\begin{array}{l} \longleftarrow 1 \\ \longleftarrow 2 \\ \longleftarrow 3 \\ \longleftarrow 4 \\ \longleftarrow 5 \end{array}
\end{array}
$$

$$\text{Node} \quad \textcircled{1} \quad \textcircled{2} \quad \textcircled{3}$$

Thus, Eq. (2.4) states that

$$
\mathbf{v} = \mathbf{A}^T\mathbf{e} = \begin{bmatrix} 1 & 0 & 0 \\ 1 & -1 & 0 \\ 0 & 1 & 0 \\ 0 & 1 & -1 \\ 0 & 0 & 1 \end{bmatrix} \begin{bmatrix} e_1 \\ e_2 \\ e_3 \end{bmatrix}
$$

or

$v_1 = e_1$

$v_2 = e_1 - e_2$

$v_3 = e_2$

$v_4 = e_2 - e_3$

$v_5 = e_3$

These five scalar equations are easily recognized as expressions of the KVL.

**Summary**   Equations (2.2) and (2.4) give

$$\mathbf{Aj} = \mathbf{0} \quad \text{(KCL)} \qquad \mathbf{v} = \mathbf{A}^T\mathbf{e} \quad \text{(KVL)}$$

and are the two basic equations of node analysis. They are obtained from the network graph and the two Kirchhoff laws, which make them independent of the nature of the elements of the network. Eq. (2.2) expresses KCL and consists of $n$ independent linear homogeneous algebraic equations in the $b$ branch currents $j_1, j_2, \ldots, j_b$. Equation (2.4) expresses KVL and expresses the $b$ branch voltages $v_1, v_2, \ldots, v_b$ in terms of the $n$ node-to-datum voltages $e_1, e_2, \ldots, e_n$.

Obviously, to solve for the $n$ network variables $e_1, e_2, \ldots, e_n$, we need to know the branch characterization of the network, i.e., the $b$ branch equations which relate the branch voltages $\mathbf{v}$ to branch currents $\mathbf{j}$. Only in these branch equations does the nature of the network elements come into the analysis. Thus, the remaining problem is to combine Eqs. (2.2) and (2.4) with the branch equations and obtain $n$ equations in $n$ unknowns, $e_1, e_2, \ldots e_n$. This requires some elimination. For nonlinear and time-varying networks the elimination problem is usually difficult, and we shall postpone its discussion until later. However, for linear time-invariant networks the branch equations can be combined with Eqs. (2.2) and (2.4), and the elimination can easily be performed. We shall therefore treat exclusively the linear time-invariant networks in Sec. 3.

| 2.3 | **Tellegen's Theorem Revisited** |

As an application of the fundamental equations (2.2) and (2.4), let us use them to give a short proof of Tellegen's theorem. Let $v_1, v_2, \ldots, v_b$ be a set of $b$ *arbitrarily chosen* branch voltages that satisfy all the constraints imposed by KVL. From these $v_k$'s, we can uniquely define node-to-datum voltages $e_1, e_2, \ldots, e_n$, and we have [from (2.4)]

$$\mathbf{v} = \mathbf{A}^T \mathbf{e}$$

Let $j_1, j_2, \ldots, j_b$ be a set of $b$ *arbitrarily chosen* branch currents that satisfy all the constraints imposed by KCL. Since we use *associated* reference directions for these currents to those of the $v_k$'s, KCL requires that [from (2.2)]

$$\mathbf{A}\mathbf{j} = \mathbf{0}$$

Now we obtain successively

$$\sum_{k=1}^{b} v_k j_k = \mathbf{v}^T \mathbf{j}$$

$$= (\mathbf{A}^T \mathbf{e})^T \mathbf{j}$$

$$= e^T (\mathbf{A}^T)^T j$$

$$= e^T \mathbf{A} \mathbf{j}$$

Hence, by (2.2)

(2.6)   $\mathbf{v}^T \mathbf{j} = \mathbf{0}$

Thus, we have shown that $\sum_{k=1}^{b} v_k j_k = 0$. This is the conclusion of the Tellegen theorem for our arbitrary network.

Let us draw some further conclusions from (2.2), (2.4), and (2.6). Consider $\mathbf{j}$ and $\mathbf{v}$ as vectors in the *same* $b$-dimensional linear space $R^b$. From (2.2) it follows that *the set of all branch-current vectors that satisfy KCL form a linear space:* call it $\mathcal{V}_I$. (See Appendix A for the definition of linear space.) To prove this, observe that if $\mathbf{j}_1$ is such that $\mathbf{A}\mathbf{j}_1 = \mathbf{0}$, then $\mathbf{A}(\alpha \mathbf{j}_1) = \alpha \mathbf{A} \mathbf{j}_1 = \mathbf{0}$ for all real numbers $\alpha$; $\mathbf{A}\mathbf{j}_1 = \mathbf{0}$ and $\mathbf{A}\mathbf{j}_2 = \mathbf{0}$ imply that $\mathbf{A}\mathbf{j}_1 + \mathbf{A}\mathbf{j}_2 = \mathbf{A}(\mathbf{j}_1 + \mathbf{j}_2) = \mathbf{0}$.

It can similarly be shown that *the set of all branch voltage vectors* $\mathbf{v}$ *that satisfy KVL form a linear space; let us call the space* $\mathcal{V}_V$.

Tellegen's theorem may be interpreted to mean that any vector in $\mathcal{V}_I$ is orthogonal to every vector of $\mathcal{V}_V$. In other words, *the subspaces* $\mathcal{V}_I$ *and* $\mathcal{V}_V$ *are orthogonal subspaces of* $R^b$.

We now show that *the direct sum of the orthogonal subspaces* $\mathcal{V}_I$ *and* $\mathcal{V}_V$ *is* $R^b$ *itself.* In other words, *any* vector in $R^b$, say $\mathbf{x}$, can be written *uniquely* as the sum of a vector in $\mathcal{V}_V$, say, $\mathbf{v}$, and a vector in $\mathcal{V}_I$, say, $-\mathbf{j}$.

To prove this, consider the graph specified by the reduced incidence

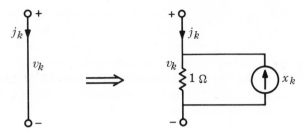

**Fig. 2.5**    Branch $k$ is replaced by a 1-ohm resistor and a constant current source $x_k$.

matrix **A**.  For $k = 1, 2, \ldots, b$, replace branch $k$ by a 1-ohm resistor in parallel with a current source of $x_k$ amp (here $\mathbf{x} = x_1, x_2, \ldots, x_b$ represents an arbitrary vector in $R^b$); this replacement is illustrated in Fig. 2.5.  Call $\mathfrak{N}$ the network resulting from the replacement.  As we shall prove later (remark 2, Sec. 3.2), this resistive network $\mathfrak{N}$ has a unique solution whatever the values of the current sources $x_1, x_2, \ldots x_b$.  Note that the branch equations read

$$\mathbf{v} = \mathbf{j} + \mathbf{x}$$

In other words, we have shown that *any* vector $\mathbf{x}$ in $R^b$ can be written in a *unique* way as the sum of a vector in $\mathfrak{V}_V$ and a vector in $\mathfrak{V}_I$.  Hence, the direct sum of $\mathfrak{V}_V$ and $\mathfrak{V}_i$ is $R^b$.

**Remark**    The subspaces $\mathfrak{V}_I$ and $\mathfrak{V}_V$ depend only on the graph.  They are completely determined by the incidence matrix, and consequently they are independent of the nature of the branches and waveforms of the sources.

**3**    **Node Analysis of Linear Time-invariant Networks**

In linear time-invariant networks all elements except the independent sources are linear and time-invariant.  We have studied in detail, but separately, the branch equations of linear time-invariant resistors, capacitors, inductors, coupled inductors, ideal transformers, and controlled sources.  The problem of general node analysis is to combine these branch equations with the two basic equations

(3.1)    $\mathbf{Aj} = \mathbf{0}$    (KCL)

and

(3.2)    $\mathbf{v} = \mathbf{A}^T\mathbf{e}$    (KVL)

The resulting equations will, in general, take the form of *linear* simultaneous differential equations or integrodifferential equations with $n$ network variables $e_1, e_2, \ldots, e_n$.  The purpose of this section is to study the formulation of these equations and to develop some important properties of the resulting equations. For simplicity we consider first the case in which only resistors and independent sources are allowed in the network. In this case the resulting equations will be linear *algebraic* equations. We next consider the sinusoidal steady-state analysis of networks using phasors and impedances. Finally, we consider the formulation of general differential and integrodifferential equations.

## 3.1   Analysis of Resistive Networks

Consider a linear time-invariant resistive network with $b$ branches, $n_t$ nodes, and one separate part. A typical branch is shown in Fig. 3.1. Note that it includes independent sources. The branch equations are of the form

(3.3a) $\quad v_k = R_k j_k + v_{sk} - R_k j_{sk} \qquad k = 1, 2, \ldots, b$

or, equivalently,

(3.3b) $\quad j_k = G_k v_k + j_{sk} - G_k v_{sk} \qquad k = 1, 2, \ldots, b$

In matrix notation, we have from Eq. (3.3b)

(3.4) $\quad \mathbf{j} = \mathbf{Gv} + \mathbf{j}_s - \mathbf{Gv}_s$

where $\mathbf{G}$ is called the **branch conductance matrix.** It is a *diagonal* matrix of order $b$; that is,

$$
\mathbf{G} = \begin{bmatrix} G_1 & 0 \cdots\cdots\cdots\cdots 0 \\ 0 & G_2 \\ \vdots & \ddots \\ & & \ddots \, 0 \\ 0 \cdots\cdots\cdots\cdots 0 & G_b \end{bmatrix}
$$

The vectors $\mathbf{j}_s$ and $\mathbf{v}_s$ are source vectors of dimension $b$; that is,

$$
\mathbf{j}_s = \begin{bmatrix} j_{s1} \\ j_{s2} \\ \vdots \\ j_{sb} \end{bmatrix} \qquad \mathbf{v}_s = \begin{bmatrix} v_{s1} \\ v_{s2} \\ \vdots \\ v_{sb} \end{bmatrix}
$$

It is only necessary to combine Eqs. (3.1) (3.2), and (3.4) to eliminate the branch variables and obtain a vector equation in terms of the vector network variable $\mathbf{e}$. Premultiplying Eq. (3.4) by the matrix $\mathbf{A}$, substituting $\mathbf{v}$ by $\mathbf{A}^T\mathbf{e}$, and using Eq. (3.1), we obtain

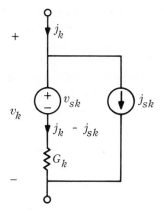

**Fig. 3.1**    The $k$th branch.

(3.5)    $\mathbf{AGA}^T\mathbf{e} + \mathbf{Aj}_s - \mathbf{AGv}_s = \mathbf{0}$

or

(3.6)    $\mathbf{AGA}^T\mathbf{e} = \mathbf{AGv}_s - \mathbf{Aj}_s$

In Eq. (3.6) $\mathbf{AGA}^T$ is an $n \times n$ square matrix, whereas $\mathbf{AGv}_s$ and $-\mathbf{Aj}_s$ are $n$-dimensional vectors.   Let us introduce the following notations:

(3.7a)    $\mathbf{Y}_n \overset{\Delta}{=} \mathbf{AGA}^T$

(3.7b)    $\mathbf{i}_s \overset{\Delta}{=} \mathbf{AGv}_s - \mathbf{Aj}_s$

then Eq. (3.6) becomes

(3.8)    $\boxed{\mathbf{Y}_n\mathbf{e} = \mathbf{i}_s}$

The set of equations (3.8) is usually called the **node equations;** $\mathbf{Y}_n$ is called the **node admittance matrix,**† and $\mathbf{i}_s$ is the **node current source vector.**
The node equations (3.8) are very important; they deserve careful examination.   First, observe that since the graph specifies the reduced incidence matrix $\mathbf{A}$ and since the branch conductances specify the branch admittance matrix $\mathbf{G}$, the node admittance matrix $\mathbf{Y}_n$ is a known matrix; indeed, $\mathbf{Y}_n \overset{\Delta}{=} \mathbf{AGA}^T$.
Similarly, the vectors $\mathbf{v}_s$ and $\mathbf{j}_s$, which specify the sources in the branches, are given; therefore, the node current source vector $\mathbf{i}_s$ is also

---

† We call $\mathbf{Y}_n$ the node *admittance* matrix rather than the node *conductance* matrix even though we are dealing with a purely resistive network.   It will be seen that in sinusoidal steady-state analysis we have exactly the same formulation; hence it is more convenient to introduce the more general term *admittance*.

known by (3.7*b*). Thus, Eq. (3.8) relates the unknown *n*-vector **e** to the known $n \times n$ matrix $\mathbf{Y}_n$ and the known *n*-vector $\mathbf{i}_s$. The vector equation (3.8) consists of a system of *n* *linear algebraic* equations in the *n* unknown node-to-datum voltages $e_1, e_2, \ldots, e_n$. Once **e** is found, it is a simple matter to find the *b* branch voltages **v** and the *b* branch currents **j**. Indeed, (3.2) gives $\mathbf{v} = \mathbf{A}^T \mathbf{e}$, and having **v**, we obtain **j** by the branch equation (3.4); that is, $\mathbf{j} = \mathbf{Gv} + \mathbf{j}_s - \mathbf{Gv}_s$.

---

**Example 1**  Let us consider the circuit in Fig. 3.2, where all element values are given. The graph of the circuit is the same one illustrated in Fig. 2.3. Let us give the detailed procedure for writing the equations and solving the branch variables.

*Step 1*  Pick a datum node, say ④, and label the remaining nodes ①, ②, and ③. Call $e_1, e_2$, and $e_3$ the voltages of nodes ①, ②, and ③, respectively, with respect to the datum node.

*Step 2*  Number the branches 1, 2, 3, 4, and 5, and assign each one a reference direction. The variable $G_i$ is the conductance of branch numbered *i*.

*Step 3*  Write the three linearly independent equations expressing KCL for nodes ①, ②, and ③; thus,

$$(3.9) \quad \mathbf{Aj} = \begin{bmatrix} 1 & 1 & 0 & 0 & 0 \\ 0 & -1 & 1 & 1 & 0 \\ 0 & 0 & 0 & -1 & 1 \end{bmatrix} \begin{bmatrix} j_1 \\ j_2 \\ j_3 \\ j_4 \\ j_5 \end{bmatrix} = \begin{bmatrix} 0 \\ 0 \\ 0 \end{bmatrix}$$

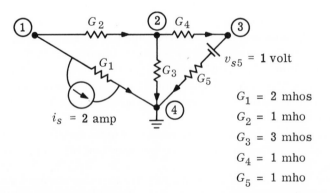

$i_s = 2$ amp

$v_{s5} = 1$ volt

$G_1 = 2$ mhos
$G_2 = 1$ mho
$G_3 = 3$ mhos
$G_4 = 1$ mho
$G_5 = 1$ mho

**Fig. 3.2**   Network for Example 1.

Note that the reduced incidence matrix **A** is the same as in Example 1 of Sec. 2.1.

*Step 4* Use KVL to express the branch voltage $v_k$ in terms of the node voltages $e_i$; thus,

$$(3.10) \quad \mathbf{v} = \mathbf{A}^T \mathbf{e} = \begin{bmatrix} 1 & 0 & 0 \\ 1 & -1 & 0 \\ 0 & 1 & 0 \\ 0 & 1 & -1 \\ 0 & 0 & 1 \end{bmatrix} \begin{bmatrix} e_1 \\ e_2 \\ e_3 \end{bmatrix}$$

*Step 5* Write the branch equations in the form

$$(3.11a) \quad \mathbf{j} = \mathbf{G}\mathbf{v} + \mathbf{j}_s - \mathbf{G}\mathbf{v}_s$$

Thus,

$$(3.11b) \quad \begin{bmatrix} j_1 \\ j_2 \\ j_3 \\ j_4 \\ j_5 \end{bmatrix} = \begin{bmatrix} 2 & 0 & 0 & 0 & 0 \\ 0 & 1 & 0 & 0 & 0 \\ 0 & 0 & 3 & 0 & 0 \\ 0 & 0 & 0 & 1 & 0 \\ 0 & 0 & 0 & 0 & 1 \end{bmatrix} \begin{bmatrix} v_1 \\ v_2 \\ v_3 \\ v_4 \\ v_5 \end{bmatrix} + \begin{bmatrix} 2 \\ 0 \\ 0 \\ 0 \\ 0 \end{bmatrix} - \begin{bmatrix} 2 & 0 & 0 & 0 & 0 \\ 0 & 1 & 0 & 0 & 0 \\ 0 & 0 & 3 & 0 & 0 \\ 0 & 0 & 0 & 1 & 0 \\ 0 & 0 & 0 & 0 & 1 \end{bmatrix} \begin{bmatrix} 0 \\ 0 \\ 0 \\ 0 \\ 1 \end{bmatrix}$$

For example, the fifth scalar equation of (3.11b) reads

$$j_5 = v_5 - 1$$

*Step 6* Substitute into (3.11) the expression for **v** given by (3.10), and multiply the result on the left by the matrix **A**; according to (3.9), the result is **0**. After reordering terms, we can put the answer in the form

$$(3.12a) \quad \mathbf{Y}_n \mathbf{e} = \mathbf{i}_s$$

where

$$\mathbf{Y}_n \triangleq \mathbf{A}\mathbf{G}\mathbf{A}^T$$

$$= \begin{bmatrix} 1 & 1 & 0 & 0 & 0 \\ 0 & -1 & 1 & 1 & 0 \\ 0 & 0 & 0 & -1 & 1 \end{bmatrix} \begin{bmatrix} 2 & 0 & 0 & 0 & 0 \\ 0 & 1 & 0 & 0 & 0 \\ 0 & 0 & 3 & 0 & 0 \\ 0 & 0 & 0 & 1 & 0 \\ 0 & 0 & 0 & 0 & 1 \end{bmatrix} \begin{bmatrix} 1 & 0 & 0 \\ 1 & -1 & 0 \\ 0 & 1 & 0 \\ 0 & 1 & -1 \\ 0 & 0 & 1 \end{bmatrix}$$

$$(3.12b) \quad = \begin{bmatrix} 3 & -1 & 0 \\ -1 & 5 & -1 \\ 0 & -1 & 2 \end{bmatrix}$$

and

$$(3.12c) \quad \mathbf{i}_s \overset{\Delta}{=} \mathbf{AGv}_s - \mathbf{Aj}_s = \begin{bmatrix} -2 \\ 0 \\ 1 \end{bmatrix}$$

Thus, the node equation

$$(3.12d) \quad \begin{bmatrix} 3 & -1 & 0 \\ -1 & 5 & -1 \\ 0 & -1 & 2 \end{bmatrix} \begin{bmatrix} e_1 \\ e_2 \\ e_3 \end{bmatrix} = \begin{bmatrix} -2 \\ 0 \\ 1 \end{bmatrix}$$

*Step 7*   Solve Eq. (3.12*d*) for **e**.  The numerical solution of such equations is done by the Gauss elimination method whenever $n > 5$.  Formally, we may express the answer in terms of the inverse matrix $\mathbf{Y}_n{}^{-1}$; thus,

$$(3.13) \quad \mathbf{e} = \mathbf{Y}_n{}^{-1} \mathbf{i}_s = \frac{1}{25} \begin{bmatrix} 9 & 2 & 1 \\ 2 & 6 & 3 \\ 1 & 3 & 14 \end{bmatrix} \begin{bmatrix} -2 \\ 0 \\ 1 \end{bmatrix} = \frac{1}{25} \begin{bmatrix} -17 \\ -1 \\ 12 \end{bmatrix}$$

Once the node voltages **e** are found, the branch voltages **v** are obtained from (3.10) as follows:

$$(3.14) \quad \mathbf{v} = \mathbf{A}^T \mathbf{e} = \frac{1}{25} \begin{bmatrix} -17 \\ -16 \\ -1 \\ -13 \\ 12 \end{bmatrix}$$

The next step is to use **v** to obtain the branch currents **j** by Eq. (3.11); then we have

$$(3.15) \quad \mathbf{j} = \mathbf{Gv} + \mathbf{j}_s - \mathbf{Gv}_s = \frac{1}{25} \begin{bmatrix} 16 \\ -16 \\ -3 \\ -13 \\ -13 \end{bmatrix}$$

This completes the analysis of the network shown in Fig. 3.2; that is, all branch voltages and currents have been determined.

| 3.2 | **Writing Node Equations by Inspection** |

The step-by-step procedure detailed above is very important for two reasons. First, it exhibits quite clearly the various facts that have to be used to analyze the network, and second, it is completely general in the sense that it works in all cases and is therefore suitable for automatic computation.

In the case of networks made only of resistors and *independent* sources (in particular, with no coupling elements such as dependent sources), the node equations can be written by inspection. Let us call $y_{ik}$ the $(i,k)$ element of the node admittance matrix $\mathbf{Y}_n$; then the vector equation

$(3.16a)$  $\mathbf{Y}_n\mathbf{e} = \mathbf{i}_s$

written in scalar form becomes

$(3.16b)$
$$\begin{bmatrix} y_{11} & y_{12} & \cdots & y_{1n} \\ y_{21} & y_{22} & \cdots & y_{2n} \\ \cdots\cdots\cdots\cdots\cdots \\ y_{n1} & y_{n2} & \cdots & y_{nn} \end{bmatrix}\begin{bmatrix} e_1 \\ e_2 \\ \vdots \\ e_n \end{bmatrix} = \begin{bmatrix} i_{s1} \\ i_{s2} \\ \vdots \\ i_{sn} \end{bmatrix}$$

The following statements are easily verified in simple examples, and can be proved for networks without coupling elements.

1. $y_{ii}$ *is the sum of the conductances of all branches connected to node* ⓘ; $y_{ii}$ *is called the* **self-admittance of node** ⓘ.

2. $y_{ik}$ *is the negative of the sum of the conductances of all branches connecting node* ⓘ *and node* ⓚ; $y_{ik}$ *is called the* **mutual admittance between node** ⓘ *and node* ⓚ.

3. *If we convert all voltage sources into current sources, then* $i_{sk}$ *is the algebraic sum of all source currents entering node* ⓚ.

**Exercise**   Prove statements 1 and 2 above.   Hint: $\mathbf{Y}_n = \mathbf{A}\mathbf{G}\mathbf{A}^T$; consequently,

$$y_{ii} = \sum_{j=1}^{b} a_{ij}G_j a_{ij} = \sum_{j=1}^{b} (a_{ij})^2 G_j$$

and

$$y_{ik} = \sum_{j=1}^{b} a_{ij}G_j a_{kj}$$

Note that $(a_{ij})^2$ can only be zero or 1; similarly, $a_{ij}a_{kj}$ can only be zero or $-1$.

**Example 1**
**(continued)**   Consider again the network of Fig. 3.2.   Let us, by inspection, write every branch current in terms of the node voltages; thus,

$$j_1 = G_1 e_1 + 2$$

$$j_2 = G_2(e_1 - e_2)$$

$$j_3 = G_3 e_2$$

$$j_4 = G_4(e_2 - e_3)$$

$$j_5 = G_5 e_3 - 1$$

In the last equation, we used the equivalent current source for branch 5, as shown in Fig. 3.3. Substituting the above into the node equations, we obtain

$$(G_1 + G_2)e_1 - G_2 e_2 = -1$$

$$-G_2 e_1 + (G_2 + G_3 + G_4)e_2 - G_4 e_3 = 0$$

$$-G_4 e_3 + (G_4 + G_5)e_3 = 1$$

By inspection, it is easily seen that statements 1, 2, and 3 above hold for the present case. Also if the numerical values for the $G_k$'s are substituted, the answer checks with (3.12a).

**Remarks**

1. For networks made of resistors and independent sources, the node admittance matrix $\mathbf{Y}_n = (y_{ik})$ in Eq. (3.8) is a *symmetric matrix;* i.e., $y_{ik} = y_{ki}$ for $i$, $k = 1, 2, \ldots, n$. Indeed, since $\mathbf{Y}_n \triangleq \mathbf{AGA}^T$, then $\mathbf{Y}_n{}^T = (\mathbf{AGA}^T)^T = \mathbf{AG}^T\mathbf{A}^T = \mathbf{AGA}^T = \mathbf{Y}_n$. In the last step we used the fact that $\mathbf{G}^T = \mathbf{G}$ because $\mathbf{G}$, the branch conductance matrix of a resistive network with *no coupling elements,* is a diagonal matrix.

2. If all the conductances of a linear resistive network are *positive,* it is easy to show that det $(\mathbf{Y}_n) > 0$.† Cramer's rule then guarantees that whatever $\mathbf{i}_s$ may be, Eq. (3.16) has a *unique solution.* The fact that det $(\mathbf{Y}_n) > 0$ also follows from $\mathbf{Y}_n = \mathbf{AGA}^T$, where $\mathbf{G}$ is an $n \times n$ diagonal matrix with *positive* elements and $\mathbf{A}$ is an $n \times b$ matrix with

† See Sec. 2.4, Appendix B.

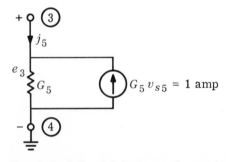

**Fig. 3.3**   Branch 5 of Fig. 3.2 in terms of current source.

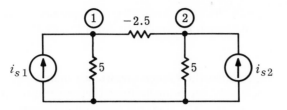

**Fig. 3.4**   A resistive network with a negative conductance.  The conductances are given in mhos.

rank $n$.  Note that the assumption that *all* conductances are positive is crucial.  For example, consider the circuit in Fig. 3.4, where one of the conductances is negative.  The node admittance matrix is singular (i.e., its determinant is zero).  Indeed, the node equations are

$$\begin{bmatrix} 2.5 & 2.5 \\ 2.5 & 2.5 \end{bmatrix} \begin{bmatrix} e_1 \\ e_2 \end{bmatrix} = \begin{bmatrix} i_{s1} \\ i_{s2} \end{bmatrix}$$

and, for example, with $i_{s1} = i_{s2} = 5$, $e_1 = 1 - \alpha$ and $e_2 = 1 + \alpha$, solutions whatever $\alpha$ may be!  Thus, this network has an infinite number of solutions!

**Exercise**   Assume that all elements of the graph of Fig. 2.3 have conductances of 5 mhos and that a current source squirts 1 amp into node ① and sucks it out of node ②.  Write the node equations by inspection.

**3.3**   **Sinusoidal Steady-state Analysis**

Consider now a linear time-invariant network containing resistors, inductors, capacitors, and independent sources; such networks are usually referred to as *RLC* networks.  Suppose that we are only interested in the sinusoidal steady-state analysis.  Let all the independent sources be sinusoidal at the *same* angular frequency $\omega$, and suppose that the branch currents and voltages have reached the steady state.  Consequently, we shall use voltage phasors, current phasors, impedances, and admittances to write the branch equations and Kirchhoff's laws.  Again we assume that each branch includes a voltage source and a current source in addition to its nonsource element.

Thus, a typical branch contains an admittance, say $Y_k$ (in the $k$th branch), which is one of the forms $G_k$, $j\omega C_k$, or $1/j\omega L_k$, depending on whether the $k$th branch is a resistor, capacitor, or inductor, respectively. The branch equation is

(3.17)   $J_k = Y_k V_k + J_{sk} - Y_k V_{sk} \qquad k = 1, 2, \ldots, b$

where $J_k$ and $V_k$ are the $k$th branch current phasor and voltage phasor, and $J_{sk}$ and $V_{sk}$ are the $k$th branch phasors representing the current and voltage sources of branch $k$.   In matrix form Eq. (3.17) can be written as

(3.18)   $\mathbf{J} = \mathbf{Y}_b\mathbf{V} + \mathbf{J}_s - \mathbf{Y}_b\mathbf{V}_s$

The matrix $\mathbf{Y}_b$ is called the **branch admittance matrix,** and the vectors $\mathbf{J}$ and $\mathbf{V}$ are, respectively, the branch-current phasor vector and the branch-voltage phasor vector.   The analysis is exactly the same as that of the resistive network in the preceding section.   The node equation is of the form

(3.19)   $\mathbf{Y}_n\mathbf{E} = \mathbf{I}_s$

where the phasor $\mathbf{E}$ represents the node-to-datum voltage vector, the phasor $\mathbf{I}_s$ represents the current-source vector, and $\mathbf{Y}_n$ is the node admittance matrix.   In terms of $\mathbf{A}$ and $\mathbf{Y}_b$, $\mathbf{Y}_n$ is given by

(3.20a)   $\mathbf{Y}_n = \mathbf{A}\mathbf{Y}_b\mathbf{A}^T$

(3.20b)   $\mathbf{I}_s = \mathbf{A}\mathbf{Y}_b\mathbf{V}_s - \mathbf{A}\mathbf{J}_s$

**Remark**   Consider on the one hand the steady-state analysis of an *RLC* network driven by sinusoidal sources having the same frequency, and, on the other hand, the analysis of a resistive network.   In both cases node analysis leads to a set of linear algebraic equations in the node voltages.   In the sinusoidal steady-state case the unknowns are node-to-datum voltage *phasors,* and the coefficients of the equations are complex numbers which depend on the frequency.   Finally, recall that the node-to-datum voltages are obtained from the phasors by

$e_k(t) = \mathrm{Re}\,(E_k\,\epsilon^{j\omega t})\qquad k = 1, 2, \ldots, n$

In the resistive network case, the equations had real numbers as coefficients and their solution gave the node-to-datum voltages directly.

If we had only networks without coupling elements, the inspection method of the preceding section would suffice.   We are going to tackle an example that has both *dependent* sources and *mutual inductances.*   It will become apparent that for this case the inspection method does not suffice, and the value of our systematic procedure will become apparent. After the example, we shall sketch out the general procedure.

**Example**   Consider the linear time-invariant network shown in Fig. 3.5.   The independent current source is sinusoidal and is represented by the phasor $I$; its waveform is $|I|\cos{(\omega t + \measuredangle I)}$.   We assume that the network is in the sinusoidal steady state, consequently, all waveforms will be represented by phasors $\mathbf{V}$, $\mathbf{J}$, $\mathbf{E}$, etc.   Note the presence of two dependent sources. The three inductors $L_3$, $L_4$, and $L_5$ are magnetically coupled, and their inductance matrix is

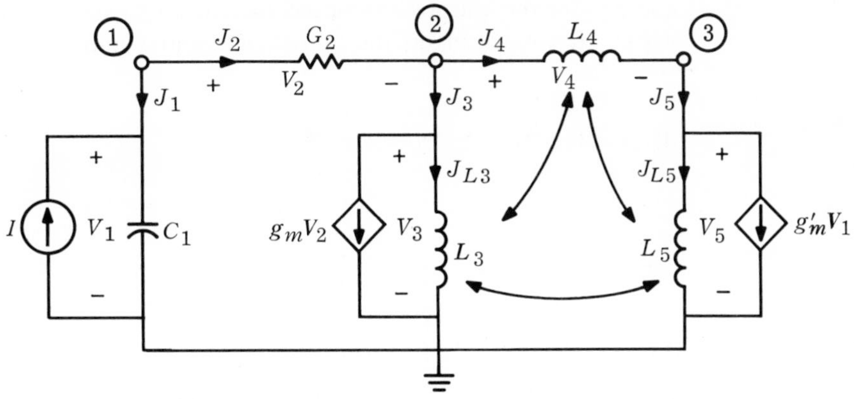

**Fig. 3.5**    Example of sinusoidal steady-state analysis when mutual magnetic coupling and dependent sources are present.

$$\mathbf{L} = \begin{bmatrix} 1 & -1 & 1 \\ -1 & \frac{5}{3} & -\frac{4}{3} \\ 1 & -\frac{4}{3} & \frac{5}{3} \end{bmatrix}$$

First, to express Kirchhoff's laws, we need the node incidence matrix $\mathbf{A}$; by inspection

$$(3.21) \quad \mathbf{A} = \begin{bmatrix} 1 & 1 & 0 & 0 & 0 \\ 0 & -1 & 1 & 1 & 0 \\ 0 & 0 & 0 & -1 & 1 \end{bmatrix}$$

Second, we need the branch equations.  We write them in terms of phasors; thus,

$$(3.22a) \quad \begin{aligned} J_1 &= j\omega C_1 V_1 - I \\ J_2 &= G V_2 \end{aligned}$$

To write $J_3$, $J_4$, and $J_5$ in terms of the $V_k$'s is not that simple; indeed the inductance matrix tells us only the relation between $V_3$, $V_4$, and $V_5$ and the *inductor currents* $J_{L3}$, $J_4$, and $J_{L5}$ (see Fig. 3.5 for the definition of $J_{L3}$ and $J_{L5}$).  Therefore,

$$\mathbf{V'} = j\omega \mathbf{L} \mathbf{J}_L$$

or

$$\begin{bmatrix} V_3 \\ V_4 \\ V_5 \end{bmatrix} = j\omega \begin{bmatrix} 1 & -1 & 1 \\ -1 & \frac{5}{3} & -\frac{4}{3} \\ 1 & -\frac{4}{3} & \frac{5}{3} \end{bmatrix} \begin{bmatrix} J_{L3} \\ J_4 \\ J_{L5} \end{bmatrix}$$

Clearly, for our purposes we need the currents as a function of the voltages. Hence, we invert the inductance matrix and get an expression of the form

$$\mathbf{J}_L = \frac{1}{j\omega}\mathbf{L}^{-1}\mathbf{V}'$$

or

$$\begin{bmatrix} J_{L3} \\ J_4 \\ J_{L5} \end{bmatrix} = \frac{1}{j\omega}\begin{bmatrix} 3 & 1 & -1 \\ 1 & 2 & 1 \\ -1 & 1 & 2 \end{bmatrix}\begin{bmatrix} V_3 \\ V_4 \\ V_5 \end{bmatrix}$$

Having obtained these relations and noting that $J_3 = J_{L3} + g_m V_2$ and $J_5 = J_{L5} + g'_m V_1$, we can write our last three branch equations as follows:

$$J_3 = g_m V_2 + \frac{3}{j\omega} V_3 + \frac{1}{j\omega} V_4 - \frac{1}{j\omega} V_5$$

(3.22b)
$$J_4 = \frac{1}{j\omega} V_3 + \frac{2}{j\omega} V_4 + \frac{1}{j\omega} V_5$$

$$J_5 = g'_m V_1 - \frac{1}{j\omega} V_3 + \frac{1}{j\omega} V_4 + \frac{2}{j\omega} V_5$$

The branch equations (3.22a) and (3.22b) constitute a system of five equations of the form, as in (3.18),

(3.23)   $\mathbf{J} = \mathbf{Y}_b\mathbf{V} + \mathbf{J}_s$

Note that the matrix $\mathbf{Y}_b$ has complex numbers as elements, is no longer diagonal (because of both mutual coupling and dependent sources), and is no longer symmetric (because of dependent sources). Let us calculate the node admittance matrix $\mathbf{Y}_n \triangleq \mathbf{A}\mathbf{Y}_b\mathbf{A}^T$ as follows:

$$\begin{bmatrix} 1 & 1 & 0 & 0 & 0 \\ 0 & -1 & 1 & 1 & 0 \\ 0 & 0 & 0 & -1 & 1 \end{bmatrix}\begin{bmatrix} j\omega C_1 & 0 & 0 & 0 & 0 \\ 0 & G_2 & 0 & 0 & 0 \\ 0 & g_m & \frac{3}{j\omega} & \frac{1}{j\omega} & -\frac{1}{j\omega} \\ 0 & 0 & \frac{1}{j\omega} & \frac{2}{j\omega} & \frac{1}{j\omega} \\ g'_m & 0 & -\frac{1}{j\omega} & \frac{1}{j\omega} & \frac{2}{j\omega} \end{bmatrix}\begin{bmatrix} 1 & 0 & 0 \\ 1 & -1 & 0 \\ 0 & 1 & 0 \\ 0 & 1 & -1 \\ 0 & 0 & 1 \end{bmatrix}$$

$$
= \begin{bmatrix} j\omega C_1 & G_2 & 0 & 0 & 0 \\ 0 & -G_2 + g_m & \dfrac{4}{j\omega} & \dfrac{3}{j\omega} & 0 \\ g_m' & 0 & \dfrac{-2}{j\omega} & \dfrac{-1}{j\omega} & \dfrac{1}{j\omega} \end{bmatrix} \begin{bmatrix} 1 & 0 & 0 \\ 1 & -1 & 0 \\ 0 & 1 & 0 \\ 0 & 1 & -1 \\ 0 & 0 & 1 \end{bmatrix}
$$

$$
= \begin{bmatrix} j\omega C_1 + G_2 & -G_2 & 0 \\ -G_2 + g_m & G_2 - g_m + \dfrac{7}{j\omega} & \dfrac{-3}{j\omega} \\ g_m' & \dfrac{-3}{j\omega} & \dfrac{2}{j\omega} \end{bmatrix} = \mathbf{Y}_n
$$

The right-hand term of the node equation, from (3.19) and (3.20b), is $-\mathbf{A}\mathbf{J}_s$ since $\mathbf{V}_s$ is identically zero. It is easily found to be a vector whose first component is $I$ (all the others are zero). Thus, the node equation is

(3.24)
$$
\begin{bmatrix} j\omega C_1 + G_2 & -G_2 & 0 \\ -G_2 + g_m & G_2 - g_m + \dfrac{7}{j\omega} & \dfrac{-3}{j\omega} \\ g_m' & \dfrac{-3}{j\omega} & \dfrac{2}{j\omega} \end{bmatrix} \begin{bmatrix} E_1 \\ E_2 \\ E_3 \end{bmatrix} = \begin{bmatrix} I \\ 0 \\ 0 \end{bmatrix}
$$

After substituting numerical values, we can solve these three equations for the phasors $E_1$, $E_2$, and $E_3$. Successively, we get $\mathbf{V}$ by

$$\mathbf{V} = \mathbf{A}^T\mathbf{E}$$

and $\mathbf{J}$ by (3.23).

*Systematic procedure*   This example exhibits clearly the value of the systematic method; obviously, it is hazardous to try to write Eq. (3.24) by inspection.

The detailed example above suggests a method for writing the sinusoidal steady-state equations of *any* linear time-invariant network driven by sinusoidal sources of the *same* frequency. Note that the network may include $R$'s, $L$'s, $C$'s, mutual inductances, dependent sources, and independent sources. The steps are as follows:

*Step 1*   Perform (if needed) source transformations as indicated in Sec. 1.

*Step 2*   Write the requirements of Kirchhoff's laws; thus,

(3.25a)   $\mathbf{AJ} = \mathbf{0}$

(3.25$b$)   $\mathbf{V} = \mathbf{A}^T\mathbf{E}$

*Step 3*   Write the branch equations [from (3.18)]

$$\mathbf{J} = \mathbf{Y}_b(j\omega)\,\mathbf{V} - \mathbf{Y}_b(j\omega)\,\mathbf{V}_s + \mathbf{J}_s$$

where $\mathbf{Y}_b(j\omega)$ is the branch admittance matrix.  Note that $\mathbf{Y}_b$ is evaluated at $j\omega$, where $\omega$ represents the angular frequency of the sinusoidal sources.

*Step 4*   Perform the substitution to obtain the node equations labeled (3.19)

$$\mathbf{Y}_n(j\omega)\mathbf{E} = \mathbf{I}_s$$

where [from (3.20$a$) and (3.20$b$)]

$$\mathbf{Y}_n(j\omega) \triangleq \mathbf{A}\mathbf{Y}_b(j\omega)\mathbf{A}^T$$
$$\mathbf{I}_s \triangleq \mathbf{A}\mathbf{Y}_b(j\omega)\,\mathbf{V}_s - \mathbf{A}\mathbf{J}_s$$

*Step 5*   Solve the node equations (3.19) for the phasor $\mathbf{E}$.

*Step 6*   Obtain $\mathbf{V}$ by (3.25$b$) and $\mathbf{J}$ by (3.18).

*Properties of the node admittance matrix $\mathbf{Y}_n(j\omega)$*   From the basic equation

$$\mathbf{Y}_n(j\omega) = \mathbf{A}\mathbf{Y}_b(j\omega)\mathbf{A}^T$$

we obtain the following useful properties:

1. Whenever there are no coupling elements (i.e., neither mutual inductances nor *dependent* sources), the $b \times b$ matrix $\mathbf{Y}_b(j\omega)$ is diagonal, and consequently the $n \times n$ matrix $\mathbf{Y}_n(j\omega)$ is *symmetric*.

2. Whenever there are no *dependent* sources (mutual inductances are allowed), both $\mathbf{Y}_b(j\omega)$ and $\mathbf{Y}_n(j\omega)$ are *symmetric*.

## 3.4   Integrodifferential Equations

In general, node analysis of linear networks leads to a set of simultaneous integrodifferential equations, i.e., equations involving unknown functions, say $e_1, e_2, \ldots$, some of their derivatives $\dot{e}_1, \dot{e}_2, \ldots$, and some of their integrals $\int_0^t e_1(t')\,dt',\ \int_0^t e_2(t')\,dt',\ldots$. We shall present a systematic method for obtaining the node integrodifferential equations of any linear time-invariant network.  These equations are necessary if we have to compute the complete response of a given network to a given input and a given initial state.  The method is perfectly general, but in order not to get bogged down in notation we shall present it by way of an example.

**Example**   We are given (1) the linear time-invariant network shown in Fig. 3.6; (2) the element values $G_1$, $G_2$, $C_3$, and $g_m$, and the reciprocal inductance matrix

$$\begin{bmatrix} \Gamma_{44} & \Gamma_{45} \\ \Gamma_{45} & \Gamma_{55} \end{bmatrix}$$

(note that this matrix corresponds to the reference directions for $j_{L4}$ and $j_5$ in Fig. 3.6); (3) the input waveform $j_{s1}(t)$ for $t \geq 0$; and (4) the initial values of initial capacitor voltage $v_3(0)$ and initial inductor currents $j_{L4}(0)$ and $j_5(0)$. We shall proceed in the same order as that used for the sinusoidal steady state.

*Step 1*   Perform (if needed) source transformations as indicated in Sec. 1.

*Step 2*   Write the requirement of Kirchhoff's laws as follows:

$$(3.26) \quad \begin{bmatrix} 1 & 0 & 1 & 0 & 0 \\ 0 & 1 & -1 & 1 & 0 \\ 0 & -1 & 0 & 0 & 1 \end{bmatrix} \begin{bmatrix} j_1 \\ j_2 \\ j_3 \\ j_4 \\ j_5 \end{bmatrix} = \begin{bmatrix} 0 \\ 0 \\ 0 \end{bmatrix} \quad \text{(KCL)}$$

$$(3.27) \quad \begin{bmatrix} v_1 \\ v_2 \\ v_3 \\ v_4 \\ v_5 \end{bmatrix} = \begin{bmatrix} 1 & 0 & 0 \\ 0 & 1 & -1 \\ 1 & -1 & 0 \\ 0 & 1 & 0 \\ 0 & 0 & 1 \end{bmatrix} \begin{bmatrix} e_1 \\ e_2 \\ e_3 \end{bmatrix} \quad \text{(KVL)}$$

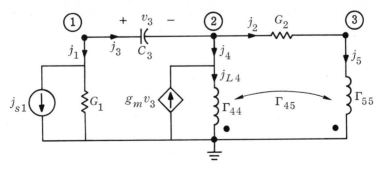

**Fig. 3.6**   Network for which the node integrodifferential equations are obtained.

Note that the $j_k$'s, $v_k$'s, and $e_i$'s are time functions and *not* phasors. To lighten the notation, we have written $j_1, j_2, \ldots, e_1, e_2, \ldots$, etc., instead of $j_1(t), j_2(t), \ldots$, etc.

*Step 3*   Write the branch equations (expressing the branch currents in terms of the branch voltages); thus,

$$j_1 = G_1 v_1 + j_{s1}$$

$$j_2 = G_2 v_2$$

$$j_3 = C_3 \dot{v}_3$$

Noting that $j_{L4}$ is the current in the fourth inductance and that $j_4 = j_{L4} - g_m v_3$, we obtain the branch equation of the inductors as follows:

$$j_4 = -g_m v_3 + \Gamma_{44} \int_0^t v_4(t') \, dt' + \Gamma_{45} \int_0^t v_5(t') \, dt' + j_{L4}(0)$$

$$j_5 = \Gamma_{45} \int_0^t v_4(t') \, dt' + \Gamma_{55} \int_0^t v_5(t') \, dt' + j_5(0)$$

It is convenient to use the notation $D$ to denote the differentiation operator with respect to time; for example,

$$D v_3 = \dot{v}_3 = \frac{dv_3}{dt}$$

and the notation $D^{-1}$ to denote the *definite* integral $\int_0^t \cdot \, ;$ for example,

$$\frac{1}{D} v_4 = \int_0^t v_4(t') \, dt'$$

With these notations the branch equations can be written in matrix form as follows:

(3.28)
$$\begin{bmatrix} j_1 \\ j_2 \\ j_3 \\ j_4 \\ j_5 \end{bmatrix} = \begin{bmatrix} G_1 & 0 & 0 & 0 & 0 \\ 0 & G_2 & 0 & 0 & 0 \\ 0 & 0 & C_3 D & 0 & 0 \\ 0 & 0 & -g_m & \Gamma_{44} D^{-1} & \Gamma_{45} D^{-1} \\ 0 & 0 & 0 & \Gamma_{45} D^{-1} & \Gamma_{55} D^{-1} \end{bmatrix} \begin{bmatrix} v_1 \\ v_2 \\ v_3 \\ v_4 \\ v_5 \end{bmatrix} + \begin{bmatrix} j_{s1} \\ 0 \\ 0 \\ j_{L4}(0) \\ j_5(0) \end{bmatrix}$$

Note that the matrix is precisely that obtained in the sinusoidal steady state if we were to replace $D$ by $j\omega$. Note also the contribution of the initial state, the terms $j_{L4}(0)$ and $j_5(0)$ in the right-hand side. In the following we shall not perform algebra with the $D$ symbols among themselves; we shall only multiply and divide them by constants.†

---

† It is *not* legitimate to treat $D$ and $D^{-1}$ as algebraic symbols analogous to real or complex numbers. Whereas it is true that $D$ and $D^{-1}$ may be added and multiplied by real numbers (constants), it is also true that $D \, D^{-1} \neq D^{-1} D$. Indeed, apply the first operator to a function $f$; thus,

*Step 4*    Eliminate the $v_k$'s and $j_k$'s from the system of (3.26), (3.27), and (3.28). If we note that they have, respectively, the form

(3.29)    $\mathbf{Aj} = \mathbf{0}$

(3.30)    $\mathbf{v} = \mathbf{A}^T\mathbf{e}$

(3.31)    $\mathbf{j} = \mathbf{Y}_b(D)\,\mathbf{v} + \mathbf{j}_s$

the result of the elimination is of the familiar form

(3.32)    $\mathbf{AY}_b(D)\,\mathbf{A}^T\mathbf{e} = -\mathbf{Aj}_s$

or

$\mathbf{Y}_n(D)\,\mathbf{e} = \mathbf{i}_s$

Let us calculate for the example the *node admittance matrix operator;* thus,

$\mathbf{Y}_n(D) \triangleq \mathbf{AY}_b(D)\,\mathbf{A}^T$

$$
= \begin{bmatrix} 1 & 0 & 1 & 0 & 0 \\ 0 & 1 & -1 & 1 & 0 \\ 0 & -1 & 0 & 0 & 1 \end{bmatrix} \begin{bmatrix} G_1 & 0 & 0 & 0 & 0 \\ 0 & G_2 & 0 & 0 & 0 \\ 0 & 0 & C_3D & 0 & 0 \\ 0 & 0 & -g_m & \Gamma_{44}D^{-1} & \Gamma_{45}D^{-1} \\ 0 & 0 & 0 & \Gamma_{45}D^{-1} & \Gamma_{55}D^{-1} \end{bmatrix} \begin{bmatrix} 1 & 0 & 0 \\ 0 & 1 & -1 \\ 1 & -1 & 0 \\ 0 & 1 & 0 \\ 0 & 0 & 1 \end{bmatrix}
$$

$$
= \begin{bmatrix} G_1 + C_3D & -C_3D & 0 \\ -C_3D - g_m & G_2 + C_3D + g_m + \Gamma_{44}D^{-1} & -G_2 + \Gamma_{45}D^{-1} \\ 0 & -G_2 + \Gamma_{45}D^{-1} & G_2 + \Gamma_{55}D^{-1} \end{bmatrix}
$$

---

$D\,D^{-1}f = \dfrac{d}{dt}\left[\displaystyle\int_0^t f(\tau)\,d\tau\right] = f(t)$

where the last step used the fundamental theorem of the calculus.    Now

$D^{-1}Df = \displaystyle\int_0^t f'(\tau)\,d\tau = f(\tau)\Big|_0^t = f(t) - f(0).$

On the other hand, $D$ and the *positive* integral powers of $D$ can be manipulated by the usual rules of algebra.    In fact, for any *positive* integers $m$ and $n$,

$D^m\,D^n = D^{m+n}$

and for any real numbers $\alpha_1, \alpha_2, \beta_1, \beta_2$

$(\alpha_1 D^n + \beta_1)(\alpha_2 D^m + \beta_2) = \alpha_1\alpha_2 D^{m+n} + \alpha_1\beta_2 D^n + \alpha_2\beta_1 D^m + \beta_1\beta_2$

Therefore, for the present example, node analysis gives the following integrodifferential equation:

$$(3.33) \quad \begin{bmatrix} G_1 + C_3 D & -C_3 D & 0 \\ -C_3 D - g_m & C_3 D + G_2 + g_m + \Gamma_{44} D^{-1} & -G_2 + \Gamma_{45} D^{-1} \\ 0 & -G_2 + \Gamma_{45} D^{-1} & G_2 + \Gamma_{55} D^{-1} \end{bmatrix} \begin{bmatrix} e_1 \\ e_2 \\ e_3 \end{bmatrix}$$

$$= \begin{bmatrix} -j_{s1} \\ -j_{L4}(0) \\ -j_5(0) \end{bmatrix}$$

The required initial conditions are easily obtained; writing the cut-set law for branches 1, 4, and 5 we have

$$(3.34) \quad e_1(0) = \frac{1}{G_1}[-j_{s1}(0) - j_{L4}(0) + g_m v_3(0) - j_5(0)]$$

where all terms in brackets are known.   Finally,

$$(3.35) \quad e_2(0) = e_1(0) - v_3(0)$$

and

$$(3.36) \quad e_3(0) = e_2(0) - G_2 j_5(0)$$

---

**Remarks**   1.   Except for the initial conditions, the writing of node equations in the integrodifferential equation form is closely related to that used in the sinusoidal steady-state analysis.   It is easily seen that if we replace $D$ by $j\omega$ in $\mathbf{Y}_n(D)$, we obtain the node admittance matrix $\mathbf{Y}_n(j\omega)$ for the sinusoidal steady-state analysis.   Therefore, in the absence of coupling elements, $\mathbf{Y}_n(D)$ can always be written by inspection.

2.   Although the equations of (3.33) correspond to the rather formidable name of "integrodifferential equations," they are in fact no different than differential equations; it is just a matter of notation!   To prove the point, introduce new variables, namely fluxes $\phi_2$ and $\phi_3$ *defined* by

$$(3.37) \quad \phi_2(t) \overset{\Delta}{=} \int_0^t e_2(t')\, dt' \qquad \phi_3(t) \overset{\Delta}{=} \int_0^t e_3(t')\, dt'$$

Clearly,

$$De_2 = D^2 \phi_2 \qquad e_2 = D\phi_2 \qquad D^{-1} e_2 = \phi_2$$

The system of (3.33) becomes

$$
\begin{bmatrix}
G_1 + C_3 D & -C_3 D^2 & 0 \\
-C_3 D - g_m & C_3 D^2 + (G_2 + g_m)D + \Gamma_{44} & -G_2 D + \Gamma_{45} \\
0 & -G_2 D + \Gamma_{45} & G_2 D + \Gamma_{55}
\end{bmatrix}
\begin{bmatrix}
e_1 \\ \phi_2 \\ \phi_3
\end{bmatrix}
$$

$$
= \begin{bmatrix}
-j_{s1} \\
-j_{L4}(0) \\
-j_5(0)
\end{bmatrix}
$$

and the initial conditions are

$e_1(0)$     given by (3.34)

$\phi_2(0) = 0$    see (3.37)      $\dot{\phi}_2(0) = e_2(0)$     given by (3.35)

$\phi_3(0) = 0$    see (3.37)      $\dot{\phi}_3(0) = e_3(0)$     given by (3.36)

### 3.5    Shortcut Method

When the network under study involves only a few dependent sources, the equations can be written by inspection if one uses the following idea: treat the *dependent* source as an independent source, and only in the last step express the source in terms of the appropriate variables.

**Example 1**    Let us write the sinusoidal steady-state equations for the network of Fig. 3.7.

*Step 1*    Replace the dependent sources by independent ones, and call them $J_{s3}$ and $J_{s5}$.

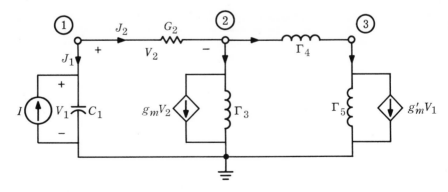

**Fig. 3.7**    Network including dependent sources.

*Step 2*    Write the equations by inspection as follows:

$$
\begin{bmatrix}
G_2 + j\omega C_1 & -G_2 & 0 \\
-G_2 & G_2 + \dfrac{\Gamma_3 + \Gamma_4}{j\omega} & -\dfrac{\Gamma_4}{j\omega} \\
0 & -\dfrac{\Gamma_4}{j\omega} & \dfrac{\Gamma_4 + \Gamma_5}{j\omega}
\end{bmatrix}
\begin{bmatrix}
E_1 \\ E_2 \\ E_3
\end{bmatrix}
=
\begin{bmatrix}
I \\ -J_{s3} \\ -J_{s5}
\end{bmatrix}
$$

*Step 3*    Express the dependent source waveforms in terms of the appropriate variables as follows:

$$ J_{s3} = g_m V_2 = g_m(E_1 - E_2) $$

and

$$ J_{s5} = g'_m V_1 = g'_m E_1 $$

*Step 4*    Substitute and rearrange terms; thus we have

$$
\begin{bmatrix}
G_2 + j\omega C_1 & -G_2 & 0 \\
g_m - G_2 & G_2 - g_m + \dfrac{\Gamma_3 + \Gamma_4}{j\omega} & -\dfrac{\Gamma_4}{j\omega} \\
g'_m & -\dfrac{\Gamma_4}{j\omega} & \dfrac{\Gamma_4 + \Gamma_5}{j\omega}
\end{bmatrix}
\begin{bmatrix}
E_1 \\ E_2 \\ E_3
\end{bmatrix}
=
\begin{bmatrix}
I \\ 0 \\ 0
\end{bmatrix}
$$

**Example 2**    Let us write by inspection the integrodifferential equations of the network shown in Fig. 3.6 under the assumption that $\Gamma_{45} = 0$. The network is redrawn in Fig. 3.8.

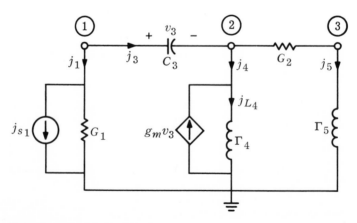

**Fig. 3.8**    Network analyzed in Example 2.

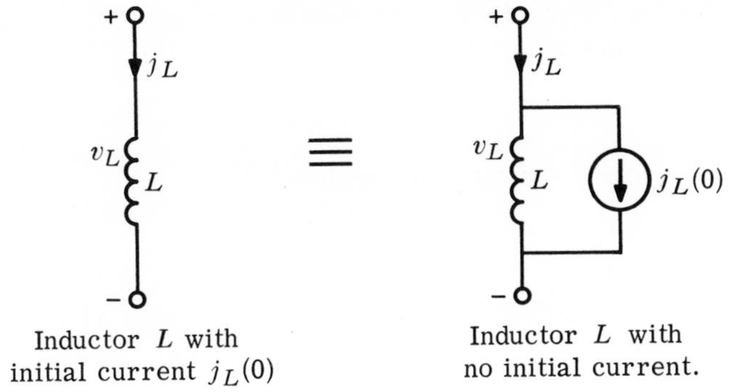

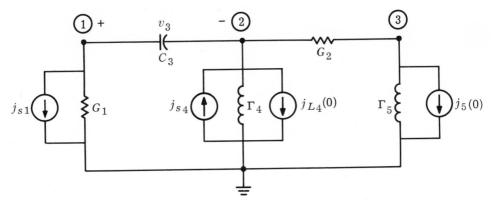

Inductor $L$ with
initial current $j_L(0)$

Inductor $L$ with
no initial current.

**Fig. 3.9**   Useful equivalence when writing equations by inspection.

*Step 1*   We must observe that the branch equations of inductors include initial currents (see Fig. 3.9); thus,

$$j_L(t) = \Gamma \int_0^t v_L(t') \, dt' + j_L(0)$$

Thus, every inductor can be replaced by an inductor without initial current in parallel with a *constant* current source $j_L(0)$. After replacing the dependent source $g_m v_3$ by an independent source $j_{s4}$, we have the network shown in Fig. 3.10.

*Step 2*   By inspection, the equations are

$$\begin{bmatrix} G_1 + C_3 D & -C_3 D & 0 \\ -C_3 D & G_2 + \Gamma_4 D^{-1} + C_3 D & -G_2 \\ 0 & -G_2 & G_2 + \Gamma_5 D^{-1} \end{bmatrix} \begin{bmatrix} e_1 \\ e_2 \\ e_3 \end{bmatrix} = \begin{bmatrix} -j_{s1} \\ j_{s4} - j_{L4}(0) \\ -j_5(0) \end{bmatrix}$$

**Fig. 3.10**   Network of Fig. 3.8 in which the initial currents have been replaced by constant current sources.

*Step 3*   Express the dependent source waveform in terms of the appropriate variable as follows:

$$j_{s4} = g_m v_3 = g_m(e_1 - e_2)$$

*Step 4*   Substitute and rearrange terms; thus,

$$\begin{bmatrix} G_1 + C_3 D & -C_3 D & 0 \\ -g_m - C_3 D & g_m + G_2 + \Gamma_4 D^{-1} + C_3 D & -G_2 \\ 0 & -G_2 & G_2 + \Gamma_5 D^{-1} \end{bmatrix} \begin{bmatrix} e_1 \\ e_2 \\ e_3 \end{bmatrix} = \begin{bmatrix} -j_{s1} \\ -j_{L4}(0) \\ -j_5(0) \end{bmatrix}$$

In summary, it is easy in many instances to write the equations by inspection. It is important to know that the systematic method of Sec. 3.1 and 3.3 always works and hence can be used as a check in case of doubt.

---

## 4   Duality

In this section we propose to develop the concept of duality which we shall use repeatedly in the remainder of this chapter and in Chap. 11.

### 4.1   Planar Graphs, Meshes, Outer Meshes

By the very definition of a *graph* that we adopted, namely, a set of nodes and a set of branches each terminated at each end into a node, it is clear that a given graph may be drawn in several different ways. For example, the three figures shown in Fig. 4.1 are representations of the same graph. Indeed, they have the same incidence matrix. Similarly, a loop is a concept that does not depend on the way the graph is drawn; for example, the branches *f*, *b*, *c*, and *e* form a loop in the three figures shown in Fig. 4.1.

If we use the term "mesh" intuitively, we would call the loop *bcef* a

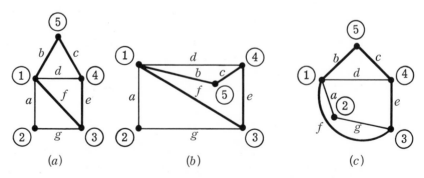

(a)                    (b)                    (c)

**Fig. 4.1**   The figures (a), (b), and (c) represent the same *graph* in the form of three different *topological graphs*.

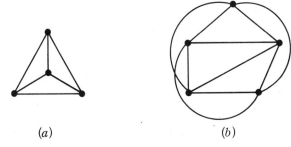

$(a)$                                                $(b)$

**Fig. 4.2**    $(a)$ Planar graph; $(b)$ nonplanar graph.

mesh in Fig. 4.1$b$, but not in Fig. 4.1$a$ or $c$.   For this reason, we need to consider graphs drawn in a specific way.   When we do consider a graph $\mathcal{G}$ drawn in a way specified by us, we refer to it as the **topological graph** $\mathcal{G}$.   For example, the three figures in Fig. 4.1 may be considered to be the same *graph* or three different *topological graphs.*

A graph $\mathcal{G}$ is said to be **planar** if it *can* be drawn on the plane in such a way that no two branches intersect at a point which is not a node.   The graph in Fig. 4.2$a$ is a planar graph, whereas the graph in Fig. 4.2$b$ is not.

Consider a topological graph $\mathcal{G}$ which is planar.   We call any loop of this graph for which there is no branch in its interior a **mesh**.   For example, for the topological graph shown in Fig. 4.1$a$ the loop *f bce* is not a mesh; for the topological graph shown in Fig. 4.1$b$ the loop *f bce* is a mesh.   In Fig. 4.1$c$ the loop *f bce* contains no branches in its exterior; it is called the **outer mesh** of the topological graph.

If you imagine the planar graph as your girl friend's hair net and if you imagine it slipped over a transparent sphere of lucite, then as you stand in the center of the sphere and look outside, you see that there is no significant difference between a mesh and the outer mesh.

We next exhibit a type of network whose graphs have certain properties that lead to simplification in analysis.   Consider the three graphs shown in Fig. 4.3$a$–$c$.   Each of these graphs has the property that it can be partitioned into *two* nondegenerate subgraphs $\mathcal{G}_1$ and $\mathcal{G}_2$ which are connected together by *one* node.†   Graphs which have this property are called **hinged graphs.**   A graph that is not hinged is called unhinged (or sometimes, nonseparable); thus, an **unhinged graph** has the property that whenever it is partitioned into two connected nondegenerate subgraphs $\mathcal{G}_1$ and $\mathcal{G}_2$, the subgraphs have at least two nodes in common.   Determining whether a graph is hinged or not is easily done by inspection (see Fig. 4.3 for examples).

From a network analysis point of view, if a network has a graph that is hinged and if there is no coupling (by mutual inductances or dependent

---

† By nondegenerate we mean that the subgraph is not an isolated node.

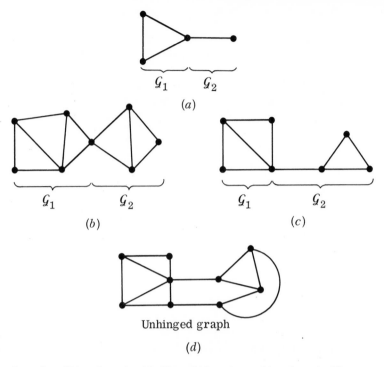

**Fig. 4.3**   Examples of hinged graphs, (a), (b), and (c), and an unhinged graph, (d).

sources) between the elements of $\mathcal{G}_1$ and $\mathcal{G}_2$, the analysis of the network reduces to the analysis of two independent subnetworks, namely the networks corresponding to the graphs $\mathcal{G}_1$ and $\mathcal{G}_2$. Since $\mathcal{G}_1$ and $\mathcal{G}_2$ are connected together by one node, KCL requires that the net current flow from $\mathcal{G}_1$ to $\mathcal{G}_2$ be zero at all times, so there is no exchange of current between the two subnetworks. Also the fact that the two subnetworks have a node in common does not impose any restriction on the branch voltages.

*Counting*
*meshes*
It is easy to see that for *a connected unhinged planar graph the number of meshes is equal to* $b - n_t + 1$ where $b$ is the number of branches and $n_t$ the number of nodes. The proof can be given by mathematical induction. Let the number of meshes be $l$; thus we want to show that

(4.1)
$$l = b - n_t + 1$$

Consider the graph in Fig. 4.4a, where $l = 1$. Here it is obvious that Eq. (4.1) is true. Next consider a graph which has $l$ meshes, and we assume that Eq. (4.1) is true. We want to show that (4.1) is still true if the graph is

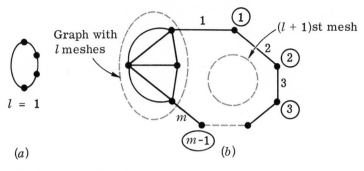

**Fig. 4.4**   Indication of proof of $l = b - n_t + 1$.

changed to have $l + 1$ meshes.   We can increase the number of meshes by 1 by adding a branch between two existing nodes or by adding $m$ branches in series which are connected to the existing graph through $m - 1$ new nodes, as shown in Fig. 4.4*b*.   For the new graph with $l + 1$ meshes Eq. (4.1) is still satisfied, because $m - 1$ nodes and $m$ branches have been added, resulting in one additional mesh.   Therefore, by induction, Eq. (4.1) is true in general.

*The matrix* $\mathbf{M}_a$   A fundamental property of a connected unhinged planar graph is that each branch of the graph belongs to exactly two meshes if we include the outer mesh.   Consider such a graph with specified branch orientation.   We shall assign by convention the following reference directions for the meshes: the clockwise direction for each mesh and the counterclockwise direction for the outer mesh.   This is illustrated in the graph of Fig. 4.5.   Thus, an oriented planar graph $\mathcal{G}$ that is connected and unhinged can be described

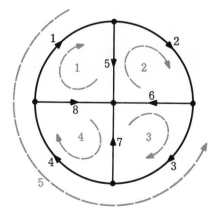

**Fig. 4.5**   An oriented planar graph with eight branches and five meshes (including the outer mesh).

analytically by a matrix $\mathbf{M}_a$. Let $\mathcal{G}$ have $b$ branches and $l + 1$ meshes (including the outer mesh); then $\mathbf{M}_a$ is defined as the rectangular matrix of $l + 1$ rows and $b$ columns whose $(i,k)$th element $m_{ik}$ is defined by

$$
m_{ik} = \begin{cases}
1 & \text{if branch } k \text{ is in mesh } i \text{ and if their reference directions } \textit{coincide} \\
-1 & \text{if branch } k \text{ is in mesh } i \text{ and if their reference directions } \textit{do not coincide} \\
0 & \text{if branch } k \text{ } \textit{does not belong} \text{ to mesh } i
\end{cases}
$$

For the graph shown in Fig. 4.5, $b = 8, l + 1 = 5$, and the matrix $\mathbf{M}_a$ is

$$
\text{Meshes}\begin{array}{c} \\ 1 \\ 2 \\ 3 \\ 4 \\ 5 \end{array}
\begin{bmatrix}
1 & 0 & 0 & 0 & 1 & 0 & 0 & -1 \\
0 & 1 & 0 & 0 & -1 & 1 & 0 & 0 \\
0 & 0 & 1 & 0 & 0 & -1 & 1 & 0 \\
0 & 0 & 0 & 1 & 0 & 0 & -1 & 1 \\
-1 & -1 & -1 & -1 & 0 & 0 & 0 & 0
\end{bmatrix}
$$

with Branches columns $1\ 2\ 3\ 4\ 5\ 6\ 7\ 8$.

Observe that the matrix $\mathbf{M}_a$ has a property in common with the incidence matrix $\mathbf{A}_a$; that is, in each column all elements are zero except for one $+1$ and one $-1$. In the succeeding subsection the concept of dual graphs will be introduced so that we may further explore the relation between these matrices.

## 4.2   Dual Graphs

Before giving a precise formulation of the concept of dual graphs and dual networks, let us consider an example. By pointing to some features of the following example, we shall provide some motivation for the later formulations.

**Example 1**   Consider the linear time-invariant networks shown in Fig. 4.6. For simplicity we assume that the sources are sinusoidal and have the same frequency, and that the networks are in the sinusoidal steady state. In the first network, say $\mathfrak{N}$, we represent the two sinusoidal node-to-datum voltages by the *phasors* $E_1$ and $E_2$. By inspection, we obtain the following node equations:

(4.2a)   $\left(j\omega C_1 + \dfrac{1}{j\omega L}\right)E_1 - \dfrac{1}{j\omega L}E_2 = I_s$

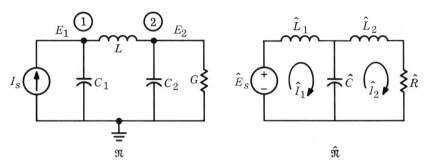

**Fig. 4.6**   Two networks used to illustrate duality; if $C_1 = \hat{L}_1$, $C_2 = \hat{L}_2$, $L = \hat{C}$, $G = \hat{R}$, and $I_s = \hat{E}_s$, then they are said to be dual networks.

$$(4.2b) \quad -\frac{1}{j\omega L} E_1 + \left( j\omega C_2 + G + \frac{1}{j\omega L} \right) E_2 = 0$$

Note that $I_s$ is the *phasor* that represents the sinusoidal current of the source.

The second network $\hat{\mathfrak{N}}$ has two meshes. We represent the two sinusoidal mesh currents by the phasors $\hat{I}_1$ and $\hat{I}_2$. Again, by inspection, we obtain the following mesh equations:

$$(4.3a) \quad \left( j\omega \hat{L}_1 + \frac{1}{j\omega \hat{C}} \right) \hat{I}_1 - \left( \frac{1}{j\omega \hat{C}} \right) \hat{I}_2 = \hat{E}_s$$

$$(4.3b) \quad -\left( \frac{1}{j\omega \hat{C}} \right) \hat{I}_1 + \left( j\omega \hat{L}_2 + \hat{R} + \frac{1}{j\omega \hat{C}} \right) \hat{I}_2 = 0$$

If the element values of the two circuits are related by

$$C_1 = \hat{L}_1 \qquad L = \hat{C} \qquad C_2 = \hat{L}_2 \qquad G = \hat{R}$$

and if the sources have the same phasor

$$I_s = \hat{E}_s$$

then Eqs. (4.2) and (4.3) are *identical*. Therefore, if we have solved one of the networks, we have solved the other. These two networks are an example of a pair of dual networks. There are some interesting relations between them. $\mathfrak{N}$ has two nodes and a datum node; $\hat{\mathfrak{N}}$ has two meshes and an outer mesh. Both have five branches. To a branch between two nodes of $\mathfrak{N}$ (say, the inductor connecting node ① and node ②) corresponds a branch of $\hat{\mathfrak{N}}$ which is common to the corresponding meshes (the capacitor common to mesh 1 and mesh 2). In $\mathfrak{N}$, the current source $I_s$ and the capacitor $C_1$ are in parallel; in $\hat{\mathfrak{N}}$, the voltage source $\hat{E}_s$ and the inductor $\hat{L}$ are in series; etc. We note that the relation between $\mathfrak{N}$ and $\hat{\mathfrak{N}}$ involves both graph-theoretic concepts (meshes and nodes) and the nature of the elements (sources, inductors, capacitors, etc.). For this reason we must proceed in two steps, first considering dual graphs and then defining dual networks.

We are ready to introduce the concept of dual graphs.  Again we start with a topological graph $\mathcal{G}$ which is assumed to be connected, unhinged, and planar.† Let $\mathcal{G}$ have $n_t = n + 1$ nodes, $b$ branches, and hence, $l = b - n$ meshes (not counting the outer mesh).  A planar topological graph $\widehat{\mathcal{G}}$ is said to be a **dual graph** of a topological graph $\mathcal{G}$ if

1.  There is a one-to-one correspondence between the *meshes* of $\mathcal{G}$ (including the outer mesh) and the *nodes* of $\widehat{\mathcal{G}}$.

2.  There is a one-to-one correspondence between the *meshes* of $\widehat{\mathcal{G}}$ (including the outer mesh) and the *nodes* of $\mathcal{G}$.

3.  There is a one-to-one correspondence between the branches of each graph in such a way that whenever two meshes of one graph have the corresponding branch in common, the corresponding nodes of the other graph have the corresponding branch connecting these nodes.

We shall use the symbol ^ to designate all terms pertaining to a dual graph.

It follows from the definition that $\widehat{\mathcal{G}}$ has $b$ branches, $l + 1$ nodes, $n$ meshes, and one outer mesh.  It is easily checked that if $\widehat{\mathcal{G}}$ is a dual graph of $\mathcal{G}$, then $\mathcal{G}$ is a dual graph of $\widehat{\mathcal{G}}$.  In other words *duality is a symmetric relation between connected, planar, unhinged topological graphs.*

**ALGORITHM**    Given a connected, planar, unhinged topological graph $\mathcal{G}$, we construct a dual graph $\widehat{\mathcal{G}}$ by proceeding as follows:

1.  To each mesh of $\mathcal{G}$, including the outer mesh, we associate a node of $\widehat{\mathcal{G}}$; thus, we associate node ① to mesh 1 and draw node ① inside mesh 1; a similar procedure is followed for nodes ②, ③, . . . , including node $\boxed{l + 1}$, which corresponds to the outer mesh.

2.  For each branch, say $k$, of $\mathcal{G}$ which is common to mesh $i$ and mesh $j$, we associate a branch $k$ of $\widehat{\mathcal{G}}$ which is connected to nodes ⓘ and ⓙ.

By its very construction, the resulting graph $\widehat{\mathcal{G}}$ is a dual of $\mathcal{G}$.

**Example 2**    The given planar graph is shown in Fig. 4.7*a*.  There are three meshes, not counting the outer mesh.  We insert nodes ①, ②, and ③, with one node inside each mesh as shown in Fig. 4.7*b*.  We place node ④ outside the graph $\mathcal{G}$ because ④ will correspond to the outer mesh.  To complete

---

† The concept of dual graph can be introduced for arbitrary planar connected graph.  However, for simplicity we rule out the case of hinged graph.

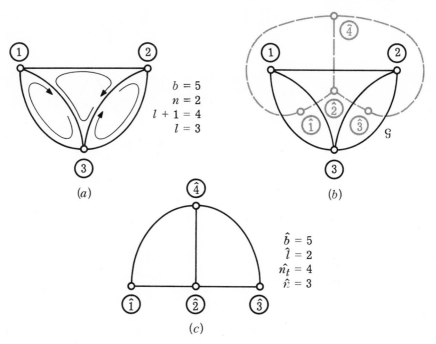

$$b = 5$$
$$n = 2$$
$$l + 1 = 4$$
$$l = 3$$

(a)

(b)

$$\hat{b} = 5$$
$$\hat{l} = 2$$
$$\hat{n}_t = 4$$
$$\hat{n} = 3$$

(c)

**Fig. 4.7**   Example 2: Illustrating the construction of a dual graph.  (a) Given graph; (b) construction step; (c) dual graph.

the dual graph $\widehat{\mathscr{G}}$, we connect two nodes with a branch whenever the corresponding meshes of $\mathscr{G}$ have a branch in common.   The dotted lines in Fig. 4.7b represent branches of the dual graph $\widehat{\mathscr{G}}$.   The dual graph $\widehat{\mathscr{G}}$ is redrawn in Fig. 4.7c.

In case the given graph $\mathscr{G}$ is oriented, i.e., in case each branch has a reference direction, the orientation of the dual graph $\widehat{\mathscr{G}}$ can be obtained by adding to the construction above a simple orientation convention.   Since the branches in both graphs are oriented, we can imagine the reference directions of the branches to be indicated by vectors which lie along the branch and point in the direction of the reference direction.   The reference direction of a branch of the dual graph $\widehat{\mathscr{G}}$ is obtained from that of the corresponding branch of the given graph $\mathscr{G}$ by rotating the vector 90° *clockwise*.   With this algorithm, given any planar oriented topological graph $\mathscr{G}$, we can obtain in a systematic fashion a dual oriented graph $\widehat{\mathscr{G}}$.

**Example 3**   Consider the oriented graph in Fig. 4.8a.   Let node ④ be the datum node. We wish to obtain the dual graph $\widehat{\mathscr{G}}$ whose outer mesh corresponds to node ④ of $\mathscr{G}$.   Following the rules for constructing a dual graph, we insert

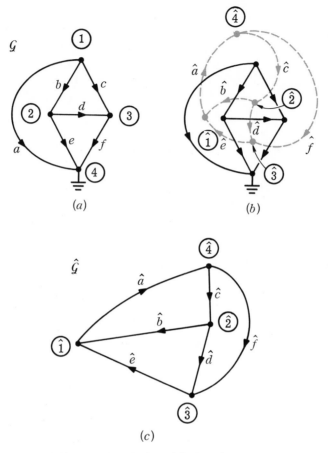

**Fig. 4.8**   Example 3: Construction of oriented dual graph.

nodes ①, ②, and ③ inside meshes 1, 2, and 3, respectively, of $\mathcal{G}$, leaving node ④ outside the mesh, as shown in Fig. 4.8*b*.   To complete the dual graph, branches are drawn in dotted lines, as shown in Fig. 4.8*b*, connecting the nodes of $\hat{\mathcal{G}}$ to correspond to the branches of $\mathcal{G}$.   The reference directions of the branches of $\hat{\mathcal{G}}$ are obtained by the method indicated above. The dual graph is redrawn as shown in Fig. 4.8*c*.   Care must be taken to ensure that the branches of $\mathcal{G}$ connecting the datum node correspond to the outer mesh of $\hat{\mathcal{G}}$.   A moment of thought will lead to the following rule, which gives the appropriate one-to-one correspondence.   Since the datum node, node ④ in the example, must stay outside all dotted lines, when placing node ④ of $\hat{\mathcal{G}}$ it is convenient to put it as far away as possible from the datum node ④, as shown in Fig. 4.8*b*.

**Remarks**

1. In general, a given topological graph $\mathcal{G}$ has many duals. However, if we specify the datum node of $\mathcal{G}$ and specify that it has to correspond to the outer mesh of $\widehat{\mathcal{G}}$, then the procedure described above defines a unique dual graph. The branches which are connected to the datum node in $\mathcal{G}$ have corresponding branches which form the outer mesh in $\widehat{\mathcal{G}}$.

2. The correspondence between the graph $\mathcal{G}$ and its dual $\widehat{\mathcal{G}}$ involves branch versus branch, node versus mesh, and datum node versus outer mesh. Furthermore, the incidence matrix $\mathbf{A}_a$ of the given graph $\mathcal{G}$ is equal to the matrix $\mathbf{M}_a$ of the dual graph $\widehat{\mathcal{G}}$.

**Exercise**

Construct a dual graph $\widehat{\mathcal{G}}$ of the oriented planar graph $\mathcal{G}$ given in Fig. 4.5. Write the KCL equations of the dual graph for all the nodes; that is,

$$\widehat{\mathbf{A}}_a\widehat{\mathbf{j}} = \mathbf{0}$$

Show that the set of equations is identical to the KVL equations of the given graph for all the meshes (including the outer mesh); that is,

$$\mathbf{M}_a\mathbf{v} = \mathbf{0}$$

## 4.3 Dual Networks

In this discussion we restrict ourselves to networks having the following properties: their graphs are *connected, planar,* and *unhinged;* and all their elements are one-port elements. In other words, we *exclude* coupled inductors, transformers, and dependent sources; we *include* independent voltage or current sources, inductors, resistors, and capacitors. It is fundamental to observe that the elements do not have to be linear and/or time-invariant.

We say that a network $\widehat{\mathcal{N}}$ is the **dual of the network** $\mathcal{N}$ if (1) the topological graph $\widehat{\mathcal{G}}$ of $\widehat{\mathcal{N}}$ is a dual of the topological graph $\mathcal{G}$ of $\mathcal{N}$, and (2) the branch equation of a branch of $\widehat{\mathcal{N}}$ is obtained from its corresponding equation of $\mathcal{N}$ by performing the following substitutions:

(4.4)
$$v \to \widehat{j} \qquad q \to \widehat{\phi}$$
$$j \to \widehat{v} \qquad \phi \to \widehat{q}$$

where $v, j, q,$ and $\phi$ are the branch voltage, current, charge, and flux variables, respectively, for $\mathcal{N}$, and $\widehat{v}, \widehat{j}, \widehat{q},$ and $\widehat{\phi}$ are the corresponding branch variables for $\widehat{\mathcal{N}}$.

Requirement 2 of the definition means that a resistor of $\mathcal{N}$ corresponds to a resistor of $\widehat{\mathcal{N}}$. Furthermore, a linear resistor of $\widehat{\mathcal{N}}$ with a resistance of $K$ ohms corresponds to a linear resistor of $\widehat{\mathcal{N}}$ with a conductance of $K$ mhos. Indeed, the branch equation of the resistor of $\mathcal{N}$ is $v = Kj$; hence, that of the corresponding resistor of $\hat{\mathfrak{N}}$ is $\hat{j} = K\hat{v}$. Similarly, an inductor of $\mathfrak{N}$ corresponds to a capacitor of $\hat{\mathfrak{N}}$. Furthermore, a nonlinear time-varying inductor of $\mathfrak{N}$ which is characterized by $\phi = f(i,t)$, where $f(\cdot, \cdot)$ is a given function of two variables, will correspond to a nonlinear time-varying capacitor of $\hat{\mathfrak{N}}$ which is characterized by $\hat{q} = f(\hat{v},t)$. A voltage source whose voltage is a function $f(\cdot)$ will correspond to a current source whose current is the same function $f(\cdot)$. Also, the dual of a short circuit is an open circuit; a short circuit is characterized by $v = 0$, hence its dual is characterized by $\hat{j} = 0$, the equation for an open circuit.

It is easily checked that if $\hat{\mathfrak{N}}$ is a dual network of $\mathfrak{N}$, then $\mathfrak{N}$ is a dual network of $\hat{\mathfrak{N}}$; in other words, *duality is a symmetric relation between networks.*

**Example 4**    Consider the nonlinear time-varying network $\mathfrak{N}$ shown in Fig. 4.9a. The inductor is nonlinear; its characteristic is $\phi_1 = \tanh j_1$. The capacitor is time-varying and linear; its characteristic is $q_3 = (1 + \epsilon^{-t^2})v_3$. The output resistor is linear and time-varying with resistance $2 + \cos t$; its characteristic is $v_4 = (2 + \cos t)j_4$. The branch orientations are indicated in the figure. Let the mesh currents be $i_1$ and $i_2$; then,

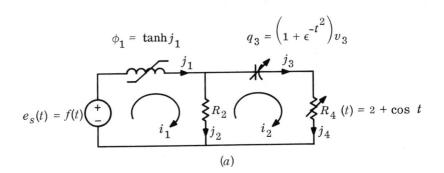

(a)

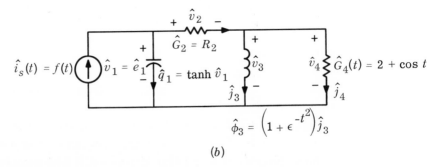

(b)

**Fig. 4.9**    Example 4: Illustrating dual networks.

$$j_1 = i_1 \qquad j_2 = i_1 - i_2 \qquad j_3 = j_4 = i_2$$

The mesh equations read

(4.5a) $\quad e_s(t) = f(t) = \dfrac{1}{\cosh^2 i_1} \dfrac{di_1}{dt} + R_2(i_1 - i_2)$

(4.5b) $\quad 0 = R_2(i_2 - i_1) + \dfrac{q_3(0)}{1 + \epsilon^{-t^2}} + \dfrac{1}{1 + \epsilon^{-t^2}} \displaystyle\int_0^t i_2(t')\, dt' + (2 + \cos t)i_2(t)$

The dual network $\widehat{\mathfrak{N}}$ is easily found. First, the dual graph $\widehat{\mathcal{G}}$ is drawn, including the orientation; then each branch is filled with the appropriate dual element as prescribed by requirement 2. The result is shown in Fig. 4.9b. Let the node-to-datum voltages be $\widehat{e}_1$ and $\widehat{e}_2$. Then the branch voltages are related to the node voltages by $\widehat{v}_1 = \widehat{e}_1$, $\widehat{v}_2 = \widehat{e}_1 - \widehat{e}_2$, and $\widehat{v}_3 = \widehat{v}_4 = \widehat{e}_2$. The node equations give

(4.6a) $\quad \widehat{i}_s(t) = f(t) = \dfrac{1}{\cosh^2 \widehat{e}_1} \dfrac{d\widehat{e}_1}{dt} + \widehat{G}_2(\widehat{e}_1 - \widehat{e}_2)$

(4.6b) $\quad 0 = \widehat{G}_2(\widehat{e}_2 - \widehat{e}_1) + \dfrac{\widehat{\phi}(0)}{1 + \epsilon^{-t^2}} + \dfrac{1}{1 + \epsilon^{-t^2}} \displaystyle\int_0^t \widehat{e}_2(t')\, dt' + (2 + \cos t)e_2(t)$

where the conductance $\widehat{G}_2$ is equal to the resistance $R_2$. Observe that Eqs. (4.6) are identical with Eqs. (4.5) except for the names of the variables.

---

*General property of dual networks*

The importance of duality cannot be overemphasized. Its power is exhibited by the following general assertion. *Consider an arbitrary planar network $\mathfrak{N}$ and its dual $\widehat{\mathfrak{N}}$. Let S be any true statement concerning the behavior of $\mathfrak{N}$. Let $\widehat{S}$ be the statement obtained from S by replacing every graph-theoretic word or phrase (node, mesh, loop, etc.) by its dual and every electrical quantity (voltage, current, impedance, etc.) by its dual. Then $\widehat{S}$ is a true statement concerning the behavior of $\widehat{\mathfrak{N}}$.* In Table 10.1 we give a tabulation of pairs of dual terms. Some of these will be illustrated in later chapters.

**Exercise 1** Consider the linear time-invariant *RLC* network $\mathfrak{N}$ (without coupled inductors) shown in Fig. 4.10. Assume that its graph is planar and unhinged. Suppose it is driven by a sinusoidal current source and is in the steady state. Consider the dual network $\widehat{\mathfrak{N}}$. Show that the driving-point impedance $Z_{in}$ of $\mathfrak{N}$ is for each $\omega$ equal to the driving-point admittance $Y_{in}$ of $\widehat{\mathfrak{N}}$.

**Exercise 2** Consider the ladder network $\mathfrak{N}$ shown in Fig. 4.11. The functions $f_1(j\omega)$, $f_2(j\omega), \ldots, f_5(j\omega)$ specify the *impedances* of the corresponding element of $\mathfrak{N}$; the function $f$ specifies the waveform of the source. Show that the dual of $\mathfrak{N}$ can be obtained (1) by replacing the current source of $f(t)$ amp

**Table 10.1**   **Dual Terms**

| Types of properties | $S$ | $\widehat{S}$ |
|---|---|---|
| Graph-theoretic properties | Node | Mesh |
| | Cut set | Loop |
| | Datum node | Outer mesh |
| | Tree branch* | Link* |
| | Fundamental cut set* | Fundamental loop* |
| | Branches in series | Branches in parallel |
| | Reduced incidence matrix | Mesh matrix* |
| | Fundamental cut-set matrix* | Fundamental loop matrix* |
| Graph-theoretic and electric properties | Node-to-datum voltages | Mesh currents |
| | Tree-branch voltages* | Link currents* |
| | KCL | KVL |
| Electric properties | Voltage | Current |
| | Charge | Flux |
| | Resistor | Resistor |
| | Inductor | Capacitor |
| | Resistance | Conductance |
| | Inductance | Capacitance |
| | Current source | Voltage source |
| | Short circuit | Open circuit |
| | Admittance | Impedance |
| | Node admittance matrix | Mesh impedance matrix* |

* The asterisk is used to indicate terms that will be encountered in this and the next chapter.

by a voltage source of $f(t)$ volts, (2) by replacing each series element of $\mathfrak{N}$ of impedance $f_i(j\omega)$ by a shunt element of admittance $f_i(j\omega)$, and (3) by replacing each shunt element of $\mathfrak{N}$ of impedance $f_i(j\omega)$ by a series element of admittance $f_i(j\omega)$.

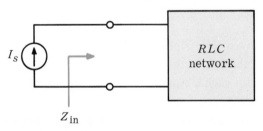

**Fig. 4.10**   Exercise 1: illustrating the dual of a driving-point impedance.

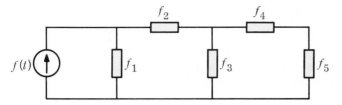

**Fig. 4.11**    A ladder network $\mathfrak{N}$.

## 5    Two Basic Facts of Mesh Analysis

Let us consider any network $\mathfrak{N}$ whose graph is connected, planar, and un-hinged. We assume that it has $n_t$ nodes and $b$ branches; consequently it has $l = b - n_t + 1$ meshes, not counting the outer mesh. We label the meshes $1, 2, \ldots, l$ and use the clockwise reference directions. The meshes are the duals of the nodes, whereas the outer mesh is the dual of the datum node. We shall employ the concept of duality to develop the two basic facts of mesh analysis. Again it should be emphasized that the two facts are independent of the nature of the network elements. Thus, the network can be linear or nonlinear, time-invariant or time-varying.

### 5.1    Implications of KVL

Let us apply KVL to meshes $1, 2, \ldots, l$ (omitting the outer mesh). As seen in the example below (Fig. 5.1) each expression is a linear homogeneous algebraic equation in the branch voltages. Thus, we have a system of $l$ linear homogeneous algebraic equations in the $b$ unknowns $v_1, v_2, \ldots, v_b$. The first basic fact of mesh analysis can be stated as follows:

*The $l$ linear homogeneous algebraic equations in $v_1, v_2, \ldots, v_l$ obtained by applying KVL to each mesh (except the outer mesh) constitute a set of $l$ linearly independent equations.*

The proof of this statement can be given in the same manner as the proof of the comparable statement in node analysis. We shall ask the reader to go through corresponding steps. It is, however, quite easy to use duality to prove the statement above: Denote by $\mathfrak{N}$ the network dual to $\mathfrak{N}$. Apply KCL to all nodes of $\widehat{\mathfrak{N}}$ except the datum node. If the basic fact above were false, the first basic fact of node analysis would also be false. Since the latter has been proved independently, it follows that the first basic fact of mesh analysis is true.

Analytically KVL may be expressed by the use of the **mesh matrix**

(5.1)   $\mathbf{Mv} = \mathbf{0}$   (KVL)

where $\mathbf{M} = (m_{ij})$ is an $l \times b$ matrix defined below by Eq. (5.2). When we write that the $i$th component of $\mathbf{Mv}$ is zero, we merely assert that the sum of all branch voltages around the $i$th mesh is zero. Since this $i$th component is of the form

$$\sum_{k=1}^{b} m_{ik}v_k = 0$$

we must have for $i = 1, 2, \ldots, l$ and $k = 1, 2, \ldots, b$,

(5.2)   $m_{ik} = \begin{cases} 1 & \text{if branch } k \text{ is in mesh } i \text{ and if their reference directions } \textit{coincide} \\ -1 & \text{if branch } k \text{ is in mesh } i \text{ and if their reference directions } \textit{do not coincide} \\ 0 & \text{if branch } k \textit{ does not belong} \text{ to mesh } i \end{cases}$

The basic fact established above implies that *the mesh matrix* $\mathbf{M}$ *has a rank equal to l.*

Note that the mesh matrix $\mathbf{M}$ is obtained from the matrix $\mathbf{M}_a$ by deleting the row of $\mathbf{M}_a$ which corresponds to the outer mesh.

**Example 1**   Consider the oriented graph of Fig. 5.1, which is the dual graph of Fig. 2.3. There are three nodes and five branches; thus, $l = 5 - 3 + 1 = 3$. The three meshes are labeled as shown. The branch voltage vector is

$$\mathbf{v} = \begin{bmatrix} v_1 \\ v_2 \\ v_3 \\ v_4 \\ v_5 \end{bmatrix}$$

The mesh matrix, obtained from Eq. (5.2), is

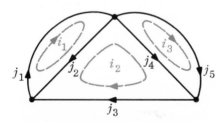

**Fig. 5.1**   An oriented graph which is the dual graph of that of Fig. 2.3.

Mesh

$$
\mathbf{M} = \begin{bmatrix} 1 & 1 & 0 & 0 & 0 \\ 0 & -1 & 1 & 1 & 0 \\ 0 & 0 & 0 & -1 & 1 \end{bmatrix} \begin{matrix} \leftarrow 1 \\ \leftarrow 2 \\ \leftarrow 3 \end{matrix}
$$

$$
\begin{matrix} \uparrow & \uparrow & \uparrow & \uparrow & \uparrow \\ \text{Branch} \quad 1 & 2 & 3 & 4 & 5 \end{matrix}
$$

The mesh equation is therefore

$$
\mathbf{Mv} = \begin{bmatrix} 1 & 1 & 0 & 0 & 0 \\ 0 & -1 & 1 & 1 & 0 \\ 0 & 0 & 0 & -1 & 1 \end{bmatrix} \begin{bmatrix} v_1 \\ v_2 \\ v_3 \\ v_4 \\ v_5 \end{bmatrix} = \begin{bmatrix} 0 \\ 0 \\ 0 \end{bmatrix}
$$

or

$$
\begin{aligned}
v_1 + v_2 &= 0 \\
-v_2 + v_3 + v_4 &= 0 \\
-v_4 + v_5 &= 0
\end{aligned}
$$

which are clearly the three mesh equations obtained from KVL for meshes 1, 2, and 3. The three equations are clearly linearly independent, since each equation includes a variable which is not present in the other two.

**Exercise**   Let $\mathbf{M}$ be the mesh matrix of an oriented graph $\mathcal{G}$. Let $\hat{\mathbf{A}}$ be the reduced incidence matrix of the dual graph $\hat{\mathcal{G}}$. Show that $\hat{\mathbf{A}} = \mathbf{M}$. Verify that the dual of definition (2.3) of $a_{ik}$ gives definition (5.2) of $m_{ik}$.

## 5.2   Implications of KCL

Let us call $i_1, i_2, \ldots, i_l$ the mesh currents. For convenience let us assign to each one a *clockwise* reference direction. First, let us observe that the mesh currents $i_1, i_2, \ldots, i_l$ are linearly independent as far as KCL is concerned. Since each mesh current runs around a loop, if mesh current $i_k$ crosses an arbitrarily chosen cut set in the positive direction, it also crosses the cut set in the negative direction and, hence, cancels out from the KCL equation applied to that cut set. In other words, if we write KCL for any cut set and express the branch currents in terms of mesh currents, everything cancels out. Thus, KCL has nothing to say about the mesh currents, which makes them linearly independent as far as KCL is concerned.

The next step is to show that the branch currents can be calculated in terms of the mesh currents by the equation

(5.3)    $\mathbf{j} = \mathbf{M}^T\mathbf{i}$    (KCL)

where $\mathbf{M}^T$ is the transpose of the mesh matrix $\mathbf{M}$. Equation (5.3) means that every branch current can be expressed as a linear combination of mesh currents, and that the matrix which specifies these linear combinations is the transpose of the mesh matrix defined previously. We shall ask the reader to apply duality to the proof that $\mathbf{v} = \mathbf{A}^T\mathbf{e}$ in order to justify Eq. (5.3).

**Example 2**    Consider the network whose graph is shown in Fig. 5.1. It is obvious that we can relate the branch currents and mesh currents as follows:

$$j_1 = i_1$$
$$j_2 = i_1 - i_2$$
$$j_3 = i_3$$
$$j_4 = i_2 - i_3$$
$$j_5 = i_3$$

or

$$\mathbf{j} = \mathbf{M}^T\mathbf{i} = \begin{bmatrix} 1 & 0 & 0 \\ 1 & -1 & 0 \\ 0 & 1 & 0 \\ 0 & 1 & -1 \\ 0 & 0 & 1 \end{bmatrix} \begin{bmatrix} i_1 \\ i_2 \\ i_3 \end{bmatrix}$$

**Exercise 1**    Express Kirchhoff's laws in the form of Eq. (5.1) and (5.3) for the graph of Fig. 4.7a.

**Exercise 2**    Suppose that a network has a graph consisting of a large square divided into 25 equal squares (five on a side). Suppose that you have a large supply of zero-impedance ammeters. What is the minimum number of ammeters required to measure all mesh currents? Where would you put them?

**Exercise 3**    Use definition (5.2) of $m_{ik}$ to prove Eq. (5.3).

**Summary**    Equation (5.1)

$$\mathbf{Mv} = \mathbf{0} \quad \text{(KVL)}$$

and Eq. (5.3)

$$\mathbf{j} = \mathbf{M}^T\mathbf{i} \quad \text{(KCL)}$$

respectively, are the two basic equations of mesh analysis. Since the two equations are obtained from the network graph (planar, connected, and unhinged) and the two Kirchhoff laws, they are independent of the nature of the elements of the network. Equation (5.1) expresses KVL and consists of a set of $l$ linearly independent equations in terms of the $b$ branch voltages $v_1, v_2, \ldots, v_b$. Equation (5.3) expresses KCL and relates the $b$ branch currents $j_1, j_2, \ldots, j_b$ to the $l$ mesh currents $i_1, i_2, \ldots, i_l$. To solve for the $l$ network variables $i_1, i_2, \ldots, i_l$, we need to know the branch characterization of the network, i.e., the $b$ branch equations which relate the branch voltages to the branch currents. Only in these branch equations does the nature of network elements come into the analysis. In the next section we shall treat exclusively linear time-invariant networks. Nonlinear and time-varying networks will be considered later.

## 6    Mesh Analysis of Linear Time-invariant Networks

The mesh analysis of a linear time-invariant network requires a sequence of steps which is the dual of that required for the node analysis of the dual network $\widehat{\mathfrak{N}}$. This will allow us to treat this material briefly.

### 6.1    Sinusoidal Steady-state Analysis

Since the analysis of resistive networks is a special case of the sinusoidal steady-state analysis, we treat only the latter.

Let $\mathfrak{N}$ be a linear time-invariant network with $b$ branches and $n_t$ nodes. Let its graph $\mathcal{G}$ be connected, planar, and unhinged. Let the sources be sinusoidal and have all the same frequency $\omega$. Call $\mathbf{J}_s$ and $\mathbf{V}_s$ the $b$-vectors whose $k$th components are the phasors representing the sinusoidal sources in the $k$th branch. Similarly, $\mathbf{V}$ and $\mathbf{J}$ are the $b$-vectors whose $k$th components are the phasors representing the branch voltage $v_k$ and the branch current $j_k$. Call $\mathbf{I}$ the $l$-vector whose components are the phasors representing the mesh currents $i_1, i_2, \ldots, i_l$. Kirchhoff's laws give

$$(6.1) \quad \mathbf{MV} = \mathbf{0} \quad \text{(KVL)}$$

$$(6.2) \quad \mathbf{J} = \mathbf{M}^T\mathbf{I} \quad \text{(KCL)}$$

The branch equations are

$$(6.3) \quad \mathbf{V} = \mathbf{Z}_b(j\omega)\mathbf{J} - \mathbf{Z}_b(j\omega)\mathbf{J}_s + \mathbf{V}_s$$

The $b \times b$ matrix $\mathbf{Z}_b(j\omega)$ is called the **branch-impedance matrix.** The substitution gives

$$(\mathbf{MZ}_b(j\omega)\mathbf{M}^T)\mathbf{I} = \mathbf{MZ}_b(j\omega)\mathbf{J}_s - \mathbf{V}_s$$

or

(6.4)    $$\boxed{\mathbf{Z}_m(j\omega)\mathbf{I} = \mathbf{E}_s}$$

where $\mathbf{Z}_m(j\omega)$ is an $l \times l$ matrix called the **mesh impedance matrix,** given by

(6.5)    $\mathbf{Z}_m(j\omega) = \mathbf{MZ}_b(j\omega)\mathbf{M}^T$

and $\mathbf{E}_s$, the **mesh voltage source vector,** is the $l$ vector given by

(6.6)    $\mathbf{E}_s = \mathbf{MZ}_b(j\omega)\mathbf{J}_s - \mathbf{MV}_s$

Equations (6.4) are called the **mesh equations** of $\mathfrak{N}$; they constitute a system of $l$ linear algebraic equations (with complex coefficients) in $l$ unknowns, the phasors representing the mesh currents $I_1, I_2, \ldots, I_l$. The solution of (6.4) specifies all mesh currents. Then the branch currents are obtained by (6.2), and the branch voltages by (6.3).

---

**Example 1**    Consider the linear time-invariant network $\mathfrak{N}$ shown in Fig. 6.1. The phasor $V_{s1}$ represents the sinusoidal voltage of the source; $v_{s1}(t) = |V_{s1}| \cos(\omega t + \angle V_{s1})$. By inspection,

(6.7)    $$\mathbf{M} = \begin{bmatrix} -1 & 1 & 1 & 0 & 0 \\ 0 & 0 & -1 & 1 & 1 \end{bmatrix}$$

Let the inductance matrix of the branches 3, 4, and 5 be

(6.8)    $$\mathbf{L} = \begin{bmatrix} 3 & 1 & -1 \\ 1 & 4 & 2 \\ -1 & 2 & 5 \end{bmatrix}$$

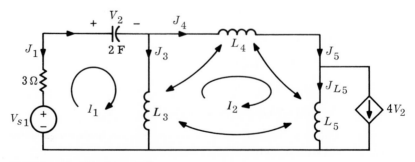

**Fig. 6.1**    Network analyzed in Example 1.

hence

$$
(6.9) \quad
\begin{bmatrix} V_3 \\ V_4 \\ V_5 \end{bmatrix}
=
\begin{bmatrix}
3j\omega & j\omega & -j\omega \\
j\omega & 4j\omega & 2j\omega \\
-j\omega & 2j\omega & 5j\omega
\end{bmatrix}
\begin{bmatrix} J_3 \\ J_4 \\ J_{L5} \end{bmatrix}
$$

If we note that the current $J_5$ is related to $J_{L5}$, the current into the inductor, by

$$
(6.10) \quad J_{L5} = J_5 - 4V_2 = J_5 - \frac{2}{j\omega} J_2
$$

we obtain the following branch equations:

$$
\begin{bmatrix} V_1 \\ V_2 \\ V_3 \\ V_4 \\ V_5 \end{bmatrix}
=
\begin{bmatrix}
3 & 0 & 0 & 0 & 0 \\
0 & \dfrac{1}{2j\omega} & 0 & 0 & 0 \\
0 & 2 & 3j\omega & j\omega & -j\omega \\
0 & -4 & j\omega & 4j\omega & 2j\omega \\
0 & -10 & -j\omega & 2j\omega & 5j\omega
\end{bmatrix}
\begin{bmatrix} J_1 \\ J_2 \\ J_3 \\ J_4 \\ J_5 \end{bmatrix}
+
\begin{bmatrix} V_{s1} \\ 0 \\ 0 \\ 0 \\ 0 \end{bmatrix}
$$

Using (6.5) and (6.6), we obtain the following mesh equations:

$$
\begin{bmatrix}
5 + 3j\omega + \dfrac{2}{j\omega} & -3j\omega \\
-16 - 3j\omega & 16j\omega
\end{bmatrix}
\begin{bmatrix} I_1 \\ I_2 \end{bmatrix}
=
\begin{bmatrix} V_{s1} \\ 0 \end{bmatrix}
$$

*Properties of the mesh impedance matrix*

1. If the network $\mathfrak{N}$ has no coupling elements, $\mathbf{Z}_b(j\omega)$ is *diagonal* and $\mathbf{Z}_m(j\omega)$ is *symmetric;* that is, $\mathbf{Z}_m(j\omega) = \mathbf{Z}_m{}^T(j\omega)$ [this follows from (6.5)].

2. Again, if $\mathfrak{N}$ has no coupling elements, the mesh impedance $\mathbf{Z}_m(j\omega)$ can be written by inspection as follows:

   a. Call $z_{ii}$ the diagonal element of $\mathbf{Z}_m$ in the $i$th row and $i$th column. $z_{ii}$ is the sum of all the impedances of branches in mesh $i$ and is called the **self-impedance of mesh** $i$.

   b. Call $z_{ik}$ the $(i,k)$th element of $\mathbf{Z}_m$. $z_{ik}$ is the negative of the sum of all the impedances of the branches which are in common with meshes $i$ and $k$† and is called the **mutual impedance between mesh $i$ and mesh** $k$.

3. If, using the Thévenin equivalent, we convert all sources into voltage sources, then $e_{sk}$ is the algebraic sum of all the source voltages in mesh $k$.

† The fact that $z_{ik}$ is the negative of the sum of the impedances common to meshes $i$ and $k$ is a consequence of the convention that *all* mesh currents are given clockwise reference directions.

4.   In the case of resistive networks, if all resistances are *positive*, then det $(\mathbf{Z}_m) > 0$.  Cramer's rule then guarantees that whatever the independent sources may be, the mesh equations (6.4) have a *unique* solution.

In case $\mathfrak{N}$ has coupling elements, the only general conclusions are that $\mathbf{Z}_b(j\omega)$ is no longer diagonal and $\mathbf{Z}_m(j\omega)$ is usually no longer symmetric.

**Exercise 1**   Prove statements 2*a* and 2*b*.  [Hint: Use (6.5) and (5.2)].

**Exercise 2**   By inspection, write the mesh equations of the network shown in Fig. 6.1, assuming that all mutual inductances and the dependent current sources are set to zero.

**Exercise 3**   Find a network with coupling elements which has a symmetric mesh impedance matrix.  (Hint: Use symmetry.)

## 6.2   Integrodifferential Equations

Let us illustrate the general procedure for writing the integrodifferential mesh equations.  We choose a simple case so that we can concentrate on the handling of initial conditions.  The method, however, is completely general.

Consider the linear time-invariant network shown in Fig. 6.2.  We are given the element values $R_3$, $C_4$, and $\alpha$ and the inductance matrix

$$\begin{bmatrix} L_1 & M \\ M & L_2 \end{bmatrix}$$

Note that $M$ is positive in view of the reference directions of $j_1$ and $j_2$, which enter the inductors through the dotted terminals.  In addition, we need the initial charge on the capacitor or, equivalently, $v_4(0)$, and the initial inductor currents $j_1(0)$ and $j_2(0)$.  Finally, we need the input waveform $v_{s3}(\cdot)$.  To obtain the mesh equations, we proceed as follows:

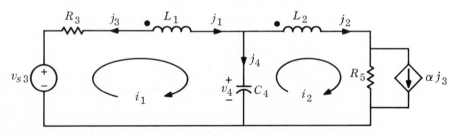

**Fig. 6.2**   Network used to illustrate the writing of integrodifferential equations.

*Step* 1   We write KVL; thus, $\mathbf{Mv} = \mathbf{0}$, and

(6.11)
$$
\begin{bmatrix} 1 & 0 & -1 & 1 & 0 \\ 0 & 1 & 0 & -1 & 1 \end{bmatrix}
\begin{bmatrix} v_1 \\ v_2 \\ v_3 \\ v_4 \\ v_5 \end{bmatrix} =
\begin{bmatrix} 0 \\ 0 \end{bmatrix}
$$

*Step* 2   We write KCL; thus, $\mathbf{j} = \mathbf{M}^T\mathbf{i}$, and

(6.12)
$$
\begin{bmatrix} j_1 \\ j_2 \\ j_3 \\ j_4 \\ j_5 \end{bmatrix} =
\begin{bmatrix} 1 & 0 \\ 0 & 1 \\ -1 & 0 \\ 1 & -1 \\ 0 & 1 \end{bmatrix}
\begin{bmatrix} i_1 \\ i_2 \end{bmatrix}
$$

*Step* 3   We write the branch equations in matrix form as follows:

(6.13)
$$
\begin{bmatrix} v_1 \\ v_2 \\ v_3 \\ v_4 \\ v_5 \end{bmatrix} =
\begin{bmatrix}
L_1D & MD & 0 & 0 & 0 \\
MD & L_2D & 0 & 0 & 0 \\
0 & 0 & R_3 & 0 & 0 \\
0 & 0 & 0 & \dfrac{1}{C_4D} & 0 \\
0 & 0 & -\alpha R_5 & 0 & R_5
\end{bmatrix}
\begin{bmatrix} j_1 \\ j_2 \\ j_3 \\ j_4 \\ j_5 \end{bmatrix} +
\begin{bmatrix} 0 \\ 0 \\ v_{s3} \\ v_4(0) \\ 0 \end{bmatrix}
$$

We may think of this equation as being in the form

$$\mathbf{v} = \mathbf{Z}_b(D)\mathbf{j} + \mathbf{v}_s$$

where, in this case, $\mathbf{Z}_b$ is a matrix whose elements involve the operators $D$ and $D^{-1}$.

*Step* 4   We use (6.12) to eliminate the $j_k$'s from (6.13); then we use (6.11) to eliminate the $v_k$'s from the resulting equation.   Rearranging terms, we obtain

(6.14)   $\mathbf{MZ}_b(D)\mathbf{M}^T\mathbf{i} = -\mathbf{Mv}_s$

or

$\mathbf{Z}_m(D)\mathbf{i} = \mathbf{e}_s$

where $\mathbf{Z}_m(D) \triangleq \mathbf{MZ}_b(D)\mathbf{M}^T$ is recognized as the *mesh impedance matrix operator*.   In this example, the matrix equation reads

$$(6.15) \quad \begin{bmatrix} L_1D + R_3 + \dfrac{1}{C_4D} & MD - \dfrac{1}{C_4D} \\[4mm] MD + \alpha R_5 - \dfrac{1}{C_4D} & L_2D + R_5 + \dfrac{1}{C_4D} \end{bmatrix} \begin{bmatrix} i_1 \\[4mm] i_2 \end{bmatrix} = \begin{bmatrix} v_{s3} - v_4(0) \\[4mm] v_4(0) \end{bmatrix}$$

or, in scalar form,

$$L_1 \frac{di_1}{dt} + R_3 i_1 + \frac{1}{C_4} \int_0^t i_1(t')\, dt' + M \frac{di_2}{dt} - \frac{1}{C_4} \int_0^t i_2(t')\, dt'$$

$$(6.16) \qquad\qquad\qquad\qquad\qquad\qquad\qquad\qquad = v_{s3}(t) - v_4(0)$$

$$M \frac{di_1}{dt} + \alpha R_5 i_1 - \frac{1}{C_4} \int_0^t i_1(t')\, dt' + L_2 \frac{di_2}{dt} + R_5 i_2(t) + \frac{1}{C_4} \int_0^t i_2(t')\, dt'$$

$$\qquad\qquad\qquad\qquad\qquad\qquad\qquad\qquad = v_4(0)$$

This system of Eqs. (6.16) is a system of two integrodifferential equations in two unknowns, $j_1(\cdot)$ and $j_2(\cdot)$. The required initial conditions are $j_1(0)$, $j_2(0)$, and $v_4(0)$.

**Remark**   Any system of integrodifferential equations such as (6.16) can always be put in the form of a system of differential equations by the introduction of appropriate variables. Define the waveforms $q_1(\cdot)$ and $q_2(\cdot)$ by

$$(6.17) \quad q_1(t) = \int_0^t i_1(t')\, dt' \qquad q_2(t) = \int_0^t i_2(t')\, dt'$$

Physically, $q_1(t)$ is the net amount of charge (in coulombs) that has passed through $R_3$ or $L_1$ in the interval $[0,t]$. In terms of $q_1(\cdot)$ and $q_2(\cdot)$, the integrodifferential system of (6.16) becomes a system of *differential* equations,

$$L_1 \ddot{q}_1 + R_3 \dot{q}_1 + \frac{1}{C_4} q_1 + M \ddot{q}_2 - \frac{1}{C_4} q_2 = v_{s3}(t) - v_4(0)$$

$$M \ddot{q}_1 + R_5 \alpha \dot{q}_1 - \frac{1}{C_4} q_1 + L_2 \ddot{q}_2 + R_5 \dot{q}_2 + \frac{1}{C_4} q_2 = v_4(0)$$

with the initial conditions

$$q_1(0) = 0 \qquad q_2(0) = 0 \qquad \text{see (6.17)}$$
$$\dot{q}_1 = j_1(0) \qquad \dot{q}_2(0) = j_2(0)$$

## Summary

- In the node analysis, the $n$ node-to-datum voltages $e_1, e_2, \ldots, e_n$ are used as network variables. By applying KCL to all the nodes except the datum node, we obtain $n$ linearly independent equations in terms of branch currents. For linear time-invariant networks, taking the branch equations into account, the $n$ equations can be written explicitly in terms of the $n$ node-to-datum voltages. In general, the resulting network equation gives in matrix form

(a)　$\mathbf{Y}_n(D)\mathbf{e} = \mathbf{i}_s$

Once $\mathbf{e}$ is determined, the $b$ branch voltages can be obtained immediately from

(b)　$\mathbf{v} = \mathbf{A}^T\mathbf{e}$

The $b$ branch currents are then obtainable from the branch equations. The writing of the node matrix equation in (a) can be done formally using the step-by-step procedure and matrix multiplications. However, for simple circuits which do not contain complicated coupling elements, the node admittance matrix $\mathbf{Y}_n(D)$ as well as the source vector $\mathbf{i}_s$ can be written by inspection. Frequently it is advisable to convert all voltage sources into current sources before starting node analysis.

■ In mesh analysis the $l$ mesh currents $i_1, i_2, \ldots, i_l$ are used as network variables, and $l$ linearly independent equations in terms of branch voltages are obtained by applying KVL to all the meshes except the outer mesh. For linear time-invariant networks, taking the branch equations into account, the $l$ equations can be written explicitly in terms of the $l$ mesh currents. In general, the resulting mesh equation gives in matrix form

(c)　$\mathbf{Z}_m(D)\mathbf{i} = \mathbf{e}_s$

Once $\mathbf{i}$ is determined, the $b$ branch currents can be obtained from

(d)　$\mathbf{j} = \mathbf{M}^T\mathbf{i}$

The $b$ branch voltages are then obtainable from the branch equations. Again, a step-by-step procedure can be used to write the mesh matrix equation of (c); or in case the network does not include complicated coupling elements, $\mathbf{Z}_m(D)$ and $\mathbf{e}_s$ can be obtained by inspection. Frequently, it is advisable to convert all current sources into voltage sources before starting mesh analysis.

■ Whereas node analysis is completely general, mesh analysis is restricted to planar networks.

■ The two methods are dual to each other in the case of planar networks. The natural question to ask is which is a better method. The answer depends on the given network. In node analysis there are $n$ network variables to be determined, whereas in mesh analysis there are $l$ network variables. Thus, it is reasonable to say that if for a given graph the number of nodes $n + 1$ is much smaller than the number of meshes $l$, the node analysis is preferable. If it is the other way around, then the mesh analysis is more advantageous. However, other factors must also be considered. One crucial point deals with the number and kind of sources in the network. If all given sources (dependent and independent) are current sources, node analysis probably is more convenient since one can often write the node equations by inspection. On the other hand, if all given sources are voltage sources, mesh analysis may be easier to use. Experience will help you to decide one way or the other.

■ Two connected, unhinged, planar topological graphs are said to be dual to each other if (1) there is a one-to-one correspondence between the meshes of one (including the outer mesh) and the nodes of the other, and vice-versa, and (2) there is a one-to-one correspondence between the branches of each graph such that whenever two meshes of one graph have a branch in common, the corresponding nodes of the other graph have the corresponding branch connecting these nodes.

■ Two networks are said to be dual to each other if (1) the topological graph of one is the dual of the topological graph of the other, and (2) the branch equations of one are obtainable from the corresponding branch equations of the other by performing the following substitutions:

$$v \to \widehat{j} \qquad q \to \widehat{\phi}$$
$$j \to \widehat{v} \qquad \phi \to \widehat{q}$$

■ The main facts of node and mesh analysis are tabulated in Table 10.2 in a form that makes the duality of the two methods readily apparent.

**Table 10.2   Summary of Node and Mesh Analyses**

|  | Node analysis | Mesh analysis |
|---|---|---|
| Network variables | **e,** node-to-datum voltages | **i,** mesh currents |
| Basic facts | $\mathbf{Aj} = \mathbf{0}$     (KCL)<br>$\mathbf{v} = \mathbf{A}^T\mathbf{e}$   (KVL) | $\mathbf{Mv} = \mathbf{0}$     (KVL)<br>$\mathbf{j} = \mathbf{M}^T\mathbf{i}$   (KCL) |
| Linear time-invariant resistive networks | Branch equations | |
|  | $\mathbf{j} = \mathbf{Gv} + \mathbf{j}_s - \mathbf{Gv}_s$ | $\mathbf{v} = \mathbf{Rj} + \mathbf{v}_s - \mathbf{Rj}_s$ |
|  | Network equations | |
|  | $\mathbf{Y}_n\mathbf{e} = \mathbf{i}_s$ | $\mathbf{Z}_m\mathbf{i} = \mathbf{e}_s$ |
|  | $\mathbf{Y}_n \overset{\Delta}{=} \mathbf{AGA}^T$<br>$\mathbf{i}_s \overset{\Delta}{=} \mathbf{AGv}_s - \mathbf{Aj}_s$ | $\mathbf{Z}_m \overset{\Delta}{=} \mathbf{MRM}^T$<br>$\mathbf{e}_s \overset{\Delta}{=} \mathbf{MRj}_s - \mathbf{Mv}_s$ |

## Problems

Rank and inverse

**1.** *a.* Give the rank of the following matrices:

$$A_1 = \begin{bmatrix} 2 & 0 & 2 & -2 & 1 & 1 \\ 1 & 1 & 0 & 1 & 0 & 1 \\ 0 & 1 & -1 & 2 & 1 & 2 \end{bmatrix} \qquad A_2 = \begin{bmatrix} 1 & 0 & 1 & 1 \\ 0 & -2 & -2 & 2 \\ -1 & 1 & 0 & -2 \end{bmatrix}$$

$$A_3 = \begin{bmatrix} 2 & 3 & 1 & 0 \\ 0 & 4 & 7 & 9 \\ 4 & 6 & 2 & 0 \\ 1 & 3 & 2 & 7 \end{bmatrix}$$

*b.* Find the inverse of the following matrices:

$$B_1 = \begin{bmatrix} -1 & 3 \\ 2 & -4 \end{bmatrix} \qquad B_2 = \begin{bmatrix} \frac{1}{2} & \frac{4}{3} \\ -\frac{1}{2} & -\frac{2}{3} \end{bmatrix} \qquad B_3 = \begin{bmatrix} 2 & -2 & 1 \\ 0 & 1 & 0 \\ -1 & 2 & 1 \end{bmatrix}$$

Incidence matrices

**2.** Are the following matrices possible reduced incidence matrices? In each case justify your answer.

$$\begin{bmatrix} -1 & 1 & 0 & 0 & 0 & 0 & 0 & 0 & 0 & 0 \\ 1 & 0 & 0 & 0 & 0 & -1 & 0 & -1 & 0 & -1 \\ 0 & -1 & 1 & 0 & 0 & 0 & 0 & 0 & 1 & 1 \\ 0 & 0 & 0 & 0 & 0 & 0 & 1 & 1 & -1 & 0 \\ 0 & 0 & 0 & 0 & -1 & 1 & -1 & 0 & 0 & 0 \\ 0 & 0 & 0 & -1 & 1 & 0 & 0 & 0 & 0 & 0 \end{bmatrix}$$

$$\begin{bmatrix} 1 & 0 & 1 & 0 \\ -1 & 1 & 0 & -1 \\ 0 & 0 & 2 & 0 \\ 0 & 0 & 0 & 1 \end{bmatrix} \quad \begin{bmatrix} 1 & 1 & 0 & 1 & 0 & 0 & 1 \\ -1 & 0 & 1 & 0 & 0 & -1 & 0 \\ 0 & 0 & 0 & 0 & 1 & 0 & -1 \\ 0 & -1 & -1 & 0 & 0 & 1 & 0 \end{bmatrix} \quad [1 \quad 0]$$

Node analysis

**3.** Consider the linear time-invariant network shown in Fig. P10.3.

*a.* Using the reference directions shown, obtain the oriented graph of the network.

*b.* Write the expression for KCL and KVL in matrix form ($\mathbf{Aj} = \mathbf{0}$, and $\mathbf{v} = \mathbf{A}^T\mathbf{e}$).

*c.* Assuming the sinusoidal steady state at frequency $\omega$ and the source phasor $J_s$ for the current source, write the node equations in matrix form.

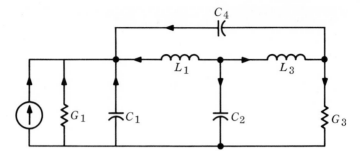

**Fig. P10.3**

d. Assuming zero initial conditions, obtain the integrodifferential equations of node analysis for the network shown.

Node analysis **4.** Consider the linear time-invariant network shown in Fig. P10.3.

a. Obtain by inspection the node equations for the sinusoidal steady state.

b. Obtain by inspection the node integrodifferential equations for the case in which all initial conditions are zero.

Node analysis **5.** Consider the linear time-invariant network shown in Fig. P10.5.

a. Obtain the node integrodifferential equations by the systematic method.

b. Assuming the sinusoidal steady state at frequency $\omega$, obtain the node equations.

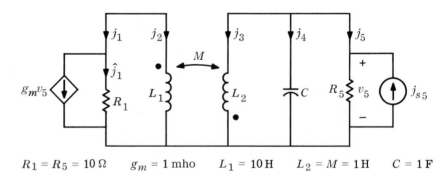

$R_1 = R_5 = 10\,\Omega$  $g_m = 1\,\text{mho}$  $L_1 = 10\,\text{H}$  $L_2 = M = 1\,\text{H}$  $C = 1\,\text{F}$

**Fig. P10.5**

Node analysis **6.** Repeat Prob. 5 with $M = 0$, using the shortcut method.

Node analysis **7.** Let the linear time-invariant network shown in Fig. P10.7 be in the sinusoidal steady state. An oriented graph of this network is shown in Fig. P10.7. Use node analysis to find all branch voltages and currents, given $e_s(t) = \cos 10^6 t$.

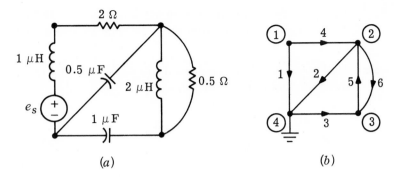

(a)                                (b)

**Fig. P10.7**

Node analysis    **8.** Consider the linear time-invariant network shown in Fig. P10.8.

    *a.* Assume it is in the sinusoidal steady state and write its node equations by inspection.

    *b.* Assume all initial conditions to be zero, and write the integrodifferential node equation.

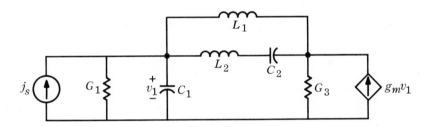

**Fig. P10.8**

Duality    **9.** Find the dual of the general resistive branch shown in Fig. P10.9. By transforming the voltage source in the dual branch, show that the dual

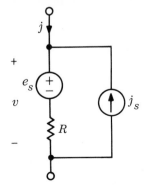

**Fig. P10.9**

takes the same form as the original with $\hat{e}_s = -j_s$ and $\hat{j}_s = -e_s$. (Hint: If a voltage source appears in series with a current source, the voltage source may be shorted out.)

Dual graphs   **10.** For each of the topological graphs shown in Fig. P10.10, give two different dual topological graphs.

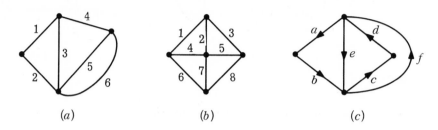

(a)          (b)          (c)

**Fig. P10.10**

Dual networks   **11.** Given the nonlinear network $\mathfrak{N}$ shown in Fig. P10.11, find a dual network $\hat{\mathfrak{N}}$. (Be sure to specify $\hat{\mathfrak{N}}$ completely.) The branch equations for $\mathfrak{N}$ are

$$v_R = 2j_R + 3j_R^2 \qquad j_L = 10^{-2}\phi_L + \tanh \phi_L$$

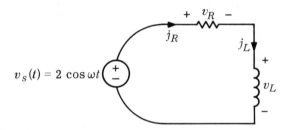

**Fig. P10.11**

Mesh analysis   **12.** The circuit shown in Fig. P10.12 is made of linear time-invariant elements. Calculate the voltage $v_1$.

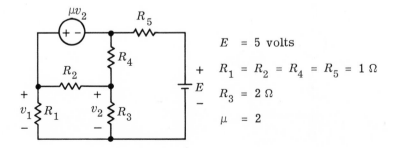

$E = 5$ volts

$R_1 = R_2 = R_4 = R_5 = 1\ \Omega$

$R_3 = 2\ \Omega$

$\mu = 2$

**Fig. P10.12**

Mesh analysis   **13.** The network shown in Fig. P10.13 is in the sinusoidal steady state with angular frequency $\omega = 2$ rad/sec.

*a.* Find the phasors $I_1$ and $I_2$.

*b.* What is the driving-point impedance seen by the voltage source?

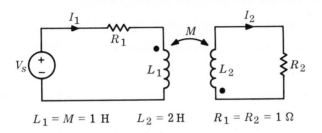

$$L_1 = M = 1 \text{ H} \qquad L_2 = 2 \text{ H} \qquad R_1 = R_2 = 1 \ \Omega$$

**Fig. P10.13**

Mesh analysis   **14.** The linear time-invariant circuit shown in Fig. P10.14 represents the Maxwell bridge, an instrument used to measure the inductance $L_x$ and resistance $R_x$ of a linear time-invariant physical inductor. The measurement is made with the circuit in sinusoidal steady state by adjusting $R_1$ and $C$ until $v_A = v_B$ (then the bridge is said to be "balanced"). It is important to note that when balance is achieved, the current through the detector $D$ is zero regardless of the value of the detector impedance.

Show that when the bridge is balanced,

$$L_x = R_2 R_3 C \qquad R_x = \frac{R_2 R_3}{R_1}$$

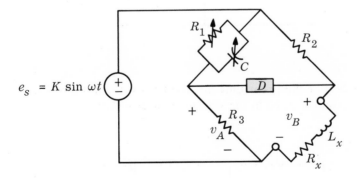

**Fig. P10.14**

Mesh analysis   **15.** Consider the linear time-invariant network shown in Fig. P10.3.

*a.* Obtain the mesh matrix for this network.

*b.* Write KVL and KCL in matrix form.

c.  Assuming the sinusoidal steady state at frequency $\omega$, use $(a)$ and $(b)$ to obtain the mesh equations.

d.  Assume the following initial conditions:

$$j_{L_1}(0) = 1 \quad j_{L_2}(0) = 0 \qquad v_{C_1}(0) = 0 \qquad v_{C_2}(0) = 1$$

Obtain the integrodifferential mesh equations.

Mesh analysis   **16.** Repeat $(c)$ and $(d)$ of Prob. 15 by inspection.

Mesh analysis   **17.** Consider the linear time-invariant network of Fig. P10.5.

a.  Obtain the mesh matrix for this network.

b.  Assuming the network is in sinusoidal steady state, write the mesh equations.

c.  Obtain, by the systematic method, the integrodifferential mesh equations for zero initial conditions.

Mesh analysis   **18.** Repeat the previous problem using the shortcut method.

Mesh analysis   **19.** Write the mesh equations of the circuit shown in Fig. P10.19 using the charge $q_1$ and $q_2$ as variables.  Indicate the necessary initial conditions.  At $t = 0$ the current in the inductor is $I_0$, and the voltage across the capacitor is $V_0$.

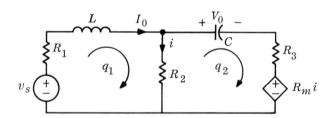

**Fig. P10.19**

Node and   **20.** For the linear time-invariant network shown in Fig. P10.20, which is
mesh analysis   assumed to be in the sinusoidal steady state,

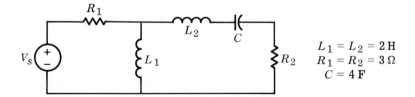

$$L_1 = L_2 = 2\,\text{H}$$
$$R_1 = R_2 = 3\,\Omega$$
$$C = 4\,\text{F}$$

**Fig. P10.20**

*a.* Draw the network graph; call it $\mathcal{G}$.

*b.* Draw the dual graph; call it $\hat{\mathcal{G}}$.

*c.* Obtain the dual network; call it $\hat{\mathcal{N}}$.

*d.* Write the mesh equations for the given network and the node equations of $\hat{\mathcal{N}}$.

*e.* Compare the equations in (*d*).

Dual networks **21.** Consider the linear time-invariant network $\mathcal{N}$ shown in Fig. P10.21.

*a.* Find a dual network of $\mathcal{N}$; call if $\hat{\mathcal{N}}$.

*b.* Assuming that the sinusoidal steady state has been reached, write by inspection the mesh equations of $\hat{\mathcal{N}}$ in terms of voltage and current phasors; that is, $\mathbf{Z}_m\mathbf{I} = \mathbf{E}_s$, where $\mathbf{Z}_m$ is the mesh impedance matrix, $\mathbf{I}$ is the phasor which represents the mesh current vector, and $\mathbf{E}_s$ is the phasor which represents the mesh voltage source vector.

*c.* Assuming that the sinusoidal steady state has been reached, write by inspection the node equations of $\hat{\mathcal{N}}$ in terms of voltage and current phasors; that is, $\mathbf{Y}_n\mathbf{E} = \mathbf{I}_s$, where $\mathbf{Y}_n$ is the node admittance matrix, $\mathbf{E}$ is the phasor which represents the node-to-datum voltage vector, and $\mathbf{I}_s$ is the phasor which represents the node current source vector.

*d.* Solve for the mesh currents of $\mathcal{N}$ and the node-to-datum voltages of $\hat{\mathcal{N}}$ Give your results in the form of real functions of time.

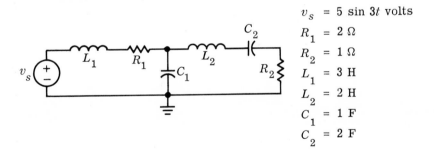

$$v_S = 5 \sin 3t \text{ volts}$$
$$R_1 = 2 \ \Omega$$
$$R_2 = 1 \ \Omega$$
$$L_1 = 3 \text{ H}$$
$$L_2 = 2 \text{ H}$$
$$C_1 = 1 \text{ F}$$
$$C_2 = 2 \text{ F}$$

**Fig. P10.21**

# 11

# Loop and Cut-set Analysis

In the previous chapter we have learned to perform systematically the node analysis of *any* linear time-invariant network. We have also learned to perform the mesh analysis for any such network provided its graph is planar. In this chapter we briefly discuss two generalizations, or perhaps variations, of these methods, namely, the cut-set analysis and the loop analysis. There are two reasons for studying loop and cut-set analysis: first, these methods are useful because they are much more flexible than mesh and node analysis, and, second, they use concepts and teach us points of view that are indispensable for writing state equations.

In Sec. 1, we introduce some new graph-theoretic concepts and prove a fundamental theorem. In Sec. 2, we study loop analysis, and in Sec. 3 we study cut-set analysis. Section 4 is devoted to comments on these methods. In Sec. 5 we establish a basic relation between the loop matrix **B** and the cut-set matrix **Q**.

## 1 Fundamental Theorem of Graph Theory

In order to develop this theorem we need to indicate precisely what we mean by a *tree*. Let $\mathcal{G}$ be a connected graph and $T$ a subgraph of $\mathcal{G}$. We say that $T$ is a **tree of the connected graph** $\mathcal{G}$ if (1) $T$ is a *connected* subgraph, (2) it contains *all the nodes* of $\mathcal{G}$, and (3) it contains *no loops*.

Given a connected graph $\mathcal{G}$ and a tree $T$, the branches of $T$ are called **tree branches,** and the branches of $\mathcal{G}$ not in $T$ are called **links.** (Some authors call them cotree branches, or chords.)

A graph has usually many trees. In Fig. 1.1 we show a few trees of a connected graph $\mathcal{G}$. To help you understand the definition, in Fig. 1.2 we show a few subgraphs (of the same graph $\mathcal{G}$) which are *not* trees of $\mathcal{G}$. To emphasize the fact that complicated graphs have many trees, remember that if a graph has $n_t$ nodes and has a single branch connecting every pair of nodes, then it has $n_t^{n_t-2}$ trees. For such graphs, when $n_t = 5$, there are 125 trees; when $n_t = 10$, there are $10^8$ trees.

**Exercise** Draw all possible trees for the graph shown in Fig. 1.3.

The following fundamental theorem relates the properties of loops, cut sets and trees.

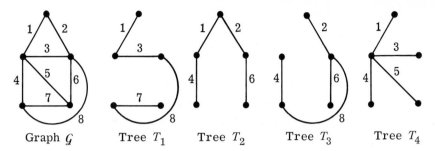

| Graph $\mathcal{G}$ | Tree $T_1$ | Tree $T_2$ | Tree $T_3$ | Tree $T_4$ |

**Fig. 1.1**   Examples of trees of graph $\mathcal{G}$.

**THEOREM**   Given a connected graph $\mathcal{G}$ of $n_t$ nodes and $b$ branches, and a tree $T$ of $\mathcal{G}$,

1. There is a unique path along the tree between any pair of nodes.

2. There are $n_t - 1$ tree branches and $b - n_t + 1$ links.

3. Every link of $T$ and the unique tree path between its nodes constitute a *unique loop* (this is called the **fundamental loop** associated with the link).

4. Every tree branch of $T$ together with some links defines a *unique* cut set of $\mathcal{G}$. This cut set is called a **fundamental cut set** associated with the tree branch.

*Proof*   1.   Suppose there were two paths along the tree between node ① and node ②. Since some branches of these two paths would constitute a loop, the tree would contain a loop. This contradicts requirement 3 of the definition of a tree.

2.   Let $T$ be a tree of $\mathcal{G}$; then $T$ is a subgraph of $\mathcal{G}$ which connects all nodes, and it therefore has $n_t$ nodes. If a node of $T$ has only one tree branch incident with it, this node is called a *terminal node* of $T$. Since $T$ is a connected subgraph which contains no loops, it has at least two terminal nodes. Let us remove from the tree one of the terminal nodes

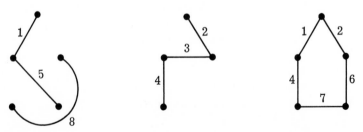

Violates property (1)   Violates property (2)   Violates property (3)

**Fig. 1.2**   Examples of subgraphs of $\mathcal{G}$ which are *not* trees.

**Fig. 1.3**    A connected graph with four nodes and six branches.

and its incident tree branch.  The remaining subgraph must still have at least two terminal nodes.  Let us continue removing terminal nodes and their incident tree branches until only one tree branch is left.  This last branch is incident with two nodes.  Thus, we have removed one tree branch for every node except for the last branch which was connected with two nodes.  Since there were $n_t$ nodes, $T$ must have had $n_t - 1$ branches.  Since all branches of $\mathcal{G}$ which are not in $T$ are called links, there are $b - (n_t - 1) = b - n_t + 1$ links.

3.   Consider a link $l_1$ which connects nodes ① and ②.  By part 1, there is a unique tree path between ① and ②.  This tree path, together with the link $l_1$, constitutes a loop.  There cannot be any other loop since the tree had no loop to start with.

4.   Consider the branch $b_1$ of $T$ as shown in Fig. 1.4.  Remove $b_1$ from $T$. What remains of $T$ is then made up of two separate (connected) parts, say $T_1$ and $T_2$.  Since every link connects a node of $T$ to another node of $T$, let us consider the set $L$ of all the links that connect a node of $T_1$ to a node of $T_2$.  It is easily verified that the links in $L$, together with the tree branch $b_1$, constitute a cut set.  All links not in $L$ cannot contribute to another cut set since each one of them with a tree path in either $T_1$ or $T_2$ constitutes a loop.

The theorem can readily be extended to the case in which the graph consists of several separate parts, as shown in the following statement.

**COROLLARY**    Suppose that $\mathcal{G}$ has $n_t$ nodes, $b$ branches, and $s$ separate parts.  Let $T_1$, $T_2, \ldots, T_s$ be trees of each separate part, respectively.  The set $\{T_1, T_2, \ldots, T_s\}$ is called **a forest of** $\mathcal{G}$.  Then the forest has $n_t - s$ branches,

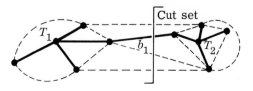

**Fig. 1.4**    Illustration of properties of a fundamental cut set.

$\mathcal{G}$ has $b - n_t + s$ links, and the remaining statements of the theorem are true.

**Exercise**    Consider the graph $\mathcal{G}$ of Fig. 1.1. List all the fundamental loops and all the fundamental cut sets corresponding to tree $T_1$. Repeat for trees $T_2$, $T_3$, and $T_4$.

## 2    Loop Analysis

### 2.1    Two Basic Facts

Consider a connected graph with $b$ branches and $n_t$ nodes. Pick an arbitrary tree $T$. There are $n = n_t - 1$ tree branches and $l = b - n$ links. Number the branches as follows: links first from 1 to $l$, tree branches next from $l + 1$ to $b$. Every link defines a fundamental loop, i.e., the loop formed by the link and the unique tree path between the nodes of that link. This is illustrated in Fig. 2.1 in terms of a simple graph with $b = 8$, $n_t = 5$, $n = 4$, and $l = 4$.

In order to apply KVL to each fundamental loop we adopt *a reference direction for the loop* which agrees with the reference direction of the link which defines that fundamental loop. This is shown in Fig. 2.1; for example, fundamental loop 1 has the same orientation as link 1, etc. The KVL equations can be written for the four fundamental loops in terms of the branch voltage as follows:

Loop 1:    $v_1 - v_5 + v_6 = 0$

Loop 2:    $v_2 + v_5 - v_6 + v_7 + v_8 = 0$

Loop 3:    $v_3 - v_6 + v_7 + v_8 = 0$

Loop 4:    $v_4 - v_6 + v_7 = 0$

In matrix form, the equation gives

$$
l \text{ loops} \left\{ \underbrace{\begin{bmatrix} 1 & 0 & 0 & 0 \\ 0 & 1 & 0 & 0 \\ 0 & 0 & 1 & 0 \\ 0 & 0 & 0 & 1 \end{bmatrix}}_{l \text{ links}} \underbrace{\begin{bmatrix} -1 & 1 & 0 & 0 \\ 1 & -1 & 1 & 1 \\ 0 & -1 & 1 & 1 \\ 0 & -1 & 1 & 0 \end{bmatrix}}_{n \text{ tree branches}} \right. \begin{bmatrix} v_1 \\ v_2 \\ v_3 \\ v_4 \\ v_5 \\ v_6 \\ v_7 \\ v_8 \end{bmatrix} = \begin{bmatrix} 0 \\ 0 \\ 0 \\ 0 \end{bmatrix}
$$

More generally, if we apply the KVL to each one of the $l$ fundamental loops, we obtain a system of $l$ linear algebraic equations in $b$ unknowns $v_1, v_2, \ldots, v_b$. The first basic fact of loop analysis is as follows:

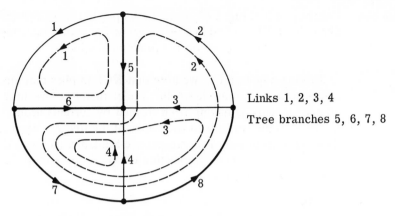

Links 1, 2, 3, 4

Tree branches 5, 6, 7, 8

**Fig. 2.1**   Fundamental loops for the chosen tree of a graph.

*The l linear homogeneous algebraic equations in $v_1, \ldots, v_b$ obtained by applying KVL to each fundamental loop constitute a set of l linearly independent equations.*

If we recall the convention that the reference direction of the loop agrees with that of the link which defines it, we see that the system of equations obtained from KVL is of the form

(2.1)   $\mathbf{Bv} = \mathbf{0}$

where $\mathbf{B}$ is an $l \times b$ matrix called the **fundamental loop matrix.** Furthermore, its $(i,k)$th element is defined as follows:

(2.2)   $b_{ik} = \begin{cases} 1 & \text{if branch } k \text{ is in loop } i \text{ and their reference directions } \textit{agree} \\ -1 & \text{if branch } k \text{ is in loop } i \text{ and their reference directions } \textit{do not} \\ & \textit{agree} \\ 0 & \text{if branch } k \textit{ is not in } \text{loop } i \end{cases}$

Since each fundamental loop includes one link only and since the orientations of the loop and the link are picked to be the same, it is clear that if we number the links $1, 2, \ldots, l$ and the tree branches $l + 1, l + 2, \ldots, b$, the matrix $\mathbf{B}$ has the form

(2.3)   $\mathbf{B} = \left[ \begin{array}{c|c} \mathbf{1}_l & \mathbf{F} \end{array} \right] \Big\}\ l \text{ loops}$

$\underbrace{\phantom{xxxx}}_{l \text{ links}}\ \underbrace{\phantom{xxxx}}_{\substack{n \text{ tree} \\ \text{branches}}}$

where $\mathbf{1}_l$ designates a unit matrix of order $l$ and $\mathbf{F}$ designates a rectangular matrix of $l$ rows and $n$ columns. It is obvious that the rank of $\mathbf{B}$ is $l$, since $\mathbf{B}$ includes the unit matrix $\mathbf{1}_l$ and has only $l$ rows. Therefore, we have established the fact that the $l$ fundamental loop equations written in terms of the branch voltages constitute a set of $l$ linearly independent equations.

**Exercise**    For the graph of Fig. 2.1, consider a loop $\ell$ that is not a fundamental loop. Show that KVL applied to loop $\ell$ gives an equation which depends linearly on the $l$ equations based on the fundamental loops.

Turning now to KCL, we note that KCL implies that any current that comes to a node must leave this node; therefore, we may think of the branch currents as having been formed by currents around loops. Call $i_1, i_2, \ldots, i_l$ the currents in the $l$ links of the tree $T$. We imagine each of these currents flowing in its respective fundamental loop; thus, each tree branch current is the superposition of one or more loop currents. More precisely, we assert that

(2.4)    $\mathbf{j} = \mathbf{B}^T\mathbf{i}$

where $\mathbf{B}^T$ is the transpose of the fundamental loop matrix. To prove Eq. (2.4), let us write it in the form $\mathbf{j} = \mathbf{Ci}$, where $\mathbf{C}$ is the appropriate matrix of $b$ rows and $l$ columns which makes the equation true. We wish to show that $\mathbf{C} = \mathbf{B}^T$. Let us consider the branch currents. For those branches which are links of the given tree, the link currents are identical to the fundamental loop currents; that is,

(2.5)    $j_k = i_k \qquad k = 1, 2, \ldots, l$

The remaining branches belong to the tree; hence they are tree branches. Each tree-branch current is a linear combination of the fundamental loop currents. More specifically, the $k$th branch current can be written as

(2.6)    $j_k = \displaystyle\sum_{i=1}^{l} c_{ki} i_i \qquad k = l + 1, l + 2, \ldots, b$

where $c_{ki}$ is given by the following equation:

(2.7)    $c_{ki} = \begin{cases} 1 & \text{if branch } k \text{ is in loop } i \text{ and their reference directions } \textit{agree} \\ -1 & \text{if branch } k \text{ is in loop } i \text{ and their reference directions } \textit{do not agree} \\ 0 & \text{if branch } k \textit{ is not in} \text{ loop } i \end{cases}$

It is obvious that Eq. (2.7) considers all branches, since for a link, branch $k$ is only in loop $k$, and their reference directions coincide; hence, as in Eq. (2.5), all $c_{kk} = 1$. Comparing Eq. (2.7) with Eq. (2.2), we conclude that $c_{ki} = b_{ik}$; hence, the matrix $\mathbf{C} = (c_{ki})$ as specified by $\mathbf{j} = \mathbf{Ci}$ is the transpose of $\mathbf{B}$; that is, $\mathbf{C} = \mathbf{B}^T$. If we partition the matrix $\mathbf{B}^T$ in Eq. (2.4) according to whether a branch is a link or a tree branch, we obtain

(2.8)    $\mathbf{j} = \mathbf{B}^T\mathbf{i} = \left[\dfrac{\mathbf{1}_l}{\mathbf{F}^T}\right]\mathbf{i}$

This equation will be useful for later applications.

Let us consider our example of Fig. 2.1. We can write the following equations according to Eq. (2.7):

$j_1 = i_1$

$j_2 = i_2$

$j_3 = i_3$

$j_4 = i_4$

$j_5 = -i_1 + i_2$

$j_6 = i_1 - i_2 - i_3 - i_4$

$j_7 = i_2 + i_3 + i_4$

$j_8 = i_2 + i_3$

In matrix form the equation is

$$
\begin{bmatrix} j_1 \\ j_2 \\ j_3 \\ j_4 \\ j_5 \\ j_6 \\ j_7 \\ j_8 \end{bmatrix} = \begin{bmatrix} 1 & 0 & 0 & 0 \\ 0 & 1 & 0 & 0 \\ 0 & 0 & 1 & 0 \\ 0 & 0 & 0 & 1 \\ -1 & 1 & 0 & 0 \\ 1 & -1 & -1 & -1 \\ 0 & 1 & 1 & 1 \\ 0 & 1 & 1 & 0 \end{bmatrix} \begin{bmatrix} i_1 \\ i_2 \\ i_3 \\ i_4 \end{bmatrix}
$$

**Summary**   KVL is expressed by $\mathbf{Bv} = \mathbf{0}$, and KCL by $\mathbf{j} = \mathbf{B}^T\mathbf{i}$ where $\mathbf{i}$ is the loop current vector. As a result of our choice of reference directions, the fundamental loop matrix $\mathbf{B}$ is of the form (2.3). These equations are valid irrespective of the nature of the branches.

**Exercise 1**   Prove Tellegen's theorem by using Eqs. (2.1) and (2.4).

**Exercise 2**   Consider the graph $\mathcal{G}$ of Fig. 1.1. Assign reference directions to each branch. Determine $\mathbf{B}$ for the tree $T_1$.

**Exercise 3**   Mesh analysis is not always a special case of loop analysis; give an example of a special case. (Hint: This will be the case if for each mesh current there is one branch that is traversed by only that mesh current.)

---

**2.2**   **Loop Analysis for Linear Time-invariant Networks**

In this section we shall restrict our consideration to linear time-invariant networks. We shall introduce branch equations and obtain by elimination

$l$ linear network equations in terms of the $l$ fundamental loop currents. For simplicity, we shall consider networks with resistors. The extension to the general case is exactly the same as the generalization discussed in Chap. 10.

The branch equations are written in matrix form as follows:

(2.9)   $\mathbf{v} = \mathbf{Rj} + \mathbf{v}_s - \mathbf{Rj}_s$

As before, $\mathbf{R}$ is a diagonal branch resistance matrix of dimension $b$, and $\mathbf{v}_s$ and $\mathbf{j}_s$ are voltage source and current source vectors, respectively. Combining Eqs. (2.1), (2.8), and (2.9), we obtain

(2.10)   $\mathbf{BRB}^T\mathbf{i} = -\mathbf{Bv}_s + \mathbf{BRj}_s$

or

(2.11)   $\mathbf{Z}_l\mathbf{i} = \mathbf{e}_s$

where

(2.12)   $\mathbf{Z}_l \triangleq \mathbf{BRB}^T \qquad \mathbf{e}_s \triangleq -\mathbf{Bv}_s + \mathbf{BRj}_s$

$\mathbf{Z}_l$ is called the **loop impedance matrix** of order $l$, and $\mathbf{e}_s$ is the **loop voltage source vector.** The loop impedance matrix has properties similar to those of the mesh impedance matrix discussed in the previous chapter. The matrix $\mathbf{Z}_l$ is symmetric. This is immediately seen once it is observed that in Eq. (2.12) $\mathbf{R}$ is a symmetric matrix. Let us rewrite Eq. (2.11) as follows:

(2.13)
$$
\begin{bmatrix}
z_{11} & z_{12} & \cdots & z_{1l} \\
z_{21} & z_{22} & \cdots & z_{2l} \\
\cdots\cdots\cdots\cdots\cdots \\
z_{l1} & z_{l2} & \cdots & z_{ll}
\end{bmatrix}
\begin{bmatrix}
i_1 \\ i_2 \\ \vdots \\ i_l
\end{bmatrix}
=
\begin{bmatrix}
e_{s1} \\ e_{s2} \\ \vdots \\ e_{sl}
\end{bmatrix}
$$

---

**Example**   Let us consider the network of Fig. 2.2. The graph of the network is that of Fig. 2.1; hence the fundamental loop matrix has been obtained before. The branch equation is

$$
\begin{bmatrix}
v_1 \\ v_2 \\ v_3 \\ v_4 \\ v_5 \\ v_6 \\ v_7 \\ v_8
\end{bmatrix}
=
\begin{bmatrix}
R_1 & & & & & & & \\
& R_2 & & & & 0 & & \\
& & R_3 & & & & & \\
& & & R_4 & & & & \\
& & & & R_5 & & & \\
& 0 & & & & R_6 & & \\
& & & & & & R_7 & \\
& & & & & & & R_8
\end{bmatrix}
\begin{bmatrix}
j_1 \\ j_2 \\ j_3 \\ j_4 \\ j_5 \\ j_6 \\ j_7 \\ j_8
\end{bmatrix}
+
\begin{bmatrix}
v_{s1} \\ 0 \\ 0 \\ 0 \\ 0 \\ 0 \\ 0 \\ 0
\end{bmatrix}
+
\begin{bmatrix}
0 \\ 0 \\ 0 \\ 0 \\ 0 \\ 0 \\ 0 \\ R_8 j_{s8}
\end{bmatrix}
$$

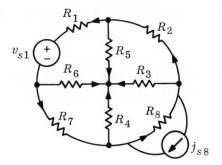

**Fig. 2.2**   Example of loop analysis.

Using Eq. (2.12), we can obtain the loop impedance matrix

$$\mathbf{Z}_l = \mathbf{BRB}^T$$

$$= \begin{bmatrix} R_1+R_5+R_6 & -R_5-R_6 & -R_6 & -R_6 \\ -R_5-R_6 & R_2+R_5+R_6+R_7+R_8 & R_6+R_7+R_8 & R_6+R_7 \\ -R_6 & R_6+R_7+R_8 & R_3+R_6+R_7+R_8 & R_6+R_7 \\ -R_6 & R_6+R_7 & R_6+R_7 & R_4+R_6+R_7 \end{bmatrix}$$

The loop equations are

$$\begin{bmatrix} R_1+R_5+R_6 & -R_5-R_6 & -R_6 & -R_6 \\ -R_5-R_6 & R_2+R_5+R_6+R_7+R_8 & R_6+R_7+R_8 & R_6+R_7 \\ -R_6 & R_6+R_7+R_8 & R_3+R_6+R_7+R_8 & R_6+R_7 \\ -R_6 & R_6+R_7 & R_6+R_7 & R_4+R_6+R_7 \end{bmatrix}$$

$$\begin{bmatrix} i_1 \\ i_2 \\ i_3 \\ i_4 \end{bmatrix} = \begin{bmatrix} -v_{s1} \\ -R_8 j_{s8} \\ -R_8 j_{s8} \\ 0 \end{bmatrix}$$

**Exercise**   Assume that the network shown in Fig. 2.2 is in the sinusoidal steady state and that its $k$th branch has an impedance $Z_k(j\omega)$. In terms of phasors, write the loop equations corresponding to the given tree.

## 2.3   Properties of the Loop Impedance Matrix

It is clear that the analysis of a resistive network and sinusoidal steady-state analysis of a similar network are very closely related. The main difference is in the appearance of phasors and impedances.

The following properties of the loop impedance matrix $\mathbf{Z}_l(j\omega)$ follow from the relation

$$\mathbf{Z}_l(j\omega) = \mathbf{B}\mathbf{Z}_b(j\omega)\mathbf{B}^T$$

1. If the network has no coupling elements, the branch impedance matrix $\mathbf{Z}_b(j\omega)$ is diagonal, and the loop impedance matrix is symmetric.

2. Also, if the network has no coupling elements, the loop impedance matrix $\mathbf{Z}_l(j\omega)$ can be written by inspection.

   a. The $i$th diagonal element of $\mathbf{Z}_l(j\omega)$, $z_{ii}$, is equal to the sum of the impedances in loop $i$; $z_{ii}$ is called the **self-impedance of loop** $i$.

   b. The $(i,k)$ element of $\mathbf{Z}_l(j\omega)$, $z_{ik}$, is equal to plus or minus the sum of the impedances of the branches common to loop $i$ and to loop $k$; the plus sign applies if, in the branches common to loop $i$ and loop $k$, the loop reference directions agree, and the minus sign applies when they are opposite.

3. If all current sources are converted, by Thévenin's theorem, into voltage sources, then the forcing term $e_{si}$ is the algebraic sum of all the source voltages in loop $i$.

4. If the network is resistive and if all its resistances are positive, then det $(\mathbf{Z}_l) > 0$.

**Exercise 1**   Write in a few sentences the circuit-theoretic consequences of property 4.

**Exercise 2**   Give an example of a linear time-invariant network made of *passive* elements such that for some tree and some frequency $\omega_0$, det $[\mathbf{Z}_l(j\omega_0)] = 0$. Can you give an example which includes a resistor?

**Exercise 3**   In the network of Fig. 2.2, pick the tree consisting of branches 1, 2, 3, and 4. Write the loop equations by inspection.

## 3   Cut-set Analysis

### 3.1   Two Basic Facts of Cut-set Analysis

Cut-set analysis is the dual of loop analysis. First, we pick a tree; call it $T$. Next we number branches; as before, the links from range 1 to $l$, and the tree branches range from $l + 1$ to $b$. We know that every tree branch defines (for the given tree) a *unique* fundamental cut set. That cut set is made up of links and of one tree branch, namely the tree branch which defines the cut set. In Fig. 3.1 we show the same graph $\mathcal{G}$ and the same tree $T$ as in the previous section. The four fundamental cut sets are also shown.

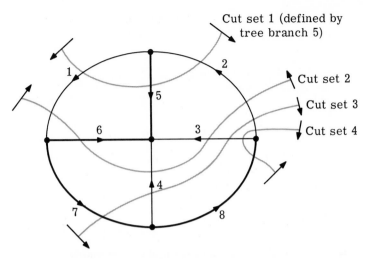

**Fig. 3.1**   Fundamental cut sets for the chosen tree of a given graph.

Let us number the cut sets as follows: cut set 1 is associated with tree branch 5, cut set 2 with tree branch 6, etc.   By analogy to previous conventions, for each fundamental cut set we adopt a *reference direction for the cut set* which agrees with that of the tree branch defining the cut set. Under these conditions, if we apply KCL to the four cut sets, we obtain

Cut set 1:   $j_1 - j_2 + j_5 = 0$

Cut set 2:   $-j_1 + j_2 + j_3 + j_4 + j_6 = 0$

Cut set 3:   $-j_2 - j_3 - j_4 + j_7 = 0$

Cut set 4:   $-j_2 - j_3 + j_8 = 0$

In matrix form, the equation is

$$\begin{bmatrix} 1 & -1 & 0 & 0 & 1 & 0 & 0 & 0 \\ -1 & 1 & 1 & 1 & 0 & 1 & 0 & 0 \\ 0 & -1 & -1 & -1 & 0 & 0 & 1 & 0 \\ 0 & -1 & -1 & 0 & 0 & 0 & 0 & 1 \end{bmatrix} \begin{bmatrix} j_1 \\ j_2 \\ j_3 \\ j_4 \\ j_5 \\ j_6 \\ j_7 \\ j_8 \end{bmatrix} = \begin{bmatrix} 0 \\ 0 \\ 0 \\ 0 \end{bmatrix}$$

More generally, if we apply KCL to each one of the $n$ fundamental cut sets, we obtain a system of $n$ linear homogeneous equations in $n$ unknowns $j_1, j_2, \ldots, j_b$.   The first basic fact of cut-set analysis is summarized in the following statement:

*The n linear homogeneous algebraic equations in $j_1, j_2, \ldots, j_b$ obtained by applying KCL to each fundamental cut set constitute a set of n linearly independent equations.*

Recalling the convention of sign for cut sets, we see that the KCL equations are of the form

(3.1)   $\mathbf{Qj} = \mathbf{0}$

where $\mathbf{Q}$ is an $n \times b$ matrix defined by

(3.2)   $q_{ik} = \begin{cases} 1 & \text{if branch } k \text{ belongs to cut set } \textcircled{i} \text{ and has the } same \text{ reference direction} \\ -1 & \text{if branch } k \text{ belongs to cut set } \textcircled{i} \text{ and has the } opposite \text{ reference direction} \\ 0 & \text{if branch } k \text{ } does \text{ } not \text{ } belong \text{ to cut set } \textcircled{i} \end{cases}$

$\mathbf{Q} = [q_{ik}]$ is called the **fundamental cut-set matrix.**   As before we note that it is of the form

(3.3)   $\mathbf{Q} = [\ \mathbf{E}\ \vdots\ \mathbf{1}_n\ ]$   $n$ cut sets

$l$ links   $n$ tree branches

where $\mathbf{E}$ is an appropriate $n \times l$ matrix with elements $-1, +1, 0$, and $\mathbf{1}_n$ is the $n \times n$ unit matrix.   Obviously, $\mathbf{Q}$ has a rank $n$ since it includes the unit matrix $\mathbf{1}_n$.   Hence, the $n$ fundamental cut-set equations in terms of the branch currents are linearly independent.

Turning now to KVL, we note that each branch voltage can be expressed as a linear combination of the tree-branch voltages.   For convenience, let us label the tree-branch voltages by $e_1, e_2, \ldots, e_n$.   For the example in Fig. 3.1, from KVL we obtain the following equations:

$v_1 = v_5 - v_6 = e_1 - e_2$

$v_2 = -v_5 + v_6 - v_7 - v_8 = -e_1 + e_2 - e_3 - e_4$

$v_3 = v_6 - v_7 - v_8 = e_2 - e_3 - e_4$

$v_4 = v_6 - v_7 = e_2 - e_3$

$v_5 = e_1$

$v_6 = e_2$

$v_7 = e_3$

$v_8 = e_4$

By following the reasoning dual to that of the loop analysis, we can prove the assertion of the second basic fact, namely

(3.4)   $\mathbf{v} = \mathbf{Q}^T\mathbf{e}$

that is, the branch voltage vector is obtained by forming the product of the cut-set matrix *transposed* and the tree-branch voltage vector.

**Summary**  KCL requires that $\mathbf{Qj} = \mathbf{0}$.  KVL is expressed by $\mathbf{v} = \mathbf{Q}^T\mathbf{e}$.  As a result of our numbering convention, the fundamental cut-set matrix $\mathbf{Q}$ is of the form of (3.3).  These equations are valid irrespective of the nature of the branches.

**Exercise 1**  Prove Tellegen's theorem by using Eqs. (3.1) and (3.4).

**Exercise 2**  Node analysis is not always a special case of cut-set analysis.  Give an example of such a non-special case.

---

**3.2**    **Cut-set Analysis for Linear Time-invariant Networks**

In cut-set analysis Kirchhoff's laws are expressed by [see (3.1) and (3.4)]

$$\mathbf{Qj} = \mathbf{0}$$

$$\mathbf{v} = \mathbf{Q}^T\mathbf{e}$$

These equations are combined with branch equations to form network equations with the $n$ tree-branch voltages $e_1, e_2, \ldots, e_n$ as network variables.

For the case of linear time-invariant resistive networks, the branch equations are easily written in matrix form.  Let us illustrate the procedure with a resistive network.  The branch equations are written in matrix form as follows:

(3.5)  $\mathbf{j} = \mathbf{Gv} + \mathbf{j}_s - \mathbf{Gv}_s$

As before, $\mathbf{G}$ is the diagonal branch conductance matrix of dimension $b$ and $\mathbf{j}_s$ and $\mathbf{v}_s$ are the source vectors.  Combining Eqs. (3.1), (3.4), and (3.5), we obtain

(3.6)  $\mathbf{QGQ}^T\mathbf{e} = \mathbf{QGv}_s - \mathbf{Qj}_s$

or

(3.7)  $\mathbf{Y}_q\mathbf{e} = \mathbf{i}_s$

where

(3.8)  $\mathbf{Y}_q \triangleq \mathbf{QGQ}^T \qquad \mathbf{i}_s \triangleq \mathbf{QGv}_s - \mathbf{Qj}_s$

$\mathbf{Y}_q$ is called the **cut-set admittance matrix,** and $\mathbf{i}_s$ is the **cut-set current source vector.**

In scalar form, the cut-set equations are

$$\begin{bmatrix} y_{11} & y_{12} & \cdots & y_{1n} \\ y_{21} & y_{22} & \cdots & y_{2n} \\ \cdots\cdots\cdots\cdots\cdots \\ y_{n1} & y_{n2} & \cdots & y_{nn} \end{bmatrix} \begin{bmatrix} e_1 \\ e_2 \\ \vdots \\ e_n \end{bmatrix} = \begin{bmatrix} i_{s1} \\ i_{s2} \\ \vdots \\ i_{sn} \end{bmatrix}$$

| **3.3** | **Properties of the Cut-set Admittance Matrix** |

As before, we note that for sinusoidal steady-state analysis the cut-set admittance matrix $\mathbf{Y}_q$ has a number of properties based on the equation

$$\mathbf{Y}_q(j\omega) = \mathbf{Q}\mathbf{Y}_b(j\omega)\mathbf{Q}^T$$

1. If the network has no coupling elements, the branch admittance matrix $\mathbf{Y}_b(j\omega)$ is diagonal, and $\mathbf{Y}_q(j\omega)$ is symmetric.

2. If there are no coupling elements,

   a. The $i$th diagonal element of $\mathbf{Y}_q(j\omega)$, $y_{ii}(j\omega)$, is equal to the sum of the admittances of the branches of the $i$th cut set.

   b. The $(i,k)$ element of $\mathbf{Y}_q(j\omega)$, $y_{ik}(j\omega)$, is equal to the sum of all the admittances of branches common to cut set $i$ and cut set $k$ when, in the branches common to their two cut sets, their reference directions agree; otherwise, $y_{ik}$ is the negative of that sum.

3. If all the voltage sources are transformed to current sources, then $i_{sk}$ is the total current-source contribution to cut set $k$.

4. If the network is resistive and if all branch resistances are positive, then det $(\mathbf{Y}_q) > 0$.

**Example**   Consider the resistive network shown in Fig. 3.2. The cut-set equations are

$$\begin{bmatrix} G_1 + G_2 + G_5 & -G_1 - G_2 & G_2 & G_2 \\ -G_1 - G_2 & G_1 + G_2 + G_3 + G_4 + G_6 & -G_2 - G_3 - G_4 & -G_2 - G_3 \\ G_2 & -G_2 - G_3 - G_4 & G_2 + G_3 + G_4 + G_7 & G_2 + G_3 \\ G_2 & -G_2 - G_3 & G_2 + G_3 & G_2 + G_3 + G_8 \end{bmatrix} \begin{bmatrix} e_1 \\ e_2 \\ e_3 \\ e_4 \end{bmatrix} = \begin{bmatrix} G_1 v_{s1} \\ -G_1 v_{s1} \\ 0 \\ j_{s8} \end{bmatrix}$$

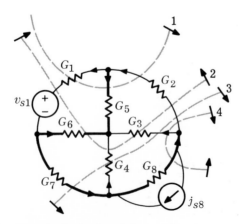

**Fig. 3.2**   Example of cut-set analysis.

## 4   Comments on Loop and Cut-set Analysis

Both the loop analysis and cut-set analysis start with choosing a tree for the given graph. Since the number of possible trees for a graph is usually large, the two methods are extremely flexible. It is obvious that they are more general than the mesh analysis and node analysis. For example, consider the graph of Fig. 4.1, where the chosen tree is shown by the emphasized branches. The fundamental loops for the particular tree coincide with the four meshes of the graph. Thus, the mesh currents are identical with the fundamental loop currents. Similarly, as shown in Fig. 4.2, the fundamental cut sets for the particular tree coincide with the sets of branches connected to nodes (1), (2), (3), and (4). If node (5) is picked as the datum node, the tree-branch voltages are identical with the node-to-datum voltages. Thus, mesh analysis and node analysis for this particular example are special cases of the loop analysis and cut-set analysis. However, it should be pointed out that for the graph of Fig. 4.3, the meshes are not special cases of fundamental loops; i.e., there exists no tree such that the five meshes are fundamental loops. Similarly, in Fig. 4.2, if node (4) is picked as datum node, there exists no tree which gives tree-branch voltages identical to the node-to-datum voltages.

As far as the relative advantages of cut-set analysis and loop analysis, the conclusion is the same as that between mesh analysis and node analysis. It depends on the graph as well as on the kind and number of sources in the network. For example, if the number of tree branches, $n$, is much smaller than the number of links, $l$, the cut-set method is usually more efficient.

It is important to keep in mind the duality among the concepts pertaining to general networks and graphs. Table 10.1 of Chap. 10 should be studied again at this juncture. Whereas in our first study, duality applied

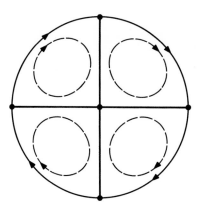

**Fig. 4.1**   Fundamental loops for the chosen tree are identical with meshes.

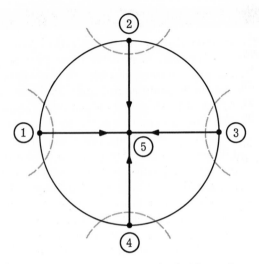

**Fig. 4.2**  The four fundamental cut sets for the chosen tree coincide with the set of branches connected to nodes ①, ②, ③, and ④.

only to planar graphs and planar networks and we thought in terms of node and mesh analysis, it is now apparent that duality extends to concepts pertaining to nonplanar graphs and networks; for example, cut sets and loops are dual concepts. The entries of Table 10.1 of Chap. 10 should be carefully considered.

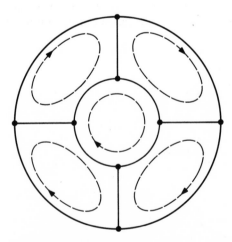

**Fig. 4.3**  A graph showing that meshes are not special cases of fundamental loops.

## 5   Relation Between B and Q

If we start with an oriented graph $\mathcal{G}$ and pick any one of its trees, say $T$, and if we write the fundamental loop matrix **B** and the fundamental cut-set matrix **Q**, we should expect to find a very close connection between these matrices. After all, **B** tells us which branch is in which fundamental loop, and **Q** tells us which branch is in which fundamental cut set. The precise relation between **B** and **Q** is stated in the following theorem.

**THEOREM**   Call **B** the fundamental loop matrix and **Q** the fundamental cut-set matrix of the *same* oriented $\mathcal{G}$, and let both matrices pertain to the *same* tree $T$; then

(5.1)   $$\mathbf{BQ}^T = \mathbf{0} \quad \text{and} \quad \mathbf{QB}^T = \mathbf{0}$$

Furthermore, if we number the links from 1 to $l$ and number the tree branches from $l + 1$ to $b$, then

(5.2)   $$\mathbf{B} = \left[\, \mathbf{1}_l \,\vdots\, \mathbf{F} \,\right] \quad \text{and} \quad \mathbf{Q} = \left[\, -\mathbf{F}^T \,\vdots\, \mathbf{1}_n \,\right]$$

Before proving these facts, let us see what the first Eq. (5.1) means. This equation tells us that the product of the $l \times b$ matrix **B** and the $b \times n$ matrix $\mathbf{Q}^T$ is the $l \times n$ zero matrix. In other words, the product of every row of **B** and every column of $\mathbf{Q}^T$ is zero. The second Eq. (5.1) is simply the first one transposed; the product of every row of **Q** by every column of $\mathbf{B}^T$ is zero.

*Proof*   Let the components of the vector $\mathbf{e} = [e_1, e_2, \ldots, e_n]^T$ be *arbitrary*. Since they are the tree-branch voltages of the tree $T$, the branch voltages of $\mathcal{G}$ are given by

$$\mathbf{v} = \mathbf{Q}^T \mathbf{e}$$

In other words, whatever the $n$-vector **e** may be, this equation gives us a set of $b$ branch voltages that satisfies KVL. On the other hand, any time a set of branch voltages $v_k$ satisfies KVL, we have

$$\mathbf{Bv} = \mathbf{0}$$

(that is, these $v_k$'s satisfy KVL along all the fundamental loops). Substituting **v**, we obtain

(5.3)   $$\mathbf{BQ}^T \mathbf{e} = \mathbf{0} \quad \text{for all } \mathbf{e}$$

Note very carefully that this equation means that given *any* $n$-vector **e**, if we multiply it on the left by $\mathbf{BQ}^T$, we get the zero vector! Observe that the product $\mathbf{BQ}^T$ is an $l \times n$ matrix. This means that whenever we multiply any $n$-vector **e** by $\mathbf{BQ}^T$, we get the zero vector. For example if we choose $\mathbf{e} = \mathbf{e}_1 \triangleq [1, 0, 0, \ldots, 0]^T$, $\mathbf{BQ}^T\mathbf{e}_1$ is easily seen to be the first column of $\mathbf{BQ}^T$; hence, the first column of $\mathbf{BQ}^T$ is a column of zeros. Similarly, if we choose $\mathbf{e} = \mathbf{e}_2 \triangleq [0, 1, 0, \ldots, 0]^T$, we see that the second column of $\mathbf{BQ}^T$ is a column of zeros, and so forth. Therefore, Eq. (5.3) implies that the matrix $\mathbf{BQ}^T$ has all its elements equal to zero. Therefore, Eqs. (5.1) are established. (The second equation is simply the first one transposed.)

To prove (5.2), let us recall that we noted that $\mathbf{Q}$ was of the form

(5.4) $$\mathbf{Q} = \left[\mathbf{E} \mid \mathbf{1}_n\right]$$

Therefore,

$$\mathbf{BQ}^t = \underbrace{\left[\mathbf{1}_l \mid \mathbf{F}\right]}_{l \quad n}\left.\left[\begin{matrix}\mathbf{E}^T \\ \mathbf{1}_n\end{matrix}\right]\right\}\begin{matrix}l \\ n\end{matrix}$$

Using the fact that a product of matrices is performed as rows by columns and noting that $\mathbf{1}_l$ has the same number of columns as $\mathbf{E}^T$ has rows, we conclude that

$$\mathbf{BQ}^T = \mathbf{1}_l\mathbf{E}^T + \mathbf{F1}_n = \mathbf{E}^T + \mathbf{F} = \mathbf{0}$$

Hence,

$$\mathbf{E}^T = -\mathbf{F}$$

and transposing,

$$\mathbf{E} = -\mathbf{F}^T$$

Using this conclusion into (5.4), we see that

$$\mathbf{Q} = \left[-\mathbf{F}^T \mid \mathbf{1}_n\right]$$

Thus, the proof is complete.

The relation between $\mathbf{B}$ and $\mathbf{Q}$ expressed by (5.2) is extremely useful since it means that whenever we know one of these matrices, we can write the other one by inspection; or, even better, both matrices $\mathbf{B}$ and $\mathbf{Q}$ are uniquely specified by the $l \times n$ matrix $\mathbf{F}$.

**Exercise 1** Verify that $\mathbf{BQ}^T = \mathbf{0}$ for the graph of Fig. 3.1.

**Exercise 2** Prove the first equation (5.1) by referring to the definitions of $\mathbf{B}$ and $\mathbf{Q}$. Note that the $(i,k)$ element of $\mathbf{BQ}^T$ is of the form

$$\sum_{j=1}^{b} q_{ij}b_{kj} = q_{ik}b_{kk} + q_{i(i+l)}b_{k(i+l)}$$

that is, the sum has two nonzero terms.

## Summary

- In both the loop analysis and the cut-set analysis we first pick a tree and number all branches. For convenience, we number the links first from 1 to $l$ and number the tree branches from $l + 1$ to $b$. Then we assign branch orientations.

- In loop analysis we use the fundamental loop currents $i_1, i_2, \ldots, i_l$ as network variables. We write $l$ linearly independent algebraic equations in terms of branch voltages by applying KVL for each fundamental loop. In linear time-invariant networks, taking the branch equations into account, the $l$ equations can be put explicitly in terms of the $l$ fundamental loop currents. In general, the resulting network equations form a system of $l$ integrodifferential equations, in matrix form,

$$\mathbf{Z}_l(D)\mathbf{i} = \mathbf{e}_s$$

The solution of this system of linear integrodifferential equations will be treated in succeeding chapters. Once the fundamental loop currents $\mathbf{i}$ are determined, the branch currents can be found immediately from

$$\mathbf{j} = \mathbf{B}^T\mathbf{i} \qquad \text{(KCL)}$$

The $b$ branch voltages are then obtainable from the $b$ branch equations.

- The cut-set analysis is the dual of the loop analysis. The $n$ tree-branch voltages $e_1, e_2, \ldots, e_n$ are used as network variables, and $n$ linearly independent equations in terms of branch currents are written by applying KCL for all the fundamental cut sets associated with the tree. In linear time-invariant networks the $n$ equations can be put explicitly in terms of the $n$ tree-branch voltages. In general, the resulting matrix equation is

$$\mathbf{Y}_q(D)\mathbf{e} = \mathbf{i}_s$$

Once $\mathbf{e}$ is determined, the $b$ branch voltages can be found immediately from

$$\mathbf{v} = \mathbf{Q}^T\mathbf{e} \qquad \text{(KVL)}$$

The $b$ branch currents are then obtainable from the $b$ branch equations.

- Given any oriented graph $\mathcal{G}$ and any of its trees, the resulting fundamental loop matrix $\mathbf{B}$ and the fundamental cut-set matrix $\mathbf{Q}$ are such that

$$\mathbf{BQ}^T = \mathbf{0} \qquad \text{and} \qquad \mathbf{QB}^T = \mathbf{0}$$

Furthermore,

$$\mathbf{B} = \left[\begin{array}{c|c} \mathbf{1}_l & \mathbf{F} \end{array}\right]\Big\}l \qquad \text{and} \qquad \mathbf{Q} = \left[\begin{array}{c|c} -\mathbf{F}^T & \mathbf{1}_n \end{array}\right]\Big\}n$$
$$\underbrace{\phantom{xxxx}}_{l}\ \underbrace{\phantom{xxxx}}_{n} \qquad\qquad\qquad \underbrace{\phantom{xxxx}}_{l}\ \underbrace{\phantom{xxxx}}_{n}$$

- The analogies between the four methods of analysis deserve to be emphasized:

$$\mathbf{Y}_n(j\omega) = \mathbf{A}\mathbf{Y}_b(j\omega)\mathbf{A}^T \qquad \text{for node analysis}$$

$$\mathbf{Z}_m(j\omega) = \mathbf{M}\mathbf{Z}_b(j\omega)\mathbf{M}^T \qquad \text{for mesh analysis}$$

$$\mathbf{Y}_q(j\omega) = \mathbf{Q}\mathbf{Y}_b(j\omega)\mathbf{Q}^T \qquad \text{for cut-set analysis}$$

$$\mathbf{Z}_l(j\omega) = \mathbf{B}\mathbf{Z}_b(j\omega)\mathbf{B}^T \qquad \text{for loop analysis}$$

- Each one of the "connection" matrices **A, M, Q,** and **B** is of full rank.

## Problems

Trees, cut
sets, and
loops

**1.** For the oriented graph shown in Fig. P11.1 and for the tree indicated,

*a.* Indicate all the fundamental loops and the fundamental cut sets.

*b.* Write all the fundamental loop and cut-set equations.

*c.* Can you find a tree such that all its fundamental loops are meshes?

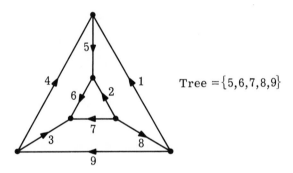

Tree $= \{5,6,7,8,9\}$

**Fig. P11.1**

Loop analysis

**2.** Your roommate analyzed a number of passive linear time-invariant *RLC* circuits. He found the loop impedance matrices given below. Which ones do you accept as correct? Give your reasons for rejecting any.

*a.*
$\begin{bmatrix} 3 & 2 \\ 2 & 5 \end{bmatrix}$
$\begin{bmatrix} -1 & 2 \\ 2 & 4 \end{bmatrix}$
$\begin{bmatrix} 1 & 2 \\ 2 & 3 \end{bmatrix}$

*b.*
$\begin{bmatrix} 3+j & -2j \\ -2j & 5+7j \end{bmatrix}$
$\begin{bmatrix} 3 & -j \\ -j & 2 \end{bmatrix}$
$\begin{bmatrix} 5 & 7j \\ 6j & 8+3j \end{bmatrix}$

Loop analysis

**3.** The linear time-invariant network of Fig. P11.3a, having a (topological) graph shown in Fig. P11.3b, is in the sinusoidal steady state. From the (topological) graph a tree is picked as shown in Fig. P11.3c.

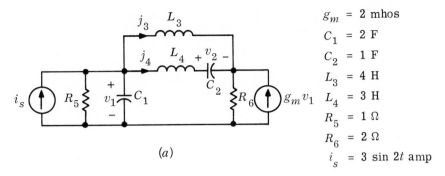

$g_m$ = 2 mhos

$C_1$ = 2 F

$C_2$ = 1 F

$L_3$ = 4 H

$L_4$ = 3 H

$R_5$ = 1 Ω

$R_6$ = 2 Ω

$i_s$ = 3 sin $2t$ amp

(a)

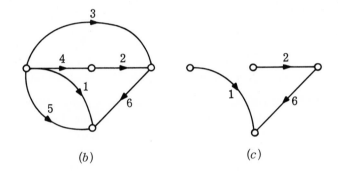

(b)  (c)

**Fig. P11.3**

a.   Write the fundamental loop matrix **B**.

b.   Calculate the loop impedance matrix $\mathbf{Z}_l$.

c.   Write the loop equations in terms of voltage and current phasors; that is, $\mathbf{Z}_l\mathbf{I} = \mathbf{E}_s$.

Loop analysis   **4.** Assume that the linear time-invariant network of Fig. P11.3 is in the sinusoidal steady state. Write the fundamental loop equations for the tree indicated by the shortcut method. (First assume that the dependent current source is independent, and introduce its dependence in the last step.)

Loop analysis   **5.** The linear time-invariant network shown in Fig. P11.5 is in the sinusoidal steady state. For the reference directions indicated on the inductors, the inductance matrix is

$$\begin{bmatrix} 4 & 2 & 1 \\ 2 & 4 & 2 \\ 1 & 2 & 4 \end{bmatrix}$$

Write the fundamental loop equation for a tree of your choice.

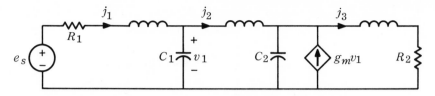

**Fig. P11.5**

Cut-set
analysis

**6.** Consider the linear time-invariant network shown in Fig. P11.3a. Suppose it is in the sinusoidal steady state. Consider the tree shown in Fig. P11.3c.

*a.* Write the fundamental cut-set matrix **Q**.

*b.* Calculate the cut-set admittance matrix $\mathbf{Y}_q$.

*c.* Write the cut-set equations in terms of the cut-set voltage and source current phasors; that is, $\mathbf{Y}_q\mathbf{E} = \mathbf{I}_s$.

Cut-set
analysis

**7.** The linear time-invariant network shown in Fig. P11.7 is in the sinusoidal steady state. It originates from delay-line designs; each inductor is coupled to its neighbor and to his neighbor(s) once removed, and the bridging capacitors compensate the coupling of the neighbors once removed. The coupling between inductors is specified by the reciprocal inductance matrix

$$\boldsymbol{\Gamma} = \begin{bmatrix} \Gamma_0 & \Gamma_1 & \Gamma_2 & 0 \\ \Gamma_1 & \Gamma_0 & \Gamma_1 & \Gamma_2 \\ \Gamma_2 & \Gamma_1 & \Gamma_0 & \Gamma_1 \\ 0 & \Gamma_2 & \Gamma_1 & \Gamma_0 \end{bmatrix}$$

Pick a tree such that the corresponding cut-set equations are easy to write. Write these cut-set equations.

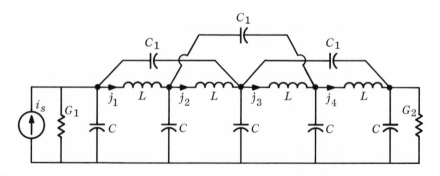

**Fig. P11.7**

Cut set and
loop matrix

**8.** For a given connected network and for a fixed tree, the fundamental loop matrix is given by

$$\mathbf{B} = \begin{bmatrix} 1 & 0 & 0 & 1 & 0 & 0 \\ 0 & 1 & 0 & 0 & 0 & -1 \\ 0 & 0 & 1 & 1 & -1 & -1 \end{bmatrix}$$

*a.* Write, by inspection, the fundamental cut-set matrix which corresponds to the same tree.

*b.* Draw the oriented graph of the network.

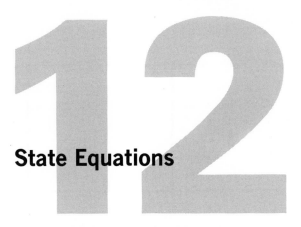

# State Equations

When the differential equations of a lumped network are written in the form

$$\dot{\mathbf{x}} = \mathbf{f}(\mathbf{x},\mathbf{w},t)$$

(where $\mathbf{x}$ is a vector, say, with $n$ components; $\mathbf{w}$ represents the set of inputs, and $t$ represents the time), we say that the equations are in the **state form** and that $\mathbf{x}$ represents the **state** of the network. There are three basic reasons for writing the equations in this form: (1) this form lends itself most easily to analog and/or digital computer programming, (2) the extension of the analysis to nonlinear and/or time-varying networks is quite easy (whereas this extension is not easy in the case of loop, mesh, cut-set, or node analysis), and (3) in this form a number of system-theoretic concepts are readily applicable to networks.

In Sec. 1 we use examples to illustrate how state equations are written for simple linear time-invariant networks. In Sec. 2 we review and extend some pertinent concepts of state. In Sec. 3 simple time-varying and nonlinear networks are treated. Finally, in Sec. 4 we give a general method of writing state equations for a broad class of linear time-invariant networks.

## 1    Linear Time-invariant Networks

Consider the linear time-invariant network shown in Fig. 1.1. It has three energy-storing elements: one capacitor $C$ and two inductors $L_1$ and $L_2$. Any response of this network is thus closely related to the behavior of the capacitor voltage $v$ and the inductor currents $i_1$ and $i_2$. If we wish to use these variables for our analysis and also wish to write equations in the state form,

$$(1.1) \quad \dot{\mathbf{x}} = \mathbf{f}(\mathbf{x},\mathbf{w},t)$$

we may choose $v$, $i_1$, and $i_2$ as the state variables; that is, we choose

$$(1.2) \quad \mathbf{x} = \begin{bmatrix} v \\ i_1 \\ i_2 \end{bmatrix}$$

as the state vector. Note that in the vector differential equation (1.1), $\dot{\mathbf{x}}$,

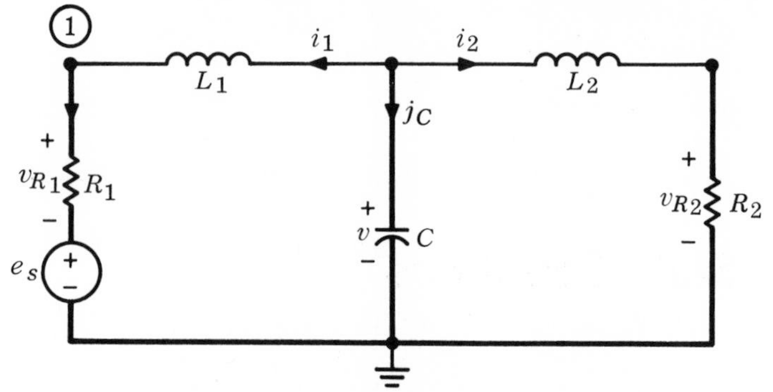

**Fig. 1.1**   Network used to illustrate the writing of state equations; the tree branches are emphasized.

the first derivative of the state vector, is expressed in terms of $\mathbf{x}$ itself and the input $\mathbf{w}$. We note that the quantities $C(dv/dt)$, $L_1(di_1/dt)$, and $L_2(di_2/dt)$ are currents in the capacitor and voltages in the inductors. To evaluate $C(dv/dt)$ in terms of the state variables and possibly the input, we must write a cut-set equation; similarly, to evaluate $L_1(di_1/dt)$ and $L_2(di_2/dt)$, we must write loop equations. This suggests that the capacitors should belong to a tree and the inductors to the links. When these suggestions are thought through and sorted out, we are led to the following procedure:

*Step* 1   Pick a tree which contains *all* the capacitors and *none* of the inductors. Such a tree is indicated in Fig. 1.1 by the emphasis on the tree branches.

*Step* 2   Use the tree-branch capacitor *voltages* and the link inductor currents as variables. In the example this means $v$, $i_1$, and $i_2$.

*Step* 3   Write a fundamental cut-set equation for each capacitor. Note that in these cut-set equations all branch currents must be expressed in terms of the variables chosen in Step 2. For the capacitor, we obtain

$$C\frac{dv}{dt} = -i_1 - i_2$$

*Step* 4   Write a fundamental loop equation for each inductor. Note that in these loop equations all branch voltages (except independent sources) must be expressed in terms of the variables chosen in Step 2. For the first inductor,

$$L_1\frac{di_1}{dt} = -v_{R_1} - e_s + v$$

and in terms of the chosen variables $i_1$, $i_2$, and $v$,

$$L_1 \frac{di_1}{dt} = -R_1 i_1 - e_s + v$$

Similarly, for the second inductor,

$$L_2 \frac{di_2}{dt} = -v_{R_2} + v$$

$$= -R_2 i_2 + v$$

Thus, we obtain the following system of state equations:

$$C \frac{dv}{dt} = -i_1 - i_2$$

$$(1.3) \qquad L_1 \frac{di_1}{dt} = -R_1 i_1 + v - e_s$$

$$L_2 \frac{di_2}{dt} = -R_2 i_2 + v$$

If we put the system in matrix form, we obtain

$$(1.4) \qquad \begin{bmatrix} \dfrac{dv}{dt} \\[2mm] \dfrac{di_1}{dt} \\[2mm] \dfrac{di_2}{dt} \end{bmatrix} = \begin{bmatrix} 0 & -\dfrac{1}{C} & -\dfrac{1}{C} \\[2mm] \dfrac{1}{L_1} & -\dfrac{R_1}{L_1} & 0 \\[2mm] \dfrac{1}{L_2} & 0 & -\dfrac{R_2}{L_2} \end{bmatrix} \begin{bmatrix} v \\[2mm] i_1 \\[2mm] i_2 \end{bmatrix} + \begin{bmatrix} 0 \\[2mm] -1 \\[2mm] 0 \end{bmatrix} e_s$$

In terms of $\mathbf{x}$, the state vector whose components are $v$, $i_1$, and $i_2$, we note that the equations have the form

$$(1.5) \qquad \dot{\mathbf{x}} = \mathbf{A}\mathbf{x} + \mathbf{b}w$$

where $\mathbf{A}$ is a constant $3 \times 3$ matrix, $\mathbf{b}$ is a constant vector, and $w = e_s$ is the scalar input. Both $\mathbf{A}$ and $\mathbf{b}$ depend only on the elements of the network and on its graph. Equation (1.5) is a special form of the general state equation (1.1) for linear time-invariant networks.

Let the initial state at $t = 0$ be specified by the three quantities $v(0)$, $i_1(0)$, and $i_2(0)$; that is,

$$(1.6) \qquad \mathbf{x}_0 \triangleq \mathbf{x}(0) = \begin{bmatrix} v(0) \\ i_1(0) \\ i_2(0) \end{bmatrix}$$

If the input waveform $e_s(\cdot)$ is given for all $t \geq 0$, then for any $t_1 > 0$, the state at time $t_1$ is uniquely specified. Indeed, the theory of differential equations teaches us that, given the initial conditions $v(0)$, $i_1(0)$, and $i_2(0)$, the forcing function $e_s(\cdot)$ uniquely determines the functions $v(\cdot)$, $i_1(\cdot)$, and $i_2(\cdot)$ for all $t \geq 0$. Also observe that each network variable can be written as a function of the state and of the input. For example, the node-to-datum voltage $e_1$ is given by

$$e_1(t) = R_1 i_1(t) + e_s(t) \qquad \text{for all } t$$

For linear time-invariant networks, if we let $y$ be any output, we can express $y$ in terms of a linear combination of the state vector and the input $w$.† Thus,

(1.7)   $y = \mathbf{c}^T \mathbf{x} + d_0 w$

where $d_0$ is a scalar and $\mathbf{c}$ is a constant vector, both depending on the network topology and element values. Thus, $e_1$ can be written in the above form as

$$e_1 = [0 \; R_1 \; 0] \begin{bmatrix} v \\ i_1 \\ i_2 \end{bmatrix} + (1) e_s$$

---

**Example**   To further illustrate the procedure, consider the linear time-invariant network shown in Fig. 1.2. It has one input (the voltage source $e_s$), two inductors $L_1$ and $L_2$, and two capacitors $C_1$ and $C_2$. This network will therefore require four state variables $i_{L_1}$, $i_{L_2}$, $v_{C_1}$, and $v_{C_2}$. We proceed as follows:

*Step 1*   We pick a tree which includes $C_1$ and $C_2$ and does not include $L_1$ and $L_2$. Let us put $R_4$ in the tree; consequently, the series connection of the voltage source $e_s$ and $R_3$ is a link.

*Step 2*   We identify the variables as the inductor currents $i_{L_1}$ and $i_{L_2}$ and the capacitor voltages $v_{C_1}$ and $v_{C_2}$.

*Step 3*   We write cut-set equations for the fundamental cut set defined by the capacitors, recalling that we must express all currents in terms of $i_{L_1}$, $i_{L_2}$, $v_{C_1}$, and $v_{C_2}$. Thus, we obtain

(1.8)   $C_1 \dfrac{dv_{C_2}}{dt} = i_{L_1}$

and

(1.9)   $C_2 \dfrac{dv_{C_2}}{dt} = i_{L_1} + i_{L_2}$

---

† In certain cases, the output may also depend on the derivatives of the input $w$. These correspond to improper systems (see Chaps. 5 and 13).

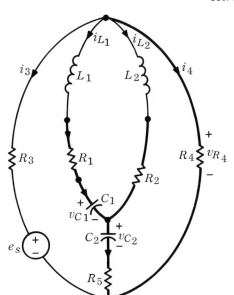

**Fig. 1.2**   Network used in the example for writing state equations.

*Step 4*   We write loop equations for the fundamental loop defined by the inductors, recalling that all the voltages must be expressed in terms of $i_{L_1}$, $i_{L_2}$, $v_{C_1}$, and $v_{C_2}$. Thus, we have

$$L_1 \frac{di_{L_1}}{dt} = -v_{R_1} - v_{C_1} - v_{C_2} - v_{R_5} + v_{R_4}$$

(1.10)
$$= -R_1 i_{L_1} - v_{C_1} - v_{C_2} - R_5(i_{L_1} + i_{L_2}) + v_{R_4}$$

In this case we notice that we are not as lucky as before, since $v_{R_4}$ in the right-hand side of the equation is not one of the state variables.

To express $v_{R_4}$ in terms of the chosen variables, we must solve the circuit shown in Fig. 1.3 (note that in calculating $v_{R_4}$, we consider that the state variables $i_{L_1}$, $i_{L_2}$, $v_{C_1}$, and $v_{C_2}$ are known). Thus,

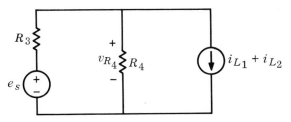

**Fig. 1.3**   Auxiliary network used to calculate $v_{R_4}$ in terms of the chosen variables.

$$v_{R_4}\left(\frac{1}{R_4} + \frac{1}{R_3}\right) - \frac{1}{R_3}e_s = -(i_{L_1} + i_{L_2})$$

or

$$(1.11) \quad v_{R_4} = e_s \frac{R_4}{R_3 + R_4} - \frac{R_3 R_4}{R_3 + R_4}(i_{L_1} + i_{L_2})$$

Hence,

$$(1.12) \quad \frac{di_{L_1}}{dt} = -\frac{v_{C_1}}{L_1} - \frac{v_{C_2}}{L_1} - \frac{(R_1 + R)}{L_1}i_{L_1} - \frac{R}{L_1}i_{L_2} + \frac{R_4}{L_1(R_3 + R_4)}e_s$$

where

$$(1.13) \quad R \triangleq R_5 + \frac{R_3 R_4}{R_3 + R_4}$$

Similarly,

$$L_2 \frac{di_{L_2}}{dt} = -R_L i_{L_2} - v_{C_2} - R_5(i_{L_1} + i_{L_2}) + v_{R_4}$$

and, after elimination of $v_{R_4}$,

$$(1.14) \quad \frac{di_{L_2}}{dt} = -\frac{v_{C_2}}{L_2} - \frac{R}{L_2}i_{L_1} - \frac{(R_2 + R)}{L_2}i_{L_2} + \frac{R_4}{L_2(R_3 + R_4)}e_s$$

Thus the state equations are

$$(1.15) \quad \begin{bmatrix} \dfrac{dv_{C_1}}{dt} \\[2mm] \dfrac{dv_{C_2}}{dt} \\[2mm] \dfrac{di_{L_1}}{dt} \\[2mm] \dfrac{di_{L_2}}{dt} \end{bmatrix} = \begin{bmatrix} 0 & 0 & \dfrac{1}{C_1} & 0 \\[2mm] 0 & 0 & \dfrac{1}{C_2} & \dfrac{1}{C_2} \\[2mm] -\dfrac{1}{L_1} & -\dfrac{1}{L_1} & -\dfrac{R_1 + R}{L_1} & -\dfrac{R}{L_1} \\[2mm] 0 & -\dfrac{1}{L_2} & -\dfrac{R}{L_2} & -\dfrac{R_2 + R}{L_2} \end{bmatrix} \begin{bmatrix} v_{C_1} \\[2mm] v_{C_2} \\[2mm] i_{L_1} \\[2mm] i_{L_2} \end{bmatrix} + \begin{bmatrix} 0 \\[2mm] 0 \\[2mm] \dfrac{1}{L_1} \\[2mm] \dfrac{1}{L_2} \end{bmatrix} \dfrac{R_4}{R_4 + R_3}e_s$$

If $v_{R_4}$ is the output, we note that it can be expressed in terms of the state and the input $e_s$; indeed, from (1.11),

$$(1.16) \quad v_{R_4}(t) = -\frac{R_3 R_4}{R_3 + R_4}i_{L_1}(t) - \frac{R_3 R_4}{R_3 + R_4}i_{L_2}(t) + \frac{R_4}{R_3 + R_4}e_s(t) \qquad \text{for all } t$$

We can express (1.16) in the standard form of Eq. (1.7) as follows:

$$(1.17) \quad v_{R_4} = \begin{bmatrix} 0 & 0 & -\dfrac{R_3 R_4}{R_3 + R_4} & \dfrac{-R_3 R_4}{R_3 + R_4} \end{bmatrix} \begin{bmatrix} v_{C_1} \\[2mm] v_{C_2} \\[2mm] i_{L_1} \\[2mm] i_{L_2} \end{bmatrix} + \frac{R_4}{R_3 + R_4}e_s$$

**Remark**   There is nothing sacred in the choice of inductor currents and capacitor voltages as state variables.   We could just as well have taken *inductor fluxes* and *capacitor charges*.   In fact, for the time-varying case there are definite advantages in doing so.   The situation is somewhat analogous to that in particle mechanics where either positions and velocities or positions and momenta may be chosen as variables.

**Example**   Consider again the example of Fig. 1.1.   We note that

(1.18)   $\phi_1(t) = L_1 i_1(t)$      and      $\phi_2(t) = L_2 i_2(t)$

and

(1.19)   $q(t) = Cv(t)$

With these variables, namely $q$, $\phi_1$, and $\phi_2$, the cut-set equation becomes

$$\dot{q} = -\frac{1}{L_1}\phi_1 - \frac{1}{L_2}\phi_2$$

and the loop equations become

$$\dot{\phi}_1 = \frac{q}{C} - R_1\frac{\phi_1}{L_1} - e_s$$

$$\dot{\phi}_2 = \frac{q}{C} - R_2\frac{\phi_2}{L_2}$$

In matrix form,

(1.20)   $$\begin{bmatrix} \dot{q} \\ \dot{\phi}_1 \\ \dot{\phi}_2 \end{bmatrix} = \begin{bmatrix} 0 & -\dfrac{1}{L_1} & -\dfrac{1}{L_2} \\ \dfrac{1}{C} & -\dfrac{R_1}{L_1} & 0 \\ \dfrac{1}{C} & 0 & -\dfrac{R_2}{L_2} \end{bmatrix} \begin{bmatrix} q \\ \phi_1 \\ \phi_2 \end{bmatrix} + \begin{bmatrix} 0 \\ -1 \\ 0 \end{bmatrix} e_s$$

Looking at these state equations, we may say that the state at $t = 0$ is specified by the initial charge $q(0)$ and the initial fluxes $\phi_1(0)$ and $\phi_2(0)$. Indeed, with these three numbers and the waveform $e_s(\cdot)$, we can integrate the differential equations and obtain (uniquely) $q(t)$, $\phi_1(t)$, and $\phi_2(t)$ for any $t > 0$.

**Exercise**   Write the state equations for the linear time-invariant networks shown in Fig. 1.4

*a.*   Using currents and voltages as state variables

*b.*   Using charges and fluxes

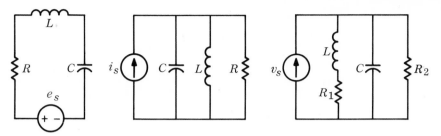

**Fig. 1.4**   Networks used to illustrate the concept of state.

<h2>2    The Concept of State</h2>

In the preceding examples, we have said "the initial state is *specified* by . . ."
or "the state is *given* by . . . ." Why didn't we say "the state is $q(0)$, $\phi_1(0)$,
and $\phi_2(0)$"? The reason is that strictly speaking the state is an abstract
concept that may be *represented* in many ways. As an analogy, think of
the number 2. It is also an abstract concept. When we think of it we may
picture the real line and the point in the middle of those with abscissa 1
and 3. When we wish to *represent* it we may write the numeral 2; if we
were Romans, we would write the symbol II. The computer, with its
binary arithmetic, insists on representing it by the symbol 10; sometimes
we may represent it by $1 + 1$, or by $(1.41421 . . .)^2$, etc. All of these *representations* evoke in our minds the concept of the number 2.

Now that we understand the idea that the state may be represented in
many ways, we shall henceforth write, for short, "the state is : . ." and
remember that more precisely we mean "the state is represented by. . . ."

Let us give a precise definition of the concept of *state*.

A set of data qualifies to be called the **state** of a network if it satisfies the
following two conditions:

1. *For any time, say $t_1$, the state at $t_1$ and the input waveforms (specified
   from time $t_1$ on) determine uniquely the state at any time $t > t_1$.*

2. *The state at time $t$ and the inputs at time $t$ (and sometimes some of their
   derivatives) determine uniquely the value at time $t$ of any network
   variable.*

We think of the state as a vector, and the components of the state are
called **state variables.**

In the case of linear time-invariant networks if the state equation can be
written in the form

$$\dot{\mathbf{x}}(t) = \mathbf{A}\mathbf{x}(t) + \mathbf{b}w(t)$$

then the vector **x** automatically satisfies property 1. Similarly, as soon as
any output $y$ can be written as

$$y(t) = \mathbf{c}^T\mathbf{x}(t) + d_0 w$$

property 2 is also automatically satisfied.

**Example 1**    Consider the parallel $RC$ circuit driven by a current source $i_s$ (see Fig. 2.1). If we use the voltage $v$ as a variable, we have

$$C\dot{v}(t) + Gv(t) = i_s(t) \qquad \text{for } t \geq 0$$

or

$$\dot{v}(t) = -\frac{G}{C}v(t) + \frac{1}{C}i_s(t) \qquad \text{for } t \geq 0$$

and

$$v(t) = v(0) + \int_0^t \frac{1}{C}\epsilon^{-(t-t')/RC}i_s(t')\,dt' \qquad \text{for } t \geq 0$$

The voltage $v(0)$ and the input $i_s(\cdot)$ uniquely specify $v(t)$ for any $t > 0$. Also, given $v(t)$, any network variable is specified; thus,

$$i_R(t) = Gv(t) \qquad q(t) = Cv(t)$$

Therefore, the voltage $v$ across the capacitor qualifies to be called the state of the $RC$ circuit.

**Exercise**    Verify that the charge $q$ of the capacitor also qualifies to be called the state of the $RC$ circuit.

**Example 2**    Consider the example of the previous section. Referring to Eq. (1.15), we see that the set of data $\{v_{C_1}(t), v_{C_2}(t), i_{L_1}(t), i_{L_2}(t)\}$ qualifies to be called the state at time $t$ of the network. Also note that if $q_1$ and $q_2$ denote the charges on the capacitors and $\phi_1$ and $\phi_2$ denote the fluxes in the inductors, then the set of data $\{q_1(t), q_2(t), \phi_1(t), \phi_2(t)\}$ also qualifies to be called the state at time $t$ of the network. Just as there are infinitely many ways of representing the number 2, so there are infinitely many ways of specifying the state of this network. For example,

$$\{q_1(t) + q_2(t), q_1(t) - q_2(t), \phi_1(t), \phi_1(t) + 2\phi_2(t)\}$$

also specifies the state of this network.

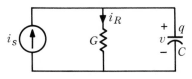

**Fig. 2.1**    Network used in Example 1.

It is worth noting that, under very general conditions, the state of any network† is specified by all the capacitor voltages (or charges) and all the inductor currents (or fluxes). By any network, we mean any interconnection of elements of the type described in Chaps. 2 and 8.

Sometimes some information in addition to the charges and fluxes is necessary to specify the state. For example, if the network includes switches, the switch position must be indicated. If there are inductors on magnetic cores that exhibit significant hysteresis (say, as in computer memories), then the condition of the magnetic core must be specified (see, for example, Chap. 2, Fig. 4.4).

The concept of state is a very basic and fundamental concept which is also found in the modern theories of control systems and sequential machines. In classical mechanics, the hamiltonian formulation of the equations of motion constitutes a way of writing the state equations for the dynamical system under consideration. The following fundamental idea is behind the concept of state in all these fields.

> *Given the state of the system at time $t_0$ and all the inputs (specified from time $t_0$ on), the behavior of the system is completely determined for all $t > t_0$.*

## 3    Nonlinear and Time-varying Networks

At present practically all the analyses of time-varying and nonlinear networks use the state equations. As we shall see for the linear time-varying case, the equations require little modification. For the nonlinear case, the state equations are the most convenient for computations.

In this section we shall show, with the aid of a few examples, how to write state equations for nonlinear time-varying networks.

### 3.1    Linear Time-varying Case

**Example 1**    Let us consider the network shown in Fig. 3.1. Note that all the elements (except the voltage source) are linear and time-varying. Their characteristics are given by

$$v_{R_1}(t) = R_1(t)j_{R_1}(t)$$
$$\phi_1(t) = L_1(t)i_1(t)$$
$$(3.1) \quad q(t) = C(t)v(t)$$
$$\phi_2(t) = L_2(t)i_2(t)$$
$$v_{R_2}(t) = R_2(t)j_{R_2}(t)$$

---

† In Chap. 13, we shall prove this fact for linear time-invariant networks.

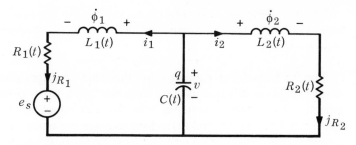

**Fig. 3.1**    Linear time-varying network used in Example 1.

To write the equations of the network of Fig. 3.1, we pick the same tree as before. We choose $q$, $\phi_1$, and $\phi_2$ as state variables and write a fundamental cut-set equation for the tree-branch capacitor $C(t)$ (remembering to express all quantities in terms of the chosen variables $q$, $\phi_1$, and $\phi_2$); the result is

(3.2)
$$\dot{q}(t) = -i_1(t) - i_2(t)$$
$$= -\frac{\phi_1(t)}{L_1(t)} - \frac{\phi_2(t)}{L_2(t)}$$

We then write fundamental loop equations for each inductor as follows:

(3.3)
$$\dot{\phi}_1(t) = v_C(t) - R_1(t)j_{R_1}(t) - e_s(t)$$
$$= \frac{q(t)}{C(t)} - \frac{R_1(t)}{L_1(t)}\phi_1(t) - e_s(t)$$

and

(3.4)
$$\dot{\phi}_2(t) = v_C(t) - R_2(t)j_{R_2}(t)$$
$$= \frac{q(t)}{C(t)} - \frac{R_2(t)}{L_2(t)}\phi_2(t)$$

Thus, in matrix form, we obtain

(3.5)
$$\begin{bmatrix} \dot{q}(t) \\ \dot{\phi}_1(t) \\ \dot{\phi}_2(t) \end{bmatrix} = \begin{bmatrix} 0 & -\dfrac{1}{L_1(t)} & -\dfrac{1}{L_2(t)} \\ \dfrac{1}{C(t)} & -\dfrac{R_1(t)}{L_1(t)} & 0 \\ \dfrac{1}{C(t)} & 0 & -\dfrac{R_2(t)}{L_2(t)} \end{bmatrix} \begin{bmatrix} q(t) \\ \phi_1(t) \\ \phi_2(t) \end{bmatrix} + \begin{bmatrix} 0 \\ -1 \\ 0 \end{bmatrix} e_s(t)$$

**Remark**    The only difference between the equations of (3.5) (which pertain to the linear *time-varying* case) and the equations of (1.20) is that the element values, $L_1$, $L_2$, $C$, $R_1$, and $R_2$ are now known functions of time rather than constants.   Otherwise, the equations are exactly the same.   This fact is true in general.   It is a consequence of our using the charges and fluxes as state variables.

For the linear time-varying case, there are two significant reasons for using charges and fluxes as state variables: (1) if we were to use inductor currents and capacitor voltages, then the derivatives $\dot{L}(t)$ and $\dot{C}(t)$ would appear in the analysis; and (2) as long as capacitor currents do not include impulses, the capacitor charge is a continuous function of time [i.e., the waveform $q(\cdot)$ versus $t$ does not have jumps].   Note that this is so even if $C(t)$ jumps from one constant value to another.   Since $q(\cdot)$ is continuous and since $q(t) = C(t)v(t)$ for all $t$, the waveform $v(\cdot)$ exhibits a jump whenever $C(\cdot)$ jumps.   Clearly, it is much simpler to solve for a $q(\cdot)$ that is continuous.

**Exercise**   Assume that all the $R$'s, $C$'s, and $L$'s of the linear networks shown in Fig. 1.4 are time-varying.   Write their state equations using charges and fluxes as state variables.

**Exercise**   For the same linear time-varying circuit as shown in Fig. 3.1 use capacitor voltage and inductor currents as the state variables.   Show that the state equation is of the form (here, for simplicity, we have dropped the explicit dependence on $t$, which should appear in each symbol)

$$\begin{bmatrix} \dfrac{dv}{dt} \\[2mm] \dfrac{di_1}{dt} \\[2mm] \dfrac{di_2}{dt} \end{bmatrix} = \begin{bmatrix} -\dfrac{\dot{C}}{C} & -\dfrac{1}{C} & -\dfrac{1}{C} \\[2mm] \dfrac{1}{L_1} & -\dfrac{R_1 + \dot{L}_1}{L_1} & 0 \\[2mm] \dfrac{1}{L_2} & 0 & -\dfrac{R_2 + \dot{L}_2}{L_2} \end{bmatrix} \begin{bmatrix} v \\[2mm] i_1 \\[2mm] i_2 \end{bmatrix} + \begin{bmatrix} 0 \\[2mm] -1 \\[2mm] 0 \end{bmatrix} e_s$$

Compare this matrix equation with that of (1.4).   [Hint: Since $L_1(t)$ and $C(t)$ depend on time $v_{L_1}(t) = \dfrac{d}{dt}[L_1(t)i_1(t)]$, etc.]

## 3.2   Nonlinear Case

We are going to write the state equations of two nonlinear networks.   For simplicity, we shall assume that these networks are time-invariant.   From Chap. 2, we recall that a nonlinear inductor is specified by its characteristic in the $\phi i_L$ plane (see Fig. 3.2a).   Since we wish to use $\phi$ as a variable, let us assume that the characteristic is described in terms of the function $f_L$ as

$$i_L = f_L(\phi)$$

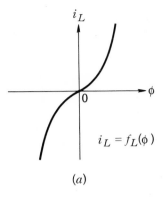

$i_L = f_L(\phi)$

(a)

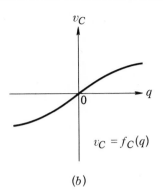

$v_C = f_C(q)$

(b)

**Fig. 3.2**   Typical characteristics of nonlinear elements.

Similarly, a nonlinear capacitor is specified by its characteristic in the $qv_C$ plane (see Fig. 3.2b). We assume that its characteristic is given by

$$v_C = f_C(q)$$

In the case of resistors, it is sometimes convenient to have the voltage expressed as a function of the current; at other times it is convenient to have the current expressed as a function of the voltage. It depends on the topology of the network.

**Example 1**   Let us assume that all elements of the network shown in Fig. 3.3 are nonlinear. The characterization of each element is given in the figure. We follow exactly the same procedure as before.

*Step 1*   Pick a tree which contains *all* the capacitors and *none* of the inductors. The tree is indicated in the figure.

*Step 2*   Choose fluxes and charges as state variables. In the present case, the chosen state variables are $q$, $\phi_1$, and $\phi_2$.

*Step 3*   Write a fundamental cut-set equation for each capacitor. In each cut-set equation, all branch currents must be expressed in terms of the chosen state variables. Thus,

$$\dot{q} = -i_1 - i_2$$
$$= -f_{L_1}(\phi_1) - f_{L_2}(\phi_2)$$

*Step 4*   Write a fundamental loop equation for each inductor. In each loop equation, all branch voltages must be expressed in terms of the chosen state variables. Therefore, we obtain

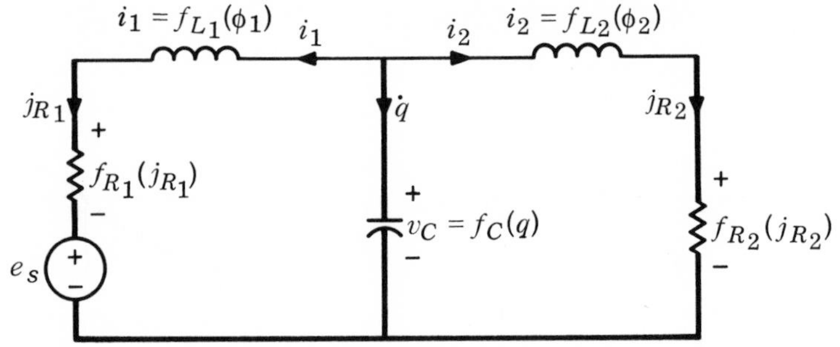

**Fig. 3.3**   In Example 1 we obtain the state equations of this nonlinear network; the tree branches are emphasized.

$$\dot{\phi}_1 = v_C - v_{R_1} - e_s$$
$$= f_C(q) - f_{R_1}(f_{L_1}(\phi_1)) - e_s$$

and

$$\dot{\phi}_2 = v_C - v_{R_2}$$
$$= f_C(q) - f_{R_2}(f_{L_2}(\phi_2))$$

Thus, if we exhibit the time dependence of the variables, the state equations have the form

$$\dot{q}(t) = -f_{L_1}(\phi_1(t)) - f_{L_2}(\phi_2(t))$$
(3.6)   $$\dot{\phi}_1(t) = f_C(q(t)) - f_{R_1}(f_{L_1}(\phi_1(t))) - e_s(t)$$
$$\dot{\phi}_2(t) = f_C(q(t)) - f_{R_2}(f_{L_2}(\phi_2(t)))$$

It is fundamental to observe that if we know the element characteristics [i.e., the functions $f_{L_1}(\cdot), f_{L_2}(\cdot), \ldots$], the state at time $t$ [i.e., the three numbers $q(t)$, $\phi_1(t)$, and $\phi_2(t)$], and the input at time $t$ [i.e., the voltage $e_s(t)$], then Eqs. (3.6) show us how to calculate the rates of change of the state variables, $\dot{q}(t)$, $\dot{\phi}_1(t)$, and $\dot{\phi}_2(t)$. It is precisely this calculation which is the key step in the numerical integration of the differential equations of (3.6).

   Let us develop another example in order to show that in some cases the input may appear in the argument of a branch characteristic.

**Example 2**   Consider the nonlinear network shown in Fig. 3.4.   The characteristics are shown in the figure.   We shall see later that it is convenient to assume that the characteristic of the first resistor is specified by

$$j_{R_1} = f_{R_1}(v_{R_1})$$

and the characteristic of the second is specified by

$$v_{R_2} = f_{R_2}(j_{R_2})$$

Steps 1 and 2 are obvious.   In fact, there is only one possible tree; the state variables are chosen to be $q_1$, $q_2$, and $\phi$.

*Step 3*   We write cut-set equations for each capacitor.   For the first capacitor,

$$\dot{q}_1 = -j_{R_1} = -f_{R_1}(v_{R_1})$$
$$= -f_{R_1}(v_{C_1} + v_{C_2} - e_s)$$
$$= -f_{R_1}[f_{C_1}(q_1) + f_{C_2}(q_2) - e_s]$$

For the second capacitor,

$$\dot{q}_2 = \dot{q}_1 - i_L$$
$$= -f_{R_1}[f_{C_1}(q_1) + f_{C_2}(q_2) - e_s] - f_L(\phi)$$

*Step 4*   We write the fundamental loop equations as follows:

$$\dot{\phi} = v_{C_2} - v_{R_2}$$
$$= f_{C_2}(q_2) - f_{R_2}(f_L(\phi))$$

Thus, the state equations have the form:

$$\dot{q}_1(t) = -f_{R_1}[f_{C_1}(q_1(t)) + f_{C_2}(q_2(t)) - e_s(t)]$$
$$\dot{q}_2(t) = -f_{R_1}[f_{C_1}(q_1(t)) + f_{C_2}(q_2(t)) - e_s(t)] - f_L(\phi(t))$$
$$\dot{\phi}(t) = f_{C_2}(q_2(t)) - f_{R_2}[f_L(\phi(t))]$$

The same fundamental remark as made in Example 1 applies. If we know the element characteristics [i.e., the functions $f_{R_1}(\cdot)$, $f_{L_1}(\cdot)$, $f_{C_1}(\cdot)$, ...], the state at time $t$ [i.e., the three numbers $q_1(t)$, $q_2(t)$, and $\phi(t)$], and the input at time $t$ [i.e., the voltage $e_s(t)$], then the equations show us how to calculate the rates of change of the state variables $\dot{q}_1(t)$, $\dot{q}_2(t)$, and $\dot{\phi}(t)$. It is customary to visualize the state as a point in the three-dimensional space with coordinates $q_1$, $q_2$, and $\phi$. This space is called the **state space**. Thus, the state at time $t$ and the input at time $t$ specify the **velocity of the state** at time $t$ in the state space. The curve traced by the state as it moves in the state space is called the **state trajectory**. For two-dimensional state spaces, the state trajectory is easily exhibited (see, for example, Fig. 6.2 of Chap. 4).

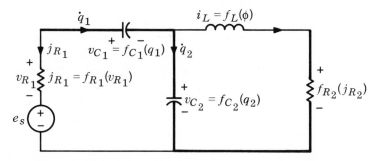

**Fig. 3.4**   In Example 2 we obtain the state equations of this nonlinear network; the tree branches are emphasized.

| **4** | **State Equations for Linear Time-invariant Networks** |

In this section we develop a systematic method for writing the state equations of linear time-invariant networks. For simplicity we shall restrict ourselves to the case in which there are no capacitor-only loops and no inductor-only cut sets. For the general case, the reader is referred to the literature.[†]

Consider a linear time-invariant network whose graph is connected. The elements are resistors, inductors, capacitors, and independent sources. As stated, we assume that capacitors do not form a loop and inductors do not constitute a cut set. The first step in the analysis is to select a tree; however, for convenience we will pick what is called a *proper tree*. A tree is called a **proper tree** if it contains all the capacitors in the network, but no inductors. For networks without capacitor-only loops and inductor-only cut sets, a proper tree can always be found. Consider the capacitors; since they do not form any loop, we can obviously include all capacitors as tree branches. Since the inductors do not form any cut set, starting from any node we can reach any other node in the graph without going through an inductor; hence we can obviously assign all inductors to the links. Once we have assigned all the capacitors to the tree, it is usually necessary to add some resistors in order to complete the tree. We shall use an example to illustrate the procedure. A possible choice for a proper tree is shown in Fig. 4.1. It includes the two capacitors (with capacitances $C_a$ and $C_b$) and four resistors (with conductances $G_a$, $G_b$, $G_d$, and $G_e$). The three links are the inductors $L_a$ and $L_b$ and the series connection of $R_c$ and $v_0$ (counted as a single branch).

In general, it is convenient to partition the branches into four subsets, namely the resistive links, the inductive links, the capacitive tree branches, and the resistive tree branches. The KVL equations for the fundamental loops are $\mathbf{Bv} = \mathbf{0}$, or

$$(4.1) \quad \left[\mathbf{1}_l \mid \mathbf{F}\right]\begin{bmatrix} \mathbf{v}_R \\ \mathbf{v}_L \\ \hline \mathbf{v}_C \\ \mathbf{v}_G \end{bmatrix} = \mathbf{0}$$

where $\mathbf{v}_R$, $\mathbf{v}_L$, $\mathbf{v}_C$, and $\mathbf{v}_G$ are subvectors representing voltages for the resistive links, inductive links, capacitive tree branches, and resistive tree branches, respectively. The matrix $\mathbf{B} = [\mathbf{1}_l \mid \mathbf{F}]$ is the fundamental loop matrix corresponding to the tree shown in Fig. 4.1. For the present problem we number the branches according to the partition of Eq. (4.1), i.e., resistive links first, inductive links next, then capacitive tree branches and

---

[†] E. S. Kuh and R. A. Rohrer, State Variable Approach to Network Analysis, *Proc. IEEE,* **53**:672–686 (1965) (bibliography included).

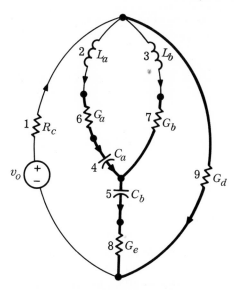

**Fig. 4.1**   The emphasized branches constitute a proper tree for the network.

resistive tree branches, as shown in Fig. 4.1 (including the orientations). Equation (4.1) becomes

$$
\begin{bmatrix}
1 & 0 & 0 & \vdots & 0 & 0 & 0 & 0 & 0 & 1 \\
0 & 1 & 0 & \vdots & 1 & 1 & 1 & 0 & 1 & -1 \\
0 & 0 & 1 & \vdots & 0 & 1 & 0 & 1 & 1 & -1
\end{bmatrix}
\begin{bmatrix}
\mathbf{v}_R \\
\hline
\mathbf{v}_L \\
\hline
\mathbf{v}_C \\
\mathbf{v}_G
\end{bmatrix}
= \mathbf{0}
$$

where

$$
\mathbf{v}_R = v_{R_c} - v_0 \qquad \mathbf{v}_L = \begin{bmatrix} v_{L_a} \\ v_{L_b} \end{bmatrix}
$$

$$
\mathbf{v}_C = \begin{bmatrix} v_{C_a} \\ v_{C_b} \end{bmatrix} \qquad \mathbf{v}_G = \begin{bmatrix} v_{R_a} \\ v_{R_b} \\ v_{R_e} \\ v_{R_d} \end{bmatrix}
$$

The KCL equations for the fundamental cut sets are $\mathbf{Qj} = \mathbf{0}$, or

$$
(4.2) \qquad [-\mathbf{F}^T, \mathbf{1}_n] \begin{bmatrix} \mathbf{j}_R \\ \mathbf{j}_L \\ \hline \mathbf{j}_C \\ \mathbf{j}_G \end{bmatrix} = \mathbf{0}
$$

where $j_R$, $j_L$, $j_C$, and $j_G$ are subvectors representing currents for the resistive links, inductive links, capacitive tree branches, and resistive tree branches, respectively. For the present example, we have

$$\mathbf{j}_R = j_{Rc} \qquad \mathbf{j}_L = \begin{bmatrix} j_{L_a} \\ j_{L_b} \end{bmatrix}$$

$$\mathbf{j}_C = \begin{bmatrix} j_{C_a} \\ j_{C_b} \end{bmatrix} \qquad \mathbf{j}_G = \begin{bmatrix} j_{R_a} \\ j_{R_b} \\ j_{R_e} \\ j_{R_d} \end{bmatrix}$$

Next we must introduce the branch equations. For convenience we assume that the independent sources located in links are voltage sources, and the independent sources located in tree branches are current sources. This certainly imposes no restriction on the method since it is easy to transform independent voltage sources into current sources, and vice versa. We write them in the following form with the new notations defined below:

(4.3a)  $\mathbf{v}_R = \mathbf{R}_R \mathbf{j}_R + \mathbf{e}_R$

(4.3b)  $\mathbf{v}_L = \mathbf{L} \dfrac{d}{dt} \mathbf{j}_L + \mathbf{e}_L$

(4.3c)  $\mathbf{j}_C = \mathbf{C} \dfrac{d}{dt} \mathbf{v}_C + \mathbf{i}_C$

(4.3d)  $\mathbf{j}_G = \mathbf{G}_G \mathbf{v}_G + \mathbf{i}_G$

The matrices $\mathbf{R}_R$, $\mathbf{L}$, $\mathbf{C}$, and $\mathbf{G}_G$ are all branch parameter matrices that denote, respectively, the link resistance matrix, link inductance matrix, tree-branch capacitance matrix, and tree-branch conductance matrix of the network. The vectors $\mathbf{e}_R$, $\mathbf{e}_L$, $\mathbf{i}_C$, and $\mathbf{i}_G$ represent the independent sources. The reference directions used in writing Eqs. (4.3) are indicated in Fig. 4.2. For the example in Fig. 4.1

$$\mathbf{R}_R = R_c \qquad \mathbf{L} = \begin{bmatrix} L_a & 0 \\ 0 & L_b \end{bmatrix}$$

$$\mathbf{C} = \begin{bmatrix} C_a & 0 \\ 0 & C_b \end{bmatrix} \qquad \mathbf{G}_G = \begin{bmatrix} G_a & & & 0 \\ & G_b & & \\ & & G_e & \\ 0 & & & G_d \end{bmatrix}$$

$$\mathbf{e}_R = -v_0 \qquad \mathbf{e}_L = \mathbf{i}_C = \mathbf{i}_G = 0$$

Clearly, the next problem is to combine the three sets of equations, i.e., the KVL equations in (4.1), the KCL equations in (4.2), and the branch equations in (4.3). We must eliminate all variables which are neither state variables nor sources.

Equations (4.1) and (4.2) can be rewritten as

$$\text{(4.4a)} \quad \begin{bmatrix} \mathbf{v}_R \\ \mathbf{v}_L \end{bmatrix} = -\mathbf{F} \begin{bmatrix} \mathbf{v}_C \\ \mathbf{v}_G \end{bmatrix} = - \begin{bmatrix} \mathbf{F}_{RC} & \mathbf{F}_{RG} \\ \mathbf{F}_{LC} & \mathbf{F}_{LG} \end{bmatrix} \begin{bmatrix} \mathbf{v}_C \\ \mathbf{v}_G \end{bmatrix}$$

$$\text{(4.4b)} \quad \begin{bmatrix} \mathbf{j}_C \\ \mathbf{j}_G \end{bmatrix} = \mathbf{F}^T \begin{bmatrix} \mathbf{j}_R \\ \mathbf{j}_L \end{bmatrix} = \begin{bmatrix} \mathbf{F}_{RC}^T & \mathbf{F}_{LC}^T \\ \mathbf{F}_{RG}^T & \mathbf{F}_{LG}^T \end{bmatrix} \begin{bmatrix} \mathbf{j}_R \\ \mathbf{j}_L \end{bmatrix}$$

where the matrix $\mathbf{F}$ has been partitioned into submatrices for convenience. Combining Eqs. (4.3) and (4.4), we obtain

$$\text{(4.5a)} \quad \mathbf{R}_R \mathbf{j}_R = -\mathbf{F}_{RC} \mathbf{v}_C - \mathbf{F}_{RG} \mathbf{v}_G - \mathbf{e}_R$$

$$\text{(4.5b)} \quad \mathbf{L} \frac{d}{dt} \mathbf{j}_L = -\mathbf{F}_{LC} \mathbf{v}_C - \mathbf{F}_{LG} \mathbf{v}_G - \mathbf{e}_L$$

$$\text{(4.5c)} \quad \mathbf{C} \frac{d}{dt} \mathbf{v}_C = \mathbf{F}_{RC}^T \mathbf{j}_R + \mathbf{F}_{LC}^T \mathbf{j}_L - \mathbf{i}_C$$

$$\text{(4.5d)} \quad \mathbf{G}_G \mathbf{v}_G = \mathbf{F}_{RG}^T \mathbf{j}_R + \mathbf{F}_{LG}^T \mathbf{j}_L - \mathbf{i}_G$$

Note that in Eq. (4.5) the only variables which are neither state variables nor sources are $\mathbf{j}_R$ and $\mathbf{v}_G$. They can be eliminated to obtain the state representation as follows (the proof is straightforward and is omitted):

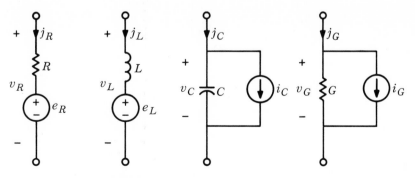

| Link resistor | Inductor | Capacitor | Tree-branch resistor |

**Fig. 4.2**    Typical branches with independent sources.

$$(4.6) \quad \frac{d}{dt}\begin{bmatrix} \mathbf{v}_C \\ \mathbf{j}_L \end{bmatrix} = \begin{bmatrix} \mathbf{C} & \mathbf{0} \\ \mathbf{0} & \mathbf{L} \end{bmatrix}^{-1} \begin{bmatrix} -\mathbf{y} & \mathbf{3C} \\ -\mathbf{3C}^T & -\mathbf{z} \end{bmatrix} \begin{bmatrix} \mathbf{v}_C \\ \mathbf{j}_L \end{bmatrix} - \begin{bmatrix} \mathbf{C} & \mathbf{0} \\ \mathbf{0} & \mathbf{L} \end{bmatrix}^{-1} \mathcal{B} \begin{bmatrix} \mathbf{i}_C \\ \mathbf{i}_G \\ \mathbf{e}_R \\ \mathbf{e}_L \end{bmatrix}$$

The terms are defined as follows:

$$(4.7a) \quad \mathbf{y} \triangleq \mathbf{F}_{RC}{}^T \mathfrak{R}^{-1} \mathbf{F}_{RC}$$

$$(4.7b) \quad \mathbf{z} \triangleq \mathbf{F}_{LG} \mathbf{G}^{-1} \mathbf{F}_{LG}{}^T$$

$$(4.7c) \quad \mathbf{3C} \triangleq \mathbf{F}_{LC}{}^T - \mathbf{F}_{RC}{}^T \mathfrak{R}^{-1} \mathbf{F}_{RG} \mathbf{R}_G \mathbf{F}_{LG}{}^T$$

$$(4.7d) \quad \mathfrak{R} \triangleq \mathbf{R}_R + \mathbf{F}_{RG} \mathbf{R}_G \mathbf{F}_{RG}{}^T \qquad \mathbf{R}_G = \mathbf{G}_G{}^{-1}$$

$$(4.7e) \quad \mathbf{G} \triangleq \mathbf{G}_G + \mathbf{F}_{RG}{}^T \mathbf{G}_R \mathbf{F}_{RG} \qquad \mathbf{G}_R = \mathbf{R}_R{}^{-1}$$

$$(4.7f) \quad \mathcal{B} \triangleq \begin{bmatrix} 1 & -\mathbf{F}_{RC}{}^T \mathfrak{R}^{-1} \mathbf{F}_{RG} \mathbf{R}_G & \mathbf{F}_{RC}{}^T \mathfrak{R}^{-1} & 0 \\ 0 & -\mathbf{F}_{LG} \mathbf{G}^{-1} & -\mathbf{F}_{LG} \mathbf{G}^{-1} \mathbf{F}_{RG}{}^T \mathbf{G}_R & 1 \end{bmatrix}$$

In our example,

$$\mathbf{F} = \begin{bmatrix} \mathbf{F}_{RC} & \mathbf{F}_{RG} \\ \mathbf{F}_{LC} & \mathbf{F}_{LG} \end{bmatrix} = \begin{bmatrix} 0 & 0 & 0 & 0 & 0 & 1 \\ 1 & 1 & 1 & 0 & 1 & -1 \\ 0 & 1 & 0 & 1 & 1 & -1 \end{bmatrix}$$

$$\mathfrak{R} = R_c + \begin{bmatrix} 0 & 0 & 0 & 1 \end{bmatrix} \begin{bmatrix} G_a & & & 0 \\ & G_b & & \\ & & G_e & \\ 0 & & & G_d \end{bmatrix}^{-1} \begin{bmatrix} 0 \\ 0 \\ 0 \\ 1 \end{bmatrix} = R_c + R_d$$

$$\mathbf{G} = \begin{bmatrix} G_a & & & 0 \\ & G_b & & \\ & & G_e & \\ 0 & & & G_d \end{bmatrix} + \begin{bmatrix} 0 \\ 0 \\ 0 \\ 1 \end{bmatrix} \frac{1}{R_c} \begin{bmatrix} 0 & 0 & 0 & 1 \end{bmatrix}$$

$$= \begin{bmatrix} G_a & & & 0 \\ & G_b & & \\ & & G_e & \\ 0 & & & G_d + G_c \end{bmatrix}$$

$$\mathbf{y} = 0$$

$$\mathbf{z} = \begin{bmatrix} 1 & 0 & 1 & -1 \\ 0 & 1 & 1 & -1 \end{bmatrix} \begin{bmatrix} G_a & & & 0 \\ & G_b & & \\ & & G_e & \\ & & & G_d + G_c \end{bmatrix}^{-1} \begin{bmatrix} 1 & 0 \\ 0 & 1 \\ 1 & 1 \\ -1 & -1 \end{bmatrix}$$

$$= \begin{bmatrix} R_a + R_e + \dfrac{R_cR_d}{R_c + R_d} & R_e + \dfrac{R_cR_d}{R_c + R_d} \\[4mm] R_e + \dfrac{R_cR_d}{R_c + R_d} & R_b + R_e + \dfrac{R_cR_d}{R_c + R_d} \end{bmatrix}$$

$$\mathcal{H} = \begin{bmatrix} 1 & 0 \\ 1 & 1 \end{bmatrix}$$

Since

$$\mathbf{e}_L = \mathbf{i}_C = \mathbf{i}_G = \mathbf{0} \qquad \mathbf{e}_R = -v_0 \qquad \mathbf{F}_{RC} = \mathbf{0}$$

for the term containing the input we only need to compute $-\mathbf{F}_{LG}{}^{-1}\mathbf{F}_{RG}{}^T\mathbf{G}_1\mathbf{e}_R$, which leads to

$$-\mathcal{B} \begin{bmatrix} \mathbf{i}_C \\ \mathbf{i}_G \\ \mathbf{e}_R \\ \mathbf{e}_L \end{bmatrix} = \begin{bmatrix} 0 \\ 0 \\ \dfrac{R_d}{R_c + R_d} \\ \dfrac{R_d}{R_c + R_d} \end{bmatrix} v_0$$

The result checks with the result obtained previously.

The writing of state equations for networks containing coupling elements can sometimes be done as in this section. For example, with coupled inductors the inductance matrix becomes symmetric and nondiagonal. For networks containing dependent sources and ideal transformers the link resistance matrix and the tree-branch conductance matrix become nondiagonal. For these cases the derivation of the closed-form representation becomes rather complicated. On the other hand, in most practical instances the intuitive approach which we presented first works fine.

## Summary

- A set of data qualifies to be called the **state of a network** if it satisfies two conditions:

1. For any time $t_1$, the state at time $t_1$ and the inputs from $t_1$ on determine uniquely the state at any time $t > t_1$.

2. For any time $t$, the state at time $t$ and the inputs at time $t$ (and sometimes some of their derivatives) determine uniquely every network variable at time $t$.

The components of the state are called **state variables.**

■ For linear time-invariant networks, the state equation is usually of the form (for a single input $w$)

$$\dot{\mathbf{x}} = \mathbf{A}\mathbf{x} + \mathbf{b}w$$

and the output equation is of the form (for a single output $y$)

$$y = \mathbf{c}^T\mathbf{x} + d_0 w$$

In some cases, derivatives of the input may appear in the state equation and/or the output equation.

■ Typically, the state variables are inductor currents and capacitor voltages. For time-varying and nonlinear networks it is often preferable to use the inductor fluxes and capacitor charges as state variables.

■ A systematic method that works in a large number of cases for $RLC$ networks is as follows: (1) choose a tree that includes *all* the capacitors and *none* of the inductors, (2) choose fluxes and charges as state variables, (3) write a fundamental cut-set equation for each capacitor (in each equation express all branch currents in terms of the chosen state variables), and (4) write a fundamental loop equation for each inductor (in each equation express all branch voltages in terms of the chosen state variables). The state equations have the form

$$\dot{\mathbf{x}}(t) = \mathbf{f}(\mathbf{x}(t),\mathbf{w}(t),t)$$

and any network variable $y$ can be expressed as

$$y(t) = g(\mathbf{x}(t),w(t),\dot{w}(t), \ldots ,t)$$

## Problems

Time-invariant networks

**1.** Write the state equations for the following linear time-invariant networks:

*a.* The network shown in Fig. P10.3
*b.* The network shown in Fig. P10.7
*c.* The network shown in Fig. P10.8
*d.* The network shown in Fig. P10.11
*e.* The network shown in Fig. P10.19
*f.* The network shown in Fig. P10.20
*g.* The network shown in Fig. P10.21

Time-varying networks

**2.** Assume that all the linear elements of the networks listed below are time-varying. Write the state equations of *a.*   The network shown in Fig. P10.3

*b.*   The network shown in Fig. P10.7

*c.*   The network shown in Fig. P10.8

*d.*   The network shown in Fig. P10.11

*e.*   The network shown in Fig. P10.19

*f.*   The network shown in Fig. P10.20

*g.*   The network shown in Fig. P10.21

Linear and nonlinear networks

**3.** Write the state equations of the network shown in Fig. P12.3 for the following situations:

*a.*   All non-source elements are linear and time-invariant (use $C_1$, $C_2$, $C_3$, $L_4$, and $L_5$).

*b.*   All non-source elements (except the resistors $R_1$, $R_3$) are linear and time-varying [these elements have characteristics specified by the functions of time $C_1(\cdot)$, $C_2(\cdot)$, $C_3(\cdot)$, $L_4(\cdot)$, and $L_5(\cdot)$].

*c.*   All non-source elements (except the resistors $R_1$, $R_3$) are nonlinear and time-invariant [the element characteristics are $v_1 = f_1(q_1)$, $v_2 = f_2(q_2)$, $v_3 = f_3(q_3)$, $i_4 = f_4(\phi_4)$, and $i_5 = f_5(\phi_5)$].

*d.*   All non-source elements (except the resistors $R_1$, $R_3$) are nonlinear and time-varying, and $v_1 = f(q_1,t), \ldots, i_5 = f_5(\phi_5,t)$.

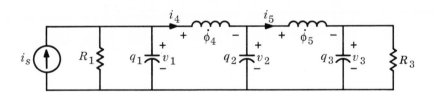

**Fig. P12.3**

Linear and nonlinear networks

**4.** Consider the two versions (linear and nonlinear) of the time-invariant circuit shown in Fig. P12.4. For each version write a set of state equations using $\phi_1$, $\phi_2$, $\phi_3$, $q_4$, and $q_5$ as state variables.

| Linear version | Nonlinear version |
|---|---|
| $L_1 = 1$ henry | $j_1 = f_1(\phi_1)$ |
| $L_2 = 2$ henrys | $j_2 = f_2(\phi_2)$ |
| $L_3 = 5$ henrys | $j_3 = f_3(\phi_3)$ |
| $C_4 = 1$ farad | $v_4 = f_4(q_4)$ |
| $C_5 = 3$ farads | $v_5 = f_5(q_5)$ |
| $R_6 = 2$ ohms | $R_6 = 2$ ohms |
| $R_7 = 3$ ohms | $R_7 = 3$ ohms |

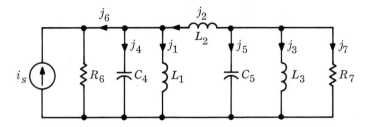

**Fig. P12.4**

The initial conditions are $j_1(0) = 2$ amp, $j_2(0) = 0$, $j_3(0) = 0$, $v_4(0) = 4$ volts, $v_5(0) = 6$ volts, and the source is $i_s = 5 \cos t$ amp.

**Nonlinear networks, proper tree**

**5.** Obtain for each of the nonlinear time-invariant networks shown in Fig. P12.5 a set of state equations. Proceed in a systematic way; i.e., pick a proper tree, write fundamental cut-set and fundamental loop equations, perform whatever algebra is necessary, and finally indicate the initial state.

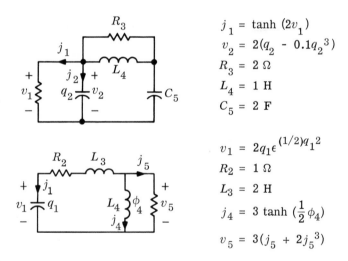

$$j_1 = \tanh (2v_1)$$
$$v_2 = 2(q_2 - 0.1q_2{}^3)$$
$$R_3 = 2 \ \Omega$$
$$L_4 = 1 \ \text{H}$$
$$C_5 = 2 \ \text{F}$$

$$v_1 = 2q_1\epsilon^{(1/2)q_1{}^2}$$
$$R_2 = 1 \ \Omega$$
$$L_3 = 2 \ \text{H}$$
$$j_4 = 3 \tanh \left(\tfrac{1}{2}\phi_4\right)$$
$$v_5 = 3(j_5 + 2j_5{}^3)$$

**Fig. P12.5**

**Networks with mutual inductances**

**6.** Write the state equations for

*a.* The network shown in Fig. P10.5

*b.* The network shown in Fig. P10.13

**Differential equations for nonlinear network**

**7.** The nonlinear elements of the time-invariant network shown in Fig. P12.7 are specified by the following relations: $i_2 = f(v_2)$, and $q_1 = \phi(v_1)$. Knowing that the remaining components are linear, write two differential equations with $v_1$ and $v_2$ as variables. [Hints: (1) Express the current in the inductor in terms of $v_2$. (2) Pick the nonlinear elements as branches of a tree. (3) Write KVL for the fundamental loop associated with the link consisting of $L$. (4) Write KCL for the fundamental cut set associated with the tree branch corresponding to the capacitor.]

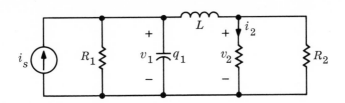

**Fig. P12.7**

# 13

# Laplace Transforms

The purpose of this chapter is to develop just enough of the method of the Laplace transform to enable us to use it in the study of linear time-invariant networks. The Laplace transform (with its close relative, the Fourier transform) is a fundamental tool for studying *linear time-invariant systems,* for example, in electromechanical systems (microphones, loudspeakers, electromagnetic transducers, etc.) or in communication systems, where we study the properties of the interconnection of networks, antennas, and propagation media. The Laplace transform is a vast subject since so many engineering problems depend on its method and properties. We shall only develop those properties which serve our present purposes; however, the student will be exposed in other courses to other aspects of the Laplace transform, and will have an opportunity to obtain a comprehensive understanding of the subject.

It should be stressed that the Laplace transform is a very important and very effective tool for studying *linear time-invariant* networks, but that it is almost useless for time-varying and/or nonlinear networks. With the increasing importance of nonlinear and time-varying networks the Laplace transform does not have the overwhelming importance it used to have, say, 10 years ago. In those days, many people had a hard time distinguishing between circuit theory and applications of the Laplace transform.

We devote the first two sections of this chapter to a concise presentation of the properties of the Laplace transform which are relevant to our purposes. In Sec. 3 we show how the Laplace transform is used to solve a single linear differential equation with constant coefficients, and we give special attention to the circuit-theoretic conclusions implied by the analysis. In Sec. 4 we develop important techniques for solving systems of *linear* differential equations with *constant* coefficients; then, in Sec. 5 applying the results of this analysis, we prove five general properties of linear time-invariant networks. We alluded to some of these properties in our study of simple circuits. These five general properties are extremely important for electrical engineers, and we must understand them thoroughly. In Sec. 6 we apply the Laplace transform to state equations. In Sec. 7 we illustrate how sloppy modeling of physical systems can produce puzzling degeneracies, and in Sec. 8 we bring out the important property of *uniqueness* of solutions for *linear time-invariant networks* composed of *passive* resistors, capacitors, and inductors.

## 1 | Definition of the Laplace Transform

Let us introduce the concept of a transform and justify its use with a very familiar example.  In the old days, when not all engineering offices had a desk and/or electronic computer, if an engineer had to multiply two numbers, say $a = 172,395$ and $b = 896,432$, he might have used logarithms; from

$$\log ab = \log a + \log b$$

he would get the logarithm of the product, and from the fact that there is a *unique* number which has log $ab$ as a logarithm, he would find the product $ab$.  The virtue of the method is that adding numbers is far easier than multiplying them.  Similarly, the Laplace transform reduces the solution of linear *differential* equations to the solution of linear *algebraic* equations.  Instead of relating a positive number $a$ to another real number log $a$, the Laplace transform associates to a time function defined on the interval $[0,\infty)$ another function which is defined in the complex frequency plane, the $s$ plane.

The importance of the Laplace transform arises also from other facts. (1) Laplace transform theory uses the concept of network function.  Since network functions can be determined experimentally (in fact, with great precision) by sinusoidal steady-state measurements, Laplace transforms help us think problems in terms of network functions, which are often more convenient to deal with than, say, impulse responses.  (2) The Laplace transform exhibits the close relation that exists between the time-domain behavior of a network (say, as waveforms on the scope) and its sinusoidal steady-state behavior.

As we said earlier, the key idea of the Laplace transform is that to a time function $f$ defined on $[0,\infty)$ it associates a function $F$ of the complex frequency $s$.  The transform is constructed as follows: $f(t)$ is multiplied by the factor $\epsilon^{-st}$, and the resulting function of $t$, $f(t)\epsilon^{-st}$, is integrated between $0-$ and $\infty$, giving

$$\int_{0-}^{\infty} f(t)\epsilon^{-st}\, dt$$

Since the integral is definite (the limits $0-$ and $\infty$ are fixed), it does not depend on $t$, but only on the parameter $s$; hence it is a function of the complex frequency $s$.  Call $F(s)$ the function defined by this integral; that is,

$$(1.1) \qquad \boxed{F(s) \triangleq \int_{0-}^{\infty} f(t)\epsilon^{-st}\, dt}$$

The integral on the right is called the **defining integral.**  We say that $F(s)$ is the **Laplace transform** of $f(t)$.  The variable $s$ is called the **complex fre-**

**quency.** To distinguish $F$ from $f$, we often refer to $f$ as the "time function" and to $F$ as the "Laplace transform." We also write Eq. (1.1) in the form

$$F(s) = \mathcal{L}[f(t)]$$

where $\mathcal{L}$ is read "Laplace transform of."

**Remarks** 1. Note that we use $0-$ as the lower limit of integration. We do this to emphasize the fact that if the time function $f$ includes an impulse at $t = 0$, the defining integral includes the impulse.

2. The integral (1.1) must be interpreted as a result of two limiting operations:

$$\int_{0-}^{\infty} f(t)\epsilon^{-st}\, dt = \lim_{\substack{\epsilon \to 0 \\ \epsilon > 0}} \lim_{T \to \infty} \int_{-\epsilon}^{T} f(t)\epsilon^{-st}\, dt$$

The limiting process $\epsilon \to 0$ usually does not cause any trouble because $f$ is either identically zero for $t < 0$ or otherwise well behaved. The limiting process $T \to \infty$ may cause trouble; as the interval of integration becomes infinitely large, the net area under the curve $f(t)\epsilon^{-st}$ may tend to infinity or tend to no particular value. The standard remedy is to restrict $s$ to having a sufficiently large positive real part so that the weighting factor $\epsilon^{-st}$ tends to zero sufficiently fast to make the integral, i.e., the area, finite.

3. We shall use throughout capital letters to denote Laplace transforms. Thus, the Laplace transforms of $v(t)$ and $j(t)$ are $V(s)$ and $J(s)$, respectively. Furthermore, since we need to deal with time function and its Laplace transform simultaneously, the functional notation for the Laplace transform, such as $V(\cdot)$, is not used because we want to know what is the independent variable of a function. The notation $(\cdot)$ does not give us that information. Thus, even for a time function the functional notation, say $v(\cdot)$, is not enforced; whenever we wish to emphasize the independent variable $t$ of a time function, we write the voltage function $v(t)$ as against its Laplace transform $V(s)$.

---

**Example 1**   Let $f(t) = u(t)$, the unit step; the defining integral then is

(1.2)   $$\int_{0-}^{\infty} u(t)\epsilon^{-st}\, dt = \int_{0-}^{\infty} \epsilon^{-st}\, dt$$

Suppose we take $s$ to be a positive real number, say $\sigma > 0$. Then the integrand of (1.2) is $\epsilon^{-\sigma t}$, a *decaying* exponential. Clearly, the area under a decaying exponential is finite. In fact,

$$\int_{0-}^{\infty} \epsilon^{-\sigma t}\, dt = \frac{\epsilon^{-\sigma t}}{-\sigma} \Big|_{0-}^{\infty} = \frac{1}{\sigma}$$

where we used the fact that (with $\sigma > 0$) $\epsilon^{-\sigma t} \to 0$ as $t \to \infty$. If $s$ were a negative real number, say $-a$ (with $a > 0$), the integrand would be $\epsilon^{at}$, an *increasing* exponential. Clearly, the area under such a curve is infinite. In other words, when $s$ is real and negative, the integral (1.2) "blows up"; i.e., the defining integral (1.2) is infinite.

If the complex frequency $s$ is of the form $s = \sigma + j\omega$ (where $\sigma$ and $\omega$ are, respectively, the real and imaginary parts of $s$) with $\sigma > 0$, the integral (1.2) may be written

$$\int_{0-}^{\infty} \epsilon^{-(\sigma+j\omega)t} \, dt = \int_{0-}^{\infty} \epsilon^{-\sigma t} \cos \omega t \, dt - j \int_{0-}^{\infty} \epsilon^{-\sigma t} \sin \omega t \, dt$$

Since $\sigma > 0$, in each integral the integrand is an exponentially damped sinusoid; hence each integral is a well-defined number. For calculation purposes it is better to use the exponential form as follows:

$$\int_{0-}^{\infty} \epsilon^{-(\sigma+j\omega)t} \, dt = \frac{\epsilon^{-(\sigma+j\omega)t}}{-(\sigma + j\omega)} \bigg|_{0-}^{\infty} = \frac{1}{\sigma + j\omega} \qquad \text{for } \sigma > 0$$

Therefore, we have shown that

(1.3)    $\mathcal{L}[u(t)] = \dfrac{1}{s} \qquad$ for Re $(s) > 0$

Strictly speaking, the Laplace transform of $u(t)$ is defined only for Re $(s) > 0$. However, we consider the Laplace transform $1/s$ to be a well-defined function for all values of $s$ except $s = 0$ (indeed, when $s = 0$, the denominator is zero, and the expression $1/s$ has no meaning). This concept of enlarging the domain of the function is usually referred to as analytic continuation.

---

**Example 2**     Let $a$ be any real or complex number. By definition, the Laplace transform of $\epsilon^{at}$ is given by

$$\mathcal{L}[\epsilon^{at}] = \int_{0-}^{\infty} \epsilon^{at} \epsilon^{-st} \, dt = \int_{0-}^{\infty} \epsilon^{-(s-a)t} \, dt$$

$$= \frac{\epsilon^{(s-a)t}}{-(s - a)} \bigg|_{0-}^{\infty}$$

Thus,

(1.4)    $\mathcal{L}[\epsilon^{at}] = \dfrac{1}{s - a} \qquad$ for Re $(s - a) > 0$

Again, we shall regard the Laplace transform of $\epsilon^{at}$, that is, the function $1/(s - a)$, as well defined for all $s$ except $s = a$.

---

**Remarks**     1.   The two Laplace transforms calculated above (Eqs. 1.3 and 1.4) are simple examples of Laplace transforms that are rational functions in $s$. We call a **rational function** in $s$ any ratio of polynomials in $s$. In these two examples the numerator polynomial is the constant polynomial with value 1. In most of our applications the Laplace transforms will turn out to be rational functions. We shall discuss rational functions further in Sec. 3 and Chap. 15.

2.   From a strictly logical point of view, the defining integral in Eq. (1.4) defines the Laplace transform $1/(s - a)$ only for $\text{Re}(s) > \text{Re}(a)$. On the other hand, the expression $1/(s - a)$ defines a function of the complex frequency $s$ for all values of $s$ (except $s = a$, where the denominator is zero). Therefore, it is intuitively reasonable to consider the Laplace transform $1/(s - a)$ as defined for all values of $s$ except $s = a$. This process of extension can be rigorously justified by the technique of analytic continuation.

Let us give an analogous example from calculus. Consider the power series $1 + x + x^2 + x^3 + \cdots + x^n + \cdots$. This is a geometric series which converges for $|x| < 1$ and whose sum is $1/(1 - x)$. Strictly, the series specifies a function only on the interval $(-1, 1)$. However, the expression $1/(1 - x)$ defines a function for all $x$ except $x = 1$.

**Example 3**    Let $f(t) = \epsilon^{t^2}$; then

$$(1.5) \quad \mathcal{L}[\epsilon^{t^2}] = \int_{0-}^{\infty} \epsilon^{-st}\epsilon^{t^2} \, dt = \int_{0-}^{\infty} \epsilon^{-st+t^2} \, dt$$

For any fixed value of $s$, however large, when $t > |s|$, the exponent $t^2 - st$ will be monotonically increasing with $t$, and the integrand $\to \infty$ as $t \to \infty$. Obviously, the integral "blows up." Thus, there is no value for $s$ for which (1.5) makes sense. We say that $\epsilon^{t^2}$ is not Laplace transformable, or that the function $\epsilon^{t^2}$ has no Laplace transform.

**Example 4**    Let $f(t) = \delta(t)$; then

$$\int_{0-}^{\infty} \epsilon^{-st} \, \delta(t) \, dt = \int_{0-}^{\infty} \delta(t)\epsilon^{-st} \, dt = \int_{0-}^{\infty} \delta(t) \, dt = 1$$

This is an example in which the defining integral has meaning for all real and all complex values of $s$.

Throughout this chapter, we shall implicitly assume that all the functions we consider are Laplace transformable. It should be stressed that this imposes no restriction on the important circuit-theoretic conclusions of Sec. 5. These conclusions apply even to inputs of the form $\epsilon^{t^2}$ because we are only interested in the properties of the response at *finite* time. Suppose we want to calculate the response to, say, $\epsilon^{t^2}$, at some time $t_1$ ($t_1$ may be as large as we wish, but it is fixed for the duration of this argument). We only need to take as input the function

$$f(t) = \begin{cases} \epsilon^{t^2} & \text{for } 0 \leq t \leq t_1 + 1 \\ 0 & \text{for } t > t_1 + 1 \end{cases}$$

Clearly, the response *at time* $t_1$ of any network to $f(t)$ will be the same as its response to $\epsilon^{t^2}$. However, $f(t)$ has a perfectly well-defined Laplace transform. In conclusion, the very general results of Sec. 5 are not invalidated by the fact that some inputs might grow so fast as $t \to \infty$ that they do not possess a Laplace transform.

## 2    Basic Properties of the Laplace Transform

We describe below only those fundamental properties of the Laplace transform that are useful to our study of linear time-invariant networks.

### 2.1    Uniqueness

Uniqueness is as fundamental as it is intuitively obvious. It means that if a given function of the complex frequency $s$, say, $F(s)$, is known to be the Laplace transform of a time function, say $f(t)$, and if some other time function $g(t)$ also has $F(s)$ as a Laplace transform, then the function $g(t)$ differs from $f(t)$ trivially.

Suppose we take the unit step for $f$; thus,

$$f(t) = \begin{cases} 1 & \text{for } t > 0 \\ 0.5 & \text{for } t = 0 \\ 0 & \text{for } t < 0 \end{cases}$$

Then $F(s) = 1/s$. Now, it is easy to verify that if we change *only* the ordinate of $f$ at $t = 0$ and thus obtain (strictly speaking) a new function $g(t)$, for example,

$$g(t) = \begin{cases} 1 & \text{for } t > 0 \\ 1 & \text{for } t = 0 \\ 0 & \text{for } t < 0 \end{cases}$$

then $\mathcal{L}[g(t)] = 1/s$. Obviously, for our purposes the difference between $f$ and $g$ is trivial.

It is an important and deep fact of analysis that, *except for those trivial differences, a time function is uniquely specified by its Laplace transform.* Of course, it is understood that for negative $t$, the time function may be arbitrary since the defining integral is for the interval $[0, \infty)$. Further discussion on this comment will be given later.

Keeping this uniqueness property in mind, we should note again the analogy with logarithms as follows:

$$f(t) \xrightarrow{\text{uniquely}} F(s) \quad \Big| \quad a \xrightarrow{\text{uniquely}} \log a$$

$$f(t) \xleftarrow{\text{uniquely}} F(s) \quad \Big| \quad a \xleftarrow{\text{uniquely}} \log a$$

This uniqueness property is fundamental to all applications of the Laplace transform. Unfortunately, its proof is very complicated and would take us too far astray.

We now know that the Laplace transform $F(s)$ of a given time function $f(t)$ is uniquely defined by the defining integral. This fact is symbolically represented by the notation

(2.1)  $F(s) = \mathcal{L}[f(t)]$

where $\mathcal{L}[\cdot]$ can be thought of as the operator $\int_{0-}^{\infty} \cdot \, \epsilon^{-st} \, dt$ which maps a time function into its Laplace transform. Conversely, given a Laplace transform $F(s)$, we know that (except for trivialities) there is a unique time function $f(t)$ over the interval $[0, \infty)$ such that (2.1) holds. This fact is symbolized by

(2.2)  $f(t) = \mathcal{L}^{-1}[F(s)]$

This equation means $f(t)$ is the *inverse Laplace transform* of $F(s)$.

| 2.2 | Linearity |
|-----|-----------|

The second most important property of the Laplace transform is that it is linear. Consider the Laplace transformation $\mathcal{L}[\cdot]$ as a transformation that can be applied to an extremely large class of time functions. To each of these time functions the Laplace transformation assigns a function of $s$, namely, the Laplace transform of the given time function. Therefore, we can consider the Laplace transformation itself as a *function* which maps time functions into their Laplace transforms. The linearity property asserts that the Laplace transformation is a *linear* function. Referring to the definition of a linear function (see Appendix A), we see that the linearity can be stated as follows.

**THEOREM**  Let $f_1$ and $f_2$ be any two time functions, and let $c_1$ and $c_2$ be two arbitrary *constants;* then

(2.3)  $\mathcal{L}[c_1 f_1(t) + c_2 f_2(t)] = c_1 \mathcal{L}[f_1(t)] + c_2 \mathcal{L}[f_2(t)]$

*Proof*  By definition the left-hand side is

$$\mathcal{L}[c_1 f_1(t) + c_2 f_2(t)] = \int_{0-}^{\infty} \epsilon^{-st}[c_1 f_1(t) + c_2 f_2(t)] \, dt$$

$$= \int_{0-}^{\infty} \epsilon^{-st} c_1 f_1(t) \, dt + \int_{0-}^{\infty} \epsilon^{-st} c_2 f_2(t) \, dt$$

because the integral of a sum is the sum of the integrals of each term.   Recalling that $c_1$ and $c_2$ are *constants,* we may pull them out of the integral sign, giving

$$\mathcal{L}[c_1 f_1(t) + c_2 f_2(t)] = c_1 \int_{0-}^{\infty} \epsilon^{-st} f_1(t) \, dt + c_2 \int_{0-}^{\infty} \epsilon^{-st} f_2(t) \, dt$$

$$= c_1 \mathcal{L}[f_1(t)] + c_2 \mathcal{L}[f_2(t)]$$

*Application*   Using the fact that $\mathcal{L}[\epsilon^{at}] = 1/(s - a)$, we obtain

$$\mathcal{L}[\cos \beta t] = \mathcal{L}\left[ \frac{1}{2} \epsilon^{j\beta t} + \frac{1}{2} \epsilon^{-j\beta t} \right] = \frac{1}{2} \frac{1}{s - j\beta} + \frac{1}{2} \frac{1}{s + j\beta}$$

hence

(2.4)   $\mathcal{L}[\cos \beta t] = \dfrac{s}{s^2 + \beta^2}$

Similarly,

$$\mathcal{L}[\sin \beta t] = \mathcal{L}\left[ \frac{1}{2j} \epsilon^{j\beta t} - \frac{1}{2j} \epsilon^{-j\beta t} \right]$$

(2.5)   $\qquad\qquad = \dfrac{\beta}{s^2 + \beta^2}$

**Exercise**   Show that

(2.6)   $\mathcal{L}[\epsilon^{\alpha t} \cos \beta t] = \dfrac{s - \alpha}{(s - \alpha)^2 + \beta^2}$

(2.7)   $\mathcal{L}[\epsilon^{\alpha t} \sin \beta t] = \dfrac{\beta}{(s - \alpha)^2 + \beta^2}$

## 2.3    Differentiation Rule

The third most important property of the Laplace transformation is the simple relation that exists between the Laplace transform of a function $f$ and the Laplace transform of its derivative $df/dt$.   In order to clarify a number of points let us consider some examples.

**Example 1**   Consider $\mathcal{L}[\sin \omega t]$.   Differentiating the time function $\sin \omega t$, we obtain $\omega \cos \omega t$ for all $t$ (see Fig. 2.1).   We note that

$$\mathcal{L}\left[ \frac{d}{dt} \sin \omega t \right] = \mathcal{L}[\omega \cos \omega t] = \omega \frac{s}{s^2 + \omega^2} = s \mathcal{L}[\sin \omega t]$$

**Example 2**   Consider $u(t) \cos \omega t$ as shown in Fig. 2.2.   Differentiating this time function, we obtain

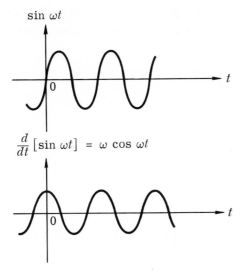

**Fig. 2.1**   Time functions for Example 1.

$$\frac{d}{dt}[u(t)\cos\omega t] = \frac{du}{dt}\cos\omega t + u(t)[-\omega\sin\omega t]$$

$$= \delta(t) - \omega u(t)\sin\omega t$$

This curve is also shown in Fig. 2.2.   Then

$$\mathcal{L}\left\{\frac{d}{dt}[u(t)\cos\omega t]\right\} = 1 - \frac{\omega^2}{s^2+\omega^2} = s\frac{s}{s^2+\omega^2} = s\,\mathcal{L}[u(t)\cos\omega t]$$

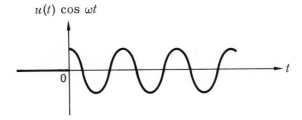

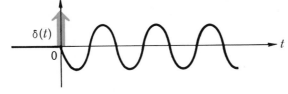

**Fig. 2.2**   Time functions for Example 2.

These examples suggest that the Laplace transform of $df/dt$ is equal to $s$ times the Laplace transform of $f$.  However, this is not the case in general, as we see by the next example.

**Example 3**   Consider the function

$$f(t) = \begin{cases} -1 & \text{for } t < 0 \\ \epsilon^{at} & \text{for } t \geq 0 \end{cases}$$

A graph of the function $f$ is given in Fig. 2.3.  The Laplace transform of $f(t)$ is

$$\mathcal{L}[f(t)] = \int_{0-}^{\infty} \epsilon^{at} \epsilon^{-st} \, dt = \frac{1}{s-a}$$

Now referring to Fig. 2.3 and observing that $f(0+) - f(0-) = 2$, we obtain

$$\frac{d}{dt} f(t) = 2\delta(t) + a\epsilon^{at}$$

and

$$\mathcal{L}\left[\frac{d}{dt} f(t)\right] = 2 + \frac{a}{s-a} = \frac{2s-a}{s-a}$$

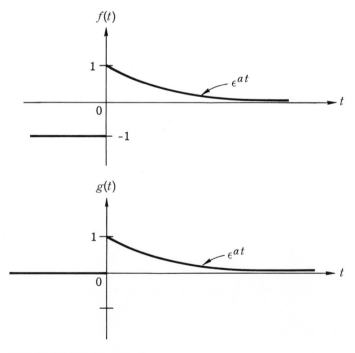

**Fig. 2.3**   Time functions for Example 3.

Let us modify the function $f$ by setting its values to zero for $t < 0$. We thus obtain a new function $g$ defined by

$$g(t) = \begin{cases} 0 & \text{for } t < 0 \\ \epsilon^{at} & \text{for } t \geq 0 \end{cases}$$

Clearly,

$$\mathcal{L}[g(t)] = \mathcal{L}[\epsilon^{at}] = \mathcal{L}[f(t)] = \frac{1}{s - a}$$

However,

$$\frac{d}{dt} g(t) = \delta(t) + a\epsilon^{at}$$

hence

$$\mathcal{L}\left[\frac{d}{dt} g(t)\right] = 1 + \frac{a}{s - a} = \frac{s}{s - a}$$

The Laplace transform of $dg/dt$ differs from that of $df/dt$ because the magnitude of the jump in $g$ at $t = 0$ is 1, whereas that of $f$ is 2.

These examples show that the *Laplace transform of $df/dt$ depends on the magnitude of the jump at $t = 0$*, that is, on $f(0+) - f(0-)$. These considerations should make the following statement appear more natural:

**DIFFEREN- TIATION THEOREM**

(2.8)   $\mathcal{L}\left[\dfrac{df}{dt}\right] = s\,\mathcal{L}[f(t)] - f(0-)$

It is crucial that when we calculate the derivative in (2.8), we insert the appropriately weighted impulse whenever $f$ has a jump; also, the defining integral of the Laplace transform starts at $0-$ and thus includes the full contribution of any impulse at the origin.

*Proof*   Integrating by parts, we have

(2.9)   $\mathcal{L}\left[\dfrac{df}{dt}\right] = \displaystyle\int_{0-}^{\infty} \dfrac{df}{dt} \epsilon^{-st}\, dt = f(t)\epsilon^{-st}\Big|_{0-}^{\infty} - \int_{0-}^{\infty} f(t)(-s)\epsilon^{-st}\, dt$

$= s\displaystyle\int_{0-}^{\infty} f(t)\epsilon^{-st}\, dt - f(0-) = s\,\mathcal{L}[f(t)] - f(0-)$

Here again we use the fact that Re $(s)$ is taken sufficiently large so that as $t \to \infty$, $f(t)\epsilon^{-st} \to 0$.

**Example 4**   From $\mathcal{L}[\delta(t)] = 1$ we obtain

(2.10)   $\mathcal{L}[\delta^{(1)}(t)] = s, \ \mathcal{L}[\delta^{(2)}(t)] = s^2, \ldots, \mathcal{L}[\delta^{(n)}(t)] = s^n$

**Example 5**     Let us calculate the impulse response of the linear time-invariant circuit shown in Fig. 2.4, where $e$ is the input and $i$ is the response. The input is $e(t) = \delta(t)$, and, by definition of the impulse response, $i(0-) = 0$, since the circuit must be in the zero state prior to the application of the impulse. Let us use our previous notation and call $h$ the impulse response. The differential equation is

$$L\frac{dh}{dt} + Rh = \delta(t) \qquad h(0-) = 0$$

Let us take the Laplace transform of both sides as follows:

$$\mathcal{L}\left[L\frac{dh}{dt} + Rh\right] = \mathcal{L}[\delta(t)] = 1$$

By linearity (note that $L$ and $R$ are *constants*) the equation becomes

$$L\mathcal{L}\left[\frac{dh}{dt}\right] + R\mathcal{L}[h] = 1$$

Using the differentiation rule and the initial condition, we obtain

$$(Ls + R)\mathcal{L}[h(t)] = 1$$

since $h(0-) = 0$. Thus,

(2.11)     $$\mathcal{L}[h(t)] = \frac{1}{Ls + R} = \frac{1}{L}\frac{1}{s + R/L}$$

and hence

(2.12)     $$h(t) = \frac{1}{L}u(t)\epsilon^{-(R/L)t} \qquad \text{for all } t$$

It should be pointed out that the Laplace transform method only gives solutions for $t \geq 0$. The fact that a step function $u(t)$ is used in (2.12) is due to the physical fact that the response is zero before the application of the impulse.

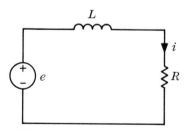

**Fig. 2.4**     Linear time-invariant circuit for Example 5.

By repeated application of the differentiation theorem, we obtain the following corollary.

**COROLLARY**

$$\mathcal{L}[f^{(2)}(t)] = s^2\mathcal{L}[f(t)] - sf(0-) - f^{(1)}(0-)$$

$$\mathcal{L}[f^{(3)}(t)] = s^3\mathcal{L}[f(t)] - s^2f(0-) - sf^{(1)}(0-) - f^{(2)}(0-)$$

and, in general,

$$\mathcal{L}[f^{(n)}(t)] = s^n\mathcal{L}[f(t)] - s^{n-1}f(0-) - \cdots - sf^{(n-2)}(0-) - f^{(n-1)}(0-)$$

## 2.4    Integration Rule

This property is the inverse of differentiation, but its precise interpretation is important. The integration rule relates the Laplace transform of a given function $f$ to the Laplace transform of its integral $\int_{0-}^{t} f(t')\, dt'$. More precisely, we have the following theorem.

**INTEGRATION THEOREM**

(2.13)    $\mathcal{L}\left[\int_{0-}^{t} f(t')\, dt'\right] = \dfrac{1}{s}\mathcal{L}[f(t)]$

Note that we start integrating the function $f$ at $0-$; hence if $f$ includes some impulses at $t = 0$, their full contribution is included in the integral.

*Proof*    Again we integrate by parts as follows:

$$\mathcal{L}\left[\int_{0-}^{t} f(t')\, dt'\right] = \int_{0-}^{\infty}\left[\int_{0-}^{t} f(t')\, dt'\right]\epsilon^{-st}\, dt$$

$$= \left[\int_{0-}^{t} f(t')\, dt'\right]\frac{\epsilon^{-st}}{-s}\Bigg|_{0-}^{\infty} - \int_{0-}^{\infty} f(t)\left(-\frac{1}{s}\right)\epsilon^{-st}\, dt$$

The first term is zero. Indeed, for $t = 0-$, the integral is zero; for $t \to \infty$ and with Re $(s)$ sufficiently large, the exponential $\epsilon^{-st} \to 0$ fast enough to make the whole term go to zero. Hence,

$$\mathcal{L}\left[\int_{0-}^{t} f(t')\, dt'\right] = \frac{1}{s}\int_{0-}^{\infty} f(t)\epsilon^{-st}\, dt = \frac{1}{s}\mathcal{L}[f(t)]$$

**Example 6**    From $\int_{0-}^{t} \delta(t')\, dt' = u(t)$ and $\mathcal{L}[\delta(t)] = 1$, we conclude that $\mathcal{L}[u(t)] = 1/s$.

**Example 7**    Integrating successively the unit step, we obtain

$$\int_{0-}^{t} u(t')\, dt' = t \qquad \int_{0-}^{t} t'\, dt' = \frac{t^2}{2} \qquad \int_{0-}^{t} \frac{t'^2}{2}\, dt' = \frac{t^3}{3!}$$

and

$$\int_{0-}^{t} \frac{t'^k}{k!} dt' = \frac{t^{k+1}}{(k+1)!} \qquad \text{for any integer } k$$

Consequently,

(2.14)   $\mathcal{L}[t] = \dfrac{1}{s^2} \qquad \mathcal{L}\left[\dfrac{t^2}{2!}\right] = \dfrac{1}{s^3}$

and

(2.15)   $\mathcal{L}\left[\dfrac{t^n}{n!}\right] = \dfrac{1}{s^{n+1}} \qquad \text{for any integer } n$

---

**Example 8**   Let $f(\cdot)$ be the rectangular pulse defined by

$$f(t) = \begin{cases} 0 & t < 1 \\ 1 & 1 \le t \le 3 \\ 0 & 3 < t \end{cases}$$

Then its Laplace transform is

$$\int_{0-}^{\infty} f(t)\epsilon^{-st}\,dt = \int_{1}^{3} \epsilon^{-st}\,dt = \frac{\epsilon^{-st}}{-s}\Big|_{1}^{3} = \frac{\epsilon^{-s} - \epsilon^{-3s}}{s}$$

---

**Example 9**   Let $f(\cdot)$ be a waveform which is identical to zero for $t < 0$. Let its Laplace transform be $F(s)$. Let us find the Laplace transform of the waveform $f_\tau(\cdot)$ which is the waveform $f(\cdot)$ *delayed* by $\tau$ sec (here $\tau > 0$). First, note that $u(t)f(t) = f(t)$ for all $t$. Then, by definition of $f_\tau(\cdot)$,

$$f_\tau(t) = u(t - \tau)f(t - \tau) \qquad \text{for all } t$$

and

$$\mathcal{L}[f_\tau(t)] = \int_{0-}^{\infty} u(t - \tau)f(t - \tau)\epsilon^{-st}\,dt$$

$$= \int_{0-}^{\infty} u(t')f(t')\epsilon^{-s(t'+\tau)}\,dt'$$

$$= \epsilon^{-s\tau}\int_{0-}^{\infty} f(t')\epsilon^{-st'}\,dt' = \epsilon^{-s\tau}F(s)$$

---

**Remark**   Since the Laplace transform suppresses the nature of the time function prior to $t = 0-$, we may take $f(t) = 0$ for $t < 0$ for any of the time functions under study. It is usually convenient to do so. With this interpretation we always have $f(0-) = 0$. In applying the differentiation rule, we must not forget the impulses that may arise at the origin when the resulting time function is differentiated.

This concludes our brief survey of the properties of the Laplace transform. Of course, it has many other properties which are very useful for some studies. However, we shall be able to get along with only the four listed above and the convolution property that we shall derive later.

In Table 13.1 we give a list of Laplace transforms of some frequently encountered time functions.

**Exercise 1**   Derive (2.13) on the basis of (2.8). [Hint: Call $g(t)$ the integral in the left-hand side of (2.13), find $g(0-)$ and $dg/dt$, and apply (2.8)].

**Exercise 2**   Verify the last four entries in Table 13.1.

**Table 13.1   Laplace Transforms of Elementary Functions**

| $f(t)$ | $F(s) \triangleq \int_{0-}^{\infty} f(t)\epsilon^{-st}\, dt$ |
|---|---|
| $\delta(t)$ | $1$ |
| $\delta^{(n)}(t)$ | $s^n \quad (n = 1, 2, \ldots)$ |
| $u(t)$ | $\dfrac{1}{s}$ |
| $\dfrac{t^n}{n!}$ | $\dfrac{1}{s^{n+1}} \quad (n = 1, 2, \ldots)$ |
| $\epsilon^{-at} \begin{pmatrix} a \text{ real or} \\ \text{complex} \end{pmatrix}$ | $\dfrac{1}{s + a}$ |
| $\dfrac{t^n}{n!} \epsilon^{-at} \begin{pmatrix} a \text{ real or} \\ \text{complex} \end{pmatrix}$ | $\dfrac{1}{(s + a)^{n+1}} \quad (n = 1, 2, \ldots)$ |
| $\cos \beta t$ | $\dfrac{s}{s^2 + \beta^2}$ |
| $\sin \beta t$ | $\dfrac{\beta}{s^2 + \beta^2}$ |
| $\epsilon^{-at} \cos \beta t$ | $\dfrac{s + \alpha}{(s + \alpha)^2 + \beta^2}$ |
| $\epsilon^{-at} \sin \beta t$ | $\dfrac{\beta}{(s + \alpha)^2 + \beta^2}$ |
| $a\epsilon^{-at} \cos \beta t + \dfrac{(b - a\alpha)}{\beta} \epsilon^{-at} \sin \beta t$ | $\dfrac{as + b}{(s + \alpha)^2 + \beta^2}$ |
| $2\,|K|\epsilon^{-at} \cos (\beta t + \angle K)$ | $\dfrac{K}{s + \alpha - j\beta} + \dfrac{\bar{K}}{s + \alpha + j\beta}$ |

## 3   Solutions of Simple Circuits

One of the principal uses of the Laplace transform is solving *linear* integro-differential equations with *constant* coefficients. We shall describe the method by way of examples.

### 3.1   Calculation of an Impulse Response

Consider the linear time-invariant circuit shown in Fig. 3.1, where $e$ is the input and $v$ is the output. Let us find its impulse response. The branch equations are

$$i = \frac{v}{R} \qquad v_C(t) = v_C(0-) + \frac{1}{C} \int_{0-}^{t} i(t')\,dt' \qquad v_L = \frac{L}{R}\frac{dv}{dt}$$

Hence, by KVL,

(3.1) $$\frac{L}{R}\frac{dv}{dt} + v + \frac{1}{RC} \int_{0-}^{t} v(t')\,dt' + v_C(0-) = e(t)$$

By definition of the impulse response, the circuit is in the zero state at $0-$, and the input is a unit impulse; that is,

$$v_C(0-) = 0 \qquad i_L(0-) = \frac{1}{R}v(0-) = 0 \qquad e(t) = \delta(t)$$

Let $h$ be the impulse response; we then have

(3.2) $$\frac{L}{R}\frac{dh}{dt} + h + \frac{1}{RC} \int_{0-}^{t} h(t')\,dt' = \delta(t)$$

with

$$h(0-) = 0$$

Let us take the Laplace transform of both sides and put

$$H(s) \overset{\Delta}{=} \mathcal{L}[h(t)]$$

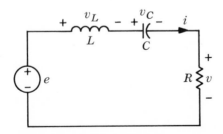

**Fig. 3.1**   Linear time-invariant RLC circuit.

Then

$$\mathcal{L}\left[\frac{L}{R}\frac{dh}{dt} + h + \frac{1}{RC}\int_{0-}^{t} h(t')\,dt'\right] = 1$$

By linearity we get

(3.3)   $$\frac{L}{R}\mathcal{L}\left[\frac{dh}{dt}\right] + H(s) + \frac{1}{RC}\mathcal{L}\left[\int_{0-}^{t} h(t')\,dt'\right] = 1$$

Using the differentiation theorem and the integration theorem, and taking into account the initial conditions, we obtain

$$\left[\frac{L}{R}s + 1 + \frac{1}{RCs}\right]H(s) = 1$$

or

(3.4)   $$H(s) = \frac{R}{L}\frac{s}{s^2 + (R/L)s + 1/LC}$$

As in Table 5.1, with $Q \triangleq \omega_0/2\alpha$ assuming $Q > \frac{1}{2}$, we put

$$\omega_0^2 \triangleq \frac{1}{LC} \qquad \alpha \triangleq \frac{R}{2L} \qquad \omega_d = \sqrt{\omega_0^2 - \alpha^2} \qquad \phi \triangleq \sin^{-1}\frac{\alpha}{\omega_0}$$

and we obtain

(3.5)   $$H(s) = \frac{R}{L}\frac{s}{(s + \alpha)^2 + \omega_d^2}$$

Referring back to Table 13.1, we conclude that

$$h(t) = \frac{R}{L}u(t)\epsilon^{-\alpha t}\left(\cos \omega_d t - \frac{\alpha}{\omega_d}\sin \omega_d t\right)$$

(3.6)   $$= \frac{\omega_0 R}{\omega_d L}u(t)\epsilon^{-\alpha t}\cos (\omega_d t + \phi)$$

This answer, of course, agrees with our results of Chap. 5.

**Remarks**   1.   Suppose we had to calculate the sinusoidal steady-state response (represented by the phasor $V$) to a sinusoidal input $e(t) = E_m \cos \omega t$ (represented by the phasor $E$). We would use the phasor method and obtain the following equation relating the unknown phasor $V$ to the phasor $E$:

$$\frac{L}{R}j\omega V + V + \frac{1}{RC}\frac{1}{j\omega}V = E$$

Hence

(3.7a) $\quad \dfrac{V}{E} = \dfrac{R}{L}\ \dfrac{j\omega}{(j\omega)^2 + (R/L)j\omega + 1/LC}$

Equation (3.7a) gives the network function relating the input phasor $E$ to the response phasor $V$. Comparing it with Eq. (3.4), we conclude

(3.7b) $\quad \dfrac{V}{E} = H(j\omega)$

2. The method we have used so far to go from the Laplace transform (3.5) to the corresponding time function is simply to look it up in the table. In other words, we use our Laplace transform table in the same way we use a table of logarithms. The procedure is perfectly legitimate because the uniqueness property guarantees that if *using any method whatsoever* we find a function $f(t)$ whose Laplace transform is the given function $F(s)$, then $f(t)$ is *the* time function we are looking for.

### 3.2   Partial-fraction Expansion

It is apparent that a more complicated problem would give a more complicated Laplace transform, i.e., a more complicated rational function in $s$. Many such transforms will not appear directly in Laplace transform tables. However, we can easily find the corresponding time function by reducing the transform to simpler elements which do appear in our table.

There is a general method (which you may have already encountered in calculus) for breaking up any rational function into simple components. The method is called *partial-fraction expansion*. Consider the following rational function

(3.8) $\quad F(s) = \dfrac{P(s)}{Q(s)} = \dfrac{b_0 s^m + b_1 s^{m-1} + \cdots + b_{m-1} s + b_m}{a_0 s^n + a_1 s^{n-1} + \cdots + a_{n-1} s + a_n}$

where $P(s)$ and $Q(s)$ are polynomials in the complex frequency variable $s$ and the coefficients $a_0, a_1, \ldots, a_n, b_0, b_1, \ldots, b_m$ are *real* numbers. A rational function is specified completely by the two sets of real coefficients which define the numerator and denominator polynomials. On the other hand, the polynomials can also be expressed in the factored form in terms of their zeros. Thus, an alternate representation of $F(s)$ is given by

(3.9) $\quad F(s) = K\ \dfrac{\displaystyle\prod_{i=1}^{m}(s - z_i)}{\displaystyle\prod_{j=1}^{n}(s - p_j)}$

where $z_i, i = 1, 2, \ldots, m$, are the zeros of the numerator polynomial $P(s)$, and $p_j, j = 1, 2, \ldots, n$, are the zeros of the denominator polynomial $Q(s)$. The $z_i$'s are called the **zeros** of the rational function. The $p_j$'s are called the

**poles** of the rational function.† If $p_j$ is a simple zero of the denominator polynomial $Q(s)$, then $p_j$ is said to be a **simple pole** of the rational function. If $p_k$ is a zero of order $r$ of the polynomial $Q(s)$, then $p_k$ is called a **multiple pole of order** $r$.

The first step in the partial-fraction expansion is to put the rational function into a proper form. We say that a rational function is **proper** if the degree of the numerator polynomial is less than the degree of the denominator polynomial. If the given rational function $F(s)$ is not proper, i.e., if the degree of $P(s)$ is greater than or equal to that of $Q(s)$, we divide $P(s)$ by $Q(s)$ and obtain

$$(3.10) \quad F(s) = \frac{P(s)}{Q(s)} = \hat{P}(s) + \frac{R(s)}{Q(s)}$$

In (3.10), $\hat{P}(s)$, the quotient, is a polynomial and $R(s)$ is the remainder; therefore, $R(s)$ has a degree less than that of $Q(s)$, and the new rational function $R(s)/Q(s)$ is proper. Since $\hat{P}(s)$ is a polynomial, the corresponding time function is a linear combination of $\delta$, $\delta^{(1)}$, $\delta^{(2)}$, etc., and can be determined directly from Table 13.1. We therefore go ahead with the new rational function $R(s)/Q(s)$ which is proper. In the remaining part of this section we assume that all rational functions are proper.

The second step of the partial-fraction expansion is to factor the denominator polynomial $Q(s)$ and obtain the poles of the rational function. We shall consider three cases, namely simple poles, multiple poles, and complex poles.

*Case 1:*
*Simple poles*

We start with a simple example as follows:

$$F(s) = \frac{P(s)}{Q(s)} = \frac{s^2 + 3s + 5}{(s + 1)(s + 2)(s + 3)}$$

We claim that there are constants $K_1$, $K_2$, and $K_3$ such that

$$(3.11) \quad \frac{s^2 + 3s + 5}{(s + 1)(s + 2)(s + 3)} = \frac{K_1}{s + 1} + \frac{K_2}{s + 2} + \frac{K_3}{s + 3}$$

for all $s$ (strictly speaking we should say for all $s$ except at the poles, that is, at $s \neq -1, \neq -2$, and $\neq -3$.) Let us clear the denominator in (3.11); then

$$s^2 + 3s + 5 = K_1(s + 2)(s + 3) + K_2(s + 1)(s + 3) + K_3(s + 1)(s + 2)$$

Since this equation must hold for all $s$, let us substitute successively $s = -1$, $s = -2$, and $s = -3$. Clearly, we get

$$K_1 = \frac{s^2 + 3s + 5}{(s + 2)(s + 3)}\bigg|_{s=-1} = \frac{3}{2} = 1.5$$

---

† We assume that the numerator and denominator polynomials have no common factors.

$$K_2 = \frac{s^2 + 3s + 5}{(s + 1)(s + 3)}\bigg|_{s=-2} = \frac{3}{-1} = -3$$

$$K_3 = \frac{s^2 + 3s + 5}{(s + 1)(s + 2)}\bigg|_{s=-3} = \frac{5}{2} = 2.5$$

Hence

$$f(t) = \mathcal{L}^{-1}\left[\frac{s^2 + 3s + 5}{(s + 1)(s + 2)(s + 3)}\right]$$

$$= \mathcal{L}^{-1}\left[\frac{1.5}{s + 1} + \frac{-3}{s + 2} + \frac{2.5}{s + 3}\right]$$

$$= 1.5\epsilon^{-t} - 3\epsilon^{-2t} + 2.5\epsilon^{-3t} \quad \text{for } t \geq 0$$

This example indicates a straightforward way of obtaining the partial-fraction expansion in the case in which all poles are simple. The coefficients $K_1$, $K_2$, and $K_3$ in Eq. (3.11) are called the **residues** of the particular poles $-1$, $-2$, and $-3$, respectively.

There is a formula for computing the residue of an arbitrary rational function $F(s)$ at a simple pole: let the denominator polynomial of $F(s)$ be

(3.12)   $$Q(s) = \prod_{j=1}^{n} (s - p_j)$$

where $p_j$ $(j = 1, 2, \ldots, n)$ are *simple* poles of $F(s)$. Then the partial-fraction expansion is of the form

(3.13a)   $$F(s) = \sum_{j=1}^{n} \frac{K_j}{s - p_j}$$

and the residue $K_j$ of the pole $p_j$ is given by

(3.13b)   $$K_j = (s - p_j)F(s)\bigg|_{s=p_j}$$

*Proof*   Multiplying on both sides of (3.13a) by the factor $s - p_j$, we obtain

(3.14)   $$(s - p_j)F(s) = K_j + (s - p_j)\sum_{\substack{i=1 \\ i \neq j}}^{n} \frac{K_i}{s - p_i}$$

Substituting $s = p_j$ in (3.14), we immediately obtain the formula in (3.13b).

Note that in the simple example of (3.11), the residues are indeed computed by means of the formula given in (3.13b).

*Case 2:*    Suppose we were given
*Multiple*
*poles*    $$F(s) = \frac{s^2 + 3s + 5}{(s + 1)^2(s + 2)}$$

We claim that there are constants $K_{12}$, $K_{11}$, and $K_2$ such that

$$\frac{s^2 + 3s + 5}{(s + 1)^2(s + 2)} = \frac{K_{12}}{(s + 1)^2} + \frac{K_{11}}{s + 1} + \frac{K_2}{s + 2}$$

for all $s$.    Clearing the denominator, we obtain

$$s^2 + 3s + 5 = K_{12}(s + 2) + K_{11}(s + 1)(s + 2) + K_2(s + 1)^2$$

This equation must hold for all $s$; thus, substituting $s = -1$ and $s = -2$, we obtain $K_{12}$ and $K_2$, respectively, as follows:

$$K_{12} = \left. \frac{s^2 + 3s + 5}{s + 2} \right|_{s=-1} = \frac{3}{1} = 3$$

$$K_2 = \left. \frac{s^2 + 3s + 5}{(s + 1)^2} \right|_{s=-2} = \frac{3}{1} = 3$$

Having found values for $K_{12}$ and $K_2$, we can take any convenient value of $s$ to get an equation for $K_{11}$.    Letting $s = 0$, we obtain

$$5 = 2K_{12} + 2K_{11} + K_2 = 9 + 2K_{11}$$

or

$$K_{11} = -2$$

Hence

$$f(t) = \mathcal{L}^{-1}\left[ \frac{s^2 + 3s + 5}{(s + 1)^2(s + 2)} \right]$$

$$= \mathcal{L}^{-1}\left[ \frac{3}{(s + 1)^2} + \frac{-2}{s + 1} + \frac{3}{s + 2} \right]$$

$$= 3t\epsilon^{-t} - 2\epsilon^{-t} + 3\epsilon^{-2t} \qquad \text{for } t \geq 0$$

Let us assume that the denominator of the rational function $F(s)$ is the polynomial

(3.15a)    $$Q(s) = (s - p_1)^{n_1}(s - p_2)^{n_2} \cdots (s - p_r)^{n_r}$$

Thus, the rational function $F(s)$ has a pole of order $n_1$ at $p_1$, a pole of order $n_2$ at $p_2, \ldots$, and a pole of order $n_r$ at $p_r$.    Clearly, if $n$ is the degree of $Q$, then

(3.15b)    $$\sum_{i=1}^{r} n_i = n$$

The rational function $F(s)$ has a partial-fraction expansion of the form

$$F(s) = \frac{K_{11}}{s - p_1} + \frac{K_{12}}{(s - p_1)^2} + \cdots + \frac{K_{1n_1}}{(s - p_1)^{n_1}}$$

$$+ \frac{K_{21}}{s - p_2} + \frac{K_{22}}{(s - p_2)^2} + \cdots + \frac{K_{2n_2}}{(s - p_2)^{n_2}}$$

. . . . . . . . . . . . . . . . . . . . . . . . . . . . . . . . .

(3.16a)
$$+ \frac{K_{r1}}{s - p_r} + \frac{K_{r2}}{(s - p_r)^2} + \cdots + \frac{K_{rn_r}}{(s - p_r)^{n_r}}$$

Note that the first index $i$ of the subscript to $K_{ij}$ corresponds to the pole $p_i$, and the second index $j$ corresponds to the order of the corresponding denominator. In order to calculate the coefficients $K_{11}, K_{12}, \ldots, K_{1n_1}$ associated with the pole $p_1$, we consider the product of $F(s)$ and $(s - p_1)^{n_1}$, namely,

$$(s - p_1)^{n_1} F(s) = K_{11}(s - p_1)^{n_1 - 1} + K_{12}(s - p_1)^{n_1 - 2} + \cdots$$

$$+ K_{1,n_1 - 1}(s - p_1) + K_{1n_1}$$

$$+ (s - p_1)^{n_1} \sum_{i=2}^{r} \sum_{j=1}^{n_r} \frac{K_{ij}}{(s - p_i)^j}$$

Because the last double sum has $(s - p_1)^{n_1}$ as a factor, if we evaluate $(s - p_1)^{n_1} F(s)$, $(d/ds)[(s - p_1)^{n_1} F(s)], \ldots, (d^{n_1-1}/ds^{n_1-1})[(s - p_1)^{n_1} F(s)]$ at $s = p_1$, the last double sum contributes zero. Consequently, we obtain successively,

$$K_{1n_1} = (s - p_1)^{n_1} F(s) \Big|_{s=p_1}$$

(3.16b)
$$K_{1,n_1 - 1} = \frac{d}{ds}[(s - p_1)^{n_1} F(s)] \Big|_{s=p_1}$$

$$K_{1,n_1 - 2} = \frac{1}{2!} \frac{d^2}{ds^2}[(s - p_1)^{n_1} F(s)] \Big|_{s=p_1}$$

. . . . . . . . . . . . . . . . . . . . . . . . . .

---

**Example**   Find the partial-fraction expansion of

$$F(s) = \frac{1}{(s + 1)^3 s^2}$$

The function has two multiple poles at $s = p_1 = -1$ (third order, $n_1 = 3$) and at $s = p_2 = 0$ (second order, $n_2 = 2$). Thus, the partial-fraction expansion is of the form

$$F(s) = \frac{K_{11}}{s + 1} + \frac{K_{12}}{(s + 1)^2} + \frac{K_{13}}{(s + 1)^3} + \frac{K_{21}}{s} + \frac{K_{22}}{s^2}$$

To calculate $K_{11}$, $K_{12}$, and $K_{13}$, we first multiply $F(s)$ by $(s + 1)^3$ to obtain

$$(s + 1)^3 F(s) = \frac{1}{s^2}$$

Using (3.16b), we find

$$K_{13} = \frac{1}{s^2}\bigg|_{s=-1} = 1$$

$$K_{12} = \frac{d}{ds}\frac{1}{s^2}\bigg|_{s=-1} = \frac{-2}{s^3}\bigg|_{s=-1} = 2$$

$$K_{11} = \frac{1}{2}\frac{d^2}{ds^2}\frac{1}{s^2}\bigg|_{s=-1} = \frac{1}{2}\frac{6}{s^4}\bigg|_{s=-1} = 3$$

Similarly, to calculate $K_{21}$ and $K_{22}$, we first multiply $F(s)$ by $s^2$ to obtain

$$s^2 F(s) = \frac{1}{(s + 1)^3}$$

Using (3.16b), we find

$$K_{22} = \frac{1}{(s + 1)^3}\bigg|_{s=0} = 1$$

$$K_{21} = \frac{d}{ds}\frac{1}{(s + 1)^3}\bigg|_{s=0} = -3$$

Therefore, the partial-fraction expansion is

$$F(s) = \frac{1}{(s + 1)^3 s^2} = \frac{3}{s + 1} + \frac{2}{(s + 1)^2} + \frac{1}{(s + 1)^3} - \frac{3}{s} + \frac{1}{s^2}$$

The corresponding time function is

$$f(t) = \mathcal{L}^{-1}\left[\frac{1}{(s + 1)^3 s^2}\right] = 3\epsilon^{-t} + 2t\epsilon^{-t} + \frac{1}{2}t^2\epsilon^{-t} - 3 + t \qquad \text{for } t \geq 0$$

*Case 3:*
*Complex*
*poles*

The two cases presented above are valid for poles which are either real or complex. However, if complex poles are present, the coefficients in the partial-fraction expansion are, in general, complex, and further simplification is possible. First, we observe that $F(s)$ in (3.8) is a ratio of polynomials with *real* coefficients; hence zeros and poles, if complex, must occur in complex conjugate pairs. More precisely, if $p_1 = \sigma + j\omega_1$ is a pole, that is, $Q_1(p_1) = 0$, then $\bar{p}_1 = \sigma_1 - j\omega_1$ is also a pole; that is, $Q_1(\bar{p}_1) = 0$. This is due to the fact that any polynomial $Q(s)$ with real coefficients has the property that $\overline{Q(s)} = Q(\bar{s})$ for all $s$.

Let us assume that the rational function has a simple pole at $s = p_1 = \alpha + j\beta$; then it must have another pole at $s = p_2 = \bar{p}_1 = \alpha - j\beta$. The partial-fraction expansion of $F(s)$ must contain the following two terms:

(3.17) $\quad \dfrac{K_1}{s - \alpha - j\beta} + \dfrac{K_2}{s - \alpha + j\beta}$

Using formula (3.13b) for simple poles, we obtain

(3.18a)
$$K_1 = (s - \alpha - j\beta)F(s)\Big|_{s=\alpha+j\beta}$$

$$K_2 = (s - \alpha + j\beta)F(s)\Big|_{s=\alpha-j\beta}$$

Since $F(s)$ is a rational function of $s$ with *real* coefficients, it follows from (3.18a) that $K_2$ is the complex conjugate of $K_1$. Let us express $K_1$ and $K_2$ in polar form; then

(3.18b)
$$K_1 = |K_1|\, \epsilon^{j\angle K_1}$$
$$K_2 = \bar{K}_1 = |K_1|\, \epsilon^{-j\angle K_1}$$

The inverse Laplace transform of (3.17) is

$$K_1\epsilon^{(\alpha+j\beta)t} + K_2\epsilon^{(\alpha-j\beta)t} = |K_1|\,\epsilon^{\alpha t}[\epsilon^{j(\beta t+\angle K_1)} + \epsilon^{-j(\beta t+\angle K_1)}]$$

(3.19)
$$= 2|K_1|\,\epsilon^{\alpha t}\cos(\beta t + \angle K_1)$$

This formula, which gives the corresponding time function for a pair of terms due to the complex conjugate poles, is extremely useful. Note that it is only necessary to find the complex residue $K_1$ using (3.18a), since the corresponding time function for both terms in (3.17) can be written immediately by means of (3.19).

---

**Example**   Find the partial-fraction expansion of

$$F(s) = \frac{s^2 + 3s + 7}{[(s + 2)^2 + 4](s + 1)}$$

$$= \frac{K_1}{s + 2 - j2} + \frac{\bar{K}_1}{s + 2 + j2} + \frac{K_3}{s + 1}$$

Using (3.18a), we have

$$K_1 = (s + 2 - j2)F(s)\Big|_{s=-2+j2}$$

$$= \frac{s^2 + 3s + 7}{(s + 2 + j2)(s + 1)}\Big|_{s=-2+j2}$$

$$= \frac{(-2 + j2)^2 + 3(-2 + j2) + 7}{j4(-1 + j2)}$$

$$= j\frac{1}{4} = \frac{1}{4}\,\epsilon^{j90°}$$

Similarly, we have for the real pole at $s = -1$

$$K_3 = \frac{s^2 + 3s + 7}{(s + 2)^2 + 4}\bigg|_{s=-1} = 1$$

The corresponding time function for $F(s)$ can be written by inspection using (3.19); thus,

$$f(t) = -\frac{1}{2}\epsilon^{-2t}\sin 2t + \epsilon^{-t} \qquad \text{for } t \geq 0$$

Note that the first term in $f(t)$ represents the time function corresponding to the pair of complex conjugate poles.

---

### 3.3   Zero-state Response

Consider the same linear time-invariant circuit shown in Fig. 3.1. Let the input $e(\,\cdot\,)$ be an arbitrary waveform whose Laplace transform is $E(s)$. Let us calculate the zero-state response to $e(\,\cdot\,)$. Repeating the previous analysis and using the fact that all initial conditions are zero, we obtain

$$\left(\frac{L}{R}s + 1 + \frac{1}{RCs}\right)V(s) = E(s)$$

We can write

$$V(s) = \left[\frac{R}{L}\frac{s}{s^2 + (R/L)s + 1/LC}\right]E(s)$$

(3.20)
$$= H(s)E(s)$$

Recall that in the sinusoidal steady-state analysis we defined the network function to be the ratio of the output phasor to the input phasor. We also noted that the factor $H(s)$ in Eq. (3.20) would be the network function if $s$ were replaced by $j\omega$. Since in Laplace transform theory we allow $s$ to take any value in the complex plane, we formally extend our definition as follows: we call the ratio of the Laplace transform of the *zero-state response* to the Laplace transform of the input the **network function.** Therefore, the interpretation of Eq. (3.20) states that the Laplace transform of the zero-state response is the product of the network function times the Laplace transform of the input. Since this fact is of great importance we restate it.

(3.21)
$$\boxed{\binom{\text{Laplace transform of}}{\text{the zero-state response}} = \binom{\text{Network}}{\text{function}}\binom{\text{Laplace transform}}{\text{of the input}}}$$

Since the Laplace transform of a unit impulse is unity and since the impulse response is a zero-state response, we obtain the following important fact.

$$\mathcal{L}[h(t)] = H(s)$$

that is, the Laplace transform of the impulse response is the network function.

**Exercise**   In Eq. (3.20) let $R/L = 6$, $1/LC = 25$, and $e(t) = u(t)\epsilon^{-t}$. Calculate the corresponding zero-state response.

---

**3.4**   **The Convolution Theorem**

From our previous knowledge of circuit theory we can obtain the basic convolution theorem of the Laplace transform. Let us observe the following three facts.

1.  We know that for linear time-invariant networks the zero-state response $v(\cdot)$ can be obtained by calculating the convolution of the input $e(\cdot)$ with the impulse response $h(\cdot)$; thus, from Eq. (4.4) of Chap. 6 with $t_0 = 0$, we have

(3.22)   $v(t) = \int_{0-}^{t+} h(t - \tau)e(\tau)\, d\tau \qquad \text{for } t \geq 0$

2.  The Laplace transform of the impulse response is the network function

(3.23)   $\mathcal{L}[h(t)] = H(s)$

3.  The Laplace transform of the zero-state response is the product of the Laplace transform of the input times the network function as follows:

(3.24)   $V(s) = H(s)E(s)$

These facts show that the Laplace transform of the convolution of two functions is the product of their Laplace transforms. More formally, we state the following theorem.

**CONVOLUTION**   Let $f_1(t)$ and $f_2(t)$ have $F_1(s)$ and $F_2(s)$, respectively, as Laplace transforms.
**THEOREM**   Let $f_3$ be the convolution of $f_1$ and $f_2$; that is,

(3.25)   $f_3(t) \triangleq \int_{0-}^{t+} f_1(t - \tau)f_2(\tau)\, d\tau \qquad \text{for } t \geq 0$

Then

(3.26)   $F_3(s) = F_1(s)F_2(s)$

We deliberately indicate $0-$ and $t+$, respectively, as limits of integration for the convolution of $f_1$ and $f_2$. We do this for the following reasons:

(1) if $f_2(\cdot)$ has an impulse at the origin, then it must be included in the computation of the convolution integral (3.25); (2) if $f_1(\cdot)$ has an impulse at the origin, $f_1(t - \tau)$, as a function of $\tau$, has an impulse at $\tau = t$, and that impulse must also be included in the computation of the convolution integral.

*Proof*   Let us prove it directly.  For convenience, let $f_1 * f_2$ denote the convolution integral.  By definition, we have

$$\mathcal{L}[f_1 * f_2] = \mathcal{L}\left[ \int_{0-}^{t+} f_1(t - \tau) f_2(\tau) \, d\tau \right]$$

$$= \int_{0-}^{\infty} \left[ \int_{0-}^{t+} f_1(t - \tau) f_2(\tau) \, d\tau \right] \epsilon^{-st} \, dt$$

If we recall that for our purposes we may take $f_1$ and $f_2$ to be identically zero when their arguments are negative, then for fixed $t$, $f(t - \tau)$ is identically zero for $\tau > t$.  Hence we may replace the upper limit of integration by $\infty$; thus,

$$\mathcal{L}[f_1 * f_2] = \int_{0-}^{\infty} \int_{0-}^{\infty} f_1(t - \tau) f_2(\tau) \, d\tau \, \epsilon^{-st} \, dt$$

Using the fact that $\epsilon^{-st} = \epsilon^{-s(t-\tau)} \epsilon^{-s\tau}$ and splitting the integrals we get

$$\mathcal{L}[f_1 * f_2] = \int_{0-}^{\infty} f_2(\tau) \epsilon^{-s\tau} \, d\tau \int_{0-}^{\infty} f_1(t - \tau) \epsilon^{-s(t-\tau)} \, dt$$

$$= \int_{0-}^{\infty} f_2(\tau) \epsilon^{-s\tau} \, d\tau \int_{0-}^{\infty} f_1(\lambda) \epsilon^{-s\lambda} \, d\lambda$$

where $\lambda = t - \tau$ is the new integration variable.  Therefore,

(3.27)   $\mathcal{L}[f_1 * f_2] = F_1(s) F_2(s)$

where $f_1 * f_2$ represents the convolution of $f_1$ and $f_2$.

**Exercise**   Work out in complete detail the change of variables in the above derivation.

### 3.5   The Complete Response

Consider the linear time-invariant circuit shown in Fig. 3.2.  Let us calculate the current $i$ given that $v_C(0-) = 1$ volt, $i_L(0-) = 5$ amp, and $e(t) = 12 \sin 5t$ for $t \geq 0$.  The differential equation is

$$L \frac{di}{dt} + Ri + \frac{1}{C} \int_{0-}^{t} i(t') \, dt' + v_C(0-) = e(t) \qquad \text{for } t \geq 0$$

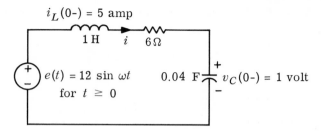

**Fig. 3.2**   Circuit used for the calculation of the complete response.

Let us take Laplace transforms of both sides; thus,

$$\frac{L}{s}\left(s^2 + \frac{R}{L}s + \frac{1}{LC}\right) I(s) = E(s) + L\, i_L(0-) - \frac{v_C(0-)}{s}$$

Using the element values of $L = 1$, $R = 6$, and $C = 0.04$, we obtain

(3.28)
$$I(s) \qquad = \frac{s}{(s+3)^2 + 4^2}\, E(s) + \frac{5s - 1}{(s+3)^2 + 4^2}$$

$$\underbrace{\qquad\qquad}_{\substack{\text{Laplace transform}\\\text{of complete}\\\text{response}}} = \underbrace{\qquad I_0(s) \qquad}_{\substack{\text{Laplace transform}\\\text{of zero-state}\\\text{response}}} + \underbrace{\qquad I_i(s) \qquad}_{\substack{\text{Laplace transform}\\\text{of zero-input}\\\text{response}}}$$

To emphasize the physical meaning of the complete response, we have separated the above into two terms, one due to the input and the other due to the initial conditions. Let us calculate the zero-state response. We first obtain $E(s)$ from the given input $e(t) = 12 \sin 5t$ for $t \geq 0$; thus,

$$E(s) = \frac{60}{s^2 + 5^2}$$

From (3.28), we have

$$I_0(s) = \frac{60s}{[(s+3)^2 + 4^2](s^2 + 5^2)}$$

Using partial-fraction expansion, we write

$$I_0(s) = \frac{K_1}{s+3-j4} + \frac{\overline{K}_1}{s+3+j4} + \frac{K_3}{s-j5} + \frac{\overline{K}_3}{s+j5}$$

where

$$K_1 = (s + 3 - j4)I_0(s)\Big|_{s=-3+j4} = j1.25 = 1.25\epsilon^{j90°}$$

and

$$K_3 = (s - j5)I_0(s)\Big|_{s=j5} = -j = \epsilon^{-j90°}$$

The inverse Laplace transform is written by inspection using (3.19) as follows:

$$i_0(t) = \mathcal{L}^{-1}[I_0(s)] = 2.5\epsilon^{-3t} \cos(4t + 90°) + 2\cos(5t - 90°)$$

$$= -2.5\epsilon^{-3t} \sin 4t + 2\sin 5t \qquad \text{for } t \geq 0$$

From (3.28), the zero-input response is

$$i_i(t) = \mathcal{L}^{-1}[I_i(s)] = 5\epsilon^{-3t} \cos 4t - 4\epsilon^{-3t} \sin 4t \qquad \text{for } t \geq 0$$

The complete response then is

$$i(t) = i_0(t) + i_i(t)$$

$$= 5\epsilon^{-3t} \cos 4t - 6.5\epsilon^{-3t} \sin 4t + 2\sin 5t \qquad \text{for } t \geq 0$$

The last term in the right-hand side is the sinusoidal steady-state response; the other terms form the transient response.

This example illustrates the general fact that the transient response is contributed by both the initial state and the sudden application of the input sinusoid $12 \sin 5t$ at $t = 0$.

**Exercise**    Consider the linear time-invariant circuit shown in Fig. 3.3. Its initial state is given by $v_C(0-) = 2$ volts and $i_L(0-) = 3$ amp. Calculate the complete response for the following inputs:

a.  $i_s(t) = 6 \cos 7t$

b.  $i_s(t) = u(t)$

c.  $i_s(t) = \delta(t)$

d.  $i_s(t) = r(t)$

Is the answer to (c) the time derivative of the answer to (b)? If not, why not?

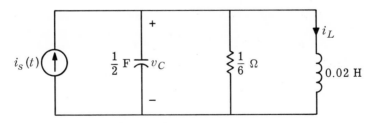

**Fig. 3.3**    Calculation of the complete response.

---

## 4 | Solution of General Networks

In this section we show how the Laplace transform is used to solve systems of *linear* integrodifferential equations with *constant* coefficients. We shall use this method to prove several important properties of linear time-invariant lumped networks.

## 4.1   Formulation of Linear Algebraic Equations

In Chap. 10, Sec. 3.4, we used node analysis to write integrodifferential equations for linear time-invariant networks. The node equations are of the form

(4.1)   $\mathbf{Y}_n(D)\mathbf{e} = \mathbf{i}$

where $\mathbf{e}$ is the node-to-datum voltage vector, $\mathbf{i}$ is the node current source vector, $\mathbf{Y}_n(D)$ is the node admittance matrix operator, $D = d/dt$ is the differentiation operator, and $D^{-1}(\,\cdot\,) = \int_0^t (\,\cdot\,)\,dt$ is the integration operator. Equation (4.1) is a set of simultaneous linear integrodifferential equations. It is easy to see that when we take the Laplace transform of Eq. (4.1), we obtain the following set of simultaneous linear algebraic equations:

(4.2)   $\mathbf{Y}_n(s)\mathbf{E}(s) = \mathbf{I}(s) + \mathbf{a}$

where $\mathbf{E}(s)$ and $\mathbf{I}(s)$ are, respectively, the Laplace transforms of $\mathbf{e}(t)$ and $\mathbf{i}(t)$; that is, the components of $\mathbf{E}(s)$ and $\mathbf{I}(s)$ are, respectively, the Laplace transforms of the corresponding components of $\mathbf{e}(t)$ and $\mathbf{i}(t)$. $\mathbf{a}$ is a vector which includes contributions due to initial conditions. $\mathbf{Y}_n(s)$ is the *node admittance matrix* in the complex frequency variable $s$. Let us pick a specific example. Although this example involves only two equations and two unknown functions, the reader should verify for himself that the methods are perfectly general.

**Example 1**   Consider the linear time-invariant circuit shown in Fig. 4.1. The input is the current $i$, and the response of interest is the voltage $e_2$. Using the node voltages $e_1$ and $e_2$ as variables, we obtain the node equations

$$(C_1 + C_3)\frac{d}{dt}e_1 + (G_1 + G_3)e_1 - \left(C_3\frac{d}{dt}e_2 + G_3e_2\right) = i$$

(4.3)   $$-\left(C_3\frac{d}{dt}e_1 + G_3e_1\right) + (C_2 + C_3)\frac{d}{dt}e_2$$

$$+ (G_2 + G_3)e_2 + \Gamma_2\int_{0-}^t e_2(t')\,dt' + j_L(0-) = 0$$

Taking the Laplace transform of both sides and applying linearity, the differentiation rule, and the integration rule, we end up with

(4.4)   $$\begin{bmatrix} (C_1 + C_3)s + (G_1 + G_3) & -(C_3s + G_3) \\[2ex] -(C_3s + G_3) & (C_2 + C_3)s + (G_2 + G_3) + \dfrac{\Gamma_2}{s} \end{bmatrix}\begin{bmatrix} E_1(s) \\[2ex] E_2(s) \end{bmatrix} = \begin{bmatrix} I(s) \\[2ex] 0 \end{bmatrix} + \begin{bmatrix} a_1 \\[2ex] a_2 \end{bmatrix}$$

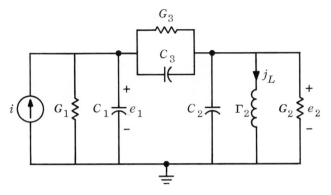

**Fig. 4.1**    Three-node linear time-invariant circuit.

where $E_1(s)$, $E_2(s)$, and $I(s)$ are the Laplace transforms of $e_1(t)$, $e_2(t)$, and $i(t)$, respectively, and

(4.5)    $a_1 = e_1(0-)(C_1 + C_3) - e_2(0-)C_3$

(4.6)    $a_2(s) = -e_1(0-)C_3 + e_2(0-)(C_2 + C_3) - \dfrac{j_L(0-)}{s}$

Note that Eq. (4.4) is in the form of Eq. (4.2).

If mesh, loop, or cut-set analysis were used, we would get mesh, loop, or cut-set equations, respectively. After the Laplace transform is taken, we again end up with a set of simultaneous linear *algebraic* equations. Thus, the remaining task is the solution of these algebraic equations.

## 4.2    The Cofactor Method

Let us consider the following set of simultaneous linear algebraic equations:

(4.7)    $\mathbf{Y}_n(s)\mathbf{E}(s) = \mathbf{F}(s)$

where $\mathbf{Y}_n(s)$ is a given $n \times n$ matrix, say, the node admittance matrix, $\mathbf{E}(s)$ is the $n$-dimensional node voltage vector which is to be determined, and $\mathbf{F}(s)$ is the given $n$-dimensional vector which represents the forcing function (including initial conditions). Obviously, the unknown vector $\mathbf{E}$ can be written immediately in terms of the inverse matrix $\mathbf{Y}_n^{-1}(s)$ as follows:

(4.8)    $\mathbf{E}(s) = \mathbf{Y}_n^{-1}(s)\mathbf{F}(s)$

Thus, the complete solution of $\mathbf{E}$ involves no more than the determination of the inverse of a given matrix. Nevertheless, to calculate the inverse of a matrix, say, anything more than a $3 \times 3$ matrix, is not a simple job. Since in most problems we are only required to find one or two of the unknowns in the vector **E,** we need not find the inverse matrix.

In discussing the cofactor method it is convenient to write (4.7) in the following expanded form:

(4.9)
$$
\begin{bmatrix}
Y_{11} & Y_{12} & \cdots & Y_{1n} \\
Y_{21} & Y_{22} & \cdots & Y_{2n} \\
\multicolumn{4}{c}{\dotfill} \\
Y_{n1} & Y_{n2} & \cdots & Y_{nn}
\end{bmatrix}
\begin{bmatrix}
E_1 \\ E_2 \\ \vdots \\ E_n
\end{bmatrix}
=
\begin{bmatrix}
F_1 \\ F_2 \\ \vdots \\ F_n
\end{bmatrix}
$$

Let us assume that we wish to determine the unknown $E_j$. We first introduce the following notation: let $\Delta_n(s) \triangleq \det \mathbf{Y}_n(s)$ be the determinant called the **network determinant on the node basis.** Let $\Delta_{ij}$ be the cofactor of $Y_{ij}$; that is, let $\Delta_{ij}$ be $(-1)^{j+i}$ times the determinant of the matrix obtained by deleting row $i$ and column $j$ of the square matrix $\mathbf{Y}_n(s)$. Now we need two facts from the theory of determinants.

1.  If we multiply each element of the $j$th column by its cofactor and add the resulting $n$ products, we obtain $\Delta_n(s)$, the network determinant. In mathematical notation,

$$
\sum_{i=1}^{n} Y_{ij}(s)\Delta_{ij}(s) = \Delta_n(s) \qquad \text{for all } s
$$

2.  If we multiply each element of the $k$th column (where $k$ is any integer different from $j$) by the cofactor of the corresponding element of the $j$th column and sum the resulting $n$ products, we obtain a polynomial which is identically zero. In mathematical notation, whenever $k \neq j$,

$$
\sum_{i=1}^{n} Y_{ik}(s)\Delta_{ij}(s) = 0 \qquad \text{for all } s
$$

These facts imply that multiplying the first scalar equation of (4.9) by $\Delta_{1j}(s)$, the second by $\Delta_{2j}(s), \ldots,$ the $n$th by $\Delta_{nj}(s)$, and by adding gives

$$
\Delta_n(s)E_j(s) = \sum_{k=1}^{n} \Delta_{kj}(s)F_k(s)
$$

Thus, the unknown $E_j(s)$ is given by

(4.10)    $$ E_j(s) = \frac{\displaystyle\sum_{k=1}^{n} \Delta_{kj}(s)F_k(s)}{\Delta_n(s)} $$

This result is often referred to as the solution obtained from Cramer's rule. To illustrate the cofactor method, let us continue with the circuit of Example 1.

---

**Example 2**    We consider the voltage $E_2(s)$ as the variable which we wish to determine. From Eq. (4.4) we have

$$\Delta_n(s) = (C_1 C_2 + C_2 C_3 + C_3 C_1) s^2$$
$$+ (G_1 C_2 + G_1 C_3 + G_3 C_2 + G_2 C_1 + G_2 C_3 + G_3 C_1) s$$
$$+ (C_1 \Gamma_2 + C_3 \Gamma_2 + G_1 G_2 + G_1 G_3 + G_2 G_3) + (G_1 + G_3) \frac{\Gamma_2}{s}$$

$$\Delta_{12}(s) = C_3 s + G_3$$

and

$$\Delta_{22}(s) = (C_1 + C_3) s + G_1 + G_3$$

The forcing functions are

$$F_1(s) = I(s) + a_1$$

and

$$F_2(s) = a_2(s)$$

Using Eq. (4.10), we can write $E_2(s)$ in terms of, separately, the part due to the input current source and the part which is contributed by the initial conditions; thus,

(4.11)   $$E_2(s) = \frac{\Delta_{12}(s)}{\Delta_n(s)} I(s) + \frac{N(s)}{\Delta_n(s)}$$

where

$$N(s) \triangleq \Delta_{12}(s) a_1 + \Delta_{22}(s) a_2(s)$$

---

**Remarks**  1. The first term on the right-hand side of (4.11) is the Laplace transform of the zero-state response, and the second term is the Laplace transform of the zero-input response; $E_2(s)$ is the Laplace transform of the complete response. The remaining problem is to obtain the time function $e_2(t)$. Observe that in (4.11), whenever $I(s)$ is a rational function of $s$, the complete response $e_2(t)$ can be obtained by partial-fraction expansion.

2. In this example the current source $i$ is the input, and the voltage $e_2$ is the output. Since the network function is defined as the ratio of the Laplace transform of the zero-state response and the Laplace transform of the input (see Eq. 3.21), we have, from (4.11), the network function

(4.12)   $$H(s) = \frac{\Delta_{22}(s)}{\Delta_n(s)}$$

which is a rational function.

---

**4.3**   **Network Functions and Sinusoidal Steady State**

Suppose we were asked to find the sinusoidal steady-state response of the network $\mathfrak{N}$ shown in Fig. 4.2. Its input is the sinusoidal current $i_1(t) = I_m$

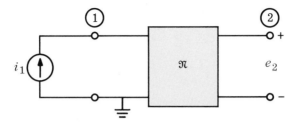

**Fig. 4.2**   A general network with input current source $i_1$ and output voltage $e_2$.

cos $\omega t$, and its output is the node-to-datum voltage $e_2$. For purposes of phasor analysis we represent the input by the phasor $I_1 = I_m$ (a real number in the present case). The (unknown) output is represented by the phasor $E_2$. The node equations have the matrix form

$$(4.13) \qquad \left[\quad \mathbf{Y}_n(j\omega) \quad\right] \begin{bmatrix} E_1 \\ E_2 \\ \vdots \\ E_n \end{bmatrix} = \begin{bmatrix} I_1 \\ 0 \\ \vdots \\ 0 \end{bmatrix}$$

If we solve this set of equations by Cramer's rule, we obtain

$$(4.14) \qquad E_2 = \frac{\Delta_{12}(j\omega)}{\Delta_n(j\omega)} I_1$$

where $\Delta_{12}(j\omega)$ is the cofactor of the (1,2) element of $\mathbf{Y}_n(j\omega)$, and $\Delta_n(j\omega)$ is the determinant of $\mathbf{Y}_n(j\omega)$. The factor $\Delta_{12}(j\omega)/\Delta_n(j\omega)$ is a rational function of $\omega$ and, since it relates the input phasor $I_1$ to the output phasor $E_2$, it is called a transfer impedance. It is a special case of a network function. If, for simplicity, we write

$$(4.15) \qquad H(j\omega) \triangleq \frac{\Delta_{12}(j\omega)}{\Delta_n(j\omega)}$$

then the relation between the waveform $e_2(\,\cdot\,)$ and $i_1(\,\cdot\,)$ is

$$(4.16) \qquad e_2(t) = \text{Re}\,[H(j\omega)I_m\epsilon^{j\omega t}] = |H(j\omega)|I_m\,\cos\,[\omega t + \sphericalangle H(j\omega)]$$

Consider a related problem. Suppose $\mathfrak{N}$ is in the zero state at $t = 0-$, and we switch on the input $i_1(t) = I_m \cos \omega t$ at $t = 0-$. The problem is to calculate the response $e_2(\,\cdot\,)$. As usual, for simplicity, we shall perform all calculations as if the input were $I_m\epsilon^{j\omega t}$, and from the response† $e_{2e}$ we take the real part. Let us use the Laplace transform and, as above, node

---

† The subscript $e$ is used to remind us that the corresponding quantity relates to the *exponential* input.

analysis. Since the network $\mathfrak{N}$ is in the zero state at $t = 0-$, the Laplace transformed node equations are of the form

$$(4.17) \quad \begin{bmatrix} & & \\ & \mathbf{Y}_n(s) & \\ & & \end{bmatrix} \begin{bmatrix} E_{1e}(s) \\ E_{2e}(s) \\ \vdots \\ E_{ne}(s) \end{bmatrix} = \begin{bmatrix} I_m/(s - j\omega) \\ 0 \\ \vdots \\ 0 \end{bmatrix}$$

Note that the matrix $\mathbf{Y}_n(s)$ is precisely the same matrix as in (4.13) except for the fact that $j\omega$ has been replaced throughout by $s$. The Laplace transform of the zero-state response is

$$(4.18) \quad E_{2e}(s) = \frac{\Delta_{12}(s)}{\Delta_n(s)} \frac{I_m}{s - j\omega}$$

Therefore, using the same previous notation, that is, $H = \Delta_{12}/\Delta_n$ [$\Delta_n$ as in (4.15)], we have

$$(4.19) \quad E_{2e}(s) = H(s) \frac{I_m}{s - j\omega}$$

Let us assume that *all* the poles of $H(s)$ are in the open left-half plane [that is, $\text{Re}(p_j) < 0, j = 1, 2, \ldots$, where $p_j$ are the poles of $H(s)$]. The partial-fraction expansion of (4.19) contains terms which can be grouped into two sets. In the first set we include all terms of the form $K_{jk}/(s - p_j)^k$ (in the case of multiple poles), and we do so for all the poles of $H(s)$. In the second set we include the single term

$$(4.20) \quad \frac{K}{s - j\omega} = \frac{H(j\omega)I_m}{s - j\omega}$$

Since $\text{Re}(p_j) < 0$ for all $j$, then for all $j$, the time functions corresponding to the terms of the first set tend exponentially to zero as $t \to \infty$. Thus, for large $t$, the response $e_{2e}$ tends exponentially to the time function which corresponds to the single term (4.20), and we have

$$e_{2e}(t) = \mathcal{L}^{-1}[E_{2e}(s)]$$
$$\approx H(j\omega)I_m \epsilon^{j\omega t} \qquad \text{for large } t$$

To obtain the response $e_2(\cdot)$ due to $i_1(t) = I_m \cos \omega t$, we take the real part of the above expression and obtain

$$(4.21) \quad e_2(t) \approx |H(j\omega)|I_m \cos[\omega t + \sphericalangle H(j\omega)] \qquad \text{for large } t$$

As $t \to \infty$, Eq. (4.21) gives the sinusoidal response, which is the same as that in (4.16) obtained from strictly sinusoidal steady-state analysis.

To summarize, we give the following two conclusions:

1. *Provided that all the poles of H(s) have negative real parts, the zero-state response to $I_m$ cos $\omega t$ tends to the sinusoidal steady-state response $|H(j\omega)|I_m$ cos $[\omega t + \measuredangle H(j\omega)]$.*

2. *The network function H(s) evaluated at $s = j\omega$ gives the ratio of the output phasor to the input phasor in the sinusoidal steady state at frequency $\omega$.*

**Remark**   This last conclusion is extremely important because it relates the network function to easily measurable quantities (amplitudes and relative phase of the sinusoidal input and output). Thus, for example, if nothing is known about the topology and element values of a linear time-invariant network and only the input and output are measurable, it is still possible to experimentally determine the network function.

**Exercise**   Show that if all the zeros of $\Delta_n(s)$ are in the open left-half plane [Re $(s_i) < 0$ for all $i$, where $s_i$ are the zeros of $\Delta_n(s)$], then whatever the initial state of $\mathfrak{N}$ may be, the response to $u(t) \, I_m$ cos $\omega t$ exponentially approaches $|H(j\omega)|$ cos $[\omega t + \measuredangle H(j\omega)]$ as $t \to \infty$.

## 5   Fundamental Properties of Linear Time-invariant Networks

In this section we restate and prove five fundamental properties of linear time-invariant networks. These properties have been encountered in the study of simple circuits earlier in the book.

**PROPERTY 1**   *For any linear time-invariant network, the Laplace transform of the complete response is the sum of the Laplace transform of the zero-state response and the Laplace transform of the zero-input response. Also, by the linearity of Laplace transforms, the same holds for the corresponding time functions.*

We have seen that this statement holds for the example of Sec. 4. To show that it holds in general, consider an arbitrary linear time-invariant network with a single input. We can always write node, mesh, loop, or cut-set equations so that the input appears in only the first equation. Specifically, let us use node analysis and let the variables be the node voltages $e_1, e_2, \ldots, e_n$, and let $e_2$ be the desired response. After taking Laplace transforms, we see that the equations are of the form

$$(5.1) \qquad \left[ \quad \mathbf{Y}_n(s) \quad \right] \left[ \mathbf{E}(s) \right] = \begin{bmatrix} I(s) \\ 0 \\ \vdots \\ 0 \end{bmatrix} + \begin{bmatrix} a_1 \\ a_2 \\ \vdots \\ a_n \end{bmatrix}$$

Using Cramer's rule, calling $\Delta_{12}(s)$ the cofactor of the (1,2) element of the node admittance matrix $\mathbf{Y}_n(s)$, and calling $\Delta_n(s)$ its determinant, we obtain

(5.2)
$$E_2(s) = \underbrace{\frac{\Delta_{12}(s)}{\Delta_n(s)} I(s)}_{} + \underbrace{\frac{\sum_{k=1}^{n} \Delta_{k2}(s)a_k}{\Delta_n(s)}}_{}$$

| Laplace transform of complete response | Laplace transform of zero-state response | Laplace transform of zero-input response |

Note that $\Delta_{12}(s)/\Delta_n(s)$ is a rational function and is (by definition) the *network function* relating the response $E_2$ to the input $I$. The first term on the right-hand side of (5.2) is the Laplace transform of the zero-state response because, when all initial conditions are zero, all $a_k$ are zero, and the second term in (5.2) drops out. Similarly, when the input is identically zero, $I(s) = 0$ for all $s$, and the first term drops out; consequently, the second term is the Laplace transform of the zero-input response.

Considering the first term of (5.2) and noting that the network function $\Delta_{12}(s)/\Delta(s)$ is a rational function with real coefficients, we state the following property.

**PROPERTY 2**   *The network function is, by definition, the function of s which, when multiplied by the Laplace transform of the input, gives the Laplace transform of the zero-state response. For any lumped linear time-invariant network, any of its network functions is a rational function with real coefficients.*

The reason for specifically requiring the network to be "lumped" is that for distributed networks network functions need not be rational functions.

In a number of examples we found that the initial conditions required to specify uniquely the solution to a network problem were the initial voltages across capacitors and the initial currents in the inductors. We are going to use the Laplace transform to prove that this is true in general. We express this property in the following statement.

**PROPERTY 3**   *For any linear time-invariant network, the initial conditions required to solve for any network variable are completely specified by the initial voltages on the capacitors and the initial currents in the inductors.*

The idea contained in this assertion can also be expressed by saying that the initial state of a linear time-invariant network is completely specified by the initial voltages on the capacitors and the initial currents in the inductors.

This is almost obvious. Consider a linear time-invariant network made of resistors, capacitors, inductors, mutual inductors, and independent and dependent sources. We are going to show that the very process of writing down the Laplace transform equations requires *all* initial inductor currents and *all* initial capacitor voltages. Suppose we perform a node or cut-set analysis; we then use voltages as variables. The initial inductor currents appear as a result of expressing the inductor current as a function of the inductor voltages; thus,

$$i_L(t) = \Gamma \int_0^t v_L(t') \, dt' + i_L(0-)$$

The initial capacitor voltages will appear when the Laplace transform of the capacitor current is taken; thus,

$$i_C(t) = C \frac{dv_C}{dt}$$

hence

$$I_C(s) = Cs V_C(s) - C v_C(0-)$$

For mesh or loop analysis, the dual relations introduce in the Laplace transform equations all the initial inductor currents and all the initial capacitor voltages. Therefore, the Laplace transform of any network variable is completely specified by the Laplace transform of the inputs and the initial capacitor voltages and initial inductor currents.

**PROPERTY 4**   *For any linear time-invariant network the network function is the Laplace transform of the corresponding impulse response.*

Indeed, suppose we wanted to calculate the impulse response, i.e., the voltage $e_2$ due to the input $i(t) = \delta(t)$, given that the network is in the zero state prior to the application of the impulse at the input. Then, all initial conditions are zero, and the right-hand side of (5.1) reduces to the single column vector $(1, 0, 0, \ldots, 0)^T$. If we call $H(s)$ the Laplace transform of the impulse response, we obtain from (5.2)

(5.3)   $$H(s) = \frac{\Delta_{12}(s)}{\Delta_n(s)}$$

The left-hand side is the Laplace transform of the impulse response, and the right-hand side is the network function. Therefore, Property 4 is established.

**PROPERTY 5**   *For any linear time-invariant network the derivative of the step response is the impulse response.*

From (5.3), if we call $s(t)$ the step response, we get

(5.4)   $$\mathcal{L}[s(t)] = \frac{\Delta_{12}(s)}{\Delta_n(s)} \frac{1}{s}$$

Since the step response $s(t)$ is, by definition, identically zero for $t < 0$, we obtain, by the differentiation rule,

(5.5)   $\mathcal{L}\left[\dfrac{d\Delta}{dt}\right] = s\mathcal{L}[\Delta] = \dfrac{\Delta_{12}(s)}{\Delta_n(s)} = H(s) = \mathcal{L}[h(t)]$

By the uniqueness of the Laplace transform we conclude that

(5.6)   $h(t) = \dfrac{d\Delta}{dt}$

When we use (5.6), it is understood that whenever $\Delta$ goes through a "jump," its derivative includes the corresponding impulse.

Property 5 is *not true* in the case of linear *time-varying* networks, or in the case of nonlinear networks.   (See Probs. 13 and 14.)

**Remark**   To be quite precise, throughout the discussion above we should have ruled out the case where the determinant $\Delta_n(s)$ is *identically* zero (i.e., zero for all values of $s$).   When this occurs, we say that we are in the *degenerate* case. In such cases, the network may have no solution, or it may have infinitely many solutions.   It is physically obvious that degenerate cases occur as a result of oversimplifications in the modeling process; certain relevant physical aspects of the problem have been overlooked.   Examples of degenerate networks will be given in Sec. 7.

---

| **6** | **State Equations** |

The Laplace transform applies equally well to state equations.

**Example**   Consider the circuit shown in Fig. 6.1.   For the element values shown, the state equations read

(6.1)   $\begin{bmatrix} \dfrac{di}{dt} \\[2mm] \dfrac{dv}{dt} \end{bmatrix} = \begin{bmatrix} 0 & -1 \\ 1 & -\frac{1}{2} \end{bmatrix} \begin{bmatrix} i \\ v \end{bmatrix} + \begin{bmatrix} 1 \\ 0 \end{bmatrix} e_s$

It is standard to write this equation in the form

(6.2)   $\dot{\mathbf{x}} = \mathbf{A}\mathbf{x} + \mathbf{b}w$

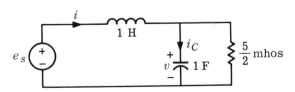

**Fig. 6.1**   Example illustrating state equations.

Thus, in the present case, $\mathbf{x} = [i \quad v]^T$ is the state vector, $w = e_s$ is the input, and

$$\mathbf{A} = \begin{bmatrix} 0 & -1 \\ 1 & -\frac{5}{2} \end{bmatrix} \qquad \mathbf{b} = \begin{bmatrix} 1 \\ 0 \end{bmatrix}$$

Taking the Laplace transform of Eq. (6.1), we obtain

$$\begin{bmatrix} s & 1 \\ -1 & s + \frac{5}{2} \end{bmatrix} \begin{bmatrix} I(s) \\ V(s) \end{bmatrix} = \begin{bmatrix} 1 \\ 0 \end{bmatrix} E(s) + \begin{bmatrix} i(0-) \\ v(0-) \end{bmatrix}$$

where $I(s)$, $V(s)$, and $E(s)$ are, respectively, the Laplace transforms of $i(t)$, $v(t)$, and $e_s(t)$. We could simply solve these two equations in two unknowns. However, to exhibit the general formalism, let us multiply on the left by the inverse matrix

$$\begin{bmatrix} s & 1 \\ -1 & s + \frac{5}{2} \end{bmatrix}^{-1} = \frac{1}{\psi(s)} \begin{bmatrix} s + \frac{5}{2} & -1 \\ 1 & s \end{bmatrix}$$

where $\psi(s)$ is the determinant of

$$\begin{bmatrix} s & 1 \\ -1 & s + \frac{5}{2} \end{bmatrix}$$

and is given by $s^2 + \frac{5}{2}s + 1$. The result is

$$(6.3) \qquad \begin{bmatrix} I(s) \\ V(s) \end{bmatrix} = \begin{bmatrix} \dfrac{s + \frac{5}{2}}{\psi(s)} & \dfrac{-1}{\psi(s)} \\ \dfrac{1}{\psi(s)} & \dfrac{s}{\psi(s)} \end{bmatrix} \begin{bmatrix} 1 \\ 0 \end{bmatrix} E(s) + \begin{bmatrix} \dfrac{s + \frac{5}{2}}{\psi(s)} & \dfrac{-1}{\psi(s)} \\ \dfrac{1}{\psi(s)} & \dfrac{s}{\psi(s)} \end{bmatrix} \begin{bmatrix} i(0-) \\ v(0-) \end{bmatrix}$$

Observe that the response is expressed as the sum of two terms; the first one is the Laplace transform of the zero-state response, and the second one is the Laplace transform of the zero-input response. For example, let the initial state be the zero state; thus, $i(0-) = v(0-) = 0$. Let $e_s(t) = \delta(t)$; thus, $E(s) = 1$.

The response is obtained from (6.3) by taking the inverse Laplace transform of

$$\begin{bmatrix} I(s) \\ V(s) \end{bmatrix} = \begin{bmatrix} \dfrac{s + \frac{5}{2}}{\psi(s)} & \dfrac{-1}{\psi(s)} \\ \dfrac{1}{\psi(s)} & \dfrac{s}{\psi(s)} \end{bmatrix} \begin{bmatrix} 1 \\ 0 \end{bmatrix} = \begin{bmatrix} \dfrac{s + \frac{5}{2}}{\psi(s)} \\ \dfrac{1}{\psi(s)} \end{bmatrix}$$

Noting that $\psi(s) = (s + 0.5)(s + 2)$ and using partial-fraction expansion, we obtain

(6.4)
$$\begin{bmatrix} i(t) \\ v(t) \end{bmatrix} = \begin{bmatrix} \frac{4}{3}\epsilon^{-0.5t} - \frac{1}{3}\epsilon^{-2t} \\ \frac{2}{3}\epsilon^{-0.5t} - \frac{2}{3}\epsilon^{-2t} \end{bmatrix}$$

In general, the output can be written as a linear combination of the components of the state vector and the input.  Let $y$ be the output; then in terms of standard notation we have

(6.5)   $y = \mathbf{c}^T \mathbf{x} + d_0 w$

where the vector $\mathbf{c}$ and the scalar $d_0$ are constant.  Suppose we say that the output variable for our example is the current $i_C$ in the capacitor; then

(6.6)   $i_C = i - \frac{5}{2}v = [1 \quad -\frac{5}{2}] \begin{bmatrix} i \\ v \end{bmatrix}$

In terms of the notation of Eq. (6.5), this means

$$\mathbf{c} = \begin{bmatrix} 1 \\ -\frac{5}{2} \end{bmatrix} \quad \text{and} \quad d_0 = 0$$

Combining (6.4) and (6.6), we have the impulse response

$$i_C(t) = -\frac{1}{3}\epsilon^{-0.5t} + \frac{4}{3}\epsilon^{-2t}$$

For general linear time-invariant networks with single input and single output, we have the standard state equations

(6.7)   $\dot{\mathbf{x}} = \mathbf{A}\mathbf{x} + \mathbf{b}w$

(6.8)   $y = \mathbf{c}^T \mathbf{x} + d_0 w$

Taking the Laplace transforms, we obtain

(6.9)   $(s\mathbf{1} - \mathbf{A})\mathbf{X}(s) = \mathbf{b}W(s) + \mathbf{x}(0-)$

(6.10)   $Y(s) = \mathbf{c}^T\mathbf{X}(s) + d_0 W(s)$

where $\mathbf{X}(s)$, $Y(s)$, and $W(s)$ are, respectively, the Laplace transforms of $\mathbf{x}(t)$, $y(t)$, and $w(t)$.  Multiplying on the left of (6.9) by $(s\mathbf{1} - \mathbf{A})^{-1}$, we obtain

(6.11)   $\mathbf{X}(s) = (s\mathbf{1} - \mathbf{A})^{-1}\mathbf{b}W(s) + (s\mathbf{1} - \mathbf{A})^{-1}\mathbf{x}(0-)$

Substituting (6.11) in (6.10), we obtain the Laplace transform of the output as follows

$$(6.12) \qquad \underbrace{Y(s)}_{\substack{\text{Laplace transform} \\ \text{of complete} \\ \text{response}}} = \underbrace{[\mathbf{c}^T(s\mathbf{1} - \mathbf{A})^{-1}\mathbf{b} + d_0]W(s)}_{\substack{\text{Laplace transform of} \\ \text{zero-state response}}} + \underbrace{\mathbf{c}^T(s\mathbf{1} - \mathbf{A})^{-1}\mathbf{x}(0-)}_{\substack{\text{Laplace transform of} \\ \text{zero-input} \\ \text{response}}}$$

**Remarks**   1.   Equation (6.12) should be compared with Eq. (5.2).   We recognize that the Laplace transform of the complete response is again the sum of two terms.   Note that if the network starts at zero state, $\mathbf{x}(0-) = \mathbf{0}$, the response is due to the input only, and thus, the first term on the right-hand side of (6.12) is the Laplace transform of the zero-state response.   Similarly, if the input is set to zero; that is, $W(s) = 0$ for all $s$, the response is due exclusively to the initial state $\mathbf{x}(0-)$, and we thus conclude that the second term on the right-hand side of (6.12) is the Laplace transform of the zero-input response.

2.   Recall that the network function is, by definition, the ratio of the Laplace transform of the zero-state response to that of the input. Thus, the network function $H(s)$ can be expressed explicitly in terms of $\mathbf{A}, \mathbf{b}, \mathbf{c},$ and $d_0$ of the state equations as

$$(6.13) \quad H(s) = \mathbf{c}^T(s\mathbf{1} - \mathbf{A})^{-1}\mathbf{b} + d_0$$

3.   To evaluate the inverse of the matrix $s\mathbf{1} - \mathbf{A},$ we need to find the determinant of $s\mathbf{1} - \mathbf{A}.$   Let us denote it by $\psi(s) \triangleq \det (s\mathbf{1} - \mathbf{A}).$   It is a polynomial in the complex frequency variable $s$ and is called the **characteristic polynomial of the matrix A.**   Then

$$(6.14) \quad (s\mathbf{1} - \mathbf{A})^{-1} = \frac{1}{\psi(s)} \mathbf{N}(s)$$

where $\mathbf{N}(s)$ is a matrix with *polynomial* elements.   Equation (6.13) can be written as

$$(6.15) \quad H(s) = \frac{1}{\psi(s)} \mathbf{c}^T\mathbf{N}(s)\mathbf{b} + d_0$$

Thus, any pole of the network function $H(s)$ is a zero of the characteristic polynomial $\psi(s).$   However, some zero of the characteristic polynomial may sometimes not be a pole because there may be a cancellation between the numerator polynomial $\mathbf{c}^T\mathbf{N}(s)\mathbf{b}$ and the denominator polynomial $\psi(s).$

**Exercise**   Determine the network function specified by the input $e_s$ and the output $i_C$ in the circuit of Fig. 6.1.

| 7 | **Degenerate Networks** |

Intuitively, we would expect that every linear time-invariant $RLC$ network has a unique solution in response to any set of initial conditions and any set of independent sources. We shall give below some examples to show that this is not the case in some limiting situations. We shall also see that when a linear time-invariant network includes some dependent sources or when it includes some negative resistance, it may happen that, for some initial conditions and/or for some source distribution, the network has no solution or more than one solution.

**Example 1**   This is the simplest example of a passive $RLCM$ network that has infinitely many solutions. Consider the closely coupled inductors shown in Fig. 7.1. The mesh equations are

$$(7.1) \quad \mathbf{L}\frac{d\mathbf{i}}{dt} = \begin{bmatrix} 4 & 6 \\ 6 & 9 \end{bmatrix} \begin{bmatrix} \dfrac{di_1}{dt} \\ \dfrac{di_2}{dt} \end{bmatrix} = \begin{bmatrix} 0 \\ 0 \end{bmatrix}$$

and suppose that

$$(7.2) \quad i_1(0) = i_2(0) = 0$$

Note that the inductance matrix $\mathbf{L}$ is singular [that is, det $(\mathbf{L}) = 0$]. This is a direct consequence of the fact that the inductors are closely coupled. It is easy to see that if we call $\mathbf{a}$ the vector $(-3, 2)^T$, then $\mathbf{La} = \mathbf{0}$. Consequently, if $f(t)$ is *any* differentiable function with $f(0) = 0$,

$$(7.3) \quad i_1(t) = -3f(t) \quad \text{and} \quad i_2(t) = 2f(t)$$

constitute a solution of (7.1) which satisfies the initial conditions of (7.2).

   This may seem an obvious contradiction to the uniqueness theorem of differential equations (Appendix C, Sec. 4). However, in order to apply the uniqueness theorem of the Appendix we must have the equations in the *normal form* $\dot{\mathbf{x}} = \mathbf{f}(\mathbf{x}, t)$. To bring Eq. (7.1) to the normal form, we would have to multiply (7.1) by the inverse of $\mathbf{L}$. However, $\mathbf{L}$ does not have an inverse since it is singular; hence Eq. (7.1) cannot be written in the normal form, and, consequently, the uniqueness theorem does not apply.

   From a physical point of view, the trouble lies, of course, in that somewhere in the modeling process the baby got thrown out with the bath water. This point is demonstrated in the following exercise.

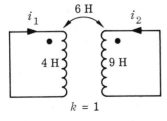

**Fig. 7.1**   Example of a degenerate network: closely coupled inductors.

**Exercise**   Verify that, if the circuit shown on Fig. 7.1 is modified in any one of the following ways, the circuit will have one and only one solution; (1) increase any of the *self*-inductances by a positive amount (this models the ever-present leakage flux), (2) insert a linear resistor in any of the meshes, or (3) insert a small capacitor in series with any of the inductors.

**Example 2**   This example exhibits the kind of difficulty which may occur when a network includes both *RLC* elements and *dependent* sources, and when the element values bear special relationships to one another.   The node equations for the circuit in Fig. 7.2 have the form

(7.4)
$$\frac{d}{dt} e_1 + 2e_1 - 4e_2 = f_1$$
$$-\frac{1}{2}\frac{d}{dt} e_1 - e_1 + 2e_2 = f_2$$

First note that if the second equation is multiplied by $-2$, the resulting left-hand side is identical with that of the first equation; consequently, the system (7.4) will have a solution if and only if

(7.5)   $f_1 = -2f_2$

In other words, unless the independent current sources have their waveforms $f_1(\cdot)$ and $f_2(\cdot)$ related by Eq. (7.5), it is impossible to find branch voltages and branch currents that satisfy the branch equations and Kirchhoff's laws.   Thus, whenever $f_1$ is not equal to $-2f_2$, Fig. 7.2 gives an example of a linear time-invariant network that has no solution.
Now suppose that

(7.6)   $f_1(t) = f_2(t) = 0$      for all $t$

and $e_1(0) = 0$.   It is easy to verify (by direct substitution) that if $g(t)$ is any differentiable function satisfying $g(0) = 0$, then

$$e_1(t) = 4g(t) \qquad \text{and} \qquad e_2(t) = \dot{g}(t) + 2g(t)$$

constitute a solution of (7.4) when (7.6) holds and $e_1(0) = 0$.   In this case

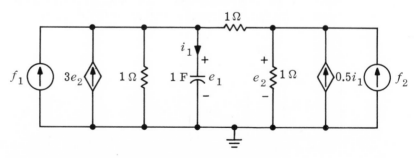

**Fig. 7.2**   Example of a degenerate network with dependent sources.

we have an infinite number of solutions. Again this is a case where the modeling has gone astray.

Now if we take the Laplace transform of both sides of Eq. (7.4) we get

$$\mathbf{Y}_n(s)\mathbf{E}(s) = \mathbf{F}(s)$$

The cause of these difficulties is that $\det [\mathbf{Y}_n(s)] = 0$ for all $s$.

The qualification "for all $s$" is of crucial importance in this case; indeed, $\det [\mathbf{Y}_n(s)]$ is a polynomial in the complex variable $s$, and, therefore, it is *always* possible to find *some* values of $s$ at which $\det [\mathbf{Y}_n(s)]$ takes the value zero. On the other hand, the only way in which $\det [\mathbf{Y}_n(s)]$ can be zero *for all s* is for *all* its coefficients to be zero.

---

**Exercise** Calculate $\det [\mathbf{Y}_n(s)]$ for each example above (verify that it is equal to zero for all $s$).

These examples justify the following definition: a linear time-invariant network will be called **degenerate** whenever the determinant of the system of Laplace transform equations obtained from node, mesh, loop, or cut-set analysis is equal to zero *for all values of s*. It is important to note that degenerate networks are a consequence of oversimplifications in modeling.

---

| **8** | **Sufficient Conditions for Uniqueness** |
|---|---|

The first fact we wish to establish is the following: any linear time-invariant *RLC* network, which has all its resistors with positive resistances, all its capacitors with positive capacitances, and all its inductors with positive inductances, has a unique solution in response to any set of independent sources and to any initial state.

Since the conditions on the *RLC* elements amount to requiring them to be passive, we can restate the result as follows:

*Given any initial state and any set of inputs, any linear time-invariant network made of passive R's, L's, and C's has a unique solution.*

*Proof* We base our proof on node analysis. In terms of Laplace transforms the node equations have the form

(8.1)  $\mathbf{Y}_n(s)\mathbf{E}(s) = \mathbf{F}(s)$

where the right-hand side vector $\mathbf{F}(s)$ includes the contributions both of the independent sources and the initial conditions. The solution will be unique if it is not the case that $\det [\mathbf{Y}_n(s)] = 0$ for all $s$. In fact, we shall show that whenever $s$ is *real* and positive, $\det [\mathbf{Y}_n(s)] \neq 0$. For purposes of our proof let $s_1$ be any *real* and *positive* number. Consider the branch admittance matrix $\mathbf{Y}_b(s)$ evaluated at $s_1$; it is a $b \times b$ matrix, $\mathbf{Y}_b(s_1)$, which is diagonal, and its elements are of the form $G$, $Cs_1$, and $1/Ls_1$, where $G$,

$C$, and $L$ are positive numbers. Therefore, the diagonal matrix $\mathbf{Y}_b(s_1)$ has all its diagonal elements positive. Consider the node admittance matrix $\mathbf{Y}_n(s)$ evaluated at $s_1$; it is related to the branch admittance matrix by

(8.2)   $\mathbf{Y}_n(s_1) = \mathbf{A}\mathbf{Y}_b(s_1)\mathbf{A}^T$

where the reduced incidence matrix $\mathbf{A}$ is an $n \times b$ matrix which is of full rank. For the purpose of a proof by contradiction, suppose that det $[\mathbf{Y}_n(s_1)] = 0$. Then there exists a *nonzero* $n$-vector $\mathbf{x}$ such that

(8.3)   $\mathbf{Y}_n(s_1)\mathbf{x} = \mathbf{0}$

Indeed, Eq. (8.3) represents a system of homogeneous linear algebraic equations in $x_1, x_2, \ldots, x_n$, and the determinant of the system is equal to zero. If we multiply Eq. (8.3) by $\mathbf{x}^T$ on the left, the result is equal to zero, and we obtain successively

(8.4)   $\begin{aligned} 0 = \mathbf{x}^T\mathbf{Y}_n(s_1)\mathbf{x} &= \mathbf{x}^T\mathbf{A}\mathbf{Y}_b(s_1)\mathbf{A}^T\mathbf{x} \\ &= (\mathbf{A}^T\mathbf{x})^T\mathbf{Y}_b(\mathbf{A}^T\mathbf{x}) \end{aligned}$

Thus, if we define the $b$-vector $\mathbf{z}$ by $\mathbf{z} \triangleq \mathbf{A}^T\mathbf{x}$, we see that

(8.5)   $\mathbf{z}^T\mathbf{Y}_b(s_1)\mathbf{z} = 0$

This new vector $\mathbf{z} \neq \mathbf{0}$ because $\mathbf{x} \neq \mathbf{0}$, and $\mathbf{A}^T$ is of full rank. Now Eq. (8.5) cannot hold because the real vector $\mathbf{z} \neq \mathbf{0}$, and $\mathbf{Y}_b(s_1)$ is a diagonal matrix with *positive* diagonal elements; in fact, (8.5) asserts that

$$\sum_{i=1}^{b} y_{ii}(s_1)z_i{}^2 = 0$$

where not all the $z_i$'s are zero and each $y_{ii}(s_1)$ is positive. This is a contradiction. Consequently, we must have det $[\mathbf{Y}_n(s_1)] \neq 0$.

The result above can be extended to the case in which there are mutual inductances. Precisely formulated, the result is as follows:

*Suppose that $\mathfrak{N}$ is a linear time-invariant RLCM network, such that all its resistors have positive resistances, all its capacitors have positive capacitances, all its inductors have positive inductances. Suppose further that every set of coupled inductors has a positive definite inductance matrix. Under these conditions, given any initial state and any set of inputs, the network $\mathfrak{N}$ has a unique solution.*

This statement can be proved by using loop analysis and by using the fact that for $s_1$ real and positive, the branch impedance matrix $\mathbf{Z}_b(s_1)$ is a real symmetric matrix that is positive definite.

**Exercise**   Consider a linear time-invariant network such that $\mathbf{Y}_b(j\omega) = \mathbf{G}_b + j\mathbf{B}_b(\omega)$, where $\mathbf{G}_b$ is a matrix of real numbers, and $\mathbf{B}_b(\omega)$ has real elements which depend on $\omega$. Establish the following facts:

*a.* If, for some fixed $\omega$, the real matrix $\mathbf{G}_b + \mathbf{B}_b(\omega)$ is symmetric and positive definite, then the node admittance matrix $\mathbf{Y}_n(j\omega)$ is nonsingular.

*b.* For any $RC$ network made of positive resistances and positive capacitances and for all $\omega > 0$, $\mathbf{Y}_n(j\omega)$ is nonsingular.

*c.* For any $RL$ network made of positive resistances and positive inductances and for all $\omega > 0$, use loop analysis to show that the loop impedance matrix $\mathbf{Z}_l(j\omega)$ is nonsingular.

## Summary

■ The basic properties of the Laplace transform are those summarized in Table 13.2. As with logarithms, the use of the Laplace transform is based on the uniqueness property, namely, to one time function corresponds only one Laplace transform, and conversely, to one Laplace transform corresponds only one time function:

**Table 13.2**  **Basic Properties of the Laplace Transform**

*Defining integral:*
$$\mathcal{L}[f(t)] \triangleq \int_{0-}^{\infty} f(t)\epsilon^{-st}\, dt$$

*Linearity:*
$$\mathcal{L}[c_1 f_1(t) + c_2 f_2(t)] = c_1 \mathcal{L}[f_1(t)] + c_2 \mathcal{L}[f_2(t)]$$
where $c_1$ and $c_2$ are arbitrary *constants*.

*Differentiation:*
$$\mathcal{L}\left[\frac{df}{dt}\right] = s\mathcal{L}[f(t)] - f(0-)$$
(If $f$ has "jumps," $df/dt$ includes corresponding impulses.)
$$\mathcal{L}\left[\frac{d^2 f}{dt^2}\right] = s^2\mathcal{L}[f(t)] - sf(0-) - f^{(1)}(0-)$$
$$\mathcal{L}\left[\frac{d^3 f}{dt^3}\right] = s^3\mathcal{L}[f(t)] - s^2 f(0-) - s f^{(1)}(0-) - f^{(2)}(0-)$$

*Integration:*
$$\mathcal{L}\left[\int_{0-}^{t+} f(t')\, dt'\right] = \frac{1}{s}\mathcal{L}[f(t)]$$

*Convolution:*
Denote the convolution of $f_1$ and $f_2$ by $f_1 * f_2$; that is,
$$(f_1 * f_2)(t) \triangleq \int_{0-}^{t+} f_1(t - \tau) f_2(\tau)\, d\tau \qquad \text{for } t \geq 0$$
$$\mathcal{L}[f_1 * f_2] = \mathcal{L}[f_1(t)]\mathcal{L}[f_2(t)]$$

- A network function of a linear time-invariant network is by definition the ratio of the Laplace transform of the zero-state response and the Laplace transform of the input.

- When it is evaluated at $s = j\omega$, the network function gives the complex number $H(j\omega)$. It is the ratio of the output phasor and the input phasor in the sinusoidal steady state at frequency $\omega$.

- The five fundamental properties of linear time-invariant networks are as follows:

  1. The complete response is the sum of the zero-state response and the zero-input response.
  2. Any network function is a rational function with real coefficients of the complex frequency variable $s$.
  3. Any network variable is completely specified by the input waveform, the initial capacitor voltages, and the initial inductor currents.
  4. Any network function is the Laplace transform of the corresponding impulse response.
  5. The derivative of the step response is the impulse response.

- A linear time-invariant network is called *degenerate* if the determinant of the system of Laplace transform equations is equal to zero for all values of $s$.

- A linear time-invariant $RLC$ network with positive resistances, inductances, and capacitances has a unique solution for any given initial state and for any set of input waveforms.

## Problems

Calculating
$F(s)$

**1.** Find the Laplace transform of

  a.  $3\epsilon^{-2t} + 4\epsilon^{-t} \cos{(3t + 4)} + t\epsilon^{-t}$ (Note: The phase angle is in radians.)

  b.  $t^3 - 2t + 1$

  c.  $\sinh bt$

  d.  $\epsilon^{-at} f(t)$, where $\mathcal{L}[f(t)] = F(s)$

Properties of
Laplace
transform

**2.** Given that $\mathcal{L}[f(t)] = F(s)$, prove that

  a.  $\mathcal{L}[f(at)] = (1/a)F(s/a)$, where $a$ is a real positive constant.

  b.  $\mathcal{L}[tf(t)] = -(d/ds)F(s)$.

  c.  $\mathcal{L}[f(t - T)u(t - T)] = \epsilon^{-Ts}F(s)$, where $T$ is a positive constant, and $f(t) = 0$ for $t < 0$. Use this result to calculate the Laplace transform of $g(t) = \epsilon^{-t} + u(t - 1)\epsilon^{-(t-1)} + \delta(t - 2)$.

Properties of
Laplace
transform

**3.** Knowing that $\mathcal{L}[\epsilon^{-at}] = 1/(s + a)$, and that $2 \sinh at = \epsilon^{at} - \epsilon^{-at}$ and $2 \cosh at = \epsilon^{at} + \epsilon^{-at}$, find

a. $\mathcal{L}[\cosh at]$

b. $\mathcal{L}[\sinh at]$

Properties of
Laplace
transform

**4.** Calculate (first by direct integration, then by using Laplace transforms)

a. $\displaystyle\int_0^t \epsilon^{-a\tau}\epsilon^{-b(t-\tau)}\,d\tau$

b. $\displaystyle\int_0^t \cos \omega_1\tau \cos \omega_2(t - \tau)\,d\tau$

Complete
response

**5.** The linear time-invariant network shown in Fig. P13.5 is in the steady state with switch $S_1$ closed. Switch $S_1$ is open at $t = 0$. Find $i_1(t)$ and $v_{L_2}(t)$ for $t \geq 0$, given that $V = 2$ volts, $L_1 = L_2 = 1$ henry, and $R_1 = R_2 = 1$ ohm.

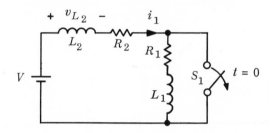

**Fig. P13.5**

Partial-fraction
expansion

**6.** Find the *inverse* Laplace transform of the following by means of the *partial-fraction expansion:*

a. $\dfrac{s^2 + 6s + 8}{s^2 + 4s + 3}$

b. $\dfrac{s^2 + s}{s^3 + 2s^2 + s + 2}$

c. $\dfrac{1}{s^3 + 2s^2 + 2s + 1}$

d. $\dfrac{1}{s^4 + 3s^3 + 4s^2 + 3s + 1}$

e. $\dfrac{(s + 1)(s + 3)}{s(s + 2)(s + 5)}$

f. $\dfrac{s(s^2 + 2)}{(s^2 + 1)(s^2 + 3)}$

g. $\dfrac{1}{s^2(s + 1)^2(s + 2)}$

h. $\dfrac{s^3 + 1}{s^2 + 2s + 2}$

Zero-input
response

**7.** Using the Laplace transform method, determine the voltage $v_2(t)$ for $t \geq 0$ for the circuit shown in Fig. P13.7, where $j_1(0) = 1$ amp, $v_2(0) = 2$ volts, and $v_3(0) = 1$ volt.

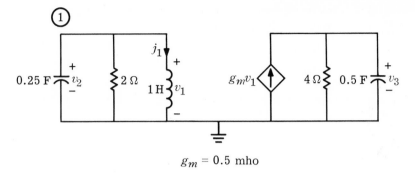

$g_m = 0.5$ mho

**Fig. P13.7**

Complete response

**8.** The network shown in Fig. P13.8 is in the steady state with switch $S$ closed. Switch $S$ is opened at $t = 0$. Find the currents $i_{L_1}(t)$ and $i_{L_2}(t)$ and the voltage $v(t)$ for $t \geq 0$. Relate the values of $i_{L_1}$ and $i_{L_2}$ at $0+$ to their values at $0-$. Explain physically.

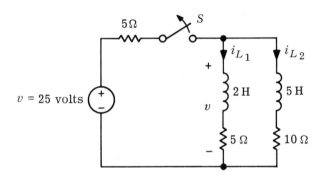

**Fig. P13.8**

Zero-input response

**9.** Find the zero-input response [that is, $v_1(t)$ and $v_2(t)$, for $t \geq 0$] of the networks in Fig. P13.9 (page 577).

Zero-state response

**10.** Calculate the zero-state response of the networks shown in Fig. P13.9 when

a. $e_1(t) = u(t) \cos \omega t$, and $e_2(t) = \delta(t)$ [use $u(t)e^{j\omega t}$ as input, and take the real part of the answer].

b. $e_1(t) = u(t)$, and $e_2(t) = \delta(t)$

c. $e_1(t) = \delta(t)$, and $e_2(t) = u(t)$

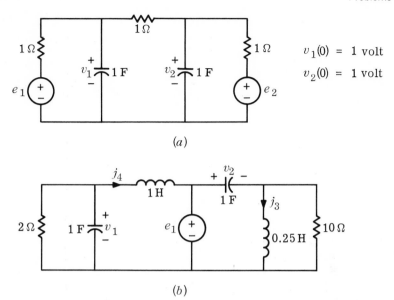

$v_1(0) = 1$ volt

$v_2(0) = 1$ volt

(a)

(b)

$v_1(0) = 0$     $v_2(0) = 2$ volts     $j_3(0) = 5$ amp     $j_4(0) = 2$ amp

**Fig. P13.9**

Complete
response

**11.** *a.* The circuit shown in Fig. P13.11 is linear and time-invariant. Write integrodifferential equations for the circuit using node-to-datum voltages $v_1$ and $v_2$ as variables and knowing that $v_1(0-) = \gamma$ and $i(0-) = \rho$.

*b.* Determine $V_2(s)$, where $V_2(s) = \mathcal{L}[v_2(t)]$.

*c.* Let $v_0(t) = u(t) \sin t$, $R_1 = R_2 = 1$ ohm, $L = 2$ henrys, $C = 2$ farads, $g_m = \frac{1}{2}$ mho, $\gamma = -2$ volts, and $\rho = 1$ amp. Calculate $v_2(t)$ for $t \geq 0$. Write the answer as the sum of the zero-input response and the zero-state response.

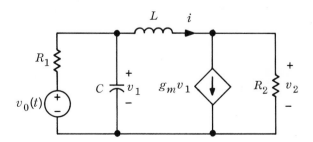

**Fig. P13.11**

Complete
response

**12.** The linear time-invariant circuit shown in Fig. P13.12 is in the steady state with switch $S$ closed.   The switch opens at $t = 0$.   Find $v_1(t)$ for $t \geq 0$, given that $R_1 = R_2 = 1$ ohm, $L = 1$ henry, $C = 1$ farad, and $v(t) = \cos t$.

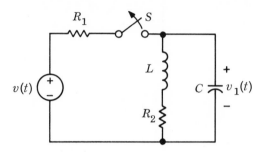

**Fig. P13.12**

Step and im-
pulse
response of
linear time-
varying
network

**13.** Consider the linear network shown in Fig. P13.13; $e$ is the input, $q$ is the response.   Show that its impulse response is $h(t) = u(t)/(t + 1)$, and its step response is $_A(t) = tu(t)/(t + 1)$.   (Hence $d_A/dt \neq h$!   This problem shows that for linear time-varying networks we do not necessarily have $d_A/dt = h$, as in the time-invariant case.)

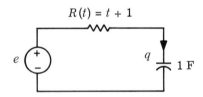

**Fig. P13.13**

Step and
impulse
response of
nonlinear
network

**14.** Consider the nonlinear time-invariant network shown in Fig. P13.14, where $e$ is the input, and $v_C$ is the response.   Show that $h(t) = (1/RC)u(t)$ and $_A(t) = (1 - \epsilon^{-t/RC})u(t)$ (hence, $d_A/dt \neq h$!).

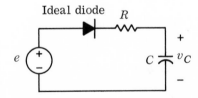

**Fig. P13.14**

**15.** Let $A(s)/B(s)$ be the network function of a linear time-invariant network. Let the input be $u(t) \cos \omega t$. Under what conditions [on the polynomials $A(s)$ and $B(s)$] is the zero-state response identical to the sinusoidal
steady-state response? (To simplify derivations, use $\epsilon^{j\omega t}$ as the input.)

**16.** The linear time-invariant circuit shown in Fig. P13.16 is a simple feedback amplifier.

a.   Determine the network function

$$H(s) = \frac{E_o(s)}{E_i(s)} = \frac{\mathcal{L}[e_o(t)]}{\mathcal{L}[e_i(t)]}$$

b.   Determine the impulse response.

c.   Determine the steady-state response for a sinusoidal input, $e_i(t) = \cos 2t$.

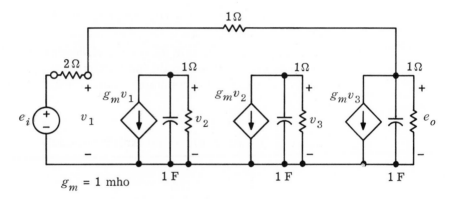

**Fig. P13.16**

**17.** a.   Calculate the Laplace transforms of $f_1(t) = u(t)\epsilon^{-at}$ and $f_2(t) = u(t)\epsilon^{-bt}$.

b.   Give an analytic expression valid for all $t$ in $(-\infty,\infty)$ of $df_1/dt$ and
$df_2/dt$; calculate their Laplace transforms, and check the differentiation rule.

c.   Calculate the convolution of $f_1$ and $f_2$.

d.   Calculate the Laplace transform of this convolution. Is it what you
expect?

**18.** For the same $f_1$ and $f_2$ as in the preceding problem,

a.   Evaluate the convolution integral of $f_1$ and $df_2/dt$.

b.   Repeat $df_1/dt$ and $f_2$.

c.   Verify these calculations using Laplace transforms.

**19.** Calculate the impulse response of the following networks:

a.   The network shown in Fig. P13.5 (input, $V$; response, $i_1$; switch $S_1$ closed at all times)

b.   The network shown in Fig. P13.8 (input, $v$; response, $i_{L_1}$; switch $S$ closed at all times)

c.   The network shown in Fig. P13.7 (input, current source connected between node ① and ground; response, $v_3$)

d.   The network shown in Fig. P13.9 (input, voltage source $e_1$; response, $v_2$; in case (a), $e_2(t) = 0$, for all $t$)

**20.** Consider the linear time-invariant network shown in Fig. P13.20, where the input voltage is $e = 2 \cos t$ volts when the switch $S$ is in position 1, and $e = 4$ volts when the switch $S$ is in position 2. The response of interest is the current $i$.

a.   Calculate the sinusoidal steady-state response when $S$ is left in position 1.

b.   Assuming the network is in the sinusoidal steady state (with $S$ in position 1) before $t = 0$, set up the mesh differential equations for $t > 0$, indicating proper initial conditions, if the switch is thrown to position 2 at $t = 0$.

c.   Calculate $i(t)$ for $t \geq 0$.

d.   Compare $i(0-)$ and $i(0+)$, and explain physically.

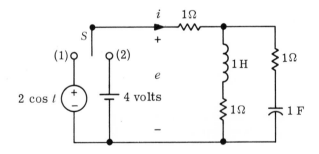

**Fig. P13.20**

**21.** Consider the linear time-invariant network shown in Fig. P13.21. The switch $S$ has been in position 1 sufficiently long so that the circuit is in the sinusoidal steady state. At the instant $t = 0$, the switch is thrown to position 2.

a.   Find the sinusoidal steady-state values of the current $i_1(t)$, $i_2(t)$, and $i_3(t)$ for $t < 0$.

b.  Find the inductor currents and capacitor voltages at $t = 0-$ (that is, immediately prior to the throwing of the switch to position 2).

c.  Find the values of $i_1(t)$, $i_2(t)$, and $i_3(t)$ both at $t = 0-$ and at $t = 0+$.

d.  Set up the network equations and solve them to obtain $i_1(t)$, $i_2(t)$, and $i_3(t)$ for $t > 0$.

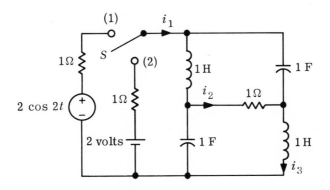

**Fig. P13.21**

# 14

# Natural Frequencies

This chapter is devoted to the study of the natural frequencies of linear time-invariant networks. In Chap. 5 we introduced the concept of natural frequency in the case of second-order circuits. The definition was given in terms of the characteristic roots of a second-order linear differential equation with constant coefficients which described the second-order circuit under consideration. In this chapter we shall give a general definition of natural frequencies that is valid for any linear time-invariant network. To be precise, we need to distinguish between two concepts: the natural frequencies of a *network variable*, to be discussed in Sec. 1, and the natural frequencies of a *network*, to be discussed in Sec. 3. Both are defined under *zero-input* conditions. These concepts are of paramount importance for understanding the behavior of linear time-invariant networks.

We start by defining physically the concept of natural frequency of a network variable. In order to calculate the natural frequencies of a particular network variable we must obtain its "minimal differential equation." For this purpose we develop the elimination method. Finally, we relate natural frequencies to state equations.

## 1    Natural Frequency of a Network Variable

We consider a linear time-invariant network $\mathfrak{N}$, for example, any one of those shown in Figs. 1.1 to 1.5. We set all its *independent* sources to zero; i.e., all the independent voltage sources are reduced to a short circuit, and all the independent current sources are reduced to an open circuit. We set our attention on one network variable of $\mathfrak{N}$; it may be a branch voltage, a node voltage, a branch current, or a loop current. In order not to prejudice our thinking we call it $x$. Given the initial state of $\mathfrak{N}$ at $t = 0$, the corresponding zero-input response is a waveform $x(\cdot)$, in general represented by

$$(1.1) \qquad x(t) = K_1 \epsilon^{s_1 t} + K_2 \epsilon^{s_2 t} + \cdots \qquad \text{for } t \geq 0$$

where the $K_i$'s and $s_i$'s are constants (possibly complex). The $s_i$'s depend on the network topology and on the element values. The $K_i$'s depend, in addition, on the initial state. It will be increasingly apparent that the numbers $s_i$ are very important, and therefore, they deserve a name. We shall say that $s_1$ is a **natural frequency of the network variable** $x$ if, for some initial state, the zero-input response $x$ includes the term $K_1\epsilon^{s_1 t}$. In other words, for some initial state, $K_1 \neq 0$, and $K_1\epsilon^{s_1 t}$ appears in the expression for the zero-input response of $x$.

Let us illustrate this definition by examples.

**Example 1**   Consider the network shown in Fig. 1.1. Let $v$ be the network variable of interest. The differential equation for $v$ is

(1.2)   $$C\frac{dv}{dt} + Gv = 0$$

To the initial state $v(0)$, the zero-input response is

$$v(t) = v(0)\epsilon^{-t/RC}$$

Thus, $-1/RC$ is the natural frequency of the network variable $v$.

If $i$ were the network variable of interest, since $i(t) = Gv(0)\epsilon^{-t/RC}$, $-1/RC$ would also be the natural frequency of $i$.

**Example 2**   Consider the parallel $RLC$ circuit shown in Fig. 1.2, where $C = 1$ farad, $R = \frac{1}{6}$ ohm, and $L = \frac{1}{25}$ henry. From the equation

(1.3)   $$C\dot{v}_C + Gv_C + \frac{1}{L}\int_0^t v_C(t')\,dt' + j_L(0-) = 0$$

and its Laplace transform

$$\left(Cs + G + \frac{1}{Ls}\right)V_C(s) = Cv_C(0-) - \frac{1}{s}j_L(0-)$$

we obtain

$$V_C(s) = \frac{sCv_C(0-) - j_L(0-)}{Cs^2 + Gs + 1/L}$$

(1.4)   $$= \frac{sv_C(0-) - j_L(0-)}{(s+3)^2 + 4^2}$$

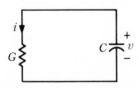

**Fig. 1.1**   The natural frequency of the network variable $v$ is $-1/RC$.

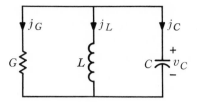

**Fig. 1.2**    *RLC* network used in the calculation of natural frequencies.

Using the partial-fraction-expansion technique, we obtain

$$v_C(t) = \frac{(-3 + j4)v_C(0-) - j_L(0-)}{j8} \epsilon^{(-3+j4)t}$$

$$+ \frac{(-3 - j4)v_C(0-) - j_L(0-)}{-j8} \epsilon^{(-3-j4)t}$$

Thus, $-3 + j4$ and $-3 - j4$ are the natural frequencies of $v_C$. It is easy to verify that $j_L, v_L, j_G, v_G, j_C$, and $v_C$ have the same natural frequencies.

**Example 3**    Consider the same parallel *RLC* circuit shown in Fig. 1.2, where $C = 1$ farad, $R = \frac{1}{6}$ ohm, and $L = \frac{1}{9}$ henry. The natural frequencies for the network variable $v_C$ again are obtained from (1.4). In this case

$$V_C(s) = \frac{sv_C(0-) - j_L(0-)}{(s + 3)^2}$$

which has a partial-fraction expansion

$$V_C(s) = \frac{-3v_C(0-) - j_L(0-)}{(s + 3)^2} + \frac{v_C(0-)}{s + 3}$$

The inverse Laplace transform is

$$v_C(t) = [-3v_C(0-) - j_L(0-)]t\epsilon^{-3t} + v_C(0-)\epsilon^{-3t}$$

Thus, $-3$ is a natural frequency of $v_C$. In view of the presence of the term proportional to $t\epsilon^{-3t}$, we say that $-3$ is a natural frequency of $v_C$ of *order* 2.

Example 3 might suggest that the above definition of natural frequency is defective. This is not the case. Indeed, if the natural frequency $s_1$ is of an order higher than 1 (say 3, for purposes of discussion), then (1.1) would become

$$x(t) = K_{11}\epsilon^{s_1 t} + K_{12}t\epsilon^{s_1 t} + K_{13}t^2\epsilon^{s_1 t} + K_2\epsilon^{s_2 t} + \cdots$$

Now it is a fact that if there is some initial state for which $K_{12}$ or $K_{13}$ can be made different from zero, then there are also initial states for which $K_{11} \neq$

0. Therefore, in all cases we need only consider the factor of the pure exponential $\epsilon^{s_1 t}$.

In the next section, we shall show that we can obtain a homogeneous differential equation for the variable $x$,

(1.5)   $Q(D)x = 0$

which has the property that *any* zero-input response $x$ of the network $\mathfrak{N}$ satisfies the differential equation (1.5), and *any* solution of (1.5) is the zero-input response $x$ corresponding to some initial state of $\mathfrak{N}$. Since no differential equation of smaller order can have this property, it is called the **minimal differential equation** of the network variable $x$. Suppose that this equation were given to us; then from our knowledge of differential equations, we would conclude that $s_1$ is a natural frequency of $x$ if and only if $s_1$ is a zero of the polynomial $Q(s)$ [that is, if and only if $Q(s_1) = 0$]. If $s_1$ is a zero of order $m$ of the polynomial $Q(s)$, then $s_1$ is said to be a **natural frequency of order** $m$ of the network variable $x$.

**Example 4**   Suppose that the network variable of interest is a branch voltage $v$ and that its minimal differential equation is

$(D^5 + 2D^4 + 2D^3 + 2D^2 + D)v = 0$

or

$(D^2 + 1)(D + 1)^2 Dv = 0$

Consequently, the zero-input response of $v$ is of the form

$v(t) = K_1 \epsilon^{jt} + K_2 \epsilon^{-jt} + (K_3 + K_4 t)\epsilon^{-t} + K_5$

The natural frequencies of the network variable $v$ are $s_1 = j1$, $s_2 = -j1$, and $s_3 = s_4 = -1$ ($s_3$ is a natural frequency of order 2), and $s_5 = 0$.

**Remarks**   1.   If $s_i = 0$, as in the above example, the zero-input response may contain a constant term. It is physically clear that this may occur in two instances.

    *a.*   The network variable of interest is an inductor current, and this inductor is in a loop made of inductors only. Since all these inductors are ideal, a constant current $I$ may flow around the loop, and no voltage will appear across any of the inductors since $L(dI/dt) = 0$. For an example, see Fig. 1.3.

    *b.*   The network variable of interest is a capacitor voltage, and this capacitor is in a cut set made of capacitors only. Since all the capacitors are ideal, a constant voltage $V$ may exist across each capacitor of the cut set, and no current will flow through them since $C(dV/dt) = 0$. Figure 1.3 also gives an example of this second possibility.

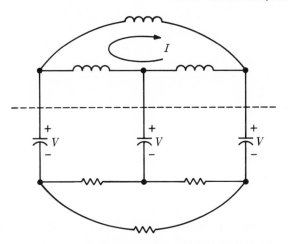

**Fig. 1.3**    An example illustrating a dc circulating current in an inductor loop and a dc voltage across a capacitor cut set.

2.  If we count a natural frequency of order $m$ as $m$ natural frequencies,
    the number of natural frequencies of a network variable is equal to the
    order of its minimal differential equation (or, equivalently, to the
    degree of the corresponding characteristic polynomial). Physically,
    the number of natural frequencies of a network variable is equal to the
    minimum number of initial conditions necessary to specify uniquely
    the response of that network variable.

In Example 1 we found that the natural frequency for the network
variable $v$, from Eq. (1.2), is $-1/RC$. We noted that this was also the
natural frequency for the current $i$, since $i = Gv$, and thus, the minimal
differential equation for the current $i$ has the same characteristic polynomial $Q(s) = s + 1/RC$. In Example 2 we also noted that the natural
frequencies for all the branch voltages and the branch currents are the
same. Is it then a general fact that for a given network, each network variable has the same set of natural frequencies? The answer is no. Let us
consider the following example.

---

**Example 5**    The purpose of this example is to show that two network variables need
not have the same set of natural frequencies. Consider the network shown
in Fig. 1.4. Note that terminals ② and ②' are left open-circuited. It
follows from our previous examples that $v_1$ has $j(1/\sqrt{L_1 C_1})$ and
$-j(1/\sqrt{L_1 C_1})$ as natural frequencies, whereas $v_2$ has $-1/R_2 C_2$ as the
natural frequency. In passing, let us observe that if we were to perform a
node analysis, we would easily verify that $-1/R_2 C_2$, $+j(1/\sqrt{L_1 C_1})$, and
$-j(1/\sqrt{L_1 C_1})$ are zeros of the network determinant, i.e., of $\det [\mathbf{Y}_n(s)]$.

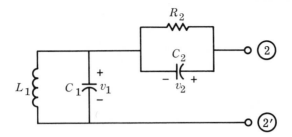

**Fig. 1.4**    Given that the terminals ② and ②′ are left open-circuited, the network variables $v_1$ and $v_2$ have no natural
frequencies in common.

This particular fact will be discussed in Sec. 3, where we introduce the concept of natural frequencies of a *network*.

**Exercise**    Consider the linear time-invariant network shown in Fig. 1.5.    Show that
the natural frequencies of the variables $e_1$ and $e_2$ are $-1$ and $-\frac{1}{3}$, whereas
$v_3$ has only $-\frac{1}{3}$ as a natural frequency.

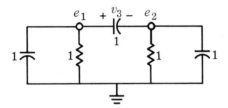

**Fig. 1.5**    The network variables $e_1$ and $e_2$ have $-1$ as a
natural frequency, whereas $v_3$ does not.

## 2    The Elimination Method

In this section we develop a systematic procedure whereby, starting from a
system of integrodifferential equations which describes a network, we obtain the minimal differential equation of a specified variable.

### 2.1    General Remarks

The analysis of linear time-invariant networks leads to simultaneous linear
integrodifferential or differential equations.    Without loss of generality, we
may assume that we have to solve a system of linear *differential* equations.
Indeed, suppose we use mesh (or loop) analysis; then whenever we encounter an integral like $(1/C)D^{-1}i(t)$, where $C$ is the capacitance, $i$ is the

mesh (or loop) current, $D = d/dt$, and $D^{-1}(\cdot) = \int_0^t (\cdot)\,dt$, we can introduce the charge as variable

$$q(t) \triangleq \int_0^t i(t')\,dt'$$

and the term $(1/C)D^{-1}i(t)$ becomes $(1/C)q(t)$. Similarly, in node (or cut-set) analysis we can introduce the flux as a variable

$$\phi(t) \triangleq \int_0^t v(t')\,dt'$$

where $v$ is the node-to-datum (or tree-branch) voltage. Thus, we end up in mesh (or loop) analysis with currents and/or charges as variables, and in node (or cut-set) analysis with voltage and/or fluxes as variables. Let us start by considering a simple example to review the writing of network equations.

**Example 1**    In Fig. 2.1 we have a two-mesh linear time-invariant circuit. The initial state is specified by the initial currents in the inductors, $j_1(0)$ and $j_2(0)$, and the initial voltage across the capacitor, $v_C(0)$. Let us use the mesh currents $i_1$ and $i_2$ as network variables. The mesh equations are

(2.1a)    $(D + 6 + 10D^{-1})i_1 - 10D^{-1}i_2 = e_s - v_C(0)$

(2.1b)    $-10D^{-1}i_1 + (10D + 10D^{-1})i_2 = v_C(0)$

Note that the initial voltage across the capacitor has already been included in the equations. Thus, we have a system of two simultaneous linear integrodifferential equations with constant coefficients in the variables $i_1$ and $i_2$. In matrix form the equations are

(2.2)    $\begin{bmatrix} \mathbf{Z}_m(D) \end{bmatrix} \begin{bmatrix} i_1 \\ i_2 \end{bmatrix} = \begin{bmatrix} e_s \\ 0 \end{bmatrix} + \begin{bmatrix} -v_C(0) \\ v_C(0) \end{bmatrix}$

where $\mathbf{Z}_m(D)$ is the mesh impedance matrix operator. If, on the other

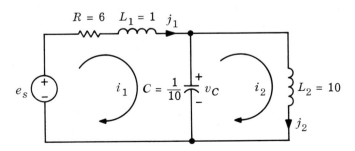

**Fig. 2.1**    Network for Example 1 with element values in ohms, farads, and henrys.

hand, we choose the mesh charges $q_1$ and $q_2$ as network variables, where

(2.3)   $q_1(t) \triangleq \int_0^t i_1(t') \, dt' \qquad q_2(t) \triangleq \int_0^t i_2(t') \, dt'$

we obtain a system of two simultaneous linear *differential* equations with variables $q_1$ and $q_2$

(2.4a)   $(D^2 + 6D + 10)q_1 - 10q_2 = e_s - v_C(0)$

(2.4b)   $-10q_1 + (10D^2 + 10)q_2 = v_C(0)$

This is a system of two linear ordinary differential equations in two un-known functions $q_1$ and $q_2$. In matrix form the equations are

(2.5)   $\begin{bmatrix} D\mathbf{Z}_m(D) \end{bmatrix} \begin{bmatrix} q_1 \\ q_2 \end{bmatrix} = \begin{bmatrix} e_s \\ 0 \end{bmatrix} + \begin{bmatrix} -v_C(0) \\ v_C(0) \end{bmatrix}$

The differences between (2.2) and (2.5) are apparent. Whereas (2.2) repre-sents a system of integrodifferential equations in terms of mesh currents, (2.5) is a system of differential equations in terms of mesh charges.

Since the changing of variables from currents to charges and from volt-ages to fluxes is readily performed, we assume from now on that we have to solve *differential* equations rather than *integrodifferential* equations. With-out prejudicing the case either in favor of loop (or mesh) or cut-set (or node) analysis, we denote the $n$ network variables by $x_1, x_2, \ldots, x_n$. The system of $n$ simultaneous linear differential equations resulting from our analysis then takes the form

$p_{11}(D)x_1 + p_{12}(D)x_2 + \cdots + p_{1n}(D)x_n = f_1$

$p_{21}(D)x_1 + p_{22}(D)x_2 + \cdots + p_{2n}(D)x_n = f_2$

(2.6)   $\cdots\cdots\cdots\cdots\cdots\cdots\cdots\cdots\cdots\cdots\cdots\cdots$

$p_{n1}(D)x_1 + p_{n2}(D)x_2 + \cdots + p_{nn}(D)x_n = f_n$

where $D = d/dt$ and the $p_{ij}(D)$ are polynomials in $D$ of degree 2 at most. The functions $f_1, f_2, \ldots, f_n$ which appear in the right-hand side are forcing functions contributed by independent sources and some of the initial conditions. It is convenient to express this system (let us call it system 1) in matrix form as

(2.7a)   $\mathbf{P}(D)\mathbf{x} = \mathbf{f}$

or

(2.7b)   $\begin{bmatrix} p_{11}(D) & p_{12}(D) & \cdots & p_{1n}(D) \\ p_{21}(D) & p_{22}(D) & \cdots & p_{2n}(D) \\ \cdots\cdots\cdots\cdots\cdots\cdots\cdots\cdots \\ p_{n1}(D) & p_{n2}(D) & \cdots & p_{nn}(D) \end{bmatrix} \begin{bmatrix} x_1 \\ x_2 \\ \vdots \\ x_n \end{bmatrix} = \begin{bmatrix} f_1 \\ f_2 \\ \vdots \\ f_n \end{bmatrix}$

where $\mathbf{x} = (x_1, x_2, \ldots, x_n)^T$ is a column vector whose components are the unknown waveforms; $\mathbf{P}(D)$, whose entries are polynomials in $D$, is the **matrix of the system;** and $\mathbf{f} = (f_1, f_2, \ldots, f_n)^T$ is the **forcing function vector.** For later reference, we write

(2.8)   $\Delta(D) = \det \mathbf{P}(D)$

and refer to $\Delta$ as the **determinant of the system.** It is important to note that $\mathbf{P}(D)$ is a matrix whose elements are *polynomials* in $D$ and that in evaluating $\det [\mathbf{P}(D)]$ we treat $D$ as an ordinary variable.

**Remark**   It is important to point out that $\mathbf{P}(D)$ is related simply to the mesh impedance matrix operator $\mathbf{Z}_m(D)$ in the case of mesh analysis. If charges instead of currents are used in mesh analysis, as in Example 1, then $\mathbf{P}(D) = D\mathbf{Z}_m(D)$. Let us consider the determinant of the system, that is, $\Delta(s) \triangleq \det [\mathbf{P}(s)]$. It is then clear that the nonzero roots of $\Delta(s) = 0$ are identical to the nonzero roots of the network determinant $\Delta_m(s) \triangleq \det [\mathbf{Z}_m(s)]$. Similarly, if node analysis is employed, and fluxes instead of voltages are used as variables, then $\mathbf{P}(D) = D\mathbf{Y}_n(D)$. Clearly, the nonzero roots of $\Delta(s) \triangleq \det [\mathbf{P}(s)]$ are identical to the nonzero roots of the network determinant $\Delta_n(s) \triangleq \det [\mathbf{Y}_n(s)]$. As a matter of fact it can be proven that for any given network, if different methods of network analysis are used, the corresponding network determinants have the same sets of nonzero roots; that is, the nonzero roots of $\Delta_m(s)$, $\Delta_n(s)$, $\Delta_l(s) \triangleq \det [\mathbf{Z}_l(s)]$ (the loop impedance matrix), and $\Delta_q(s) \triangleq \det [\mathbf{Y}_q(s)]$ (the cut-set admittance matrix) are all identical and are equal to the nonzero roots of $\Delta(s)$, the determinant of the system of differential equations $\mathbf{P}(D)\mathbf{x} = \mathbf{f}$.

## 2.2   Equivalent Systems

We wish to perform some operations on the system of (2.6) or (2.7a) in order to obtain the minimal differential equation pertaining to a particular network variable, say $x_n$. In doing so, we wish to be sure we do not introduce spurious solutions. In other words, if we start with system 1 of (2.7a); that is,

$\mathbf{P}(D)\mathbf{x} = \mathbf{f}$

and we end up with a system 2,

(2.9)   $\widehat{\mathbf{P}}(D)\mathbf{x} = \widehat{\mathbf{f}}$

we want to be sure that they have the same solutions. To be precise, we introduce the notion of equivalent systems: the system of differential equations, 1, is said to be **equivalent** to system 2 if every solution of system 1 is a solution of system 2, and if every solution of system 2 is a solution of system 1.

Let us make an observation which will be useful later. It follows directly from this definition that if system 1 is equivalent to system 2 and if system 2 is equivalent to some system 3, then system 1 is equivalent to system 3.

In order to make further progress we propose to transform the given system 1 of (2.7a) into an *equivalent system* of the following form, called system *T*:

(2.10)
$$\hat{p}_{11}(D)x_1 + \hat{p}_{12}(D)x_2 + \cdots\cdots\cdots\cdots\cdots\cdots + \hat{p}_{1n}(D)x_n = \hat{f}_1$$
$$\hat{p}_{22}(D)x_2 + \cdots\cdots\cdots\cdots\cdots + \hat{p}_{2n}(D)x_n = \hat{f}_2$$
$$\cdots\cdots\cdots\cdots\cdots\cdots\cdots\cdots\cdots\cdots$$
$$\hat{p}_{n-1,n-1}(D)x_{n-1} + \hat{p}_{n-1,n}(D)x_n = \hat{f}_{n-1}$$
$$\hat{p}_{nn}(D)x_n = \hat{f}_n$$

In matrix notation, we have

(2.11a)   $$\hat{\mathbf{P}}(D)\mathbf{x} = \hat{\mathbf{f}}$$

or

(2.11b)
$$
\begin{bmatrix}
\hat{p}_{11}(D) & \hat{p}_{12}(D) & \cdots & \hat{p}_{1n}(D) \\
0 & \hat{p}_{22}(D) & \cdots & \hat{p}_{2n}(D) \\
\cdots & \cdots & \cdots & \cdots \\
0 & 0 & \hat{p}_{n-1,n-1}(D) & \hat{p}_{n-1,n}(D) \\
0 & \cdots & 0 & \hat{p}_{nn}(D)
\end{bmatrix}
\begin{bmatrix}
x_1 \\
x_2 \\
\vdots \\
\\
x_n
\end{bmatrix}
=
\begin{bmatrix}
\hat{f}_1 \\
\hat{f}_2 \\
\vdots \\
\\
\hat{f}_n
\end{bmatrix}
$$

As before, $\hat{\mathbf{P}}(D)$ is the matrix (in the operator $D$) of the new system; its $ij$th element, $\hat{p}_{ij}(D)$, is a polynomial in $D$ whose degree may be greater than 2, whereas $\hat{\mathbf{f}} = (\hat{f}_1, \hat{f}_2, \ldots, \hat{f}_n)^T$ is the forcing function vector of the new system. Again we define

(2.12)   $$\hat{\Delta}(D) = \det [\hat{\mathbf{P}}(D)]$$

as the determinant (in the operator $D$) of the new system. For convenience we shall refer to the original system as system 1 and the new system as system $T$ ($T$, for triangular). For obvious reasons, system $T$ is said to be in the *triangular form*. Note that in the last equation of system $T$ all variables except $x_n$ have been eliminated. Thus, the last equation in system $T$ represents a single differential equation in the variable $x_n$ and is the minimal differential equation of the variable $x_n$.

**Remarks**   1.   Systems 1 and $T$ above are equivalent. This means that every solution of 1 is a solution of $T$, and every solution of $T$ is a solution of 1. In particular, the last component of any solution $\mathbf{x}$ of system 1 is a solution of $\hat{p}_{nn}(D)x_n = \hat{f}_n$, and conversely any solution of the last equation of system $T$, namely, $\hat{p}_{nn}(D)x_n = \hat{f}_n$, is the last component of a solution of system 1. Thus,

(2.13)   $\hat{p}_{nn}(D)x_n = \hat{f}_n$

is the differential equation of *least* order in the variable $x_n$ which is satisfied by the $n$th component of all solutions of system 1. For this reason it is *the minimal differential equation of the network variable $x_n$.*

2.  It is important to note that the polynomial $\hat{p}_{nn}(D)$ and the function $\hat{f}_n$, which appear in *the* minimal differential equation of the variable $x_n$, are uniquely determined (except possibly for a constant nonzero multiplier). In other words, the polynomial $\hat{p}_{nn}$ is independent of the method used to analyze the network; it therefore describes some basic properties of the variables $x_n$. To see this, assume that in system 1 all the inputs are equal to zero. The minimal differential equation of $x_n$ predicts all possible zero-input responses of $x_n$ (due to any initial state) and none other. For example, if the equation has a solution $Ke^{s_1 t}$ (where $K$ and $s_1$ are suitable constants), then there is an initial state of the network which will have $Ke^{s_1 t}$ as a response. Since all methods of analysis predict the same responses, they must all lead, after elimination, to the same minimal differential equation (except possibly for a *constant nonzero factor*).

3.  The characteristic polynomial of the minimal differential equation of the network variable $x_n$ in (2.13) is $\hat{p}_{nn}(s)$. Consequently, $s_1$ is a natural frequency of the network variable $x_n$ if and only if $\hat{p}_{nn}(s_1) = 0$.

**Exercise**   Using the notation of Eqs. (2.11$a$), (2.11$b$), and (2.12), show that $\Delta(D) = \hat{p}_{11}(D)\hat{p}_{22}(D) \cdots \hat{p}_{nn}(D)$.

A systematic procedure for obtaining the minimal differential equation is based on the following theorem.

**THEOREM**   Let the given system $A$ be a set of $n$ linear differential equations with constant coefficients,

(2.14$a$)   $U_j(\mathbf{x}) = 0 \qquad j = 1, 2, \ldots, n$

where

(2.14$b$)   $U_j(\mathbf{x}) \triangleq p_{j1}(D)x_1 + p_{j2}(D)x_2 + \cdots + p_{jn}(D)x_n - f_j$

and $\mathbf{x} = (x_1, x_2, \ldots, x_n)^T$ is the unknown. Let

(2.15)   $U_i^*(\mathbf{x}) \triangleq mU_i(\mathbf{x}) + M(D)U_k(\mathbf{x}) \qquad k \neq i$

where $m$ is a *nonzero constant*, $M(D)$ an *arbitrary polynomial* in $D$, and $k$ is any index different from $i$. Consider the system $B$ obtained from the system $A$ by replacing *only* the $i$th equation of $A$, namely $U_i = 0$, by $U_i^* = 0$. Under these conditions, the system $A$ is equivalent to the system $B$.

*Proof*   We shall give a direct proof based on the definition of equivalent systems. For convenience we denote the equations of system $B$ as follows:

(2.16)   $U_j^*(\mathbf{x}^*) = 0 \qquad j = 1, 2, \ldots, n$

where we use the symbol $\mathbf{x}^*$ to denote a solution of system $B$. We show first that any solution, say $\mathbf{x}$, of system $A$ is a solution of system $B$. Since all equations of $B$ are identical with those of $A$, except the $i$th equation, if we substitute $\mathbf{x}$ in $B$, we obtain

(2.17)   $U_j^*(\mathbf{x}) = U_j(\mathbf{x}) = 0 \qquad j = 1, 2, \ldots, i - 1, i + 1, \ldots, n$

Now we substitute $\mathbf{x}$ in the $i$th equation of system $B$; then by (2.15) we obtain

$$U_i^*(\mathbf{x}) = mU_i(\mathbf{x}) + M(D)U_k(\mathbf{x}) = 0$$

where the last equality follows from (2.14). Thus, we have

$$U_j^*(\mathbf{x}) = 0 \qquad \text{for all } j$$

In other words, we have shown that every solution of system $A$ is a solution of system $B$.

Next we must show the converse: every solution $\mathbf{x}^*$ of system $B$ is a solution of system $A$. As before, we find immediately that

(2.18)   $U_j(\mathbf{x}^*) = 0 \qquad j = 1, 2, \ldots, i - 1, i + 1, \ldots, n$

Therefore $U_k(\mathbf{x}^*) = 0$ because $k \neq i$. From (2.15) we obtain

$$mU_i(\mathbf{x}^*) = U_i^*(\mathbf{x}^*) - M(D)U_k(\mathbf{x}^*) = 0$$

where the last equality follows from (2.16) and (2.18). Since $m$ is a *nonzero constant*, $U_i(\mathbf{x}^*) = 0$. Thus, every solution to system $B$ is a solution to system $A$. Therefore, the two systems $A$ and $B$ are equivalent.

**Remarks**   1.   It is important to note that $m$ of Eq. (2.15) in the transformation is a *nonzero constant;* otherwise the theorem does not hold.

2.   Consider the matrix descriptions of the two systems

(2.19a)   $\mathbf{P}(D)\mathbf{x} - \mathbf{f} = \mathbf{0}$

(2.19b)   $\mathbf{P}^*(D)\mathbf{x}^* - \mathbf{f}^* = \mathbf{0}$

where

(2.20)   $\mathbf{P}^*(D) = \begin{bmatrix} p_{11}^*(D) & p_{12}^*(D) & \cdots & p_{1n}^*(D) \\ p_{21}^*(D) & p_{22}^*(D) & \cdots & p_{2n}^*(D) \\ \multicolumn{4}{c}{\dotfill} \\ p_{n1}^*(D) & p_{n2}^*(D) & \cdots & p_{nn}^*(D) \end{bmatrix}$

The transformation stated in the theorem is usually referred to as an *elementary row transformation* of the matrix of the given system.   The transformation can be expressed in the following matrix form:

(2.21a)   $\mathbf{P}^*(D) = \mathbf{T}(D)\mathbf{P}(D)$

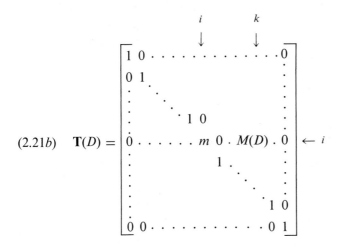

(2.21b)   $\mathbf{T}(D) =$

Since a product of matrices is performed row by column, it is clear that (2.21a) and (2.21b) predict that the matrices $\mathbf{P}(D)$ and $\mathbf{P}^*(D)$ differ only in the *i*th row.   The operation $\mathbf{T}(D)$ amounts to multiplying the *i*th row of $\mathbf{P}(D)$ by *m* and adding to it the *k*th row after it has been operated by $M(D)$.   From the theory of determinants,

det $[\mathbf{P}^*(D)] = $ det $[\mathbf{T}(D)]$ det $[\mathbf{P}(D)]$

By expanding det $[\mathbf{T}(D)]$ successively by rows, we see that det $[\mathbf{T}(D)] = m$; thus, det $[\mathbf{P}^*(D)] = m$ det $[\mathbf{P}(D)]$, where *m* is a nonzero constant.

The following example illustrates the use of elementary row transformations.

**Example 2**   Consider Eqs. (2.4a) and (2.4b) of Example 1.   Suppose we wish to eliminate $q_2$ and obtain a single differential equation in $q_1$.   Let us denote the two original equations by $\mathcal{E}_1$ and $\mathcal{E}_2$ as follows:

(2.22a)   $\mathcal{E}_1$:   $(D^2 + 6D + 10)q_1 - 10q_2 = e_s - v_C(0)$

(2.22b)   $\mathcal{E}_2$:   $-10q_1 + (10D^2 + 10)q_2 = v_C(0)$

By the theorem just proved we can obtain an equivalent system as follows:

(2.23a)   $\mathcal{E}_1^* = \mathcal{E}_1$:   $(D^2 + 6D + 10)q_1 - 10q_2 = e_s - v_C(0)$

(2.23b)   $\mathcal{E}_2^* = \mathcal{E}_2 + (D^2 + 1)\mathcal{E}_1$:   $[-10 + (D^2 + 1)(D^2 + 6D + 10)]q_1$
$$= v_C(0) + (D^2 + 1)[e_s - v_C(0)]$$

The system of (2.23) is in triangular form.  The second equation contains a single variable $q_1$ and is *the* minimal differential equation of the variable $q_1$; thus,

(2.24)   $(D^4 + 6D^3 + 11D^2 + 6D)q_1 = (D^2 + 1)e_s$

The characteristic polynomial of this equation is of fourth degree, and the natural frequencies of $q_1$ are $0$, $-1$, $-2$, and $-3$.  If we let $q_p$ be of any particular solution of (2.24), then the solution for $q_1$ is of the form

(2.25)   $q_1(t) = K_1 + K_2\epsilon^{-t} + K_3\epsilon^{-2t} + K_4\epsilon^{-3t} + q_p(t)$

where $K_1, K_2, K_3,$ and $K_4$ are arbitrary constants to be determined from the initial conditions.  Note that the current $i_1$ is related to $q_1$ by

(2.26)   $i_1 = Dq_1$

From (2.24) we obtain the following minimal differential equation for the variable $i_1$:

(2.27)   $(D^3 + 6D^2 + 11D + 6)i_1 = (D^2 + 1)e_s$

The natural frequencies for the network variable $i_1$ are therefore $-1$, $-2$, and $-3$; note that the number of natural frequencies for $i_1$ is equal to the number of energy-storing elements in the circuit.

---

**Remark**   It is customary to use the term "natural frequency" with a network variable which is either a current or a voltage.  If the charge or the flux is used as a network variable, the zero-input response for that charge or flux can be obtained by integrating the corresponding zero-input response for the current or voltage, respectively.  Now if $s_1$ is a nonzero natural frequency of a current $i$, then for some initial conditions the zero-input response $i$ will contain a term of the form $K\epsilon^{s_1 t}$.  The charge corresponding to that current,

$$q(t) = \int_0^t i(t')\, dt'$$

will contain a term of the form $(K/s_1)\epsilon^{s_1 t}$.  Thus, if $s_1$ is a nonzero natural frequency of the current $i$, it is also a natural frequency of the charge $q$.  Similarly, if $s_1$ is a nonzero natural frequency of a voltage, $s_1$ is also a natural frequency of the corresponding flux.  Now since $Dq = i$, the differential equation of the charge is one order higher than that of the current; the characteristic polynomial of the charge includes an extra zero at the origin.  From a physical point of view, no extra physical initial conditions are required because the definition of $q$ itself specifies the initial condition, namely, $q(0) = 0$ (see Eq. 2.3).  Thus, the increased order of the differential equation has no particular physical significance.

## 2.3    The Elimination Algorithm

We shall present an algorithm for obtaining an equivalent triangular system from a given system. The algorithm constitutes a proof that the transformation to triangular form can always be achieved by means of successive elementary row transformations.

For convenience let us redefine the two systems again. We denote the $n$ variables by $x_1, x_2, \ldots, x_n$ and the $n$ equations by $\mathscr{E}_k$, $k = 1, 2, \ldots, n$. The given system 1 of linear differential equations is

$$\mathscr{E}_1: \quad p_{11}(D)x_1 + p_{12}(D)x_2 + \cdots + p_{1n}(D)x_n = f_1$$

$$\mathscr{E}_2: \quad p_{21}(D)x_1 + p_{22}(D)x_2 + \cdots + p_{2n}(D)x_n = f_2$$

$$\vdots \quad \ldots\ldots\ldots\ldots\ldots\ldots\ldots\ldots\ldots\ldots$$

$$\mathscr{E}_k: \quad p_{k1}(D)x_1 + p_{k2}(D)x_k + \cdots + p_{kn}(D)x_n = f_k$$

$$\vdots \quad \ldots\ldots\ldots\ldots\ldots\ldots\ldots\ldots\ldots\ldots$$

$$\mathscr{E}_n: \quad p_{n1}(D)x_1 + p_{n2}(D)x_2 + \cdots + p_{nn}(D)x_n = f_n$$

where $D$ denotes $d/dt$, the $p_{ij}$ are polynomials of degree 2 at most in $D$ with constant coefficients, and the $f_i$ include contributions from the independent sources and from some of the initial conditions. Let us denote the equations of the equivalent triangular system by $\widehat{\mathscr{E}}_k$, $k = 1, 2, \ldots, n$. The triangular system $T$ of linear differential equations is

$$\widehat{\mathscr{E}}_1: \quad \widehat{p}_{11}(D)x_1 + \widehat{p}_{12}(D)x_2 + \widehat{p}_{13}(D)x_3 + \cdots + \widehat{p}_{1n}(D)x_n = \widehat{f}_1$$

$$\widehat{\mathscr{E}}_2: \qquad\qquad \widehat{p}_{22}(D)x_2 + \widehat{p}_{23}(D)x_3 + \cdots + \widehat{p}_{2n}(D)x_n = \widehat{f}_2$$

$$\vdots \qquad\qquad \ldots\ldots\ldots\ldots\ldots\ldots\ldots\ldots\ldots\ldots$$

$$\widehat{\mathscr{E}}_{n-1}: \qquad\qquad\qquad \widehat{p}_{n-1,n-1}(D)x_{n-1} + \widehat{p}_{n-1,n}(D)x_n = \widehat{f}_{n-1}$$

$$\widehat{\mathscr{E}}_n: \qquad\qquad\qquad\qquad \widehat{p}_{nn}(D)x_n = \widehat{f}_n$$

**ALGORITHM**

*Step 1*    In case one needs to calculate only one network variable, reorder, if necessary, the unknown functions $x_1, \ldots, x_n$, so that $x_n$ is the network variable of interest.

*Step 2*    Examine the first column of the matrix $\mathbf{P}(D)$. If there is only one polynomial that is not identically zero, say, $p_{k1}$, then interchange the first and the $k$th equation, and go to Step 7; if not, go to Step 3.

*Step 3*    Pick out of the first column the polynomial of least degree which is not identically zero. Call $k$ the number of the row of this polynomial; then the name of the polynomial of least degree is $p_{k1}$.

*Step 4*    Divide $p_{k1}(D)$ into each polynomial $p_{i1}(D)$ for all $i \neq k$; call $q_{i1}$ the quotient of the division of $p_{i1}$ by $p_{k1}$, and call the remainder $r_{i1}$. Hence,

$$p_{i1}(D) - q_{i1}(D)p_{k1}(D) = r_{i1}(D) \qquad \text{for all } i \neq k$$

(Observe that the degree of each polynomial $r_{i1}$ is at least 1 less than the degree of $p_{k1}$.)

*Step 5*  Write the following equivalent system:

$$\mathcal{E}_1' \triangleq \mathcal{E}_1 - q_{11}(D)\mathcal{E}_k$$
$$\mathcal{E}_2' \triangleq \mathcal{E}_2 - q_{21}(D)\mathcal{E}_k$$
. . . . . . . . . . . . . . . . . . . .
$$\mathcal{E}_k' \triangleq \mathcal{E}_k$$
$$\mathcal{E}_{k+1}' \triangleq \mathcal{E}_{k+1} - q_{k+1,1}(D)\mathcal{E}_k$$
. . . . . . . . . . . . . . . . . . . .
$$\mathcal{E}_n' \triangleq \mathcal{E}_n - q_{n1}(D)\mathcal{E}_k$$

The first column of coefficients of the system $(\mathcal{E}_1', \mathcal{E}_2', \ldots, \mathcal{E}_n')$ is $r_{11}(D)$, $r_{21}(D), \ldots, p_{k1}(D), \ldots, r_{n1}(D)$.

*Step 6*  Repeat Steps 2 through 5 for this equivalent system.

*Step 7*  After at most three such cycles, the system (now system 1′) is in the following form (we write $p_{ij}^*$ because the present polynomial coefficients are possibly different from those of system 1):

$$\mathcal{E}_1^*: \quad p_{11}^*(D)x_1 + p_{21}^*(D)x_2 + \cdots + p_{1n}^*(D)x_n = f_1^*$$
$$\mathcal{E}_2^*: \qquad\qquad\quad p_{22}^*(D)x_2 + \cdots + p_{2n}^*(D)x_n = f_2^*$$
$$\vdots$$
$$\mathcal{E}_n^*: \qquad\qquad\quad p_{n2}^*(D)x_2 + \cdots + p_{nn}^*(D)x_n = f_3^*$$

(System 1′ is equivalent to system 1; that is, every solution of one is a solution of the other, and vice versa.)  Now, disregard the *first* equation. If there is only one equation left, stop; if there is more than one equation left, go to Step 2, and repeat Steps 2 through 7 for the remaining equations.

This algorithm reduces system 1 step by step to the triangular form and can be readily programmed in a computer.

---

**Example 3**  Consider the linear time-invariant network shown in Fig. 2.2.  The initial inductor currents $j_1(0)$ and $j_2(0)$ and the initial capacitor voltage $v_C(0)$ are given.  We wish to find the minimal differential equation of the network variable $j_2$, the current in the inductor $L_2$.  In view of the graph of the network under consideration we shall use mesh analysis.  The mesh currents $i_1$ and $i_2$ are indicated in Fig. 2.2.  Since there is a capacitor in mesh 1, we shall use the capacitor charge $q_1$ as the mesh variable instead of $i_1$.  We have

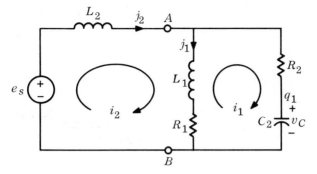

**Fig. 2.2**   With $R_1 = R_2 = 1$ ohm, $C_2 = 1$ farad, $L_1 = 1$ henry, and $L_2 = 2$ henrys, the natural frequencies of $j_2$ are $-1$ and $-\frac{1}{2}$.

$$i_1(t) = \frac{dq_1}{dt} \qquad \text{and} \qquad q_1(0) = Cv_C(0)$$

The system of differential equations is

$$\mathcal{E}_1: \quad (D^2 + 2D + 1)q_1 - (D + 1)i_2 = 0$$
$$\mathcal{E}_2: \quad -D(D + 1)q_1 + (3D + 1)i_2 = e_s$$

*Step 1*   No reordering of the variables is necessary since the network variable of interest is already in the last column.

Here $k = 1$, and $p_{11}(D) = D^2 + 2D + 1$.

*Step 4*   $p_{21}(D) - (-1)p_{11}(D) = D + 1$

*Steps 2 and 3*   Here $q_{21}(D) = -1$ and $r_{21}(D) = D + 1$.

*Step 5*   $\mathcal{E}_1' = \mathcal{E}_1:$ $\qquad\qquad\qquad (D^2 + 2D + 1)q_1 - (D + 1)i_2 = 0$
$\mathcal{E}_2' = \mathcal{E}_2 - q_{21}(D)\mathcal{E}_1 = \mathcal{E}_2 + \mathcal{E}_1: \quad (D + 1)q_1 + \qquad 2Di_2 = e_s$

Since the first column is not yet in the desired form, we go back to Step 2. We obtain successively [we call $p'_{ij}(D)$ the coefficients of the system $\mathcal{E}_1', \mathcal{E}_2', \ldots$]

$k = 2 \qquad p'_{21}(D) = D + 1$

$p'_{11}(D) - (D + 1)(D + 1) = 0 \qquad \text{or} \qquad \begin{aligned} q'_{11}(D) &= D + 1 \\ r'_{11}(D) &= 0 \end{aligned}$

The new system is

$\mathcal{E}_1'' = \mathcal{E}_1' - (D + 1)\mathcal{E}_2': \qquad 0 \qquad - (D + 1)(2D + 1)i_2 = -(D + 1)e_s$
$\mathcal{E}_2'' = \mathcal{E}_2': \qquad\qquad\qquad (D + 1)q_1 + 2D\, i_2 \qquad\qquad = e_s$

By interchanging the order of the equations, we obtain the triangular form

$$(D + 1)q_1 \quad\quad + 2D\, i_2 \quad\quad = e_s$$
$$(2D^2 + 3D + 1)i_2 = (D + 1)e_s$$

Therefore, the minimal polynomial of the network variable $i_2$ is $2D^2 + 3D + 1$, and the natural frequencies of $i_2$ are $-1$ and $-\frac{1}{2}$.

**Exercise** Consider the network shown in Fig. 2.2. Calculate the natural frequencies of $j_2$ for the case in which $L_2 = 1$ henry, $L_1 = 4$ henrys, $C_1 = 1$ farad, $R_1 = R_2 = 2$ ohms.

In conclusion, the elimination algorithm provides us with a systematic technique for obtaining the minimal differential equation and the natural frequencies of any network variable that we wish. Indeed, the zeros of the polynomial $\hat{p}_{nn}(s)$ are the natural frequencies of $x_n$.

## 3 Natural Frequencies of a Network

We now broaden our point of view and consider the linear time-invariant network $\mathfrak{N}$ as a whole. The main concept we wish to consider is that of natural frequency of a network. We say that a number $s_k$ is **a natural frequency of the network** $\mathfrak{N}$ if $s_k$ is a natural frequency of some *voltage* or a natural frequency of some *current* in the network $\mathfrak{N}$. It will turn out that in order to find the natural frequencies of a network we need not find all the natural frequencies of each voltage and of each current in the network.

**Example** The natural frequencies of the network of Example 5 in Sec. 1 are the natural frequencies of the voltage $v_1$, namely $j(1/\sqrt{L_1C_1})$ and $-j(1/\sqrt{L_1C_1})$, and the natural frequency of the voltage $v_2$, namely $-1/R_2C_2$.

Let us start by making two observations.

1. If $s_1 \neq 0$ and if $s_1$ is a natural frequency of a branch current, then it is also a natural frequency of the corresponding branch voltage. The reason is as follows: by assumption, for some initial state, the branch current $j$ will include the term $K_1\epsilon^{s_1t}$, where $K_1 \neq 0$. Depending on the nature of the branches, the corresponding branch voltages will behave as follows:

   *a.* For a resistor, $v = Rj$, and $v$ includes the term $RK_1\epsilon^{s_1t}$.

   *b.* For an inductor, $v = L(dj/dt)$, and $v$ includes the term $LK_1s_1\epsilon^{s_1t}$.

   *c.* For a capacitor, $v = v(0-) + (1/C)\int_0^t j(t')\,dt'$, and $v$ includes the term $CK_1(1/s_1)\epsilon^{s_1t}$.

2.  If $s_1 \neq 0$, and if $s_1$ is a natural frequency of a branch voltage, then it is also a natural frequency of the corresponding branch current.

The reason for the qualification $s_1 \neq 0$ is physically obvious. If a constant current flows through an inductor, the voltage across the inductor is identically zero; dually, if a constant voltage exists across a capacitor, the current through the capacitor is identically zero. Therefore, the number 0 may be a natural frequency of a branch current without being a natural frequency of the corresponding branch voltage, and vice versa.

One important consequence of observations 1 and 2 is that in order to find the *nonzero* natural frequencies of a network we may use any method of network analysis that we like. Indeed, if $s_1 \neq 0$ and if $s_1$ is a natural frequency of, say, some loop currents, it is necessarily a natural frequency of any branch voltages of branches in that loop, and vice versa.

Let us now state the main result of this section.

**THEOREM**    The nonzero natural frequencies of any linear time-invariant network are identical to the nonzero roots of the polynomial $\Delta(s) \triangleq \det [\mathbf{P}(s)] = 0$; here, $\mathbf{P}(s)$ is the matrix of *any* system of differential equations which describe the network.

**Remarks**    1.    Furthermore, since the nonzero roots of $\Delta(s)$ are identical to the nonzero roots of various network determinants $\Delta_n(s)$, $\Delta_m(s)$, etc., we conclude that nonzero natural frequencies of any linear time-invariant network are identical to nonzero roots of any network determinant.

2.    It is important to note that although $s_1$ is a natural frequency of a network, it is possible that some network variables do not have $s_1$ as a natural frequency. This is shown by Example 5, Sec. 1.

*Proof*    We have to show that the set of nonzero roots of $\det [\mathbf{P}(s)] = 0$ is identical with the set of nonzero natural frequencies of the network. We proceed in two steps.

1.    We show that if $s_1 \neq 0$ and if $\det [\mathbf{P}(s_1)] = 0$, then $s_1$ is a natural frequency of the network.

Since we are interested in zero-input responses, the network is described by a system of differential equations of the form

(3.1)    $\mathbf{P}(D)\mathbf{x} = \mathbf{0}$

If we replace the differential operator $D$ by the complex variable $s$, the determinant of the resulting matrix $\mathbf{P}(s)$ is a polynomial in $s$. Suppose that $s_1$ is a zero of this determinant; that is,

$\det [\mathbf{P}(s_1)] = 0$

This means that the matrix $\mathbf{P}(s_1)$ (whose elements are real or possibly complex numbers) is singular. Consequently, the system of $n$ linear homogeneous algebraic equations in $n$ unknowns $z_1, z_2, \ldots, z_n$

$$\mathbf{P}(s_1)\mathbf{z} = \mathbf{0}$$

has at least one nonzero solution.  Let $\mathbf{y}$ be one such nonzero solution. $\mathbf{y}$ is a vector whose components are (real or complex) numbers, and $\mathbf{P}(s_1)\mathbf{y} = \mathbf{0}.$

Consider now the waveforms specified by

$$\mathbf{y}\epsilon^{s_1 t} = \begin{bmatrix} y_1 \epsilon^{s_1 t} \\ y_2 \epsilon^{s_1 t} \\ \vdots \\ y_n \epsilon^{s_1 t} \end{bmatrix} \quad \text{for } t \geq 0$$

From the fact that $D\epsilon^{s_1 t} = s_1 \epsilon^{s_1 t}$, $D^2 \epsilon^{s_1 t} = s_1{}^2 \epsilon^{s_1 t}, \ldots$, it follows that $\mathbf{P}(D)\mathbf{y}\epsilon^{s_1 t} = \mathbf{P}(s_1)\mathbf{y}\epsilon^{s_1 t} = \mathbf{0}$ for all $t \geq 0$.  In other words, the waveforms $\mathbf{y}\epsilon^{s_1 t}$ are solutions of the system of (3.1).  Therefore, since $\mathbf{y}$ is a nonzero vector, $s_1$ is a natural frequency of all the network variables $y_i$, $i = 1, 2, \ldots, n$ for which $y_i \neq 0$.  If $y_i$ represents a voltage or a current, $s_1$, by definition, is a natural frequency of the network.  If $y_i$ represents a charge or flux, since $s_1 \neq 0$, $s_1$ is also a natural frequency of the corresponding current or voltage, and hence a natural frequency of the network.

2.  We have to show that, conversely, if $s_1 \neq 0$ and if $s_1$ is a natural frequency, then det $[\mathbf{P}(s_1)] = 0$.

If we take the Laplace transform of Eq. (3.1), we obtain

(3.2)   $\mathbf{P}(s)\mathbf{X}(s) = \mathbf{F}(s)$

where the components of $\mathbf{F}(s)$ are polynomials in $s$ whose coefficients involve the initial conditions.  If we solve (3.2) by Cramer's rule, we see that each component of $\mathbf{X}(s)$ is a ratio of polynomials and that the denominator polynomial is det $[\mathbf{P}(s)]$.  Therefore, by the partial-fraction expansion, in order that for some initial state, some component of $\mathbf{x}(t)$ includes a term like $K_1 \epsilon^{s_1 t}$ we must have det $[\mathbf{P}(s_1)] = 0$.  Therefore, whenever $s_1$ is a natural frequency, det $[\mathbf{P}(s_1)] = 0$.

We conclude this section by commenting that the number of natural frequencies of a network (including the origin) is no larger than the number of energy-storing elements in the network.  We give a physical argument to justify this statement.  The initial state of a network is specified by the initial currents of all the inductors and the initial voltages across all the capacitors.  The initial state determines all the arbitrary constants in the solution.  Now, in the initial-state specifications there may be some constraints imposed by Kirchhoff's laws; for example, the initial voltages on capacitors which form a loop or the initial currents in inductors which form a cut set must sum to zero.  Thus, the number of *independent* specifications may be less than the number of energy-storing elements.  Therefore, the number of natural frequencies in a network, or the number of independent initial conditions needed to determine uniquely all the voltages and currents in the network, is no larger than the number of energy-storing elements in the network.

**Exercise** Perform a node and a mesh analysis of the network shown in Fig. 3.1. Obtain for each case a system of differential equations. Compare the behavior of their determinants at $s = 0$. Give a physical interpretation.

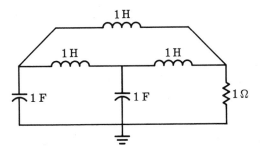

**Fig. 3.1** Network to be analyzed by node and mesh analysis; note the loop of inductors.

## 4 | Natural Frequencies and State Equations

Let us consider the question of the natural frequencies of a network in the light of the state equations. To be specific, consider the linear time-invariant network shown in Fig. 4.1. Following the standard technique, we find the following state equations

$$(4.1) \quad \begin{bmatrix} \dot{v}_1 \\ \dot{v}_2 \\ \dot{i}_3 \end{bmatrix} = \begin{bmatrix} -1 & 0 & -1 \\ 0 & -1 & 1 \\ 1 & -1 & -4 \end{bmatrix} \begin{bmatrix} v_1 \\ v_2 \\ i_3 \end{bmatrix}$$

of the general form

$$(4.2) \quad \dot{\mathbf{x}}(t) = \mathbf{A}\mathbf{x}(t)$$

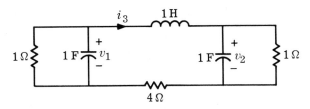

**Fig. 4.1** The state equations of this network are given by Eq. (4.1).

The state equations in (4.2) are actually a special case of the system of the linear differential equation

(4.3)   $\mathbf{P}(D)\mathbf{x} = \mathbf{f}$

where the matrix operator is

(4.4)   $\mathbf{P}(D) = D\mathbf{1} - \mathbf{A}$

The determinant of the system is

(4.5)   $\Delta(s) = \det [\mathbf{P}(s)] = \det [s\mathbf{1} - \mathbf{A}]$

Therefore, the natural frequencies of the network are the zeros of the polynomial $\Delta(s)$. In our example,

$$s\mathbf{1} - \mathbf{A} = \begin{bmatrix} s+1 & 0 & 1 \\ 0 & s+1 & -1 \\ -1 & 1 & s+4 \end{bmatrix}$$

and

$$\Delta(s) = (s+1)(s+2)(s+3)$$

Therefore, the natural frequencies of the network are $-1$, $-2$, and $-3$.

*Relation to eigenvectors*   Recalling basic facts from linear algebra, we see that each natural frequency, say, $s_i$, is an eigenvalue of $\mathbf{A}$ since

(4.6)   $\det [\mathbf{A} - s_i\mathbf{1}] = 0$

Now to each eigenvalue $s_i$ there is a (nonzero) eigenvector $\mathbf{u}_i$, that is, a vector $\mathbf{u}_i$ which satisfies the homogeneous algebraic equations

(4.7)   $\mathbf{A}\mathbf{u}_i = s_i\mathbf{u}_i$

In our example, it is easy to verify that we can associate to each eigenvalue the following eigenvectors:

$$\text{to } s_1 = -1 \quad \mathbf{u}_1 = \begin{bmatrix} 1 \\ 1 \\ 0 \end{bmatrix}$$

$$\text{to } s_1 = -2 \quad \mathbf{u}_2 = \begin{bmatrix} 1 \\ -1 \\ 1 \end{bmatrix}$$

$$\text{to } s_1 = -3 \quad \mathbf{u}_3 = \begin{bmatrix} 1 \\ -1 \\ 2 \end{bmatrix}$$

To physically interpret the eigenvectors, suppose that the initial state is $\mathbf{x}(0-) = \mathbf{u}_2$. We assert that the zero-input response is then

$$\mathbf{x}(t) = \mathbf{u}_2 \epsilon^{-2t}$$

Indeed, by substitution in Eq. (4.2), we find

$$-2\mathbf{u}_2\epsilon^{-2t} = (\mathbf{A}\mathbf{u}_2)\epsilon^{-2t} = (-2)\mathbf{u}_2\epsilon^{-2t}$$

where we used (4.7). We therefore obtain the following interesting conclusion:

*If the initial state lies along the eigenvector $\mathbf{u}_i$, then (1) the state trajectory remains along this eigenvector and (2) all network variables are proportional to $\epsilon^{-s_it}$.*

**Exercise 1** For the network shown in Fig. 4.1 find the waveforms when the initial state is

    *a.*   $\mathbf{x}(0) = \mathbf{u}_1$

    *b.*   $\mathbf{x}(0) = \mathbf{u}_2$

    *c.*   $\mathbf{x}(0) = \mathbf{u}_3$

    *d.*   $\mathbf{x}(0) = \alpha_1\mathbf{u}_1 + \alpha_2\mathbf{u}_2 + \alpha_3\mathbf{u}_3$

(where $\alpha_1$, $\alpha_2$, and $\alpha_3$ are prescribed numbers).

**Exercise 2** For the network shown in Fig. 1.5, find an initial state such that all branch voltages and all branch currents are proportional to

    *a.*   $\epsilon^{-t}$

    *b.*   $\epsilon^{-(1/3)t}$

## Summary

- Let $x$ be a network variable of a linear time-invariant network. The number $s_1$ is said to be a natural frequency of the network variable $x$ if, for some initial state, the zero-input response $x$ includes the term $K_1\epsilon^{s_1t}$.

- The minimal differential equation of the network variable $x_n$, $\hat{p}_{nn}(D)x_n = \hat{f}_n$, predicts all possible responses $x_n$ to any input waveform and to any initial state.

- The number $s_1$ is a natural frequency of $x_n$ if and only if $\hat{p}_{nn}(s_1) = 0$, that is, if $s_1$ is a zero of the characteristic polynomial of its minimal differential equation.

- Given any system of differential equations describing the network in terms of network variables $x_1, x_2, \ldots, x_n$, the elimination method is a systematic procedure for obtaining the minimal differential equation of any of these variables.

- We say that $s_k$ is a natural frequency of the linear time-invariant network $\mathfrak{N}$ if $s_k$ is a natural frequency of some voltage or of some current in $\mathfrak{N}$. $s_k$

may be a natural frequency of $\mathfrak{N}$ but not a natural frequency of some particular network variable.

■ Let $\mathbf{P}(D)$ be the polynomial matrix of a system of differential equations describing $\mathfrak{N}$, and let $s_1$ be a nonzero number. Then $s_1$ is a natural frequency of $\mathfrak{N}$ if and only if $\Delta(s_1) = 0$, where $\Delta(s) = \det [\mathbf{P}(s)]$. The set of nonzero roots of $\Delta(s) = 0$ is identical to the set of nonzero natural frequencies of $\mathfrak{N}$.

## Problems

Natural frequencies

**1.** Find the natural frequencies of the network variables indicated on the networks shown in Fig. P14.1.

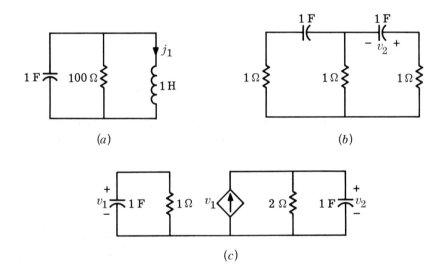

(a)       (b)

(c)

**Fig. P14.1**

Natural frequencies

**2.** Find the natural frequencies of the networks shown in Fig. P14.2.

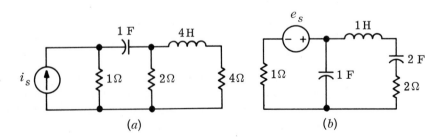

(a)       (b)

**Fig. P14.2**

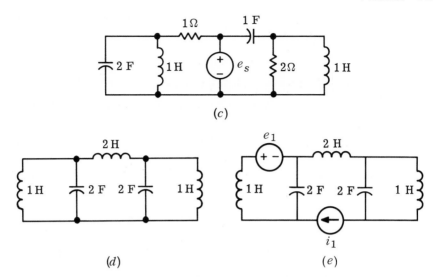

(c)

(d)                                      (e)

**Fig. P14.2**
(*continued*)

Equivalent systems

**3.** Given below are the differential equations representing two systems:

1: $(D^2 + 5D + 6)x = \epsilon^{-2t}$

2: $(D^3 + 6D^2 + 11D + 6)x = -\epsilon^{-2t}$

*a.*  Show that *any* solution of system 1 is a solution of system 2.

*b.*  Given $x(0) = 1$, and $\dot{x}(0) = 2$ as initial conditions for system 1, find suitable initial conditions for system 2 so that the solution of system 2 is also a solution of system 1.

*c.*  Find a solution of system 2 which is not a solution of system 1.

Elimination algorithm

**4.** Reduce the following systems to the triangular form:

*a.*  $\begin{bmatrix} D^2 + 5D + 7 & D + 5 & 3 \\ D^2 + 3D & D^2 + D + 1 & D + 1 \\ D + 2 & D^2 + 1 & 5D + 2 \end{bmatrix} \begin{bmatrix} x_1 \\ x_2 \\ x_3 \end{bmatrix} = \begin{bmatrix} \epsilon^{-t} \\ 5 \\ 1 - \epsilon^{-t} \end{bmatrix}$

*b.*  $\begin{bmatrix} D + 2 & -(D^2 + D) \\ -(D + 1) & D^2 + 2D + 1 \end{bmatrix} \begin{bmatrix} x_1 \\ x_2 \end{bmatrix} = \begin{bmatrix} u(t) \\ 0 \end{bmatrix}$

*c.*  $\begin{bmatrix} 2D + 1 & 1 \\ -1 & 2D + 1 \end{bmatrix} \begin{bmatrix} x_1 \\ x_2 \end{bmatrix} = \begin{bmatrix} \epsilon^t \\ \epsilon^t \end{bmatrix}$

Minimal equation

**5.** Consider the linear time-invariant circuit shown in Fig. P13.20.

*a.*  Find the minimal differential equation for the network variable *i*.

*b.*  Compare the characteristic polynomial of its minimal differential equation with the determinant of the system of Laplace transformed equations.

**6.** The network shown in Fig. P14.6 is linear and time-invariant.

*a.*    Find its natural frequencies.

*b.*    Find an initial state such that only the smallest natural frequency is excited.

*c.*    Find how to locate (at $t = 0$) 1 joule of energy in the network so that the resulting (zero-input) response includes only the largest natural frequency.   (Specify the required initial voltages on each capacitor.)

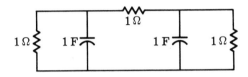

**Fig. P14.6**

**7.** Repeat Prob. 6 for the network shown in Fig. P14.7.

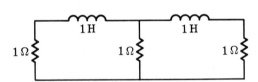

**Fig. P14.7**

# 15

# Network Functions

In this chapter we consider, exclusively, *linear time-invariant networks.* Whereas in Chap. 14 we were concerned primarily with zero-input responses, now we concentrate on *zero-state responses.* Our purpose is to consider the main properties of network functions. We shall show how a network function is related to the sinusoidal steady-state response and to the impulse response. We shall then relate the poles and the zeros of the network function to the network's frequency response and to its impulse response. We shall physically interpret the poles and the zeros and show how they can be used in oscillator design. Finally, we shall derive the symmetry properties of network functions.

## 1    Definition, Examples, and General Property

In Chap. 7 we introduced the definition of network function exclusively for sinusoidal steady-state responses. In Chap. 13 we extended this definition to the general case in terms of the Laplace transform. Let us review the extended definition. Consider a linear time-invariant network which contains a single independent voltage or current source as *input* with an arbitrary waveform $a(\cdot)$. Let the *zero-state response* be $b(\cdot)$, where the response is either a voltage across any two nodes of the network or a current in any branch of the network. We denote the Laplace transforms of the input and zero-state response by

(1.1a)   $A(s) = \mathcal{L}[a(t)]$

and

(1.1b)   $B(s) = \mathcal{L}[b(t)]$

Then the **network function** $H(s)$ is defined as follows:

(1.2a)
$$\text{Network function} \triangleq \frac{\mathcal{L}[\text{zero-state response}]}{\mathcal{L}[\text{input}]}$$

or

$$(1.2b) \quad H(s) = \frac{B(s)}{A(s)}$$

where $s$ is the complex frequency

$$(1.3) \quad s = \sigma + j\omega$$

and $\omega$ is the real frequency (in radians per second).

This extended concept of the network function is much more general than the one introduced for performing phasor calculations. First, the extended concept considers the network function as a function of the complex variable $s$ (rather than the purely imaginary variable $j\omega$). Second, it relates the *zero-state* response to *any* input (rather than relating the output phasor to the input phasor). It is, however, also important to remember that the network function only gives us a way to calculate the *zero-state* response. In general, it is not sufficient to calculate the complete response.

There are many kinds of network functions. In Chap. 7 we introduced the definition of the driving-point impedance of a one-port as the ratio of the phasor which represents the sinusoidal steady-state output voltage to the phasor which represents the sinusoidal input current. Since the driving-point impedance is a special case of a network function, we can similarly extend the definition of the **driving-point impedance** function of a one-port as the ratio of the Laplace transform of the zero-state voltage response to the Laplace transform of the driving current. Clearly, the driving-point impedances of the resistor, inductor, and capacitor are then $R$, $sL$, and $1/sC$, respectively, as shown in Fig. 1.1. We can similarly define the **driving-point admittance** function of a one-port as the ratio of the Laplace transform of the zero-state current response to the Laplace transform of the input voltage. Obviously, the driving-point admittances for the $R$, $L$, and $C$ elements in Fig. 1.1 are $1/R$, $1/sL$, and $sC$, respectively. It is easy to show that the rules for combining impedances and admittances are identical with those applicable in the sinusoidal steady state. For a series connection of elements, we add the impedances of individual elements to obtain the overall driving-point impedance. For a parallel connection of elements, we add the admittances of individual elements to obtain the overall driving-point admittance. Often, by simple series and parallel reduction we can compute the driving-point impedance of a complicated one-port. An important practical example is the ladder network. The general methods of network analysis (mesh, loop, node, and cut-set) can also be used directly in terms of impedances, admittances, and the transformed voltage and current variables to yield a set of linear *algebraic* equations. These equations can then be solved for the desired variables to obtain the desired network functions. Thus, if we only wish to determine the network function, we can bypass the writing of integrodifferential equations. Such analysis, which uses directly the transformed variables,

$$Z = \frac{\mathcal{L}[v]}{\mathcal{L}[i]} = \frac{V}{I} = R$$

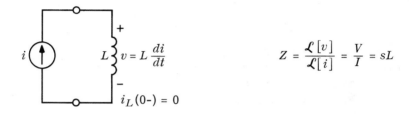

$$Z = \frac{\mathcal{L}[v]}{\mathcal{L}[i]} = \frac{V}{I} = sL$$

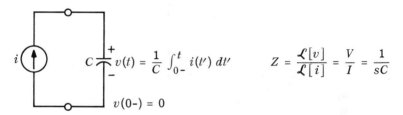

$$Z = \frac{\mathcal{L}[v]}{\mathcal{L}[i]} = \frac{V}{I} = \frac{1}{sC}$$

**Fig. 1.1**   The driving-point impedances of linear time-invariant $R$, $L$, and $C$ elements; note that the inductor and the capacitor are in the zero state prior to the application of the input $i$ at $t = 0$.

is often referred to as analysis in the "frequency domain," in contrast to the analysis with integrodifferential equations, which is called analysis in the "time domain."

We shall encounter different types of network functions, depending on the types of inputs and responses. Since the input and the response may either be a current or a voltage, the network function may be a driving-point impedance, a driving-point admittance, a transfer impedance, a transfer admittance, a transfer voltage ratio, or a transfer current ratio. In the examples to follow we shall illustrate the ways of determining these network functions. Even though each of the network functions mentioned above has distinct properties, our concern in this chapter is to obtain some general and broad properties of network functions. We leave the special properties to an advanced course in network theory.

**Example 1**  Consider the usual parallel $RC$ circuit driven by an independent current source, as shown in Fig. 1.2. Let the voltage $v$ be the zero-state response, and let $V(s)$ and $I(s)$ be the Laplace transforms of $v(t)$ and $i(t)$, respectively. Since $G + sC$ is the driving-point admittance of the parallel $RC$ circuit, we find

$$V(s) = \frac{1}{G + sC} I(s)$$

The network function is the driving-point impedance; thus,

(1.4)  $$H(s) = \frac{V(s)}{I(s)} = \frac{1}{G + sC}$$

**Example 2**  Consider the low-pass filter shown in Fig. 1.3. The input is the voltage source $e_0$, and the zero-state response is the current $i_2$ in the output resistor. The network function in this case is a transfer admittance, $H(s) = I_2/E_0$, where $I_2$ and $E_0$ are, respectively, the Laplace transforms of $i_2(t)$ and $e_0(t)$; the transfer admittance can be obtained easily by means of mesh analysis. The mesh equations in terms of the transformed network variables are

$$\left(L_1 s + \frac{1}{Cs}\right) I_1 - \frac{1}{Cs} I_2 = E_0$$

$$- \frac{1}{Cs} I_1 + \left(\frac{1}{Cs} + L_2 s + R\right) I_2 = 0$$

Note that in writing these equations we used the fact that the network was in the zero state at time $t = 0-$. Solving for $I_2$, we obtain

$$I_2 = \frac{E_0/Cs}{\begin{vmatrix} L_1 s + \dfrac{1}{Cs} & -\dfrac{1}{Cs} \\[2ex] -\dfrac{1}{Cs} & \dfrac{1}{Cs} + L_2 s + R \end{vmatrix}}$$

$$= \frac{E_0}{L_1 L_2 Cs^3 + RL_1 Cs^2 + (L_1 + L_2)s + R}$$

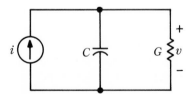

**Fig. 1.2**  Network for Example 1: a parallel $RC$ circuit.

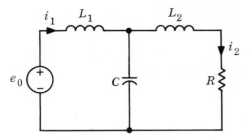

**Fig. 1.3**   Network for Example 2: a low-pass filter.

Therefore, the network function is the transfer admittance

(1.5)   $$H(s) = \frac{I_2}{E_0} = \frac{1}{L_1 L_2 C s^3 + R L_1 C s^2 + (L_1 + L_2)s + R}$$

**Example 3**   Consider the transistor amplifier shown in Fig. 1.4; its small-signal equivalent circuit is also shown on the figure. The network function of interest is a transfer voltage ratio, $H(s) = V_2/V_0$. The node equations, in terms of the transformed network variables, are

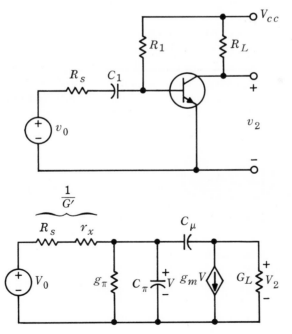

**Fig. 1.4**   Network for Example 3: a transistor amplifier. With $r_\pi = R_L = 1/G' = 1,000$ ohms, $g_m = 0.04$ mho, $C_\mu = 10^{-11}$ farad, and $C_\pi = 10^{-9}$ farad; the blocking capacitor $C_1$ and the biasing resistor $R_1$ are disregarded in the equivalent circuit.

$$\begin{bmatrix} (G' + g_\pi) + (C_\pi + C_\mu)s & -C_\mu s \\[2mm] g_m - C_\mu s & C_\mu s + G_L \end{bmatrix} \begin{bmatrix} V \\[2mm] V_2 \end{bmatrix} = \begin{bmatrix} G'V_0 \\[2mm] 0 \end{bmatrix}$$

The network function is found to be

$$H(s) = \frac{V_2}{V_0}$$

$$= \frac{(G'/C_\pi)[s - g_m/C_\mu]}{s^2 + [G_L/C_\mu + (G' + g_\pi + G_L + g_m)/C_\pi]s + G_L(G' + g_\pi)/C_\pi C_\mu}$$

For the typical set of element values shown in the figure,

(1.6) $\qquad H(s) = 10^6 \dfrac{s - 4 \times 10^9}{s^2 + 1.43 \times 10^8 s + 2 \times 10^{14}}$

---

*General
property*

We have shown in Chap. 13, Sec. 5, that network functions are rational functions of the complex frequency variable $s$ with *real* coefficients. We have checked this fact in the three examples above. This fact is true for any *lumped* linear time-invariant network. Thus, we write, in general,

(1.7) $\qquad H(s) = \dfrac{P(s)}{Q(s)} = \dfrac{b_0 s^m + b_1 s^{m-1} + \cdots + b_{m-1} s + b_m}{a_0 s^n + a_1 s^{n-1} + \cdots + a_{n-1} s + a_n}$

where $P(s)$ and $Q(s)$ are polynomials in the variable $s$ and the coefficients $a_0, a_1, \ldots, a_n, b_0, b_1, \ldots, b_m$ are *real* numbers. These coefficients are real because each one is a sum of products of element values of resistors, inductors, and capacitors, etc., and these element values are specified by real numbers. Thus, a network function is specified completely by the two sets of real coefficients which define the numerator and the denominator polynomials. Alternatively, Eq. (1.7) can be written in the factored form as

(1.8) $\qquad H(s) = K \dfrac{\displaystyle\prod_{i=1}^{m} (s - z_i)}{\displaystyle\prod_{j=1}^{n} (s - p_j)}$

where $K$ is a (real) scale factor, $z_i$, $i = 1, 2, \ldots, m$ are called the **zeros** of the network function, and $p_j$, $j = 1, 2, \ldots, n$ are called the **poles** of the network function. Thus, an alternate complete specification of a network function is given by the $m$ zeros $(z_1, z_2, \ldots, z_m)$, the $n$ poles $(p_1, p_2, \ldots, p_n)$, and the scale factor $K$. Since the numerator polynomial $P(s)$ and the denominator polynomial $Q(s)$ have real coefficients, zeros and poles must be real or occur in complex conjugate pairs. More precisely, if $p_1 = \sigma_1 + j\omega_1$ is a pole; that is, $Q(p_1) = 0$, then $\bar{p}_1 = \sigma_1 - j\omega_1$ is also a pole, that is, $Q(\bar{p}_1) = 0$. Similarly, if $z_2 = \sigma_2 + j\omega_2$ is a zero; that is, $P(z_2) = 0$, then

$\bar{z}_2 = \sigma_2 - j\omega_2$ is also a zero; that is, $P(\bar{z}_2) = 0$. These are immediate consequences of the fact that any polynomial $F(s)$ with *real* coefficients has the property that

$$\overline{F(s)} = F(\bar{s}) \qquad \text{for all } s$$

or

$$\overline{F(\sigma + j\omega)} = F(\sigma - j\omega) \qquad \text{for all real } \sigma \text{ and } \omega$$

**Exercise**  Let $H(s)$ have one zero and three poles.  Let the zero be at $s = 2$ and the poles at $s = -3$, $s = -1 \pm j2$.  Given also that $H(0) = 1$, express $H(s)$ as a rational function in the two forms shown in Eqs. (1.7) and (1.8).

**Exercise**  Write the following rational function in factored form [i.e., as in Eq. (1.8) above]:

$$H(s) = \frac{2s^2 - 12s + 16}{s^3 + 4s^2 + 6s + 3}$$

## 2    Poles, Zeros, and Frequency Response

In Chap. 13 we showed that substituting $s$ for $j\omega$ in the network function $H(s)$ gives $H(j\omega)$, defined as the ratio of the phasors representing the sinusoidal steady-state response and the corresponding sinusoidal input. Thus, it is of particular importance to understand the behavior of a network function when $s = j\omega$ and $\omega$ varies from 0 to $\infty$.  In this way we can visualize the sinusoidal steady-state properties from very low frequencies to very high frequencies.  Since for each fixed frequency $\omega$, $H(j\omega)$ is usually a complex number, we represent it in its polar form,

$$(2.1) \quad H(j\omega) = |H(j\omega)|\epsilon^{j\angle H(j\omega)}$$

where $|H(j\omega)|$ is called the **magnitude** and $\angle H(j\omega)$ is called the **phase** of the network function at the frequency $\omega$.  Whenever a network function represents a transfer function, it is convenient to introduce the logarithmic measure as follows:

$$(2.2) \quad \Theta(j\omega) \triangleq \ln H(j\omega) = \ln |H(j\omega)| + j\angle H(j\omega)$$

The real part of the above expression is usually called the **gain** and is measured in units called nepers.  We denote the gain by $\alpha(\omega)$; that is,

$$(2.3) \quad \alpha(\omega) \triangleq \ln |H(j\omega)| \qquad \text{nepers}$$

Note that in Eqs. (2.2) and (2.3) we use *natural* logarithms.  If we measure the gain in decibels, we calculate

$$20 \log |H(j\omega)| \qquad \text{db}$$

It is useful to remember that

1 neper $\approx$ 8.686 decibels

The magnitude and phase of a network function are of paramount importance, for they not only give the sinusoidal steady-state response at any frequency, but also contain all the information for calculating the zero-state response due to an *arbitrary* input. From the practical point of view, the curves of magnitude and phase vs. frequency are easily measured in the laboratory; furthermore, they can be measured with great precision. The combined information of magnitude and phase of a network function for all $\omega$ is usually referred to as the **frequency response.** In this section we shall investigate the relation between poles, zeros, and the frequency response.

**Example 1**   The *RC* circuit of Fig. 1.2 has a network function (see Eq. 1.4)

$$H(s) = \frac{1}{G + sC} = \frac{1/C}{s + 1/RC}$$

The numerator polynomial is equal to a constant for all $s$. Hence we say that $H(s)$ has no *finite* zeros.† The denominator polynomial is zero when $s = -1/RC$; that is, $H(s)$ has a pole at $s = -1/RC$.

In order to consider the magnitude and the phase we write

$$H(j\omega) = \frac{1}{C} \frac{1}{j\omega - (-1/RC)}$$

(2.4)   $$|H(j\omega)| = \frac{1}{C} \frac{1}{|j\omega - (-1/RC)|} = \frac{1}{C} \frac{1}{\sqrt{\omega^2 + (1/RC)^2}}$$

(2.5)   $$\angle H(j\omega) = -\tan^{-1} \omega RC$$

From Eq. (2.4), we see that as $\omega$ increases, $|H(j\omega)|$ decreases monotonically. At direct current, $H(0) = R$, and for large $\omega$, $|H(j\omega)| \approx 1/(\omega C)$. From Eq. (2.5), $\angle H(0) = 0$; then $\angle H(j\omega)$ decreases monotonically, and $\angle H(j\omega) \to -90°$ as $\omega \to \infty$. The frequency response is shown in Fig. 2.1a. Let us now obtain these conclusions from $s$-plane considerations. Refer to Fig. 2.1b, which shows the pole located at $s = -1/RC$; the complex number $j\omega - (-1/RC)$ represents the vector whose *tip* is at the point $s = j\omega$ and whose *origin* is at the pole $-1/RC$. Hence

$$\left| j\omega - \left( \frac{-1}{RC} \right) \right| = d_1$$

is the length of the vector, and

† Some authors find it convenient to introduce the concept of "zero at infinity." If $H(s)$ behaves for large $|s|$ like $K/s^k$ (where $K$ is a constant), they say that $H(s)$ has *a zero at infinity of order k.*

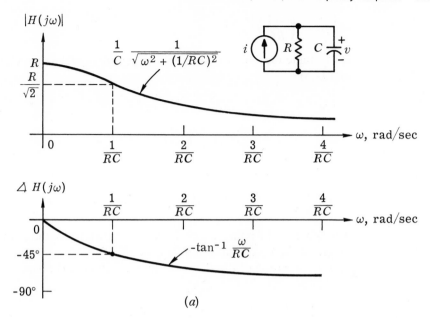

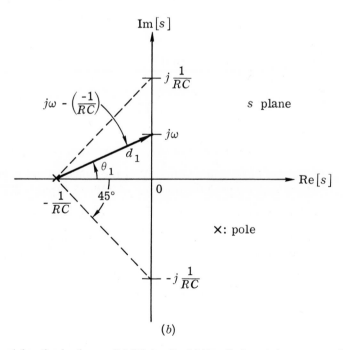

**Fig. 2.1**   Network function for the parallel $RC$ circuit.   (*a*) Magnitude and phase curves; (*b*) interpretation in terms of distances and angles on the $s$ plane.

$$\angle\left[j\omega - \left(\frac{-1}{RC}\right)\right] = \theta_1$$

is the angle the vector makes with the positive real axis. Now from (2.4) and (2.5)

$$|H(j\omega)| = \frac{1}{Cd_1} \quad \text{and} \quad \angle H(j\omega) = -\theta_1$$

Clearly, at $\omega = 0$, $H(0) = R$, and $\angle H(0) = 0$; as $\omega \to \infty$, $|H(j\omega)| \to 0$, and $\angle H(j\omega) \to -90°$. Also, at $\omega = 1/RC$, the vector $j\omega - (-1/RC)$ makes an angle of $45°$ with the real axis; consequently,

$$\left|H\left(j\frac{1}{RC}\right)\right| = \frac{R}{\sqrt{2}} \quad \angle H\left(j\frac{1}{RC}\right) = -45°$$

Therefore, as the point $j\omega$ travels along the Im $(s)$ axis, the changes in $d_1$, the length of the vector, and in $\theta_1$, the angle of the vector, allow us to obtain an interpretation of the shape of the magnitude and phase of $H$ shown on Fig. 2.1a.

---

**Example 2**  Consider the usual parallel $RLC$ tuned circuit driven by a current source. Let $v$, the voltage across the elements, be the zero-state response. Clearly, from the circuit in Fig. 2.2 the admittance of the parallel connection is $Cs + G + 1/Ls$; hence

$$(2.6) \quad V(s) = \frac{1}{Cs + G + 1/Ls} I(s)$$

$$H(s) = \frac{1}{C} \frac{s}{s^2 + (G/C)s + 1/LC}$$

$$(2.7) \quad = \frac{1}{C} \frac{s - 0}{[s - (-\alpha + j\omega_d)][s - (-\alpha - j\omega_d)]}$$

where we have assumed that $Q$ of the tuned circuit is larger than $\frac{1}{2}$, and where we have used the notation of Table 5.1 (pages 222 and 223); that is,

$$\alpha \triangleq G/2C \qquad \omega_d \triangleq \sqrt{\omega_0^2 - \alpha^2} \qquad \omega_0^2 \triangleq 1/LC$$

The network function has one zero at $s = 0$ and has a pair of complex conjugate poles at $-\alpha \pm j\omega_d$. From Eq. (2.7) we obtain

$$(2.8) \quad |H(j\omega)| = \frac{1}{C} \frac{|j\omega - 0|}{|j\omega - (-\alpha + j\omega_d)||j\omega - (-\alpha - j\omega_d)|}$$

and

$$(2.9) \quad \angle H(j\omega) = \angle(j\omega - 0) - \angle[j\omega - (-\alpha + j\omega_d)] - \angle[j\omega - (-\alpha - j\omega_d)]$$

If we interpret (2.8) and (2.9) in terms of the lengths and angles of the vectors shown in Fig. 2.2, we obtain

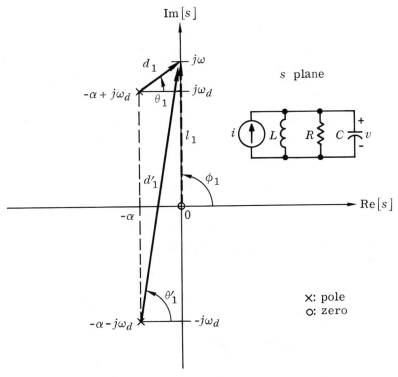

**Fig. 2.2**    Frequency response interpretation in terms of poles and zeros of the network function of a parallel tuned circuit.

(2.10)    $|H(j\omega)| = \dfrac{1}{C}\dfrac{l_1}{d_1 d'_1}$

(2.11)    $\measuredangle H(j\omega) = \phi_1 - \theta_1 - \theta'_1$

It is important to develop our intuition for relating the pole and zero locations and the gain and phase curves. For example, by analyzing Eqs. (2.10) and (2.11) and Fig. 2.2, we should be able to deduce the following facts:

1.  $H(0) = 0$.
2.  For $\omega > 0$ and $\omega \ll \omega_d$, $\measuredangle H(j\omega) \approx 90°$ (because the angles of the vectors represented by $d_1$, $d'_1$ will approximately cancel out).
3.  $|H(j\omega)|$ will increase as $\omega$ increases and reach a maximum at about $\omega \approx \omega_d$ (because the denominator factor $d_1$ is the minimum).
4.  Using the fact that for the case of Fig. 2.2 $\alpha \ll \omega_d$, hence $d'_1 \approx 2\omega_d$ when $j\omega = j\omega_d$,

$$|H(j\omega_d)| \simeq \frac{1}{C}\frac{\omega_d}{\alpha 2\omega_d} = \frac{1}{2\alpha C} = \frac{1}{G} = R$$

5. $\measuredangle H(j\omega_d) \approx 0°$ (because $\theta_1' \approx 90°$).

6. As $\omega$ increases beyond $\omega_d$, $|H(j\omega)|$ decreases and tends to zero as $\omega \to \infty$; also $\measuredangle H(j\omega) \to -90°$ as $\omega \to \infty$. Indeed, when $\omega \gg \omega_d$, the two denominator factors $d_1$ and $d_1'$ are proportional to $\omega$, and the numerator factor is equal to $\omega$; hence $|H(j\omega)|$ is proportional to $1/\omega$.

It is intuitively clear that the shape of the curves depends on the ratio $\omega_0/\alpha$, that is, on the $Q$ of the tuned circuit, since $Q \triangleq \omega_0/2\alpha$. For exact curves see Figs. 3.2 and 3.3 of the next section.

**Example 3**   The transistor amplifier in Fig. 1.4 has a network function (see Eq. 1.6)

$$H(s) = 10^6 \frac{s - 4 \times 10^9}{s^2 + 1.43 \times 10^8 s + 2 \times 10^{14}}$$

There are two poles which are located on the negative $\sigma$ axis at

$$p_1 = -0.14 \times 10^7 \qquad p_2 = -14.16 \times 10^7$$

There is a finite zero on the positive $\sigma$ axis at

$$z_1 = 4 \times 10^9$$

The magnitude and the phase are

$$|H(j\omega)| \approx \left[ \frac{1600 + 10^{-16}\omega^2}{4 + 2 \times 10^{-12}\omega^2 + 10^{-28}\omega^4} \right]^{1/2}$$

$$\measuredangle H(j\omega) = \tan^{-1}\frac{10^{-9}\omega}{-4} - \tan^{-1}\frac{14.3 \times 10^{-7}\omega}{2 - 10^{-14}\omega^2}$$

The magnitude and phase curves are plotted in Fig. 2.3, where we used a logarithmic scale for the frequency. The magnitude curve indicates that the transistor circuit is a low-pass amplifier with a midband magnitude of 20 (or 20 log 20 = 26 db gain). The 3-db cutoff frequency is approximately at $\omega = 2 \times 10^6$ rad/sec (or 320 kHz), where the magnitude is $20/\sqrt{2}$db.

*General case*   In general, we have a network function in the form of a ratio of polynomials with real coefficients as follows:

$$H(s) = \frac{b_0 s^m + \cdots + b_{m-1}s + b_m}{a_0 s^n + \cdots + a_{n-1}s + a_n}$$

To obtain the frequency response, we proceed as follows:

*Step 1*   Find the poles and the zeros. For convenience we call them $p_1, p_2, \ldots,$ and $z_1, z_2, \ldots,$ respectively.

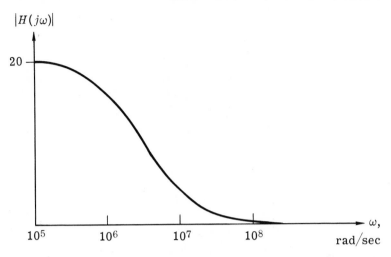

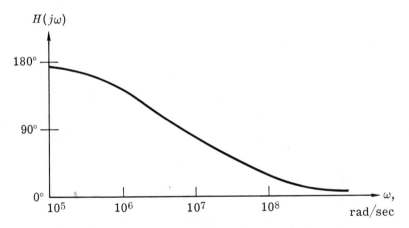

**Fig. 2.3**   Frequency response of the transistor amplifier in Fig. 1.4.

*Step 2*   Express each polynomial as a product of first-order factors. For example, assume we have three zeros and four poles; the zeros $z_2$ and $\bar{z}_2$ and the poles $p_3$ and $\bar{p}_3$ form complex conjugate pairs (see Fig. 2.4); we then obtain

$$(2.12) \quad H(s) = \frac{b_0}{a_0} \frac{(s - z_1)(s - z_2)(s - \bar{z}_2)}{(s - p_1)(s - p_2)(s - p_3)(s - \bar{p}_3)}$$

Note the presence of the scale factor $b_0/a_0$.

*Step 3*   Putting $s = j\omega$, and taking absolute values of both sides of Eq. (2.12), we obtain the gain

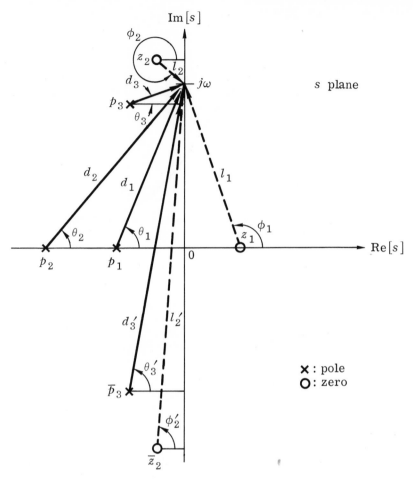

**Fig. 2.4**    Interpretation of the magnitude and phase of the network function $H(j\omega)$ [specified by Eq. (2.12)] in terms of distances and angles on the $s$ plane.

(2.13a)    $$|H(j\omega)| = \left|\frac{b_0}{a_0}\right| \frac{|j\omega - z_1||j\omega - z_2||j\omega - \bar{z}_2|}{|j\omega - p_1||j\omega - p_2||j\omega - p_3||j\omega - \bar{p}_3|}$$

or, using the notations defined in Fig. 2.4,

(2.13b)    $$|H(j\omega)| = \left|\frac{b_0}{a_0}\right| \frac{l_1 l_2 l_2'}{d_1 d_2 d_3 d_3'}$$

If we take the logarithms of both sides of (2.13a) and (2.13b), we obtain

$$\alpha(\omega) = \ln|H(j\omega)|$$

$$= \ln\left|\frac{b_0}{a_0}\right| + \ln|j\omega - z_1| + \ln|j\omega - z_2| + \ln|j\omega - \bar{z}_2|$$

$$-\ln |j\omega - p_1| - \ln |j\omega - p_2| - \ln |j\omega - p_3| - \ln |j\omega - \bar{p}_3|$$

$$= \ln \left| \frac{b_0}{a_0} \right| + \ln l_1 + \ln l_2 + \ln l_2' - \ln d_1 - \ln d_2 - \ln d_3 - \ln d_3'$$

The advantage of using logarithms is that the gain for the network function in decibels (or nepers) can be obtained by taking the sums and differences of the gains in decibels (or nepers) contributed by individual zeros and poles.

*Step 4*  Putting $s = j\omega$ and taking the argument of both sides of Eq. (2.12), we have

$$\sphericalangle H(j\omega) = \sphericalangle \frac{b_0}{a_0} + \sphericalangle (j\omega - z_1) + \sphericalangle (j\omega - z_2) + \sphericalangle (j\omega - \bar{z}_2)$$

$$- \sphericalangle (j\omega - p_1) - \sphericalangle (j\omega - p_2) - \sphericalangle (j\omega - p_3) - \sphericalangle (j\omega - \bar{p}_3)$$

or, using the notations of Fig. 2.4,

(2.14)  $$\sphericalangle H(j\omega) = \sphericalangle \frac{b_0}{a_0} + (\phi_1 + \phi_2 + \phi_2') - (\theta_1 + \theta_2 + \theta_3 + \theta_3')$$

Equations (2.13b) and (2.14) show that, given the poles and zeros and the scale factor $b_0/a_0$, it is possible to calculate graphically the curve $|H(j\omega)|$ and the curve $\sphericalangle H(j\omega)$ on the basis of measurements made on the s plane.

These graphical interpretations of the magnitude and phase curves lead to some general ideas connecting pole-zero locations with gain and phase: (1) in the neighborhood of a pole close to the $j\omega$ axis, we expect the magnitude to have a local maximum and the phase to vary rapidly; (2) in the neighborhood of a zero close to the $j\omega$ axis, we expect the magnitude to be small and the phase to vary rapidly. Hence we obtain the useful design ideas of piling up the poles where the magnitude has to be large and piling up the zeros where the magnitude has to be small. Formulas (2.13) and (2.14) are extremely important for design purposes; they relate the poles and zeros to the magnitude and phase of the network function. We shall see in the next section that the behavior of the impulse response of the network is also closely related to the location of the poles and the zeros.

## 3  Poles, Zeros, and Impulse Response

In the previous section we saw how the knowledge of the pole and zero locations allowed us by graphical means to deduce the shape of the magnitude and the phase curves of any network function. We propose now to explore further these ideas and tie them to the behavior of the impulse response.

We start by recalling that the inverse Laplace transform of a network function is the corresponding impulse response

$$\mathcal{L}^{-1}[H(s)] = h(t)$$

We propose to consider the relation between pole and zero locations and the impulse response. The best way to get hold of these ideas is to consider simple examples.

---

**Example 1**   We again consider the parallel $RC$ circuit. As before,

$$(3.1) \quad H(s) = \frac{1}{C} \frac{1}{s + 1/RC}$$

Consequently,

$$(3.2) \quad h(t) = \frac{1}{C} u(t) \epsilon^{-t/RC}$$

The magnitude curve and the impulse response are shown for $RC = 1$ and $RC = 0.5$ in Fig. 3.1. From the figure we conclude that (1) the closer the pole is to the $j\omega$ axis the narrower the 3-db bandwidth, and (2) the closer the pole is to the $j\omega$ axis, the longer it takes the impulse response to decay to zero. Equivalently, the closer the pole is to the $j\omega$ axis, the longer the memory of the circuit.

---

**Example 2**   We turn now to the parallel $RLC$ circuit driven by a current source; as usual, we take the voltage across the elements to be the response. Then

$$(3.3) \quad H(s) = \frac{1}{C} \frac{s}{s^2 + (G/C)s + 1/LC} = \frac{1}{C} \frac{s}{s^2 + 2\alpha s + \omega_0^2}$$

and, for $Q > \frac{1}{2}$, we have

$$(3.4) \quad h(t) = \frac{1}{C} u(t) \epsilon^{-\alpha t} \cos (\omega_d t + \phi)$$

where as usual

$$\omega_d^2 \overset{\Delta}{=} \omega_0^2 - \alpha^2 \qquad \phi \overset{\Delta}{=} \sin^{-1} \frac{\alpha}{\omega_0}$$

From Eq. (3.4) we note two facts. (1) The distance between the pole and the $j\omega$ axis, $\alpha$, determines completely the rate of decay of the impulse response. The closer the pole is to the $j\omega$ axis, the smaller the rate of decay; if the pole were on the $j\omega$ axis, there would be no decay, and if it were to the *right* of the $j\omega$ axis, the impulse response would be *exponentially increasing*! (2) The ordinate of the pole, $\omega_d$, determines $2\pi/\omega_d$, the distance between successive zero-crossings of the impulse response; the larger $\omega_d$, the higher the "frequency" of the impulse response. These facts are illustrated by Figs. 3.2 and 3.3, where $\alpha = 0.3$ and $\omega_d = 1$, and $\alpha = 0.1$ and

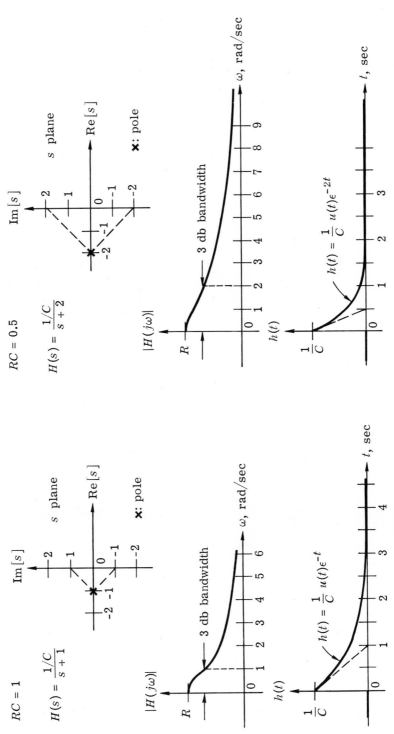

**Fig. 3.1** Network functions for parallel *RC* network with different time constants; the relations between the pole location and the magnitude curves and the impulse responses are illustrated.

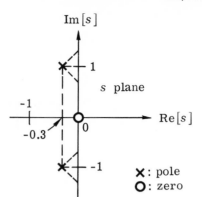

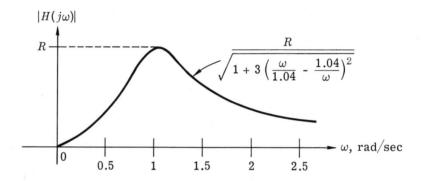

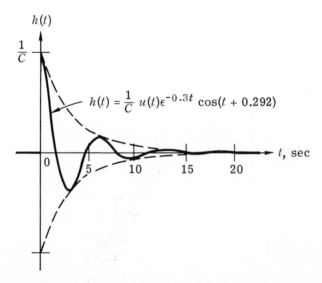

**Fig. 3.2**    Network function for a parallel $RLC$ tuned circuit; magnitude curve and impulse response with $Q = 1.74$.

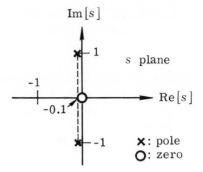

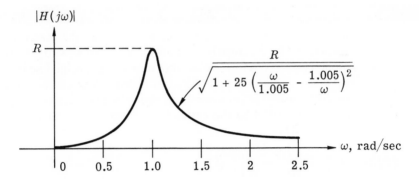

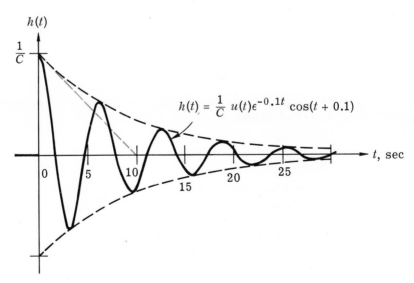

**Fig. 3.3**    Network function for a parallel *RLC* tuned circuit; magnitude curve and impulse response with $Q = 5.025$.

$\omega_d = 1$, respectively.   Note also that the closer the pole is to the $j\omega$ axis, the sharper the "peak" in the magnitude curve.   When $Q$ is larger than 5, the 3-db bandwidth is approximately $2\alpha$ rad/sec or $\alpha/\pi$ Hz.

In general, we may state the following.   Isolated poles close to the $j\omega$ axis tend to produce sharp peaks in the magnitude curve.

The ordinate of the pole determines the location of the peak as well as the "ringing frequency" of the impulse response.   The distance of the pole to the $j\omega$ axis can be estimated by the fact that $2\alpha \simeq$ 3-db bandwidth (in radians per second).   This distance $\alpha$ also determines the rate of decay of the impulse response.

## 4 | Physical Interpretation of Poles and Zeros

So far we have related the poles and zeros to the magnitude and phase curves, and to the impulse response.   We wish now to give physical interpretations.   As before we assume we have a specific linear time-invariant network, a specific input (say, an independent current source $i$), and a specific zero-state response (say, the voltage $v_L$ across a designated node pair).†   Let $H(s)$ be the network function relating this input and this response.   Then

$(4.1a)$ $\qquad H(s) = \dfrac{V_L(s)}{I(s)} = \dfrac{\mathcal{L}[v_L(t)]}{\mathcal{L}[i(t)]}$

where

$(4.1b)$ $\qquad H(s) = \dfrac{P(s)}{Q(s)} = K\dfrac{\displaystyle\prod_{i=1}^{m}(s - z_i)}{\displaystyle\prod_{j=1}^{n}(s - p_j)}$

Thus, $z_1, z_2, \ldots, z_m$ are the zeros, and $p_1, p_2, \ldots, p_n$ are the poles of the network function.

### 4.1 | Poles

In this subsection we wish to derive an important relation between the poles of a network function and the natural frequencies of the corresponding network variable.   In fact, we are going to show that *any pole of a network function is a natural frequency of the corresponding (output) network variable*.   For example, Eq. (4.1b) shows that $p_1$ is a pole of $H(s)$; the assertion above states that $p_1$ is a natural frequency of the node-pair voltage

---

† We use the subscript $L$ to signify that $v_L$, the output, is the voltage across the load on the network.

$v_L$. To prove the assertion, suppose that the network is in the zero state at time $0-$ and that the input current is $\delta(t)$. Since $\mathcal{L}[\delta] = 1$, the Laplace transform of the output voltage is

$$v_L(t) = \mathcal{L}^{-1}[H(s)]$$

By Eq. (4.1b), the partial-fraction expansion of $H(s)$ is of the form

$$(4.2) \quad H(s) = \sum_{i=1}^{n} \frac{K_i}{s - p_i}$$

where $K_i$ is the residue at the pole $p_i$. Consequently,

$$(4.3) \quad v_L(t) = \sum_{i=1}^{n} K_i \epsilon^{p_i t} \qquad t \geq 0$$

Now, observe that for $t > 0$, the input is identically zero; therefore, the response waveform $v_L(\cdot)$ for $t > 0$ given by Eq. (4.3) can be considered to be a zero-input response. In fact, it can be interpreted as being the zero-input response due to the state at time $0+$ in which the network found itself as a result of the impulse. Since, for $t > 0$, Eq. (4.3) is the expression of a zero-input response and since $K_1 \neq 0$, $p_1$ is a natural frequency of $v_L$.

In fact, it can be shown that if the network starts in the zero state at time $0-$ and is driven by an input waveform $i(\cdot)$ over the interval $[0,T]$ (with the understanding that $i$ is switched off for $t > T$), then the input waveform may be so chosen that for $t > T$, $v_L(t) = K\epsilon^{p_1 t}$, that is, for $t > T$, the output is purely exponential. We say that the natural frequency $p_1$, and only $p_1$, has been excited by this particular input. We simply illustrate it by an example.

**Example 1**    Consider the linear time-invariant network shown in Fig. 4.1. Let us determine an input $i_s$ over $[0,1]$ such that the zero-state response is, for $t \geq 1$, $e_2(t) = \epsilon^{p_1 t}$ with $p_1$ being the pole nearest the origin. First, let us find the transfer function $H(s) = E_2/I_s$. From the node equations

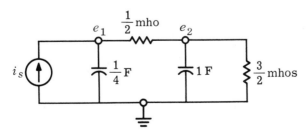

**Fig. 4.1**    Linear time-invariant network used in Example 1.

$$\begin{bmatrix} \frac{1}{4}s + \frac{1}{2} & -\frac{1}{2} \\ -\frac{1}{2} & s + 2 \end{bmatrix} \begin{bmatrix} E_1(s) \\ E_2(s) \end{bmatrix} = \begin{bmatrix} 1 \\ 0 \end{bmatrix}$$

where the input $i_s$ is taken to be a unit impulse, we obtain

$$H(s) = E_2(s) = \frac{2}{s^2 + 4s + s} = \frac{1}{s+1} - \frac{1}{s+3}$$

Hence,

$$h(t) = u(t)[\epsilon^{-t} - \epsilon^{-3t}]$$

Second, let us take $i_s(t) = C_1\epsilon^{-t} + C_2\epsilon^{-3t}$ over [0,1] and zero elsewhere. Then, using the convolution, we find that the zero-state response is, for $t \geq 1$,

$$e_2(t) = \epsilon^{-t} \int_0^1 \epsilon^\tau i_s(\tau)\, d\tau + \epsilon^{-3t} \int_0^1 \epsilon^{3\tau} i_s(\tau)\, d\tau$$

Our condition requires that the first integral be equal to 1 and the second be zero. Upon substituting the assumed form for $i_s$, we have

$$3.194C_1 + 13.40C_2 = 1$$
$$13.40C_1 + 67.07C_2 = 0$$

Therefore,

$$C_1 = 1.945 \qquad C_2 = -0.389$$

In conclusion, if

$$i_s(t) = \begin{cases} 1.945\epsilon^{-t} - 0.389\epsilon^{-3t} & 0 \leq t \leq 1 \\ 0 & t > 1 \end{cases}$$

then the zero-state response is

$$e_2(t) = \epsilon^{-t} \qquad \text{for } t \geq 1$$

---

**Exercise**   Consider the network shown in Fig. 4.1.

a.   Find the initial state such that the zero-input response is $e_2(t) = \epsilon^{-t}$ for $t \geq 0$.

b.   Repeat with $e_2(t) = \epsilon^{-3t}$ for $t \geq 0$.

---

**Example 2**   The purpose of this example is to show that a node-pair voltage $e_1$ may have $s_1$ as a natural frequency, but that the driving-point impedance at this node pair does not have a pole at $s_1$. Consider the network shown in Fig. 4.2. In order to write the state equations we pick a tree which includes all capacitors and no inductors. The tree is shown in the figure.

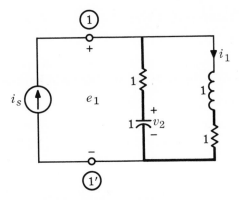

**Fig. 4.2**   Linear time-invariant network analyzed in Example 2; the driving-point impedance seen at terminals ① ① is equal to 1 ohm for all $s$.

Writing the fundamental loop equation for the inductor and the fundamental cut-set equation for the capacitor, we obtain

$$\begin{bmatrix} \dfrac{di_1}{dt} \\[2mm] \dfrac{dv_2}{dt} \end{bmatrix} = \begin{bmatrix} -2 & 1 \\ -1 & 0 \end{bmatrix} \begin{bmatrix} i_1 \\ v_2 \end{bmatrix} + \begin{bmatrix} 1 \\ 1 \end{bmatrix} i_s$$

Taking the Laplace transform, we obtain

(4.4)   $$\begin{bmatrix} s + 2 & -1 \\ 1 & s \end{bmatrix} \begin{bmatrix} I_1(s) \\ V_2(s) \end{bmatrix} = \begin{bmatrix} 1 \\ 1 \end{bmatrix} I_s(s) + \begin{bmatrix} i_1(0-) \\ v_2(0-) \end{bmatrix}$$

*a.*   Let us calculate the *zero-state response.*   By Cramer's rule,

(4.5)   $$I_1(s) = \frac{s + 1}{(s + 1)^2} I_s(s) = \frac{1}{s + 1} I_s(s)$$

(4.6)   $$V_2(s) = \frac{s + 1}{(s + 1)^2} I_s(s) = \frac{1}{s + 1} I_s(s)$$

In order to calculate the zero-state response $e_1$ we observe that

$$e_1 = \frac{di_1}{dt} + i_1$$

or, in Laplace transforms,

$$E_1(s) = (s + 1)I_1(s)$$

Consequently, by Eq. (4.5),

(4.7)   $$E_1(s) = \frac{s + 1}{s + 1} I_s(s) = I_s(s)$$

If we think of this equation in terms of the impedance $Z_1(s)$ seen at terminals ①①', we note that (for zero-state responses)

$$E_1(s) = Z_1(s)I_s(s)$$

Hence, on comparison with (4.7), the driving-point impedance

$$Z_1(s) = 1 \quad \text{for all } s$$

We note the very interesting fact that $Z_1(s)$, *the driving-point impedance seen at terminals* ①①', *is equal to* 1 *ohm for all s*. Algebraically, this is a consequence of the cancellations which occurred in Eqs. (4.5), (4.6), and (4.7). Physically, at very low frequencies, the capacitor is an open circuit, and all the current flows through the *RL* combination. Since at low frequencies the impedance of the inductor is very small, the impedance seen at ①①' should be close to 1 ohm. At very high frequencies the impedance of the capacitor is very low, the impedance of the inductor is very high, and the same conclusion follows.

*b.* Let us calculate the *zero-input response*. Now $I_s(s) \equiv 0$. By Cramer's rule applied to (4.4), we obtain

$$I_1(s) = \frac{si_1(0-) + v_2(0-)}{(s+1)^2}$$

$$V_2(s) = \frac{-i_1(0-) + (s+2)v_2(0-)}{(s+1)^2}$$

By partial-fraction expansion, we obtain

$$i_1(t) = [-i_1(0-) + v_2(0-)]te^{-t} + i_1(0-)\epsilon^{-t} \quad \text{for } t \geq 0$$
$$v_2(t) = [-i_1(0-) + v_2(0-)]te^{-t} + v_2(0-)\epsilon^{-t} \quad \text{for } t \geq 0$$

We see that the network variables $i_1$ and $v_2$ have $-1$ as a natural frequency.

Consider now $e_1$. Since, in the present case, the network does not start from the zero state, we have

$$E_1(s) = (s+1)I_1(s) - i_1(0-)$$

and, in fact,

$$E_1(s) = \frac{-i_1(0-) + v_2(0-)}{s+1}$$

Thus

$$e_1(t) = [-i_1(0-) + v_2(0-)]\epsilon^{-t} \quad \text{for } t \geq 0$$

Thus $-1$ is the natural frequency of $e_1$.

Let us stress that $-1$ is the natural frequency of $e_1$ and $-1$ is not a pole of $Z_1(s)$. This fact need not seem paradoxial, since natural frequencies are properties of *zero-input responses* (i.e., responses to arbitrary initial states) and network functions predict only *zero-state responses*.

In summary, we state that *any pole of a network function is a natural frequency of the corresponding (output) network variable, but any natural frequency of a network variable need not be a pole of a given network function which has this network variable as output.*

## 4.2    Natural Frequencies of a Network

In Chap. 14 we defined two concepts: natural frequencies of a network and natural frequencies of a network variable. Recall that, by definition, the natural frequencies of a network are the natural frequencies of any voltage or any current in the network. We showed that the (nonzero) natural frequencies of a network are the zeros of the determinant of the system of equations obtained by any general method of analysis. Conversely, we showed that each zero of the determinant is a natural frequency of the network. The concept of natural frequencies is of paramount importance in understanding linear time-invariant networks; for this reason, we shall explore further the relations between the natural frequencies of a network and the poles of some typical network functions appropriately defined for the network.

Since the natural frequencies of a network depend *only* on the *network topology* and the *element values,* but not on the input, we can set all *independent sources* to zero when we consider the natural frequencies of a network. Note that when we set a voltage source to zero, we replace it by a short circuit; when we set a current surce to zero, we replace it by an open circuit. We use the term **unforced network** to designate the network obtained from the given network by setting all its *independent* sources to zero.

Starting with the unforced network, we can apply appropriate independent sources to introduce various network functions. For example, if an independent current source $i_0$ is applied to the unforced network shown in Fig. 4.3a, the driving-point impedance $Z$ faced by the current source is defined by

$$(4.8) \quad Z(s) = \frac{V(s)}{I_0(s)} = \frac{\mathcal{L}[v(t)]}{\mathcal{L}[i_0(t)]}$$

The poles of $Z(s)$ are natural frequencies for the variable $v$, and hence are natural frequencies of the network. Note that in applying the current source we attach the terminals of the source to two nodes of the unforced network. Thus, when the current source is set to zero by opening its connection, we recover the unforced network; i.e., the application of the current source does not alter the topology nor the element values of the unforced network, and hence it does not alter its natural behavior. Similarly, if an independent voltage source $v_0$ is applied to the basic network shown in Fig. 4.3b, the driving-point admittance $Y$ faced by the voltage source is defined by

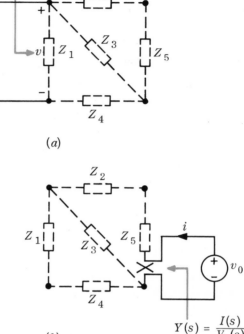

$$Z(s) = \frac{V(s)}{I_0(s)}$$

$(a)$

$$Y(s) = \frac{I(s)}{V_0(s)}$$

$(b)$

**Fig. 4.3**    Illustrations of $(a)$ soldering-iron entry and $(b)$ pliers entry to an unforced network.

$(4.9)$    $Y(s) = \dfrac{I(s)}{V_0(s)} = \dfrac{\mathcal{L}[i(t)]}{\mathcal{L}[v_0(t)]}$

The poles of $Y(s)$ are natural frequencies for the variable $i$ and hence are natural frequencies of the network. Note that in applying the voltage source we break a branch of the unforced network and insert the voltage source. Again, when the voltage source is set to zero by replacing it with a short circuit, we recover the unforced network. To emphasize the two distinct ways of applying independent sources to any unforced network, we introduce the following descriptive terminology: soldering-iron entry and pliers entry. By **soldering-iron entry** we mean that we enter the network by connecting the two terminals of a source to any two nodes of the network. By **pliers entry** we mean that we enter the network by cutting any branch of the network and connecting the two terminals of a source to the terminals created by the cut. Clearly, we apply a current source only by a soldering-iron entry and a voltage source only by a pliers entry. In this way we maintain the natural behavior of the unforced network.

We have shown that the set of natural frequencies of a network includes all the poles of the appropriately defined driving-point impedances and admittances of the network. In fact, we can also show that it is possible to obtain all the natural frequencies of a network by calculating the poles of all the impedance and admittance functions of the network. This fact has to be related to the fact that all the (nonzero) natural frequencies of the network can also be obtained by calculating the zeros of its network determinant.

### 4.3   Zeros

To interpret the physical meaning of a zero in a network function, let us start with a simple case. Consider the ladder network shown in Fig. 4.4; the ladder is driven by a current source, and $v_2$ is the zero-state response. Let the series-tuned circuit $L_2C_2$ resonate at $\omega_2$, and let the parallel-tuned circuit $L_3C_3$ resonate at $\omega_3$. We assert that the network function $H(s) = V_2(s)/I(s)$ relating the response $v_2$ to the input $i$ has a zero at $s = j\omega_2$ and another zero at $s = j\omega_3$.† Indeed, suppose we are in the sinusoidal steady state at the frequency $\omega_2$. Then the series resonant circuit $L_2C_2$ has a zero impedance at that frequency; i.e., in the sinusoidal steady state at frequency $\omega_2$, the voltage across $AB$ is zero. Furthermore, since the impedance to the right of $A'B'$ is not zero at frequency $\omega_2$, all the current will go through the $L_2C_2$ series-tuned circuit, and none will reach the resistor $R$. Hence $v_2$ is also zero. Using phasor representation, we write

$$v_2(t) = \text{Re}\,(\hat{V}_2\epsilon^{j\omega_2 t}) \qquad i(t) = \text{Re}\,(\hat{I}\epsilon^{j\omega_2 t})$$

where $v_2$ is the sinusoidal steady-state response and $i$ is the sinusoidal input. Consequently, using the relation between the input phasor $\hat{I}$, the response phasor $\hat{V}_2$, and the network function $H(j\omega_2)$, we obtain

† If $\omega_2 = \omega_3$, then $H(s)$ has a double zero at $s = j\omega_2$.

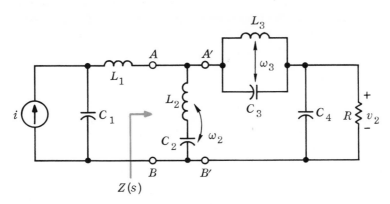

**Fig. 4.4**   Example illustrating the zeros of a network function.

$\widehat{V}_2 = H(j\omega_2)\widehat{I} = 0$

Since the phasor $\widehat{I}$ is not zero, we conclude that

$H(j\omega_2) = 0$

Suppose now that we are in the sinusoidal steady state at frequency $\omega_3$. The impedance of the parallel circuit $L_3C_3$ is infinite at that frequency, and the impedance of the series circuit $L_2C_2$ is finite. Hence, again all the current that flows through $L_1$ will go through the $L_2C_2$ series-tuned circuit, and none will enter the $L_3C_3$ parallel-tuned circuit. Clearly, $\widehat{V}_2 = 0$, and, as before,

$\widehat{V}_2 = H(j\omega_3)\widehat{I} = 0$

Consequently,

$H(j\omega_3) = 0$

The reasoning above exhibits an extremely important idea which is basic to filter design. To repeat it for emphasis, a parallel resonant circuit in a *series* arm of the ladder inserts a zero of transmission at its resonant frequency; a series resonant circuit in a *shunt* arm of the ladder inserts a zero of transmission at its resonant frequency.

In practice, the designer often feels he cannot afford two reactive elements for each series branch and for each shunt branch. In such cases he sometimes uses a ladder network made of series inductors and shunt capacitors. It is important to intuitively understand that the resulting network shown in Fig. 4.5 is a low-pass filter. Indeed, at direct current ($\omega = 0$), all the current goes through the inductors directly to the load resistor $R$. No current "leaks" through the capacitors, since they have infinite impedance at $\omega = 0$, and the inductors have zero impedance at $\omega = 0$. As long as $\omega$ is small enough so that the impedance of each inductor is very small compared with that of the preceding capacitor (i.e., at each node, $\omega L \ll 1/\omega C$), most of the current will still go through the inductors to the load resistor $R$. When $\omega$ is very large, so that the impedance of each inductor is very large compared with that of the preceding capacitor (i.e., at each node $\omega L \gg 1/\omega C$) then, at each node, most of the current goes through the capacitor, and only a very small fraction (of the

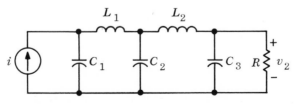

**Fig. 4.5**   A low-pass filter which has a network function without finite zeros.

order of $1/\omega^2 LC$) goes through the series inductor. In conclusion, low-frequency signals go through the network with little attenuation and very-high-frequency signals suffer considerable attenuation before reaching the load resistor. For these reasons such a network is called a low-pass filter. Note also that, in contrast to the case of Fig. 4.3, the network with series inductors and shunt capacitors does not block completely any purely sinusoidal input; its transfer function has no finite zeros.

**Exercise**   Show intuitively that a ladder network made of series inductors and shunt capacitors has a transfer function of the form

$$H(s) = \frac{V_2}{I} = \frac{K}{s^n + a_1 s^{n-1} + \cdots + a_n}$$

where $n$ is the number of reactive elements, and $K^{-1}$ is the product of all inductances and all capacitances. (Hint: Consider $|H(j\omega)|$ when $\omega$ is very large; see how the current divides at every node between the series and the shunt element.)

**Exercise**   Give an intuitive explanation for the fact that a ladder network made of *series capacitors* and *shunt inductors* is called a high-pass filter.

Generalizing this reasoning a little, we see that if a zero of the network function is on the $j\omega$ axis, it can be interpreted to mean that, in the sinusoidal steady state at that frequency, the response is identically zero. It turns out that if $z_1$ is *any* zero of the network function $H(s)$ (not necessarily on the $j\omega$ axis) and if the input is $u(t)\epsilon^{z_1 t}$, then there exists an initial state of the network such that the complete response (to both that initial state and that exponential input) is identically zero. In other words, with suitable interpretation, it is correct to associate to a *zero* of a network function the idea of *zero transmission*. This interpretation is illustrated by the following exercise.

**Exercise**   Consider the ladder network shown in Fig. 4.6. Since the parallel-tuned circuit resonates at $\omega = \frac{1}{3}$ rad/sec, the network function $H(s) =$

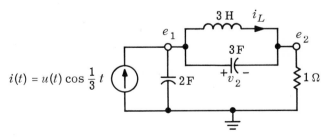

**Fig. 4.6**   Dynamic interpretation of a zero of a network function; if $e_1(0) = v_2(0) = 0$ and $i_L(0) = -\frac{3}{2}$, then for the input shown above, $e_2(t) = 0$ for all $t \geq 0$.

$E_2(s)/I(s)$ has a zero at $\omega = \frac{1}{3}$ rad/sec.  Show that if $e_1(0) = 0$, $v_2(0) = 0$, and $i_L(0) = -\frac{3}{2}$ amp, then the response to $i(t) = u(t) \cos (t/3)$ amp is, for all $t \geq 0$, $e_1(t) = \frac{3}{2} \sin (t/3)$ volts and $e_2(t) = 0$.

| **5** | **Application to Oscillator Design** |
|---|---|

In the design of oscillators we usually proceed as follows: first, we model the active element (transistor, tunnel diode, etc.) by a small-signal equivalent circuit; second, we embed the active element in a linear time-invariant network which we select either by experience or intuition; third, we check whether the element values selected are going to sustain oscillations. It is this last question that we shall examine in detail.

Suppose we propose to design a tunnel-diode oscillator based on the configuration shown in Fig. 5.1a.  This configuration does not show the

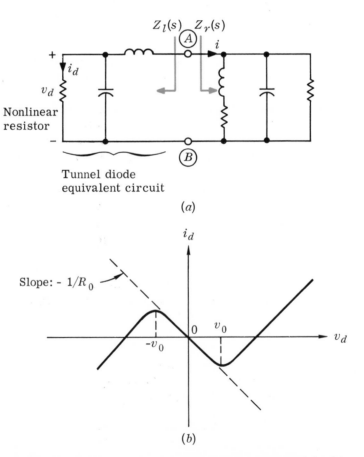

(a)

(b)

**Fig. 5.1**    A tunnel-diode oscillator.    (a) Equivalent circuit; (b) the $vi$ characteristic of the nonlinear resistor.

biasing; the only elements shown are those essential in the analysis. In the equivalent circuit representation of the tunnel diode, the resistor is nonlinear and has the *vi* characteristic shown in Fig. 5.1*b*. For the purpose of a small-signal analysis we may assume that the nonlinear resistor is replaced by a linear resistor with negative resistance $-R_0$. However, if the voltage $v_d$ is beyond $\pm v_0$ as shown in the figure, our small-signal approximation is no longer valid, and we must use the non linear characteristic shown in Fig. 5.1*b*. In the following discussion, since we want to find out whether the oscillations will build up, we use small-signal analysis. Let us consider ways in which we could check whether this small-signal equivalent circuit can oscillate.

*Method* 1   The straightforward method consists of writing the node equations in their transform form as follows:

(5.1)   $\mathbf{Y}(s)\mathbf{E}(s) = \mathbf{0}$

where $\mathbf{Y}(s)$ is the node admittance matrix and $\mathbf{E}(s)$ is the column vector whose elements are the Laplace transforms of the node voltages. Next we calculate the zeros of det $[\mathbf{Y}(s)]$ to obtain the natural frequencies of the network. The zero-input response of the network depends on the locations of the natural frequencies of the network. Thus, if det $[\mathbf{Y}(s_1)] = 0$, $s_1$ is a natural frequency, and the node voltage vector contains a typical term $\mathbf{K}_1 \epsilon^{s_1 t}$, where $\mathbf{K}_1$ is a vector with constant components which depend on the initial state of the network. For sustained sinusoidal oscillation it is necessary that the network have at least one pair of natural frequencies on the $j\omega$ axis at $j\omega_0$. Indeed, if all the other natural frequencies are in the open left-half $s$ plane, the transient will die out, and the zero-input response eventually becomes a sinusoid with angular frequency $\omega_0$.

   An alternate way for obtaining an oscillatory response is to have one or more natural frequencies in the right-half $s$ plane, that is, Re $(s_1) > 0$. We know that under this situation the node voltages grow exponentially. In particular, when the voltage $v_d$ across the nonlinear resistor is beyond $\pm v_0$, the small-signal approximation is no longer valid. We must then use a nonlinear analysis of the circuit. The characteristic in Fig. 5.1*b* indicates that the voltage $v_d$ will saturate and will form a limit cycle; i.e., an oscillation will result. In conclusion, the straightforward method consists of checking the determinant of (5.1) to see whether it has any zeros in the closed right-half $s$ plane. However, if the answer turns out to be negative, this method gives little suggestion as to how to modify the circuit to get some of the natural frequencies to move into the right-half $s$ plane. For this reason, another method is also of interest.

*Method* 2   Consider again Fig. 5.1 and use small-signal analysis. In essence, we want the tunnel diode to produce an oscillation in the tank circuit on the right of terminals Ⓐ and Ⓑ. This will require a current $i$ to flow from the

left part of the circuit to the right part.   Calling $Z_l$ and $Z_r$ the corresponding impedances, we are led to think of the circuit of Fig. 5.1 as if it were
the circuit shown in Fig. 5.2.   Let us apply the method of Sec. 4.2 to calculate the natural frequencies; thus, we shall derive conditions for oscillation.   Using a pliers entry in the loop, we obtain the driving-point admittance

(5.2)   $Y(s) = \dfrac{1}{Z_l(s) + Z_r(s)}$

The poles of $Y(s)$ are natural frequencies of the network.   Obviously, if $s_1$
is a zero of the denominator of $Y(s)$ in (5.2), $s_1$ is a pole of the network
function $Y(s)$ and hence is a natural frequency of the network.   Thus, a
sufficient condition for oscillation is the following: *given an appropriate
initial state, the current i in the linear equivalent circuit shown in Fig. 5.1 will
be oscillatory or growing exponentially if the algebraic equation*

(5.3)   $Z_l(s) + Z_r(s) = 0$

*has one or more solutions in the closed right-half s plane, in other words, if the
loop impedance $Z_l(s) + Z_r(s)$ has one or more zeros in the closed right-half
s plane.*

From a design point of view, Eq. (5.3) is very informative.   Indeed,
suppose we would like to have an oscillation at frequency $\omega_0$; we would
then split $Z_l(j\omega_0)$ and $Z_r(j\omega_0)$ into their real and imaginary parts and
obtain

$$[R_l(j\omega_0) + R_r(j\omega_0)] + j[X_l(j\omega_0) + X_r(j\omega_0)] = 0$$

Hence we must have simultaneously

(5.4)   $R_l(j\omega_0) + R_r(j\omega_0) = 0$

(5.5)   $X_l(j\omega_0) + X_r(j\omega_0) = 0$

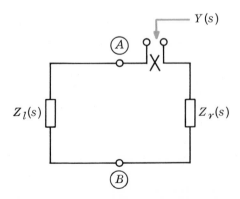

**Fig. 5.2**   Circuit illustrating a design method for linear
oscillators.

The second condition can always be satisfied. If $X_r(j\omega_0)$ is too small for (5.5) to hold, add an appropriate inductor in series with $Z_r$; conversely, if $X_r(j\omega_0)$ is too large for (5.5) to hold, add a capacitor in series with $Z_r$. Equation (5.4) indicates directly whether or not the active device has a sufficiently negative resistance to allow oscillations to occur with the particular circuit chosen. Thus, this approach clearly indicates how the embedding network has to be modified to obtain oscillations.

## 6   Symmetry Properties

We return to the $j\omega$-axis behavior of a network function. We have seen that any network function of any lumped linear time-invariant network is a rational function; hence we write

$$(6.1) \quad H(s) = \frac{b_0 s^m + b_1 s^{m-1} + \cdots + b_{m-1} s + b_m}{a_0 s^n + a_1 s^{n-1} + \cdots + a_{n-1} s + a_n}$$

where the coefficients $a_0, a_1, \ldots, a_n, b_0, b_1, \ldots, b_m$ are *real* numbers. In order to evaluate $H$ at the frequency $\omega$, we put $s = j\omega$ in (6.1) and collect terms in the following way:

$$(6.2) \quad H(j\omega) = \frac{(b_m - b_{m-2}\omega^2 + b_{m-4}\omega^4 + \cdots) + j\omega(b_{m-1} - b_{m-3}\omega^2 + \cdots)}{(a_n - a_{n-2}\omega^2 + a_{n-4}\omega^4 + \cdots) + j\omega(a_{n-1} - a_{n-3}\omega^2 + \cdots)}$$

It is important to note that $H(j\omega)$ is of the form

$$(6.3) \quad H(j\omega) = \frac{[\text{polynomial in } \omega^2] + j\omega[\text{polynomial in } \omega^2]}{[\text{polynomial in } \omega^2] + j\omega[\text{polynomial in } \omega^2]}$$

Since all the coefficients of the polynomials involved are real, the numerator and denominator of (6.3) are neatly split according to their real and imaginary parts.

Let us take the complex conjugate of (6.3); if we observe that the polynomials take real values whenever $\omega$ is real, we obtain

$$\overline{H(j\omega)} = \frac{[\text{polynomial in } \omega^2] - j\omega[\text{polynomial in } \omega^2]}{[\text{polynomial in } \omega^2] - j\omega[\text{polynomial in } \omega^2]}$$

The interesting observation is that we would have obtained precisely the same result if in (6.3) we replaced the variable $\omega$ by $-\omega$. Indeed, a polynomial in $\omega^2$ does not change its value if $\omega$ is changed into $-\omega$ because $\omega^{2k} = (-\omega)^{2k}$. Therefore, we have shown that, for all real $\omega$

$$(6.4) \quad \overline{H(j\omega)} = H(-j\omega)$$

Now, referring to the definition of the complex conjugate, we conclude that for all real $\omega$

$$(6.5) \quad \begin{aligned} \text{Re}\,[\overline{H(j\omega)}] &= \text{Re}\,[H(j\omega)] & |\overline{H(j\omega)}| &= |H(j\omega)| \\ \text{Im}\,[\overline{H(j\omega)}] &= -\text{Im}\,[H(j\omega)] & \sphericalangle\overline{H(j\omega)} &= -\sphericalangle H(j\omega) \end{aligned}$$

Consequently from Eq. (6.4), for all real $\omega$

$$(6.6) \quad \begin{aligned} \text{Re}\,[H(j\omega)] &= \text{Re}\,[H(-j\omega)] & |H(j\omega)| &= |H(-j\omega)| \\ \text{Im}\,[H(j\omega)] &= -\text{Im}\,[H(-j\omega)] & \sphericalangle H(j\omega) &= -\sphericalangle H(-j\omega) \end{aligned}$$

These important conclusions may be stated as follows: *the real part* Re $[H(j\omega)]$ *and the magnitude of a network function* $|H(j\omega)|$ *are* **even** *functions of $\omega$; the imaginary part* Im $[H(j\omega)]$ *and the phase of a network function* $\sphericalangle H(j\omega)$ *are* **odd** *functions of $\omega$.*

This leads to the important practical conclusion that in order to display the frequency response [i.e., the behavior of $H(j\omega)$ on the whole $j\omega$ axis] we only need to plot Re $[H(j\omega)]$ and Im $[H(j\omega)]$, or $|H(j\omega)|$ and $\sphericalangle H(j\omega)$, for $\omega \geq 0$. The plots for $\omega < 0$ are readily obtained by using the symmetry properties just established [see Eq. (6.6)].

**Exercise**   Derive the symmetry properties given by Eq. (6.6) from $H(s) = \int_{0-}^{\infty} h(t)\epsilon^{-st}\,dt$. [Hint: Put $s = j\omega$; use the fact that the values of the impulse response $h(\,\cdot\,)$ are real numbers.]

## Summary

- Any network function of *any* lumped linear time-invariant network is a rational function of $s$ with *real* coefficients. As a consequence, whenever zeros and poles of network functions are complex, they occur in complex conjugate pairs.

- A network function can either be specified by its numerator and denominator polynomials or by its poles and zeros together with the scale factor.

- There is a close relation between the pole and zero locations and the frequency response on the one hand, and the impulse response on the other hand. For example, an isolated pole close to the $j\omega$ axis produces a "peak" in the magnitude curve; the 3-db width of the peak is twice the distance of the pole to the $j\omega$ axis. The 3-db width also determines the rate of decay of its contribution to the impulse response. The location of the pole along the $j\omega$ axis determines the "frequency" of the oscillations in the impulse response. Poles or zeros close to the $j\omega$ axis cause the phase to vary rapidly in their neighborhood.

- Poles and zeros have interpretations in terms of the dynamics of the network. If $p_1, p_2, \ldots, p_n$ are the poles of the network function $H(s)$, then they are natural frequencies of the corresponding response variable, and the corresponding impulse response is a linear combination of the exponentials $\epsilon^{p_i t}, i = 1, 2, \ldots, n$. For some initial states, the network variable corresponding to $H$ has a zero-input response which is purely exponential, namely, $\epsilon^{p_k t}$, where $k$ may be $1, 2, 3, \ldots, n$. There are some special cases where, for some initial state, the network variable has a zero-input response which is purely exponential, say, $\epsilon^{p_0 t}$, where $p_0$ is *not* a pole of the network function.

■  Any natural frequency of a network is a pole of some network function. In particular, starting with the unforced network, we can apply a current source by means of a soldering-iron entry at any node pair of the network and define a driving-point impedance and transfer impedances; we can also apply a voltage source by means of a pliers entry in any branch of the network and define a driving-point admittance and transfer admittances. The poles of all of these network functions are the natural frequencies of the network.

■  If $z_1$ is a zero of a network function, then initial states may be selected so that the input $u(t)\epsilon^{z_1 t}$ produces a response which is *identically zero*. It is impossible to find such initial states if $z_1$ is not a zero of the network function.

■  Given an appropriate initial state, some network variables of a linear time-invariant network will be oscillatory or will grow exponentially if and only if the network has one or more natural frequencies in the closed right-half $s$ plane. In particular, if we consider a loop made of impedances $Z_l(s)$ and $Z_r(s)$, the loop current will be oscillatory or will grow exponentially if the loop impedance $Z_l(s) + Z_r(s)$ has a zero in the closed right-half $s$ plane. We showed that the latter condition gives much design information.

■  The symmetry properties of $H(j\omega)$ are as follows: $\text{Re}\,[H(j\omega)]$ and $|H(j\omega)|$ are *even* functions of the real variable $\omega$, whereas $\text{Im}\,[H(j\omega)]$ and $\sphericalangle H(j\omega)$ are *odd* functions of $\omega$.

## Problems

Poles and zeros

**1.** Calculate the driving-point impedances of the linear time-invariant circuits shown in Fig. P15.1, and plot their poles and zeros on the complex plane.

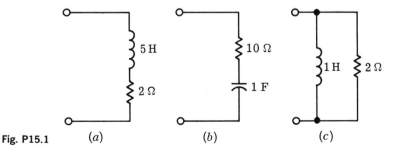

**Fig. P15.1**    (a)    (b)    (c)

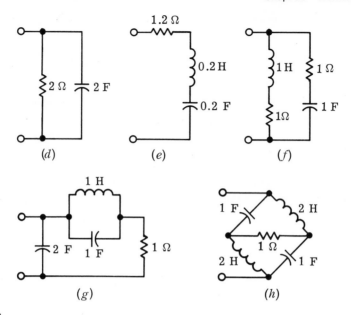

**Fig. P15.1**
(*continued*)

Poles and zeros

**2.** Obtain expressions for the driving-point impedances $Z(s)$ of the linear time-invariant networks shown in Fig. P15.2, and plot the finite poles and

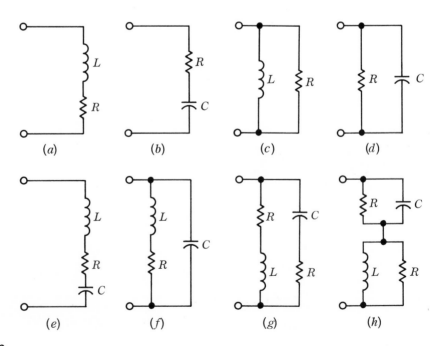

**Fig. P15.2**

zeros on the complex $s$ plane. Show that for networks (g) and (h), if $R^2 = L/C$, $Z(s) = R$ for all $s$.

**3.** Consider the linear time-invariant circuit shown in Fig. P15.3 with input $i_s$.

*a.*   Calculate the driving-point impedance $Z(s) \triangleq V_1/I_s$.

*b.*   Calculate the transfer impedance $H(s) \triangleq V_2/I_s$.

*c.*   Plot the poles and zeros of $H(s)$ on the complex plane.

*d.*   Evaluate $|H(j4)|$ and $\sphericalangle H(j4)$ graphically.

*e.*   Suppose that $i_s(t) = 15 \cos (4t - 30°)$ and that the network is in the sinusoidal steady state. Write the expression for $v_2(t)$ as a real-valued function of time.

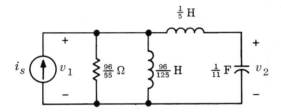

**Fig. P15.3**

**4.** For the networks shown in Fig. P15.4, obtain expressions for the voltage transfer functions $G(s) = V_2/V_1$. Plot the magnitude and phase of $G(j\omega)$ against $\omega$. Plot the locus of $G(j\omega)$ as $\omega$ varies from 0 to $\infty$ on the complex $G$ plane. For network (b), show that if $R_1C_1 = R_2C_2$,

$$G(s) = \frac{R_2}{R_1 + R_2} = \frac{C_1}{C_1 + C_2} \qquad \text{for all } s$$

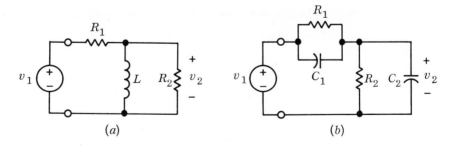

**Fig. P15.4**

**5.** The transfer function $H(s)$ has the pole-zero configuration illustrated in Fig. P15.5. Given $|H(j2)| = 7.7$, and $0 < \sphericalangle H(j2) < 180°$,

*a.*   Evaluate graphically $H(j4)$.

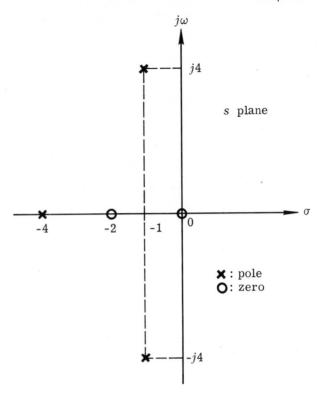

**Fig. P15.5**

b.  Give $|H(j4)|$ and $\measuredangle H(j4)$ in decibels and degrees, respectively.

Network function and impulse response

**6.** *a.*  The impulse response of a linear time-invariant network is

$$h(t) = u(t)(\epsilon^{-t} + 2\epsilon^{-2t})$$

Calculate the corresponding transfer function.

*b.*  The transfer function of a linear time-invariant network is

$$H(j\omega) = \frac{1}{1 + j(\omega/\omega_0)}$$

Calculate its impulse response.

Zero-state response

**7.** For each case below, $H(s)$ is the network function, and $e(\cdot)$ is the input; calculate the corresponding zero-state response.

*a.*  $H(s) = \dfrac{s + 1}{s^2 + 5s + 6}$

$e(t) = 3u(t) \sin 5t$

b. $H(s) = \dfrac{s + 1}{s^2 + s - 2}$

$e(t) = (1 + t)u(t)$

c. $H(s) = \dfrac{s}{(s^2 + 25)}$

$e(t) = u(t) \cos 5t$

**Zero-state and steady-state response**

**8.** For the network shown in Fig. P15.8, let $i_1$, $i_2$, and $i_3$ be mesh currents. Let $e_s$ and $i_3$ be the input and response of the network, respectively.

a. Find the network function $H(s) = I_3/E_s$.

b. Find the zero-state response when $e_s(t) = 3\epsilon^{-t} \cos 6t$.

c. Find the steady-state response when $e_s(t) = 2 + \cos 2t$.

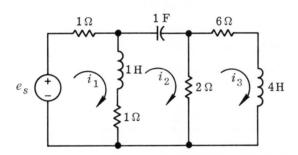

**Fig. P15.8**

**Impulse response and network functions**

**9.** The impulse response of four networks can be written as

a. $h(t) = \delta(t) + \epsilon^{-at} \sin(\omega t + \theta)u(t)$

b. $h(t) = \epsilon^{-at}\left(1 - \dfrac{t^2}{2!} + \dfrac{t^4}{4!} - \dfrac{t^6}{6!} + \cdots\right)u(t)$

c. $h(t) = (\frac{3}{5}\epsilon^{-t} - \frac{7}{5}t\epsilon^{-3t} + 3t)u(t)$

d. $h(t) = \delta'(t) + \delta(t)$

Find the corresponding network functions.

**Frequency and time response**

**10.** Consider the filter circuit shown in Fig. P15.10, where $C_1 = 1.73$ farads, $C_2 = C_3 = 0.27$ farad, $L = 1$ henry, $R = 1$ ohm.

a. Find the network function $H(s) = V_2(s)/I_1(s)$.

b. Plot the poles and zeros of this network function.

c. Plot $|H(j\omega)|$ versus $\omega$ (magnitude plot).

d. Plot $\angle H(j\omega)$ versus $\omega$ (phase plot).

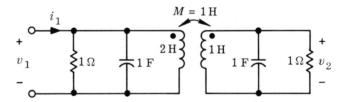

**Fig. P15.10**

e.   Determine the impulse response of this filter.

f.   Using impedance and frequency normalization, determine a filter network which cuts off at about 4 MHz and has a load resistor $R$ of 600 ohms (refer to Chap. 7, Sec. 8).

Network functions

**11.** Consider the coupled circuit shown in Fig. P15.11. Determine

a.   The driving-point impedance $V_1(s)/I_1(s)$.

b.   The transfer impedance $V_2(s)/I_1(s)$.

c.   The transfer voltage ratio $V_2(s)/V_1(s)$.

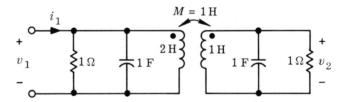

**Fig. P15.11**

Input, zero-state response

**12.** For three different networks, the following $x(t)$, $y(t)$ combinations are obtained, where $x$ is input, $y$ is the output, and the network is initially in the zero state.

a.   $x(t) = \delta(t)$       $y(t) = [\sinh t + 2(\cosh t - 1)]u(t)$

b.   $x(t) = u(t)$       $y(t) = (2\epsilon^{-t} + \sin 2t - 2)u(t)$

c.   $x(t) = u(t)\epsilon^{-t}$     $y(t) = (\epsilon^{-t} \sin t - \epsilon^{-2t} \cos 3t)u(t)$

Find the transfer function of each network.

Operational amplifier and integrator

**13.** In the circuit diagrams shown in Fig. P15.13, $A$ is a high-gain amplifier. For purposes of a preliminary analysis, assume that the output voltage of the amplifier, $e_0$, is related to the input voltage $e_g$ by $e_0 = -Ke_g$, where $K$ is a constant.   Assume also that the input impedance of the amplifier is so large that $i_0(t) = 0$ for all $t$.   Under these conditions, assuming that the output terminals are open-circuited,

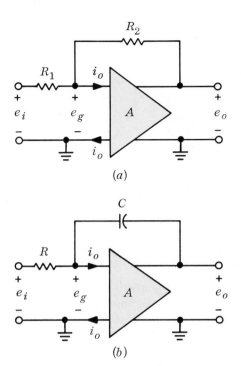

**Fig. P15.13**

a.  Find the transfer voltage ratio $E_0(s)/E_i(s)$ for the two configurations shown in Fig. P15.13.

b.  Assuming $K$ to be very large, simplify the expression for $E_0(s)/E_i(s)$ obtained in ($a$).

Frequency and time response
**14.** The magnitude frequency response for a network is sketched in Fig. P15.14.   The function $H(s)$ is assumed to be rational.   The sketch is not necessarily to scale, but the following specifics should be noted:

$|H(j\omega)|$ is an even function of $\omega$

$|H(j0)| = 1 \qquad |H(j)| = 0 \qquad |H(j\infty)| = 0$

a.  In order to be compatible with the above data, what is the minimum number of zeros that $H(s)$ can have?   What is the minimum number of poles?   Find an $H(s)$ compatible with the above data.

b.  Suppose that the network function $H(s)$ has the minimum number of zeros and simple poles at $s = -1, -1 \pm j$ and no other poles.   Suppose that the zero-state response $y$ due to some input $x$ and corresponding to this $H(s)$ is

$$y(t) = u(t)\epsilon^{-t} - u(t)\epsilon^{-t} \cos t$$

Find this input $x(t)$.

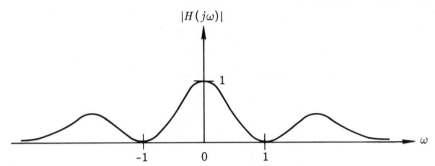

Fig. P15.14

**15.** Consider a linear time-invariant network with two accessible terminals ① and ①′ (see Fig. P15.15). Call $Y(s)$ and $Z(s)$ its driving-point admittance and driving-point impedance, respectively.

*a.* Suppose that the terminals ① and ①′ are short-circuited and that we observe the current $i$ through the short circuit. Is it true that if $s_1$ is a pole of $Y(s)$, then for some initial state we may observe $i(t) = K\epsilon^{s_1 t}$? Justify.

*b.* Suppose that the terminals ① and ①′ are open-circuited and that we observe the voltage $v$ across the open-circuited terminals. Is it true that if $s_2$ is a pole of $Z(s)$, then for some initial state we may observe $v(t) = K\epsilon^{s_2 t}$? Justify.

*c.* Work out the details for the circuit shown in Fig. P15.15c.

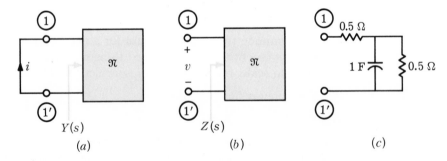

Fig. P15.15

**16.** Suppose that a linear time-invariant network $\mathfrak{N}$ is described by the following state equations

$$\mathbf{x} = \mathbf{A}\mathbf{x} + \mathbf{b}u$$

$$y = \mathbf{c}\mathbf{x}$$

Call $H(s)$ the network function which relates $u$ and $y$. Let $z$ be a zero of $H(s)$. Denote by $y(t,\mathbf{x}_0,\epsilon^{zt})$ the output at time $t$ due to the input $\epsilon^{zt}$ applied at $t = 0$, the network starting from state $\mathbf{x}_0$ at time 0. Calculate in terms of $\mathbf{A}, \mathbf{b}, \mathbf{c}$ the initial state $\mathbf{x}_0$ such that $y(t,\mathbf{x}_0,\epsilon^{zt}) = 0$ for all $t \geq 0$. [Hint: $\mathbf{x}_0 = (z\mathbf{I} - \mathbf{A})^{-1}\mathbf{b}$.]

# 16

# Network Theorems

In this chapter we shall study four very general and useful network theorems: *the substitution theorem, the superposition theorem, the Thévenin-Norton equivalent network theorem, and the reciprocity theorem.* The usefulness of these theorems lies in that (1) they are applicable to very large classes of networks encountered in practice and (2) their conclusions are very simple. This generality and simplicity may be deceiving; often people do not perceive the breadth of application of these theorems, or know precisely what they mean.

The principal assumption underlying all these network theorems is the *uniqueness* of the solution for the network under consideration. We discussed, in Secs. 7 and 8 of Chap. 13, the fact that for linear networks (with the exception of degenerate cases) the solution is always unique for any given set of inputs and for any given initial state. The substitution theorem is a very simple but general theorem which holds for all networks with a unique solution. It can be applied to linear and nonlinear networks, time-invariant and time-varying networks. The other three theorems apply only to linear networks. Recall that a linear network, by definition, consists of elements which are either linear or independent sources. These independent sources are the inputs to the network. The superposition theorem and the Thévenin-Norton equivalent network theorem apply to all linear networks (time-invariant or time-varying). Thus, these networks may include linear resistors, linear inductors, linear coupled inductors, linear capacitors, linear transformers, linear dependent sources and independent sources. The reciprocity theorem applies to a more restricted class of linear networks; the elements must be time-invariant, and they can not include dependent sources, independent sources, and gyrators. A gyrator is a linear time-invariant two-port which we shall define later.

In presenting the four theorems we shall give first some motivation, then state the theorem, and follow it with examples, remarks, and corollaries. We shall give proofs at the end, so the reader can thoroughly digest and understand the precise meaning of the theorem before worrying about its proof.

## 1 The Substitution Theorem

### 1.1 Theorem, Examples, and Application

The substitution theorem allows us to replace any particular branch of a network by a suitably chosen independent source without changing any branch current or any branch voltage.  The primary reason for the substitution is that the substitute network is easier to solve than the original one.  With this idea in mind we are ready to state the theorem.

**SUBSTITUTION THEOREM**  Consider an *arbitrary* network which contains a number of independent sources.  Suppose that for these sources and for the given initial conditions the network has a unique solution for *all* its branch voltages and branch currents.  Consider a particular branch, say branch $k$, which is not coupled to other branches of the network.†  Let $j_k(\cdot)$ and $v_k(\cdot)$ be the current and voltage waveforms of branch $k$.  Suppose that branch $k$ is replaced by either an independent current source with waveform $j_k(\cdot)$ or an independent voltage source with waveform $v_k(\cdot)$.  If the modified network has a unique solution for *all* its branch currents and branch voltages, then these branch currents and branch voltages are identical with those of the original network.

**Remark on uniqueness**  Applying this theorem requires in principle that we check that the modified network has a unique solution, i.e., that the specified initial conditions and the inputs determine uniquely all branch voltages and all branch currents.  For linear networks (time-invariant or time-varying) this is always the case except for some degenerate cases (see Secs. 7 and 8 of Chap. 13).  In fact, we have shown that if a network consists of resistors, inductors, and capacitors that are linear, time-invariant, and passive, then the initial currents in the inductors, the initial voltages on the capacitors, and the inputs determine uniquely all branch voltages and all branch currents.  The statement is still true when coupled inductors are present, provided every set of mutually coupled inductors has an inductance matrix which is positive definite.

**Example 1**  Consider the circuit shown in Fig. 1.1*a*.  The problem is to find the voltage $V$ and current $I$ of the tunnel diode given $E$, $R$, and the tunnel-diode characteristic (shown in Fig. 1.1*b*).  The solution is obtained graphically in Fig. 1.1*b*.  According to the substitution theorem we may replace the tunnel diode either by a current source $I$ or by a voltage source $V$ as shown in Fig. 1.2.  It is easy to see that in both cases the solutions are the same as that obtained originally.

To illustrate the importance of the condition that the modified network must have a *unique* solution, let us consider the voltage source $E$ and resistor $R$ of Fig. 1.1 as forming the branch $k$ of the theorem.  Let us replace it by a current source $I$ as shown in Fig. 1.3*a*.  The possible solutions of the modified network are located at the intersection of the diode characteristic and the current source characteristic.  Figure 1.3*b* shows that there are three possible solutions, namely $(V_1,I)$, $(V_2,I)$, and $(V_3,I)$.  Clearly,

---

† That is, branch $k$ may not be a branch of a coupled inductor nor a branch of a dependent source.

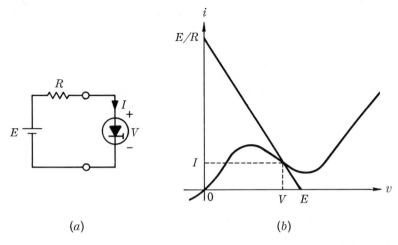

$(a)$ $(b)$

**Fig. 1.1**    A tunnel-diode circuit and its characteristic; the point $(V,I)$ is the solution of the circuit.

only one of these solutions can coincide with that of the given network. Hence, as a tool for solving networks, the substitution theorem is effective only when the modified network has a unique solution.

The substitution theorem has many applications.  In Sec. 3 we shall use it to prove the Thévenin-Norton equivalent network theorem.  The substitution theorem is particularly useful in the analysis of networks which contain a single nonlinear or time-varying element.  We shall now give a simple application to illustrate the power of the theorem.

**Example 2**    Consider any linear time-invariant network which is in the zero state at time zero and has no independent sources (see Fig. 1.4).  We shall use the frequency-domain characterizations of the network, i.e., network functions in terms of the complex frequency variable $s$.  Assume that there are two accessible terminals to the network which form a port.  Let us apply at time zero a voltage source $e_0$ to the port and let the response be

$(a)$ $(b)$

**Fig. 1.2**    The modified networks of Fig. 1.1$a$ obtained by the substitution theorem. $(a)$ With a current source $I$; $(b)$ with a voltage source $V$.

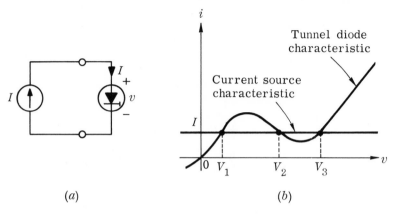

(a)                                                              (b)

**Fig. 1.3**   Example showing the importance of requiring that the modified network have a unique solution; in the circuit of (a), $v$ has three solutions $V_1$, $V_2$, and $V_3$.

the current $i$ entering the port.   Call $E_0(s)$ and $I(s)$ the Laplace transforms of $e_0(t)$ and $i(t)$, respectively, as shown in Fig. 1.4a.   The network function, by definition the ratio of the Laplace transform of the response to that of the input, is in this case the driving-point admittance

(1.1)   $$Y(s) = \frac{I(s)}{E_0(s)}$$

Next let us apply a current source $i_0$ to the port and let the response be the voltage $v$ across the port.   Call $I_0(s)$ and $V(s)$ the Laplace transforms of $i_0(t)$ and $v(t)$, respectively, as shown in Fig. 1.4b.   The network function is then the driving-point impedance

(1.2)   $$Z(s) = \frac{V(s)}{I_0(s)}$$

The driving-point impedance is equal to the reciprocal of the driving-point admittance; i.e.,

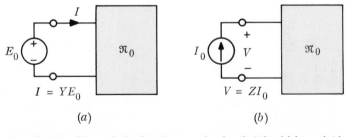

(a)                                                              (b)

**Fig. 1.4**   An application of the substitution theorem showing that the driving-point impedance of a linear time-invariant network $\mathfrak{N}_0$ is the reciprocal of the driving-point admittance of the same network; $\mathfrak{N}_0$ is in the zero state at $t = 0-$ and contains no independent sources.

$$Z(s) = \frac{1}{Y(s)}$$

The proof of this relation can be made by standard loop and node analyses; however, it is lengthy. On the other hand, if we use the substitution theorem, the conclusion is obvious. Let us consider the voltage source $E_0$ in Fig. 1.4a as the specific branch $k$ of the theorem. Applying the theorem, we replace the voltage source by a current source $I_0(s) = I(s)$, where $I(s)$ is the current entering the port in Fig. 1.4a. The network after substitution is shown in Fig. 1.4b. The substitution theorem means that the voltage across the port in Fig. 1.4b must be the same as the voltage across the port in Fig. 1.4a; that is, $V(s) = E_0(s)$. Hence we conclude from Eqs. (1.1) and (1.2) that $Z(s) = 1/Y(s)$.

---

| 1.2 | **Proof of the Substitution Theorem** |

Let $v_1(\cdot)$, $v_2(\cdot)$, ..., $v_b(\cdot)$ and $j_1(\cdot)$, $j_2(\cdot)$, ..., $j_b(\cdot)$ be the branch-voltage waveforms and branch-current waveforms of the given network; by assumption, these waveforms are the unique waveforms that satisfy the KVL constraints, the KCL constraints, the initial conditions, and the branch equations of the given network. Consider the case in which the modified network is obtained by replacing the $k$th branch by a voltage source which has the waveform $v_k(\cdot)$. We assert that these same branch voltages and branch currents $v_1(\cdot)$, $v_2(\cdot)$, ..., $v_b(\cdot)$ and $j_1(\cdot)$, $j_2(\cdot)$, ..., $j_b(\cdot)$ constitute the solution (unique by assumption) of the *modified* network. Since the topology of the given network and the modified network are the same, the Kirchhoff constraints are the same. The initial conditions are also the same for both networks. All branch equations are the same except those of the $k$th branch. In the modified network the $k$th branch is a voltage source; hence the only constraint is that the branch voltage be equal to the waveform of the source. However, the latter has been selected to be $v_k(\cdot)$ itself. Hence the set of branch voltages and branch currents $\{v_i(\cdot), j_i(\cdot), i = 1, 2, \ldots, b\}$ satisfies all the conditions of the modified network, and hence it is *the* unique solution of the modified network. The proof for the case in which the $k$th branch is replaced by a current source which has the waveform $j_k(\cdot)$ is entirely similar. The reader should construct the argument for himself.

**Exercise**   The ladder network shown in Fig. 1.5 is in the sinusoidal steady state. The

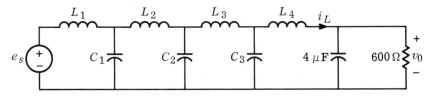

**Fig. 1.5**   A ladder network.

current $i_L$ has been measured to be $10 \cos 377t$ mA.   Calculate the output voltage $v_0$.   (If you use phasors, be sure to give your answer as a real-valued function of time.)

## 2 | The Superposition Theorem

The importance of the superposition theorem is hard to exaggerate, for it is the foundation of many engineering systems in daily use, such as high-fidelity audio systems, telephone systems, broadcasting systems, analog computer components, and numerous measuring instruments and techniques.

Roughly speaking, the superposition theorem means that, for a linear network, the zero-state response caused by several independent sources is the sum of the zero-state responses due to each independent source acting alone.   Let us illustrate this concept by a high-fidelity microphone-amplifier-loudspeaker system.   If we concern ourselves only with the electric circuit aspects of it, we may think of the microphone as a voltage source in series with an impedance; the output of the circuit is the current in the driving coil of the loudspeaker.   Suppose we want to amplify the music produced by a violin and a piano.   If superposition applies, the response when both the violin and the piano play simultaneously is the sum of the responses due to each one of them acting alone.   If superposition did not apply, we would hear the sum of their respective responses plus some "interaction."   Imagine the effects that would then appear in the case of a 140-piece orchestra!   Because high-fidelity enthusiasts demand that a violin sound like a violin whether or not the piano is played simultaneously, the designer of high-fidelity systems must make sure he ends up with a *linear* system, for then he is assured that superposition applies.

### 2.1 | Theorem, Remarks, Examples, and Corollaries

With this general picture in mind, we give the theorem's precise formulation as follows.

**SUPER-POSITION THEOREM**   Let $\mathfrak{N}$ be a linear network; i.e., let each of its elements be either an independent source or a linear element (linear resistor, linear inductor, linear capacitor, linear transformer, or linear dependent source).   The elements may be time-varying.   We further assume that $\mathfrak{N}$ has a unique zero-state response to the independent source waveforms, whatever they may be.   Let the response of $\mathfrak{N}$ be either the current in a specific branch of $\mathfrak{N}$, or the voltage across any specific node pair of $\mathfrak{N}$, or more generally any linear combination of currents and voltages.   Under these conditions, the *zero-state response* of $\mathfrak{N}$ due to all the independent sources acting simultaneously is equal to the sum of the *zero-state responses* due to each independent source acting one at a time.†

---

**Example 1**    Consider the linear network shown in Fig. 2.1.   The network is in the zero state; that is, $v(0-) = 0$.   Let the voltage across the capacitor be the response.   The independent sources are $i_s(t) = Iu(t)$ and $e_s(t) = E\delta(t)$, where $I$ and $E$ are constants.   Let $v_i(\cdot)$ be the zero-state response due to $i_s$ acting alone on the network (i.e., with $e_s = 0$, hence a short circuit); $v_i$ is given by

$$v_i(t) = IR(1 - \epsilon^{-t/RC}) \qquad \text{for } t \geq 0$$

Let $v_e(\cdot)$ be the zero-state response due to $e_s$ acting alone (i.e., with $i_s = 0$, hence an open circuit); it is given by

$$v_e(t) = \frac{E}{RC}\epsilon^{-t/RC} \qquad \text{for } t \geq 0$$

Consider the differential equation of the network with both sources present as follows:

$$i_s = \frac{v - e_s}{R} + C\frac{dv}{dt}$$

or

$$C\frac{dv}{dt} + \frac{v}{R} = i_s + \frac{e_s}{R} = Iu(t) + \frac{E}{R}\delta(t)$$

It is easy to verify that the zero-state response due to $i_s$ and $e_s$ acting simultaneously is

$$v(t) = IR + \left(\frac{E}{RC} - IR\right)\epsilon^{-t/RC} \qquad \text{for } t \geq 0$$

Clearly, we have, as the superposition theorem requires,

$$v(t) = v_i(t) + v_e(t) \qquad \text{for } t \geq 0$$

---

† In other words, we set successively all independent sources to zero except one.

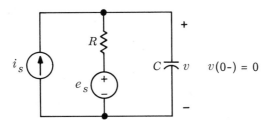

**Fig. 2.1**    Example 1: A linear network with two independent sources.

**Remarks**  1.  The superposition theorem is extremely general; it applies to all linear networks, time-invariant or time-varying.  There is no restriction on the nature, on the waveforms, or on the location of the independent sources.  It is important to note that when a voltage source is set equal to zero, it becomes a short circuit, and when a current source is set equal to zero, it becomes an open circuit.

2.  The superposition theorem can be expressed in terms of the concept of the linear function.  (Reread the definition of a linear function in Appendix A, Sec. 2.3.)  As in the statement of the theorem, we consider an arbitrary linear network $\mathfrak{N}$; its topology, its element values, and its independent sources are specified.  We assume that the waveforms of the independent sources determine uniquely the waveform of the zero-state response.  Consider now the waveforms of the independent sources to be the components of a vector which we call the *vector-input waveform*.  Then the superposition theorem simply asserts that the waveform of the zero-state response of a linear network is a linear function of the vector-input waveform.

For example, suppose that there are only two independent sources; a voltage source $e_s$ and a current source $i_s$.  The vector-input waveform is then

$$\begin{bmatrix} e_s(\,\cdot\,) \\ i_s(\,\cdot\,) \end{bmatrix}$$

Note that our notation emphasizes that the first component of the vector is the waveform $e_s(\,\cdot\,)$, and the second component is the waveform $i_s(\,\cdot\,)$.  Suppose that the response of interest is the branch-voltage waveform $v(\,\cdot\,)$.  Then the zero-state response $v(\,\cdot\,)$ is a function of the vector-input waveform

$$v(\,\cdot\,) = f\left(\begin{bmatrix} e_s(\,\cdot\,) \\ i_s(\,\cdot\,) \end{bmatrix}\right)$$

Let us show that if the function $f$ is linear, then the superposition theorem holds.  We first note that

$$\begin{bmatrix} e_s(\,\cdot\,) \\ i_s(\,\cdot\,) \end{bmatrix} = \begin{bmatrix} e_s(\,\cdot\,) \\ 0 \end{bmatrix} + \begin{bmatrix} 0 \\ i_s(\,\cdot\,) \end{bmatrix}$$

Now, $f$ is linear; therefore it is additive, and

$$\underbrace{f\left(\begin{bmatrix} e_s(\,\cdot\,) \\ i_s(\,\cdot\,) \end{bmatrix}\right)}_{\substack{\text{Zero-state} \\ \text{response when} \\ e_s \text{ and } i_s \text{ are} \\ \text{turned on}}} = \underbrace{f\left(\begin{bmatrix} e_s(\,\cdot\,) \\ 0 \end{bmatrix}\right)}_{\substack{\text{Zero-state} \\ \text{response when} \\ e_s \text{ acts alone}}} + \underbrace{f\left(\begin{bmatrix} 0 \\ i_s(\,\cdot\,) \end{bmatrix}\right)}_{\substack{\text{Zero-state} \\ \text{response when} \\ i_s \text{ acts alone}}}$$

**Exercise** For the example in Fig. 2.1 determine the linear function $f$ which relates the zero-state response $v(\cdot)$ to the vector input $[e_s(\cdot), i_s(\cdot)]^T$. (In this connection see Example 4, Appendix A.)

3. One often sees the following statement: "*In general, superposition does not apply to nonlinear networks.*" Properly interpreted, this statement is true. If we consider all networks consisting of linear and nonlinear elements, then the superposition theorem does not hold for *all* possible locations, types, and waveforms of independent sources. However, if there is a single nonlinear element in a network, we can often choose two independent sources with appropriate waveforms and locations so that superposition does hold. In other words, if a particular network has all linear elements except for a few nonlinear ones, it is possible to make superposition hold by careful selection of element values, source location, source waveform, and response. The point is that we cannot guarantee superposition for all network topologies, and all choices of response. Let us illustrate these facts by the following examples.

**Example 2** Consider the nonlinear circuit shown in Fig. 2.2. Let $i_s$ be a dc current source of 4 amp and $e_s$ be a dc voltage source of 10 volts. Let the response be the voltage $v$. For $i_s$ acting alone, the response is $v_i = 4$ volts. For $e_s$ acting alone, the response is $v_e = 0$ (the ideal diode is reverse-biased). For $i_s$ and $e_s$ acting simultaneously, the response due to both sources is $v = 0$ (it is easy to check that with both $i_s$ and $e_s$ turned on the diode is still reverse-biased). Hence, for these element values, source location, source waveform, etc., the superposition theorem does not apply.

Consider next the *same* circuit, but with $i_s = 10$ amp and $e_s = -10$ volts. Then it is easy to show that $v_i = 10$ volts, $v_e = 5$ volts, and $v = 15$ volts. For these element values, source location, source waveform, and choice of response, the superposition theorem applies. In the present situation, the source waveforms are such that the diode is, in all three cases, forward-biased; thus, the nonlinearity of the diode does not enter the picture.

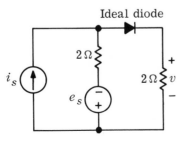

**Fig. 2.2** Example 2: A nonlinear circuit with an ideal diode.

**Example 3**    Balanced networks often furnish examples of nonlinear networks in which a nonlinear element does not affect the application of the superposition theorem. Consider the balanced bridge shown in Fig. 2.3. Let $v$ be the response. Since the bridge is balanced, it is clear that neither $i_s$ nor $e_s$ can cause current to flow through the diode. Therefore, if $v$ is the response, the superposition theorem applies (as far as the sources $e_s$ and $i_s$ are concerned). It should be stressed, however, that if the voltage source were placed in series with the diode and the current source in parallel with it, then the superposition theorem could not possibly hold for all source waveforms. For one waveform the diode might be conducting; for the other waveform and for their sum it might be reverse-biased.

**Exercise**    Consider a series $RLC$ circuit whose reactive elements are linear and time-invariant but whose resistor has a characteristic $v_R = \alpha i + \beta i^2$, where $\alpha$ and $\beta$ are real constants. Let $i_1$ be the zero-state response due to the voltage source $e_{s1}$. Similarly, let $i_2$ be the zero-state response due to $e_{s2}$. Show that $i_1 + i_2$ is not the zero-state response due to the input $e_{s1} + e_{s2}$. (Hint: Write the mesh equation; express that $i_1$ ($i_2$, respectively) satisfies it when $e_{s1}$ ($e_{s2}$, respectively) is the source. Add these equations and compare with the source voltage required to have $i_1 + i_2$ flow through the mesh.)

Up to now the superposition theorem has been stated exclusively in terms of the zero-state response of a linear network. Since *the sinusoidal steady state is the limiting condition (as $t \to \infty$) of the zero-state response to a sinusoidal input,* it follows that the superposition theorem applies in particular to the sinusoidal steady state. More formally, we state the following corollary.

**COROLLARY 1**    Let $\mathfrak{N}$ be a linear time-invariant network; i.e., let each of its elements be either an independent source or a linear element. Suppose that all the independent sources are sinusoidal (not necessarily of the same frequency). Then the steady-state response of $\mathfrak{N}$ due to all the independent sources

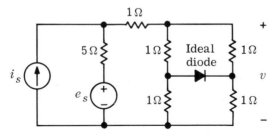

**Fig. 2.3**    Example 3: A balanced bridge.

acting simultaneously is equal to the sum of the *sinusoidal steady-state responses* due to each independent source acting one at a time.

For linear time-invariant networks it is usually more convenient to use the frequency-domain characterization. The superposition theorem can then be stated in terms of network functions.

**Example 4**   Consider the linear time-invariant network shown in Fig. 2.4. The inputs are the two independent sources $e_1$ and $i_2$, and the output is taken as the voltage $v$ across the resistor with resistance $R_2$. More precisely, $v$ is the zero-state response to $e_1(\cdot)$ and $i_2(\cdot)$ acting simultaneously on the network. Let $V(s)$, $E_1(s)$, and $I_2(s)$ be the Laplace transforms of the waveforms $v$, $e_1$, and $i_2$, respectively. Then, the superposition theorem gives

$$V(s) = H_1(s)E_1(s) + H_2(s)I_2(s)$$

The network function $H_1(s)$ is given by

$$H_1(s) = \left. \frac{V(s)}{E_1(s)} \right|_{I_2 \equiv 0} = \frac{1}{1 + (R_1 + Ls)(G_2 + Cs)}$$

Note that in calculating $H_1(s)$, the current source $i_2$ is set to zero. Similarly,

$$H_2(s) = \left. \frac{V(s)}{I_2(s)} \right|_{E_1 \equiv 0} = \frac{R_1 + Ls}{1 + (R_1 + Ls)(G_2 + Cs)}$$

Again in calculating $H_2(s)$, the voltage source $e_1$ is set to zero.

This idea can be generalized to any number of inputs. Since it is important in practice we state it formally as a corollary.

**COROLLARY 2**   Let $\mathfrak{N}$ be a linear time-invariant network. Let the response be the voltage across any node pair or the current through any branch of $\mathfrak{N}$. More spe-

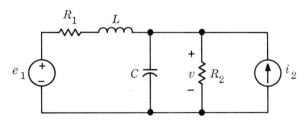

**Fig. 2.4**   Example 4: Illustration of the superposition theorem in terms of transfer functions.

cifically, call $X(s)$ the Laplace transform of the *zero-state* response due to all the independent sources acting simultaneously. Then

$$X(s) = \sum_{k=1}^{m} H_k(s)I_k(s)$$

where $I_k(s)$, $k = 1, 2, \ldots, m$, are the Laplace transforms of the $m$ inputs and $H_k(s)$, $k = 1, 2, \ldots, m$, are the respective network functions from the $m$ inputs to the specified output.

---

**2.2   Proof of the Superposition Theorem**

Although the statement of the theorem allows any number and any kind of independent sources, we shall for simplicity consider only *two* sources: one voltage source $e_s$ and one current source $i_s$. We assume that $e_s$ and $i_s$ are applied at $t = 0$. Imagine then a network $\mathfrak{N}$ consisting of *two* independent sources $e_s$ and $i_s$ as inputs and linear (possibly time-varying) resistors, capacitors, inductors, and dependent sources. In proving the theorem it is only necessary to consider as responses all the branch currents and all the branch voltages of these linear elements, since any node-pair voltage of $\mathfrak{N}$ is a linear combination of branch voltages.

We shall consider the network $\mathfrak{N}$ under three different conditions.

*Condition 1*    *The voltage source $e_s$ acts alone.* In other words, the current source is set to zero (i.e., replaced by an open circuit). To determine all branch voltages and branch currents, let us write the loop equations and solve for the zero-state loop currents (due to the unique input $e_s$). Since the network starts from the zero state and since the input $e_s(\cdot)$ is specified, then by the uniqueness assumption, all the loop currents are uniquely determined. Given the loop currents, we can calculate *all* the branch currents by forming suitable linear combinations of loop currents. The branch equations will give us all the branch voltages. Let us disregard the voltage source $e_s$ and call $v_1'(\cdot), v_2'(\cdot), \ldots, v_b'(\cdot)$ and $j_1'(\cdot), j_2'(\cdot), \ldots, j_b'(\cdot)$ the branch voltages and branch currents thus calculated. These are branch voltages and branch currents of all branches except the source. Note that (1) the symbols $v_k'(\cdot)$ and $j_k'(\cdot)$, $k = 1, 2, \ldots, b$, denote the whole *waveform* of the voltage $v_k'$ and current $j_k'$, respectively; (2) each one of these waveforms are zero-state responses to $e_s$ acting alone.

*Condition 2*    *The current source $i_s$ acts alone.* Now the voltage source is set to zero (i.e., it is replaced by a short circuit). To determine all branch voltages and all branch currents, let us write the node equations and solve them for the node voltages. From these we obtain successively all branch voltages and all branch currents. Denote them by $v_1''(\cdot), v_2''(\cdot), \ldots, v_b''(\cdot)$ and $j_1''(\cdot)$, $j_2''(\cdot), \ldots, j_b''(\cdot)$. Note that these symbols denote *waveforms* that are zero-state responses to $i_s$ acting alone.   Again by the uniqueness assumption, these waveforms are uniquely determined by the initial conditions and the input $i_s$.

*Condition 3*   *The voltage source $e_s$ and the current source $i_s$ act together.*   We have to show that when $e_s$ and $i_s$ act simultaneously the resulting zero-state branch voltages and branch currents are precisely the sum of the preceding ones. Instead of calculating the response when both sources act on the network we shall verify the following: (*a*) the branch-voltage waveforms $v'_k(\cdot) + v''_k(\cdot)$ satisfy all KVL constraints; (*b*) the branch-current waveforms $j'_k(\cdot) + j''_k(\cdot)$ satisfy all the KCL constraints; (*c*) all the branch equations are satisfied; and (*d*) the voltage waveforms $v'_k(\cdot) + v''_k(\cdot)$ and the current waveforms $j'_k(\cdot) + j''_k(\cdot)$ are equal to zero just prior to the application of the inputs.   Once these facts are established, it follows, by uniqueness, that the waveforms $v'_k(\cdot) + v''_k(\cdot)$, $j'_k(\cdot) + j''_k(\cdot)$ are the zero-state responses when both sources act simultaneously on the network.

*Proof*   *a.*   First, consider the voltages.   Around any loop, say loop $i$, which does not include $e_s$, writing KVL around that loop gives an algebraic sum of voltages of the form

$$\sum_k b_{ik}v'_k = 0 \qquad \text{for condition 1}$$

and

$$\sum_k b_{ik}v''_k = 0 \qquad \text{for condition 2}$$

where $b_{ik}$ is the $(i,k)$ element of the loop matrix.   Hence, by addition, we conclude that, for all such loops,

$$(2.1) \quad \sum_k b_{ik}(v'_k + v''_k) = 0$$

For any loop, say loop $l$, that includes the voltage source $e_s$ we have

$$\sum_k b_{lk}v'_k = e_s \qquad \text{for condition 1}$$

$$\sum_k b_{lk}v''_k = 0 \qquad \text{for condition 2}$$

Hence by addition, for all loops that include the voltage source $e_s$,

$$(2.2) \quad \sum_k b_{lk}(v'_k + v''_k) = e_s$$

This equation expresses KVL for condition 3, since the voltage source is turned on in condition 3.   Equations (2.1) and (2.2) assert that *the branch-voltage waveforms $v'_k(\cdot) + v''_k(\cdot)$ satisfy the KVL constraints around any loop of $\mathfrak{N}$.*

*b.*   Second, consider the branch currents.   By analogous reasoning we see that for any node that is *not* connected to the current source $i_s$,

(2.3)   $\displaystyle\sum_k a_{ik}(j'_k + j''_k) = 0$

where $a_{ik}$ is the $(i,k)$ element of the reduced incidence matrix. For the two nodes that are connected to the current source

(2.4)   $\displaystyle\sum_k a_{lk}(j'_k + j''_k) = i_s$

Thus, (2.3) and (2.4) assert that *the branch-current waveforms* $j'_k(\,\cdot\,) + j''_k(\,\cdot\,)$ *satisfy the KCL constraints about any node of* $\mathfrak{N}$.

$c.$   Third, consider the branch equations. If the $k$th branch is a resistor (linear and, possibly, time-varying by assumption), we have

$$v'_k(t) = R_k(t)j'_k(t)$$
$$v''_k(t) = R_k(t)j''_k(t)$$

for all $t$

and, by addition,

(2.5)   $v'_k(t) + v''_k(t) = R_k(t)[j'_k(t) + j''_k(t)]$      for all $t$

For inductors,

$$v'_k(t) = \frac{d}{dt}[L_k(t)j'_k(t)]$$
$$v''_k(t) = \frac{d}{dt}[L_k(t)j''_k(t)]$$

for all $t$

and, by addition,

(2.6)   $v'_k(t) + v''_k(t) = \dfrac{d}{dt}\{L_k(t)[j'_k(t) + j''_k(t)]\}$      for all $t$

(If there were coupled inductors, the same manipulation would go through except that the right-hand side would involve a sum over all the coupled inductors.) For capacitors,

$$j'_k(t) = \frac{d}{dt}[C_k(t)v'_k(t)]$$
$$j''_k(t) = \frac{d}{dt}[C_k(t)v''_k(t)]$$

for all $t$

and, by addition,

(2.7)   $j'_k(t) + j''_k(t) = \dfrac{d}{dt}\{C_k(t)[v'_k(t) + v''_k(t)]\}$      for all $t$

Consider now the case of linear *dependent* sources. Suppose that branch 3 consists of a current source whose current depends on the branch voltage $v_1$. By assumption, all dependent sources are linear (though possibly time-varying), hence, for conditions 1 and 2, we have, respectively,

$$j'_3(t) = g_m(t)v'_1(t)$$
$$j''_3(t) = g_m(t)v''_1(t)$$

for all $t$

By addition,

(2.8)   $j_3'(t) + j_3''(t) = g_m(t)[v_1'(t) + v_1''(t)]$      for all $t$

In other words, the current waveform $j_3'(\cdot) + j_3''(\cdot)$ and the voltage wave-form $v_1'(\cdot) + v_1''(\cdot)$ satisfy the constraint imposed by the dependent source. A similar reasoning obviously applies to dependent voltage sources. Equating (2.5) and (2.6), and (2.7) and (2.8) shows that *the branch-voltage waveforms $v_k'(\cdot) + v_k''(\cdot)$ and the branch-current waveforms $j_k'(\cdot) + j_k''(\cdot)$ satisfy all the branch equations.*

*d.* Finally, since the $v_k'$, $v_k''$, $j_k'$ and $j_k''$ are zero-state responses, we have

$$v_k'(0-) = v_k''(0-) = j_k'(0-) = j_k''(0-) = 0$$

hence

$$v_k'(0-) + v_k''(0-) = 0 \qquad j_k'(0-) + j_k''(0-) = 0$$

In other words, *the waveforms $v_k'(\cdot) + v_k''(\cdot)$ and $j_k'(\cdot) + j_k''(\cdot)$ are also zero-state responses.*

Thus, if we put together all the partial conclusions, we conclude that the branch-voltage waveforms $v_k'(\cdot) + v_k''(\cdot)$ and the branch-current waveforms $j_k'(\cdot) + j_k''(\cdot)$ satisfy (*a*) all KVL constraints, (*b*) all KCL constraints, (*c*) all the branch equations, and (*d*) the initial conditions required from zero-state responses. Hence by the uniqueness assumption, these waveforms must be the zero-state responses to the sources $e_s$ and $i_s$ acting simultaneously.

**Remark**   This proof illustrates the fact that the superposition theorem rests on four fundamental facts.

1. KVL is expressed by *linear homogeneous* algebraic equations relating branch voltages.

2. KCL is expressed by *linear homogeneous* algebraic equations relating branch currents.

3. The linearity of the network elements implies that the branch currents and the branch voltages are related by a *linear function.*

4. The initial conditions corresponding to the zero state are expressed by *linear homogeneous* equations

$$v_k(0-) = 0 \qquad j_k(0-) = 0 \qquad k = 1, 2, \ldots, b$$

---

**3**   **Thévenin-Norton Equivalent Network Theorem**

The Thévenin-Norton theorem is a powerful tool in calculating the response of complicated networks. It has added importance because it gives a mental picture of any linear network as seen from any two of its terminals. It is a very general theorem; it applies to an extremely broad class

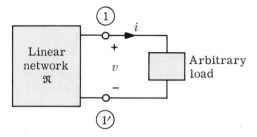

**Fig. 3.1**   Circuit illustrating the conditions for application of the Thévenin-Norton equivalent network theorem.

of networks, and whatever the network, it gives an equivalent network of the same form.

## 3.1   Theorem, Examples, Remarks, and Corollary

The Thévenin-Norton network theorem considers the following situation: a linear network $\mathfrak{N}$ is connected to an arbitrary load by two of its terminals, $(1)$ and $(1')$, as shown in Fig. 3.1. We assume that the only interaction between $\mathfrak{N}$ and the load comes from the current flowing through terminals $(1)$ and $(1')$. In particular no other coupling (e.g., magnetic or through dependent sources) is allowed between $\mathfrak{N}$ and the load. *It is important to stress the fact that we make no assumptions concerning the load;* it may be nonlinear and/or time-varying. The network $\mathfrak{N}$ is only required to be linear; it may include both dependent and independent sources. In broad terms, the Thévenin-Norton theorem asserts that the terminal current waveform $i(\cdot)$ and the terminal voltage waveform $v(\cdot)$ will not be affected if $\mathfrak{N}$ is replaced by either a *"Thévenin equivalent network"* or by a *"Norton equivalent network."*

The **Thévenin equivalent network** is shown in Fig. 3.2a. It consists of a two-terminal network $\mathfrak{N}_0$ in series with a voltage source $e_{oc}$. The waveform $e_{oc}(\cdot)$ of the voltage source is the open-circuit voltage of $\mathfrak{N}$, that is, the voltage across the terminals $(1)$ and $(1')$ when the load is disconnected, as shown in Fig. 3.2b. The voltage $e_{oc}$ is caused by all the independent sources of $\mathfrak{N}$ *and* the initial state; it is measured with the polarity indicated on the figure. The network $\mathfrak{N}_0$ is obtained from the network $\mathfrak{N}$ by setting all *independent* sources to zero (i.e., by replacing every independent voltage source by a short circuit and every independent current source by an open circuit) and by setting all the initial conditions to zero. Note that *dependent* sources are left unchanged.

The **Norton equivalent network** is shown in Fig. 3.3a. It consists of the same two-terminal network $\mathfrak{N}_0$ placed in parallel with a current source $i_{sc}$. The waveform $i_{sc}(\cdot)$ of the current source is the short-circuit current of

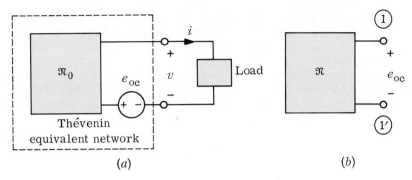

**Fig. 3.2**   (a) Thévenin equivalent network; (b) circuit defining the voltage source $e_{oc}$ of the Thévenin equivalent network; the subscript "oc" emphasizes that $e_{oc}$ is an open-circuit voltage.

$\mathfrak{N}$, that is, the current flowing in the short circuit which connects terminals ①  and ①′, as shown in Fig. 3.3$b$. The current is caused by all the independent sources of $\mathfrak{N}$ and the initial state; it is measured with the polarity indicated in the figure.

**THÉVENIN-NORTON EQUIVALENT NETWORK**   Let the linear network $\mathfrak{N}$ be connected by two of its terminals ① and ①′ to an arbitrary load. Let $\mathfrak{N}$ consist of independent sources and linear resistors, linear capacitors, linear inductors, linear transformers, and linear dependent sources. The elements may be time-varying. We further assume that $\mathfrak{N}$ has a unique solution when it is terminated by the load, and when the load is replaced by an independent source. Let $\mathfrak{N}_0$ be the network obtained from $\mathfrak{N}$ by setting all independent sources to zero and all initial conditions to zero. Let $e_{oc}$ be the open-circuit voltage of $\mathfrak{N}$ observed at terminals ① and ①′, as shown in Fig. 3.2$b$. Let $i_{sc}$ be the short-circuit current of $\mathfrak{N}$ flowing out of ① into ①′ as shown in Fig. 3.3$b$. Under these conditions, whatever the load may be, the voltage waveform

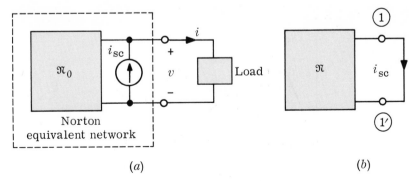

**Fig. 3.3**   (a) Norton equivalent network; (b) the circuit defining current source $i_{sc}$ of the Norton equivalent network; the subscript "sc" emphasizes that $i_{sc}$ is a short-circuit current.

$v(\,\cdot\,)$ across ① and ①′ and the current waveform $i(\,\cdot\,)$ through ① and ①′ remain unchanged when the network $\mathfrak{N}$ is replaced by either its Thévenin equivalent or by its Norton equivalent network.

**Example 1**    Consider the resistive circuit shown in Fig. 3.4a.  We want to determine the voltage across the tunnel diode.  We can use the Thévenin theorem and consider the tunnel diode as the load.  First we determine the Thévenin equivalent network of the one-port faced by the tunnel diode.  By inspection, the open-circuit voltage is given by

$$e_{oc} = \frac{R_2 E}{R_1 + R_2}$$

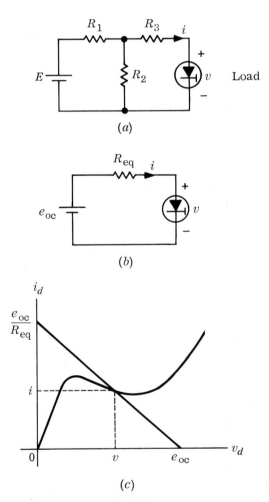

(a)

(b)

(c)

**Fig. 3.4**    Example 1: a tunnel-diode circuit.

The network $\mathfrak{N}_0$ is obtained by shorting out the battery; it is thus the series combination of $R_3$ with the parallel combination of $R_1$ and $R_2$. The equivalent resistance is therefore

$$R_{\text{eq}} = R_3 + \frac{R_1 R_2}{R_1 + R_2}$$

Given $e_{\text{oc}}$ and $R_{\text{eq}}$, the Thévenin theorem asserts that $v$ and $i$ will remain unchanged if we consider the circuit shown in Fig. 3.4$b$. With the notations defined on the figure, the terminal characteristic of the Thévenin equivalent network is

(3.1)  $v = e_{\text{oc}} - R_{\text{eq}}i$

The circuit of Fig. 3.4$b$ can be solved by plotting the characteristic (3.1) on the same graph as the tunnel-diode characteristic. Any intersection of the characteristics will give one solution of the problem. The plot is shown in Fig. 3.4$c$, and the solution is obvious.

**Remarks**   1.   The Thévenin-Norton theorem is extremely general. No restriction whatsoever has been imposed on the load except that its interaction with $\mathfrak{N}$ occur exclusively through the current and voltage at terminals ① and ①′, and that $\mathfrak{N}$ terminated by that load have a unique solution. The network $\mathfrak{N}$ must satisfy the same restrictions as in the superposition theorem; this is natural since we shall derive the Thévenin-Norton theorem from the superposition theorem.

2.   The Thévenin-Norton theorem is important not only because it saves steps in some calculations but also because it is a powerful intellectual tool. Given any linear (possibly time-varying) network $\mathfrak{N}$, as far as any one of its terminal pairs ①, ①′ is concerned it behaves as if it consisted of one voltage source $e_{\text{oc}}$ in series with a "relaxed" network $\mathfrak{N}_0$, or equivalently, as a "relaxed" network $\mathfrak{N}_0$ in parallel with a current source $i_{\text{sc}}$.

3.   Since the "relaxed" network $\mathfrak{N}_0$ is obtained from $\mathfrak{N}$ by reducing all its *independent* sources to zero and setting all initial conditions to zero, $\mathfrak{N}_0$ could legitimately be referred to as the *zero-input and zero-state equivalent network*. (The term "zero-input" refers to the fact that all independent sources are set equal to zero, and "zero state" refers to the fact that all initial conditions are set equal to zero.) It is equivalent in the sense that $\mathfrak{N}_0$, as seen from terminals ① and ①′, is indistinguishable from $\mathfrak{N}$ provided all the independent sources and all initial conditions are set equal to zero.

**3.2**   **Special Cases**

If the given linear network $\mathfrak{N}$ is time-invariant it is more convenient to use network function concepts. The Thévenin-Norton equivalent network theorem can be stated in terms of the driving-point impedance or admittance of the "relaxed" network $\mathcal{N}_0$.

**COROLLARY**    Let the linear *time-invariant* network $\mathcal{N}$ be connected by two of its terminals, $\textcircled{1}$ and $\textcircled{1'}$, to an *arbitrary* load. Let $E_{oc}(s)$ be the Laplace transform of the open-circuit voltage $e_{oc}(t)$ observed at terminals $\textcircled{1}$ and $\textcircled{1'}$, that is, the voltage when no current flows into $\mathcal{N}$ through $\textcircled{1}$ and $\textcircled{1'}$. Let $I_{sc}(s)$ be the Laplace transform of the current $i_{sc}(t)$ flowing out of $\textcircled{1}$ and into $\textcircled{1'}$ when the load is shorted. Let $Z_{eq} = 1/Y_{eq}$ be the impedance (seen between terminals $\textcircled{1}$ and $\textcircled{1'}$) of the network obtained from $\mathcal{N}$ by setting all independent sources to zero and all initial conditions to zero. Under these conditions, whatever the load may be, the voltage $V(s)$ across $\textcircled{1}$ and $\textcircled{1'}$ and the current $I(s)$ through $\textcircled{1}$ and $\textcircled{1'}$ remain unchanged when the network $\mathcal{N}$ is replaced by either its Thévenin equivalent network or its Norton equivalent network, as shown in Fig. 3.5. Furthermore,

$$(3.2)\quad E_{oc} = Z_{eq}I_{sc}$$

Formula (3.2) is easily verified by referring to the definitions of $E_{oc}$ and $I_{sc}$. Indeed, if we replace the load in Fig. 3.5$a$ by a short circuit, the current $I$ becomes the short-circuit current $I_{sc}$ and (3.2) follows by KVL.

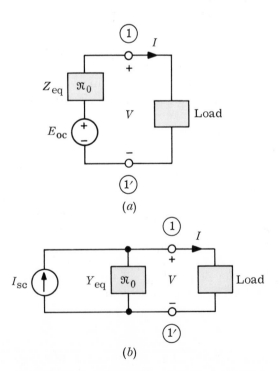

(a)

(b)

**Fig. 3.5**    The Thévenin and Norton equivalent networks for linear time-invariant networks.

Dually, if we replace the load in Fig. 3.5*b* by an open circuit, the voltage $V$ becomes the open-circuit voltage $E_{oc}$ and KCL gives

$$I_{sc} = Y_{eq}E_{oc}$$

Multiplying this equation by $Z_{eq}$ and noting that $Z_{eq}(s)Y_{eq}(s) = 1$, we obtain (3.2) again.  The formula (3.2) is especially useful in computing the Thévenin equivalent network in circuits with dependent sources.

**Example 2**   Consider the simple transistor amplifier shown in Fig. 3.6*a*.  Its small-signal equivalent circuit is shown in Fig. 3.6*b*.  We want to determine the amplifier voltage ratio $V_L/V_0$ and the output impedance (i.e., the impedance faced by the load resistor $R_L$ at terminals ① and ①'). Let us obtain the Thévenin equivalent network faced by $R_L$.  We shall calculate the open-circuit voltage $E_{oc}$ and the short-circuit current $I_{sc}$ at terminals ① and ①'.  First, we convert the series connection of the voltage source $V_0$ and resistor $r_x$ into the parallel connection of a current source $V_0/r_x$ and resistor $r_x$.  The open-circuit voltage $E_{oc}$ can be found from the circuit, as shown in Fig. 3.7*a*.  Using node analysis, we obtain

$$\begin{bmatrix} g_t + s(C_\mu + C_\pi) & -sC_\mu \\ -sC_\mu + g_m & sC_\mu \end{bmatrix}\begin{bmatrix} V_1 \\ V_2 \end{bmatrix} = \begin{bmatrix} \dfrac{V_0}{r_x} \\ 0 \end{bmatrix}$$

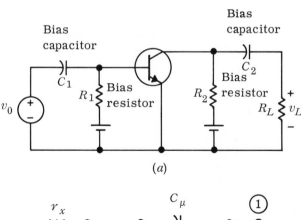

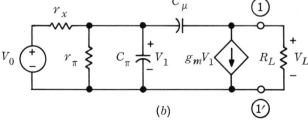

**Fig. 3.6**   A simple transistor amplifier and its small-signal equivalent circuit; the bias elements $R_1$, $R_1$, $C_1$, and $C_2$ (being large by design) are neglected in the small-signal equivalent circuit.

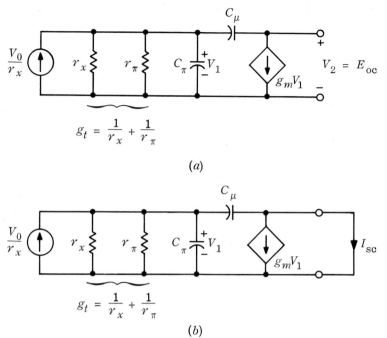

$$g_t = \frac{1}{r_x} + \frac{1}{r_\pi}$$

(a)

$$g_t = \frac{1}{r_x} + \frac{1}{r_\pi}$$

(b)

**Fig. 3.7**   Example 2: derivation of the Thévenin and Norton equivalent networks.

Thus

$$
E_{oc} = V_2 = \frac{\begin{vmatrix} g_t + s(C_\mu + C_\pi) & V_0/r_x \\ g_m - sC_\mu & 0 \end{vmatrix}}{\begin{vmatrix} g_t + s(C_\mu + C_\pi) & -sC_\mu \\ g_m - sC_\mu & sC_\mu \end{vmatrix}}
$$

(3.3)

$$
= \frac{(sC_\mu - g_m)V_0/r_x}{s^2 C_\mu C_\pi + sC_\mu(g_t + g_m)}
$$

Next we compute $I_{sc}$ from the circuit in Fig. 3.7b. We first compute $V_1$ and obtain

$$
V_1 = \frac{V_0}{r_x} \frac{1}{g_t + s(C_\mu + C_\pi)}
$$

The short-circuit current is therefore

$$
I_{sc} = (sC_\mu - g_m)V_1 = \frac{(sC_\mu - g_m)V_0/r_x}{s(C_\pi + C_\mu) + g_t}
$$

Thus the equivalent impedance in the Thévenin equivalent network is

(3.4a)   $$Z_{eq} = \frac{E_{oc}}{I_{sc}} = \frac{s(C_\pi + C_\mu) + g_t}{s^2 C_\mu C_\pi + sC_\mu(g_t + g_m)}$$

The voltage ratio of the amplifier is obtained next as follows:

(3.4b)  $$V_L = \frac{E_{oc}}{Z_{eq} + R_L} R_L$$

Combining (3.3), (3.4a), and (3.4b), we have

$$\frac{V_L}{V_0} = \frac{[(sC_\mu - g_m)/r_x]/[s^2 C_\mu C_\pi + sC_\mu(g_t + g_m)]}{[s(C_\pi + C_\mu) + g_t]/[s^2 C_\mu C_\pi + sC_\mu(g_t + g_m)] + R_L} R_L$$

$$= \frac{(sC_\mu - g_m)/r_x}{s^2 C_\mu C_\pi R_L + s(C_\pi + C_\mu + C_\mu g_t R_L + C_\mu g_m R_L) + g_t} R_L$$

It is interesting to note that the network function relating $V_L$ to $V_0$ has a zero in the right-half plane; the network function is zero when $s = g_m/C_\mu$.

*Alternate method*   We obtained the impedance $Z_{eq}$ in Eq. (3.4a) by observing that it was the ratio of $E_{oc}$ to $I_{sc}$, and we had already easily computed them. In general, the Thévenin equivalent impedance can always be obtained by (1) setting to zero all the independent sources of $\mathfrak{N}$, (2) connecting a "test" current source $I_t(s)$ to terminals ① and ①', and (3) using node analysis to calculate $V_t(s)$, the Laplace transform of the zero-state response to $I_t(s)$. Then

$$V_t(s) = Z_{eq}(s)I_t(s)$$

For this case the node equations are

$$\begin{bmatrix} g_t + s(C_\mu + C_\pi) & -sC_\mu \\ -sC_\mu + g_m & sC_\mu \end{bmatrix} \begin{bmatrix} E_1 \\ V_t \end{bmatrix} = \begin{bmatrix} 0 \\ I_t \end{bmatrix}$$

By Cramer's rule,

$$V_t(s) = \frac{g_t + s(C_\mu + C_\pi)}{s^2 C_\mu C_\pi + sC_\mu(g_t + g_m)} I_t(s)$$

The ratio $V_t(s)/I_t(s)$ is $Z_{eq}(s)$. It gives the same result for $Z_{eq}$ as before.

---

**3.3    Proof of Thévenin Theorem**

We shall prove only the Thévenin equivalent network theorem. The Norton equivalent network theorem can be proved using the dual argument. Referring to Fig. 3.1, let us first replace all initial conditions by independent sources; we replace initial voltages across capacitors by constant independent voltage sources placed in series with the capacitor, and we replace initial currents through inductors by constant independent current sources placed in parallel with the inductor. Hence we shall be concerned only with the zero-state response of the network. Let $v(\cdot)$ and $i(\cdot)$ be the actual voltage and current at the terminals of the load as shown in Fig. 3.1. The proof consists of three steps.

1.  First, we substitute the load with a current source $i(\cdot)$, as shown in Fig. 3.8. The resulting network is linear, and its solution is unique by assumption. By the substitution theorem, the solution for all branch voltages and branch currents in the resulting network are the same as those of the original network. In particular, the voltage $v(\cdot)$ across the terminals is the same as that of the original network.

2.  Let us look at the problem from the superposition theorem point of view; the network $\mathfrak{N}$ of Fig. 3.8 is driven by the load current source $i$ and all the independent sources of $\mathfrak{N}$. (Recall that these independent sources include all the original independent sources of $\mathfrak{N}$ and the independent sources which replace the initial conditions.) Thus, the terminal voltage $v(\cdot)$ is the zero-state response to two sets of sources: the independent current source $i$ which replaces the load and all the independent sources in $\mathfrak{N}$.

    Let us call $v_1$ the *zero-state response* of $\mathfrak{N}$ caused by the independent current source $i$ acting alone and $e_{oc}$ the *zero-state response* when the independent sources of $\mathfrak{N}$ are acting alone. To calculate $v_1$, we drive $\mathfrak{N}$, with all its independent sources turned off (i.e., we drive the "relaxed" network $\mathfrak{N}_0$), by the current source $i$, as shown in Fig. 3.9a. Thus,

$$v_1(t) = \int_0^t h(t,\tau)i(\tau)\,d\tau$$

where $h(t,\tau)$ represents the impulse response of the relaxed network $\mathfrak{N}_0$ at time $t$ due to a unit impulse applied at time $\tau$. To calculate $e_{oc}$, we turn off the load current source (hence $\mathfrak{N}$ becomes open-circuited) and observe the terminal voltage of $\mathfrak{N}$ due to the independent sources of $\mathfrak{N}$, as shown in Fig. 3.9b. Since the terminal current is zero, $e_{oc}$ is actually the open-circuit voltage of $\mathfrak{N}$. Now the superposition theorem asserts that

(3.5)    $$v(t) = e_{oc}(t) + v_1(t) = e_{oc}(t) + \int_0^t h(t,\tau)i(\tau)\,d\tau$$

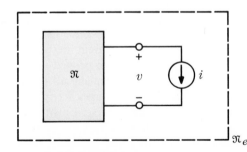

**Fig. 3.8**    By the substitution theorem, the load is replaced by a current source $i$.

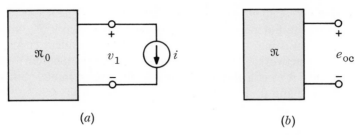

**Fig. 3.9**    The response to the circuit in Fig. 3.8 is, by the superposition theorem, the sum of two responses, $v_1$, due to the load current-source $i$ only, and $e_{oc}$, due to the sources in $\mathfrak{N}$.

3.   Now consider the *Thévenin equivalent network* shown in Fig. 3.10. Writing the KVL equation for the mesh, we obtain

$$v(t) = e_{oc}(t) + v_1(t) = e_{oc} + \int_0^t h(t,\tau)i(\tau)\,d\tau$$

which is exactly the equation in (3.5). Thus, we have shown that the Thévenin equivalent network has the same terminal voltage $v$ and terminal current $i$ as the given network $\mathfrak{N}$. Since at no point in the derivation did we make any assumptions about the properties of the load, the network $\mathfrak{N}$ and its Thévenin equivalent network have the same terminal voltage and currents for all possible loads.

**Remark**    The Thévenin-Norton theorem has just been derived from the superposition theorem. We have previously remarked that the Thévenin-Norton theorem stated that the zero-state response is a linear function of the vector input. Let us then interpret the Thévenin-Norton theorem in terms of linear functions. Assume, for simplicity, that all initial conditions have been replaced by (constant) independent sources. Thus, the waveform $v(\cdot)$ becomes a zero-state response to the (enlarged) set of independent

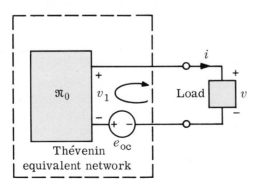

**Fig. 3.10**    The Thévenin equivalent network of $\mathfrak{N}$.

sources. Call $\mathbf{s}(\cdot)$ the vector whose $i$th component $s_i(\cdot)$ is the waveform of the $i$th independent source of $\mathfrak{N}$. The network $\mathfrak{N}_e$, shown in Fig. 3.8 and consisting of $\mathfrak{N}$ terminated by the independent current source $i$, is a *linear* network. Hence (by the superposition theorem) its zero-state response $v(\cdot)$ is a linear function of the vector input. Since $\mathfrak{N}_e$ is driven by $\mathbf{s}(\cdot)$ and $i(\cdot)$, we have

$$v(\cdot) = f\left(\begin{bmatrix} \mathbf{s}(\cdot) \\ i(\cdot) \end{bmatrix}\right)$$

where $f$ is a *linear* function. This equation asserts that the waveform $v(\cdot)$ is uniquely determined by the waveforms $\mathbf{s}(\cdot)$ and $i(\cdot)$, that is, by the vector input whose components are $s_1(\cdot), s_2(\cdot), \ldots, s_n(\cdot), i(\cdot)$. The function $f$ is linear, and therefore obeys the additivity property; hence, from the vector equality

$$\begin{bmatrix} \mathbf{s}(\cdot) \\ i(\cdot) \end{bmatrix} = \begin{bmatrix} \mathbf{s}(\cdot) \\ 0 \end{bmatrix} + \begin{bmatrix} 0 \\ i(\cdot) \end{bmatrix}$$

we conclude

$$v(\cdot) = f\left(\begin{bmatrix} \mathbf{s}(\cdot) \\ 0 \end{bmatrix}\right) + f\left(\begin{bmatrix} 0 \\ i(\cdot) \end{bmatrix}\right)$$

Now, $f\left(\begin{bmatrix} \mathbf{s}(\cdot) \\ 0 \end{bmatrix}\right)$ is the zero-state response of $\mathfrak{N}_e$ when $i(\cdot) = 0$; hence it is $e_{oc}(\cdot)$. The second term $f\left(\begin{bmatrix} 0 \\ i(\cdot) \end{bmatrix}\right)$ is the zero-state response of $\mathfrak{N}_e$ when $\mathbf{s}(\cdot) = 0$. Hence it is the voltage waveform that $i(\cdot)$ would develop across $\mathfrak{N}_0$; the voltage waveform is labeled $v_1(\cdot)$ in Fig. 3.9$a$. We conclude that

$$v(\cdot) = v_1(\cdot) + e_{oc}(\cdot)$$

which is precisely what the Thévenin equivalent network (shown in Fig. 3.10) predicts.

### 3.4   An Application of the Thévenin Equivalent Network Theorem

One important practical problem in system or circuit design is the study of sensitivity of the system or circuit with respect to changes in its components. Such changes may be due to manufacturing deviation, temperature effects, or aging. The Thévenin theorem can be used conveniently for such a study. For simplicity, let us consider a linear time-invariant network which is in the sinusoidal steady state. Suppose that we know the solution, and we wish to determine the change of the current in the $k$th branch due to a small change in the impedance of the $k$th branch, as shown in Fig. 3.11$a$. Let us denote the impedance of the $k$th branch by $Z$.

The Thévenin equivalent network is shown in Fig. 3.11$b$, with the $k$th branch as the load. $E_0$ and $Z_{eq}$ are, respectively, the open-circuit voltage phasor and the equivalent impedance. The sinusoidal current of branch $k$ is represented by the phasor $I$; this phasor is given by

(3.6)   $I(Z_{eq} + Z) = E_0$

Let the $k$th branch be changed slightly so that the new impedance is $Z + \delta Z$, and let $I' = I + \delta I$ be the new current, as shown in Fig. 3.11$c$. Since the remainder of the network remains unchanged, we have the same Thévenin equivalent network as before. Thus, the new current is given by

(3.7)   $I'(Z_{eq} + Z + \delta Z) = (I + \delta I)(Z_{eq} + Z + \delta Z) = E_0$

Substituting (3.6) in (3.7), we obtain

(3.8)   $I\,\delta Z + \delta I(Z_{eq} + Z) + \delta I\,\delta Z = 0$

Since $\delta Z$ is a small change, $\delta I\,\delta Z$ represents a second-order term in (3.8) and can be neglected. We therefore obtain

(3.9$a$)   $\delta I \approx \dfrac{-I\,\delta Z}{Z_{eq} + Z}$

or

(3.9$b$)   $\dfrac{\delta I}{I} \approx \dfrac{-\delta Z}{Z_{eq} + Z}$

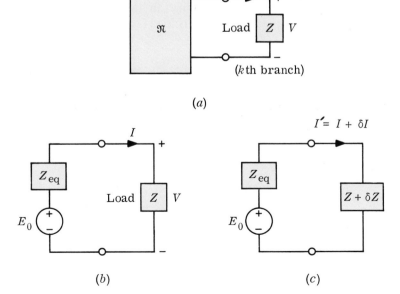

$(a)$

$(b)$          $(c)$

**Fig. 3.11**   Sensitivity study of a network using the Thévenin theorem; the impedance of branch $k$ changes from $Z$ to $Z + \delta Z$.

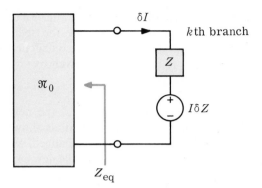

**Fig. 3.12**   A proposed measurement procedure to study the effect of a change in impedance $\delta Z$ of branch $k$.

Thus, we have obtained a simple relation between $\delta Z$, the change in the impedance $Z$, and $\delta I$, the change in the branch current $I$. More significantly, this relation leads to a simple interpretation of $\delta I$; the change in current, $\delta I$, is equal to the current that would flow if the *only* independent source in the circuit were a voltage source $I\,\delta Z$ with the polarity indicated in Fig. 3.12 ($\delta Z$ is the change in the impedance and $I$ is the current that used to flow through $Z$). A moment of thought will indicate that under this situation the measured voltages and currents in other branches of the network are precisely the changes of voltages and currents caused in the original network by the change $\delta Z$ in the $k$th branch.

**Exercise 1**   Suppose that a generator has an internal impedance $Z_{eq} = 1{,}250\epsilon^{j30°}$ ohms and $E_0 = 100$ volts. Its load is an impedance $Z$ of 100 ohms. Cal-

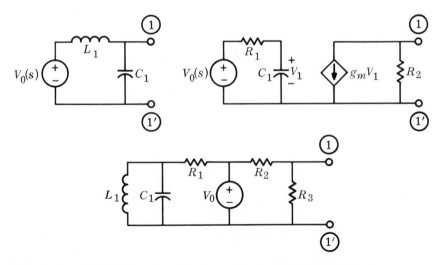

**Fig. 3.13**   Networks for which the Thévenin and Norton equivalent networks must be found.

culate (approximately) $\delta I$ and $\delta I / I$ when $\delta Z = 10$ ohms, $\delta Z = 50 \epsilon^{-j30°}$, and $\delta Z = 50 \epsilon^{j30°}$.

**Exercise 2**   Find the Thévenin and Norton equivalent networks of the networks in Fig. 3.13 (all the networks are in the zero state at $t = 0$).

## 4   The Reciprocity Theorem

Reciprocity is a property that is encountered very often in physics; it occurs in electrostatics, in mechanics, in acoustics, etc. Therefore, we should not be astonished that it comes up in circuits. Roughly speaking, when reciprocity applies to a physical system, the input and the output can be interchanged without altering the response of the system to a given input waveform. It is extremely important not only in the analysis and design of systems but also in measurement techniques.

In electric circuits reciprocity applies to a subset of all *linear time-invariant* networks. Applicable networks may have resistors, inductors, coupled inductors, capacitors, and transformers; however, *gyrators*,† *dependent sources, and independent sources* are ruled out. We shall use the symbol $\mathfrak{N}_R$ to designate networks satisfying these conditions (the subscript $R$ stands for reciprocity).

As an example, consider a telephone link between two points $A$ and $B$. Suppose that the circuit includes only elements from the allowed list. Note that since resistors are included in the allowed list, the circuit may include negative-resistance amplifiers. On the basis of only this information, the reciprocity theorem allows us to conclude that the transmission from $A$ to $B$ is identical with the transmission from $B$ to $A$. It is obvious that this fact greatly simplifies the design and the testing of the telephone link.

Let $\mathfrak{N}_R$ be any network made of elements from the allowed list. The reciprocity theorem concerns the zero-state response of $\mathfrak{N}_R$ to either an independent current source or an independent voltage source. This theorem allows considerable freedom in the way the independent source is applied and in the way the response is measured. A simple way to visualize this freedom is the following: let us connect to $\mathfrak{N}_R$ two pairs of wires. The first pair of wires will give us terminals ① and ①′, and the second pair will give us terminals ② and ②′. We may choose to connect either or both pairs of wires to existing nodes of $\mathfrak{N}_R$. This connection is called, for obvious reasons, a soldering-iron entry. It is illustrated in Fig. 4.1a. We may also choose to connect either or both pairs of wires as follows: we cut the lead of a branch and solder the wires to the terminals created by the cut. This connection is called, for obvious reasons, a pliers entry. It is illustrated in Fig. 4.1b. In stating the theorem, we shall connect a source

---

† The gyrator is a network element that will be defined in Example 5 below.

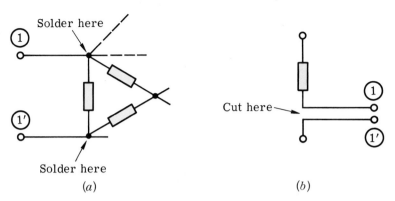

Solder here

Solder here

Cut here

(a)

(b)

**Fig. 4.1** (a) Illustration of a soldering-iron entry—the wires are connected to existing nodes of the network: (b) illustration of a pliers entry—the lead is cut, and the wires are soldered to the open leads thus created.

to ① ①′ and measure either the open-circuit voltage across ② ②′ or the current through a short circuit connected to ② ②′.

**Theorem, Examples, and Remarks**

Since in the following we shall have to consider the current waveform in a wire under two different sets of inputs and since the same reference directions are used in both cases, we shall use, say, $j_2$ for the current response to the first set of inputs and $\hat{j}_2$ (to be read "$j_2$ hat") for the current response to the second set of inputs.   For voltages we use $v_1$ and $\hat{v}_1$, etc.   In short, the network variables associated with the second set of inputs are distinguished by the "hat" symbol ⌢.   We state the reciprocity theorem as follows.

**RECIPROCITY THEOREM**   Consider a linear time-invariant network $\mathfrak{N}_R$ which consists of resistors, inductors, coupled inductors, capacitors, and transformers only.   $\mathfrak{N}_R$ is in the zero state and is not degenerate.   Connect four wires to $\mathfrak{N}_R$ thus obtaining two pairs of terminals ① ①′ and ② ②′.

*Statement 1*   Connect a voltage source $e_0(\cdot)$ to terminals ① ①′ and observe the zero-state current response $j_2(\cdot)$ in a short circuit connected to ② ②′ (see Fig. 4.2a).   Next, connect the same voltage source $e_0(\cdot)$ to terminals ② ②′ and observe the zero-state current response $\hat{j}_1(\cdot)$ in a short circuit connected to ① ①′.   The reciprocity theorem asserts that *whatever the topology and the element values of the network $\mathfrak{N}_R$ and whatever the waveform $e_0(\cdot)$ of the source,*

$$j_2(t) = \hat{j}_1(t) \qquad \text{for all } t$$

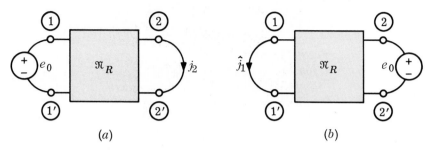

**Fig. 4.2**   Statement 1 of the reciprocity theorem asserts that the waveforms $j_2(\,\cdot\,)$ and $\hat{j}_1(\,\cdot\,)$ are equal; observe that the currents $j_2$ and $\hat{j}_1$ are short-circuit currents and note the reference directions.

*Statement 2*   Connect a current source $i_0$ to terminals ① ①' and observe the zero-state voltage response $v_2(\,\cdot\,)$ across the open-circuited terminals ② ②' (see Fig. 4.3$a$). Next, connect the same current source $i_0$ to terminals ② ②' and observe the zero-state voltage response $\hat{v}_1(\,\cdot\,)$ across the open-circuited terminals ① ①' (see Fig. 4.3$b$). The reciprocity theorem asserts that *whatever the topology and the element values of the network* $\mathfrak{N}_R$ *and whatever the waveform* $i_0(\,\cdot\,)$ *of the source,*

$$v_2(t) = \hat{v}_1(t) \quad \text{for all } t$$

*Statement 3*   Connect a current source $i_0$ to terminals ① ①' and observe the zero-state current response $j_2(\,\cdot\,)$ in a short circuit connected to ② ②' (see Fig. 4.4$a$). Next, connect a voltage source $e_0$ to terminals ② ②' and observe the zero-state response $\hat{v}_1(\,\cdot\,)$ across the open-circuited terminals ① ①' (see Fig. 4.4$b$). The reciprocity theorem asserts that *whatever the topology and the element values of the network* $\mathfrak{N}_R$, *and whatever the waveform of the source, if* $i_0(t)$ *and* $e_0(t)$ *are equal for all* $t$, *then*

$$\hat{v}_1(t) = j_2(t) \quad \text{for all } t$$

**Remarks**   1.   In Statement 1, we observe short-circuit currents. The assertion says

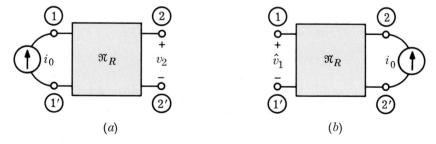

**Fig. 4.3**   Statement 2 of the reciprocity theorem asserts that the waveforms $v_2(\,\cdot\,)$ and $\hat{v}_1(\,\cdot\,)$ are equal; note that $v_2$ and $\hat{v}_1$ are open-circuit voltages and note also the reference directions.

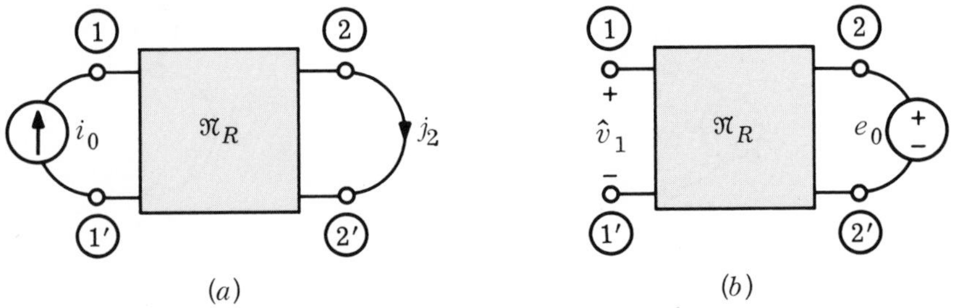

**Fig. 4.4**   Statement 3 of the reciprocity theorem asserts that if the source waveforms $i_0(\cdot)$ and $e_0(\cdot)$ are equal, then the zero-state responses $j_2(\cdot)$ and $\hat{v}_1(\cdot)$ are also equal; note the reference directions.

that if the voltage source $e_0$ is interchanged for a zero-impedance ammeter, the reading of the ammeter will not change.

2.   In Statement 2, we observe open-circuit voltages. The assertion says that if the current source $i_0$ is interchanged for an infinite-impedance voltmeter, the readings of the voltmeter will not change.

3.   In Statement 1, the source and the meter are "zero-impedance" devices. In Statement 2, the source and the meter are "infinite-impedance" devices. In Statement 3, for both measurements, there is an "infinite impedance" connected to ①①' and a "zero impedance" connected to ②②'.

*Reciprocity in terms of network functions*   Since the reciprocity theorem deals exclusively with the zero-state response (including the steady-state response as $t \to \infty$) of a linear time-invariant network, it is convenient to describe it in terms of network functions. The equivalent statements which correspond to those stated in the theorem in terms of network functions are given below.

*Statement 1*   Consider the two networks shown in Fig. 4.2. In Fig. 4.2a the input is a voltage source $e_0$ connected to terminal pair ①①', and the response is the short-circuit current $j_2$. Call $E_0(s)$ and $J_2(s)$ the Laplace transforms of $e_0$ and $j_2$, respectively. We define the transfer admittance from ①①' to ②②' as

$$y_{21}(s) \triangleq \frac{J_2(s)}{E_0(s)}$$

In Fig. 4.2b, the input is the same voltage source $e_0$ connected to terminal pair ②②', and the response is the short-circuit current $\hat{j}_1$. Using obvious notations, we define the transfer admittance from ②②' to ①①' as

$$y_{12}(s) \triangleq \frac{\hat{J}_1(s)}{E_0(s)}$$

The reciprocity theorem asserts that

$$y_{21}(s) = y_{12}(s) \qquad \text{for all } s$$

*Statement 2*   Consider the two networks shown in Fig. 4.3. In Fig. 4.3a the input is a current source $i_0$ applied to terminal pair ①①′, and the response is the open-circuit voltage $v_2$ across terminal ②②′. Let $I_0(s)$ and $V_2(s)$ be the Laplace transforms of $i_0$ and $v_2$, respectively. We define the transfer impedance from ①①′ to ②②′ as

$$z_{21}(s) \triangleq \frac{V_2(s)}{I_0(s)}$$

In Fig. 4.3b, the input is the same current source $i_0$ applied to terminal pair ②②′, and the response is the open-circuit voltage $\hat{v}_1$ across terminal pair ①①′. We define the transfer impedance from ②②′ to ①①′ as

$$z_{12}(s) \triangleq \frac{\hat{V}_1(s)}{I_0(s)}$$

The reciprocity theorem asserts that

$$z_{12}(s) = z_{21}(s) \qquad \text{for all } s$$

*Statement 3*   Consider the two networks in Fig. 4.4. In Fig. 4.4a the input is a current source $i_0$ applied to the terminal pair ①①′, and the response is the short-circuit current $j_2$. We define the transfer current ratio

$$H_I(s) \triangleq \frac{J_2(s)}{I_0(s)}$$

In Fig. 4.4b the input is a voltage source $e_0$ applied to branch 2, and the response is the voltage $\hat{v}_1$ across the node pair ①①′. We define the transfer voltage ratio as

$$H_V(s) \triangleq \frac{\hat{V}_1(s)}{E_0(s)}$$

The reciprocity theorem asserts that

$$H_I(s) = H_V(s) \qquad \text{for all } s$$

---

**Example 1**   The purpose of this example is to illustrate the full meaning of the reciprocity theorem and, in particular, that it applies to dc conditions, sinusoidal steady state, and transients. We consider the network $\mathfrak{N}_R$ shown in Fig. 4.5.

*Statement 1*   Terminal pairs ①①′ and ②②′ are obtained by performing pliers entries in the resistive branches with resistances of 5 and 1 ohm, respec-

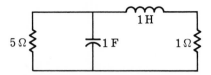

**Fig. 4.5**   Network $\mathfrak{N}_R$ used in Example 1.

tively.   Let us apply a *constant* voltage source of 1 volt at ① ①' and measure the current $j_2$ in the short circuit joining ② ②' (see Fig. 4.6*a*). Suppose that we wish only to investigate dc conditions.   Since, at direct current, the capacitor is an open circuit and the inductor is a short circuit, we obtain, by inspection, that $j_2$ is a constant current of ⅙ amp.   Next, as shown in Fig. 4.6*b*, we apply the same voltage source across terminals ② ②' and calculate the current $\hat{j}_1$ in the short circuit joining ① ①'.   We find $\hat{j}_1 = $ ⅙ amp, as required by Statement 1 of the theorem.

*Statement 2*   For the sake of variety, let us now pick soldering-iron entries.   The terminals ①, ①', ②, and ②' are shown in Fig. 4.7.   In terms of network

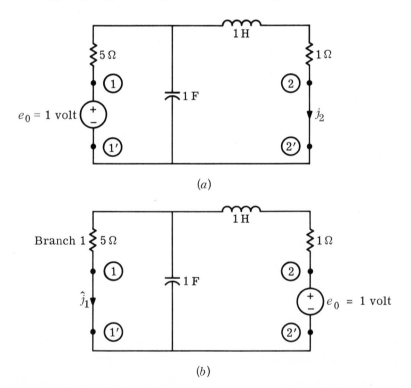

**Fig. 4.6**   Network $\mathfrak{N}_R$ is used to illustrate Statement 1 of the reciprocity theorem under dc conditions.

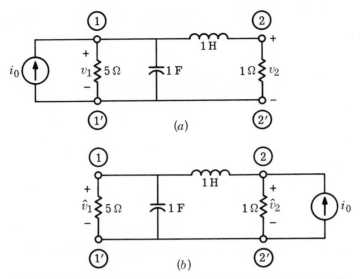

**Fig. 4.7**   Network $\mathfrak{N}_R$ is used to illustrate Statement 2 of the reciprocity theorem under sinusoidal steady-state conditions.

functions, Statement 2 states that the transfer impedance from node pair ①①' to ②②' is equal to the transfer impedance from node pair ②②' to ①①'. This means in particular that the corresponding sinusoidal steady-state responses to the same sinusoidal current source will be equal. Let us verify this fact for the network shown in Fig. 4.5.

First, as shown in Fig. 4.7a, we apply a sinusoidal current source $i_0(t) = 2 \cos (2t + \pi/6)$ at node pair ①①' and calculate the open-circuit voltage at node pair ②②'. Using phasor analysis, we write the node equations as follows:

$$
\begin{bmatrix} 0.2 + j2 + \dfrac{1}{j2} & -\dfrac{1}{j2} \\[2mm] -\dfrac{1}{j2} & \dfrac{1}{j2} + 1 \end{bmatrix} \begin{bmatrix} V_1 \\[2mm] V_2 \end{bmatrix} = \begin{bmatrix} 2\epsilon^{j(\pi/6)} \\[2mm] 0 \end{bmatrix}
$$

We obtain successively

$$
V_2 = \frac{-j0.5(2\epsilon^{j(\pi/6)})}{(0.2 + j1.5)(1 - j0.5) - (j0.5)^2} = \frac{\epsilon^{-j(\pi/3)}}{0.95 + j1.15}
$$

$$
= 0.685\epsilon^{-j9.6°}
$$

Thus, the sinusoidal steady-state response is

$$
v_2(t) = 0.685 \cos (2t - 9.6°)
$$

For the second experiment we apply the same current source to node pair ②②′ and calculate the open-circuit voltage at node pair ①①′ (see Fig. 4.7b). The node equations are

$$\begin{bmatrix} 0.2 + j2 + \dfrac{1}{j2} & -\dfrac{1}{j2} \\[2ex] -\dfrac{1}{j2} & 1 + \dfrac{1}{j2} \end{bmatrix} \begin{bmatrix} \hat{V}_1 \\[2ex] \hat{V}_2 \end{bmatrix} = \begin{bmatrix} 0 \\[2ex] 2\epsilon^{-(\pi/6)} \end{bmatrix}$$

Then we obtain

$$\hat{V}_1 = \frac{-j0.5(2\epsilon^{j(\pi/6)})}{(0.2 + j1.5)(1 - j0.5) - (j0.5)^2} = 0.685\epsilon^{-j9.6°}$$

as before.   The sinusoidal steady-state response is

$$\hat{v}_1(t) = 0.685 \cos{(2t - 9.6°)}$$

The waveforms $v_2(\cdot)$ and $\hat{v}_1(\cdot)$ are equal, as predicted by the theorem.

*Statement 3*   For variety, let us pick ①①′ to be defined by a soldering-iron entry at the terminals of the 5-ohm resistor, and pick ②②′ to be defined as a pliers entry in the 1-ohm branch (see Fig. 4.8).   Let the source waveform be a unit impulse applied at $t = 0$.   To calculate the required zero-state responses, we use Laplace transform.

First, as shown in Fig. 4.8a, we apply a current source at node pair ①①′ and calculate the current $j_2$ in the short circuit joining ②②′.   Since branch 2 is a 1-ohm resistor, $j_2(t) = v_2(t)$, so we need only calculate $v_2$. By node analysis and Laplace transform, we obtain

$$\begin{bmatrix} 0.2 + s + \dfrac{1}{s} & -\dfrac{1}{s} \\[2ex] -\dfrac{1}{s} & 1 + \dfrac{1}{s} \end{bmatrix} \begin{bmatrix} V_1(s) \\[2ex] V_2(s) \end{bmatrix} = \begin{bmatrix} 1 \\[2ex] 0 \end{bmatrix}$$

Hence

$$V_2(s) = \frac{1/s}{(0.2 + s + 1/s)(1 + 1/s) - (1/s)^2} = \frac{1}{s^2 + 1.2s + 1.2}$$

Taking the inverse Laplace transform and noting that $j_2(t) = v_2(t)$, we obtain

$$j_2(t) = \mathcal{L}^{-1}\left[\frac{1}{s^2 + 1.2s + 1.2}\right] = \mathcal{L}^{-1}\left[\frac{1}{(s + 0.6)^2 + (0.916)^2}\right]$$

$$= \frac{1}{0.916}\epsilon^{-0.6t} \sin 0.916t$$

where we used Table 13.1.   Our conclusion is that

$$j_2(t) = 1.09\epsilon^{-0.6t} \sin 0.916t \qquad \text{for } t \geq 0$$

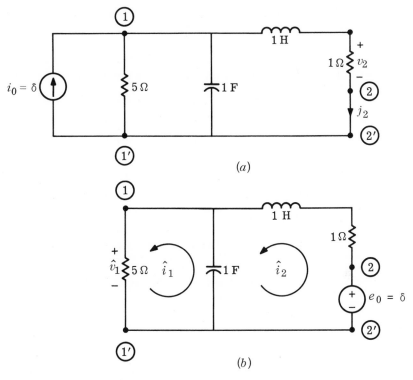

**Fig. 4.8**   Network $\mathfrak{N}_R$ is used to illustrate Statement 3 of the reciprocity theorem under transient conditions.

For the second experiment, as shown in Fig. 4.8b, we apply a voltage source $e_0$ with the *same* waveform (hence $e_0 = \delta$) across ②②′, and we calculate the open-circuit voltage across node pair ①①′. Using mesh analysis, we obtain

$$
\begin{bmatrix} 5 + \dfrac{1}{s} & -\dfrac{1}{s} \\[2mm] -\dfrac{1}{s} & 1 + s + \dfrac{1}{s} \end{bmatrix}
\begin{bmatrix} \widehat{I}_1(s) \\[2mm] \widehat{I}_2(s) \end{bmatrix}
= \begin{bmatrix} 0 \\[2mm] 1 \end{bmatrix}
$$

and

$$
\widehat{I}_1(s) = \frac{1/s}{(5 + 1/s)(s + 1 + 1/s) - (1/s)^2}
$$

Since $\widehat{v}_1(t) = 5\widehat{i}_1(t)$, we have

$$
\widehat{V}_1(s) = \frac{5}{(5s + 1)(s + 1 + 1/s) - 1/s} = \frac{5}{5s^2 + 6s + 6}
$$

$$
= \frac{1}{s^2 + 1.2s + 1.2}
$$

Recognizing this function of $s$ to be the transform of $j_2(t)$, we use previous calculations and conclude that

$$\hat{v}_1(t) = 1.09\epsilon^{-0.6t} \sin 0.916t \qquad \text{for } t \geq 0$$

Thus, the two responses are equal, as required by the theorem.

**Remark**   It is important to observe that the class of *linear* networks allowed in the reciprocity theorem is much more restricted than the corresponding class allowed in the superposition and Thévenin-Norton theorems.   Indeed, *no sources* (either dependent or independent), *no time-varying elements,* and *no gyrators* are allowed for the reciprocity theorem.   Let us show by counter examples that these restrictions are necessary.

**Example 2**   Consider the circuit shown in Fig. 4.9 with the notations defined on the figure.   The current source $i_0$ is applied across node pair ①①′, and the open-circuit voltage $v_2$ is observed across node pair ②②′.   For $i_0(t) = Iu(t)$, the zero-state response is

$$v_2(t) = -R_1 g_m R_2 I (1 - \epsilon^{-t/R_2 C_2}) \qquad \text{for } t \geq 0$$

If we interchange the source and the response, i.e., if we apply $i_0$ across

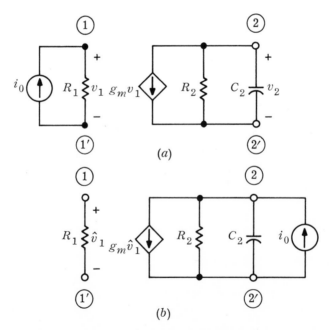

**Fig. 4.9**   Example 2: circuit containing a dependent source.

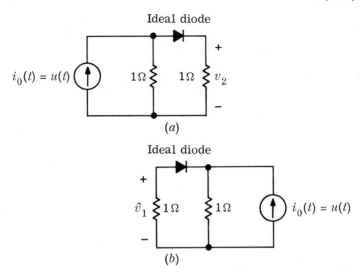

Fig. 4.10   Example 3: circuit with a nonlinear element.

node pair $\textcircled{2}\,\textcircled{2'}$, and observe the open-circuit voltage across node pair $\textcircled{1}\,\textcircled{1'}$, we see that

$$\hat{v}_1(t) = 0 \qquad \text{for } t \geq 0$$

This example shows that when dependent sources are present, reciprocity does not hold, in general.

**Example 3**   Consider the circuit shown in Fig. 4.10, which includes a nonlinear element. Clearly, in Fig. 4.10a we have $v_2(t) = 0.5u(t)$. If we interchange the input and the response as shown in Fig. 4.10b, we get $\hat{v}_1(t) = 0$. Thus, when nonlinear elements are present, reciprocity does not hold, in general.

**Example 4**   We shall discuss heuristically the case of a linear circuit which includes a time-varying resistor, as shown in Fig. 4.11a. We shall show that reciprocity does not hold. Consider only the steady-state response of the circuit to a sinusoidal current input $i_0(t) = \cos 10t$. Note that the time-varying resistor is in parallel with the current source and its resistance varies sinusoidally; that is, $R(t) = 1 + 0.1 \cos t$. The angular frequency of the source is 10 rad/sec, and that of the resistor is 1 rad/sec. The effect of this time variation is to create a steady-state voltage response $v_1$ which contains both the sum and the difference of the two frequencies;† that is, $\omega_1 = 10 + 1 = 11$ rad/sec, and $\omega_2 = 10 - 1 = 9$ rad/sec. In the circuit, we see that there are two lossless tuned circuits tuned at $\omega_1 = 11$ rad/sec

---

† Recall that $2 \cos \omega_1 t \cos \omega_2 t = \cos(\omega_1 + \omega_2)t + \cos(\omega_1 - \omega_2)t$.

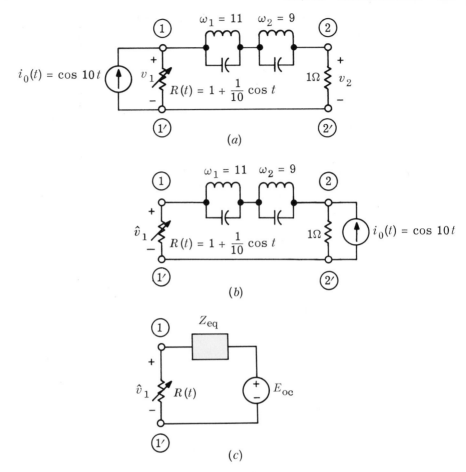

**Fig. 4.11**   Example 4: circuit with a linear time-varying resistor.

and $\omega_2 = 9$ rad/sec, respectively.   Thus, at these two frequencies the impedance faced by the parallel combination of the source and the time-varying resistor is infinite.   Clearly, the voltage response $v_2$ across the output 1-ohm resistor does not contain sinusoids at these two frequencies.

Next let us interchange the input and the response, as shown in Fig. 4.11b.   The input is now applied across the 1-ohm resistor and the response is taken as the voltage $\hat{v}_1$ across the time-varying resistor.   Let us take advantage of the Thévenin equivalent network theorem and consider the time-varying resistor as the load, as shown in Fig. 4.11c.   Obviously, the equivalent source $E_{\text{oc}}$ has a steady-state component at $\omega_s = 10$ rad/sec, and $Z_{\text{eq}}$ is finite at $\omega_s = 10$ rad/sec.   This equivalent source combined with the effect of the time-varying resistor will produce a voltage $\hat{v}_1$ which contains sinusoids at $\omega_1 = 11$ rad/sec and $\omega_2 = 9$ rad/sec.   Thus, $\hat{v}_1$ cannot be identical to $v_2$ of the previous case.   This example shows that when time-varying elements are present, reciprocity does not hold, in general.

**Remark**    Any network which satisfies the reciprocity theorem is called a **reciprocal network.**  Using this concept, we see that the reciprocity theorem guarantees that any network made of linear time-invariant resistors, capacitors, inductors, coupled inductors, and transformers is a reciprocal network. It is also a fact that *some* linear time-invariant networks that contain dependent sources are reciprocal, whereas others are not.

**Example 5**    Some texts suggest that all networks made of *passive* linear time-invariant elements are reciprocal.  This is false.  To show this, we introduce a new two-port element called a **gyrator.**  By definition, a gyrator is a two-port element (shown in Fig. 4.12) described by the equations

$$v_1(t) = \alpha i_2(t)$$

$$v_2(t) = -\alpha i_1(t)$$

or, in matrix form,

$$\begin{bmatrix} v_1(t) \\ v_2(t) \end{bmatrix} = \begin{bmatrix} 0 & \alpha \\ -\alpha & 0 \end{bmatrix} \begin{bmatrix} i_1(t) \\ i_2(t) \end{bmatrix}$$

In both equations, the number $\alpha$ is a constant and is called the **gyration ratio.**  From these equations it follows that the gyrator is a *linear time-invariant* element.  It is linear because its branch voltages $v_1$ and $v_2$ are given by linear functions of its branch currents.  Furthermore, from the defining equations, it follows that the power delivered by the external world to the gyrator is, for all $t$,

$$v_1(t)i_1(t) + v_2(t)i_2(t) = 0$$

Thus, the gyrator *neither absorbs nor delivers energy to the outside world.* It is worth noting that the ideal transformer also has these three properties.

However, in contrast to the ideal transformer, the gyrator as a network element does not obey the reciprocity theorem.  Hence, we say that

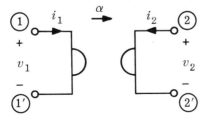

**Fig. 4.12**    Symbolic representation of the gyrator; the constant $\alpha$ is called the gyration ratio.

*the gyrator is not reciprocal.* We check this fact as follows: we apply Statement 2 of the reciprocity theorem to the terminals ①, ①', ②, and ②' shown in Fig. 4.12. With $i_1 = 1$ amp and $i_2 = 0$ (a 1-amp current source at ①①' and an open circuit at ②②'), we have $v_2 = -\alpha$ volts; with $i_2 = 1$ amp and $i_1 = 0$, we have $v_1 = \alpha$ volts. If the gyrator were reciprocal, we would have $v_1 = v_2$. In conclusion, as far as linear time-invariant networks are concerned, passivity of every network element is not sufficient to guarantee that the network is reciprocal.

**Exercise**   There has been a proposal to replace inductors by gyrators terminated by capacitors. To understand it, suppose that the gyrator shown in Fig. 4.12 is loaded by a linear time-invariant capacitor $C$. Using the current and voltage reference directions defined in Fig. 4.12, we have $i_2(t) = -C(dv_2/dt)$. Show that the equations relating $v_1(t)$ to $i_1(t)$ imply that the loaded gyrator looks like a linear time-invariant inductor. What is the value of its inductance?

## 4.2   Proof of the Reciprocity Theorem

We shall use Tellegen's theorem to prove the three statements of the reciprocity theorem. Observe that each of the three statements asserts that two waveforms are equal. The first waveform is observed when the network is in the first condition, that is, the independent source is connected to ①①', and the observation is performed at ②②'. The second waveform is observed when the network is in the second condition, that is, the independent source is at ②②', and the observation is performed at ①①'. Let us call $\alpha$ and $\beta$ the branches joining ①①' and ②②'. Thus, in the first condition, $\alpha$ is an independent source, and $\beta$ is either an open circuit or a short circuit. In the second condition, $\alpha$ is either an open circuit or a short circuit, and $\beta$ is an independent source. Altogether there will be $b + 2$ branches, where $b$ denotes the number of branches of the given network $\mathfrak{N}_R$. For the first condition, let us denote the Laplace transforms of the branch voltages by $V_\alpha(s)$, $V_\beta(s)$, $V_1(s)$, $V_2(s)$, ..., $V_b(s)$, and those of the branch currents by $J_\alpha(s)$, $J_\beta(s)$, $J_1(s)$, $J_2(s)$, ..., $J_b(s)$. Similarly, for the second condition, the Laplace transforms are, for the branch voltages, $\hat{V}_\alpha(s)$, $\hat{V}_\beta(s)$, $\hat{V}_1(s)$, $\hat{V}_2(s)$, ..., $\hat{V}_b(s)$, and for the branch currents, $\hat{J}_\alpha(s)$, $\hat{J}_\beta(s)$, $\hat{J}_1(s)$, $\hat{J}_2(s)$, ..., $\hat{J}_b(s)$. Since, in each condition, the Laplace transforms of the branch voltages and the branch currents satisfy Kirchhoff's laws, Tellegen's theorem applies.† In fact, it asserts that

---

† Tellegen's theorem was stated in Chap. 9 in terms of instantaneous voltages and currents. Obviously, it also holds for the Laplace transforms of the voltages and currents. Indeed, Tellegen's theorem follows directly from Kirchhoff's laws, and the Laplace transforms of the branch voltages and branch currents also satisfy Kirchhoff's laws.

(4.1)   $V_\alpha(s)\widehat{J}_\alpha(s) + V_\beta(s)\widehat{J}_\beta(s) + \sum_{k=1}^{b} V_k(s)\widehat{J}_k(s) = 0$

and

(4.2)   $\widehat{V}_\alpha(s)J_\alpha(s) + \widehat{V}_\beta(s)J_\beta(s) + \sum_{k=1}^{b} \widehat{V}_k(s)J_k(s) = 0$

Note that for all branches we use associated reference directions.   Let us consider the terms under the summation signs.   They are products of voltages and currents of branches of the given network $\mathfrak{N}_R$.   By assumption, these branches are resistors, capacitors, inductors, coupled inductors, and ideal transformers.   If branch $k$ is an inductor, capacitor, or resistor, we have

(4.3)   $V_k(s)\widehat{J}_k(s) = Z_k(s)J_k(s)\widehat{J}_k(s) = J_k(s)[Z_k(s)\widehat{J}_k(s)] = \widehat{V}_k(s)J_k(s)$

where $Z_k(s)$ is the impedance of branch $k$.   If branches $m$ and $n$ are two coupled inductors or are the branches of an ideal transformer, then it is easy to show, by means of the branch equations of two coupled inductors or those of an ideal transformer, that

(4.4)   $V_m(s)\widehat{J}_m(s) + V_n(s)\widehat{J}_n(s) = \widehat{V}_m(s)J_m(s) + \widehat{V}_n(s)J_n(s)$

From Eqs. (4.3) and (4.4) we conclude that the indicated sums in Eqs. (4.1) and (4.2) are equal; therefore,

(4.5)   $V_\alpha(s)\widehat{J}_\alpha(s) + V_\beta(s)\widehat{J}_\beta(s) = \widehat{V}_\alpha(s)J_\alpha(s) + \widehat{V}_\beta(s)J_\beta(s)$

This equation is very useful.   Indeed, the three statements of the reciprocity theorem follow directly from it.

*Statement 1*   The two conditions corresponding to Statement 1 are shown in Fig. 4.13. In the first condition, $V_\alpha(s) = E_0(s)$, and $V_\beta(s) = 0$.   In the second condition, $\widehat{V}_\beta(s) = E_0(s)$, and $\widehat{V}_\alpha(s) = 0$.   Thus, from (4.5) we obtain

$\widehat{J}_\alpha(s) = J_\beta(s)$

or, in terms of network functions,

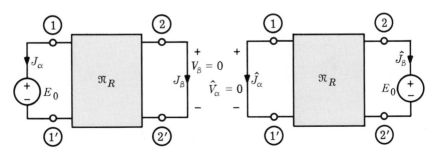

**Fig. 4.13**   Conditions corresponding to Statement 1 of the reciprocity theorem.

$$\frac{\widehat{J}_\alpha(s)}{E_0(s)} = \frac{J_\beta(s)}{E_0(s)}$$

that is,

$$y_{12}(s) = y_{21}(s)$$

*Statement 2*   We refer to the conditions shown in Fig. 4.14.   In the first condition, $J_\alpha(s) = -I_0(s)$, and $J_\beta(s) = 0$.   In the second condition $\widehat{J}_\beta(s) = -I_0(s)$, and $\widehat{J}_\alpha(s) = 0$.   Thus, from (4.5) we obtain

$$\widehat{V}_\alpha(s) = V_\beta(s)$$

or, in terms of network functions,

$$\frac{\widehat{V}_\alpha(s)}{I_0(s)} = \frac{V_\beta(s)}{I_0(s)}$$

that is,

$$z_{12}(s) = z_{21}(s)$$

*Statement 3*   We refer to the conditions shown in Fig. 4.15.   In the first condition, $J_\alpha(s) = -I_0(s)$, and $V_\beta(s) = 0$.   In the second, $\widehat{V}_\beta(s) = E(s)$, and $\widehat{J}_\alpha(s) = 0$.   Thus, from (4.5)

$$\frac{\widehat{V}_\alpha(s)}{E_0(s)} = \frac{J_\beta(s)}{I_0(s)}$$

Consequently, if the source waveforms are the same, that is, $E_0(s) = I_0(s)$, then the output waveforms are equal; that is,

$$\widehat{V}_\alpha(s) = J_\beta(s)$$

Interpreting the previous equation in terms of the network functions, we conclude that

$$H_V(s) = H_I(s)$$

This concludes the proof of the reciprocity theorem.

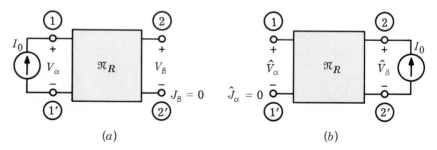

**Fig. 4.14**   Conditions corresponding to Statement 2 of the reciprocity theorem.

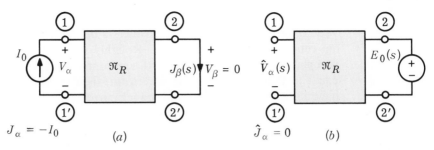

**Fig. 4.15**    Conditions corresponding to Statement 3 of the reciprocity theorem.

**Remark**    If we examine the proof of the reciprocity theorem, we note that the nature of the branches entered the proof only when we wrote

$$V_k(s) = Z_k(s)J_k(s) \qquad k = 1, 2, \ldots, b$$

Therefore, the reciprocity theorem applies to any interconnection of "black box" one-ports whose port voltage and port current are related by an equation of the form $V_k(s) = Z_k(s)J_k(s)$. The nature of the elements inside the black boxes is irrelevant.

**Exercise 1**    Let $\mathbf{V}(s) \triangleq [V_1(s), V_2(s), \ldots, V_b(s)]^T$, and $\mathbf{J}(s) \triangleq [J_1(s), J_2(s), \ldots, J_b(s)]^T$. Define $\hat{\mathbf{V}}(s)$ and $\hat{\mathbf{J}}(s)$ similarly. Let $\mathbf{Z}_b(s)$ be the branch impedance matrix of the network $\mathfrak{N}_R$. Note that

$$\sum_{k=1}^{b} V_k(s)\hat{J}_k(s) = \mathbf{V}^T(s)\hat{\mathbf{J}}(s) = [\mathbf{Z}_b(s)\mathbf{J}(s)]^T\hat{\mathbf{J}}(s) = \mathbf{J}^T(s)\mathbf{Z}_b{}^T(s)\hat{\mathbf{J}}(s)$$

Show that the reciprocity theorem would still apply to $\mathfrak{N}_R$ if its branch impedance matrix $\mathbf{Z}_b(s)$ is a symmetric matrix.

**Exercise 2**    Show that the reciprocity theorem would still apply to $\mathfrak{N}_R$ if it has *time-varying* resistors only. Hint: Follow the steps of the proof above, but start with

$$v_\alpha(t)\hat{j}_\alpha(t) + v_\beta(t)\hat{j}_\beta(t) + \sum_{k=1}^{b} v_k(t)\hat{j}_k(t) = 0$$

and note that $v_k(t) = R_k(t)j_k(t)$.

## Summary

- The four network theorems that have been presented are very important because of their great usefulness. To help you keep their main features in mind, let us describe their main conclusions.

- The *substitution theorem* applies to *any* network, linear or nonlinear, time-varying or time-invariant. It considers a branch which is not coupled to any other branch of the network. Roughly, it says that if any such branch is replaced by an independent voltage source with the source waveform equal to that branch-voltage waveform, then the resulting network will have all its branch voltages and branch currents equal to those of the given network. The same conclusion holds if the branch is replaced by a current source whose source waveform is equal to that branch-current waveform. There is one condition which must be satisfied (and this condition is particularly important in the case of nonlinear networks): both networks must have a *unique* solution. If this is not the case, the only conclusion possible is that one of the solutions of the modified network is equal to one of the solutions of the given network; the \$64,000 question then is, which is which?

- The *superposition theorem* applies to any *linear* network, time-invariant or time-varying. It states that if a linear network is driven simultaneously by a number of independent sources (called inputs), then its *zero-state response* to these simultaneous inputs is equal to the sum of the zero-state responses due to each input acting alone on the network. Let us note that if in a particular problem the network does not start from the zero state, one can always replace these initial conditions by appropriate independent sources.

- The *Thévenin-Norton theorem* applies to any *linear* network $\mathfrak{N}$, time-invariant or time-varying. It says that if any *linear* network is connected by two of its terminals to an *arbitrary* load, then the voltage across the load and the current into it are not changed if either of the following replacements are made:
  1. (Thévenin form) $\mathfrak{N}$ is replaced by the series connection of a voltage source $e_{oc}(\cdot)$ and a one-port $\mathfrak{N}_0$; $e_{oc}(\cdot)$ is the open-circuit voltage of $\mathfrak{N}$ (for the reference direction see Fig. 3.2b), and $\mathfrak{N}_0$ is obtained from $\mathfrak{N}$ by setting all its *independent* sources and all its *initial conditions* to zero (the *dependent* sources are left undisturbed!).
  2. (Norton form) $\mathfrak{N}$ is replaced by the parallel connection of a current source $i_{sc}(\cdot)$ and a one-port $\mathfrak{N}_0$; $i_{sc}(\cdot)$ is the short-circuit current of $\mathfrak{N}$ (for reference direction see Fig. 3.3b), and $\mathfrak{N}_0$ is obtained from $\mathfrak{N}$ by setting all its *independent* sources and all its *initial conditions* to zero.

- The *reciprocity theorem* applies to any *linear time-invariant* network made of resistors, inductors (including mutual coupling), capacitors, and transformers. Let ①①' and ②②' be terminal pairs obtained from $\mathfrak{N}_R$ either by pliers entry or by soldering-iron entry.

The reciprocity theorem can be stated in three ways.

1.  Let a voltage source $e_0(\cdot)$ be connected to ①①', and let the zero-state current response $j_2(\cdot)$ be observed in a short circuit connected to ②②'. Let the *same* voltage source $e_0(\cdot)$ be connected to ②②', and let the zero-state current response $\widehat{j}_1(\cdot)$ be observed in a short circuit connected to ① ①' (for reference directions, refer to Fig. 4.2). Then, $j_2(t) = \widehat{j}_1(t)$ for all $t$.

2.  Let a current source $i_0(\cdot)$ be connected to ①①', and let the zero-state voltage response $v_2(\cdot)$ be observed across ②②'. Let the *same* current source $i_0(\cdot)$ be connected to ②②', and let the zero-state voltage response $\widehat{v}_1(\cdot)$ be observed across ①①' (for reference directions refer to Fig. 4.3). Then $v_2(t) = \widehat{v}_1(t)$ for all $t$.

3.  Let a current source $i_0(\cdot)$ be connected to ①①', and let the zero-state current $j_2(\cdot)$ be observed in a short circuit connected to ②②'. Let a voltage source $e_0(\cdot)$ be connected to ②②', and let the zero-state voltage $\widehat{v}_1(\cdot)$ be observed across ①①' (for reference directions refer to Fig. 4.4). Under these conditions, if the waveforms $i_0(\cdot)$ and $e_0(\cdot)$ are equal, then $j_2(t) = \widehat{v}_1(t)$ for all $t$.

▪  A network that obeys the reciprocity theorem is called *reciprocal*. Passivity has nothing to do with reciprocity. There are linear time-invariant passive networks that are not reciprocal. Any linear time-invariant *RLCM* network (some elements of which may be active) is reciprocal. There are special linear time-invariant networks which include dependent sources which are reciprocal.

▪  A gyrator is a linear, time-invariant two-port element that neither absorbs nor delivers energy. The gyrator is an example of a passive element that is not reciprocal.

## Problems

Zero-state response and superposition

**1.** Two sets of observations are made on a lumped linear time-invariant network $\mathfrak{N}$, with terminal pairs $AB$, $CD$, and $EF$ as indicated in Fig. P16.1. In both instances, $\mathfrak{N}$ is in the zero state at time $t = 0$. From the waveforms shown in the figure, sketch to scale the voltage $v_3$ appearing at the terminal pair $EF$ when the sources $e_1$ and $e_2$ are applied simultaneously (as shown in the figure) where $e_1(t) = u(t) - u(t - 2)$, and $e_2(t) = u(t - 1) \sin [(\pi/2)(t - 1)]$.

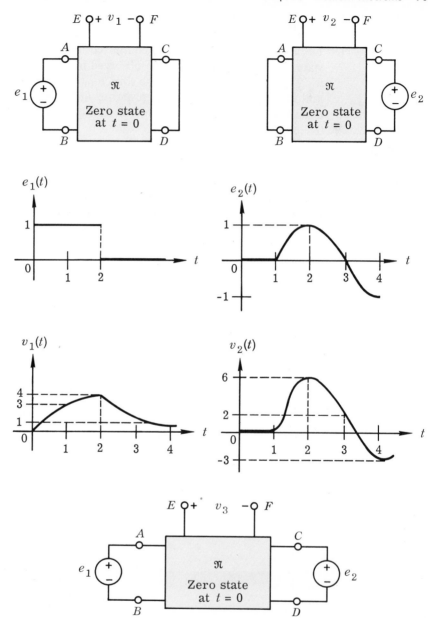

**Fig. P16.1**

Steady-state response and superposition

**2.** For the circuit shown in Fig. P16.2, find the steady-state voltage $v$ across the capacitor.   Express your answer as a real-valued function of time.

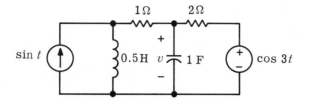

**Fig. P16.2**

Nonlinear
network

**3.** In the network shown in Fig. P16.3, $i_L(0) = v_{C_1}(0) = v_{C_2}(0) = 0, i_1(t) = u(t)\epsilon^{-t}$, and $e_1(t) = 3u(t)$ (note that there is an ideal diode in series with the current source). Find $v(t)$ for $t \geq 0$. Can you use superposition in this case? If so, why? The elements of the circuit have the values $L = \frac{5}{17}$ henry, $C_1 = \frac{17}{3}$ farads, $C_2 = 1$ farad, $R = R_2 = 1$ ohm, and $R_1 = 15$ ohms.

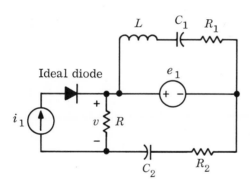

**Fig. P16.3**

Operational
amplifier

**4.** The operational amplifier given in Fig. P16.4 is a multiple-input $(e_1, e_2, \ldots, e_n)$, single-output $(e_o)$ circuit. The amplifier $A$ is characterized by the relation $e_o = -Ke_g$, where $K$ is a constant. Assume $i_A(t) = 0$ for all $t$.

  a.  Use the superposition theorem to find $E_o(s)$, the Laplace transform of the output in terms of $E_1(s), E_2(s), \ldots, E_n(s)$ and $R_1, R_2, \ldots, R_n, R_f$ and $K$, the amplifier gain.

  b.  Simplify the expression obtained in (a) for the case in which $K$ is very large.

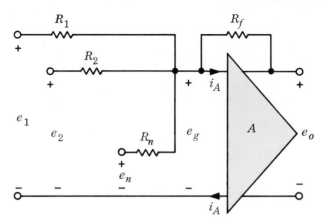

**Fig. P16.4**

Thévenin-
Norton
**5.** The network shown in Fig. P16.5 is in the sinusoidal steady state, $e_s(t) = 9\cos 10t$, and $i_s(t) = 2\cos(10t - \pi/3)$. For the network on the left of the terminals $AB$, obtain

a.   The Thévenin equivalent network.

b.   The Norton equivalent network.

c.   Calculate $v$ for $R = 1$ ohm and for $R = 10$ ohms (express your answer as a real-valued function of time).

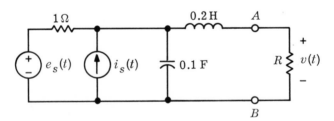

**Fig. P16.5**

Thévenin-
Norton
**6.** In the network shown in Fig. P16.6, the switch $S$ is closed at $t = 0$, a steady state having previously existed. Find $v(t)$ for $t \geq 0$, using

a.   Thévenin's theorem.

b.   Norton's theorem.

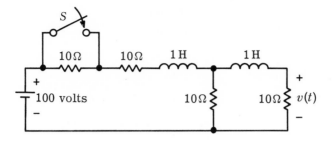

**Fig. P16.6**

**7.** The network shown in Fig. P16.7 is the equivalent circuit of a common-emitter transistor amplifier driving a nonlinear resistive load.   Find

*a.*   The Thévenin equivalent network of the amplifier.

*b.*   The output voltage $v$.

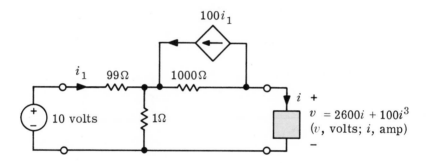

**Fig. P16.7**

**8.** Consider the small-signal equivalent network of a common-emitter transistor shown in Fig. P16.8, where $i_s(t) = 10^{-5} \cos \omega t$ amp and

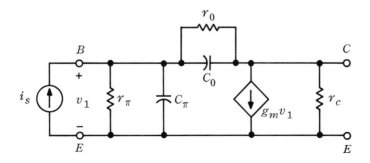

**Fig. P16.8**

$$r_\pi = 10^3 \text{ ohm} \qquad g_m = 10^{-1} \text{ mho}$$
$$C_\pi = 10^{-10} \text{ farad} \qquad r_c = 10^4 \text{ ohms}$$
$$r_0 = 10^6 \text{ ohms} \qquad \omega = 10^7 \text{ rad/sec}$$
$$C_0 = 10^{-11} \text{ farad}$$

Assuming a sinusoidal steady state exists, find

a.  The Thévenin equivalent of the network which lies to the left of terminals $C$ and $E$.

b.  The Norton equivalent.

Thévenin-  **9.** The linear time-invariant circuit shown in Fig. P16.9 corresponds to a
Norton  simplified equivalent network for a transistor. Obtain the Thévenin and
Norton equivalent circuits with respect to terminals ②②′, given $R_1 = 500$ ohms, $R_2 = 10$ kilohms, $\gamma = 10^{-4}$, and $\beta = 20$.

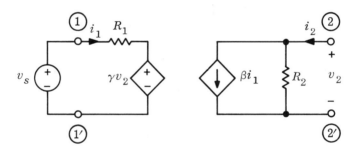

**Fig. P16.9**

Norton  **10.** The network shown in Fig. P16.10 is in the zero state prior to the closing of switch $S_1$ at $t = 0$. Using the Norton equivalent network for the

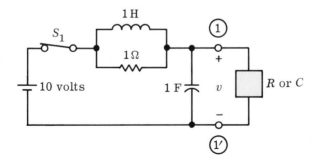

**Fig. P16.10**

circuit to the left of terminals ① and ①', calculate the voltage $v(t)$ across the terminals ① and ①' when the element in the box is

*a.* A resistor $R = 1$ ohm only.

*b.* A capacitor $C = 1$ farad only.

Thévenin **11.** The linear time-invariant circuit shown in Fig. P16.11 is in the steady state. In order to determine the steady-state inductor current $i$ by means of Thévenin's theorem,

*a.* Determine the open-circuit voltage $v_{oc}$ at terminals ① and ①' when the inductor is open-circuited.

*b.* Determine $Z_{eq}$, the equivalent impedance which is faced by the inductor.

*c.* Determine the steady-state current $i$.

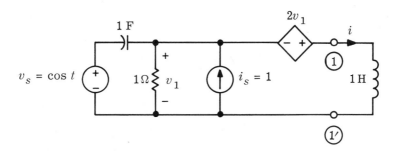

**Fig. P16.11**

Thévenin **12.** The network $\mathfrak{N}$ shown in Fig. P16.12 is made of linear time-invariant resistors and batteries, where $e_{oc} = 2$ volts. When terminated by a resistor of 10 ohms, the voltage $v$ at its terminals drops to 1 volt. Find the Thévenin equivalent circuit for $\mathfrak{N}$.

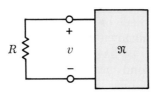

**Fig. P16.12**

Norton **13.** The network $\mathfrak{N}$ is made of linear time-invariant elements and sinusoidal sources at a frequency of 60 Hz. All measurements reported are sinusoidal steady-state measurements. The current under short-circuit conditions is $10\angle 0°$ amp. When terminated by an impedance of $20 + j5$ ohms (at 60 Hz), the terminal voltage is $50\angle -10°$ volts. Find the Norton equivalent circuit of $\mathfrak{N}$.

Nonlinear network
**14.** Consider the network shown in Fig. P16.14. All the elements of the network are linear and time-invariant except for the 20-volt battery and the nonlinear device $\mathfrak{D}$ which is modeled as a time-invariant resistor whose characteristic is given in the figure. Determine all the possible steady-state values of $(v, i)$.

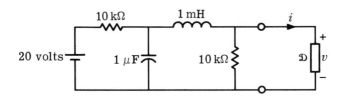

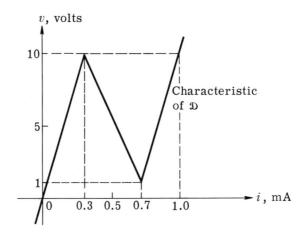

**Fig. P16.14**

Reciprocity and superposition
**15.** Two sets of measurements are made on a resistive network consisting of one known resistor and four unknown resistors, as indicated in Fig. P16.15. In the first measurement we have $i_1 = 0.6i_s$ and $i_1' = 0.3i_s$, as shown in Fig. P16.15a; in the second measurement, we have $i_2 = 0.2i_s$ and $i_2' = 0.5i_s$, as shown in Fig. P16.15b.

   *a.* Use the reciprocity theorem to calculate $R_1$.

   *b.* Consider the configuration of sources shown in Fig. P16.15c, where $k$ has been adjusted so that no voltage appears across $R_3$ (that is, $i_3 = i_3'$). Use the superposition theorem to determine this value of $k$.

    *c.* From the value of $k$ obtained above, calculate $i_3$ ($= i_3'$) in terms of $i_s$, and hence determine $R_2$ and $R_4$.

    *d.* Determine $R_3$, using either measurement.

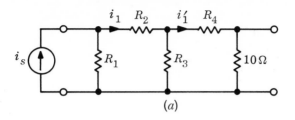

(a)

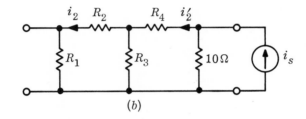

(b)

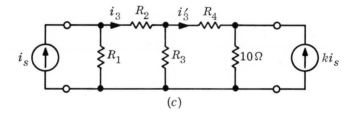

(c)

**Fig. P16.15**

Reciprocity and superposition **16.** Two sets of measurements are made on a resistive network consisting of one known resistor and four unknown resistors, as indicated in Figs. P16.16a and b. Given $R_5 = 10$ ohms, $V_1 = 0.9E$, $V_1' = 0.5E$, $V_2 = 0.3E$, and $V_2' = 0.5E$.

    *a.* Use the reciprocity theorem to determine $R_1$.

    *b.* Consider the network in the configuration shown in P16.16c. Use the superposition theorem to calculate the value of $k$ for which no current flows through $R_3$ (i.e., for which $V_3 = V_3'$).

    *c.* Determine the values of $R_2$, $R_3$, and $R_4$. (Hint: Use the condition set up in Fig. P16.16c to calculate $R_2$ and $R_4$; then go back to either of the two earlier sets of measurements to find $R_3$.)

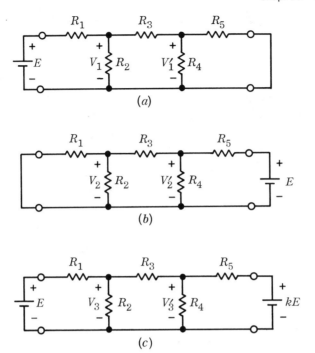

Fig. P16.16

Review   **17.** The two-port shown on Fig. P16.17 contains linear time-invariant resistors, capacitors, and inductors only.   For the connection shown in

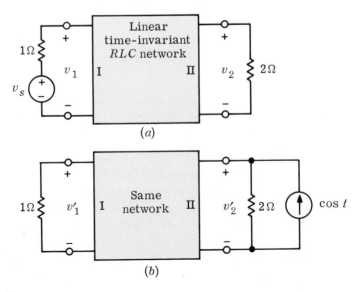

Fig. P16.17

Fig. P16.17*a*, if the voltage source $v_s$ is a unit impulse, the impulse responses for $t \geq 0$ are

$$v_1(t) = \delta(t) + \epsilon^{-t} + \epsilon^{-2t}$$
$$v_2(t) = 2\epsilon^{-t} - \epsilon^{-2t}$$

*a.* What do you know about the natural frequencies of the network variables $v_1$ and $v_2$?

*b.* What do you know about the natural frequencies of the network?

*c.* Consider the connection shown in Fig. P16.17*b*. The voltage source $v_s$ has been set to zero, and a current source has been connected to the output port. Is it possible to determine the steady-state voltages $v_1'$ and/or $v_2'$? If so, determine $v_1'$ and/or $v_2'$; if not, give your reasoning.

Review    **18.** The two-port $\mathfrak{N}$ shown in Fig. P16.18 is made of linear time-invariant resistors, inductors, and capacitors. We observe the following zero-state response (see Fig. P16.18*a*):

Input—current source:    $i_1(t) = u(t)$

Response—short-circuit current:    $i_2(t) = f(t)$

where $f(\cdot)$ is a given waveform. Find, in terms of $f(\cdot)$, the zero-state response under the following conditions (see Fig. P16.18*b*):

Input—voltage source:    $\hat{v}_2(t) = \delta(t)$

Response—open-circuit voltage:    $\hat{v}_1(t) = ?$

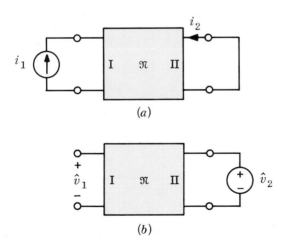

(*a*)

(*b*)

**Fig. P16.18**

Review    **19.** The two-port $\mathfrak{N}$ shown in Fig. P16.19 is made of linear time-invariant resistors, capacitors, inductors, and transformers. The following zero-state response is given:

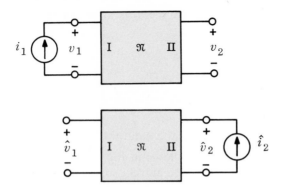

**Fig. P16.19**

Input—current source:   $i_1(t) = \delta(t)$

Response—open-circuit voltage:   $v_2(t) = 3\epsilon^{-t} + 5\epsilon^{-t} \cos{(500t - 30°)}$

a.   What do you know about the natural frequencies of the network varia-bles $v_1$, $v_2$, $\hat{v}_1$, and $\hat{v}_2$?

b.   If $\hat{i}_2$ is a unit step, find the corresponding zero-state response $\hat{v}_1$.

c.   If $\hat{i}_2(t) = \cos{500t}$, find the sinusoidal steady-state voltage $\hat{v}_1$ as a real-valued function of time.

Reciprocity   **20.** Consider the $RC$ circuit shown on Fig. P16.20.   The sliding contact $S$ moves so that the resistance between $S$ and ground is, for all $t$, given by

$R(t) = 0.5 + 0.1 \cos{2t}$ ohms

Do the conclusions of the reciprocity theorem hold true for voltages and currents at the ports ①①' and ②②'?

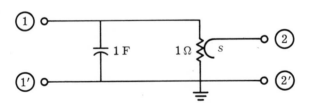

**Fig. P16.20**

# Two-ports

In the foregoing chapters we concerned ourselves with the following type of question: given a network and some inputs, how do we calculate the voltage or the current in one or more specified branches? In the process we obtained standard ways of describing networks and derived general properties of specific classes of network. With this as a background let us think in terms of design problems. For example, we might want to design a hi-fi system, a transcontinental telephone link, or some measuring instrument. Let us take the hi-fi system as a specific case. Suppose we have purchased an amplifier and the speakers; we have to design the filter which will direct the low frequencies to the woofer and the high frequencies to the tweeter. Suppose we have decided on a filter configuration, and we are ready to calculate the voltages and currents in the filter. Clearly, to do so, we must take into account the loading effects of the amplifier and the speakers on the filter. At this stage of the problem we are not interested in what happens inside the amplifier or the speakers but only what happens at their terminals. More precisely, the only data we need is the Thévenin equivalent circuit of the amplifier and the input impedance of the speakers. When we think in these terms, we mean that we are only interested in the amplifier and the speakers as *one-ports*, because we are only considering their port properties. When engineers adopt this point of view they often say, "For the filter design we treat the amplifier and the speakers as *black boxes.*"

A two-port is simply a network inside a black box, and the network has only *two pairs* of accessible terminals; usually one represents the input and the other represents the output.

A good example of the use of two-port concepts in describing a fairly complicated subsystem is given by the repeater of the blastproof transcontinental coaxial cable communication system. The repeater itself can be considered as a two-port connected at each end to the coaxial cables. The repeater is obtained by interconnection of smaller two-ports, as shown in Fig. 0.1. An overall system analysis puts requirements on the repeater. Then by further analysis, experience, and common sense, detailed requirements are obtained for each individual two-port. On the basis of these requirements the design of each two-port begins. It is for these reasons that the present chapter is devoted to two-ports.

We start by reviewing the characterization of various one-ports. By means of

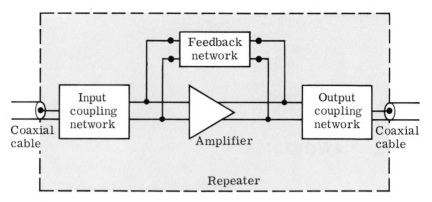

**Fig. 0.1**   A typical repeater configuration for the transcontinental coaxial-cable communication system.

nonlinear resistive networks, we introduce the two-port problem. The small-signal analysis is used as an intermediate step to specialize from the characterization of nonlinear two-ports to linear time-invariant two-ports.

## 1   Review of One-ports

When we use the expression "one-port" rather than "two-terminal net-work," we implicitly indicate that we are only interested in the external characteristics of a network. In other words, the one-port is a black box, meaning that when we think of one-ports, the only variables of interest are the port voltage and the port current. Thus, a **one-port** is completely described by all possible currents $i(\cdot)$ through the port and the corre-sponding voltages $v(\cdot)$ across the port, as indicated in Fig. 1.1$a$.

Given a linear network (i.e., by definition, a network whose elements are either linear elements or independent sources) and a pair of terminals of this network, we obtain a one-port. Clearly, such a one-port is char-acterized by the Thévenin equivalent network shown in Fig. 1.1$b$. The effect of independent sources and initial conditions is taken into account

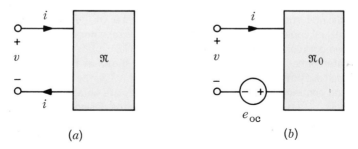

$$(a) \qquad\qquad\qquad (b)$$

**Fig. 1.1**   ($a$) Linear one-port; ($b$) its Thévenin equivalent network.

by the voltage source $e_{oc}$. The network $\mathfrak{N}_0$ is the relaxed network obtained from $\mathfrak{N}$ by setting all initial conditions and independent sources to zero. Let $h(t,\tau)$ be the impulse voltage response of the network $\mathfrak{N}_0$ for a current input; the one-port network $\mathfrak{N}$ can then be characterized by the equation

$$(1.1) \quad v(t) = e_{oc}(t) + \int_0^t h(t,\tau)i(\tau)\,d\tau \qquad t \geq 0$$

If the linear network is, in addition, time-invariant, the value of the zero-state response at time $t$ to a unit impulse applied at time $\tau$ depends only on $t - \tau$; it is denoted by $h(t - \tau)$. Thus, for the time-invariant case, Eq. (1.1) becomes

$$(1.2) \quad v(t) = e_{oc}(t) + \int_0^t h(t - \tau)i(\tau)\,d\tau \qquad t \geq 0$$

and we can more conveniently use the frequency-domain characterization. Let $V(s)$, $I(s)$, $E_{oc}(s)$, and $Z(s)$ be the Laplace transforms of $v(t)$, $i(t)$, $e_{oc}(t)$, and $h(t)$, respectively. Note that $Z(s)$ is the driving-point impedance of the one-port $\mathfrak{N}_0$. Then (1.2) becomes

$$(1.3) \quad V(s) = E_{oc}(s) + Z(s)I(s)$$

Further simplification is possible if the network $\mathfrak{N}$ does not contain independent sources and starts from the zero state. Under these conditions $E_{oc}(s)$ is identically zero, and the one-port is characterized by the driving-point impedance function $Z(s)$, or in the time domain, by the impulse response $h(t)$. Recall that this impulse response is the zero-state voltage response to an impulse of current. Equivalently, we can say that the linear time-invariant relaxed one-port is characterized by its driving-point admittance function $Y(s)$, where

$$(1.4) \quad Y(s) = \frac{1}{Z(s)} = \frac{I(s)}{V(s)}$$

The treatment of general nonlinear one-ports is very difficult and is beyond the scope of this course. Simple nonlinear one-ports, such as nonlinear resistors and nonlinear inductors, have been discussed in Chap. 2. In order to pursue the study of nonlinear resistive two-ports, we wish to review briefly some nonlinear resistive one-ports. A nonlinear resistor is completely described by its characteristic in the voltage-current plane. If the resistor is voltage-controlled, the characteristic is described by a functional relation

$$(1.5) \quad i = g(v)$$

and if the resistor is current-controlled, the characteristic is described by

$$(1.6) \quad v = r(i)$$

where $g(\cdot)$ and $r(\cdot)$ represent single-valued nonlinear functions. For example, a tunnel diode is a voltage-controlled nonlinear resistor, whereas a *pn*-junction diode is a monotonic resistor and hence is both voltage-controlled and current-controlled. Their characteristics are shown in Fig. 1.2. A linear resistor is a special case of a nonlinear monotonic resistor whose characteristic has a constant slope and passes through the origin in the *vi* plane. A linear resistive one-port which includes independent sources can be represented by the Thévenin or Norton equivalent network, as shown in Fig. 1.3. Its characteristic is given by

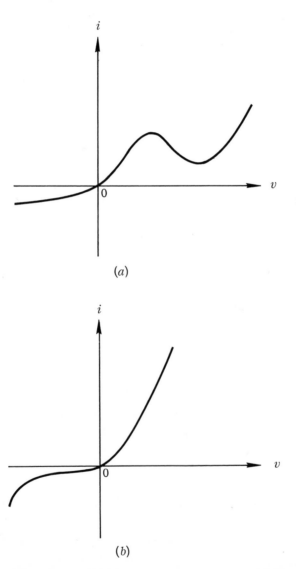

(a)

(b)

**Fig. 1.2**   Nonlinear resistive one-port characteristics.   (a) Tunnel diode; (b) *pn* junction diode.

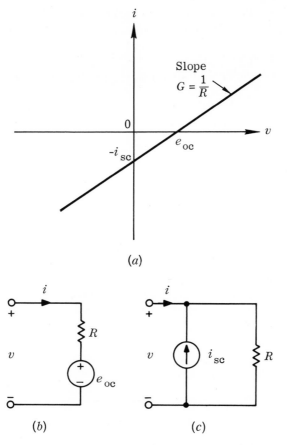

**Fig. 1.3**    Linear resistive one-port.    (*a*) Characteristic; (*b*) Thévenin
equivalent network; (*c*) Norton equivalent network.

(1.7*a*)    $v = Ri + e_{\text{oc}}$

(1.7*b*)    $i = Gv - i_{\text{sc}}$

where

(1.7*c*)    $e_{\text{oc}} = Ri_{\text{sc}}$

---

## 2    Resistive Two-ports

The two-port concept is an extension of the one-port concept. We call a
**two-port** any four-terminal network in which the four terminals have been
paired into ports ① ①′ and ② ②′. This means that for all *t* and for
all possible external connections at these terminals, the current entering
the network by terminal ① is equal to that leaving the network by ter-

minal $\textcircled{1'}$, and the current entering the network by terminal $\textcircled{2}$ is equal to that leaving the network by terminal $\textcircled{2'}$. It is important to point out that a *four-terminal network* is more general than a two-port. For example, any network with terminals $\textcircled{1}$, $\textcircled{1'}$, $\textcircled{2}$, and $\textcircled{2'}$ is a four-terminal network. Clearly, KVL and KCL will require that (see Fig. 2.1)

$$i_1(t) + i_{1'}(t) + i_2(t) + i_{2'}(t) = 0$$
$$v_{11'}(t) + v_{21}(t) - v_{22'}(t) - v_{2'1'}(t) = 0$$

for all $t$ and all possible connections of the four-terminal network. This four-terminal network will become a two-port only if we impose the *additional* constraints that for all $t$

(2.1)
$$i_{1'}(t) = -i_1(t)$$
$$i_{2'}(t) = -i_2(t)$$

It is important to stress the distinction between a *network* and a *two-port*. By a *network* (or circuit) we mean an interconnection of network elements; in a network we are free to measure any branch voltage or branch current, and we may connect any current source to any node pair or insert any voltage source in series with any branch. To solve a network means to calculate all the branch voltages and all the branch currents. When we adopt the *two-port* point of view, the only variables of interest are the port variables; the only places we may connect independent sources are the ports. In short, when we think of two-ports, we restrict our interest to the four port variables $v_1, v_2, i_1,$ and $i_2$. Roughly speaking, if we consider an amplifier, we would say that the amplifier designer considers it as a *network*, whereas the engineer who uses it on his test bench considers it as a *two-port*.

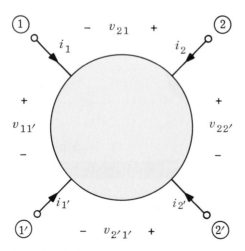

**Fig. 2.1**   A four-terminal network.

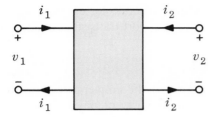

**Fig. 2.2**   A two-port.

Figure 2.2 shows a two-port. The port at the left is usually referred to as the *input port* and has voltage $v_1$ and current $i_1$, whereas the port at the right is usually referred to as the *output port* and has voltage $v_2$ and current $i_2$. Two-ports are completely characterized by all possible waveforms $i_1(\cdot)$, $v_1(\cdot)$, $i_2(\cdot)$, and $v_2(\cdot)$. The purpose of this chapter is to study the various methods of characterizing and analyzing two-ports. Many concepts and techniques are direct extensions of those used for one-ports; in particular, we have to go from scalar variables to vector variables, from scalar functions to vector-valued functions, and from numbers to matrices. However, some additional concepts will also be required.

Besides the familiar two-port devices already mentioned in earlier chapters, such as transformers, coupled inductors, and gyrators, there is the important class of three-terminal devices which may be treated as two-ports, such as transistors and vacuum tubes. Consider, for example, the common-base transistor shown in Fig. 2.3. The base terminal (node ③) is connected to the input and the output by the short circuit ① ②. At node ③, the base current $i_b$ is separated into $i_1$ and $i_2$, so that the two terminals ① and ①' at the input form the input port, and the two terminals ② and ②' at the output form the output port; that is, the current entering terminal ① is equal to the current leaving terminal ①', and the current entering terminal ② is equal to the current leaving terminal ②'. Following the same scheme, we can form three different but related two-port networks for a transistor by splitting, for example, the emitter current into two parts and obtaining a common-emitter two-port tran-

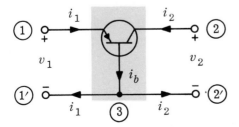

**Fig. 2.3**   A common-base transistor.

sistor configuration, or splitting the collector current into two parts and obtaining a common-collector two-port transistor configuration.

## 2.1    Various Two-port Descriptions

We start our study of two-ports with the simpler case of resistive two-ports. It should be recalled that the majority of physical devices behave like resistive two-ports when operated at low frequency. A resistive two-port differs from a general two-port in that for all time $t$ its voltages at $t$ are related to its currents at $t$ only, but are not related to either the derivatives at $t$ or the integrals up to $t$; in other words, a resistive two-port has no memory.

A resistive two-port is characterized by its *characteristic surface*. This characteristic surface is the generalization of the characteristic curve of the resistive one-port. Whereas the one-port had only two pertinent variables, $v$ and $i$, the two-port has *four* pertinent variables, $v_1$, $v_2$, $i_1$, and $i_2$. Consequently, the characteristic surface is a two-dimensional surface in a four-dimensional space whose coordinates are $v_1$, $v_2$, $i_1$, and $i_2$.

In some cases, the two-port is voltage-controlled; i.e., given the port voltages, the port currents are uniquely specified. We can then write

$(2.2a)$    $\mathbf{i} = \mathbf{g}(\mathbf{v})$

or

$(2.2b)$    $\begin{aligned} i_1 &= g_1(v_1,v_2) \\ i_2 &= g_2(v_1,v_2) \end{aligned}$

These equations are similar to those of the voltage-controlled nonlinear one-port of Eq. (1.5). In the current-controlled case, the port voltages can be expressed in terms of the port currents, and we write

$(2.3a)$    $\mathbf{v} = \mathbf{r}(\mathbf{i})$

or

$(2.3b)$    $v_1 = r_1(i_1,i_2)$
$v_2 = r_2(i_1,i_2)$

These equations should be compared to Eq. (1.6). Sometimes we can use a voltage and a current as dependent variables; for example,

$(2.4)$    $v_1 = h_1(i_1,v_2)$
$i_2 = h_2(i_1,v_2)$

These equations are referred to as a hybrid representation. For linear two-port problems, it is a simple task to derive the relations among these various characterizations and their corresponding equivalent circuit models. The details will be treated in later sections.

**2.2**     **Terminated Nonlinear Two-ports**

The first problem we encounter in electronic circuits is determining the "operating point" (also called "equilibrium point"). Typically, the problem is as follows: given the battery voltages, the load, and bias resistors, find the terminal voltages and currents of the transistor. Conceptually, the problem can be visualized as follows: the transistor is considered to be a nonlinear two-port, and the battery-resistor combinations are viewed as linear resistive one-ports. The circuit is shown in Fig. 2.4a. We want to determine the four unknown variables $i_1$, $i_2$, $v_1$, and $v_2$. Clearly, the two (battery-resistor) linear one-ports can be viewed as the Thévenin equivalent circuits of any linear one-ports which are connected to the input and output of the nonlinear two-port, as shown in Fig. 2.4b. Two simple mesh equations written for the input and output ports give

(2.5a)   $v_1 = E_1 - R_1 i_1$

(2.5b)   $v_2 = E_2 - R_2 i_2$

Let us assume that the nonlinear two-port is current-controlled; then

(2.6a)   $v_1 = r_1(i_1, i_2)$

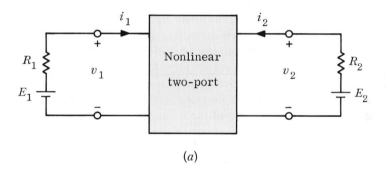

(a)

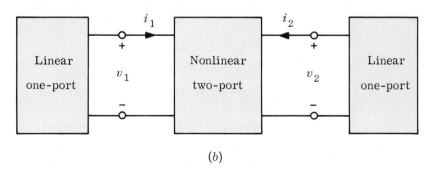

(b)

**Fig. 2.4**   (a) A nonlinear two-port terminated into battery-resistor one-ports; (b) a nonlinear two-port terminated into general linear one-ports.

(2.6b)   $v_2 = r_2(i_1, i_2)$

The four equations (2.5a), (2.5b), (2.6a), and (2.6b) must be solved to obtain the solution, i.e., the operating point. Unlike linear simultaneous equations, nonlinear equations may have no solution, a single solution, or many solutions. The condition for the existence and uniqueness of the solution depends on the properties of the nonlinear functions $r_1$ and $r_2$. Simple numerical and graphical means can often be used to obtain an approximate solution. Fortunately, in most electronic circuit problems, one of the nonlinear functions in Eqs. (2.6a) and (2.6b) contains only one argument in its approximated form, and as a result, the solution can easily be obtained graphically.

The problem that we have just outlined is usually referred to as the analysis of the bias circuit.

## 2.3   Incremental Model and Small-signal Analysis

In many applications of nonlinear devices, we are interested in the situation where a "small" signal $e_1(\cdot)$ is superimposed with the battery $E_1$ [that is, $|e_1(t)| \ll E_1$ for all $t$], and we wish to calculate the currents due to $e_1$. The assumption that $e_1$ is small leads to simplification, and the resulting analysis is called the *small-signal analysis*. In Chap. 3 we illustrated the small-signal analysis with a tunnel-diode amplifier circuit. The main difference between the tunnel-diode amplifier example and the following treatment is that we are now dealing with two scalar functions in two variables, or equivalently, a vector-valued function.

It is helpful to consider first that the nonlinear problem of Fig. 2.4a has been solved without the inclusion of the small-signal voltage $e_1$. Let $(V_1, V_2, I_1, I_2)$ be the solution of the problem with $e_1 = 0$; that is, $(V_1, V_2, I_1, I_2)$ satisfies (2.5) and (2.6). For convenience we call the solution $(V_1, V_2, I_1, I_2)$ the "operating point" $Q$. With a small signal $e_1$ inserted in series with the battery $E_1$, the new solution is, in effect, a small perturbation from the old solution $(V_1, V_2, I_1, I_2)$. Let $(v_1, v_2, i_1, i_2)$ be the solution of the problem with $e_1 \neq 0$. Then

$$v_1 = V_1 + \delta v_1 \qquad i_1 = I_1 + \delta i_1$$
$$v_2 = V_2 + \delta v_2 \qquad i_2 = I_2 + \delta i_2$$

where $(\delta v_1, \delta v_2, \delta i_1, \delta i_2)$ represents the small perturbation or the *incremental* voltages and currents.

The mesh equations corresponding to (2.5) become

(2.7a)   $V_1 + \delta v_1 = E_1 + e_1 - R_1(I_1 + \delta i_1)$

(2.7b)   $V_2 + \delta v_2 = E_2 - R_2(I_2 + \delta i_2)$

Subtracting (2.5a) from (2.7a) and (2.5b) from (2.7b), we obtain the equations for the incremental voltages and currents as follows:

(2.8a)   $\delta v_1 = e_1 - R_1 \, \delta i_1$

(2.8b)   $\delta v_2 = -R_2 \, \delta i_2$

These two equations represent the constraints imposed by the battery-resistor one-port terminations.

Similarly, the two-port characterization [see Eqs. (2.6a) and (2.6b)] becomes

(2.9a)   $v_1 = V_1 + \delta v_1 = r_1(I_1 + \delta i_1, I_2 + \delta i_2)$

(2.9b)   $v_2 = V_2 + \delta v_2 = r_2(I_1 + \delta i_1, I_2 + \delta i_2)$

Writing out the Taylor expansions of the right-hand sides, dropping all terms higher than the first, and remembering that $V_1 = r_1(I_1, I_2)$, $V_2 = r_2(I_1, I_2)$, we obtain, to a high degree of approximation,

(2.10a)   $\delta v_1 = \dfrac{\partial r_1}{\partial i_1}\bigg|_Q \delta i_1 + \dfrac{\partial r_1}{\partial i_2}\bigg|_Q \delta i_2$

(2.10b)   $\delta v_2 = \dfrac{\partial r_2}{\partial i_1}\bigg|_Q \delta i_1 + \dfrac{\partial r_2}{\partial i_2}\bigg|_Q \delta i_2$

Note that the partial derivatives must be evaluated at the fixed operating point $Q$. For instance, we differentiate the function $r_1$ with respect to its first argument $i_1$ and then evaluate the derivative at $(I_1, I_2)$.

In vector notation, Eq. (2.10) becomes

(2.11a)   $\delta \mathbf{v} = \dfrac{\partial \mathbf{r}}{\partial \mathbf{i}} \, \delta \mathbf{i}$

where

(2.11b)   $\dfrac{\partial \mathbf{r}}{\partial \mathbf{i}} \triangleq \begin{bmatrix} \dfrac{\partial r_1}{\partial i_1} & \dfrac{\partial r_1}{\partial i_2} \\[2mm] \dfrac{\partial r_2}{\partial i_1} & \dfrac{\partial r_2}{\partial i_2} \end{bmatrix}$

is called the **jacobian matrix**.

The reader should pause and verify that what we have done here is a generalization to two variables of the procedure we carried out in Chap. 3 for the case of the tunnel diode. In that case we approximated the tunnel-diode characteristic in the neighborhood of the operating point by its tangent to the operating point [see Chap. 3, Eq. (4.10)]. Here if we imagine (as in Fig. 2.5) the surface $v_1 = r_1(i_1, i_2)$ of (2.6a) in the three-dimensional space with axes $i_1$, $i_2$, and $v_1$, we see that we replace the surface in the neighborhood of the operating point $Q$ [with coordinates $(I_1, I_2, V_1)$] by its *tangent plane;* indeed, Eq. (2.10a) gives the change in $\delta v_1$ in terms of the changes of the other two coordinates, $\delta i_1$ and $\delta i_2$. It is intuitively clear that, provided $e_1$ is sufficiently small, $\delta i_1$ and $\delta i_2$ will also be small and the

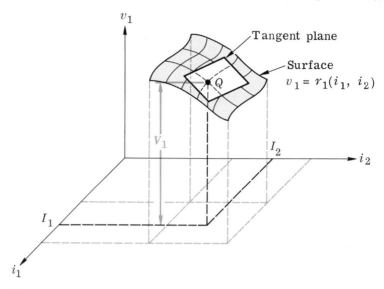

**Fig. 2.5**   Illustration of the equation $v_1 = r_1 (i_1, i_2)$; the operating point $i_1 = I_1$, $i_2 = I_2$, and $v_1 = V_1$; and the incremental resistance parameters $r_{11} = \partial r_1/\partial i_1$ and $r_{12} = \partial r_1/\partial i_2$.

higher-order terms that we dropped from the Taylor expansions are, in fact, negligible. Geometrically, in neighborhoods sufficiently close to the operating point $Q$, the tangent plane is almost indistinguishable from the surface.

It is convenient to introduce "small-signal resistance parameters" $r_{11}$, $r_{12}$, $r_{21}$, and $r_{22}$. They are defined as

(2.12)   $$r_{kj} = \left.\frac{\partial r_k}{\partial i_j}\right|_{I_1, I_2} \qquad k = 1, 2 \qquad j = 1, 2$$

Since the values of the function $r_k$ are in volts and the values of the function $i_j$ are in amperes, the $r_{kj}$'s are given in ohms. Since each partial derivative is evaluated at $(I_1, I_2)$, to be completely precise, we should say the $r_{kj}$'s are the "small-signal resistance parameters at the operating point $Q$." With Eq. (2.12), Eq. (2.11) becomes

(2.13)   $$\delta v_1 = r_{11}\, \delta i_1 + r_{12}\, \delta i_2$$
$$\delta v_2 = r_{21}\, \delta i_1 + r_{22}\, \delta i_2$$

or, in matrix notation,

(2.14)   $$\delta \mathbf{v} = \mathbf{R}\, \delta \mathbf{i}$$

The four resistance parameters (or, equivalently, the resistance matrix $\mathbf{R}$) describe completely the behavior of the *resistive two-port in the neighbor-*

*hood of the operating point Q.*   The matrix **R** is called **the incremental re-sistance matrix** of the two-port about the operating point $Q$.   In the next section we shall illustrate these concepts with a transistor.

**Example**   Suppose that a nonlinear resistive two-port is characterized by (in volts and amperes)

$$v_1 = i_1(i_2 + 0.1i_2{}^3)$$

$$v_2 = i_1 + (2 + 0.1i_1)\epsilon^{i_2}$$

Let us calculate the incremental resistance matrix about the operating point $Q$; $I_1 = 1$, and $I_2 = 3$.

$$r_{11} = \frac{\partial}{\partial i_1}[i_1(i_2 + 0.1i_2{}^3)]\Big|_{I_1,I_2} = I_2 + 0.1I_2{}^3$$

$$= 5.7 \text{ ohms}$$

$$r_{12} = \frac{\partial}{\partial i_2}[i_1(i_2 + 0.1i_2{}^3)]\Big|_{I_1,I_2} = I_1 + I_1 0.1(3I_2{}^2)$$

$$= 3.7 \text{ ohms}$$

Similarly,

$$r_{21} = 1 + 0.1\epsilon^{I_2} = 3.0 \text{ ohms}$$

$$r_{22} = (2 + 0.1I_1)\epsilon^{I_2} = 42 \text{ ohms}$$

The problem of determining the solution under the small-signal input $e_1$ is now reduced to solving the four linear simultaneous equations in (2.8) and (2.13).   These can be solved immediately.   Eliminating $\delta v_1$ and $\delta v_2$, we obtain

(2.15)
$$(R_1 + r_{11})\,\delta i_1 + r_{12}\,\delta i_2 = e_1$$
$$r_{21}\,\delta i_1 + (R_2 + r_{22})\,\delta i_2 = 0$$

From the solution of these equations, the (small-signal) output current and voltage are readily obtained as follows:

(2.16)
$$\delta i_2 = \frac{-r_{21}e_1}{(R_1 + r_{11})(R_2 + r_{22}) - r_{12}r_{21}}$$

$$\delta v_2 = \frac{R_2 r_{21}e_1}{(R_1 + r_{11})(R_2 + r_{22}) - r_{12}r_{21}}$$

The incremental input variables $\delta i_1$ and $\delta v_1$ can be obtained from (2.8). As far as the incremental voltages and currents are concerned, we can draw a linearized circuit as shown in Fig. 2.6.

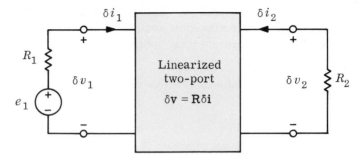

**Fig. 2.6**   Linearized circuit in terms of the incremental voltages and currents and the small-signal input $e_1$.

## 3   Transistor Examples

In this section we will use a *pnp* transistor to illustrate various useful concepts and standard techniques in characterizing a nonlinear resistive two-port.

### 3.1   Common-base Configuration

Consider the common-base *pnp* transistor shown in Fig. 3.1. The physical device can be analyzed using a one-dimensional diffusion model to yield the Ebers-Moll equations†

$$(3.1a) \quad i_e = a_{11}(\epsilon^{v_{eb}/v_T} - 1) + a_{12}(\epsilon^{v_{cb}/v_T} - 1)$$

$$(3.1b) \quad i_c = a_{21}(\epsilon^{v_{eb}/v_T} - 1) + a_{22}(\epsilon^{v_{cb}/v_T} - 1)$$

Here the $a_{ij}$'s are constants, and $v_T$ is a parameter.‡ From the form of the Ebers-Moll equations, the transistor is a voltage-controlled two-port,

† See D. O. Pederson, J. J. Studer, and J. R. Whinnery, "Introduction to Electronics Systems, Circuits and Devices," McGraw-Hill Book Company, New York, 1966.

‡ The parameter $v_T$ is often called the thermal voltage. At room temperature it is approximately 25 mV. In our analysis, we assume that $v_T$ is a constant.

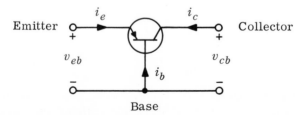

**Fig. 3.1**   Common-base *pnp* transistor.

since the port currents can be expressed as (single-valued) functions of the port voltages.  (See Eq. 2.2.)

Recall that in the one-port case the equation could be represented by a curve in the $vi$ plane, as in Fig. 1.2.  As mentioned in the previous section, if a corresponding representation for a nonlinear resistive two-port were desired, we would have a characteristic surface in a four-dimensional space whose coordinates were the port variables $v_{eb}$, $v_{cb}$, $i_e$, and $i_c$.  This is clearly impractical.  A practical way, which contains all the equivalent information of the two-port, is to use two graphs, each consisting of a set of characteristic curves.  For example, one graph would have coordinates $i_e$ and $v_{eb}$ and use $v_{cb}$ as a parameter; the other graph would have coordinates $i_c$ and $v_{cb}$ and use $v_{eb}$ as a parameter.  The transistor characteristics given in handbooks are simply variations of the above and are plotted in the most revealing way to indicate the useful ranges of voltages and currents.

We now give an alternate to the Ebers-Moll equation which is used more commonly in describing a transistor.  Eliminating $v_{cb}$ in (3.1$a$) and (3.1$b$), we obtain

(3.2$a$)   $i_e + \alpha_R i_c = I_{e0}(\epsilon^{v_{eb}/v_T} - 1)$

Similarly, eliminating $v_{eb}$, we obtain

(3.2$b$)   $\alpha_F i_e + i_c = I_{c0}(\epsilon^{v_{cb}/v_T} - 1)$

Note that the terms on the right-hand sides of the equations represent currents in $pn$-junction diodes.  $I_{e0}$ and $I_{c0}$ are, respectively, the reverse saturation current of the emitter-base diode with the collector open-circuited and the reverse saturation current of the collector-base diode with the emitter open-circuited.  The factor $\alpha_R$ (a positive number less than unity) is the reverse short-circuit current ratio,† and $\alpha_F$ (a positive number usually in the range of 0.9 and 0.995) is the forward short-circuit current ratio.  It is a simple task to express the parameters $I_{e0}$, $I_{c0}$, $\alpha_R$, and $\alpha_F$ in terms of the constants $a_{ij}$ of the Ebers-Moll equations in (3.1).  Note that the two equations can be put in a form such that $v_{eb}$ and $v_{cb}$ are expressed explicitly in terms of logarithmic functions of $i_e$ and $i_c$.  Thus, the two-port is current-controlled.  However, the particular form in Eq. (3.2) has the advantage that the equations have simple physical interpretations.  First, from the equations we can easily draw an equivalent circuit of the common-base $pnp$ transistor using $pn$-junction diodes and dependent current sources, as shown in Fig. 3.2.  Second, the most common description of a common-base transistor follows immediately from the equations in (3.2).  The two graphs are shown in Fig. 3.3.  In the $v_{eb}i_e$ plane, $i_c$ is used as a parameter.  In particular, when $i_c$ is zero, the curve represents the characteristic of a $pn$-junction diode.  When $i_c$ is positive, $i_e$ is decreased; thus, the collector current $i_c$ controls the emitter current $i_e$.  Similarly, in

---

† The term "short-circuit current ratio" will be explained in Sec. 6.1.

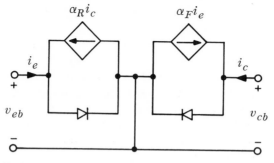

**Fig. 3.2**   Equivalent circuit of a common-base *pnp* transistor.

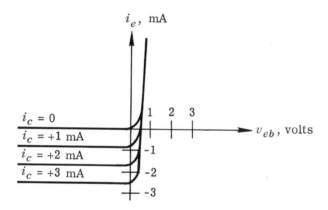

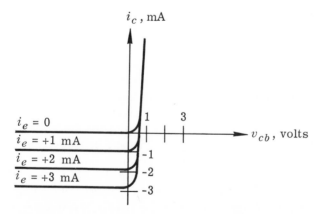

**Fig. 3.3**   Common-base *pnp* transistor characteristics.   (*After Searle et al., "Elementary Circuit Properties of Transistors," John Wiley & Sons, Inc., New York, 1964.*)

the $v_{cb}i_c$ plane, $i_e$ is used as a parameter, and its value controls the collector current.

Finally, we wish to mention briefly the piecewise linear model of the transistor using ideal diodes. The piecewise linear model is often useful for analyzing the approximate behavior of a transistor circuit. If we assume that the diodes in Fig. 3.2 are ideal diodes, i.e., instead of having exponential properties, they behave as a short circuit when forward-biased and an open circuit when reverse-biased, the graphical plots are then in the form of straight-line segments, as shown in Fig. 3.4.

We have indicated various characterizations of the common-base transistor configuration. All of them are useful for different reasons. The

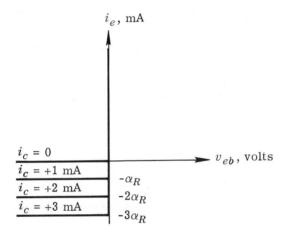

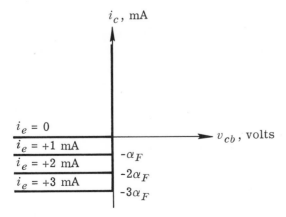

**Fig. 3.4**   Ideal piecewise linear diode *pnp* transistor characteristic in the common-base configuration.

equivalent circuit helps us know the physical behavior of the device.   The nonlinear characterization equations are essential for precise analysis. The graphical characteristics are important for numerical calculation and for design purposes.   The piecewise linear representation is useful for a quick analysis and enables us to reduce a nonlinear circuit problem to a simpler problem, one for which linear analysis suffices.

| 3.2 | **Common-emitter Configuration** |

In most circuit applications, transistors are used in the common-emitter configuration, as shown in Fig. 3.5.   It is important to know that the characterizations of a common-emitter configuration can be readily obtained from those of the common-base configuration.   The port currents in the common-emitter two-port of Fig. 3.5 are $i_b$ and $i_c$; they are related to the port currents of the common-base two-port according to

$(3.3a)$   $i_b = -(i_e + i_c)$

$(3.3b)$   $i_c = i_c$

The port voltages are $v_{be}$ and $v_{ce}$; they are related to the port voltages of the common-base two-port by

$(3.4a)$   $v_{be} = -v_{eb}$

$(3.4b)$   $v_{ce} = v_{cb} - v_{eb}$

If we substitute Eqs. (3.3) and (3.4) in (3.2), we obtain the relations

$(3.5a)$   $i_b + (1 - \alpha_R)i_c = I_{e0}(1 - \epsilon^{-v_{be}/v_T})$

$(3.5b)$   $-\alpha_F i_b + (1 - \alpha_F)i_c = I_{c0}(\epsilon^{(v_{ce}-v_{be})/v_T} - 1)$

These two equations, even though they contain all the information of the two-port, are not convenient to work with; indeed, Eq. (3.5b) contains both port currents and both port voltages.   We shall not attempt to simplify these equations any further since the derivation is tedious and the result is not particularly useful.

The most useful description for a common-emitter transistor two-port is given by the following hybrid form:

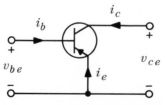

**Fig. 3.5**   Common-emitter *pnp* transistor.

(3.6a)    $v_{be} = h_1(i_b, v_{ce})$

(3.6b)    $i_c = h_2(i_b, v_{ce})$

The most commonly used graphical characteristics, given in Fig. 3.6, are compatible with the equations in the hybrid form. The main reason for using the hybrid representation is the convenience in measurement. Let us take a look at the curves in the $v_{be}i_b$ plane. Note that for $v_{ce}$ negative, the curves are in the third quadrant, and the input current $i_b$ and voltage $v_{be}$ are virtually independent of the output voltage $v_{ce}$. In the normal operating condition of a *pnp* common-emitter transistor, both $v_{ce}$ and $v_{be}$ are negative, and thus only the third quadrants of both graphs need to be considered. In the following discussion we shall use the two graphs to

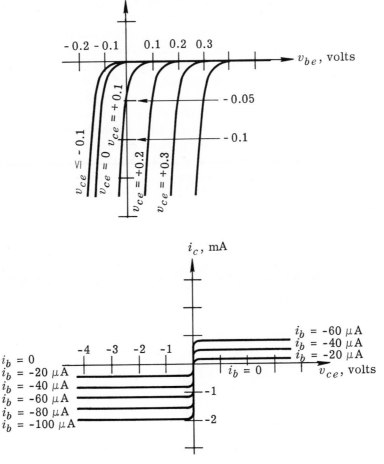

**Fig. 3.6**    Common-emitter *pnp* transistor characteristics.

illustrate the determination of the operating point and the meanings of the incremental hybrid parameters of a common-emitter transistor.

Consider the biased transistor circuit shown in Fig. 3.7. At the input and the output two mesh equations describe the terminations of the two-port as follows:

(3.7a)   $v_{be} = -V_{bb} - R_b i_b$

(3.7b)   $v_{ce} = -V_{cc} - R_c i_c$

These equations can be superimposed on the curves of the common-emitter characteristics of Fig. 3.6 as straight lines (called load lines), as shown in Fig. 3.8. The problem is to determine the operating point $Q$, that is, the solution $(V_{be}, V_{ce}, I_b, I_c)$ of the problem in Fig. 3.7. Since the input characteristic of the common-emitter two-port is relatively insensitive to the output port voltage $v_{ce}$ when $v_{be}$ is negative, the intersection of any characteristic curve with the load line gives approximately the values of $V_{be}$ and $I_b$, as shown in the figure. These values are then used to determine $V_{ce}$ and $I_c$ by means of the output characteristics in the $v_{ce} i_c$ plane. An accurate value of $V_{be}$ and $I_b$ can be found if we simply use the cut-and-try method, i.e., if we go back and forth successively a few times in the two planes. For example, we may assume a value of $v_{ce}$ and find the intersection of the characteristic in the input graph with the load line. Using the value corresponding to this intersection, we locate a point in the output graph and find that it is not quite on the load line. We next change the assumed value of $v_{ce}$ slightly and repeat the process. Finally, we obtain the solution; i.e., we have a solution $(V_{be}, V_{ce}, I_b, I_c)$ which sits on the load lines for both the input graph and the output graph.

In the neighborhood of the operating point $Q$, $(V_{be}, V_{ce}, I_b, I_c)$ we can use the small-signal incremental equations to describe the behavior of the transistor two-port. Taking the incremental voltages and currents and using Eqs. (3.6a) and (3.6b), we obtain the following linear equations:

(3.8a)   $\delta v_{be} = \dfrac{\partial h_1}{\partial i_b}\bigg|_Q \delta i_b + \dfrac{\partial h_1}{\partial v_{ce}}\bigg|_Q \delta v_{ce}$

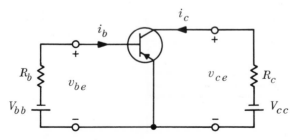

**Fig. 3.7**   A common-emitter transistor with biasing.

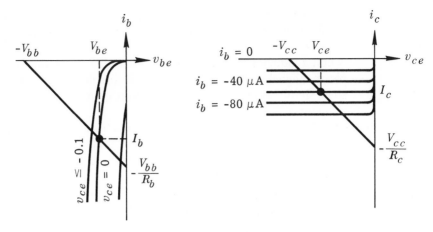

**Fig. 3.8**   Load lines for the biasing circuit of Fig. 3.7.

$$(3.8b) \quad \delta i_c = \left.\frac{\partial h_2}{\partial i_b}\right|_Q \delta i_b + \left.\frac{\partial h_2}{\partial v_{ce}}\right|_Q \delta v_{ce}$$

The partial derivatives above are called the **hybrid parameters** of the incremental two-port model. They are denoted by $h_{ij}$ and evaluated at the operating point according to the following equations:

$$(3.9) \quad \begin{aligned} h_{11} &= \left.\frac{\partial h_1}{\partial i_b}\right|_Q & h_{12} &= \left.\frac{\partial h_1}{\partial v_{ce}}\right|_Q \\[2ex] h_{21} &= \left.\frac{\partial h_2}{\partial i_b}\right|_Q & h_{22} &= \left.\frac{\partial h_2}{\partial v_{ce}}\right|_Q \end{aligned}$$

More detailed discussion of the significance of these incremental two-port parameters and their relations to other parameters will be given later in the treatment of linear time-invariant two-ports.

---

**4**   | **Coupled Inductors**

Up to this point we have considered only nonlinear resistive two-ports and their small-signal analysis. General nonlinear two-ports will not be treated in this book because they are too complicated. However, we wish to introduce the characterization of a special nonlinear two-port which stores energy, namely, the coupled inductor. It is well known that the flux of an iron core "saturates" when the magnetomotive force becomes large; therefore, an iron-core inductor is basically a nonlinear element. Iron-core or ferrite-core coupled inductors are used in numerous applications, for example, as transformers in power supplies, in Variacs, audio amplifiers, IF amplifiers, and magnetic amplifiers. For most applications they may be treated as linear elements, since the applied signal remains in the linear region, and the linear coupled-inductor analysis introduced in Chap. 8 is sufficient. However, in some instances the saturation of the coupled inductors is the key to the operation of the device, as, for example, in magnetic amplifiers or magnetic-core memories. In this section we shall simply study the characterization of a pair of nonlinear coupled inductors.

Consider the nonlinear coupled inductors shown in Fig. 4.1, where the reference directions of port voltages and port currents are indicated. The physical characteristics of the coupled inductors are usually given in terms of fluxes and currents. The relation between fluxes and currents is analogous to the relation between voltages and currents of the resistive two-ports in that the relation is described by algebraic or transcendental nonlinear equations. A pair of coupled inductors is described by a two-dimensional characteristic surface in the $(\phi_1,\phi_2,i_1,i_2)$ space. Let us denote $\phi = (\phi_1,\phi_2)^T$ as the flux vector; if the currents can be expressed as functions of the flux, we have the more special representation

(4.1)   $\mathbf{i} = \widehat{\mathbf{i}}(\phi)$

or

(4.2)   $\begin{aligned}i_1 &= \widehat{i}_1(\phi_1,\phi_2)\\ i_2 &= \widehat{i}_2(\phi_1,\phi_2)\end{aligned}$

The coupled inductors are then said to be flux-controlled. Similarly, if the fluxes can be expressed as functions of the currents, we have the representation

(4.3a)   $\phi = \widehat{\phi}\,(\mathbf{i})$

or

(4.3b)   $\begin{aligned}\phi_1 &= \widehat{\phi}_1(i_1,i_2)\\ \phi_2 &= \widehat{\phi}_2(i_1,i_2)\end{aligned}$

The coupled inductors are then said to be current-controlled. The relation between voltages and currents can be obtained immediately from Faraday's law; thus,

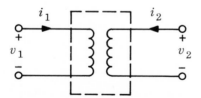

**Fig. 4.1**   Nonlinear coupled inductors with $\mathbf{v} = d\phi/dt$.

(4.4)   $\mathbf{v} = \dfrac{d\boldsymbol{\phi}}{dt}$

For the flux-controlled case, the instantaneous currents $\mathbf{i}(t)$ are related to the instantaneous fluxes $\boldsymbol{\phi}(t)$ by $\mathbf{i}(t) = \hat{\mathbf{i}}[\boldsymbol{\phi}(t)]$.   Consequently,

$$\frac{d\mathbf{i}}{dt} = \frac{d}{dt}\{\hat{\mathbf{i}}[\boldsymbol{\phi}(t)]\} = \frac{\partial \hat{\mathbf{i}}}{\partial \boldsymbol{\phi}}\frac{d\boldsymbol{\phi}}{dt}$$

(4.5)   $\qquad = \boldsymbol{\Gamma}[\boldsymbol{\phi}(t)]\mathbf{v}$

where

(4.6)   $\boldsymbol{\Gamma} = \dfrac{\partial \hat{\mathbf{i}}}{\partial \boldsymbol{\phi}} = \begin{bmatrix} \dfrac{\partial \hat{i_1}}{\partial \phi_1} & \dfrac{\partial \hat{i_1}}{\partial \phi_2} \\[2mm] \dfrac{\partial \hat{i_2}}{\partial \phi_1} & \dfrac{\partial \hat{i_2}}{\partial \phi_2} \end{bmatrix} = \begin{bmatrix} \Gamma_{11} & \Gamma_{12} \\[2mm] \Gamma_{21} & \Gamma_{22} \end{bmatrix}$

is the usual jacobian matrix and is called the **incremental reciprocal inductance matrix** of the nonlinear coupled inductors.   The scalar form of (4.5) is

(4.7a)   $\dfrac{di_1}{dt} = \Gamma_{11}v_1 + \Gamma_{12}v_2$

(4.7b)   $\dfrac{di_2}{dt} = \Gamma_{21}v_1 + \Gamma_{22}v_2$

For the current-controlled case [see Eq. (4.3)] we have

(4.8)   $\mathbf{v} = \dfrac{d\boldsymbol{\phi}}{dt} = \dfrac{\partial \hat{\boldsymbol{\phi}}}{\partial \mathbf{i}}\dfrac{d\mathbf{i}}{dt} = \mathbf{L}(\mathbf{i})\dfrac{d\mathbf{i}}{dt}$

where

(4.9)   $\mathbf{L} = \dfrac{\partial \hat{\boldsymbol{\phi}}}{\partial \mathbf{i}} = \begin{bmatrix} \dfrac{\partial \hat{\phi_1}}{\partial i_1} & \dfrac{\partial \hat{\phi_1}}{\partial i_2} \\[2mm] \dfrac{\partial \hat{\phi_2}}{\partial i_1} & \dfrac{\partial \hat{\phi_2}}{\partial i_2} \end{bmatrix} = \begin{bmatrix} L_{11} & L_{12} \\[2mm] L_{21} & L_{22} \end{bmatrix}$

is called the **incremental inductance matrix** of the nonlinear coupled inductors.   The scalar form of (4.8) is

(4.10a)   $v_1 = L_{11}\dfrac{di_1}{dt} + L_{12}\dfrac{di_2}{dt}$

(4.10b)   $v_2 = L_{21}\dfrac{di_1}{dt} + L_{22}\dfrac{di_2}{dt}$

It should be emphasized that in both Eq. (4.7) and Eq. (4.10), the parameters $\Gamma_{ij}$ and $L_{ij}$ are usually nonlinear functions of $(\phi_1,\phi_2)$ and $(i_1,i_2)$, respectively.

---

**Example**   Suppose the characteristics $\hat{\phi}_1$ and $\hat{\phi}_2$ in (4.3b) have the form

$$\phi_1 = \tanh(i_1 + i_2) + 10^{-2}i_1$$
$$\phi_2 = \tanh(i_1 + i_2) + 10^{-2}i_2$$

The second term of the right-hand side represents the effects of the leakage flux through the air. Suppose that the waveforms $i_1(\cdot)$ and $i_2(\cdot)$ are known; then the port voltages are given by

$$v_1(t) = \frac{d\phi_1}{dt} = \left\{\frac{1}{\cosh^2[i_1(t) + i_2(t)]} + 10^{-2}\right\}\frac{di_1}{dt} + \frac{1}{\cosh^2[i_1(t) + i_2(t)]}\frac{di_2}{dt}$$

$$v_2(t) = \frac{d\phi_2}{dt} = \frac{1}{\cosh^2[i_1(t) + i_2(t)]}\frac{di_1}{dt} + \left\{\frac{1}{\cosh^2[i_1(t) + i_2(t)]} + 10^{-2}\right\}\frac{di_2}{dt}$$

---

## 5   Impedance and Admittance Matrices of Two-ports

In the present section and the next section, we consider two-ports which may include energy-storing elements and dependent sources. We require, however, all elements to be *linear* and *time-invariant;* for this reason, we shall use Laplace transforms. We shall adopt throughout the reference directions shown in Fig. 5.1. Thus, the waveforms $v_1(\cdot)$ and $i_1(\cdot)$ are the port-voltage and port-current waveforms at port 1, and the reference directions are such that $v_1(t)i_1(t)$ is the power (at time $t$) entering the two-port through port 1. The notation and reference directions pertaining to port 2 are similar. The Laplace transforms are denoted as usual by capital letters, thus; we have $V_1(s)$, $V_2(s)$, $I_1(s)$, and $I_2(s)$. Throughout Secs. 5 and 6 we assume that

1. *All* the elements in the two-ports are *linear* and *time-invariant* (in particular, this excludes *independent* sources).

2. We calculate only *zero-state* responses.

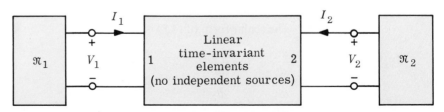

**Fig. 5.1**   The two-port is connected to networks $\mathfrak{N}_1$ and $\mathfrak{N}_2$; the variables of interest are $V_1$, $V_2$, $I_1$, and $I_2$.

Let us consider the situation shown in Fig. 5.1. We note that there are four unknowns of interest: $V_1$, $V_2$, $I_1$, and $I_2$. We assert that the two-port can impose only two (linear) constraints on these four variables. Indeed, we can, by the substitution theorem, replace $\mathfrak{N}_1$ and $\mathfrak{N}_2$ by appropriate independent sources. If both sources are voltage sources, then $V_1$ and $V_2$ are specified, and the corresponding $I_1$ and $I_2$ are determined by the given two-port. Similarly, if both sources are current sources, then $I_1$ and $I_2$ are specified, and the corresponding $V_1$ and $V_2$ are determined by the given two-port. Looking at the question from a different angle, we note that since there are six ways of picking two elements out of a set of four elements, there are six ways of characterizing the zero-state response of a two-port composed of linear time-invariant elements. We shall start with the impedance matrix characterization.

### 5.1    The (Open-circuit) Impedance Matrix

Let us imagine the two-port driven by current sources, as shown in Fig. 5.2. In other words, we use the Laplace transforms $I_1$ and $I_2$ as independent variables, and we wish to calculate the Laplace transforms of the port voltages $V_1$ and $V_2$ in terms of those of the currents. Since, by assumption, the elements of the two-port are *linear* (and time-invariant) and since we seek only the *zero-state* response, the superposition theorem asserts that

$$V_1(s) = \begin{pmatrix} \text{Laplace transform of} \\ \text{voltage at port 1 due} \\ \text{to } I_1 \text{ acting alone} \end{pmatrix} + \begin{pmatrix} \text{Laplace transform of} \\ \text{voltage at port 1 due} \\ \text{to } I_2 \text{ acting alone} \end{pmatrix}$$

By the definition of a network function, the first term is the product of a suitable network function and $I_1(s)$. Call this network function $z_{11}(s)$. Similarly, the second term may be put in the form $z_{12}(s)I_2(s)$. Thus, we may write

(5.1a)    $V_1(s) = z_{11}(s)I_1(s) + z_{12}(s)I_2(s)$

An analogous reasoning applied to port 2 introduces the network functions $z_{21}(s)$ and $z_{22}(s)$ and allows us to write

(5.1b)    $V_2(s) = z_{21}(s)I_1(s) + z_{22}(s)I_2(s)$

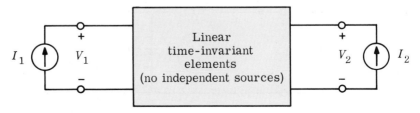

**Fig. 5.2**    A two-port driven by two current sources.

*In most of the equations below we shall not exhibit the dependence on s of the voltages, the currents, and the network functions.*

Rewriting the two equations above in matrix form, we obtain

(5.2a)   $\mathbf{V} = \mathbf{ZI}$

where the matrix

(5.2b)   $\mathbf{Z} = \begin{bmatrix} z_{11} & z_{12} \\ z_{21} & z_{22} \end{bmatrix}$

is called the **open-circuit impedance matrix** of the two-port. The impedances $z_{ij}$ are called the **open-circuit impedance parameters.**

It is important to note that as a consequence of the substitution theorem, Eqs. (5.1a) and (5.1b) would still be valid if the two-port were connected to arbitrary networks (as shown in Fig. 5.1) which supplied its ports with the currents $I_1$ and $I_2$.

---

**Example 1**   Let us calculate the open-circuit impedance matrix of the two-port shown in Fig. 5.3. Since there are only two loops and since the two loop currents $I_1$ and $I_2$ are specified, we obtain, by inspection

$$\begin{bmatrix} V_1(s) \\ V_2(s) \end{bmatrix} = \begin{bmatrix} L_1 s + \dfrac{1}{Cs} & \dfrac{1}{Cs} \\ \dfrac{1}{Cs} & L_2(s) + \dfrac{1}{Cs} \end{bmatrix} \begin{bmatrix} I_1(s) \\ I_2(s) \end{bmatrix}$$

Note that in writing these equations, we had to take into account the fact that the initial currents in each inductor and the initial voltage across the capacitor were zero.

---

*Physical interpretation*   In the description of $\mathbf{Z}$ we use the qualification "open-circuit" because each one of the $z_{ij}$'s may be interpreted as an open-circuit impedance. Indeed, suppose port 2 is left open-circuited (hence $I_2 = 0$); then the input impedance seen at port 1 is [see Eq. (5.1a)]

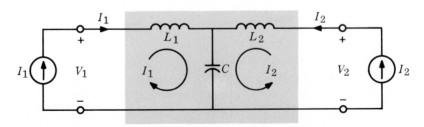

**Fig. 5.3**   A two-port for Example 1.

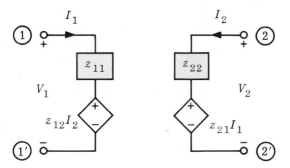

**Fig. 5.4**   An equivalent circuit of a two-port in terms of the open-circuit impedance parameters.

$$(5.3a) \quad z_{11} = \frac{V_1}{I_1}\bigg|_{I_2=0}$$

Thus, $z_{11}$ is the driving-point impedance at port 1 with port 2 open-circuited.   Under the same conditions, we conclude from Eq. (5.1b) that

$$(5.3b) \quad z_{21} = \frac{V_2}{I_1}\bigg|_{I_2=0}$$

Thus, $z_{21}$ is the forward open-circuit transfer impedance since it is the ratio of the Laplace transform of the *open-circuit* voltage at port 2 to the Laplace transform of the input current at port 1.   Let us now have an open circuit at port 1; then

$$(5.3c) \quad z_{22} = \frac{V_2}{I_2}\bigg|_{I_1=0}$$

$$(5.3d) \quad z_{12} = \frac{V_1}{I_2}\bigg|_{I_1=0}$$

The impedance parameters $z_{ij}$ may be used in many ways.   It is possible to draw several two-ports that obey Eq. (5.1).   For example, Fig. 5.4 shows such a two-port.   The interpretation of the $z_{ij}$'s leads to an immediate verification of the fact that this two-port obeys Eq. (5.1).   Another equivalent circuit is the so-called *T equivalent circuit;* it is shown in Fig. 5.5. This equivalent circuit uses only one dependent source, but it is not as general as the previous one since it implicitly assumes that terminals (1') and (2') are at the same potential.   In particular, if $z_{12} = z_{21}$, the dependent voltage source vanishes; the two-port network is then representable in terms of two-terminal impedances only.   When $z_{12} = z_{21}$, the **two-port** is said to be **reciprocal.**   The reciprocity theorem guarantees that if the two-port is made of resistors, inductors, capacitors, and transformers, it is a reciprocal two-port.

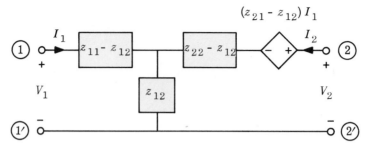

Fig. 5.5   The T equivalent circuit of a two-port.

**Example 2**   As far as small signals are concerned, a grounded-base transistor at low frequencies has the T equivalent circuit representation shown in Fig. 5.6a. Using a Norton equivalent network, we obtain an alternate equivalent circuit which uses a dependent current source. It is shown in Fig. 5.6b.

## 5.2   The (Short-circuit) Admittance Matrix

Let us consider the two-port from a point of view which is the dual of the one used previously. Now the port voltages $V_1$ and $V_2$ are the independent variables, and the port currents $I_1$ and $I_2$ are the dependent variables.

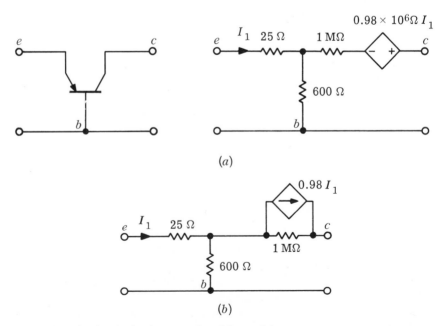

Fig. 5.6   T equivalent circuits of a low-frequency transistor model.

Thus, we consider the situation in which the two-port is driven by voltage sources, as shown in Fig. 5.7. Since the elements of the two-port are linear and since we seek only the zero-state response, the superposition theorem guarantees that each port current is the sum of the contributions due to each voltage source acting alone; therefore, in terms of Laplace transforms, the equations have the form

(5.4a)   $I_1(s) = y_{11}(s)V_1(s) + y_{12}(s)V_2(s)$

(5.4b)   $I_2(s) = y_{21}(s)V_1(s) + y_{22}(s)V_2(s)$

Using the matrix form and suppressing the dependence on $s$, we write

(5.5a)   $\mathbf{I} = \mathbf{YV}$

where

(5.5b)   $\mathbf{Y} = \begin{bmatrix} y_{11} & y_{12} \\ y_{21} & y_{22} \end{bmatrix}$

is called the **short-circuit admittance matrix** of the two-port, and the admittances $y_{ij}$ are called the **short-circuit admittance parameters.**
From Eq. (5.4a) the admittance $y_{11}$ can be interpreted by

(5.6a)   $y_{11} = \dfrac{I_1}{V_1}\bigg|_{V_2=0}$

Thus, $y_{11}$ is the driving-point admittance at port 1 when port 2 is short-circuited. Similarly,

(5.6b)   $y_{21} = \dfrac{I_2}{V_1}\bigg|_{V_2=0}$

is the forward transfer admittance, which is defined under the condition that port 2 is short-circuited. In applying (5.6b) note the reference directions for $I_2$ and $V_1$; thus, for networks made of passive resistors, $y_{21}$ is a negative number. From Eq. (5.4b) we obtain the following interpretations:

(5.6c)   $y_{22} = \dfrac{I_2}{V_2}\bigg|_{V_1=0}$

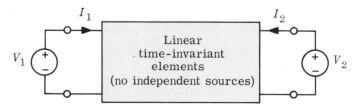

**Fig. 5.7**   A two-port driven by two voltage sources.

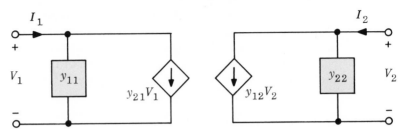

**Fig. 5.8**   An equivalent circuit of a two-port in terms of the short-circuit admittance parameters.

and

$$(5.6d) \quad y_{12} = \frac{I_1}{V_2}\bigg|_{V_1=0}$$

The parameters $y_{22}$ and $y_{12}$ are, respectively, the driving-point admittance at port 2 and the reverse transfer admittance under the condition that port 1 is short-circuited.

The equivalent circuit based on the short-circuit two-port parameters is shown in Fig. 5.8. Note that two dependent current sources are used in the figure. An alternate equivalent circuit using one dependent current source is shown in Fig. 5.9. It is called the $\pi$ *equivalent circuit.* However, it assumes implicitly that the terminals ① and ② are at the same potential. In particular, if $y_{12} = y_{21}$, the dependent current source vanishes; the **two-port** is then said to be **reciprocal.**

**Example 3**   A pentode in the common-cathode connection has the equivalent circuit shown in Fig. 5.10. The short-circuit admittance parameters are easily obtainable from their physical interpretations [Eqs. (5.6a) to (5-6d)] as follows:

$$y_{11} = s(C_{gk} + C_{gp})$$
$$y_{22} = s(C_{pk} + C_{gp}) + g_p$$

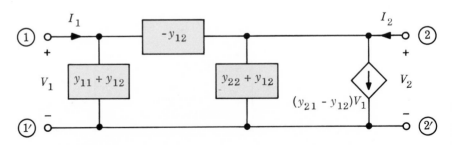

**Fig. 5.9**   The $\pi$ equivalent circuit of a two-port.

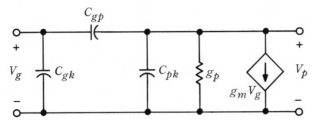

**Fig. 5.10**   The equivalent circuit of a pentode.

$$y_{12} = -sC_{gp}$$
$$y_{21} = -sC_{gp} + g_m$$

*Relation between* **Z** *and* **Y**   It is easy to derive the relation between the open-circuit impedance parameters and the short-circuit admittance parameters by means of the two matrix equations in (5.2) and (5.5).   Clearly,

(5.7)   $\mathbf{Z} = \mathbf{Y}^{-1}$   and   $\mathbf{Y} = \mathbf{Z}^{-1}$

From the inverse relations of a nonsingular matrix, we obtain

(5.8)   $z_{11} = \dfrac{y_{22}}{\Delta_Y}$     $z_{12} = \dfrac{-y_{12}}{\Delta_Y}$     $z_{21} = \dfrac{-y_{21}}{\Delta_Y}$     $z_{22} = \dfrac{y_{11}}{\Delta_Y}$

(5.9)   $y_{11} = \dfrac{z_{22}}{\Delta_Z}$     $y_{12} = \dfrac{-z_{12}}{\Delta_Z}$     $y_{21} = \dfrac{-z_{21}}{\Delta_Z}$     $y_{22} = \dfrac{z_{11}}{\Delta_Z}$

where

(5.10)   $\Delta_Y = \det [\mathbf{Y}]$   and   $\Delta_Z = \det [\mathbf{Z}]$

### 5.3   A Terminated Two-port

In many applications a two-port is connected to a generator and to a load, as shown in Fig. 5.11.   The generator is represented by the series connec-

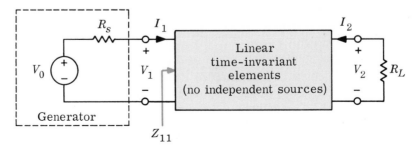

**Fig. 5.11**   A terminated two-port.

tion of a voltage source and a linear resistor with resistance $R_s$. The load is a linear resistor with resistance $R_L$. We are interested in obtaining (1) the driving-point impedance at the input port, i.e., the relation between the current $I_1$ and the voltage $V_1$, and (2) the overall transfer properties of the terminated two-port, i.e., the relation between the output voltage $V_2$ and the source voltage $V_0$. Let the two-port be described by its open-circuit impedance parameters; i.e.,

(5.11a)   $V_1 = z_{11}I_1 + z_{12}I_2$

(5.11b)   $V_2 = z_{21}I_1 + z_{22}I_2$

The source and the load impose the following additional constraints on the port voltages and currents:

(5.11c)   $V_1 = V_0 - R_sI_1$

(5.11d)   $V_2 = -R_LI_2$

It is a simple exercise to eliminate the variables $I_2$ and $V_2$ to obtain the following useful formula of the driving-point impedance of the *terminated* two-port:

(5.12)   $Z_{11} = \dfrac{V_1}{I_1} = z_{11} - \dfrac{z_{12}z_{21}}{z_{22} + R_L}$

Similarly, eliminating $I_1$, $V_1$, and $I_2$ from the four equations in (5.11), we obtain an equation relating $V_2$ and $V_0$. The ratio $V_2/V_0$ is called the **transfer voltage ratio** and is given by

(5.13)   $\dfrac{V_2}{V_0} = \dfrac{z_{21}R_L}{(z_{11} + R_s)(z_{22} + R_L) - z_{12}z_{21}}$

Thus, given the open-circuit impedance parameters of a two-port and the source and load impedances, we can calculate the driving-point impedance and the overall transfer voltage ratio by using Eqs. (5.12) and (5.13).

**Exercise 1**   Find the impedance matrix of the two-ports shown in Fig. 5.12. (Hint: For Fig. 5.12c and d, assume that the ports are terminated by voltage sources $V_1$ and $V_2$, respectively; write three loop equations, use the third loop equation to calculate $I_3$ in terms of $I_1$ and $I_2$, and substitute in the first two.)

**Exercise 2**   Calculate the admittance matrix of the two-ports shown in Fig. 5.13. (Hint: For Fig. 5.13c, assume the ports are connected to current sources $I_1$ and $I_2$; write three node equations, solve the third one for $V_3$, and substitute the result in the first two.)

**Exercise 3**   Which of the two-ports (in Figs. 5.12 and 5.13) are reciprocal?

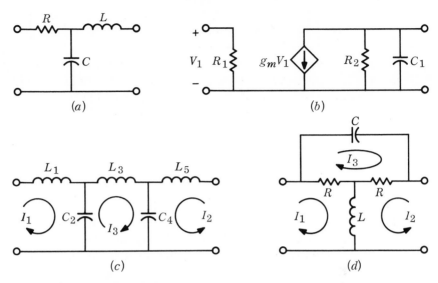

**Fig. 5.12**   Two-ports whose impedance matrix must be found.

**Exercise 4**   A two-port has the following impedance matrix

$$\begin{bmatrix} \dfrac{2}{s+1} & \dfrac{1}{s+1} \\[2ex] \dfrac{1}{s+1} & \dfrac{6}{s+1} \end{bmatrix}$$

It is terminated as shown in Fig. 5.11, with $R_s = 2$ ohms and $R_L = 1$ ohm.

a.   Calculate the zero-state response $v_2$ to a unit step.

b.   If $v_0(t) = 10 \cos (3t + 60°)$, find the sinusoidal steady-state output voltage $v_2$ (express your answer as a real-valued function of time).

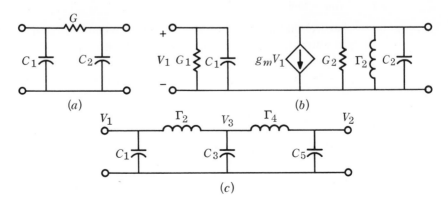

**Fig. 5.13**   Two-ports whose admittance matrix must be found.

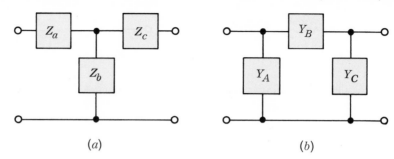

*(a)*                                    *(b)*

**Fig. 5.14**    Exercise to derive the T-$\pi$ equivalence.

**Exercise 5**    Let a reciprocal two-port network be represented by its T equivalent, as shown in Fig. 5.14$a$. Derive its $\pi$ equivalent, as shown in Fig. 5.14$b$; that is, calculate the admittances $Y_A$, $Y_B$, and $Y_C$ in terms of $Z_a$, $Z_b$, and $Z_c$. Similarly, assuming the $\pi$ equivalent is given, obtain its T equivalent.

## 6    Other Two-port Parameter Matrices

The open-circuit impedance matrix and the short-circuit admittance matrix relate the current vector and the voltage vector of a two-port. In the former the currents $I_1$ and $I_2$ are the independent variables, whereas the voltages $V_1$ and $V_2$ are the dependent variables. In the latter the voltages $V_1$ and $V_2$ are the independent variables, whereas the currents $I_1$ and $I_2$ are the dependent variables. It is possible and often more convenient to use a mixed set consisting of one current and one voltage as the independent variables and the remaining parameters as the dependent variables. In this section we shall give four descriptions of this kind. Each leads to a new matrix which describes the two-port.

### 6.1    The Hybrid Matrices

The hybrid matrices describe a two-port when the voltage of one port and the current of the other port are taken as the independent variables. Two such descriptions are possible for a two-port. For the first one we use $I_1$ and $V_2$ as independent variables. As before, superposition allows us to write

(6.1$a$)    $V_1 = h_{11}I_1 + h_{12}V_2$

(6.1$b$)    $I_2 = h_{21}I_1 + h_{22}V_2$

or, in matrix notation

(6.2$a$)    $\begin{bmatrix} V_1 \\ I_2 \end{bmatrix} = \begin{bmatrix} h_{11} & h_{12} \\ h_{21} & h_{22} \end{bmatrix} \begin{bmatrix} I_1 \\ V_2 \end{bmatrix}$

We define **H** by

$$(6.2b) \quad \mathbf{H} \triangleq \begin{bmatrix} h_{11} & h_{12} \\ h_{21} & h_{22} \end{bmatrix}$$

The matrix **H** is called a **hybrid matrix.** It is straightforward to derive the physical interpretations of the parameters $h_{ij}$ as in the impedance and admittance cases. The following results, which give both the physical interpretations and the relations with the impedance and admittance parameters, are easily derived:

$$(6.3a) \quad h_{11} = \left.\frac{V_1}{I_1}\right|_{V_2=0} = \frac{1}{y_{11}}$$

$$(6.3b) \quad h_{12} = \left.\frac{V_1}{V_2}\right|_{I_1=0} = \frac{z_{12}}{z_{22}}$$

$$(6.3c) \quad h_{21} = \left.\frac{I_2}{I_1}\right|_{V_2=0} = \frac{y_{21}}{y_{11}}$$

$$(6.3d) \quad h_{22} = \left.\frac{I_2}{V_2}\right|_{I_1=0} = \frac{1}{z_{22}}$$

Note that $h_{11}$ has the dimension of an impedance, and $h_{22}$ has the dimension of an admittance. $h_{12}$ is a reverse transfer voltage ratio, whereas $h_{21}$ is a forward transfer current ratio. Since $h_{21}$ is obtained with $V_2 = 0$, it is usually called the **short-circuit current ratio.** Transistors are often characterized in terms of these hybrid parameters.

An equivalent circuit based on the hybrid matrix **H** is shown in Fig. 6.1.

From Eqs. (6.3b) and (6.3c) we can easily see that the condition for the two-port to be reciprocal is that $h_{21} = -h_{12}$.

A second hybrid matrix can be defined in a similar way as follows:

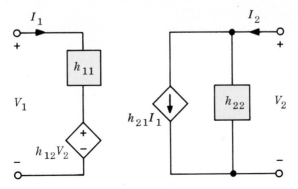

**Fig. 6.1**    An equivalent circuit of a two-port in terms of the hybrid parameters.

(6.4a)   $I_1 = g_{11}V_1 + g_{12}I_2$

(6.4b)   $V_2 = g_{21}V_1 + g_{22}I_2$

or, in matrix notation,

(6.5)
(6.6)
$$\begin{bmatrix} I_1 \\ V_2 \end{bmatrix} = \mathbf{G} \begin{bmatrix} V_1 \\ I_2 \end{bmatrix} \begin{bmatrix} g_{11} & g_{12} \\ g_{21} & g_{22} \end{bmatrix} \begin{bmatrix} V_1 \\ I_2 \end{bmatrix}$$

It can be checked that

$$\mathbf{G} = \mathbf{H}^{-1}$$

## 6.2   The Transmission Matrices

The two other conventional two-port parameters are called the transmission parameters. The transmission matrix relates the input variables $V_1$ and $I_1$ to the output variables $V_2$ and $-I_2$. Note that the variable used is $-I_2$ rather than $I_2$. Again, superposition allows us to write the input variables $(V_1, I_1)$ in terms of the output variables $(V_2, -I_2)$ by equations of the form

(6.7)
$$\begin{bmatrix} V_1 \\ I_1 \end{bmatrix} = \begin{bmatrix} A & B \\ C & D \end{bmatrix} \begin{bmatrix} V_2 \\ -I_2 \end{bmatrix}$$

We define

$$\mathbf{T} \triangleq \begin{bmatrix} A & B \\ C & D \end{bmatrix}$$

The minus sign is used with $I_2$ for historical reasons. With a minus sign in (6.7), the current at port 2 which *leaves* the two-port is designated as positive. The matrix **T** is called the **transmission matrix.** The parameters $A$, $B$, $C$, and $D$ are called the **transmission parameters.**† They are simply related to the open-circuit impedance and short-circuit admittance parameters. (See Table 17.1).

The main use of the transmission matrix is in dealing with a cascade connection of two-ports, as shown in Fig. 6.2. Assume that the transmission matrices of the two-ports $\mathfrak{N}^{(1)}$ and $\mathfrak{N}^{(2)}$ are known. As a result of the

† The matrix **T** is also called the *chain* matrix.

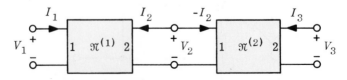

**Fig. 6.2**   The cascade connection of two two-ports.

**Table 17.1   Conversion Chart of Two-port Matrices***

| | Z | Y | T | T' | H | G |
|---|---|---|---|---|---|---|
| **Z** | $\begin{matrix} z_{11} & z_{12} \\ z_{21} & z_{22} \end{matrix}$ | $\begin{matrix} \dfrac{y_{22}}{\Delta_Y} & -\dfrac{y_{12}}{\Delta_Y} \\[2mm] -\dfrac{y_{21}}{\Delta_Y} & \dfrac{y_{11}}{\Delta_Y} \end{matrix}$ | $\begin{matrix} \dfrac{A}{C} & \dfrac{\Delta_T}{C} \\[2mm] \dfrac{1}{C} & \dfrac{D}{C} \end{matrix}$ | $\begin{matrix} \dfrac{D'}{C'} & \dfrac{1}{C'} \\[2mm] \dfrac{\Delta_{T'}}{C'} & \dfrac{A'}{C'} \end{matrix}$ | $\begin{matrix} \dfrac{\Delta_H}{h_{22}} & \dfrac{h_{12}}{h_{22}} \\[2mm] -\dfrac{h_{21}}{h_{22}} & \dfrac{1}{h_{22}} \end{matrix}$ | $\begin{matrix} \dfrac{1}{g_{11}} & -\dfrac{g_{12}}{g_{11}} \\[2mm] \dfrac{g_{21}}{g_{11}} & \dfrac{\Delta_G}{g_{11}} \end{matrix}$ |
| **Y** | $\begin{matrix} \dfrac{z_{22}}{\Delta_Z} & -\dfrac{z_{12}}{\Delta_Z} \\[2mm] -\dfrac{z_{21}}{\Delta_Z} & \dfrac{z_{11}}{\Delta_Z} \end{matrix}$ | $\begin{matrix} y_{11} & y_{12} \\ y_{21} & y_{22} \end{matrix}$ | $\begin{matrix} \dfrac{D}{B} & -\dfrac{\Delta_T}{B} \\[2mm] \dfrac{-1}{B} & \dfrac{A}{B} \end{matrix}$ | $\begin{matrix} \dfrac{A'}{B'} & -\dfrac{1}{B'} \\[2mm] \dfrac{-\Delta_{T'}}{B'} & \dfrac{D'}{B'} \end{matrix}$ | $\begin{matrix} \dfrac{1}{h_{11}} & -\dfrac{h_{12}}{h_{11}} \\[2mm] \dfrac{h_{21}}{h_{11}} & \dfrac{\Delta_H}{h_{11}} \end{matrix}$ | $\begin{matrix} \dfrac{\Delta_G}{g_{22}} & \dfrac{g_{12}}{g_{22}} \\[2mm] -\dfrac{g_{21}}{g_{22}} & \dfrac{1}{g_{22}} \end{matrix}$ |
| **T** | $\begin{matrix} \dfrac{z_{11}}{z_{21}} & \dfrac{\Delta_Z}{z_{21}} \\[2mm] \dfrac{1}{z_{21}} & \dfrac{z_{22}}{z_{21}} \end{matrix}$ | $\begin{matrix} -\dfrac{y_{22}}{y_{21}} & -\dfrac{1}{y_{21}} \\[2mm] -\dfrac{\Delta_Y}{y_{21}} & -\dfrac{y_{11}}{y_{21}} \end{matrix}$ | $\begin{matrix} A & B \\ C & D \end{matrix}$ | $\begin{matrix} \dfrac{D'}{\Delta_{T'}} & \dfrac{B'}{\Delta_{T'}} \\[2mm] \dfrac{C'}{\Delta_{T'}} & \dfrac{A'}{\Delta_{T'}} \end{matrix}$ | $\begin{matrix} -\dfrac{\Delta_H}{h_{21}} & -\dfrac{h_{11}}{h_{21}} \\[2mm] \dfrac{h_{22}}{h_{21}} & -\dfrac{1}{h_{21}} \end{matrix}$ | $\begin{matrix} \dfrac{1}{g_{21}} & \dfrac{g_{22}}{g_{21}} \\[2mm] \dfrac{g_{11}}{g_{21}} & \dfrac{\Delta_G}{g_{21}} \end{matrix}$ |
| **T'** | $\begin{matrix} \dfrac{z_{22}}{z_{12}} & \dfrac{\Delta_Z}{z_{12}} \\[2mm] \dfrac{1}{z_{12}} & \dfrac{z_{11}}{z_{12}} \end{matrix}$ | $\begin{matrix} -\dfrac{y_{11}}{y_{12}} & -\dfrac{1}{y_{12}} \\[2mm] -\dfrac{\Delta_Y}{y_{12}} & -\dfrac{y_{22}}{y_{12}} \end{matrix}$ | $\begin{matrix} \dfrac{D}{\Delta_T} & \dfrac{B}{\Delta_T} \\[2mm] \dfrac{C}{\Delta_T} & \dfrac{A}{\Delta_T} \end{matrix}$ | $\begin{matrix} A' & B' \\ C' & D' \end{matrix}$ | $\begin{matrix} \dfrac{1}{h_{12}} & \dfrac{h_{11}}{h_{12}} \\[2mm] \dfrac{h_{22}}{h_{12}} & \dfrac{\Delta_H}{h_{12}} \end{matrix}$ | $\begin{matrix} -\dfrac{\Delta_G}{g_{12}} & -\dfrac{g_{22}}{g_{12}} \\[2mm] -\dfrac{g_{11}}{g_{12}} & -\dfrac{1}{g_{12}} \end{matrix}$ |
| **H** | $\begin{matrix} \dfrac{\Delta_Z}{z_{22}} & \dfrac{z_{12}}{z_{22}} \\[2mm] -\dfrac{z_{21}}{z_{22}} & \dfrac{1}{z_{22}} \end{matrix}$ | $\begin{matrix} \dfrac{1}{y_{11}} & -\dfrac{y_{12}}{y_{11}} \\[2mm] \dfrac{y_{21}}{y_{11}} & \dfrac{\Delta_Y}{y_{11}} \end{matrix}$ | $\begin{matrix} \dfrac{B}{D} & \dfrac{\Delta_T}{D} \\[2mm] -\dfrac{1}{D} & \dfrac{C}{D} \end{matrix}$ | $\begin{matrix} \dfrac{B'}{A'} & \dfrac{1}{A'} \\[2mm] -\dfrac{\Delta_{T'}}{A'} & \dfrac{C'}{A'} \end{matrix}$ | $\begin{matrix} h_{11} & h_{12} \\ h_{21} & h_{22} \end{matrix}$ | $\begin{matrix} \dfrac{g_{22}}{\Delta_G} & -\dfrac{g_{12}}{\Delta_G} \\[2mm] -\dfrac{g_{21}}{\Delta_G} & \dfrac{g_{11}}{\Delta_G} \end{matrix}$ |
| **G** | $\begin{matrix} \dfrac{1}{z_{11}} & -\dfrac{z_{12}}{z_{11}} \\[2mm] \dfrac{z_{21}}{z_{11}} & \dfrac{\Delta_Z}{z_{11}} \end{matrix}$ | $\begin{matrix} \dfrac{\Delta_Y}{y_{22}} & \dfrac{y_{12}}{y_{22}} \\[2mm] -\dfrac{y_{21}}{y_{22}} & \dfrac{1}{y_{22}} \end{matrix}$ | $\begin{matrix} \dfrac{C}{A} & -\dfrac{\Delta_T}{A} \\[2mm] \dfrac{1}{A} & \dfrac{B}{A} \end{matrix}$ | $\begin{matrix} \dfrac{C'}{D'} & -\dfrac{1}{D'} \\[2mm] \dfrac{\Delta_{T'}}{D'} & \dfrac{B'}{D'} \end{matrix}$ | $\begin{matrix} \dfrac{h_{22}}{\Delta_H} & -\dfrac{h_{12}}{\Delta_H} \\[2mm] -\dfrac{h_{21}}{\Delta_H} & \dfrac{h_{11}}{\Delta_H} \end{matrix}$ | $\begin{matrix} g_{11} & g_{12} \\ g_{21} & g_{22} \end{matrix}$ |
| The two-port is reciprocal if | $z_{12} = z_{21}$ | $y_{12} = y_{21}$ | $\Delta_T = 1$ | $\Delta_{T'} = 1$ | $h_{12} = -h_{21}$ | $g_{12} = -g_{21}$ |

* All matrices in the same row are equal.

connection, $-I_2$ is both the current *leaving* $\mathfrak{N}^{(1)}$ and the current *entering* $\mathfrak{N}^{(2)}$, and $V_2$ is the output voltage of $\mathfrak{N}^{(1)}$ as well as the input voltage of $\mathfrak{N}^{(2)}$; thus, with the notations of Fig. 6.2,

$$(6.8a) \qquad \begin{bmatrix} V_1 \\ I_1 \end{bmatrix} = \begin{bmatrix} A^{(1)} & B^{(1)} \\ C^{(1)} & D^{(1)} \end{bmatrix} \begin{bmatrix} V_2 \\ -I_2 \end{bmatrix}$$

$$(6.8b) \qquad \begin{bmatrix} V_2 \\ -I_2 \end{bmatrix} = \begin{bmatrix} A^{(2)} & B^{(2)} \\ C^{(2)} & D^{(2)} \end{bmatrix} \begin{bmatrix} V_3 \\ -I_3 \end{bmatrix}$$

Combining (6.8*a*) and (6.8*b*), we obtain a relation of the form

(6.9*a*)
$$\begin{bmatrix} V_1 \\ I_1 \end{bmatrix} = \begin{bmatrix} A & B \\ C & D \end{bmatrix} \begin{bmatrix} V_3 \\ -I_3 \end{bmatrix}$$

where

(6.9*b*)
$$\begin{bmatrix} A & B \\ C & D \end{bmatrix} = \begin{bmatrix} A^{(1)} & B^{(1)} \\ C^{(1)} & D^{(1)} \end{bmatrix} \begin{bmatrix} A^{(2)} & B^{(2)} \\ C^{(2)} & D^{(2)} \end{bmatrix}$$

Thus, *the transmission matrix of a cascade of two two-ports is the product of the transmission matrices of the individual two-ports.* It is this property which makes the transmission matrix so useful. This property is used in the design of telephone systems, microwave networks, radars, and, with the help of a fairly obvious analogy, in the design of optical instruments.

Finally, if we express the output variables $V_2$ and $-I_2$ in terms of the input variables $V_1$ and $I_1$, we obtain

(6.10)
$$\begin{bmatrix} V_2 \\ -I_2 \end{bmatrix} = \begin{bmatrix} A' & B' \\ C' & D' \end{bmatrix} \begin{bmatrix} V_1 \\ I_1 \end{bmatrix}$$

By comparing Eq. (6.10) with Eq. (6.7), we conclude that

(6.11)
$$\mathbf{T}' \triangleq \begin{bmatrix} A' & B' \\ C' & D' \end{bmatrix} = \begin{bmatrix} A & B \\ C & D \end{bmatrix}^{-1}$$

This matrix is the last of the six possible characterizations of a two-port made of linear time-invariant elements. Table 17.1 shows how, given any one of these characterizations, one can calculate any one of the others.

**Remark**   Not all two-ports have a particular two-port representation. For example, the two-port shown in Fig. 6.3*a* does not have an impedance matrix representation; indeed, all the elements $z_{ij}$ are infinite! However, it has an admittance matrix representation; in fact, $y_{11} = y_{22} = Z^{-1}$ and $y_{12} = y_{21} = -Z^{-1}$. Another way of grasping the fact that its $\mathbf{Z}$ matrix does not exist is to observe that det $(\mathbf{Y}) = 0$. Thus, its admittance matrix $\mathbf{Y}$ is singular,

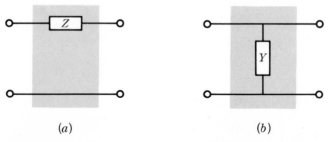

(*a*)                               (*b*)

**Fig. 6.3**   Examples of two-ports that have no impedance and no admittance matrix, respectively.

and hence the matrix **Y** does not have an inverse.   By Eq. (5.7) if such an inverse existed, it would be the **Z** matrix.

Consider now the two-port shown in Fig. 6.3b; this two-port does not have an admittance matrix representation, but it has an impedance matrix representation.

Finally, consider the ideal transformer; its defining equations are (in terms of Laplace transforms)

$$V_1 = nV_2$$

$$I_1 = -\frac{1}{n}I_2$$

Since it is impossible to express **V** as a function of **I** and vice versa, the ideal transformer has neither an impedance matrix nor an admittance matrix representation.   It clearly has a hybrid matrix representation, namely,

$$\begin{bmatrix} V_1 \\ I_2 \end{bmatrix} = \begin{bmatrix} 0 & n \\ -n & 0 \end{bmatrix} \begin{bmatrix} I_1 \\ V_2 \end{bmatrix}$$

Given this background the following fact is important.   Suppose that we are given any linear time-invariant network made of passive $R$'s, $L$'s, $C$'s, transformers, and gyrators, and suppose that we pick out two pairs of terminals ①① and ②②; then the resulting two-port has a hybrid matrix representation.

**Exercise 1**   Find the transmission matrices and hybrid matrices of the two-ports shown in Fig. 6.3a and b.

**Exercise 2**   Use the result of Exercise 1 to calculate the transmission matrix of the two-port shown in Fig. 6.4.

**Exercise 3**   Use Eq. (6.5) to derive the physical interpretation of the $g$ parameters.

**Exercise 4**   Show that if a two-port is reciprocal (that is, $z_{12} = z_{21}$), then $h_{21} = -h_{12}$ and $AD - BC = 1$.

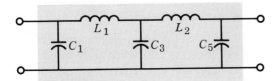

**Fig. 6.4**    Ladder network.

**Exercise 5**    If port 2 is loaded by the impedance $Z_L$, show that the driving-point impedance at port 1 is given by

$$Z_1 = \frac{AZ_L + B}{CZ_L + D}$$

## Summary

- The subject of this chapter is the characterization of two-ports. A *two-port* is a four-terminal network subject to the restriction that the current entering the two-port by terminal ① is always equal to the current leaving it by terminal ①′, and the current entering it by terminal ② is always equal to that leaving it by terminal ②′. Terminals ① and ①′ constitute the input port, and its corresponding port variables are $v_1$ and $i_1$. Terminals ② and ②′ constitute the output port, and its corresponding variables are $v_2$ and $i_2$. For reference directions, refer to Fig. 2.2. A two-port is completely specified by the set of all its possible port-voltage and port-current waveforms, that is, $v_1(\cdot)$, $i_1(\cdot)$, $v_2(\cdot)$, and $i_2(\cdot)$. Among the four port variables $v_1$, $i_1$, $v_2$, and $i_2$, the two-port imposes two constraining equations, and the terminations impose two more equations.

- Typical characterizations of nonlinear resistive two-ports are $\mathbf{v} = \mathbf{r(i)}$ and $\mathbf{i} = \mathbf{g(v)}$. Such two-ports exhibit no memory. When a nonlinear resistive two-port is terminated at both ports by dc sources and linear resistors, the four port variables are constant and determine the operating point $Q$. In practice the operating point is found graphically. In the neighborhood of the operating point, the nonlinear two-port can be approximated by using only the first-order terms in the Taylor expansions of the two-port equations. This yields a "small-signal" linear model of the two-port; this model is only valid in the neighborhood of the operating point. Thus if a small-signal input is superimposed on the battery at the input port, the port variables can be obtained by determining first the operating point and then the incremental port voltages and currents. These incremental variables are related by incremental impedance, admittance, or hybrid matrices which are defined for this operating point. In this discussion all networks were purely resistive.

- Two nonlinear coupled inductors constitute a two-port which is characterized by equations of the form $\mathbf{i} = \hat{\mathbf{i}}(\boldsymbol{\phi})$ for the flux-controlled case, and $\boldsymbol{\phi} = \hat{\boldsymbol{\phi}}(\mathbf{i})$ for the current-controlled case. For both cases we showed how the port variables $\mathbf{v}$ and $\mathbf{i}$ are related.

- When a two-port is made of linear time-invariant elements (in particular, with no independent sources) and when only the zero-state response is sought, the two-port may be characterized by a two-port matrix. The elements of such matrices are network functions. Each element is thus a rational function of $s$. The matrix $\mathbf{Z}$, the open-circuit impedance matrix, and $\mathbf{Y}$, the short-circuit admittance matrix, relate the port-voltage vector $\mathbf{V}$ to the port-current vector $\mathbf{I}$. The elements $z_{ij}$ of the impedance matrix are open-circuit driving-point and open-circuit transfer impedances of the two-port. The elements $y_{ij}$ of the admittance matrix are short-circuit driving-point and transfer admittances of the two-port. The $z_{ij}$ and $y_{ij}$ are related because $\mathbf{Z} = \mathbf{Y}^{-1}$. The hybrid matrices relate a vector whose components are the voltage at one port and the current at the other port to the vector whose components are the other port variables. The transmission matrices relate the input port variables and the output port variables. The transmission matrices are particularly useful for cascade connections of two-ports. The relations between the six two-port matrices are given in Table 17-1, page 747.

- A two-port is said to be *reciprocal* when $z_{12} = z_{21}$. To express this condition in terms of the other parameters, refer to Table 17.1.

## Problems

One-port
(small-signal
equivalent)

**1.** For the current-controlled nonlinear resistor shown in Fig. P16.14 find the small-signal equivalent resistance when the bias current is

*a.*  00.2 mA

*b.*  00.5 mA

*c.*  1.0 mA

Two-port
(small-signal
equivalent)

**2.** The nonlinear resistive two-port shown in Fig. P17.2 is characterized by the equations

$$v_1 = -i_1 + i_1^3$$
$$v_2 = 2i_1 + 3i_1 i_2$$

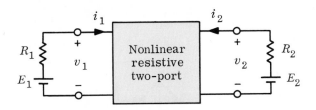

**Fig. P17.2**

The two-port is terminated on each side by series connections of batteries and linear resistors, as shown in the figure. Let $E_1 = 8$ volts, $E_2 = 14$ volts, $R_1 = 1$ ohm, and $R_2 = 4$ ohms.

*a.* Determine the operating point of the nonlinear circuit; i.e., determine the port voltages $V_1$ and $V_2$ and the port currents $I_1$ and $I_2$ which satisfy Kirchhoff's laws, the branch equations, and the two-port characterization. Do you think the circuit has a unique operating point? If so, why?

*b.* Determine the incremental impedance matrix of the two-port at the operating point obtained in (*a*). Draw a small-signal equivalent circuit of the two-port using resistors and dependent sources.

*c.* If a small-signal voltage $e_1$ is superimposed with the dc battery $E_1$ at the input, compute the small-signal voltage ratio, i.e., the ratio of the incremental voltage at the output port to the input incremental voltage $e_1$.

Transistor amplifier

**3.** The circuit shown in Fig. P17.3*b* is the incremental model corresponding to the common-emitter transistor amplifier shown in Fig. P17.3*a*.

*a.* Determine the operating point $Q$ of the amplifier shown in Fig. P17.3*a*. Write KCL for the path consisting of $R_L$, collector, emitter, and $R_e$,

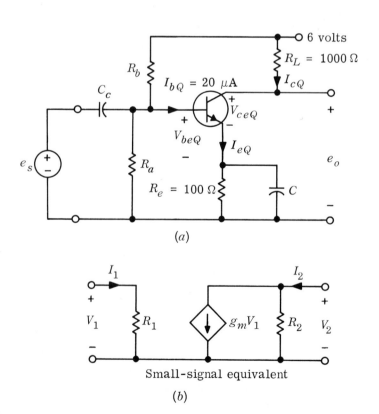

(*a*)

Small-signal equivalent

(*b*)

Fig. P17.3

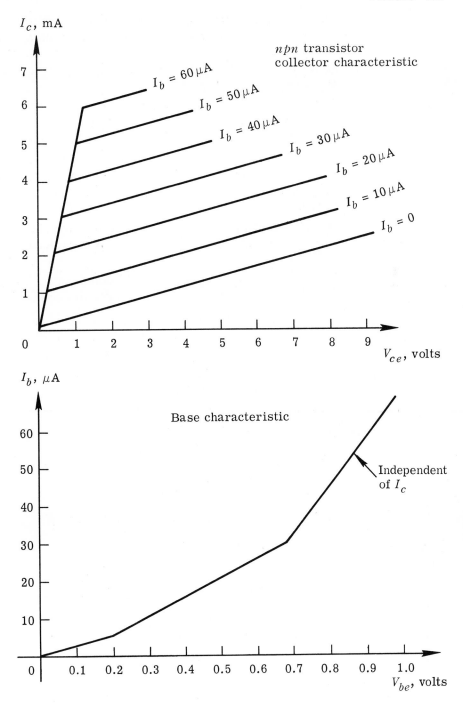

**Fig. P17.3**
(*Continued*)

and use the accompanying transistor characteristics to obtain $I_{cQ}$, $V_{ceQ}$, and $V_{beQ}$. (Observe that for determining the operating point, $C_c$ and $C$ are open circuits, and we set $e_s = 0$. Note also $I_{bQ} \ll I_{cQ}$.)

b.  Write the two-port equations for the incremental circuit shown in Fig. P17.3b in the form

$$V_1 = h_{11}I_1 + h_{12}V_2$$
$$I_2 = h_{21}I_1 + h_{22}V_2$$

c.  From the accompanying curves and the dc operating point calculated in (a), determine the small-signal parameters $R_1$, $g_m$, and $R_2$.

d.  Suppose that the small-signal voltage source $e_s$ operates at such frequencies that $C$ and $C_c$ are short circuits. Write KVL for the $R_L$-collector-emitter path and draw a line representing the possible values $I_c$-$V_{ce}$ for various values of $I_b$; that is, draw the ac load line. (Assume $I_b \ll I_c$.)

e.  If the source phasor $E_s(j\omega)$ has a magnitude of 0.1 volt, determine $K = E_o(j\omega)/E_s(j\omega)$ with respect to the operating point of $I_{bQ} = 20~\mu A$.

Impedance
matrix

**4.** a.  Obtain the two-port admittance matrix of the circuit shown in Fig. P16.8 (the ports are $BE$ and $CE$) in terms of the frequency $\omega$.

b.  Calculate the two-port impedance matrix of the network shown in Fig. P17.4, between ports $AE$ and $DE$.

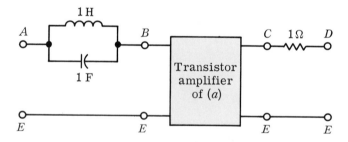

**Fig. P17.4**

Two-port
matrices

**5.** Consider the linear time-invariant two-ports shown in Fig. P17.5. For *each* of these two-ports, write *one* of the following two-port matrices:

a.  Impedance matrix

b.  Admittance matrix

c.  Hybrid matrix

d.  Transmission matrix

Choose in each case the matrix easiest to calculate.

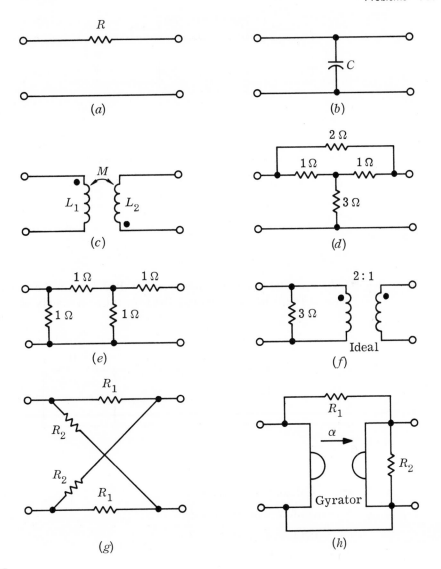

**Fig. P17.5**

**6.** Consider a linear time-invariant *RLC* two-port shown in Fig. P17.6. Zero-state responses to a step-function current source at port 1 are measured in two cases: (*a*) when port 2 is short-circuited and (*b*) when port 2 is terminated by a 4-ohm resistor. From the measurements given below, determine the short-circuit admittance matrix $\mathbf{Y}(s)$ of the two-port.

$$v_{1a} = \tfrac{2}{3}u(t)[1 - \epsilon^{-3t/2}] \qquad v_{1b} = \tfrac{6}{7}u(t)[1 - \epsilon^{-7t/6}]$$
$$i_{2a} = \tfrac{1}{2}u(t)[1 - \epsilon^{-3t/2}] \qquad i_{2b} = \tfrac{1}{14}u(t)[1 - \epsilon^{-7t/6}]$$

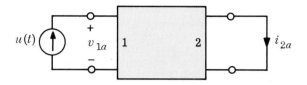

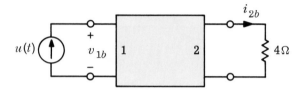

**Fig. P17.6**

Transmission
matrix **7.** Assume the two-ports shown in Fig. P17.7 are in the sinusoidal steady state at frequency $\omega$; calculate the transmission matrix of each two-port.

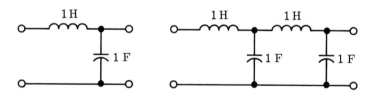

**Fig. P17.7**

Relations
between
matrices **8.** *a.*  Calculate **Z** in terms of **Y**.

    *b.*  Calculate **Y** in terms of **Z**.

    *c.*  Calculate **T** in terms of **Z**.

    *d.*  Calculate **H** in terms of **T**.

Refer to Table 17.1 to check your answers.

Cascade
connection **9.** Prove that a two-port obtained by the cascade connection of two ideal gyrators is equivalent to an ideal transformer.  Give the turns ratio of the ideal transformer in terms of the gyration ratios of the gyrators.

Impedance
matrix **10.** Calculate the impedance matrix and the admittance matrix of the two-port shown in Fig. P17.10 (such a two-port is called a **symmetric lattice**).

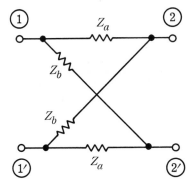

**Fig. P17.10**

Input imped-
ance of a
terminated
two-port **11.** The two-port $\mathfrak{N}$ is driven by the generator $G$ and is terminated by the impedance $Z_L$; $V_1, I_1$, and $I_L$ are the Laplace transforms of the corresponding zero-state responses.   The reference direction of $I_L$ is shown in Fig. P17.11.   The two-port $\mathfrak{N}$ is made of linear time-invariant elements.

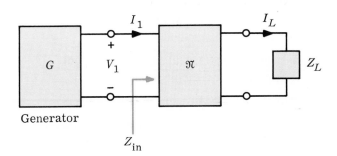

**Fig. P17.11**

*a.* Show that if $\mathfrak{N}$ is a reciprocal two-port, then

$$\left(\frac{I_L}{I_1}\right)^2 = \frac{\partial Z_{\text{in}}}{\partial Z_L}$$

(Hint: Formulate the two-port equations using the transmission matrix.)

*b.* Under the same conditions show that

$$\frac{\partial^2 Z_{\text{in}}}{\partial Z_L{}^2} = -2C\left(\frac{I_L}{I_1}\right)^3$$

where $C$ is the $(2,1)$ element of the transmission matrix of $\mathfrak{N}$.

*c.* Your boss suggests that you design a two-port made of linear time-invariant resistors such that when it is loaded by $R_L$, its input resistance $R_{in}$ varies according to the following schedule:

| $R_L$, ohms | $R_{in}$, ohms |
|---|---|
| 100 | 300 |
| 150 | 340 |
| 200 | 420 |

Can you do it? If yes, design the network; if not, say why it cannot be done.

**Network with independent sources**

**12.** Let $\mathfrak{N}$ be a two-port made of linear time-invariant elements *and* independent sources.

a. Show that the relations between $V_1$, $V_2$, $I_1$, and $I_2$ can be written in the form

$$V_1 = z_{11}I_1 + z_{12}I_2 + E_1$$
$$V_2 = z_{21}I_1 + z_{22}I_2 + E_2$$

b. Give a physical interpretation for $E_1$ and $E_2$.

**Impedance ratios**

**13.** Let $\mathfrak{N}$ be a two-port made of linear time-invariant elements. Show that

$$\frac{Z(\text{at port 1; port 2 oc})}{Z(\text{at port 1; port 2 sc})} = \frac{Z(\text{at port 2; port 1 oc})}{Z(\text{at port 2; port 1 sc})}$$

(Hint: Relate these ratios to $z_{11}y_{11} = z_{22}y_{22}$.)

**Reciprocal two-ports**

**14.** The purpose of this problem is to justify a method for checking whether a given two-port is a reciprocal two-port at frequency $\omega_0$. Consider the two situations shown in Fig. P17.14. All measurements are sinusoidal

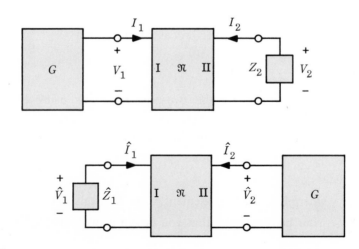

**Fig. P17.14**

steady-state measurements made at frequency $\omega_0$; consequently $V_1, V_2, I_1,$ $I_2, \hat{V}_1, \hat{V}_2, \hat{I}_1,$ and $\hat{I}_2$ are the *phasors* representing the sinusoidal waveforms. The impedances $\hat{Z}_1$ and $Z_2$ and the internal impedance of the generator are arbitrary, except that

$$\frac{I_1}{I_2} \neq \frac{\hat{I}_1}{\hat{I}_2}$$

Show that the two-port is reciprocal at frequency $\omega_0$ if and only if $V_1\hat{I}_1 + V_2\hat{I}_2 = \hat{V}_1 I_1 + \hat{V}_2 I_2$.

*Y-Δ trans-*
*formation*

**15.** Show that the three-terminal circuits shown in Fig. P17.15 are equivalent if and only if

$$Y_{ab} = \frac{Y_a Y_b}{Y_a + Y_b + Y_c}$$

and two similar formulas obtained by permuting the subscripts, viz.,

$$Y_{bc} = \frac{Y_b Y_c}{Y_a + Y_b + Y_c}$$

$$Y_{ca} = \frac{Y_c Y_a}{Y_a + Y_b + Y_c}$$

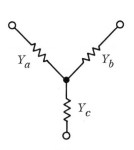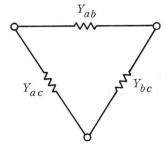

**Fig. P17.15**

(Rule: The admittance of an arm of the Δ is equal to the product of the admittance of the *adjacent* arms of the Y divided by the sum of Y admittances.)

*Δ-Y trans-*
*formation*

**16.** Refer to Fig. P17.15. Show that the conditions for equivalence can also be expressed as follows:

$$Y_a = \frac{Y_{ab}Y_{bc} + Y_{bc}Y_{ca} + Y_{ca}Y_{ab}}{Y_{bc}}$$

and two similar formulas obtained by permuting subscripts. (Rule: The admittance of an arm of the Y is equal to the sum of the products of the Δ admittances taken two at a time divided by the admittance of the opposite arm of the Δ.)

# 18

# Resistive Networks

In this chapter we shall discuss some important and interesting properties of lumped resistive networks. We shall restrict our treatment to resistive networks but we shall consider nonlinear as well as linear networks.† We shall present three fundamental concepts. The first one is related to the relationships between physical networks and network models and faces the question of the existence and uniqueness of solution of a network. The second deals with the nature of the solution of a resistive network and, in particular, its relation to power dissipation. The third concept is the basic property of passive resistive networks which states that the voltage gain and the current gain of such a network cannot exceed 1 in absolute value.

## 1  Physical Networks and Network Models

In Chap. 2 we discussed briefly the problem of modeling. Given a *physical* network, we use a network *model* to represent (approximately) its behavior. Under proper physical environment, within the allowed tolerance of engineering specifications, and assuming good judgment on the part of the engineer who selects the network model, the model serves as a good approximation for prediction and design. As an example, we usually think of an attenuator pad (made of physical resistors) as a resistive network. The function of an attenuator pad is, of course, to reduce the signal level but maintain its waveform. However, strictly speaking, an attenuator pad is not a resistive network. First, whenever there is a voltage difference, there is an electric field and some electrostatic energy is stored; second, the presence of a current implies some magnetic energy stored. From a modeling point of view, these energies should be modeled by capacitive and inductive elements. Thus, an attenuator pad is, strictly speaking, not a purely resistive network. On the other hand, since the capacitive and inductive effects in the usual operating frequencies are so small, we can comfortably choose a resistive network model for analysis and synthesis of the attenuator pad. When approximations are wisely

† General nonlinear networks are beyond the scope of this introductory book.

made and when the physical network is judiciously built, the errors involved are so small that they have no practical consequences.

In nonlinear networks, because of the fact that the description of a single nonlinear element involves a whole curve rather than a single number (as in linear time-invariant elements), the modeling is far more difficult. Often we find that a network model behaves much differently than the physical network. This means that some important aspect of the physical network is missing in the model.

To illustrate the point, consider the situation shown in Fig. 1.1. Suppose the internal resistance $R$ of the source is small so that the voltage drop across the source resistor is negligible compared to the source voltage $E$ for the values of the current $i$ which we expect. Then, the $vi$ characteristic of the source is shown in Fig. 1.2$a$ as a solid line and its approximate characteristic is shown as a dashed line. Suppose that measurements of the nonlinear load resistor yield a characteristic that saturates, and that as $i$ increases the slope becomes very small, making it seem reasonable to take the slope to be zero for current values larger than, say, $I_0$, as shown by the dotted curve in Fig. 1.2$b$. Now if the source is connected to the nonlinear resistor, as shown in Fig. 1.1, the same current will flow through both elements, and their terminal voltage will be the same. Therefore, an easy way of finding the actual voltage $v$ and current $i$ in the circuit is to plot both characteristics on the same set of axes and find their intersection. If we were to do this using the two *approximate* characteristics, as Fig. 1.3 shows, we would find that there is no intersection! A mathematician would say, "The problem does not have a solution." Obviously, this is physically absurd. We know that there is a solution to the physical problem; the point is that *our model (resulting from the approximations) has no solution.*

This example is, of course, trivial, and the remedy is obvious: fix up the models of the source and the nonlinear resistor. On the other hand, when we deal with problems so complicated that they have to be solved on computers, the diagnosis is not so easy.† For these reasons, it is useful to know in advance whether or not a particular network *model* will have a solution, and in case it has a solution, whether it is unique.

This example indicates that even though physical networks always have at least one solution, there are *nonlinear resistive network models which have no solutions* (i.e., it is impossible to find a set of branch voltages and branch currents such that the two Kirchhoff laws and the branch equations are satisfied). Furthermore, we have seen examples of *nonlinear resistive networks that have several solutions* (i.e., the solution is not unique, or, equivalently, there are several sets of branch voltages and branch currents

---

† Computers operate on the basis of "garbage in, garbage out." When we face a computer output which is obviously garbage, it is often difficult to decide whether the error lies in the programming (i.e., the computer did not receive correct instructions), or in the data fed to the computer (the computer worked on the wrong problem), or finally, in the model (this one hurts the most, since the engineer started the garbage production!).

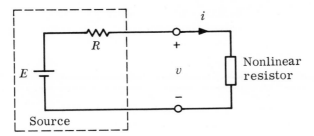

**Fig. 1.1**   A nonlinear resistor is connected to a source.

that satisfy the two Kirchhoff laws and the branch equations). A simple example is that of a suitably biased tunnel diode. The sets $(v_1,i_3)$, $(v_2,i_2)$, and $(v_3,i_1)$ shown in Fig. 1.4 are solutions. We should add that the model used to represent a tunnel diode is rather primitive in that it neglects to consider the combined diffusion and depletion-layer capacitance of the diode which makes the solution $(v_2,i_2)$ unstable. This, however, is another

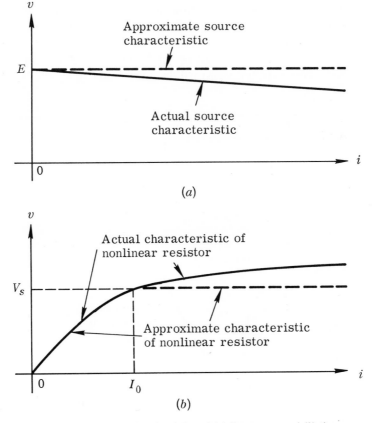

**Fig. 1.2**   Approximate and actual characteristics of (a) the source and (b) the non-linear resistor.

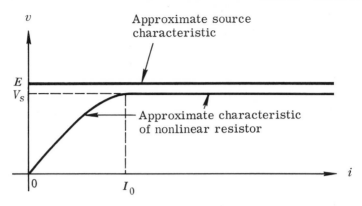

**Fig. 1.3**   The approximate characteristics of the source and the nonlinear resistor superimposed indicate that there is no solution.

story.   Recall also that in Chap. 13 we encountered some linear time-invariant network models that were degenerate.   Some had no solution, whereas others had an infinite number of solutions.   Thus, we see that both linear and nonlinear network models may have *no solution,* or some may have *more than one solution.*   We therefore have a definite interest in knowing some conditions under which a solution exists and is unique.   Some results to this effect will be given in the next section.

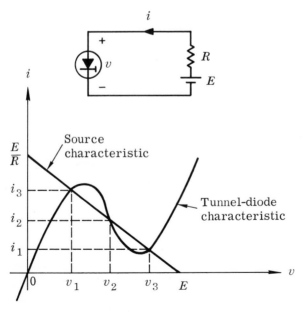

**Fig. 1.4**   A tunnel-diode circuit with three solutions.

**2** **Analysis of Resistive Networks from a Power Point of View**

In this section we would like to bring out some general facts concerning linear and nonlinear resistive networks. The common features of the presentation are its generality (there is no bound on the number of elements constituting the network, nor is there any condition on the graph of the network) and the key role played by the *power dissipation.* We begin our discussion with linear networks; we comment on direct methods of solution of the network equations and then calculate the power dissipated in a linear resistive network in terms of the loop resistance matrix and the node conductance matrix. We use these results to establish the *minimum property.*

**2.1**   **Linear Networks Made of Passive Resistors**

As a first step we would like to review several aspects of the problem of solving a linear resistive network. Mathematically, the problem is to solve a set of linear algebraic equations.

Consider a linear resistive network which has $b$ resistors and $l$ independent loops, and which is driven by a number of independent sources. By loop analysis, we obtain loop equations of the form

(2.1)   $\mathbf{R}\mathbf{i} = \mathbf{e}_s$

where $\mathbf{i} = (i_1, i_2, \ldots, i_l)^T$ is the loop current vector, $\mathbf{R}$ is the $l \times l$ loop resistance matrix, $\mathbf{e}_s = (e_{s1}, e_{s2}, \ldots, e_{sl})^T$ is the loop voltage source vector, and its $k$th component $e_{sk}$ is the sum of the source voltages in loop $k$. We know that $\mathbf{R}$ is a real symmetric matrix. The system of equations (2.1) is system of linear algebraic equations; therefore, by Cramer's rule the solution can be written in terms of determinants. Then

$$i_1 = \frac{\Delta_1}{\Delta}, \; i_2 = \frac{\Delta_2}{\Delta}, \ldots, \; i_l = \frac{\Delta_l}{\Delta}$$

where $\Delta = \det(\mathbf{R})$ and $\Delta_k$ is the determinant of the matrix obtained by replacing the $k$th column of $\mathbf{R}$ by the right-hand side of (2.1). We shall see later that if all the resistors are passive (i.e., all resistances are positive), then $\Delta$ is always a positive number.

**Example 1**   For the circuit shown in Fig. 2.1, the equations are

(2.2)   $$\begin{bmatrix} R_1 + R_2 & -R_2 & -R_1 \\ -R_2 & R_2 + R_4 & -R_4 \\ -R_1 & -R_4 & R_1 + R_3 + R_4 \end{bmatrix} \begin{bmatrix} i_1 \\ i_2 \\ i_3 \end{bmatrix} = \begin{bmatrix} e_{s1} \\ 0 \\ 0 \end{bmatrix}$$

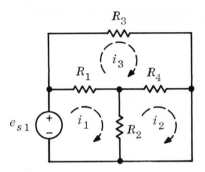

**Fig. 2.1**   Example of a linear resistive network.

$$i_2 = \frac{\begin{vmatrix} R_1 + R_2 & e_{s1} & -R_1 \\ -R_2 & 0 & -R_4 \\ -R_1 & 0 & R_1 + R_3 + R_4 \end{vmatrix}}{\begin{vmatrix} R_1 + R_2 & -R_2 & -R_1 \\ -R_2 & R_2 + R_4 & -R_4 \\ -R_1 & -R_4 & R_1 + R_3 + R_4 \end{vmatrix}}$$

$$= e_{s1} \frac{R_2R_1 + R_2R_3 + R_2R_4 + R_1R_4}{R_1R_2R_3 + R_1R_3R_4 + R_2R_3R_4}$$

In principle, the equations of (2.1) can always be solved by determinants. In practice however, whenever the number of equations is larger than four or five, the Gauss elimination method or any of its variants is always used. Indeed, solving for all the currents by the Gauss elimination method requires about $l^3/3$ multiplication operations, whereas evaluating $\Delta$ requires at least $l \times l!$ multiplications. For $l = 10$, the numbers turn out to be about 350 and 36,300,000, respectively.

**Remark**   In the remainder of this section, in order to simplify notation, we shall assume that all voltage sources are constant voltage sources (batteries). Consequently, the loop voltage source vector $\mathbf{e}_s$ is constant, and so are the loop currents and the branch currents. Note that if $\mathbf{e}_s$ were a function of time, then the system of equations of (2.1) would have to be solved for each $t$. All the results that follow are applicable to this case provided one keeps track of the time.

*Power dissipated*   Let us next calculate the power dissipated in the resistors. Let $\mathbf{j} = (j_1, j_2, \ldots, j_b)^T$ be the branch current vector of the $b$ resistors. The power dissipated by the resistors is

(2.3)   $p = \sum_{k=1}^{b} R_k j_k^2$

where $R_k$ is the resistance of the $k$th resistor.   We may write (2.3) in matrix notation as follows:

$$p = [j_1 \ j_2 \ \cdots \ j_b] \begin{bmatrix} R_1 & 0 & \cdots & 0 \\ 0 & R_2 & \cdots & 0 \\ \cdots\cdots\cdots\cdots\cdots \\ 0 & 0 & \cdots & R_b \end{bmatrix} \begin{bmatrix} j_1 \\ j_2 \\ \vdots \\ j_b \end{bmatrix}$$

Let $\mathbf{R}_b$ be the branch resistance matrix; it is diagonal and $\mathbf{R}_b \triangleq$ diag $(R_1, R_2, \ldots, R_b)$.   In terms of $\mathbf{R}_b$, Eq. (2.3) may be written compactly as

(2.4)   $p = \mathbf{j}^T \mathbf{R}_b \mathbf{j}$

If now we recall the relation between the branch currents and loop currents ($\mathbf{j} = \mathbf{B}^T \mathbf{i}$), we may rewrite (2.4) as

(2.5)   $p = \mathbf{i}^T \mathbf{B} \mathbf{R}_b \mathbf{B}^T \mathbf{i}$

where $\mathbf{B}$ is the $l \times b$ fundamental loop matrix.   Observing that $\mathbf{B R}_b \mathbf{B}^T \triangleq \mathbf{R}$ is the loop resistance matrix which appears naturally in the loop analysis, we obtain

(2.6)   $p = \mathbf{i}^T \mathbf{R} \mathbf{i}$

---

**Example 2**   Using the circuit of Fig. 2.1, we have

$$p = \sum_{k=1}^{4} R_k j_k^2 = [i_1 \ i_2 \ i_3] \begin{bmatrix} R_1 + R_2 & -R_2 & -R_1 \\ -R_2 & R_2 + R_4 & -R_4 \\ -R_1 & -R_4 & R_1 + R_3 + R_4 \end{bmatrix} \begin{bmatrix} i_1 \\ i_2 \\ i_3 \end{bmatrix}$$

Completing the calculations, we obtain

$$p = (R_1 + R_2)i_1{}^2 + (R_2 + R_4)i_2{}^2 + (R_1 + R_2 + R_4)i_3{}^2$$
$$- 2R_2 i_1 i_2 - 2R_1 i_1 i_3 - 2R_4 i_2 i_3$$

---

A dual analysis in terms of branch voltages and node-to-datum voltages would give us the dual of (2.4) and (2.6).   Let $\mathbf{v} = (v_1, v_2, \ldots, v_b)^T$ be the branch voltage vector and $\mathbf{e} = (e_1, e_2, \ldots, e_n)^T$ be the node-to-datum voltage vector.   Note that $n$ denotes the number of node-to-datum voltages. The power dissipated is then

(2.7)   $p = \mathbf{v}^T \mathbf{G}_b \mathbf{v} = \mathbf{e}^T \mathbf{G} \mathbf{e}$

where $\mathbf{G}_b \triangleq$ diag $(G_1, G_2, \ldots, G_b)$ is the branch conductance matrix and $\mathbf{G} \triangleq \mathbf{A G}_b \mathbf{A}^T$ is the node conductance matrix which appears in the node analysis, where $\mathbf{A}$ is the reduced incidence matrix.   We are now in a position to establish an important property.

**THEOREM**   The loop resistance matrix **R** and the node conductance matrix **G** of a linear network made of passive resistors (i.e., all resistances are positive) are positive definite matrices.†

*Proof*   The demonstration of the above property is immediate.   We wish to prove that for a passive resistive network, $\mathbf{i} \neq \mathbf{0}$ implies that $p = \mathbf{i}^T\mathbf{R}\mathbf{i} > 0$. Since **i** is a fundamental loop current vector, $\mathbf{i} \neq \mathbf{0}$ implies that some branch current, say $j_k$, is nonzero, since by assumption $R_k$ is positive, $R_k j_k^2 > 0$; hence $p = \sum_{k=1}^{b} R_k j_k^2 = \mathbf{i}^T\mathbf{R}\mathbf{i} > 0$. The proof for the node conductance matrix follows from similar arguments.

**COROLLARY 1**   The loop resistance matrix **R** and the node conductance matrix **G** of a linear passive resistive network are nonsingular matrices; equivalently,

(2.8)   $\det \mathbf{R} \neq 0$   and   $\det \mathbf{G} \neq 0$

*Proof*   Suppose $\det \mathbf{R} = 0$; then there would be a nonzero set of loop currents $\hat{\mathbf{i}}$ such that $\mathbf{R}\hat{\mathbf{i}} = \mathbf{0}$, and consequently,

$$p = \hat{\mathbf{i}}^T\mathbf{R}\hat{\mathbf{i}} = 0$$

which contradicts the positive definite property of **R.**

Using well-known facts concerning systems of linear algebraic equations (Cramer's rule), we obtain immediately an important consequence of (2.8).

**COROLLARY 2**   Whatever the voltage sources may be, the loop equations

$$\mathbf{R}\mathbf{i} = \mathbf{e}_s$$

of a linear network made of passive resistors have one and only one solution.

**COROLLARY 3**   Whatever the current sources may be, the node or cut-set equations

$$\mathbf{G}\mathbf{e} = \mathbf{i}_s$$

of a linear network made of passive resistors have one and only one solution; where the vector $\mathbf{i}_s$ is the node or cut-set current source vector, and the $k$th component of $\mathbf{i}_s$ represents the total source current at the $k$th node or cut set.

---

† By definition, **a positive definite matrix** is a real symmetric matrix **S** such that for *all* vectors $\mathbf{x} \neq \mathbf{0}$, the quadratic form $\mathbf{x}^T\mathbf{S}\mathbf{x} > 0$.   For example, $\mathbf{S} \triangleq \mathrm{diag}\,(1,1)$ is positive definite because $\mathbf{x}^T\mathbf{S}\mathbf{x} = x_1^2 + x_2^2$, whereas $\mathbf{S}' \triangleq \mathrm{diag}\,(1,-1)$ is not because $\mathbf{x}^T\mathbf{S}'\mathbf{x} = x_1^2 - x_2^2$; this expression is negative whenever $x_1 = 0$ and $x_2 \neq 0$.

We have seen previously that nonlinear resistive networks may have several solutions or none at all. The example below has the purpose of showing that if *active* resistors (negative resistances) are present in a *linear* network, then the solution need not be unique. In fact, for some values of the source voltages the system of equations may have no solution.

**Example 3**  Consider the resistive network shown in Fig. 2.2: all its resistors are linear, and all of them except one are passive. If all the resistors were passive, we would know that $i_1 = i_2 = i_3 = 0$. Writing mesh equations, we obtain

$$
\begin{bmatrix}
2 & -1 & -1 \\
-1 & 0.5 & 0.5 \\
-1 & 0.5 & 1.5
\end{bmatrix}
\begin{bmatrix}
i_1 \\
i_2 \\
i_3
\end{bmatrix}
=
\begin{bmatrix}
0 \\
0 \\
0
\end{bmatrix}
$$

Let $\alpha$ be an arbitrary number; then it is easy to check that

$$i_1 = \alpha \qquad i_2 = 2\alpha \qquad i_3 = 0$$

is a solution of this set of equations. Thus, this network has an infinite number of solutions! Mathematically, this is related to the fact that the matrix of coefficients is singular; it is a $3 \times 3$ matrix of rank 2.

Suppose now that we insert a 1-volt battery in the first mesh. The equations become

$$
\begin{bmatrix}
2 & -1 & -1 \\
-1 & 0.5 & 0.5 \\
-1 & 0.5 & 1.5
\end{bmatrix}
\begin{bmatrix}
i_1 \\
i_2 \\
i_3
\end{bmatrix}
=
\begin{bmatrix}
1 \\
0 \\
0
\end{bmatrix}
$$

The augmented matrix is

$$
\begin{bmatrix}
2 & -1 & -1 & 1 \\
-1 & 0.5 & 0.5 & 0 \\
-1 & 0.5 & 1.5 & 0
\end{bmatrix}
$$

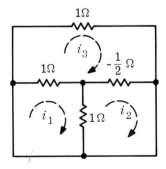

**Fig. 2.2**  An active resistive network.

Since the first, third, and fourth column are linearly independent, it follows that the matrix is of rank 3. However, we know that whenever the augmented matrix and the matrix of coefficients have different rank, the system of equations has no solution. Therefore, there is no set of branch voltages and branch currents that satisfy the Kirchhoff laws and the branch equations!

---

| 2.2 | **Minimum Property of the Dissipated Power** |

This section has two purposes: first, to derive an interesting property of resistive networks and second, to show that the problem of solving a linear resistive network is equivalent to minimizing an appropriately chosen function.

The property we are going to establish is often stated informally as follows: *the current in a resistive network made of linear passive resistors distributes itself in such a way that the power dissipated is minimum.* As stated this sentence suggests the following absurdity: if we were asked to minimize the power $i^T R i$, we would answer that the minimum occurs at $i = 0$, because, $R$ being positive definite, $i^T R i > 0$ whenever $i \neq 0$. This is obviously not what is meant! A more precise statement is the following theorem.

**THEOREM**   Consider a linear network $\mathcal{G}$ made of passive resistors and independent voltage sources. The minimum of the power dissipated in the resistors of $\mathcal{G}$, $i^T R i$, subject to the condition that the source currents remain unchanged, occurs at $\hat{i}$, the only solution of the network equations.

**Remarks**   1. To visualize the meaning of this statement, consider the network $\mathcal{G}$ analyzed by the loop method. Suppose the voltage sources have been included only in the links; then only one loop current flows through each voltage source (since fundamental loops are chosen). Now, the power dissipated in $\mathcal{G}$ depends on *all* the loop currents. The statement means that if the loop currents through sources are held constant and if the other loop currents were allowed to change, then the power dissipated in the resistors of $\mathcal{G}$ is minimum when the currents in the loops without sources have the values specified by the two Kirchhoff laws and Ohm's law.

2. This property of the power dissipated is very interesting from a philosophical point of view. However, in order to know the current through the voltage sources we need to solve the network equations. For this reason we shall reformulate the minimization problem in a more useful fashion later on.

*Proof*   Consider a linear network made of passive resistors driven by independent voltage sources. We use the same notation as in the previous sections.

Suppose that in the loop analysis we choose the tree so that the voltage sources are included only in the links of that tree, and suppose that we number the loop currents starting with the fundamental loops which contain no sources. The loop current vector $\mathbf{i}$ may be partitioned as $(\mathbf{i}_1^T, \mathbf{i}_2^T)$ where $\mathbf{i}_1$ represents loop currents that traverse fundamental loops without sources and $\mathbf{i}_2$ represents loop currents that traverse fundamental loops with sources. Since the sources are in the links of the tree, each component of $\mathbf{i}_2$ is the current that goes through the corresponding voltage source. The loop equations written for the fundamental loops have the following form:

(2.9)
$$\begin{bmatrix} \mathbf{R}_{11} & \mathbf{R}_{12} \\ \mathbf{R}_{21} & \mathbf{R}_{22} \end{bmatrix} \begin{bmatrix} \mathbf{i}_1 \\ \mathbf{i}_2 \end{bmatrix} = \begin{bmatrix} \mathbf{0} \\ \mathbf{e}_s \end{bmatrix}$$

where

$$\mathbf{R} = \begin{bmatrix} \mathbf{R}_{11} & \mathbf{R}_{12} \\ \mathbf{R}_{21} & \mathbf{R}_{22} \end{bmatrix}$$

is the loop resistance matrix. Since $\mathbf{R}$ is symmetric, $\mathbf{R}_{11}$ and $\mathbf{R}_{22}$ are symmetric, and $\mathbf{R}_{12} = \mathbf{R}_{21}^T$. Call $(\hat{\mathbf{i}}_1^T, \hat{\mathbf{i}}_2^T)$ the solution of (2.9). The power dissipated in the resistors is given by

$$p(\hat{\mathbf{i}}_1, \hat{\mathbf{i}}_2) = [\hat{\mathbf{i}}_1^T \hat{\mathbf{i}}_2^T] \begin{bmatrix} \mathbf{R}_{11} & \mathbf{R}_{12} \\ \mathbf{R}_{12}^T & \mathbf{R}_{22} \end{bmatrix} \begin{bmatrix} \hat{\mathbf{i}}_1 \\ \hat{\mathbf{i}}_2 \end{bmatrix}$$

$$= \hat{\mathbf{i}}_1^T \mathbf{R}_{11} \hat{\mathbf{i}}_1 + \hat{\mathbf{i}}_1^T \mathbf{R}_{12} \hat{\mathbf{i}}_2 + \hat{\mathbf{i}}_2^T \mathbf{R}_{12}^T \hat{\mathbf{i}}_1 + \hat{\mathbf{i}}_2^T \mathbf{R}_{22} \hat{\mathbf{i}}_2$$

$$= \hat{\mathbf{i}}_1^T \mathbf{R}_{11} \hat{\mathbf{i}}_1 + 2\hat{\mathbf{i}}_1^T \mathbf{R}_{12} \hat{\mathbf{i}}_2 + \hat{\mathbf{i}}_2^T \mathbf{R}_{22} \hat{\mathbf{i}}_2$$

Suppose now that we vary the loop currents in an arbitrary fashion but subject to the condition that the source currents remain unchanged; that is, $\hat{\mathbf{i}}_1$ becomes $\hat{\mathbf{i}}_1 + \delta\mathbf{i}_1$, and $\hat{\mathbf{i}}_2$ remains as is. Note that $\delta\mathbf{i}_1$ is arbitrary; it is not required to be small. The power equation becomes

$$p(\hat{\mathbf{i}}_1 + \delta\mathbf{i}_1, \hat{\mathbf{i}}_2) = (\hat{\mathbf{i}}_1 + \delta\mathbf{i}_1)^T \mathbf{R}_{11}(\hat{\mathbf{i}}_1 + \delta\mathbf{i}_1) + 2(\hat{\mathbf{i}}_1 + \delta\mathbf{i}_1)^T \mathbf{R}_{12}\hat{\mathbf{i}}_2 + \hat{\mathbf{i}}_2^T \mathbf{R}_{22}\hat{\mathbf{i}}_2$$

or

(2.10)
$$p(\hat{\mathbf{i}}_1 + \delta\mathbf{i}_1, \hat{\mathbf{i}}_2) = \hat{\mathbf{i}}_1^T \mathbf{R}_{11}\hat{\mathbf{i}}_1 + 2\hat{\mathbf{i}}_1^T \mathbf{R}_{12}\hat{\mathbf{i}}_2 + \hat{\mathbf{i}}_2^T \mathbf{R}_{22}\hat{\mathbf{i}}_2$$
$$+ 2\delta\mathbf{i}_1^T \mathbf{R}_{11}\hat{\mathbf{i}}_1 + 2\delta\mathbf{i}_1^T \mathbf{R}_{12}\hat{\mathbf{i}}_2 + \delta\mathbf{i}_1^T \mathbf{R}_{11}\,\delta\mathbf{i}_1$$

We recognize the first three terms of the right-hand side of (2.10) to be $p(\hat{\mathbf{i}}_1, \hat{\mathbf{i}}_2)$. The next two terms are

$$2\delta\mathbf{i}_1^T(\mathbf{R}_{11}\hat{\mathbf{i}}_1 + \mathbf{R}_{12}\hat{\mathbf{i}}_2) = 0$$

where the equals sign follows from the first equation of (2.9). Hence (2.10) becomes

$$p(\hat{\mathbf{i}}_1 + \delta\mathbf{i}_1, \hat{\mathbf{i}}_2) = p(\hat{\mathbf{i}}_1, \hat{\mathbf{i}}_2) + \delta\mathbf{i}_1^T \mathbf{R}_{11}\,\delta\mathbf{i}_1$$

Since the network is made of passive resistors, $\mathbf{R}_{11}$ is positive definite. Hence, for all $\delta\mathbf{i}_1 \neq \mathbf{0}$, $\delta\mathbf{i}_1{}^T \mathbf{R}_{11}\, \delta\mathbf{i}_1 > 0$. Consequently,

$$p(\hat{\mathbf{i}}_1 + \delta\mathbf{i}_1, \hat{\mathbf{i}}_2) > p(\hat{\mathbf{i}}_1, \hat{\mathbf{i}}_2) \qquad \text{for all } \delta\mathbf{i}_1 \neq \mathbf{0}$$

that is, the power dissipated in the resistors of $\mathscr{G}$ for any loop currents which are not the solution of the network is always larger than that for the loop currents which are the solution of the network; hence, the stated minimum property is established.

---

### 2.3   Minimizing Appropriate Functions

We now turn to the second task, namely, to show that the problem of solving the network equations is equivalent to that of minimizing a well-selected function. For the sake of variety we use node equations in the development; of course, we could just as well use loop equations. Since we use node equations, it is more natural to assume that the network is driven by current sources. This poses no problem, since a given network driven by voltage sources can be changed into a network with equivalent current sources. Let us illustrate the idea by a simple example.

**Example 4**   Refer to Fig. 2.3*a*. The unknown node voltage $e$ can be found graphically by locating the intersection between the current source characteristic $i = i_s$, a constant, and the resistor characteristic $i = Ge$; these characteristics are shown by the dashed lines in Fig. 2.3*b*. If, on the other hand, we were to calculate the minimum of the function of $e$ defined by

$$\phi(e) \triangleq Ge^2 - 2i_s e$$

we would find first

$$\frac{d\phi}{de} = 2Ge - 2i_s = 0$$

Hence $e = i_s/G$ is the abscissa of the minimum, as shown in Fig. 2.3*b*. Thus, the solution of the resistive circuit occurs at $\hat{e} = Ri_s$, which also gives the minimum of the function $\phi(\cdot)$.

---

Turning now to the general case, we consider as before a linear resistive network with passive resistors driven by current sources. The node voltages satisfy the equation

(2.11)   $\mathbf{Ge} = \mathbf{i}_s$

where $\mathbf{G}$ is the $n \times n$ node conductance matrix and $\mathbf{i}_s = (i_{s1}, i_{s2}, \ldots, i_{sn})^T$ is such that $i_{sk}$ is the source current entering the $k$th node, $k = 1, 2, \ldots, n$. Call $\hat{\mathbf{e}}$ the unique solution of (2.11). Note the function $\phi(\cdot)$ defined by

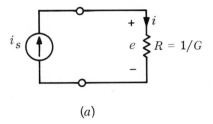

$(a)$

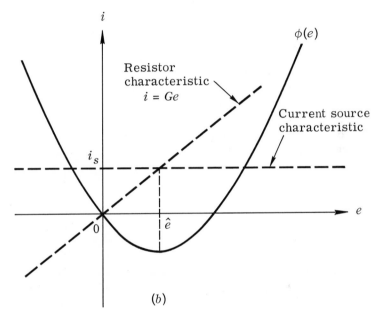

$(b)$

**Fig. 2.3**    A simple example illustrating that the solution of a resistive circuit occurs at a voltage $\hat{e} = Ri_s$, where the function $\phi(\,\cdot\,)$ is the minimum.

(2.12)    $$\phi(\mathbf{e}) \triangleq \mathbf{e}^T \mathbf{Ge} - 2\mathbf{e}^T \mathbf{i}_s = \sum_{k=1}^{n}\sum_{i=1}^{n} G_{ki} e_k e_i - 2\sum_{k=1}^{n} e_k i_{sk}$$

For each value of $\mathbf{e}$, (2.12) specifies the value of $\phi(\mathbf{e})$. Suppose we wished to minimize $\phi(\mathbf{e})$. The first step would be to set all the partial derivatives $\partial\phi/\partial e_k$ to zero; thus,

$$\frac{\partial\phi}{\partial e_k} = 2\sum_{i=1}^{n} G_{ki} e_i - 2i_{sk} = 0 \qquad k = 1, 2, \ldots, n$$

The equality above merely means that

$$\mathbf{Ge} = \mathbf{i}_s$$

which is the network equation (2.11). In other words, all the partial derivatives of $\phi$ are equal to zero if and only if $\mathbf{e}$ satisfies the network equations. Since we know that (2.11) has only one solution $\hat{\mathbf{e}}$, we conclude that $\phi$ is stationary at the point $\hat{\mathbf{e}}$ (that is, all its partial derivatives are zero at that point). However, at that point, $\phi$ may have a minimum, a maximum, or a saddle point. Let us verify that it is a minimum. Consider an arbitrary† set of node voltages designated $\hat{\mathbf{e}} + \delta\mathbf{e}$; then

$$\phi(\hat{\mathbf{e}} + \delta\mathbf{e}) = (\hat{\mathbf{e}} + \delta\mathbf{e})^T\mathbf{G}(\hat{\mathbf{e}} + \delta\mathbf{e}) - 2(\hat{\mathbf{e}} + \delta\mathbf{e})^T\mathbf{i}_s$$

or

(2.13)   $$\phi(\hat{\mathbf{e}} + \delta\mathbf{e}) = \hat{\mathbf{e}}^T\mathbf{G}\hat{\mathbf{e}} - 2\hat{\mathbf{e}}^T\mathbf{i}_s + 2\delta\mathbf{e}^T\mathbf{G}\hat{\mathbf{e}} - 2\delta\mathbf{e}^T\mathbf{i}_s + \delta\mathbf{e}^T\mathbf{G}\,\delta\mathbf{e}$$

We recognize the first two terms on the right-hand side of (2.13) to be $\phi(\hat{\mathbf{e}})$, as can be checked by referring to (2.12). Equation (2.11) implies that the next two terms cancel out; that is,

$$2\delta\mathbf{e}^T(\mathbf{G}\hat{\mathbf{e}} - \mathbf{i}_s) = 0$$

Thus, (2.13) becomes

(2.14)   $$\phi(\hat{\mathbf{e}} + \delta\mathbf{e}) = \phi(\hat{\mathbf{e}}) + \delta\mathbf{e}^T\mathbf{G}\,\delta\mathbf{e}$$

Since all the resistors of the network have positive resistances, $\mathbf{G}$ is positive definite, and the quadratic form $\delta\mathbf{e}^T\mathbf{G}\,\delta\mathbf{e}$ is positive for *all* $\delta\mathbf{e} \neq \mathbf{0}$; consequently,

$$\phi(\hat{\mathbf{e}} + \delta\mathbf{e}) > \phi(\hat{\mathbf{e}})$$

whenever $\delta\mathbf{e} \neq \mathbf{0}$. This means that $\phi$ is minimum at $\hat{\mathbf{e}}$. Note that it is not just a local minimum; in fact, it is an absolute minimum. The above inequality means that whenever $\mathbf{e} \neq \hat{\mathbf{e}}$, $\phi(\mathbf{e}) > \phi(\hat{\mathbf{e}})$. Finally, observe that

$$\phi(\hat{\mathbf{e}}) = \hat{\mathbf{e}}^T\mathbf{G}\hat{\mathbf{e}} - 2\hat{\mathbf{e}}^T\mathbf{i}_s = -\hat{\mathbf{e}}^T\mathbf{G}\hat{\mathbf{e}}$$

Thus,

$$-\phi(\hat{\mathbf{e}}) = \hat{\mathbf{e}}^T\mathbf{G}\hat{\mathbf{e}} = \text{power dissipated in the network}$$

**Remarks**   1.   The dual derivation, based on the loop equations $\mathbf{Ri} = \mathbf{e}_s$ and on a function $\psi(\mathbf{i}) \triangleq \mathbf{i}^T\mathbf{Ri} - 2\mathbf{i}^T\mathbf{e}_s$, follows precisely the same steps. It is left as an exercise to the reader.

2.   It should be mentioned that the minimum property presented here is an elementary example of many of the minimum principles in physics and engineering.

We summarize the results in this section by the following statement.

**MINIMUM PROPERTY**   Consider a linear network made of passive resistors and independent sources. The problem of solving the network equations, either in terms of node voltages,

---

† It is important to note that $\delta\mathbf{e}$ is *not* required to be small.

(2.15)   $\mathbf{Ge} = \mathbf{i}_s$

or in terms of loop currents,

(2.16)   $\mathbf{Ri} = \mathbf{e}_s$

is equivalent to that of minimizing the function

(2.17)   $\phi(\mathbf{e}) \overset{\Delta}{=} \mathbf{e}^T\mathbf{Ge} - 2\mathbf{e}^T\mathbf{i}_s$

or the function

(2.18)   $\psi(\mathbf{i}) \overset{\Delta}{=} \mathbf{i}^T\mathbf{Ri} - 2\mathbf{i}^T\mathbf{e}_s$

respectively.   More precisely, $\phi(\cdot)$ and $\psi(\cdot)$ attain *only one minimum* which occurs at the *unique* solution $\hat{\mathbf{e}}$ and $\hat{\mathbf{i}}$, respectively, of the network equations; furthermore,

(2.19)   $\min_{\mathbf{e}} \phi(\mathbf{e}) = \phi(\hat{\mathbf{e}}) = -\hat{\mathbf{e}}^T\mathbf{G}\hat{\mathbf{e}}$

= the negative of the power dissipated in the resistors

and

(2.20)   $\min_{\mathbf{i}} \psi(\mathbf{i}) = \psi(\hat{\mathbf{i}}) = -\hat{\mathbf{i}}^T\mathbf{R}\hat{\mathbf{i}}$

= the negative of the power dissipated in the resistors

## 2.4   Nonlinear Resistive Networks

In the last three subsections we learned that, given any linear network made of passive resistors, if an arbitrary set of voltage sources is inserted in series with the branches and another arbitrary set of current sources is inserted between node pairs, then the resulting network has one and only one solution.   The voltages and currents of the solution may be obtained either by solving the equations directly or by minimizing a suitably chosen function.   We also learned in Sec. 1 that some nonlinear resistive networks may have no solution or several solutions for some values of the applied voltages or currents.   We would like to consider now more carefully the problem of solving nonlinear resistive networks.

**Example 5**   To get a feel for the problem of solving nonlinear resistive networks, consider the simple circuit illustrated in Fig. 2.4.   Suppose the nonlinear resistors $\mathcal{R}_2$ and $\mathcal{R}_3$ are voltage-controlled and have characteristics given by the functions $j_2 = g_2(v_2)$ and $j_3 = g_3(v_3)$.   The node equations, using the node-to-datum voltages $e_1$ and $e_2$ as variables, are

(2.21a)   $G_1 e_1 + g_2(e_1 - e_2) = i_s$

(2.21b)   $-g_2(e_1 - e_2) + g_3(e_2) = 0$

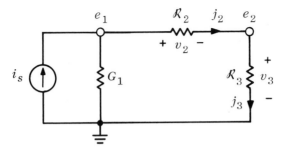

**Fig. 2.4**   A simple nonlinear resistive network.

For example, the characteristics $g_2$ and $g_3$ might be

$$g_2(e) = 5e + \epsilon^e \qquad g_3(e) = e + 0.02e^3$$

The degree of difficulty of actually solving these equations depends on the properties of the characteristics $g_2(\,\cdot\,)$ and $g_3(\,\cdot\,)$. Usually we must resort to numerical methods because there is no direct method for solving equations like (2.21a) and (2.21b). However, it is possible to reformulate the problem as a minimization problem. Indeed, let $p_2$ and $p_3$ be

$$p_2(e) \triangleq \int_0^e g_2(e') \, de' \qquad p_3(e) \triangleq \int_0^e g_3(e') \, de'$$

and let

$$\phi(\mathbf{e}) = G_1{}^2 e_1 + 2p_2(e_1 - e_2) + 2p_3(e_2) - 2e_1 i_s$$

Then we immediately verify that the necessary conditions for the existence of an extremum for $\phi$ (namely $\partial\phi/\partial e_1 = 0$ and $\partial\phi/\partial e_2 = 0$) turn out to be Eqs. (2.21a) and (2.21b). Thus, the problem of solving the equations in (2.21) is equivalent to finding the extrema (if any) of $\phi$.

It is not too difficult to show† that if we impose mild conditions on the characteristics of the nonlinear resistors, the function $\phi(\,\cdot\,)$ will have one and only one extremum which, in fact, will be a minimum. Thus, when these conditions are fulfilled, the solution of (2.21) is equivalent to finding the *unique* minimum of $\phi(\,\cdot\,)$. The latter problem can be solved numerically on computers by the many varieties of the steepest descent method. A set of sufficient conditions may be described as follows.

*If all the resistors of the network have strictly increasing characteristics with the property that $|v| \to \infty$ whenever $|j| \to \infty$, then, for all current sources connected between any pair of nodes (soldering-iron entry) and for all voltage sources connected in series with resistors (pliers entry) there is one and only*

† C. A. Desoer and J. Katzenelson, Nonlinear RLC Circuits, *Bell System Tech. J.* Jan., 1965.

*one set of branch voltages and branch currents that satisfy the two laws of Kirchhoff and the branch equations.*

Note that the characteristics of the resistors need not go through the origin; this allows series combination of batteries and monotonically increasing resistors. As a special case, the statement above means that any network made of batteries, linear passive resistors, and ideal diodes will have one and only one solution provided there is a resistor in series with each ideal diode and each battery.

| 3 | The Voltage Gain and the Current Gain of a Resistive Network |
|---|---|

| 3.1 | Voltage Gain |
|---|---|

We wish to establish the intuitively obvious fact that the voltage gain† of any network made of linear passive resistors cannot exceed unity.‡ We state it more precisely as follows.

**THEOREM**    Consider a linear network made of only passive resistors; i.e., all resistances are *positive*. Suppose that a voltage source $e_s$ is connected between two nodes ① and ①′. Then given any two nodes, say ② and ②′, we have

(3.1)    $$|v_2 - v_2'| \leq |e_s|$$

where $v_2$ and $v_2'$ denote the node-to-datum voltages of nodes ② and ②′, respectively.

*Proof*    For simplicity, let $e_s$ be positive, and let node ① be taken as datum node, as shown in Fig. 3.1. Under these conditions we first show that, for any node ⓚ, $v_k \leq e_s$.

Consider the node-to-datum voltages of all the nodes. One of these voltages is the largest one; call it $v_{max}$. If the node with maximum voltage $v_{max}$ is node ①′, the assertion is proved. If not, suppose the node-to-datum voltage of node ⓝ is equal to $v_{max}$. Consider the equation resulting from applying KCL to node ⓝ. Since all resistors have positive resistances, the currents in all the branches connected to node ⓝ must flow away from node ⓝ. KCL requires that all branch currents for the branches connected to node ⓝ must be zero. Hence all nodes connected to ⓝ have the same voltage $v_{max}$ as their node-to-datum voltage. Repeat the argument for any one of these nodes and then to the nodes connected

---

† By "voltage gain" we mean here the ratio of an output voltage to an input voltage due to a voltage source.

‡ R. J. Schwartz, "A Note on the Transfer Voltage Ratio of Resistive Networks with Positive Elements," *Proc. IRE,* **43**(11):1670 (1955).

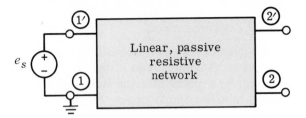

**Fig. 3.1**   An arbitrary linear, passive resistive network is connected to a voltage source at nodes ① and ①'.

to them, and so forth. Clearly, we must end up with node ①' and conclude that $v_{max} = e_s$. Thus, we have established that $v_k \leq e_s$, for all $k$.

Consider now the minimum node voltage $v_{min}$. A similar argument shows that $v_{min} = 0$, since the node-to-datum voltage of node ① is 0. Thus,

$$0 \leq v_k \leq e_s \qquad \text{for all } k$$

and

$$-e_s \leq v_2' - v_2 \leq e_s$$

which is equivalent to

$$|v_2 - v_2'| \leq |e_s|$$

Now each side of this last inequality is a voltage difference between two nodes; hence each side is unaffected by the choice of datum node. Consequently, (3.1) is established.

**Remark**   It is crucial to note the importance of the assumption that all resistances are positive. Consider the circuit shown in Fig. 3.2 where a $-1$-ohm resistor is in series with a 2-ohm resistor. The voltages across the resistors are, respectively, $-1$ volt and 2 volts for a 1-volt source!

If we examine the proof of the theorem we note that we never used the *linearity* of the resistor characteristics. We used only two facts: (1) if the

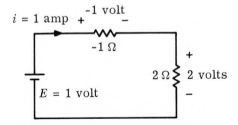

**Fig. 3.2**   An example of active resistive circuit.

current is positive, the voltage drop across the resistor is positive; and
(2) if the current is zero, the voltage drop is zero. Hence, we may assert
that *the voltage gain of a resistive network cannot exceed unity even if the
resistors are nonlinear, provided that their characteristics satisfy the condition that*

$$v = 0 \qquad \text{for } j = 0$$

and

(3.2)   $vj > 0 \qquad \text{for } j \neq 0$

A little thought will convince the reader that these conditions on the resistor characteristics are equivalent to saying that the characteristics must
go through the origin and must be inside the first and third quadrants.
Equivalently, the resistors must be *passive;* indeed, the power delivered
at time $t$ to the resistor is $v(t)j(t)$, which is always $\geq 0$.

Figure 3.3 shows typical nonlinear characteristics which satisfy these
conditions. The first one has the typical shape of a tunnel-diode characteristic and the second one that of a *pn*-junction diode. On the other hand,
if a constant voltage source is connected in series with a diode, the conditions of (3.2) will not be satisfied; hence biased nonlinear resistors may
contribute voltage gain.

| 3.2 | **Current Gain** |
|---|---|

Intuitively, we expect that the current gain of a linear passive resistive network cannot exceed unity.† In fact, this is still true when the network is
made of nonlinear resistors of the type considered above. These ideas
are stated precisely as follows.

† By "current gain" we mean here the ratio of an output current to an input current due to a current source.

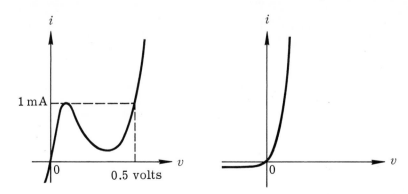

**Fig. 3.3**    Characteristics of tunnel-diode and *pn*-junction diodes; both are passive resistors.

**THEOREM**   Consider a network made of $b$ nonlinear passive resistors. (The characteristic of each branch satisfies the condition that, for $j = 0, v = 0$, and for
$j \neq 0, vj > 0$.) Suppose a current source $i_s$ is connected to the nodes ①
and ①'; then none of the branch currents can exceed the source current.
More precisely,

$$|j_k| \leq |i_s| \qquad k = 1, 2, \ldots, b$$

*Proof*   If the network is planar, this theorem follows from the preceding one by
duality. Indeed, the dual of a branch voltage is a branch current. For
*linear* resistors, this theorem also follows from the preceding one by the
reciprocity theorem. Consider the case of nonlinear resistors. Let us,
for the purpose of a proof by contradiction, assume that the current
through one branch is larger than $i_s$, say, $j_1 > i_s$. Consider this branch
and the operating point on its characteristic, as shown in Fig. 3.4. If we
were to replace this nonlinear resistor by a linear resistor whose characteristic goes through the operating point, all the branch voltages and branch
currents of the circuit would remain unchanged. Repeat the operation
with all nonlinear resistors. Then our assumption would mean that there
would exist a *linear* resistive network with positive resistances which would
have a current gain larger than 1. However, we have just shown that this
is impossible. Hence, in the nonlinear circuit, we must have

$$|j_k| \leq |i_s| \qquad k = 1, 2, \ldots, b$$

We may condense all the results of this section by the statement that
*neither the voltage gain nor the current gain of a resistive network made of*
*passive resistors (linear or nonlinear) can exceed unity.*
This completes our study of nonlinear resistive networks. It is fasci-

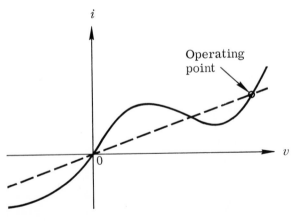

**Fig. 3.4**   Characteristic of a nonlinear resistor and characteristic of a
linear resistor going through the same operating point.

nating to observe how the requirements of the Kirchhoff laws and some simple requirements on the resistor characteristics lead to very general properties.

## Summary

■ An engineer must constantly keep in mind that his analyses and design techniques are based on *network models* of *physical networks*. Sometimes, as a result of unwarranted simplifications, the network model has properties vastly different from that of the physical network.

■ As shown by examples, in some cases a nonlinear network model may have no solution; in other cases it may have several solutions and even an infinite number of solutions, even though the physical network has only one solution.

■ Let $\mathfrak{N}$ be a *linear* network made of *passive* resistors (positive resistances); then $\mathfrak{N}$ has the following properties.

1. Its loop impedance matrix $\mathbf{R}$ and its node admittance matrix $\mathbf{G}$ are positive definite matrices (i.e., for all $\mathbf{i} \neq \mathbf{0}$, $\mathbf{i}^T \mathbf{R} \mathbf{i} > 0$, and for $\mathbf{e} \neq \mathbf{0}$, $\mathbf{e}^T \mathbf{G} \mathbf{e} > 0$). This fact implies that $\det \mathbf{R} \neq 0$, and $\det \mathbf{G} \neq 0$ (equivalently, $\mathbf{R}$ and $\mathbf{G}$ are nonsingular; that is, $\mathbf{R}^{-1}$ and $\mathbf{G}^{-1}$ are well defined).

2. If $\hat{\mathbf{i}}$ represents the actual loop currents and $\hat{\mathbf{e}}$ the actual node voltages, then the power dissipated in the network $\mathfrak{N}$ is given by

$$p = \hat{\mathbf{i}}^T \mathbf{R} \hat{\mathbf{i}} = \hat{\mathbf{e}}^T \mathbf{G} \hat{\mathbf{e}}$$

3. Whatever the sources may be, the loop equations

$$\mathbf{R} \mathbf{i} = \mathbf{e}_s$$

and the node equations

$$\mathbf{G} \mathbf{e} = \mathbf{i}_s$$

have one and only one solution (this is obvious, since $\mathbf{R}$ and $\mathbf{G}$ are nonsingular).

4. Suppose $\mathfrak{N}$ is driven by voltage sources only. Then the loop currents can be found by minimizing the power dissipated, $\mathbf{i}^T \mathbf{R} \mathbf{i}$, *subject to the condition* that the currents of the voltage sources remain what they are.

5. The solution of the node equations of $\mathfrak{N}$, $\mathbf{G} \mathbf{e} = \mathbf{i}_s$, can be obtained by minimizing the function

$$\phi(\mathbf{e}) = \mathbf{e}^T \mathbf{G} \mathbf{e} - 2\mathbf{e}^T \mathbf{i}_s$$

The function $\phi(\cdot)$ has a unique minimum at $\hat{\mathbf{e}}$, and

$$\phi(\hat{\mathbf{e}}) = -\hat{\mathbf{e}}^T \mathbf{G} \hat{\mathbf{e}} = -\text{power dissipated in } \mathfrak{N}$$

6.   The solution of the loop equations of $\mathfrak{N}$, $\mathbf{Ri} = \mathbf{e}_s$, can be obtained by minimizing the function

$$\psi(\mathbf{i}) = \mathbf{i}^T \mathbf{Ri} - 2\mathbf{i}^T \mathbf{e}_s$$

The function $\psi(\,\cdot\,)$ has a unique minimum at $\hat{\mathbf{i}}$, and

$$\psi(\hat{\mathbf{i}}) = -\hat{\mathbf{i}}^T \mathbf{R}\hat{\mathbf{i}} = -\text{power dissipated in } \mathfrak{N}$$

- For a *nonlinear* resistive network, if all resistors have strictly increasing characteristics with the property that $|v| \to \infty$ whenever $|j| \to \infty$, then, for all current sources connected between any pair of nodes (soldering-iron entry) and for all voltage sources connected in series with resistors (pliers entry), the network has one and only one solution; more precisely, there is one and only one set of branch voltages and branch currents that satisfy Kirchhoff's laws and the branch equations.

- If a resistive network (linear or nonlinear) is made of *passive* resistors (i.e., the *vi* characteristic of each resistor lies in the first and third quadrants), then any voltage gain and any current gain cannot be larger than 1 in absolute value.

## Problems

**1.** The middle resistor of the network shown in Fig. P18.1 is nonlinear; its characteristic is

$$v_3 = \frac{i_3}{\sqrt{1 + i_3{}^2}}$$

Calculate $i_1$ and $i_2$.   (Hint: Replace the two resistor-battery combinations by a Thévenin or Norton equivalent, and solve the resulting network graphically.)

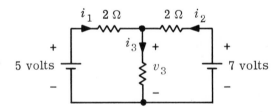

**Fig. P18.1**

**2.** For the network shown in Fig. P18.2, the characteristic of the nonlinear resistor is given by

$$v_3 = \tanh i_3$$

Calculate $e_1$.

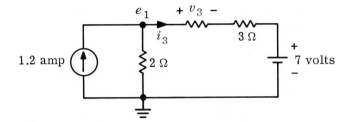

**Fig. P18.2**

**3.** Propose a resistor characteristic such that the circuit shown in Fig. P18.3 has no solution (specify the characteristic by a graph).

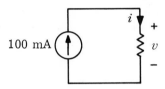

**Fig. P18.3**

**4.** Propose a resistor characteristic such that the current shown in Fig. P18.4 has

*a.*   A unique solution

*b.*   Two solutions

*c.*   An infinite number of solutions

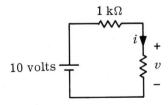

**Fig. P18.4**

**5.** The characteristic of the nonlinear resistor $\mathcal{R}$ is given by (see Fig. P18.5)

$$i = 0.527 - (v - 1.20) + (v - 1.20)^3$$

Find the largest resistance $R$ such that, whatever the battery voltage $E$ may be, the circuit has only one operating point.

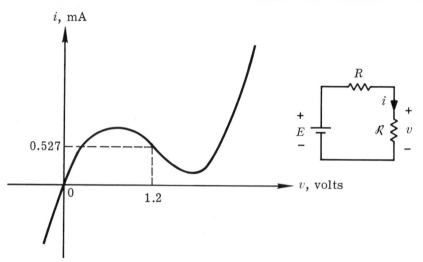

**Fig. P18.5**

Linear resis-
tive network,
power calcula-
tion, and the
minimum
property

**6.** For the linear resistive network shown in Fig. P18.6, determine the node conductance matrix and show that it is positive definite.   Determine all branch currents and calculate the total power dissipation.   Form the function $\phi(\cdot)$ and show that it attains the minimum at the solution.

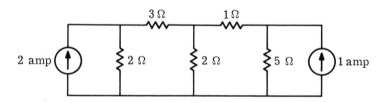

**Fig. P18.6**

Minimum
property

**7.** For the nonlinear network in Prob. 1, determine the power dissipation. Form the function $\psi(\cdot)$ and show that it attains the minimum at the solution.

Voltage gain
property

**8.** For the nonlinear circuit in Prob. 5, where the battery $E$ is the input and $v$ is the output, demonstrate that the voltage gain is always less than unity for any positive resistance $R$.

Dual circuit
and current
gain

**9.** What is the dual of the nonlinear circuit of Prob. 5? Demonstrate that the current gain is less than unity for the dual circuit.

Active resistive
network

**10.** The linear resistive network in Fig. P18.10 contains a dependent current source.

a.   Determine the value of $g_m$ for which the conductance matrix is singular.

b.   Under the above situation, what can you say about the voltages $v_1$ and $v_2$?

c.   Consider the 1-ohm resistor as the load; determine the power delivered to the load and the power delivered by the source in terms of $g_m$. What must be the values of $g_m$ such that there is a power gain?

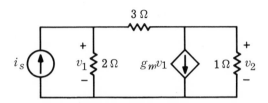

**Fig. P18.10**

# 19

# Energy and Passivity

The concept of *energy* is one of the most important concepts of science and engineering. We first encountered it in Chap. 2 when we calculated the power delivered to a two-terminal element and the energy stored in linear time-invariant inductors and capacitors. In Chap. 7 we calculated, for the sinusoidal steady state, the instantaneous power and the average power, and related the $Q$ of a resonant $RLC$ circuit to the average energy stored and dissipated at resonance. Also, in Chap. 9 we used Tellegen's theorem and sinusoidal steady-state analysis to relate Re $[Z(j\omega)]$ to the average power dissipated and Im $[Z(j\omega)]$ to $2\omega$ times $\mathcal{E}_M - \mathcal{E}_E$, the difference between the average magnetic energy stored and the average electric energy stored. Thus, we have had many opportunities to observe the interrelation between energy and other circuit-theoretic concepts.

So far, we have not discussed energy storage or energy balance of circuits with time-varying elements. It is, however, a very important subject. Suppose that we think of an electric motor as a network element and that we select for it a very simple network model. For example, the model involves two inductors, one representing the stator windings and the other representing the rotor windings; the self- and mutual inductances of these inductors depend on the shaft position $\theta$ (see the highly idealized diagram of Fig. 0.1). Thus, as the shaft rotates at a given speed, we have a pair of time-varying inductors. We know that when the shaft angle $\theta$ changes, either mechanical work is done to the outside world (in which case we have a motor), or electric energy is transferred to the circuit from the outside world (in which case we have a generator). Therefore, it is important to study the behavior of the energy in time-varying networks.

In Sec. 1 we shall consider an $LC$ circuit whose capacitor is time-varying. In this simple situation we can easily analyze the energy transfer and understand how such a circuit can become unstable. In Sec. 2 we shall study the energy stored in nonlinear time-varying elements. Section 3 is devoted to the study of passive one-ports with, in particular, the characterization of passive resistors, capacitors, and inductors. Section 4 is a preparation section in which we show that a linear time-invariant $RLCM$ network driven by a single source $\epsilon^{s_0 t}$ can be made (by suitable choice of the initial state) to have a *complete* response proportional to $\epsilon^{s_0 t}$. In Sec. 5 we shall use this result to show that the driving-point impedance of any passive one-port made of linear time-invariant elements must be a *positive real* function. This result is the basis of practically all network synthesis. In Sec. 6 we shall use energy ideas to show that passive networks must be stable. Finally, in Sec. 7 we shall use a parallel $RLC$ circuit to show how a parametric amplifier can exhibit current gain.

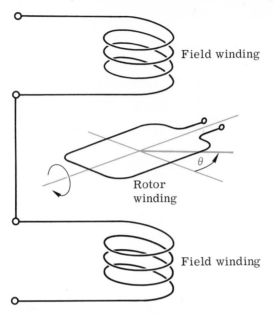

**Fig. 0.1**   Schematic diagram of an electric motor or generator.

## 1 | Linear Time-varying Capacitor

The calculation of the energy stored in a time-varying capacitor and in a time-varying inductor is definitely more subtle than that of the time-invariant case.   The reason is that in order to change the capacitance of a time-varying capacitor we move the plates of the capacitor; since there is an attractive electrostatic force between the charges on the two plates, any relative motion of the plates requires mechanical work. Depending on the direction of motion, either the capacitor is receiving electric energy from mechanical work or it transforms electric energy into mechanical work.†   Thus, a time-varying capacitor may receive energy both from the sources in the electric circuit to which it is connected and from the agent which moves the plates.

### 1.1 | Description of the Circuit

To make these ideas more concrete, we consider a specific circuit, namely the linear time-varying $LC$ circuit shown in Fig. 1.1.   The inductor $L$ is time-invariant.   The parallel-plate capacitor has a fixed plate and a mov-

---

† Some electric motors have beeen built on this principle.   However, they are very bulky compared with those using magnetic fields.

able plate; therefore, the distance $x$ between the plates can be varied. The capacitance is

(1.1)    $C = \dfrac{\varepsilon A}{x}$

where $A$ is the effective area of the plates and $\varepsilon$ is the dielectric constant of the medium. (We use the following units: $A$, in square meters; for free space, $\varepsilon_0 = 8.85 \times 10^{-12}$ coul²/newton-m²; $x$, in meters; $C$, in farads.)

If the plates are fixed and if, at time $t = 0$, a charge $q_0$ is on the right plate and no current flows in the inductor, then for $t \geq 0$ the charge on the right plate is given by

(1.2)    $q(t) = q_0 \cos \omega_0 t$

where $\omega_0 = 1/\sqrt{LC}$. Whenever the charge is nonzero, there is an *attractive* electrostatic force between the plates. This force is given in newtons by

(1.3)    $f = \dfrac{q^2}{2\varepsilon A}$

To keep the moving plate from slamming into the fixed one, we must balance this force by applying to the moving plate an external force $f_e$ of the same magnitude but opposite direction, as shown in Fig. 1.1. Note that, irrespective of the sign of $q$, we must apply a force to the moving plate in the direction of increasing $x$. Hence whenever we increase $x$ (while $q \neq 0$), we perform some mechanical work against the electrostatic forces; this work appears as energy in the circuit. Suppose for simplicity that at time

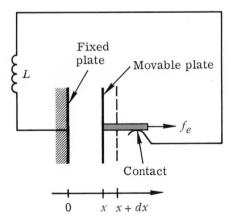

**Fig. 1.1**    *LC* circuit whose capacitor has a movable plate; the outside force $f_e$ must be applied to the movable plate to prevent it from slamming into the fixed plate.

$t_0$ we suddenly increase the separation from $x_0$ to $x_1$; assume, in fact, that the displacement of the plate is so fast that the charge $q(t_0)$ does not change. The mechanical work done by the force applied to the moving plate against the electrostatic attractive forces is

$$(1.4) \quad W_m = \int_{x_0}^{x_1} f(x)\, dx = \int_{x_0}^{x_1} \frac{q^2(t_0)}{2\varepsilon A}\, dx = \frac{q^2(t_0)}{2\varepsilon A}(x_1 - x_0)$$

since the charge $q(t_0)$ remains constant during the (instantaneous) displacement. The increase in electric energy stored is

$$\mathcal{E}_E(x_1) - \mathcal{E}_E(x_0) = \frac{1}{2}\frac{q^2(t_0)}{C(x_1)} - \frac{1}{2}\frac{q^2(t_0)}{C(x_0)}$$

$$(1.5) \qquad\qquad = \frac{q^2(t_0)}{2\varepsilon A}(x_1 - x_0)$$

Hence, as expected, the mechanical work $W_m$ done against the attractive electrostatic forces is equal to the increased energy stored.

**Exercise**   Derive Eq. (1.3) from the basic laws of electrostatics. (Neglect fringing, assume infinite planar plates, and calculate the force per unit area. Consider a test charge on one of the plates. The coulomb forces due to other charges on that plate cancel out, and the resultant force on the test charge is due exclusively to charges on the other plate.)

## 1.2   Pumping Energy into the Circuit

Now that we have some feel for the mechanism whereby energy can be pumped into the circuit, let us try to pump as much energy as possible into the circuit per unit time. Clearly, to pump the maximum energy for a given displacement, we should wait for $q^2(t)$ to reach a maximum; i.e., we should wait for the maxima and the minima of $q(t)$. In Fig. 1.2a we show the time variation of $q$; the first maximum occurs at $t_1$. Let us then pull the plates apart from the separation $x_0$ to the separation $x_1$. As we have seen, this process delivers energy to the circuit. Since we cannot pull the plates apart indefinitely, we have to bring the movable plate back; we ask ourselves how can we bring the plates back to separation $x_0$ and pump the *least* energy *out* of the circuit? Clearly, we should wait for $q$ to be zero; then the attractive electrostatic forces are zero, and we can bring back the plates without receiving any work from the electrostatic forces. Thus, we pull the plates back at time $t_2$ since, as shown in Fig. 1.2a, $q(t_2) = 0$. Clearly, we can repeat the process indefinitely, as suggested by Fig. 1.2. In Fig. 1.2c we show the value of $C(t)$ versus $t$, and in Fig. 1.2d we show the *total* energy $\mathcal{E}$ stored in the circuit ($\mathcal{E}$ is all electrostatic when $q^2$ is maximum, for then $i = 0$ and the magnetic energy stored is zero. When $q = 0$,

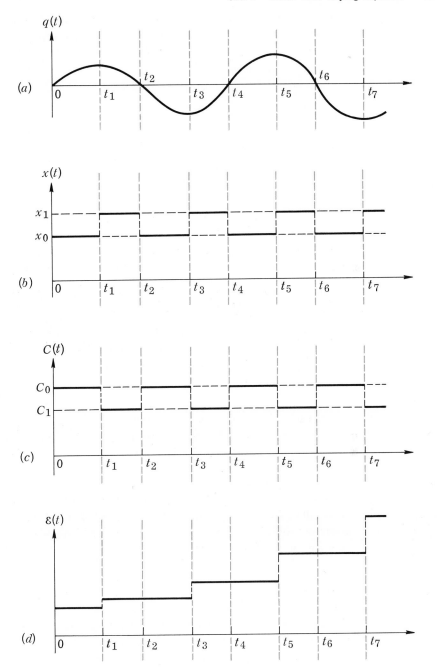

**Fig. 1.2**   Pumping energy into the $LC$ circuit of Fig. 1.1; the waveform of charge, displacement, capacitance, and energy stored are plotted to scale.   Note that at $t_1, t_3, t_5, t_7, \ldots$, the stored energy $\mathcal{E}(t)$ increases by 44 percent.

$\mathscr{E}$ is all magnetic). Since the circuit has no dissipation, $\mathscr{E}$ is constant, except at times $t_1, t_3, t_5, \ldots$ when energy is pumped into the circuit.

Let us note in passing that at $t_1, t_3, t_5, \ldots$ the voltage across the capacitor jumps; indeed, for all $t$,

(1.6) $\quad q(t) = C(t)v(t)$

Therefore, just before $t_3$

$$q(t_3-) = C(t_3-)v(t_3-) = C_0 v(t_3-)$$

and just after $t_3$ [since the charge has had no chance to leak through the inductor, and hence $q(t_3+) = q(t_3-)$],

$$q(t_3+) = q(t_3-) = C(t_3+)v_3(t_3+) = C_1 v(t_3+)$$

Hence

(1.7) $\quad \dfrac{v(t_3+)}{v(t_3-)} = \dfrac{C_0}{C_1} > 1$

In other words, as shown in Fig. 1.2a the curve of $q$ versus $t$ is *continuous;* however, the curve of $v$ versus $t$ exhibits jumps at $t_1, t_3, t_5, \ldots$, according to the rule of (1.7). The charge waveform in Fig. 1.2a is not a sinusoid; the amplitude grows from cycle to cycle since greater energy is obtained at the same capacitance values.

**Exercise** Show that $\mathscr{E}(t_3+)/\mathscr{E}(t_3-) = C_0/C_1$.

**Remark** Parallel-plate capacitors are used extensively for pumping energy in practical electric circuits. These circuits are usually referred to as parametric circuits. In practice, the capacitor is not mechanically varied but is made of the depletion layer of a reverse-biased varactor diode; the pump changes the bias and hence the width of the depletion layer. The varactor diode is reverse-biased by a large dc voltage with a small ac voltage supplied by an oscillator. The capacitor appears in the small-signal analysis as a linear periodically varying capacitor. The analysis of parametric amplifiers will be treated briefly in Sec. 7.

## 1.3  State-space Interpretation

From past experience we know that we might pick either the capacitor voltage and inductor current or the charge and flux as state variables. In the case of *time-varying* circuits whose elements vary discontinuously (as the example shown in Fig. 1.2c), the charge-flux choice is superior because these state variables change continuously [see Eq. (1.7) for the voltage discontinuity]. The reference directions are shown in Fig. 1.3; by inspection, the state equations are

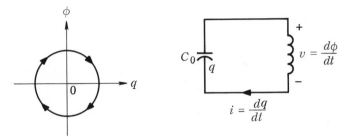

**Fig. 1.3**   State-space trajectory of the time-invariant $LC$ circuit shown; scales are so selected that the trajectory is circular.

(1.8) $\dfrac{dq}{dt} = \dfrac{\phi}{L}$

(1.9) $\dfrac{d\phi}{dt} = -\dfrac{q}{C(t)}$

*Case 1*   $C(t)$ is constant and equal to $C_0$. We choose to plot the state trajectory on the $q\phi$ plane with scales so chosen that the trajectory is circular (see Fig. 1.3). The state moves along this trajectory in the *clockwise* direction. Indeed, consider an arbitrary point on the trajectory in the *first* quadrant ($q > 0$, $\phi > 0$); then by (1.8) and (1.9), the flux is decreasing whereas the charge is increasing.

*Case 2*   $C(t)$ is piecewise constant. Whenever $q^2(t)$ reaches a maximum, $C(t)$ drops from $C_0$ to $C_1$, and whenever $q(t) = 0$, $C(t)$ jumps up from $C_1$ to $C_0$.

   In the $q\phi$ plane (shown in Fig. 1.4), the state-space trajectory consists of arcs of circles in the first and third quadrants [because then $C(t) = C_0$] and of arcs of ellipses in the second and fourth quadrants [then $C(t) = C_1$]. In theory, if these capacitance variations are maintained indefinitely, the amplitudes of oscillation of $q$ and $\phi$ increase indefinitely to infinity; hence the circuit is unstable. In practice, the process ends by either dielectric breakdown in the capacitor or melting of the coil wires.

**Exercise**   Calculate $t_2 - t_1$ and $t_3 - t_2$ (use data from Table 5.1).

### 1.4   Energy Balance

The energy stored in a linear capacitor at any time $t$ is

(1.10) $\mathcal{E}_E(t) = \tfrac{1}{2}C(t)v^2(t)$

We can easily check this in two ways. First, we can calculate directly the energy stored in the electrostatic field $\vec{E}$ using

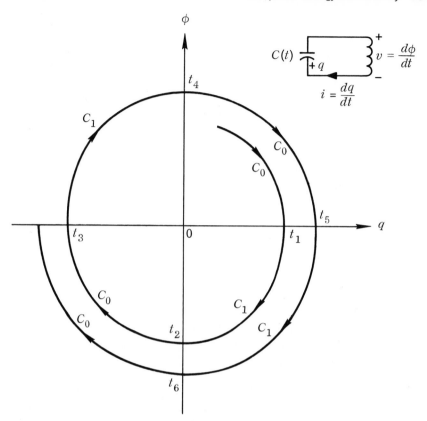

**Fig. 1.4**   State-space trajectory for the time-varying $LC$ circuit for which $C(t)$ is given by Fig. 1.2c; when $C(t) = C_1$, the trajectory is an ellipse, and when $C(t) = C_0$, the trajectory is circular (because of the choice of scales).

(1.11)   $\mathcal{E}_E = \iiint \frac{1}{2}\varepsilon|\vec{E}|^2 \, dv$

Second, by keeping $C$ fixed for all times after time $t$ and connecting a linear time-invariant resistor $R$ in parallel with $C$, we can calculate the energy delivered to $R$ as follows:

(1.12)   $\mathcal{E}_E = \displaystyle\int_t^\infty Ri^2(t') \, dt'$

The rate of change of the energy stored is obtained from Eq. (1.10) as

(1.13)   $\dfrac{d\mathcal{E}_E}{dt} = \dfrac{1}{2}\dot{C}(t)v^2(t) + C(t)v(t)\dot{v}(t)$

where we use the dot over $C$ to denote the derivative of $C$. From a circuit point of view, the capacitor voltage and current are related by

$$i(t) = \frac{d}{dt}[C(t)v(t)] = \dot{C}(t)v(t) + C(t)\dot{v}(t)$$

Thus, the circuit delivers energy *to* the capacitor at the rate

(1.14)    $p_e(t) = i(t)v(t) = \dot{C}(t)v^2(t) + C(t)v(t)\dot{v}(t)$

Hence from (1.13) and (1.14) we see that the capacitor receives energy at the rate $p_e(t)$ and stores energy at the rate $d\mathcal{E}_E/dt$, and that

(1.15)    $p_e(t) - \dfrac{d\mathcal{E}_E}{dt} = \dfrac{1}{2}\dot{C}(t)v^2(t)$

We assert that this difference is the mechanical work done by the electrostatic forces in the capacitor. Indeed, since these forces are attractive (hence their reference direction is opposite to that of the velocity $\dot{x}$), they perform work at the rate

$$-f\dot{x} = -\frac{q^2}{2\varepsilon A}\dot{x} = -\frac{1}{2}\frac{C^2v^2}{2\varepsilon A}\dot{x}$$

For reference directions, refer to Fig. 1.5. Now since $C = \varepsilon A/x$ and $\dot{C}/C = -\dot{x}/x$, we obtain successively

(1.16)    $-f\dot{x} = -\dfrac{1}{2}Cv^2\dfrac{\dot{x}}{x} = \dfrac{1}{2}\dot{C}v^2$

The result checks with (1.15). Therefore, we have verified in detail that

(1.17)    $p_e(t) = \dfrac{d}{dt}\mathcal{E}_E + p_m(t)$

where $p_m(t)$ is the rate at which mechanical work is done by the electrostatic forces against the outside world. Thus the electric power delivered to the capacitor is equal to the sum of the rate at which energy is stored and the rate at which mechanical work is done against the outside world.

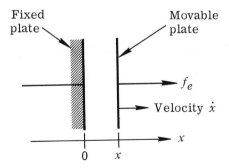

**Fig. 1.5**    $f_e$ is the force applied to the movable plate by the outside world to prevent it from slamming into the fixed plate; with the velocity $\dot{x}$ measured with the reference direction indicated, the attractive electrostatic forces deliver energy to the outside world at the rate $-f\dot{x}$.

| **2** | **Energy Stored in Nonlinear Time-varying Elements** |

The previous section has shown us why calculating the energy stored in circuits which include time-varying elements requires care. Indeed, not only do the voltage sources and the current sources deliver energy to the circuit but also the mechanical forces that cause the elements to change their values. We start by some general considerations.

Consider a one-port which may include nonlinear and/or time-varying elements. Let us drive the one-port by a generator, as shown in Fig. 2.1. From basic physics we know that the instantaneous power entering the one-port at time $t$ is

(2.1)   $p(t) = v(t)i(t)$

and the energy delivered to the one-port from time $t_0$ to $t$ is

(2.2)   $W(t_0,t) = \int_{t_0}^{t} p(t')\,dt' = \int_{t_0}^{t} v(t')i(t')\,dt'$

The energy delivered to the port may be dissipated into heat if the one-port is a passive resistor. It may also be stored in the elements if the one-port is an inductor or capacitor.

In Chap. 2 we used Eq. (2.2) to derive the energy stored in a nonlinear *time-invariant* inductor. Let us review briefly the derivation. Consider a nonlinear inductor whose characteristic is represented by the function $\hat{i}$ as follows:

(2.3)   $i = \hat{i}(\phi)$

Note that $\hat{i}$ is the function which describes the inductor characteristic. Since the inductor is time-invariant, the characteristic is a fixed curve; consequently, $\hat{i}$ does *not* depend explicitly on $t$. The voltage across the inductor is given by Faraday's law as follows:

(2.4)   $v = \dfrac{d\phi}{dt}$

The energy delivered by the generator to the inductor from $t_0$ to $t$ is then

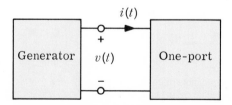

**Fig. 2.1**   A one-port driven by a generator; the one-port receives energy from the generator at the rate $v(t)i(t)$.

(2.5)   $W(t_0,t) = \int_{t_0}^{t} i(t')v(t')\,dt' = \int_{t_0}^{t} \hat{\imath}[\phi(t')]\dfrac{d\phi'}{dt'}\,dt' = \int_{\phi(t_0)}^{\phi(t)} \hat{\imath}(\phi')\,d\phi'$

Note that $W(t_0,t)$ is a function of the flux at the starting time $t_0$ and at the observing time $t$. If we assume that initially the flux is zero, that is, $\phi(t_0) = 0$, and if we choose the state of zero flux to correspond to zero stored energy, then, recalling that an inductor stores energy but does not dissipate energy, we see that the energy stored $\mathcal{E}$ must be equal to the energy delivered by the generator from $t_0$ to $t$, namely, $W(t_0,t)$, must be equal to the energy stored; that is,

(2.6)   $\mathcal{E}[\phi(t)] = W(t_0,t) = \int_{0}^{\phi(t)} \hat{\imath}(\phi')\,d\phi'$

Thus, if the characteristic of the inductor is given as in Fig. 2.2, the shaded area gives the energy stored at time $t$.

**2.1   Energy Stored in a Nonlinear Time-varying Inductor**

The characteristic of a nonlinear time-varying inductor is given *for each t* by a curve similar to that shown in Fig. 2.2, except that now the curve changes as $t$ varies. For simplicity, let us assume that at all times $t$ the characteristic curve goes through the origin; thus, the inductor is in the zero state when the flux (or equivalently the current) is zero. We also assume that at all times $t$ the inductor is flux-controlled; consequently, we may represent the nonlinear *time-varying* inductor by

(2.7)   $i = \hat{\imath}(\phi,t)$

Note that $\hat{\imath}$ is now an explicit function of both $\phi$ and $t$. By analogy with (2.6) we define $\mathcal{E}[\phi(t),t]$ by

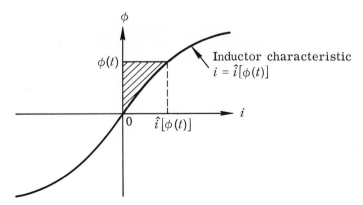

**Fig. 2.2**   The energy stored in the nonlinear time-invariant inductor is equal to the area indicated.

(2.8)   $\mathscr{E}[\phi(t),t] \triangleq \int_0^{\phi(t)} \hat{\imath}(\phi',t)\,d\phi'$

Note that $\phi'$ is the variable of integration and that $t$ is considered a fixed parameter during the integration process. The expression $\mathscr{E}[\phi(t),t]$ has the same interpretation as in Fig. 2.2, provided the characteristic used is the characteristic at time $t$.

We now show that $\mathscr{E}$ defined in (2.8) is indeed the energy stored in the inductor at time $t$. For this purpose we conduct a "thought" experiment: let us first freeze the characteristic of the inductor at time $t$; that is, for all times later than $t$, the characteristic is the same as that at time $t$. Next we connect a linear time-invariant resistor $R$ across the inductor, as shown in Fig. 2.3. Let us calculate the energy delivered by the inductor to the resistor during $[t,\infty)$. From KCL we have, for $t' \geq t$,

(2.9a)   $i_R(t') = -i_L(t') = -\hat{\imath}[\phi(t'),t]$

Note that $t$ in (2.9a) is a *fixed* parameter, and $t'$ in the interval $[t,\infty)$ is the time variable. From KVL we have

(2.9b)   $v_R(t') = v_L(t') = \dot{\phi}(t')$

As time $t'$ approaches infinity, the circuit reaches its zero state; that is, $\lim_{t'\to\infty} \phi(t') = 0$. For brevity, we denote this limiting value by $\phi(\infty)$. Furthermore, as $t' \to \infty$, $i(t')$ as well as $\phi(t') \to 0$. The energy dissipated in the resistor is, from (2.9a) and (2.9b),

(2.10)   $\displaystyle\int_t^\infty i_R(t')v_R(t')\,dt' = \int_t^\infty -\hat{\imath}[\phi(t'),t]\dot{\phi}(t')\,dt'$

$\displaystyle = -\int_{\phi(t)}^{\phi(\infty)} \hat{\imath}(\phi',t)\,d\phi'$

(2.11)   $\displaystyle = \int_0^{\phi(t)} \hat{\imath}(\phi',t)\,d\phi'$

Note that in (2.10) the time $t$ in $\hat{\imath}[\phi(t'),t]$ is a *fixed* parameter. The result in (2.11) is the integral that we labeled $\mathscr{E}$ in (2.8). Since all the energy stored in the inductor at time $t$ has been dissipated in the resistor, we have shown

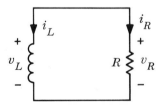

**Fig. 2.3**   The resistor $R$ dissipates, during the interval $(t,\infty)$, the energy stored at time $t$ in the nonlinear time-varying inductor.

that Eq. (2.8) is the expression for the energy stored in a nonlinear time-varying inductor at time $t$.

## 2.2   Energy Balance in a Nonlinear Time-varying Inductor

Let us consider the nonlinear time-varying inductor as a one-port and connect it to a generator, as shown in Fig. 2.4. Let us calculate the energy delivered to the inductor by the generator. From Eq. (2.2) we obtain

$$W(t_0,t) = \int_{t_0}^{t} v(t')i(t')\,dt'$$

$$(2.12) \qquad = \int_{t_0}^{t} \dot{\phi}(t')\,\hat{\imath}\,[\phi(t'),t']\,dt'$$

We are going to show that this expression can be rewritten in the form

$$(2.13) \quad W(t_0,t) = \mathscr{E}[\phi(t),t] - \mathscr{E}[\phi(t_0),t_0] - \int_{t_0}^{t} \frac{\partial}{\partial t'} \mathscr{E}[\phi(t'),t']\,dt'$$

The first two terms give the difference between the energy stored at time $t$ and the energy stored at time $t_0$. The third term

$$-\int_{t_0}^{t} \frac{\partial}{\partial t'} \mathscr{E}[\phi(t'),t']\,dt'$$

is the energy delivered by the circuit to the agent which changes the characteristic of the inductor; thus, it is the work done by the electrodynamic forces during the changes of the configuration of the inductor. To prove the equality of (2.13), we first note that both sides are equal to zero when $t = t_0$. Therefore, if we show that their rates of change with respect to $t$ are the same, they will necessarily be equal for all $t \geq t_0$. From (2.12)

$$(2.14) \qquad \frac{dW}{dt} = \dot{\phi}(t)\hat{\imath}\,[\phi(t),t]$$

To differentiate the first term in the right-hand side of (2.13), observe that $\mathscr{E}$ is a function of $\phi$ and $t$, but that $\phi$ itself is a function of $t$; hence by (2.8)

$$\frac{d}{dt}\mathscr{E}[\phi(t),t] = \frac{\partial \mathscr{E}}{\partial \phi}\dot{\phi}(t) + \frac{\partial \mathscr{E}}{\partial t}$$

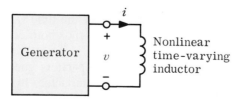

**Fig. 2.4**   A nonlinear time-varying inductor driven by a generator.

(2.15) $$= \hat{i}[\phi(t),t]\dot{\phi}(t) + \frac{\partial}{\partial t} \mathcal{E}[\phi(t),t]$$

The derivative of the second term in the right-hand side of (2.13) is zero. The derivative of the last term is

$$- \frac{\partial}{\partial t} \mathcal{E}[\phi(t),t]$$

Combining the above with (2.15), we see that the derivative of the right-hand side of (2.13) is precisely that of the left-hand side as given by (2.14). Therefore, we have shown that (2.13) is valid for all $t$.

For time-invariant inductors the energy stored does not depend explicitly on time, and the last term in (2.13) is zero; hence the energy delivered to a nonlinear *time-invariant* inductor from $t_0$ to $t$ is equal to the difference between the energy stored in the inductor at time $t$ and that stored at time $t_0$.

For a *linear time-varying inductor,* we use a time-varying inductance $L(t)$ to describe its time-varying characteristic. Thus,

(2.16) $\quad i = \hat{i}(\phi) = \dfrac{\phi}{L(t)}$

From (2.8), the energy stored is

(2.17) $\quad \mathcal{E}[\phi(t),t] = \displaystyle\int_0^{\phi(t)} \frac{\phi'}{L(t)} d\phi' = \frac{\phi^2(t)}{2L(t)}$

and

$$\frac{\partial}{\partial t} \mathcal{E}[\phi(t),t] = -\frac{1}{2} \frac{\phi^2(t)}{L^2(t)} \dot{L}(t)$$

Therefore, the energy delivered to the one-port by a generator from $t_0$ to $t$ is

(2.18) $\quad W(t_0,t) = \dfrac{1}{2} \dfrac{\phi^2(t)}{L(t)} - \dfrac{1}{2} \dfrac{\phi^2(t_0)}{L(t_0)} + \displaystyle\int_{t_0}^t \dfrac{1}{2} \dfrac{\phi^2(t')}{L^2(t')} \dot{L}(t') \, dt'$

or, in terms of current $i(\cdot)$,

(2.19) $\quad W(t_0,t) = \dfrac{1}{2}L(t)i^2(t) - \dfrac{1}{2}L(t_0)i^2(t_0) + \displaystyle\int_{t_0}^t \dfrac{1}{2}\dot{L}(t')i^2(t') \, dt'$

where the term $\int_{t_0}^t \dfrac{1}{2}\dot{L}(t')i^2(t') \, dt'$ is the energy delivered by the circuit to

the agent which changes the characteristic of the time-varying inductor. It is important to note that this last term depends both on the waveform $\dot{L}(\cdot)$ and on the waveform $i(\cdot)$.

The derivation of energy relations for capacitors goes along similar lines. In fact, it follows the derivation for inductors by duality and is therefore omitted. The results for both inductors and capacitors are summarized in Table 19.1.

**Table 19.1  Summary of Energy Relations for Inductors and Capacitors**

| | Linear time-varying | Nonlinear time-invariant | Nonlinear time-varying |
|---|---|---|---|
| Inductors, characterization $v = \dot\phi$ | $i = \dfrac{\phi}{L(t)}$ | $i = \hat{i}(\phi)$ | $i = \hat{i}(\phi,t)$ |
| Energy stored $\mathscr{E}[\phi(t),t]$ | $\mathscr{E}(t) = \dfrac{\phi^2(t)}{2L(t)} = \dfrac{1}{2}L(t)i^2(t)$ | $\mathscr{E}(t) = \displaystyle\int_0^{\phi(t)} \hat{i}(\phi')\,d\phi'$ | $\mathscr{E}[\phi(t),t] = \displaystyle\int_0^{\phi(t)} \hat{i}(\phi',t)\,d\phi'$  ($t$ is a fixed parameter) |
| Energy delivered from $t_0$ to $t$, $W(t_0,t)$ | $\mathscr{E}(t) - \mathscr{E}(t_0) + \dfrac{1}{2}\displaystyle\int_{t_0}^t \dot{L}(t')i^2(t')\,dt'$ | $\mathscr{E}(t) - \mathscr{E}(t_0)$ | $\mathscr{E}[\phi(t),t] - \mathscr{E}[\phi(t_0),t_0] - \displaystyle\int_{t_0}^t \dfrac{\partial}{\partial t'}\,\mathscr{E}[\phi(t'),t']\,dt'$ |
| Capacitors, characterization $i = \dot q$ | $v = \dfrac{q}{C(t)}$ | $v = \hat{v}(q)$ | $v = \hat{v}(q,t)$ |
| Energy stored $\mathscr{E}[q(t),t]$ | $\mathscr{E}(t) = \dfrac{q^2(t)}{2C(t)} = \dfrac{1}{2}C(t)v^2(t)$ | $\mathscr{E}(t) = \displaystyle\int_0^{q(t)} \hat{v}(q')\,dq'$ | $\mathscr{E}[q(t),t] = \displaystyle\int_0^{q(t)} \hat{v}(q',t)\,dq'$  ($t$ is a fixed parameter) |
| Energy delivered from $t_0$ to $t$, $W(t_0,t)$ | $\mathscr{E}(t) - \mathscr{E}(t_0) + \dfrac{1}{2}\displaystyle\int_{t_0}^t \dot{C}(t')v^2(t')\,dt'$ | $\mathscr{E}(t) - \mathscr{E}(t_0)$ | $\mathscr{E}[q(t),t] - \mathscr{E}[q(t_0),t_0] - \displaystyle\int_{t_0}^t \dfrac{\partial}{\partial t'}\,\mathscr{E}[q(t'),t']\,dt'$ |

Note: All characteristics are assumed to go through the origin at all times.

| 3 | **Passive One-ports** |
|---|---|

In Chap. 2 we introduced the terms "passive resistor," "passive inductor," and "passive capacitor." The term "passive element" implies roughly that the element absorbs energy. We now turn to one-ports. Recall that when we think of a one-port, we have in mind a network made of an arbitrary interconnection of lumped elements, and this network is put (symbolically) in a black box with two of its terminals sticking out. These two terminals constitute the *port,* and the one-port is the black box and its two terminals. The point is that when we talk of one-ports, the only measurements allowed are those made at the port; hence, we can only talk about the port voltage and the port current. In this section we give a formal definition of a passive *one-port* and examine the characterization of passivity for some typical elements.

Consider the one-port shown in Fig. 2.1, where $v(\cdot)$ and $i(\cdot)$ are the port voltage and current, respectively. Either the voltage or the current may be considered as the input of the one-port. Let us denote by $\mathcal{E}(t_0)$ the energy stored in the one-port at time $t_0$. As before we denote by $W(t_0,t)$ the energy delivered by the generator to the one-port from time $t_0$ to time $t$; then

$$(3.1) \quad W(t_0,t) = \int_{t_0}^{t} v(t')i(t')\,dt'$$

A one-port $\mathfrak{N}$ is said to be **passive** if the energy

$$(3.2) \quad W(t_0,t) + \mathcal{E}(t_0) \geq 0$$

for *all* initial time $t_0$, for *all* time $t \geq t_0$, and for *all* possible input waveforms. A one-port is said to be **active** if it is not passive.

From Eq. (3.2) passivity requires that the sum of the stored energy at time $t_0$ and the energy delivered to the one-port from $t_0$ to $t$ be nonnegative under all circumstances. The inequality must hold for all possible applied inputs. In particular, it must hold for step functions, sinusoids, exponentials, or any arbitrary waveform. The inequality must hold for all time $t$ and for any starting time $t_0$. Let us illustrate the significance of this definition with some familiar two-terminal elements.

| 3.1 | **Resistors** |
|---|---|

Consider a nonlinear time-varying resistor which is characterized for each $t$ by a curve in the $vi$ plane. Since a resistor does not store energy, we have $\mathcal{E}(t_0) = 0$ for all $t_0$; the condition for passivity is reduced to

$$(3.3) \quad W(t_0,t) = \int_{t_0}^{t} v(t')i(t')\,dt' \geq 0$$

Obviously, if the characteristic curve of a resistor stays in the first and third quadrants of the $vi$ plane for all times $t$, the integrand in (3.3), that is, the instantaneous power entering the one-port, is always nonnegative; then the energy delivered to the one-port is nonnegative at all $t$, for all $t_0$, and for all inputs. Thus, if the characteristic of a resistor lies in the first and third quadrants of the $vi$ plane for all times $t$, the resistor is passive. It is easily shown that if the characteristic is in the second or fourth quadrant for any time interval, however small, the resistor is active. Assume that during the interval $(t_1, t_1 + \Delta t)$ part of the characteristic lies in the second or fourth quadrant. To be specific, suppose that during the interval $(t_1, t_1 + \Delta t)$ the operating points corresponding to some fixed current $i_0$ lie inside the second or fourth quadrant. Then if during this time interval we connect to the resistor a constant current source $i_0$, the energy delivered *by* the current source to the resistor is

(3.4)   $$\int_{t_1}^{t_1 + dt} i_0 v(t') \, dt' < 0$$

where the inequality follows from the fact that the operating points are inside the second or fourth quadrant. We have shown that if part of the characteristic of a time-varying resistor lies in the second or fourth quadrant, this resistor is not passive. Thus, if the resistor is passive, its characteristic must for all time be in the first and third quadrants. Collecting the two conclusions, we state that *a nonlinear time-varying resistor is passive if and only if its characteristic is for all time in the first and third quadrants.*

It follows that independent voltage and current sources are active resistors, but germanium diodes and tunnel diodes, which have their characteristics passing through the origin of the $vi$ plane and lying completely in the first and third quadrants, are passive resistors. It is also obvious that a resistive one-port formed by an arbitrary interconnection of passive resistors is passive. [To prove this we only need to apply Tellegen's theorem (see Sec. 3.3).] A passive resistive one-port can only dissipate energy, whereas an active resistive one-port may supply energy to the device connected to it. Active resistive one-ports can be used as amplifiers. For example, the series connection of a tunnel diode and a battery, as shown in Fig. 3.1a, forms an active one-port whose characteristics are shown in Fig. 3.1b. In Chap. 3, Sec. 4, we demonstrated that if we connect to this active one-port a small-signal ac voltage source in series with a suitable linear resistor, the average power delivered to the resistor is larger than the average power delivered by the ac voltage source. Thus, the particular active one-port of Fig. 3.1a has the ability to *amplify* ac power under small-signal excitation. Not all active one-ports have this property; a constant voltage source is an active resistor, yet it cannot be used to amplify an ac signal. To distinguish between active one-ports with and without this amplification property we introduce the concepts of local activity and local passivity.

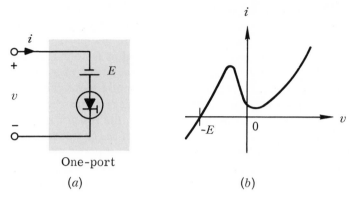

**Fig. 3.1**    An active resistive one-port and its characteristic.

A resistive one-port is said to be **locally passive** at an operating point if the slope of its characteristic in the $vi$ plane is nonnegative at that point. A resistive one-port is said to be **locally active** at an operating point if the slope of its characteristic in the $vi$ plane is negative at the point.

A tunnel diode is a passive resistor, but it is locally active over a certain interval of voltages. An independent voltage source is an active device, but it is locally passive. The interest of the concept of local activity is that if a *passive nonlinear* resistor is *locally active,* then it is possible to design an amplifier around this resistor. More specifically, given a passive nonlinear resistor which is locally active, it is possible to bias it and insert it in a suitably designed network so that this network exhibits *power gain;* by this we mean that, in the sinusoidal steady state, the average power delivered to the output is larger than the average power delivered by the signal source.† A thorough discussion of this fact would, of course, lead us into a course on amplifier design. The fundamental idea is that a *passive nonlinear resistor, which is locally active and embedded by a biasing circuit, can be the basic element for an amplifier.*

## 3.2    Inductors and Capacitors

As discussed in Sec. 2, the energy stored in a nonlinear *time-invariant* inductor at time $t$ is equal to the sum of the initially stored energy at $t_0$ and the energy delivered to the inductor by the outside world from $t_0$ to $t$. Thus, *the passivity condition requires that for all time $t$, and for all possible inductor current and inductor voltage waveforms,*

(3.5)    $$W(t_0,t) + \mathcal{E}(t_0) = \mathcal{E}(t) = \int_0^{\phi(t)} \hat{\imath}(\phi')\, d\phi' \geq 0$$

† This difference in these average powers, together with the average power dissipated in the resistor, is supplied by the biasing battery.

The condition can be represented in terms of the area in the shaded region of Fig. 2.2. Thus, for a nonlinear time-invariant inductor to be passive, its characteristic (in the $i\phi$ plane) must pass through the origin and lie in the first and third quadrants in the neighborhood of the origin. Also, if the characteristic of a nonlinear time-invariant inductor is monotonically increasing and lies in the first and third quadrants, it is passive. Unlike nonlinear passive resistors, this is a sufficient but not necessary condition. Indeed, an inductor with a characteristic like the one shown in Fig. 3.2 is passive as long as the net area (the difference of the shaded area and the crosshatched area) is positive for all $\phi_1$.

Let us now consider *linear* time-varying inductors. From Table 19.1 we know that

(3.6)   $\mathcal{E}(t) = \tfrac{1}{2}L(t)i^2(t)$

and

(3.7)   $W(t_0,t) + \mathcal{E}(t_0) = \mathcal{E}(t) + \dfrac{1}{2}\displaystyle\int_{t_0}^{t} \dot{L}(t')i^2(t')\, dt'$

Thus, the condition for passivity is

(3.8)   $\dfrac{1}{2}L(t)i^2(t) + \dfrac{1}{2}\displaystyle\int_{t_0}^{t} \dot{L}(t')i^2(t')\, dt' \ge 0$

for all times $t$, for all starting times $t_0$, and for all possible currents $i(\cdot)$. We assert that *a linear time-varying inductor is passive if and only if*

(3.9)   $L(t) \ge 0$     and     $\dot{L}(t) \ge 0$     for all $t$

First, if the two conditions in (3.9) are satisfied, the passivity condition of (3.8) will hold for all $t$ and all $i(\cdot)$. Conversely, passivity implies both

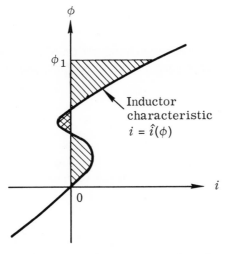

**Fig. 3.2**    Characteristic of a time-invariant nonlinear inductor which is passive.

$L(t) \geq 0$ and $\dot{L}(t) \geq 0$.  Indeed, suppose for some time $t_0$, $L(t_0) < 0$; then putting $t = t_0$ in (3.8), any $i(t_0) \neq 0$ would make the left-hand side of (3.8) negative, and hence would contradict the assumption of passivity.  Similarly, if for some interval $(t_0, t_1)$, $\dot{L} < 0$, and we choose a pulse-like current waveform such that $i(t_0) = i(t_1) = 0$ but is nonzero in the interval $(t_0, t_1)$, then the left-hand side of (3.8) with $t = t_1$ would again be negative, which contradicts the passivity assumption.

Another way of demonstrating the fact that $\dot{L}(t) \geq 0$ is needed for passivity is by considering the equation which characterizes a linear time-varying inductor, namely,

$$v = \dot{\phi} = \frac{d}{dt}[L(t)i(t)]$$

(3.10)
$$= L(t)\frac{di}{dt} + \dot{L}(t)i(t)$$

We may interpret the right-hand side of (3.10) as follows.  The term $L(t)(di/dt)$ gives the voltage at time $t$ across a *time-invariant* inductor whose inductance is equal to the number $L(t)$.  In the second term of (3.10) the voltage at time $t$ is proportional to the current at time $t$; hence it can be interpreted as the voltage across a linear time-varying *resistor* whose resistance is equal to $\dot{L}(t)$.  The series connection of these elements describes the voltage-current relation of a linear time-varying inductor.  Clearly, if $\dot{L}(t)$ is negative, we have an active resistor; thus, it is necessary that $\dot{L}(t)$ be nonnegative for the time-varying inductor to be passive.

The conditions for passivity of capacitors can be derived in a dual manner and are omitted.  We shall simply state that *a nonlinear time-invariant capacitor is passive if and only if for all q*

(3.11)   $\int_0^q \hat{v}(q')\, dq' \geq 0$

(Here the function $\hat{v}$ describes the characteristic of the nonlinear capacitor.)  *A linear time-varying capacitor is passive if and only if*

(3.12)   $C(t) \geq 0$   and   $\dot{C}(t) \geq 0$   for all $t$

---

| 3.3 | **Passive One-ports** |
|---|---|

Consider a one-port formed by an arbitrary interconnection of passive elements.  Is this one-port passive?  Intuitively, we expect an affirmative answer.  Let us prove it by means of Tellegen's theorem.

Consider the one-port shown in Fig. 3.3.  Let the one-port be driven by a current source $i_s$, and let the voltage across the port be $v$.  Note that as far as the current source is concerned the voltage $v$ is not in its associated reference direction.  Let there be $b$ branches inside the one-port, and let us assign to each of them associated reference directions; then Tellegen's theorem says that for all $t$

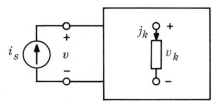

**Fig. 3.3**   A one-port which contains $b$ branches is driven by a current source $i_s$.

(3.13)   $$i_s(t)v(t) = \sum_{k=1}^{b} j_k(t)v_k(t)$$

Let $\mathcal{E}_k(t_0)$ be the energy stored in the $k$th element at $t_0$; then obviously, the energy stored in the one-port at $t_0$ is

(3.14)   $$\mathcal{E}(t_0) = \sum_{k=1}^{b} \mathcal{E}_k(t_0)$$

By assumption, we know that every one of the $b$ elements in the one-port is passive, i.e., that for all possible branch currents $j_k$ and branch voltages $v_k$ and for all $t, t_0$

(3.15)   $$\int_{t_0}^{t} j_k(t')v_k(t')\,dt' + \mathcal{E}_k(t_0) \geq 0 \qquad k = 1, 2, \ldots, b$$

Hence, adding the $b$ inequalities of (3.15), we obtain

(3.16)   $$\sum_{k=1}^{b} \int_{t_0}^{t} j_k(t')v_k(t')\,dt' + \sum_{k=1}^{b} \mathcal{E}_k(t_0) \geq 0$$

By Tellegen's theorem the first term of the above equation is equal to

(3.17)   $$\int_{t_0}^{t} i_s(t')v(t')\,dt' = W(t_0,t)$$

where $W(t_0,t)$ designates the energy delivered to the one-port by the current source during the interval $[t_0,t]$. Substituting Eq. (3.14) and (3.17) in (3.16), we obtain

$$W(t_0,t) + \mathcal{E}(t_0) \geq 0$$

which is the condition for passivity for the one-port. Thus, we have shown that *a one-port made of an arbitrary interconnection of passive elements is a passive one-port.*

---

| 4 | **Exponential Input and Exponential Response** |

Suppose we are given a linear time-invariant network driven by a single independent current source $i(\cdot)$. The source is connected at terminals ① and ①′, and the port voltage is called $e_1$ (see Fig. 4.1). Suppose

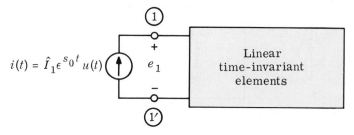

**Fig. 4.1**   A one-port with linear time-invariant elements is driven by an exponential current source.

$i(t) = \hat{I}_1 \epsilon^{s_0 t} u(t)$.   We want to find out whether it is possible to set up initial conditions in the network such that the port voltage $e_1$ is of the form $\hat{E}_1 \epsilon^{s_0 t}$ for $t \geq 0$. (In these expressions, $s_0$, $\hat{I}$, and $\hat{E}_1$ are constants, possibly complex.)   In other words, we want to find out whether we can suddenly switch in, at $t = 0$, the exponential current $\hat{I}_1 \epsilon^{s_0 t}$ and get a voltage $e_1$ that is purely exponential.   Since the response of the network will usually be of the form

$$e_1(t) = \hat{E}_1 \epsilon^{s_0 t} + \sum_i k_i \epsilon^{s_i t}$$

(where the $s_i$ are the natural frequencies of the network and the $k_i$ are constants dependent on the initial conditions), the question amounts to asking whether it is possible to set up initial conditions in the network such that *all* the constants $k_i$ are zero.

Note that we do not want to put any restrictions on $s_0$. For example, $s_0$ may be purely imaginary.   If $i(t) = A \cos(\omega_0 t - \phi)$, we would solve the problem for $\hat{I}_1 = A \epsilon^{-j\phi}$ and $s_0 = j\omega_0$, and at the end take the real part of the answer. (To justify this procedure, see Chap. 7.)   Also, $s_0$ may be a growing exponential [then Re $(s_0) > 0$] or a decaying exponential [then Re $(s_0) < 0$].   In the following, we take $s_0$ to be possibly complex.

To answer the question, we must start by analyzing the problem.   Let us write the node equations, taking terminal ① as datum node.   From Sec. 4 of Chap. 10 the equations are of the form

(4.1)   $\mathbf{Y}_n(D)\mathbf{e}(t) = \mathbf{i}(t) + \mathbf{i}_0$

where $\mathbf{Y}_n(D)$ is the node admittance matrix with $D = d/dt$, $\mathbf{e}(t) = [e_1(t), e_2(t), \ldots, e_n(t)]^T$ is the node-to-datum voltage vector, $\mathbf{i}(t) = (\hat{I}\epsilon^{s_0 t}, 0, 0, \ldots, 0)^T$ is the input, and $\mathbf{i}_0$ is a constant column vector which represents the contributions of the initial currents in inductors.   Initial voltages across capacitors are specified separately as initial conditions.   Our problem is to solve for the complete solution of the node-to-datum voltage $e_1(t)$ and show that with a suitable choice of initial inductor currents and capacitor voltages, $e_1(t)$ is equal to $\hat{E}_1 \epsilon^{s_0 t}$; that is, as far as the input port is concerned, we can find initial conditions such that the sudden application of $i(t) = \hat{I}_1 \epsilon^{s_0 t}$ does not create a "transient."

We shall solve our problem by first calculating the constant vector $\hat{\mathbf{E}}$ such that $\mathbf{e}(t) = \hat{\mathbf{E}} \epsilon^{s_0 t}$ is a solution of (4.1). Then by setting $t = 0$, we obtain all the node-to-datum voltages at $t = 0$. These will determine the required initial capacitor voltages. The initial inductor currents will then be determined by referring to the branch equations and the network topology.

We start with Eq. (4.1). For convenience, let us differentiate all the scalar equations which have a nonzero component of $\mathbf{i}_0$ in their right-hand sides. For simplicity, let us assume that the first component of $\mathbf{i}_0$ is zero, so that the first equation is not differentiated in this process.† We then get a set of equations of the form

(4.2)    $\mathbf{M}(D)\mathbf{e}(t) = \mathbf{i}(t)$

where $\mathbf{M}(D)$ is a matrix whose elements are polynomials in the operator $D$.

As an example, for the circuit shown in Fig. 4.2, Eqs. (4.1) and (4.2) have the forms

(4.1′)
$$
\begin{bmatrix} (C_1 + C_2)D + G_1 & -C_2 D \\ -C_2 D & \dfrac{1}{L_2 D} + C_2 D \end{bmatrix}
\begin{bmatrix} e_1(t) \\ e_2(t) \end{bmatrix}
=
\begin{bmatrix} \hat{I}_1 \epsilon^{s_0 t} \\ 0 \end{bmatrix}
+
\begin{bmatrix} 0 \\ -i_{L_2}(0-) \end{bmatrix}
$$

(4.2′)
$$
\begin{bmatrix} (C_1 + C_2)D + G_1 & -C_2 D \\ -C_2 D^2 & \dfrac{1}{L_2} + C_2 D^2 \end{bmatrix}
\begin{bmatrix} e_1(t) \\ e_2(t) \end{bmatrix}
=
\begin{bmatrix} \hat{I} \epsilon^{s_0 t} \\ 0 \end{bmatrix}
$$

Note that in (4.2′) we had to differentiate the second equation in order to get rid of the initial current $-i_{L_2}(0-)$ which appeared in the right-hand side of the original node equation.

---

† If this assumption does not hold, we simply have to replace $\hat{I}_1$ by $s_0 \hat{I}_1$ in all the following equations.

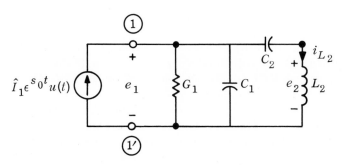

**Fig. 4.2**    An example to illustrate an operation on the equations.

Let us go back to (4.2), and let us write $\mathbf{i}(t) = \hat{\mathbf{I}}\epsilon^{s_0 t}$ where $\hat{\mathbf{I}} = (\hat{I}_1, 0, \ldots, 0)^T$. We try a solution of the form $\mathbf{e}(t) = \hat{\mathbf{E}}\epsilon^{s_0 t}$ where $\hat{\mathbf{E}}$ is a constant vector which we have to determine. Substituting and observing that $\mathbf{M}(D)\hat{\mathbf{E}}\epsilon^{s_0 t} = \mathbf{M}(s_0)\hat{\mathbf{E}}\epsilon^{s_0 t}$, we obtain

(4.3)    $\mathbf{M}(s_0)\hat{\mathbf{E}} = \hat{\mathbf{I}}$

If det $[\mathbf{M}(s_0)] \neq 0$,† the matrix $\mathbf{M}(s_0)$, whose elements are complex numbers, is nonsingular, has an inverse, and

(4.4)    $\hat{\mathbf{E}} = [\mathbf{M}(s_0)]^{-1}\hat{\mathbf{I}}$

In particular,

(4.5)    $\hat{E}_1 = \dfrac{\Delta_{11}(s_0)}{\Delta(s_0)}\hat{I}_1$

where $\Delta(s_0) = \det [\mathbf{M}(s_0)]$, and $\Delta_{11}(s_0)$ is the cofactor of the (1,1) element of $\mathbf{M}(s_0)$. Observe that if we had performed a sinusoidal steady-state analysis, we would have performed the same calculation (except that $s_0$ would have been replaced by $j\omega_0$), and we would have recognized $\Delta_{11}(j\omega_0)/\Delta(j\omega_0)$ to be $Z(j\omega_0)$, the driving-point impedance at terminals ① and ①'. Thus, using the extended definition of network functions, we put

(4.6)    $\dfrac{\Delta_{11}(s_0)}{\Delta(s_0)} = Z(s_0)$

So far we have shown that any exponential solution of (4.2) is of the form $\hat{\mathbf{E}}\epsilon^{s_0 t}$, where $\hat{\mathbf{E}}$ is given by (4.4). This solution specifies all node voltages at $t = 0$. In particular, it specifies all initial capacitor voltages. It also specifies all other branch variables except possibly some inductor currents; indeed, since the node voltages are $\hat{\mathbf{E}}\epsilon^{s_0 t}$ for $t \geq 0$, all branch voltages are specified, and the branch equations specify all resistor and capacitor currents. Only the inductor currents are subject to some arbitrariness, since

$$j_L(t) = \frac{1}{L}\int_0^t v_L(t')\, dt' + j_L(0)$$

To specify the $j_L(0)$'s, we proceed as follows. (1) If there are no loops composed only of inductors, we pick a tree that contains all the inductors. Then each KCL equation applied to a cut set defined by an inductive tree branch is one equation in one unknown, namely, $j_L(0)$ (since all other branch currents are specified). This procedure specifies all the inductor currents. (2) If there are loops composed only of inductors, we pick a tree which includes as many inductors as possible and select all the link inductor currents to be zero at $t = 0$. We determine the other initial induc-

---

† This requires that the network be nondegenerate.

tor currents at $t = 0$ by the method used in (1). This again determines all initial inductor currents.

Suppose we have calculated all these initial capacitor voltages and initial inductor currents. [For ease of reference call them $\mathbf{v}_C(0)$ and $\mathbf{j}_L(0)$, respectively]. Consider the following problem: at $t = 0-$ the initial conditions are given by $\mathbf{v}_C(0-)$ and $\mathbf{j}_L(0-)$; at $t = 0$ the current $\hat{\mathbf{I}}\epsilon^{s_0t}u(t)$ is switched on. We assert that

$$\mathbf{e}(t) = \hat{\mathbf{E}}\epsilon^{s_0t} = [\mathbf{M}(s_0)]^{-1}\hat{\mathbf{I}}\epsilon^{s_0t} \qquad t \geq 0$$

is the resulting set of node voltages. Indeed, this set of node voltages satisfies the differential equation (4.2), and by the process used to calculate $\mathbf{v}_C(0)$ and $\mathbf{j}_L(0)$ it satisfies all the initial conditions. Therefore, we conclude that *given a nondegenerate linear time-invariant network (see Fig. 4.1) driven by a single independent current source $i(t) = \hat{I}_1\epsilon^{s_0t}u(t)$, there is a set of initial capacitor voltages and initial inductor currents such that the port voltage is*

$$e_1(t) = Z(s_0)\hat{I}_1\epsilon^{s_0t} \qquad \text{for } t \geq 0$$

*provided $s_0$ is not a (complex) frequency at which the system determinant is zero.*

Let us make two observations. (1) Any complex frequency $s_i$ at which the system determinant is zero is a natural frequency of the network; hence, *provided $s_0$ is not a natural frequency of the network, the node voltages are*

$$\mathbf{e}(t) = \hat{\mathbf{E}}\epsilon^{s_0t} = [\mathbf{M}(s_0)]^{-1}\mathbf{I}\epsilon^{s_0t} \qquad \text{for } t \geq 0$$

(2) The driving-point impedance $Z(s)$ is a rational function of $s$, and its poles (i.e., the zeros of its denominator) are natural frequencies of the network. It may happen that $s_0$ is a natural frequency of the network, but not a pole of $Z(s)$. Our proof may be extended to show that *provided $s_0$ is not a pole of the driving-point impedance $Z(s)$, the input voltage is*

$$e_1(t) = Z(s_0)\hat{I}_1\epsilon^{s_0t} \qquad \text{for } t \geq 0$$

This last result is so important that it is worth repeating in complete detail.

*Consider a nondegenerate linear time-invariant one-port which has no internal independent sources and which is driven by the current source $i(t) = \hat{I}_1\epsilon^{s_0t}u(t)$. If $s_0$ is any (complex) frequency which is not a pole of the driving-point impedance $Z(s)$ of the one-port, then there is an initial state of the one-port such that the port voltage $e_1$ (in response to the current $i$ and that initial state) is of the form*

$$e_1(t) = Z(s_0)\hat{I}_1\epsilon^{s_0t} \qquad \text{for } t \geq 0$$

**Remark**   It is important to note that the same conclusion holds for the case in which the one-port is driven by a *voltage source $e(t) = E\epsilon^{s_0t}$* and $s_0$ is not a pole of the *driving-point admittance $Y(s)$* of the one-port.

| **5** | **One-ports Made of Passive Linear Time-invariant Elements** |

In this section we investigate a special class of passive one-ports, namely those consisting of an arbitrary interconnection of passive linear time-invariant elements. We have just shown that if all the elements of a one-port are passive, the one-port itself is passive. We wish now to take advantage of the fact that each element is, in addition, linear and time-invariant. Clearly then, we can use network functions. We propose to characterize the one-ports under consideration in terms of their driving-point impedances. To proceed in an orderly fashion, we must introduce the concept of a *positive real function*. It is a very general concept with applications in many fields. We need only concern ourselves with rational functions that are positive real. We formulate, therefore, our definition for rational functions only.

A function $F(\cdot)$ of the complex variable $s$ is said to be **positive real** if

(5.1)    $F(\cdot)$ is a rational function of $s$ with *real* coefficients

(5.2)    Re $(s) > 0$ implies that Re $[F(s)] \geq 0$

**Example 1**    $Z_1(s) = 1/(1 + s)$ is positive real. Indeed, $Z_1(s)$ is a rational function with real coefficients. Now let $s = \sigma + j\omega$; then

$$Z_1(s) = \frac{1}{(1 + \sigma) + j\omega} = \frac{1 + \sigma}{(1 + \sigma)^2 + \omega^2} - j\frac{\omega}{(1 + \sigma)^2 + \omega^2}$$

Thus,

$$\text{Re}\,[Z_1(s)] = \frac{1 + \sigma}{(1 + \sigma)^2 + \omega^2} > 0 \qquad \text{for } \sigma > 0$$

**Example 2**    $Z_2(s) = \frac{1}{s} = \frac{\sigma}{\sigma^2 + \omega_2} - j\frac{\omega}{\sigma^2 + \omega^2}$

It is easily shown that $Z_2$ is positive real.

**Example 3**    $Z_3(s) = \frac{1}{1 - s} = \frac{1 - \sigma}{(1 - \sigma)^2 + \omega^2} - j\frac{\omega}{(1 - \sigma)^2 + \omega^2}$

Clearly, for $\sigma = 2$, Re $[Z_3(s)] < 0$, hence $Z_3$ is not positive real.

**Example 4**    Let $Z_4(s) = 1/(1 + \bar{s})$, where $\bar{s}$ is the complex conjugate of $s$. As far as we are concerned, $Z_4$ is not a positive real function because its denominator is not a polynomial in $s$.

**Remark**    It is important to note that the fact that $F(s)$ is a rational function allows us to state that if (5.2) holds, that is, Re $[F(s)] \geq 0$ for all Re $(s) > 0$, then, by continuity, Re $[F(j\omega)] \geq 0$ for all real $\omega$ at which $F(j\omega)$ is defined.

Given the concept of positive real functions, we can state the following theorem.

**THEOREM**    If a one-port is made of lumped passive linear time-invariant elements and if it has a driving-point impedance $Z(s)$, then $Z(s)$ is a positive real function.

Note that the second assumption is necessary because some passive linear time-invariant networks do not have a driving-point impedance, e.g., a two-winding ideal transformer whose secondary is open-circuited.

*Proof*    To start the proof, we note that the driving-point impedance satisfies condition (5.1); indeed, the driving-point impedance of such a one-port is a rational function with *real* coefficients. It remains to prove that the driving-point impedance satisfies (5.2). Let us restate the definition of a passive one-port. Let $\mathcal{E}(t_0)$ be the energy stored in the one-port at time $t_0$; then

(5.3)    $$W(t_0,t) + \mathcal{E}(t_0) = \int_{t_0}^{t} v(t')i(t')\, dt' + \mathcal{E}(t_0) \geq 0$$

for all time $t$, for all initial time $t_0$, and for all input waveforms. Here $W(t_0,t)$ is the energy delivered to the one-port from $t_0$ to $t$. We pick the port current to be the input and the port voltage to be the response. Let us choose a particular current input of the form

(5.4)    $$i(t) = \mathrm{Re}\,(I\epsilon^{s_0 t})u(t - t_0)$$

where $I$ and $s_0$ are any fixed complex numbers and $u(t)$ is the unit step function. We have shown in the previous section that we can always choose a suitable set of initial conditions for the one-port such that the *complete* response (port voltage) is given by

(5.5)    $$v(t) = \mathrm{Re}\,(V\epsilon^{s_0 t})    \qquad t \geq t_0$$

where the complex numbers $V$ and $I$ are related by

(5.6)    $$V = Z(s_0)I$$

provided that $s_0$ is not a natural frequency of the one-port. In (5.6), $Z(s_0)$ is the driving-point impedance of the one-port evaluated at $s_0$. Let us denote the initially stored energy associated with the set of suitable initial conditions by $\mathcal{E}(t_0)$. Now we proceed to evaluate the integral $W(t_0,t)$.

First, we calculate the power entering the one-port for the current and voltage, as given by (5.4) and (5.5). For $t \geq 0$

(5.7)    $$p(t) = v(t)i(t) = |V||I|\epsilon^{2\sigma_0 t} \cos(\omega_0 t + \psi_1) \cos(\omega_0 t + \psi_2)$$

where

(5.8)    $$V = |V|\epsilon^{j\psi_1}    \qquad I = |I|\epsilon^{j\psi_2}$$

and

(5.9)   $s_0 = \sigma_0 + j\omega_0$

Using standard trigonometric identities, we can put Eq. (5.7) in the form

$$p(t) = \tfrac{1}{2}|V||I|\epsilon^{2\sigma_0 t}\,[\cos\,(\psi_1 - \psi_2) + \cos\,(2\omega_0 t + \psi_1 + \psi_2)]$$

(5.10)   $$= \tfrac{1}{2}\,\mathrm{Re}\;(\epsilon^{2\sigma_0 t}V\bar{I} + \epsilon^{2s_0 t}VI)$$

Using (5.6), we can express the instantaneous power in terms of the impedance $Z(s_0)$.   Let

(5.11)   $Z(s_0) = |Z(s_0)|e^{j\phi}$

Consequently by (5.6),

$$\measuredangle V = \psi_1 = \measuredangle Z(s_0) + \measuredangle I = \phi + \psi_2$$

Then (5.10) becomes

(5.12)   $p(t) = \tfrac{1}{2}\,\mathrm{Re}\;[\epsilon^{2\sigma_0 t}|I|^2 Z(s_0) + \epsilon^{2s_0 t}I^2 Z(s_0)]$

The energy delivered to the one-port from $t_0$ to $t$ is

$$W(t_0,t) = \int_{t_0}^{t} p(t')\,dt'$$

$$= \frac{1}{4}\,\mathrm{Re}\left[\epsilon^{2\sigma_0 t}|I|^2\,\frac{Z(s_0)}{\sigma_0} + \epsilon^{2s_0 t}I^2\,\frac{Z(s_0)}{s_0}\right]_{t_0}^{t}$$

(5.13)   $$= \frac{1}{4}|I|^2\epsilon^{2\sigma_0 t}\Big[a + b\,\cos\,(2\omega_0 t + \gamma)\Big]_{t_0}^{t} + c$$

where

(5.14)   $$a = \frac{\mathrm{Re}\,[Z(s_0)]}{\sigma_0} \qquad b = \frac{|Z(s_0)|}{|s_0|}$$

$\gamma$ in (5.13) represents the phase angle which is contributed by $Z(s_0)$, $I$, and $s_0$; finally, the *constant* $c$ is the preceding term evaluated at $t_0$.

Let us consider the special case in which $s_0$ is in the open right-half complex frequency plane; that is, $\sigma_0 > 0$.   Then the right-hand side of (5.13) consists of two terms: one that is exponentially increasing and one that is constant term.   As $t$ increases indefinitely, the exponential term will dominate any constant term.   Thus, for $W(t,t_0) + \mathcal{E}(t_0)$ to be nonnegative for all $t \geq 0$, the coefficient of the exponential term must be nonnegative; that is,

$$a - b \geq 0$$

To see why we have $a - b$, consider large values of $t$ for which $\cos\,(2\omega_0 t + \gamma) = -1$ [typically, $t_k = (2k\pi - \gamma)/2\omega_0$ where $k$ is an integer].   Going back to (5.14), we see that the condition $a - b \geq 0$, together with $\sigma_0 > 0$, implies

(5.15)   $\mathrm{Re}\,[Z(s_0)] \geq \sigma_0 \dfrac{|Z(s_0)|}{|s_0|} \geq 0$

Therefore, condition (5.2) is satisfied by the driving-point impedance. Thus, we have shown that the driving-point impedance is positive real.

**Remarks**   1.   We have shown that the driving-point impedance of any one-port made of passive linear time-invariant elements is positive real. It turns out that the converse is true. Given $F(s)$, a rational function of $s$ which is positive real, then there exists a one-port, made of passive linear time-invariant resistors, capacitors, and inductors, which has $F(s)$ as a driving-point impedance. In fact there are many ways of designing such one-ports. The consideration of these matters would lead us too far astray. However, it should be stressed that the above theorem is the basic result of passive network synthesis; it is basic to the design of filters, equalizers, delay networks, etc.

2.   Let us note that the theorem implies by continuity the familiar result

$\mathrm{Re}\,[Z(j\omega)] \geq 0$      for all $\omega$ at which $Z(j\omega)$ is defined

which had been established in Chaps. 7 and 9.

3.   In the course of the derivation, we have obtained an important property of positive real functions. Indeed, if we rearrange (5.15), we obtain

(5.16)   $\dfrac{\mathrm{Re}\,[Z(s_0)]}{|Z(s_0)|} \geq \dfrac{\sigma_0}{|s_0|} > 0$

where the last inequality follows from the assumption that $\sigma_0 > 0$. Interpreting inequality (5.16) geometrically [by considering the complex numbers $s_0$ and $Z(s_0)$ plotted in the complex plane], we conclude that for all $s_0$ in the right-half plane

$|\measuredangle s_0| \geq |\measuredangle Z(s_0)|$

In other words, *at any point $s_0$ in the right-half plane, the angle of a positive real function is in absolute value no larger than the angle of $s_0$.*

4.   Strictly speaking, we have only established (5.15) in the case where $\sigma_0 > 0$ and $s_0$ *is not a natural frequency of the one-port.* Now if $s_0$ is a natural frequency of the one-port, $s_0$ is a pole of the rational function $Z(s)$. By checking the behavior of $Z$ in the neighborhood of the pole, we can conclude that it is impossible for $Z(s)$ to have a pole at $s_0$ and satisfy (5.15) at all points around this pole. Consequently, (5.15) forbids $Z(s)$ from having poles in the open right-half $s$ plane. Let us remark that $Z(s)$ may have simple poles on the $j\omega$ axis. For example, the driving-point impedance of the parallel $LC$ circuit has simple poles at $\pm j1/\sqrt{LC}.$

## 6   Stability of Passive Networks

In the discussion of the preceding sections we concentrated on the behavior of one-ports. When we think in terms of one-ports, we do not concern ourselves with what is inside the one-port, provided the elements are lumped. For example, it is perfectly possible to have active elements in a network and still have the one-port of interest be passive, i.e., satisfying the passivity condition [see Eq. (3.2)]. A typical example is the balanced bridge circuit of Fig. 6.1. This one-port has an input impedance equal to 1 ohm for all complex frequencies no matter what the two-terminal element denoted by $Z$ may be; in particular, $Z$ may be a $-1$ ohm resistor, an active element.

### 6.1   Passive Networks and Stable Networks

In this section we consider the behavior of a complete network, rather than its behavior at a particular port. We say that a *network* is **passive** if all the elements of the network are passive. We shall study a general property of passive networks in terms of the energy stored. In many engineering problems it is possible to derive a number of properties of a system exclusively in terms of its energy. In particular, it is often possible to detect oscillation and instability in a system.

**Exercise**   Restate the definitions of a passive *network* and a passive *one-port*. Give a new example of a passive *one-port* which is *not* a passive network.

In Chap. 5 we used a linear time-invariant *RLC* tuned circuit to introduce the intuitive notions of stability, oscillation, and instability. We

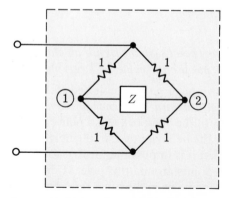

**Fig. 6.1**   A one-port which behaves as a 1-ohm resistor independently of the nature of the element $Z$ in the box between node ① and node ②.

calculated the zero-input responses of an ordinary passive $RLC$ tuned circuit, a lossless $LC$ tuned circuit, and an active negative-resistance tuned circuit. We found the state-space trajectories for the three cases. In the passive case, the trajectory reaches the origin as $t$ tends to infinity; we then called this circuit asymptotically stable. In the lossless tuned circuit, the trajectory is a closed path; we called this circuit oscillatory. In the active case, the trajectory may become unbounded and approach infinity as $t \to \infty$; we called this circuit unstable.

Stability is a vast subject. For our present purpose we wish to deal exclusively with the stability of zero-input responses; in other words, the networks under consideration contain no *independent* sources. Equivalently, we consider exclusively the *unforced network*, i.e., the network with all *independent* sources set to zero. We shall treat passive networks, linear as well as nonlinear, and we shall show that the energy stored in passive networks in general has interesting properties; in particular, *passive networks cannot be unstable*. Intuitively, this statement may be fairly obvious. However, we wish to go through the derivation in order to bring out the detailed assumptions and their consequences. For simplicity we will restrict our treatment to time-invariant networks, even though the extension to the time-varying case is not too difficult.

First let us introduce the following definition: an unforced network is said to be **stable** if for all initial states at time $t_0$ the state trajectory is bounded in the interval $[t_0, \infty)$.

**Remarks**
1. Let us insist on the following exact meaning of the word *bounded:* that the state trajectory is bounded in the interval $[t_0, \infty)$ means that, given the trajectory in state space, we can find a *fixed finite* number $M$ such that each component of the state vector is smaller than $M$ during the *whole* interval $[t_0, \infty)$; equivalently, for all $t$ in $[t_0, \infty)$

$$|x_i(t)| < M \qquad i = 1, 2, \ldots, n$$

where $x_i(t)$ is the $i$th component of the state vector $\mathbf{x}(t)$. For example, on $[0, \infty)$ the functions $\epsilon^{-t}$ and $\sin \omega t$ are bounded, but $\epsilon^t$ is not. Note however that $\epsilon^t$ is finite; indeed, for each instant of time $t$, the number $\epsilon^t$ is finite. Of course, as $t$ increases, $\epsilon^t$ increases without limit; this is the reason $\epsilon^t$ is not a bounded function on $[0, \infty)$.

2. The definition of stability allows the oscillatory responses (as in a lossless tuned circuit).

| 6.2 | **Passivity and Stability** |
|---|---|

We consider a general nonlinear *time-invariant RLC* network (see Fig. 6.2). There are no independent sources in the network. We assume that altogether there are $m$ *passive* nonlinear inductors and $n$ *passive* nonlinear

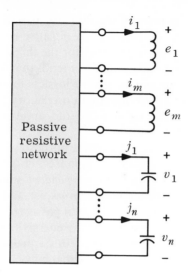

**Fig. 6.2**    A general nonlinear time-invariant network with $m$ inductors and $n$ capacitors interconnected by non-linear resistors.

capacitors.  They are connected to an arbitrary *passive* resistive network, as shown in Fig. 6.2.  The $m$ passive inductors are assumed to be flux-controlled and are characterized by

(6.1)    $i_k = \hat{i}_k(\phi_k) \qquad e_k = \dfrac{d\phi_k}{dt} \qquad k = 1, 2, \dots, m$

The $n$ passive capacitors are assumed to be charge-controlled and are characterized by

(6.2)    $v_l = \hat{v}_l(q_l) \qquad j_l = \dfrac{dq_l}{dt} \qquad l = 1, 2, \dots, n$

Let us calculate the energy stored in these elements.  The magnetic energy stored in the $m$ inductors is

(6.3)    $\mathscr{E}_M = \displaystyle\sum_{k=1}^{m} \int_0^{\phi_k} \hat{i}_k(\phi_k') \, d\phi_k'$

The electric energy stored in the $n$ capacitors is

(6.4)    $\mathscr{E}_E = \displaystyle\sum_{l=1}^{n} \int_0^{q_l} \hat{v}_l(q_l') \, dq_l'$

The energy stored in the network is then

(6.5)    $\mathscr{E} = \mathscr{E}_M + \mathscr{E}_E$

We assume further that all these elements have the following properties:

(6.6)

$\mathcal{E}_M \to \infty$       whenever one or more of the $\phi_k$'s tend to infinity

$\mathcal{E}_E \to \infty$       whenever one or more of the $q_l$'s tend to infinity

These two properties are obviously satisfied in the case of linear inductors and capacitors.

We shall now introduce the following vector notations:

Inductor-current vector:      $\mathbf{i} = (i_1, i_2, \ldots, i_m)^T$

Flux vector:      $\boldsymbol{\phi} = (\phi_1, \phi_2, \ldots, \phi_m)^T$

Inductor-voltage vector:      $\mathbf{e} = (e_1, e_2, \ldots, e_m)^T$

Capacitor-voltage vector:      $\mathbf{v} = (v_1, v_2, \ldots, v_n)^T$

Charge vector:      $\mathbf{q} = (q_1, q_2, \ldots, q_n)^T$

Capacitor current vector:      $\mathbf{j} = (j_1, j_2, \ldots, j_n)^T$

In vector notation the branch equations of the inductors and capacitors are

(6.7)    $\mathbf{i} = \hat{\mathbf{i}}(\boldsymbol{\phi}) \qquad \mathbf{e} = \dfrac{d\boldsymbol{\phi}}{dt}$

(6.8)    $\mathbf{v} = \hat{\mathbf{v}}(\mathbf{q}) \qquad \mathbf{j} = \dfrac{d\mathbf{q}}{dt}$

We will choose as the state vector of the $RLC$ network

(6.9)    $\mathbf{x} = \begin{bmatrix} \boldsymbol{\phi} \\ \mathbf{q} \end{bmatrix}$

and the state trajectory is then in an $(m + n)$-dimensional space with $\phi_k$, $k = 1, 2, \ldots, m$, and $q_l, l = 1, 2, \ldots, n$ as the state variables. The energy stored can then be written as

(6.10)

$\mathcal{E}[\mathbf{x}(t)] = \mathcal{E}_M[\boldsymbol{\phi}(t)] + \mathcal{E}_E[\mathbf{q}(t)]$

$= \displaystyle\int_0^{\phi(t)} \hat{\mathbf{i}}(\boldsymbol{\phi}') \cdot d\boldsymbol{\phi}' + \int_0^{q(t)} \hat{\mathbf{v}}(\mathbf{q}') \cdot d\mathbf{q}'$

Equations (6.3) and (6.4) imply that if $\mathbf{x} = \mathbf{0}$, then

(6.11)    $\mathcal{E}[\mathbf{0}] = \mathcal{E}_M[\mathbf{0}] + \mathcal{E}_E[\mathbf{0}] = 0$

The passivity requirement on the elements implies that for all $t$, for all $\boldsymbol{\phi}(t)$ and for all $\mathbf{q}(t)$,

(6.12)    $\mathcal{E}[\mathbf{x}(t)] = \mathcal{E}_M[\boldsymbol{\phi}(t)] + \mathcal{E}_E[\mathbf{q}(t)] \geq 0$

Taking the derivative with respect to time $t$, we obtain

$$\frac{d\mathcal{E}}{dt} = \sum_{k=1}^{m} \frac{\partial \mathcal{E}_M}{\partial \phi_k} \frac{d\phi_k}{dt} + \sum_{l=1}^{n} \frac{\partial \mathcal{E}_E}{\partial q_l} \frac{dq_l}{dt}$$

(6.13)           $= \sum\limits_{k=1}^{m} i_k e_k + \sum\limits_{l=1}^{n} j_l v_l$

Since the resistive network is passive, the instantaneous power entering it is nonnegative, that is (refer to Fig. 6.2),

$$\sum_{k=1}^{m} (-i_k)e_k + \sum_{l=1}^{n} (-j_l)v_l \geq 0$$

or

(6.14)     $\dfrac{d\mathscr{E}}{dt} = \sum\limits_{k=1}^{m} i_k e_k + \sum\limits_{l=1}^{n} j_l v_l \leq 0$

Thus, the energy stored in the network is a nonincreasing function of time. In other words,

(6.15)   $\mathscr{E}(t) \leq \mathscr{E}(t_0)$     for all $t \geq t_0$

Let us consider the consequences of the nonincreasing character of the energy stored as far as a state trajectory is concerned. In view of (6.6), it is impossible for any of the $\phi_k$'s or the $q_l$'s to become arbitrarily large. Otherwise, either $\mathscr{E}_M$ or $\mathscr{E}_E$ (or both) would become arbitrarily large, and the stored energy $\mathscr{E}(t)$ would become larger than $\mathscr{E}(t_0)$. This contradicts (6.15). Consequently, *all* the $\phi_k$'s *and all* the $q_l$'s remain bounded. Thus, we have established the following general fact:

*Any nonlinear passive time-invariant RLC network made of flux-controlled inductors and charge-controlled capacitors is a stable network.*

In many cases we can conclude that the state tends to the origin of the state space as $t \to \infty$. Intuition and Eq. (6.14) say that the energy stored $\mathscr{E}$ will decrease as long as the resistive network absorbs energy; in many cases this process will force $\mathscr{E}$ to go to 0 as $t \to \infty$. This will necessarily be the case if each inductor is in series with a passive linear resistor and each capacitor is in parallel with a passive linear resistor. The exercise below suggests situations in which the state does not necessarily go to zero as $t \to \infty$.

**Exercise**   The networks shown in Fig. 6.3 are passive, linear, and time-invariant. In each case, find capacitor voltages and inductor currents (that are not all identically zero) which satisfy the network equations. [Hints: (a) Is there a capacitor-only cut set? If so, what happens when all the capacitors of the cut set have the same constant voltage? (b) Have you ever thought of the dual of a capacitor-only cut set? (c) Observe that $Z(j) = 0$ and that the series resonant circuit (in parallel with the 1-ohm resistor) also has a zero impedance at $\omega = 1$; find then if the current $i$ can be of the form $\cos t$.]

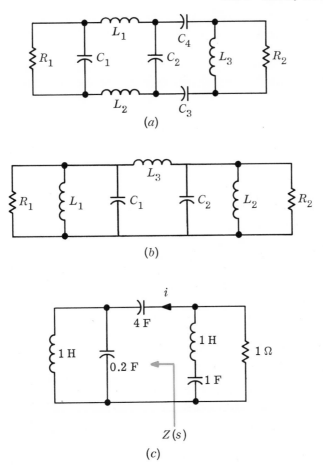

$(a)$

$(b)$

$(c)$

**Fig. 6.3**   Examples of passive linear time-invariant networks in which the energy stored does not necessarily reach zero as $t \to \infty$.

### 6.3   Passivity and Network Functions

Let us consider an important consequence of the fact that passivity implies stability. In the particular case of *passive linear time-invariant RLC* networks the fact that the stored energy $\mathcal{E}$ cannot be an increasing function of time has implications on *all* network functions. To be specific, suppose we consider a transfer impedance $Z_T(s)$; for example, we imagine the *RLC* network driven by a current source at some node pair ①①′, and the response is the voltage across another node pair ②②′. If the current source applies an impulse at $t = 0$ and if the network is in the zero state at $t = 0-$, the voltage which appears across ②②′ will be equal to $h(t)$, where $h = \mathcal{L}^{-1}[Z_T(s)]$. At $0+$, just after the impulse, $\mathcal{E}(0+)$, the energy stored, is finite; hence for all later $t$'s, $\mathcal{E}(t) \leq \mathcal{E}(0+)$. If $Z_T(s)$ has a pole, say $s_1$, in the open right-half plane [that is, Re $(s_1) > 0$], then $h(t)$ would contain a term of the form $k_1 \epsilon^{s_1 t}$, that is, an exponentially increasing voltage. Intuitively, this exponentially increasing voltage can only appear if the energy in the network is also exponentially increasing. Hence we have a contradiction, since passivity requires that $\mathcal{E}$ be nonincreasing. Therefore, the transfer impedance $Z_T(s)$ cannot have a pole in the open right-half plane.

Similarly, if we had assumed that $Z_T(s)$ had a multiple pole on the $j\omega$ axis, say, a pole of order $k \geq 2$ at $j\omega_0$, then $h(t)$ would contain a term of the form $At^{k-1} \cos(\omega_0 t - \phi)$. Clearly, this term takes arbitrarily large values as $t$ increases, and again we have a contradiction, since passivity requires that $\mathcal{E}$ be nonincreasing. Thus, we have shown that the transfer function $Z_T(s)$ cannot have poles in the open right-half plane and that if it has a pole on the $j\omega$ axis, this pole must be simple. Clearly, this reasoning applies to any network function, driving-point impedance or admittance, transfer impedance or admittance, voltage or current ratio. Then we conclude by making the following statement:

*For any **passive** linear time-invariant RLC network, if $s_1$ is a pole of any of its network functions, then* Re $(s_1) \leq 0$; *and if* Re $(s_1) = 0$, *the pole $s_1$ is simple.*

A similar reasoning shows that *any* natural frequency of a passive, linear, time-invariant network must lie in the closed left-half plane, and any $j\omega$-axis natural frequency must be simple.

---

| 7 | **Parametric Amplifier** |

In Sec. 1 we showed that by doing mechanical work against the electrostatic forces acting on the plates of a charged capacitor, it is possible to transfer energy from the outside world to the circuit. Intuitively, it would seem that the same mechanism could be used to boost the power of a sinusoidal signal. In fact, this can be done, and such amplifiers are called *parametric amplifiers*. In this section we shall give an approximate analysis of a very simple parametric amplifier. Our purpose is to illustrate the approach to the problem and to show how our methods of circuit analysis can be applied to practical problems.

The amplifier we are considering consists of a parallel *RLC* circuit (see Fig. 7.1) driven by a sinusoidal current source at angular frequency $\omega_0$. The inductor $L$ and the resistor of conductance $G$ are linear and time-invariant. The capacitor is linear and *time-varying*. Although, in practice, it consists of a reverse-biased varactor diode, we represent it as a fixed capacitor $C_0$ in parallel with a sinusoidally varying capacitor of capacitance

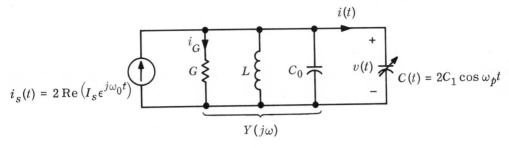

$i_s(t) = 2\,\mathrm{Re}\left(I_s\epsilon^{j\omega_0 t}\right)$

$C(t) = 2C_1 \cos \omega_p t$

$Y(j\omega)$

**Fig. 7.1**  A simple parametric amplifier circuit.

$$C(t) = 2C_1 \cos \omega_p t = C_1 \epsilon^{j\omega_p t} + C_1 \epsilon^{-j\omega_p t}$$

where $2C_1$ is the amplitude of the capacitance variation and $\omega_p$ is the angular frequency of the variation. It is called the *pump frequency*. We assume that $\omega_p > \omega_0$. In the steady state the current $i(\,\cdot\,)$ and the voltage $v(\,\cdot\,)$ of the time-varying capacitor have components at frequencies $\omega_0$, $\omega_0 \pm \omega_p$, $\omega_0 \pm 2\omega_p$, $\omega_0 \pm 3\omega_p, \ldots$. Now if the inductance $L$ is so chosen that it resonates with $C_0$ at the signal frequency $\omega_0$, then with a suitable choice of the pump frequency $\omega_p$ and the $Q$ of the tuned circuit, it turns out that $v(\,\cdot\,)$ and $i(\,\cdot\,)$ consist mainly of a linear combination of two sinusoids at $\omega_0$ and $\omega_p - \omega_0$.† A typical plot of the relative magnitude of the ordinates of the curve $|Z(j\omega)|$ at $\omega_0$, $\omega_p$, $\omega_p - \omega_0$, and $\omega_p + \omega_0$ is shown on Fig. 7.2. For convenience we denote $\omega_1$ as the difference frequency (usually called the idler frequency); thus, $\omega_1 = \omega_p - \omega_0$. The approximation that $v(\,\cdot\,)$ and $i(\,\cdot\,)$ contain only frequencies $\omega_0$ and $\omega_1$ simplifies considerably the derivation to follow. Let us represent the steady-state voltage $v(\,\cdot\,)$ by its phasors $V_0$ and $V_1$ at frequencies $\omega_0$ and $\omega_1$, respectively. Then

(7.1) $\quad v(t) = V_0 \epsilon^{j\omega_0 t} + \overline{V}_0 \epsilon^{-j\omega_0 t} + V_1 \epsilon^{j\omega_1 t} + \overline{V}_1 \epsilon^{-j\omega_1 t}$

The current $i(\,\cdot\,)$ in the time-varying capacitor is then

(7.2) $\quad i(t) = I_0 \epsilon^{j\omega_0 t} + \overline{I}_0 \epsilon^{-j\omega_0 t} + I_1 \epsilon^{j\omega_1 t} + \overline{I}_1 \epsilon^{-j\omega_1 t}$

This voltage and current must satisfy the branch equation

(7.3) $\quad i(t) = \dfrac{d}{dt}[C(t)v(t)]$

By using Eqs. (7.1) and (7.2) in (7.3) and equating the phasors at frequencies $\omega_0$ and $\omega_1$, we find that the phasors $V_0$, $V_1$, $I_0$, and $I_1$ must satisfy the equations

(7.4a) $\quad I_0 = j\omega_0 C_1 \overline{V}_1$

(7.4b) $\quad I_1 = j\omega_1 C_1 \overline{V}_0$

† The intuitive justification is that, at other frequencies, there is not enough impedance to develop substantial voltages across (hence current through) the time-varying capacitors.

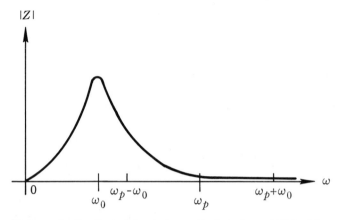

**Fig. 7.2**    Diagram showing the relation between the impedance $Z(j\omega)$ of the $LC_0G$ tuned circuit at $\omega_0$, $\omega_p$, $\omega_p - \omega_0$, and $\omega_p + \omega_0$.

Call $Y(j\omega)$ the admittance at frequency $\omega$ of the time-invariant circuit made of $G$, $L$, and $C_0$. Let also

$$(7.5) \quad Y_0 \overset{\Delta}{=} Y(j\omega_0) \qquad Y_1 \overset{\Delta}{=} Y(j\omega_1)$$

From KCL we obtain (see Fig. 7.1),

$$(7.6a) \quad I_s = Y_0 V_0 + I_0 \qquad \text{at } \omega_0$$

$$(7.6b) \quad 0 = Y_1 V_1 + I_1 \qquad \text{at } \omega_1$$

The four equations in (7.4a), (7.4b), (7.6a), and (7.6b) have to be solved for the four unknown phasors $I_0$, $I_1$, $V_0$, and $V_1$. Eliminating $I_1$ and $V_1$ in (7.4a) and (7.4b) and using (7.6b), we obtain

$$I_0 = j\omega_0 C_1 j\omega_1 C_1 \frac{V_0}{Y_1}$$

or

$$(7.7) \quad \frac{I_0}{V_0} = \frac{-\omega_0 \omega_1 C_1{}^2}{Y_1} = \frac{-\omega_0 \omega_1 C_1{}^2}{Y(-j\omega_1)}$$

Now $I_0/V_0$ is the apparent admittance of the time-varying capacitor at frequency $\omega_0$; call it $Y_C(j\omega_0)$. The equation above indicates that as far as the signal frequency $\omega_0$ is concerned the time-varying capacitance $C(t)$ has an equivalent admittance $Y_C(j\omega_0)$ which is related to the admittance $Y$ of the time-invariant $RLC$ circuit evaluated at frequency $-\omega_1$. Moreover, the negative sign in Eq. (7.7) indicates that this equivalent admittance represents an active element. Thus, the total admittance faced by the current source is

(7.8)   $Y_t = Y_C(j\omega_0) + Y_0 = Y_0 - \dfrac{\omega_0\omega_1 C_1{}^2}{\overline{Y}_1}$

If the output current is taken as $I_G$ through the resistor, we have the current gain of the parametric amplifier as

(7.9)   $\dfrac{I_G}{I_s} = \dfrac{G}{Y_0 - \omega_0\omega_1 C_1{}^2/\overline{Y}_1}$

Thus, the current gain depends on the values of the signal frequency $\omega_0$, the idler frequency $\omega_1$, the amplitude of the capacitance variation $2C_1$, and the admittance (at $\omega_0$ and $-\omega_1$) of the time-invariant tuned circuit. If the $LGC_0$ circuit resonates exactly at $\omega_0$, then $Y_0 \triangleq Y(j\omega_0) = G$. Since $\overline{Y}_1$ represents the admittance of a passive one-port at $-\omega_1$, its real part is positive. It is clear from (7.9) that an arbitrarily large current gain can be obtained with suitable design of the tuned circuit.

As mentioned earlier, a practical parametric amplifier employs a varactor diode which is actually a nonlinear capacitor. The value of $C_0$ and $C_1$ depends on the pump voltage. A typical parametric amplifier may have $C_1 = 1$ pF, $C_0 = 4$ pF, $G = 10^{-4}$ mho, $\omega_0 = 10^9$ rad/sec, $\omega_1 = 5 \times 10^9$ rad/sec, and a power gain of 20 db. The principal virtue of parametric amplifiers is that they add very little noise to the signal they amplify. In this respect, they are superior to vacuum-tube and transistor amplifiers.

Thus, we have demonstrated that an active time-varying circuit can be used to amplify signals.

## Summary

- The concept of passivity is naturally related to the basic physical concepts of energy and time. For a one-port, whether it is linear or nonlinear, time-varying or time-invariant, we can define passivity in terms of the energy delivered to the one-port. Let $\mathcal{E}(t_0)$ be the energy stored in the one-port at time $t_0$, and let $W(t_0,t)$ be the energy delivered to the one-port from time $t_0$ to $t$; then we say that the *one-port is passive* if $\mathcal{E}(t_0) + W(t_0,t)$ is nonnegative for all possible input waveforms, for all initial time $t_0$, and for all time $t \geq t_0$. Based on this definition, we obtain the following important conclusions.

  1. A resistor is passive if and only if its characteristic in the $vi$ plane is, for all time, in the first and third quadrants.

  2. A nonlinear time-invariant inductor is passive if and only if, for all $\phi$, the stored energy is nonnegative; that is,

  $$\mathcal{E} = \int_0^\phi \hat{\imath}(\phi') \, d\phi' \geq 0 \qquad \text{for all } \phi$$

A linear time-varying inductor is passive if and only if $L(t) \geq 0$ and $\dot{L}(t) \geq 0$ for all time $t$.

3.  A nonlinear time-invariant capacitor is passive if and only if, for all $q$, the stored energy is nonnegative; that is,

$$\mathcal{E} = \int_0^q \hat{v}(q') \, dq' \geq 0 \qquad \text{for all } q$$

A linear time-varying capacitor is passive if and only if $C(t) \geq 0$ and $\dot{C}(t) \geq 0$ for all time $t$.

4.  A one-port made of an arbitrary interconnection of passive elements is a passive one-port.

5.  If a one-port is made of lumped passive linear time-invariant elements and if it has a driving-point impedance $Z(s)$, then $Z(s)$ is a positive real function; that is, $Z(s)$ is a rational function of $s$ with *real* coefficients, and if Re $(s) > 0$, then Re $[Z(s)] \geq 0$.

■   Passivity is closely related to stability. An unforced network (a network with all independent sources set to zero) is said to be *stable* if, for all initial states at time $t_0$, the state trajectory is bounded in the interval $[t_0, \infty)$. A *network* is said to be *passive* if all its elements are passive. Based on these definitions, we showed that passive networks are stable. In the case of linear time-invariant networks, this means that any natural frequency of a *passive* network must lie in the closed left-half plane, and any $j\omega$-axis natural frequency must be simple.

■   Time-varying networks and time-varying circuit elements are useful in many practical problems. If we assume that the time variation of a circuit element is originated by mechanical means, we can study the energy balance of the whole network and obtain simple relations between energy dissipation, energy stored in the network, electric energy delivered, and mechanical energy delivered by the network. In particular, for a parallel-plate time-varying capacitor, the electric energy stored at time $t$ is

$$\mathcal{E}_E(t) = \tfrac{1}{2}C(t)v^2(t)$$

The electric power delivered to the capacitor by the network is

$$p_e(t) = i(t)v(t) = \dot{C}v^2(t) + C(t)v(t)\dot{v}(t)$$

Mechanical work (against the outside world) is done by the electrostatic force at a rate

$$p_m(t) = -f\dot{x} = \tfrac{1}{2}\dot{C}(t)v^2(t)$$

The energy balance yields

$$p_e(t) = \frac{d}{dt}\mathcal{E}_E + p_m(t)$$

■ We can design a simple parametric amplifier using a *RLC* tuned circuit connected in parallel with a sinusoidally varying capacitor. A sinusoidally varying capacitor is an active element. We demonstrated that an active time-varying element can be used to amplify signals.

## Problems

Linear time-varying capacitor, energy

**1.** A voltage source $v(t) = \cos \omega t$ is connected to a linear time-varying capacitor. The capacitance is $C(t) = 2 + \cos(\omega_p t + \phi)$, where the pump frequency $\omega_p$ and the phase $\phi$ can be adjusted. Assume that the ratio $\omega/\omega_p$ is a rational number.

   *a.* Find $\omega_p$ and $\phi$ so that the average power delivered by the capacitor to the source is maximum.

   *b.* Where does the energy received by the source come from?

Linear time-varying capacitor, energy

**2.** A voltage source $v(t) = \cos \omega t$ is connected to a linear time-varying capacitor. This capacitor may take any value subject to the condition that $1 \le C(t) \le 3$. How should one vary the capacitance periodically so that this constraint is satisfied *and* the maximum amount of energy is supplied by the capacitor to the source over one period. Compare this energy with that found in Prob. 1.

Linear time-varying inductor

**3.** A current source $i(t) = \sin \omega t$ is connected to a linear time-varying inductor with $L(t) = 2 + \cos(\omega_p t + \phi)$. Find the pump frequency $\omega_p$ and the phase $\phi$ so that the average power delivered by the inductor to the source is maximum. Assume that the ratio $\omega/\omega_p$ is a rational number.

Linear time-varying elements, passivity

**4.** In the expressions below, the symbols $a$ and $b$ denote constants that may be positive or negative. State what conditions $a$ and $b$ must satisfy in order that the corresponding element be active.

   *a.* A resistor with $R(t) = a + b \cos \omega_p t$.

   *b.* A capacitor with $C(t) = a + b \cos \omega_p t$.

   *c.* An inductor with $L(t) = a + b \cos \omega_p t$.

Nonlinear elements, passivity

**5.** Discuss whether the following nonlinear elements are passive or active:

   *a.* A resistor specified by $i = v^2$.

   *b.* An inductor specified by $i = -\phi + \phi^3$.

   *c.* A capacitor specified by $q = v^3$.

   *d.* A resistor specified by $i = v - 2v^2 + v^3$.

Passive network vs. passive one-port

**6.** *a.* The network shown in Fig. P19.6 is linear and time-invariant. Is it a passive *network*?

   *b.* If we define terminals ① and ①′ as a port and consider the network as a one-port, is the resulting *one-port* passive?

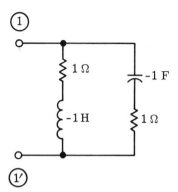

**Fig. P19.6**

Appropriate
initial state for
exponential
response

**7.** For the two circuits shown in Fig. P19.7, find the initial state at $t = 0$ which has the property that if $i(t) = u(t)\epsilon^t$, then all the voltages and all the currents of the network are proportional to $\epsilon^t$. [Hint: Assume $v_2(t) = A\epsilon^t$, and calculate back to $i(t)$, thereby adjusting the constant $A$.]

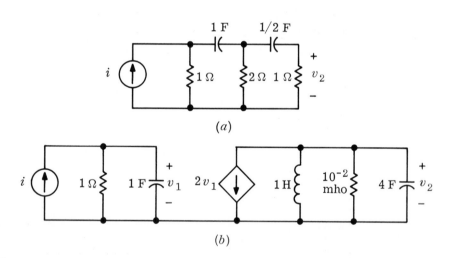

(*a*)

(*b*)

**Fig. P19.7**

Properties of
positive real
function

**8.** Given a linear time-invariant *passive* network and two of its nodes ①
and ①′. Call $Z(s)$ the impedance seen at these nodes. Justify the following properties of $Z(s)$:

*a.* $Z(s)$ cannot have poles in the open right-half plane.

*b.* If $Z(s)$ has a pole on the $j\omega$ axis, that pole must be simple.

*c.* For all $\omega$ where it is defined (i.e., except at $j\omega$-axis poles of $Z$), $\text{Re}\,[Z(j\omega)] \geq 0$.

Test of positive real functions

**9.** Are the following functions positive real?  If not, say why.

*a.* $\dfrac{s^3 + 1}{s}$    *c.* $\dfrac{s}{s - 1}$

*b.* $\dfrac{s^2 + s + 1}{s^2}$    *d.* $\dfrac{s}{s + 1}$

Properties of positive real functions

**10.** Prove the following facts:

*a.* If $Z_1(s)$ and $Z_2(s)$ are positive real, so is $Z_1(s) + Z_2(s)$.

*b.* If $Z(s)$ is positive real, so is $1/Z(s)$.

Properties of positive real functions

**11.** Give examples to show that if $Z_1(s)$, $Z_2(s)$, $Z_3(s)$, $Z_4(s)$, $Z_5(s)$, and $Z_6(s)$ are positive real, then

*a.* $Z_1(s) - Z_2(s)$ need not be positive real.

*b.* $Z_3(s)Z_4(s)$ need not be positive real.

*c.* $Z_5(s)/Z_6(s)$ need not be positive real.

Passivity and stability

**12.** The network shown in Fig. P19.12 contains a linear capacitor and a nonlinear resistor.  Is the network passive?  Is the circuit asymptotically stable?  Justify your answer.

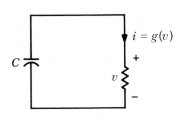 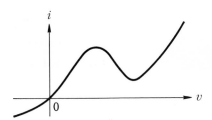

**Fig. P19.12**

Stability

**13.** Two lossless tuned networks are coupled together by means of a dependent source as shown in Fig. P19.13.  Is the network stable or unstable? If we remove the dependent source and we couple the two tuned circuits by an ideal transformer, is the new network stable or unstable?

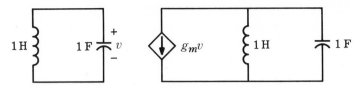

**Fig. P19.13**

Parametric
amplifier **14.** A parallel $RLC$ circuit, where the resistor is time-varying with conductance $G(t) = G_0 + 2G_1 \cos \omega_p t$ (where $G_0$, $G_1$, and $\omega_p$ are constants) is driven by a sinusoidal current source.

   *a.*  Show that power amplification is possible only if $2G_1 > G_0$. Justify this conclusion by physical argument.
   *b.*  Carry out an approximate calculation similar to that in Sec. 7 to obtain a sufficient condition for power amplification.

# Appendix: Functions and Linearity

In the course of our study of circuit theory we shall find ourselves using expressions such as "linear element," "linear space," "linear circuit," "linear function," etc. In order to understand these concepts we start by reviewing the most fundamental ones, the concept of *function* and then the concept of *linear function*.

There is a lot of clarity to be gained in engineering problems if one is familiar with these more fundamental concepts. Once the concepts of *function* and *linear function* are understood (and this requires hard thinking), the remaining ones are straightforward.

## 1 Functions

### 1.1 Introduction to the Concept of Function

Most of your work up to the present has been with a special class of functions like polynomials, sines, cosines, and exponentials, which assign to any real number another real number. The idea of a function is, however, far more general. To develop this more general concept, let us examine the following examples.

**Example 1**  The simplest and most familiar function is one defined by a polynomial. Let

$$y = x + x^2 + x^3$$

For each real number $x$ this equation defines a unique real number $y$. We say that this equation defines a function which maps real numbers into real numbers.

**Example 2**  Consider the three equations

$$y_1 = x_1 + 3x_2 + 5x_3$$

$$y_2 = 3x_1 + 2x_2 + 7x_3$$

$$y_3 = x_1 + x_2 + 5x_3$$

or, in vector form,

(1)   $\mathbf{y} = \mathbf{Ax}$

$$\mathbf{x} = \begin{bmatrix} x_1 \\ x_2 \\ x_3 \end{bmatrix} \qquad \mathbf{y} = \begin{bmatrix} y_1 \\ y_2 \\ y_3 \end{bmatrix} \qquad \text{and} \qquad \mathbf{A} = \begin{bmatrix} 1 & 3 & 5 \\ 3 & 2 & 7 \\ 1 & 1 & 5 \end{bmatrix}$$

Equation (1) assigns to each vector $\mathbf{x}$ a unique vector $\mathbf{y}$. We say that Eq. (1) defines a function which maps vectors into vectors.

**Example 3**   Consider a periodic function with period $T$; suppose it is smooth enough to have a Fourier series representation as follows:

$$F(t) = \sum_{n=-\infty}^{+\infty} c_n \epsilon^{jn2\pi t/T}$$

The Fourier coefficients $(\ldots, c_{-1}, c_0, c_1, \ldots)$ are calculated in terms of the function $F$ by

(2)   $c_n = \dfrac{1}{T} \displaystyle\int_0^T F(t) \epsilon^{-jn2\pi t/T}\, dt \qquad n = \ldots, -1, 0, 1, \ldots$

Equation (2) associates to the function $F$ and to each integer $n$ a complex number $c_n$. Considering all the Fourier coefficients as a whole, we see that Eq. (2) is a recipe that calculates, for any smooth periodic function $F$, the sequence $(\ldots, c_{-1}, c_0, c_1, \ldots)$ of its Fourier coefficients. We say that Eq. (2) defines a new function that maps any smooth periodic function $F$ into the set of all its Fourier coefficients $(\ldots, c_{-1}, c_0, c_1, \ldots)$.

Let us extract from these examples the common features of the idea of a function

1.   *There is a set (call it $X$) of elements.*   In the case of polynomials, $X$ was the set of all real numbers; in Example 2, $X$ was the set of all vectors of "space" geometry; in Example 3, $X$ was the set of all smooth periodic functions with period $T$.

2.   *There is another set (call it $Y$) of elements.*   For polynomials, $Y$ was the set of all real numbers; in Example 2, $Y$ was the set of all vectors of "space" geometry; in Example 3, $Y$ was the set of all infinite sequences of complex numbers $(\ldots, c_{-1}, c_0, c_1, \ldots)$.

3.   The function $f$ assigns to each element $x$ of $X$ a *unique* element $y$ of $Y$. This particular element $y$ is denoted by $f(x)$, hence the notation

$y = f(x)$.   We say that the function $f$ *maps* the element $x$ in $X$ into the element $f(x)$ in $Y$, and $f(x)$ is called *the value of f at x*.

4.   The set $X$ is called the *domain* of the function $f$.

It should be stressed that the generality and importance of the concept of function lie in the fact that the sets $X$ and $Y$ are arbitrary.   For instance, the elements of $X$ and $Y$ may be functions themselves, as shown by the following example from physics.

**Example 4**   Consider the following problem in electrostatics.   On an infinitely long ribbon of nylon there is a charge distribution specified by the charge density $\rho(\xi)$ coul/m.   We know that the electrostatic potential at the point of abscissa $x$ is given by

$$(3) \quad \phi(x) = \frac{1}{4\pi\epsilon_0} \int_{-\infty}^{+\infty} \frac{\rho(\xi)}{|x - \xi|}\, d\xi$$

The equation above defines a *function;* its domain is the set of all possible charge densities $\rho$, and its range is the set of all potential distributions $\phi$.   Indeed Eq. (3) gives a rule for computing the *unique* $\phi$ associated to any given $\rho$.

We are now ready to state the formal definition of a function.

## 1.2   Formal Definition

Let $f$ be a collection of ordered pairs $(x, y)$ where the first element $x$ belongs to some arbitrary set $X$ and the second element $y$ belongs to another arbitrary set $Y$.   The collection $f$ of ordered pairs is called a **function** if it has the property that for each $x$ in $X$ there is a *unique y* in $Y$ such that the pair $(x, y)$ belongs to the collection $f$.

The set $X$ of all possible first elements is called the **domain** of the function $f$.   We say that the function $f$ is *defined on X* and *takes its values in Y*.

In other words, if $f$ is a function and if the ordered pairs $(x, y)$ and $(x,z)$ belong to the collection defining $f$, then $y = z$.

**Remarks**   1.   It is important to perceive the difference between a function $f$ and $f(x)$, its value at a point $x$.   For example, consider the logarithm (base 10). To know the function logarithm, we need a whole table consisting of the collection of all ordered pairs $(x, \log x)$.   However, to know $\log 2$ we need only to remember the number 0.30103.   In short, it is best to recall that the definition of a function specifies a collection of ordered pairs $(x, f(x))$ which we may think of as listed in a table.   In the first column of the table we list all the elements $x$ of the domain; next to each $x$ we list in the second column $f(x)$, the value of the function $f$ at that $x$.   *The function f is the whole table.*   The symbol $f(x)$ denotes only the entry next to $x$.

2.   Expressing the same idea in geometric terms for the case in which the sets $X$ and $Y$ are the real line, we think of the function $f$ as being the *whole graph;* $f(x)$ is simply the ordinate corresponding to abscissa $x$. In circuit theory, we encounter voltages (or currents, charges, etc.) that change with time.   We then think of them as functions of time. When we want to emphasize that we are interested in the *function,* we shall say the function $v(\,\cdot\,)$ or the **waveform** $v(\,\cdot\,)$ to emphasize that we mean the whole graph (see also Example 4, Sec. 2, below).

3.   Let $f$ be a function which *maps $X$ into $Y$,* or $f$ is a *mapping* (or transformation or operator) of $X$ into $Y$.   Thus "function," "mapping," "transformation," and "operator" are synonymous terms.   Which term to use is a matter of tradition.   It is often convenient to write $f(\,\cdot\,)$ instead of $f$ to emphasize the difference between $f(x)$ and $f(\,\cdot\,)$. Stated again, $f(x)$ is the value of the function $f$ at $x$, whereas $f(\,\cdot\,)$—or $f$— denotes the function itself, i.e., the collection of *all ordered pairs* $(x, f(x))$.

4.   In some applications, such as nonlinear circuits, we encounter what used to be called "multivalued functions," which are rules that assign several values to some or all of the elements in the function's domain. In modern parlance, the term "function" always means *single-valued function,* and the term "multivalued function" is slowly being replaced by "multivalued relation," or simply "relation."

## 2    Linear Functions

This section is mainly review, for it ties together in modern language a lot of facts that you already know.   However, in order to understand the concept of *linear function* we must agree on the meaning of *linear space.* Roughly, a linear space is obtained by combining a set of *scalars* with a set of *vectors* and two operations, the *addition of vectors* and the *multiplication of vectors by scalars.*

### 2.1    Scalars

Technically, the set of scalars is required to be a *field* (here the word "field" is a technical term of algebra).   In engineering we encounter four fields: (1) the set of all real numbers; (2) the set of all complex numbers; (3) the set of all rational functions with real (or complex) coefficients; for example,

$$\frac{1 + 2s + \sqrt{3}s^2}{1 + s^4}$$

(4) the set of binary numbers $\{0,1\}$.   Over each of these sets the operations of addition and multiplication (and their inverses, subtraction and division) are defined in a well-known way.   However, because each of these sets constitutes a field, the rules for combining these operations are identical.

Technically, if we wish to check whether or not a set of elements over which two operations are defined qualifies as a set of scalars for some vector space, we need only check the following nine axioms of a field.

Let $F$ be a set of elements $\alpha$, $\beta$, $\gamma$, .... For the set $F$ to be called a **field** it must satisfy the following axioms:

1.   *Associative law of addition.*   For all $\alpha$, $\beta$, $\gamma$ in $F$, $\alpha + (\beta + \gamma) = (\alpha + \beta) + \gamma$.

2. .  *Commutative law of addition.*   For all $\alpha$, $\beta$ in $F$, $\alpha + \beta = \beta + \alpha$.

3.   There is an element $0$ in $F$ such that $0 + \alpha = \alpha$, for every $\alpha$ in $F$.   This element $0$ is called the *additive identity*.

4.   *Additive inverse.*   For each element $\alpha$ in $F$ there is an element in $F$ called $-\alpha$, such that $\alpha + (-\alpha) = 0$.

5.   *Associative law of multiplication.*   For all $\alpha$, $\beta$, $\gamma$ in $F$, $\alpha(\beta\gamma) = (\alpha\beta)\gamma$.

6.   *Commutative law of multiplication.*   For all $\alpha$, $\beta$ in $F$, $\alpha\beta = \beta\alpha$.

7.   There is an element $1 \neq 0$ in $F$ such that $1\alpha = \alpha$ for every $\alpha$ in $F$. This element $1$ is called the *multiplicative identity*.

8.   *Multiplicative inverse.*   For each $\alpha \neq 0$ in $F$, there is an element in $F$, called $\alpha^{-1}$, such that $\alpha\alpha^{-1} = \alpha^{-1}\alpha = 1$.

9.   *Distributive law.*   For all $\alpha$, $\beta$, $\gamma$ in $F$, $\alpha(\beta + \gamma) = \alpha\beta + \alpha\gamma$.

In circuit theory we find use of the following fields of scalars: the real numbers, the complex numbers, and the rational functions.   In the study of computers and automata we shall encounter the field of binary numbers.

## 2.2   Linear Spaces

A linear space is a collection of elements called vectors.   The basic idea associated with vectors is that they may always be combined linearly; i.e., they may be added to one another and multiplied by scalars.   Some authors use the redundant term *linear vector space*, or just *vector space*, for the term *linear space*.   We have encountered many linear spaces:

1.   The set of all vectors $\mathbf{x} = (x_1, x_2, x_3)^T$ of "space" geometry.

2.   The set of all $n$-tuples $\mathbf{x} = (x_1, x_2, \ldots, x_n)^T$ where the $x_i$'s are real numbers; this linear space is called $R^n$.

3. The set of all $n$-tuples $\mathbf{x} = (x_1, x_2, \ldots, x_n)^T$ where the $x_i$'s are complex numbers; this linear space is called $C^n$.

4. The set of all periodic functions of period $T$.

5. The set of all solutions of the homogeneous differential equation $d^2y/dt^2 + 3(dy/dt) + 2y = 0$. Recall that any such solution is of the form

$$c_1\epsilon^{-t} + c_2\epsilon^{-2t}$$

where $c_1$ and $c_2$ are any fixed real numbers.

6. Suppose that we excite a circuit with a voltage source and that we consider the effect on the circuit of all possible waveforms of the source. For such an investigation the set of all possible waveforms $e(\cdot)$ constitutes a linear space; they can be added together or multiplied by a scalar to form a new waveform.

Let us now introduce the formal definition of a linear space. $V$ is called a **linear space over the field** $F$ and the elements of $V$ are called **vectors,** if the following axioms are satisfied:

1. To every $\mathbf{x, y}$ in $V$, the operation of addition defines a unique element $\mathbf{x} + \mathbf{y}$ in $V$.

2. For all $\mathbf{x, y, z}$ in $V$, $\mathbf{x} + (\mathbf{y} + \mathbf{z}) = (\mathbf{x} + \mathbf{y}) + \mathbf{z}$.

3. For all $\mathbf{x, y}$ in $V$, $\mathbf{x} + \mathbf{y} = \mathbf{y} + \mathbf{x}$.

4. There is an element $\mathbf{0}$ in $V$ such that $\mathbf{x} + \mathbf{0} = \mathbf{x}$ for all $\mathbf{x}$ in $V$. This element $\mathbf{0}$ is called the *zero vector* or the *origin* of the linear space $V$.

5. For each element $\mathbf{x}$ in $V$ there is an element $-\mathbf{x}$ in $V$ such that $\mathbf{x} + (-\mathbf{x}) = \mathbf{0}$.

6. For each scalar $\alpha$ in $F$ and each vector $\mathbf{x}$ in $V$, the operation of scalar multiplication defines a unique element $\alpha\mathbf{x}$ in $V$.

7. For all $\alpha, \beta$ in $F$ and for all $\mathbf{x}$ in $V$, $(\alpha\beta)\mathbf{x} = \alpha(\beta\mathbf{x})$.

8. For all $\alpha$ in $F$ and for $\mathbf{x, y}$ in $V$, $\alpha(\mathbf{x} + \mathbf{y}) = \alpha\mathbf{x} + \alpha\mathbf{y}$.

9. For all $\alpha, \beta$ in $F$ and for all $\mathbf{x}$ in $V$, $(\alpha + \beta)\mathbf{x} = \alpha\mathbf{x} + \beta\mathbf{x}$.

10. With 1 being the identity element of multiplication in $F$, for all $\mathbf{x}$ in $V$, $1\mathbf{x} = \mathbf{x}$.

It is easy to verify that the 10 axioms hold for all the linear spaces indicated in the beginning of the section.

The fact that the first four axioms of a field are identical with axioms 2 to 5 of a linear space is not an accident; it is a reflection of the fact that addition over a field and addition over a linear space are both required to be commutative groups, which precisely means that they must satisfy the axioms 2 to 5.

### 2.3   Linear Functions

In order to define the concept of linear function we need to say what we mean by the *range* of a function.  Let the function $f$ be defined by all its ordered pairs $(x, y)$, where $x$ ranges over $X$, the domain of $f$.  We call **range** of $f$ the set of all $y$ in $Y$ for which there is an $x$ such that $(x, y)$ belongs to the collection defining $f$.  If we think of $f$ as a table of all $(x, f(x))$, the domain of $f$ is the set of all first entries of the table, and the range of $f$ is the set of all second entries of the table.

We shall have to impose the requirement that the functions under consideration have a domain and a range which are linear spaces.  This is only a technical requirement so that given any $x_1$, $x_2$ in the domain and scalars $\alpha_1$, $\alpha_2$, the expressions $\alpha_1 x_1 + \alpha_2 x_2$, $\alpha_1 f(x_1) + \alpha_2 f(x_2)$ are meaningful; indeed, once the above requirements are satisfied, these expressions amount to taking linear combinations of vectors.

A function $f$ is said to be **linear** if

1.  Its domain and its range are linear spaces over the same scalar field.
2.  *Homogeneity property.*  For every $x$ in the domain and for every scalar $\alpha$, $f(\alpha x) = \alpha f(x)$.
3.  *Additivity property.*  For every pair of elements $x_1$, $x_2$ of its domain, $f(x_1 + x_2) = f(x_1) + f(x_2)$.

**Remark**  Conditions 2 and 3 are equivalent to the single condition

2'.  *Superposition property.*  For every pair of scalars $\alpha_1$, $\alpha_2$ and for every pair of vectors $x_1$, $x_2$ in the domain,

$$f(\alpha_1 x_1 + \alpha_2 x_2) = \alpha_1 f(x_1) + \alpha_2 f(x_2)$$

Let us prove that conditions 2 and 3 are equivalent to condition 2'.

*a.*  Suppose 2 and 3 hold; let us show that 2' follows.  We calculate

$$f(\alpha_1 x_1 + \alpha_2 x_2) = f(\alpha_1 x_1) + f(\alpha_2 x_2)$$

where the equality follows by 3 once we observe that, since $x_1$ and $x_2$ are elements of the linear vector space, so are $\alpha_1 x_1$ and $\alpha_2 x_2$.  By 2, the right-hand side can easily be rewritten to obtain

$$f(\alpha_1 x_1 + \alpha_2 x_2) = \alpha_1 f(x_1) + \alpha_2 f(x_2)$$

*b.*  Suppose that 2' holds; let us show that 2 and 3 follow.  Let $\alpha_1 = \alpha_2 = 1$; then 2' becomes 2.  Let $\alpha_2 = 0$; then 2' becomes 3.  This completes the proof.

Thus, to show that a function is linear, we may either establish that it satisfies the superposition property (in the precise sense indicated above) or that it satisfies both the additivity property and the homogeneity property.

**Exercise**   Let $f(\cdot)$ be a linear function, mapping real numbers into real numbers; show that $f(0) = 0$.   (Hint: Use the homogeneity property.)

**Example 1**   Consider the function $f$ whose graph on the $xy$ plane is the straight line through the origin defined by

$$y = f(x) = 2x$$

We assert that $f$ is a linear function.   Indeed,

$$f(\alpha x) = 2\alpha x = \alpha(2x) = \alpha f(x) \qquad \text{for all } \alpha \text{ and all } x$$

and

$$f(x_1 + x_2) = 2(x_1 + x_2) = 2x_1 + 2x_2 = f(x_1) + f(x_2) \qquad \text{for all } x_1, x_2$$

Slightly generalizing these calculations we observe that a function whose domain and range are the set of all real numbers is a *linear* function if and only if it is of the form $y = kx$, where $k$ is a constant (that is, where $k$ is independent of $x$).   Equivalently, a function is *linear* if and only if its graph is a straight line through the origin of the $xy$ plane.   Note that a function of the form $y = kx + b$, where $b$ is a *nonzero* constant, is not linear, since when $x = 0$, $y = b \neq 0$.

**Example 2**   Suppose that the sliding contact of the rheostat shown on Fig. 1 is moved back and forth so that the resistance between terminals $A$ and $B$ at time $t$ is $(5 + \cos t)$ ohms.   The equation relating the instantaneous voltage $v(t)$ to the instantaneous current $i(t)$ is

$$v(t) = (5 + \cos t)i(t)$$

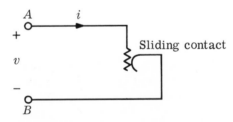

**Fig. 1**   A linear time-varying resistor which is obtained by moving the sliding contact of the rheostat back and forth.

This equation may be viewed as defining a function

(4)   $v(t) = f[i(t),t] = (5 + \cos t)i(t)$

Note that $f$ has two arguments: the instantaneous value of the current $i(t)$, and the time $t$.  For each fixed $t$, Eq. (4) may be regarded as defining a function associating to any real number $i(t)$ the real number $v(t)$; we denote this function by $f(\,\cdot\,, t)$.  We assert that $f(\,\cdot\,, t)$ *is a linear function;* i.e., for each fixed value of its second argument $t$, the function $f$ in (4) is linear with respect to its first argument.  Additivity follows from the following calculation:

$$f[i_1(t) + i_2(t), t] = (5 + \cos t)[i_1(t) + i_2(t)]$$
$$= (5 + \cos t)i_1(t) + (5 + \cos t)i_2(t)$$
$$= f[i_1(t),t] + f[i_2(t),t]$$

Since for each fixed $t$ this holds for all real numbers $i_1(t)$, $i_2(t)$, then $f(\,\cdot\,, t)$ is additive.  Homogeneity is verified as follows:

$$f[\alpha i_1(t),t] = (5 + \cos t)\alpha i_1(t) = \alpha(5 + \cos t)i_1(t) = \alpha f[i_1(t),t]$$

We shall see later that (4) describes the characteristic of a *linear time-varying resistor.*

---

**Example 3**   Consider the function **f** whose domain and range is the vector space $R^n$ (the set of all real $n$-tuples) defined by

$$\mathbf{f}(\mathbf{x}) = \mathbf{A}\mathbf{x}$$

where **A** is a given $n \times n$ matrix whose elements are real numbers.  It is immediate that **f** is a linear function, since

$$\mathbf{A}(\alpha_1 \mathbf{x}_1 + \alpha_2 \mathbf{x}_2) = \alpha_1 \mathbf{A}\mathbf{x}_1 + \alpha_2 \mathbf{A}\mathbf{x}_2$$

for all real numbers $\alpha_1$, $\alpha_2$ and for all $n$-tuples $\mathbf{x}_1$ and $\mathbf{x}_2$.

---

**Example 4**   This example shows precisely why we needed the general concept of *function* and the general concept of *linear function*.  Consider the circuit of Fig. 2, where a voltage generator drives a linear time-invariant series $RL$ circuit.  Suppose that at time $t = 0$, the current through the inductor (of inductance $L$) is zero and that from $t = 0$ on, the voltage generator applies a voltage $e_s$ across the terminals $A$ and $B$ of the circuit.  Physically, it is obvious that to each voltage waveform $e_s(\,\cdot\,)$ there results a unique current waveform $i(\,\cdot\,)$.  It is crucial to observe that by the waveform $e_s(\,\cdot\,)$ we

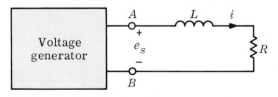

**Fig. 2**   A series $RL$ circuit driven by a voltage generator.

mean the whole "curve" specifying the voltages $e_s(t)$ for all $t \geq 0$; similarly, by the waveform $i(\cdot)$ we mean the whole "curve" specifying the current $i(t)$ for all $t \geq 0$.  Technically, the waveforms $e_s(\cdot)$ and $i(\cdot)$ are real-valued functions whose domains are the time interval $t \in [0,\infty)$.†
Thus, we may think of the $RL$ circuit shown as a device defining a function $f$ which assigns a (unique) current waveform $i(\cdot)$ to each voltage waveform $e_s(\cdot)$; therefore, we write $i = f(e_s)$.  Later on in the book we shall show that the value of $i$ at time $t$, which is also the value at time $t$ of the waveform $f(e_s)$, and is hence denoted by the clumsy notation $[f(e_s)](t)$, is given by

$$i(t) = [f(e_s)](t) = \int_0^t \frac{1}{L} \epsilon^{-(R/L)(t-t')} e_s(t') \, dt' \qquad \text{for all } t \geq 0$$

The function $f$ is completely defined since this equation gives for each $t$ the value of $f(e_s)$ for any $e_s$.

We assert that the *function f is linear*.  First, let us verify *homogeneity*. Let $\alpha$ be any real number.  Let $i$ be the response to the voltage waveform $e_s$; that is, $i = f(e_s)$.  Consider now the voltage waveform $\alpha e_s$; then

$$[f(\alpha e_s)](t) = \int_0^t \frac{1}{L} \epsilon^{-(R/L)(t-t')} \alpha e_s(t') \, dt'$$

By an obvious property of the integral—if all ordinates of a curve are multiplied by $\alpha$, the area under the curve is multiplied by $\alpha$—the right-hand side is

$$[f(\alpha e_s)](t) = \alpha \int_0^t \frac{1}{L} \epsilon^{-(R/L)(t-t')} e_s(t') \, dt'$$

$$= \alpha[f(e_s)](t)$$

Since the last equality holds for all $t$, we conclude that

$$f(\alpha e_s) = \alpha f(e_s)$$

for all real numbers $\alpha$ and waveforms $e_s$; that is, $f$ is homogeneous.  Physically, this equation means that the current waveform due to the voltage waveform $\alpha e_s$ is $\alpha$ times the current waveform due to the voltage waveform $e_s$.

Let us next verify *additivity*.  Call $i_1$ the current waveform produced by some voltage waveform $e_1$; then $i_1 = f(e_1)$.  More explicitly,

$$i_1(t) = [f(e_1)](t) = \int_0^t \frac{1}{L} \epsilon^{-(R/L)(t-t')} e_1(t') \, dt' \qquad \text{for } t \geq 0$$

Call $i_2$ the current waveform produced by some voltage waveform $e_2$; then $i_2 = f(e_2)$.  More explicitly,

---

† The symbol $[a, b]$ means the set of all real numbers $t$ such that $a \leq t \leq b$, whereas $(a, b)$ means the set of all $t$ such that $a < t < b$.  Thus $[0,\infty)$ means the set of all $t$ such that $0 \leq t < \infty$.

$$i_2(t) = [f(e_2)](t) = \int_0^t \frac{1}{L} \epsilon^{-(R/L)(t-t')} e_2(t') \, dt' \qquad \text{for } t \geq 0$$

Consider now the current $i_{1+2}$ produced by the voltage waveform $e_1 + e_2$; in this case $i_{1+2} = f(e_1 + e_2)$.   More explicitly,

$$i_{1+2}(t) = [f(e_1 + e_2)](t) = \int_0^t \frac{1}{L} \epsilon^{-(R/L)(t-t')} [e_1(t') + e_2(t')] \, dt' \qquad \text{for } t \geq 0$$

Now writing this last integral as a sum of two integrals and using the two previous equations, we obtain successively

$$i_{1+2}(t) = \int_0^t \frac{1}{L} \epsilon^{-(R/L)(t-t')} e_1(t') \, dt' + \int_0^t \frac{1}{L} \epsilon^{-(R/L)(t-t')} e_2(t') \, dt'$$

$$= [f(e_1)](t) + [f(e_2)](t)$$

$$= i_1(t) + i_2(t)$$

Since the equations hold for all $t \geq 0$, we have

$$i_{1+2} = f(e_1 + e_2) = f(e_1) + f(e_2) = i_1 + i_2 \qquad \text{for all } e_1, e_2$$

Thus, the function $f$ is additive.   Physically, we observe that if $i_1(\cdot)$ and $i_2(\cdot)$ are, respectively, the currents produced by the voltage waveform $e_1(\cdot)$ and $e_2(\cdot)$ acting alone on the circuit, then the current produced by the voltage waveform $e_1(\cdot) + e_2(\cdot)$ is $i_1(\cdot) + i_2(\cdot)$.

Since $f$ is homogeneous and additive, it is, by definition, a *linear* function.   In physical terms, provided $i(0) = 0$, the current waveform $i(\cdot)$ resulting from an applied voltage waveform $e_s(\cdot)$ is a *linear function* of $e_s(\cdot)$.

# B

# Appendix: Matrices and Determinants

The purpose of this appendix is to recall the basic definitions and the basic facts concerning matrices and determinants.

## 1 Matrices

### 1.1 Definitions

A rectangular array of *scalars* is called a **matrix.** These scalars must, by definition, be elements of a field. (Here the word *field* refers to the algebraic concept defined in Appendix A, Sec. 2. This concept of field has nothing to do with the fields of physics.) In engineering we most often encounter four fields of scalars:

1. The field of real numbers.

2. The field of complex numbers.

3. The field of rational functions with real or complex coefficients [for example, $(s + 1)/(s^2 + 2s + 4)$].

4. The field of binary numbers, more technically, the integers modulo 2.

Since scalars are by definition elements of a field, their rules of algebra are precisely the same as the rules of the algebra of real numbers.

The rectangular array

$$\begin{bmatrix} a_{11} & a_{12} & \cdots & a_{1n} \\ a_{21} & a_{22} & \cdots & a_{2n} \\ \cdots & \cdots & \cdots & \cdots \\ a_{m1} & a_{m2} & \cdots & a_{mn} \end{bmatrix}$$

is called a **matrix of $m$ rows and $n$ columns.** The scalar $a_{ij}$ is called the *element* of the $i$th row and $j$th column or the $ij$th **element** of the matrix. We shall use capital boldface letters, such as **A,** to denote matrices and the corresponding lowercase letters with subscripts, such as $a_{ij}$, to denote its elements. Sometimes it is convenient to write $\mathbf{A} = (a_{ij})$ to indicate that the $ij$th element of $\mathbf{A}$ is $a_{ij}$.

A matrix with $m$ rows and $n$ columns is referred to as an $m \times n$ matrix (read "$m$ by $n$ matrix"). If a matrix has $n$ rows and $n$ columns, it is said to be a **square matrix of order** $n$. A matrix with one row and $n$ columns is called a **row vector**. A matrix with $m$ rows and one column is called a **column vector**. Two matrices are said to be of the **same size** if and only if they have the same number of rows and the same number of columns.

| 1.2 | **Operations** |
|---|---|

*Multiplication of a matrix by a scalar*   If $\mathbf{A}$ is a matrix and $c$ is a scalar (for example, a real or a complex number), then $c\mathbf{A}$ is the matrix obtained by multiplying *each* element of $\mathbf{A}$ by the scalar $c$.

*Addition of matrices*   If $\mathbf{A}$ and $\mathbf{B}$ are matrices of the same size, we define their *sum* $\mathbf{A} + \mathbf{B}$ as the matrix $\mathbf{C}$ with elements $c_{ij}$, such that $c_{ij} = a_{ij} + b_{ij}$ for each $i$ and $j$.

*Equality of matrices*   Two matrices $\mathbf{A}$ and $\mathbf{B}$ are equal if and only if they have the same size and $a_{ij} = b_{ij}$ for each $i$ and $j$.

*Multiplication of matrices*   If $\mathbf{A}$ is an $m \times n$ matrix, and $\mathbf{B}$ is an $n \times p$ matrix, then the product $\mathbf{AB}$ is defined to be the $m \times p$ matrix $\mathbf{C}$ with elements $c_{ij}$ given by

$$c_{ij} = \sum_{k=1}^{n} a_{ik}b_{kj} \qquad i = 1, 2, \ldots, m \qquad j = 1, 2, \ldots, p$$

**Remarks**   1.   The sum $\mathbf{A} + \mathbf{B}$ is not defined unless $\mathbf{A}$ and $\mathbf{B}$ have the same number of rows and the same number of columns.

2.   The product $\mathbf{BA}$ is *not defined* unless $m = p$.

3.   Usually $\mathbf{BA} \neq \mathbf{AB}$.

| 1.3 | **More Definitions** |
|---|---|

If $\mathbf{A}$ is a square matrix, then the elements of the form $a_{ii}$ are called **diagonal elements**. A matrix in which all elements except the diagonal elements are zero is called a **diagonal matrix**. If $\mathbf{A}$ and $\mathbf{B}$ are $n \times n$ diagonal matrices, then $\mathbf{AB} = \mathbf{BA}$.

If, in a *diagonal* matrix $\mathbf{A}$, $a_{ii} = 1$ for all $i$, the matrix is called the **identity (or unit) matrix**; it is denoted by $\mathbf{1}$. If $\mathbf{A}$ is any $n \times n$ matrix and $\mathbf{1}$ is the $n \times n$ unit matrix, then $\mathbf{A1} = \mathbf{1A}$.

If $a_{ij} = 0$ for all $i$ and all $j$, the matrix is called the **zero matrix**; it is de-

noted by **0**.  If **A** is any $n \times n$ matrix and **0** the zero matrix of order $n$, then
**A** + **0** = **A**.

If **A** is a square $n \times n$ matrix, and if there exists a square $n \times n$ matrix,
**B** such that

**AB** = **BA** = **1**

**B** is called the **inverse** of **A** and is denoted by **A**$^{-1}$.  If **A** has an inverse, **A** is
said to be **nonsingular**.  It can be shown that if **A** is nonsingular, then its
inverse is unique.  It can also be shown that **A** is nonsingular if and only
if det **A** $\neq$ 0.  The calculation of **A**$^{-1}$ will be detailed in Sec. 2.3.

**THEOREM**   1.   If **A** and **B** are nonsingular $n \times n$ matrices, then

(**AB**)$^{-1}$ = **B**$^{-1}$**A**$^{-1}$

2.   If **A, B,** and **C** are nonsingular $n \times n$ matrices, then

(**ABC**)$^{-1}$ = **C**$^{-1}$**B**$^{-1}$**A**$^{-1}$

If **A** is an $m \times n$ matrix, then the $n \times m$ matrix obtained by interchanging the rows and columns of **A** is called the **transpose** of **A** and is denoted
by the symbol **A**$^T$.

**THEOREM**   1.   If **A** is an $m \times n$ matrix and **B** is an $n \times p$ matrix, then

(**AB**)$^T$ = **B**$^T$**A**$^T$

2.   If matrices **A, B, C** have dimensions such that the product **ABC** is
defined, then

(**ABC**)$^T$ = **C**$^T$**B**$^T$**A**$^T$

## 1.4   The Algebra of $n \times n$ Matrices

In the following, **A, B, C,** ... denote $n \times n$ matrices with elements in the
same field.

1.   **A** = **B** if and only if $a_{ij} = b_{ij}$ for $i = 1, 2, \ldots, n$ and $j = 1, 2, \ldots, n$.
2.   **A** + **B** = **B** + **A**; addition is commutative.
3.   *Distributive property.*  **A**(**B** + **C**) = **AB** + **AC**; (**A** + **B**)**C** = **AC** + **BC**.
4.   Usually, **AB** $\neq$ **BA**; multiplication of matrices is *not* a commutative
operation.
5.   **A** + **0** = **0** + **A** = **A**, for all **A**.
6.   **A1** = **1A** = **A**, for all **A**.

*The algebra* of $n \times n$ *matrices is substantially different from that of real
or complex numbers.*  The reason is that the class of $n \times n$ matrices does
not constitute a field.  Consider the following facts:

1. In general, multiplication is *not* commutative; for example,

$$\begin{bmatrix} 1 & 1 \\ 0 & 1 \end{bmatrix}\begin{bmatrix} 1 & 0 \\ 0 & 2 \end{bmatrix} = \begin{bmatrix} 1 & 2 \\ 0 & 2 \end{bmatrix}$$

$$\begin{bmatrix} 1 & 0 \\ 0 & 2 \end{bmatrix}\begin{bmatrix} 1 & 1 \\ 0 & 1 \end{bmatrix} = \begin{bmatrix} 1 & 1 \\ 0 & 2 \end{bmatrix}$$

2. If $\mathbf{A} \neq \mathbf{0}$ (i.e., the matrix $\mathbf{A}$ is not the zero matrix; equivalently, not all $a_{ij}$ are zero) then $\mathbf{A}$ does not necessarily have an inverse (in addition, det $\mathbf{A} \neq \mathbf{0}$ is required). For example,

$$\mathbf{A} = \begin{bmatrix} 1 & 0 \\ 0 & 0 \end{bmatrix}$$

does not have an inverse. Because there is no $2 \times 2$ matrix of real numbers which when multiplied by $\mathbf{A}$ gives the identity matrix $\mathbf{I}$.

3. In general, the cancellation law does not apply. This means that if $\mathbf{AB} = \mathbf{AC}$, it **does not** follow that $\mathbf{B} = \mathbf{C}$. (In order for this to follow, $\mathbf{A}$ must be nonsingular.) For example,

$$\begin{bmatrix} 1 & 0 \\ 0 & 0 \end{bmatrix}\begin{bmatrix} 1 & 2 \\ 0 & 1 \end{bmatrix} = \begin{bmatrix} 1 & 0 \\ 0 & 0 \end{bmatrix}\begin{bmatrix} 1 & 2 \\ 5 & 7 \end{bmatrix} = \begin{bmatrix} 1 & 2 \\ 0 & 0 \end{bmatrix}$$

## 2    Determinants

### 2.1    Definitions

We call a **determinant** a function whose domain is the collection of *square* matrices (whose elements are scalars, i.e., elements of a field) and whose range is the field of scalars. The determinant associated with the square matrix $\mathbf{A}$ is denoted by det $\mathbf{A}$. Thus, if $\mathbf{A}$ has real numbers as elements, det $\mathbf{A}$ is a real number; if $\mathbf{A}$ has rational functions as elements, det $\mathbf{A}$ is a rational function. We shall give below an inductive procedure for determining the value of det $\mathbf{A}$ for any square matrix $\mathbf{A}$.

The determinant associated with the submatrix of $\mathbf{A}$ obtained by deleting its $i$th row and $j$th column is called the **minor** of the element $a_{ij}$ and is denoted by $M_{ij}$.

The **cofactor** $A_{ij}$ of the element $a_{ij}$ is defined by the relation

$$A_{ij} = (-1)^{i+j}M_{ij}$$

*Evaluation of a determinant*    Let $\mathbf{A}$ be an $n \times n$ matrix with $n \geq 2$, then

1. $\det \mathbf{A} = \sum_{i=1}^{n} a_{ij}A_{ij}$ for any $j = 1, 2, \ldots, n$.

This formula is called *the expansion of det $\mathbf{A}$ along the $j$th column.* Clearly, rule 1 defines inductively the determinant of an $n \times n$ matrix;

indeed, it expresses the determinant of an $n \times n$ matrix as a linear combination of determinants of $(n-1) \times (n-1)$ matrices. It can be shown that rule 1 is equivalent to rules 2 and 3 below.

2. $\det \mathbf{A} = \sum\limits_{j=1}^{n} a_{ij}A_{ij}$ for any $i = 1, 2, \ldots, n$. This formula is called *the expansion of det **A** along the ith row.*

3. $\det \mathbf{A} = \Sigma \varepsilon_i\, a_{1i_1}\, a_{2i_2} \ldots a_{ni_n}$ for all permutations $i_1, i_2, \ldots, i_n$, where $\varepsilon_i$ is $+1$ or $-1$ according to whether the permutation $i_1, i_2, \ldots, i_n$ is even or odd.

## 2.2   Properties of Determinants

1. If $\mathbf{A}, \mathbf{B}$ are $n \times n$ matrices, then

   *a.*   $\det \mathbf{A}^T = \det \mathbf{A}$.

   *b.*   $\det (\mathbf{AB}) = \det \mathbf{A} \cdot \det \mathbf{B}$.

2. Let $k$ be an integer with $1 \le k \le n$. If *all* the elements of the $k$th row or of the $k$th column of the matrix $\mathbf{A}$ are zero, then $\det \mathbf{A} = 0$.

3. If the matrix $\mathbf{A}'$ is obtained from $\mathbf{A}$ by multiplying all elements of the $k$th row (or of the $k$th column) by a scalar $c$ and leaving all the other elements unchanged, then

$$\det \mathbf{A}' = c \det \mathbf{A}$$

**Remark**   A consequence of property 3 and the rule of multiplication of a matrix by a scalar is that if $\mathbf{A}$ is an $n \times n$ matrix and $c$ is a scalar, then

$$\det (c\mathbf{A}) = c^n \det \mathbf{A}$$

4. Let $\mathbf{A}$ be an $n \times n$ matrix. Suppose that the $k$th row of $\mathbf{A}$ is written as a sum

$$a_{kj} = a'_{kj} + a''_{kj} \qquad j = 1, 2, \ldots, n$$

   Call $\mathbf{A}'$ and $\mathbf{A}''$ the matrices identical with $\mathbf{A}$ except that their $k$th rows consist of $a'_{k1}, a'_{k2}, \ldots, a'_{kn}$ and $a''_{k1}, a''_{k2}, \ldots, a''_{kn}$, respectively. Then

$$\det \mathbf{A} = \det \mathbf{A}' + \det \mathbf{A}''$$

   For example,

$$\det \begin{bmatrix} 1 & 2+5 & 5 \\ 1 & 3+6 & 5 \\ 1 & 4+7 & 5 \end{bmatrix} = \det \begin{bmatrix} 1 & 2 & 5 \\ 1 & 3 & 5 \\ 1 & 4 & 5 \end{bmatrix} + \det \begin{bmatrix} 1 & 5 & 5 \\ 1 & 6 & 5 \\ 1 & 7 & 5 \end{bmatrix}$$

5. The same result holds for the $k$th column.

6. If $\mathbf{A}'$ is obtained from $\mathbf{A}$ by interchanging two rows (or two columns), then

$$\det \mathbf{A}' = -\det \mathbf{A}$$

7. If $\mathbf{A}'$ is obtained from $\mathbf{A}$ by multiplying the $k$th row by the scalar $c$ and adding the result to the $i$th row where $i \neq k$, then

$$\det \mathbf{A}' = \det \mathbf{A}$$

8. The same result holds for two columns.

9.   $a.$ $\displaystyle\sum_{i=1}^{n} a_{ij}A_{ik} = \begin{cases} 0 & \text{if } k \neq j \\ \det \mathbf{A} & \text{if } k = j \end{cases}$

     $b.$ $\displaystyle\sum_{k=1}^{n} a_{ik}A_{jk} = \begin{cases} 0 & \text{if } i \neq j \\ \det \mathbf{A} & \text{if } i = j \end{cases}$

In the definition of the determinant of a square matrix and in the statement of the properties of the determinant, the assumption was made that all the elements of the square matrix were scalars, i.e., elements of a field. In circuit theory this is the case in the analysis of linear resistive circuits (where the matrix elements are real numbers because they are resistances or linear combinations thereof) and in the sinusoidal steady-state analysis of linear time-invariant circuits (where the matrix elements are complex numbers because they are impedances or linear combinations thereof). In writing the differential equations of linear time-invariant circuits, we encounter matrices whose elements are polynomials in the differential operator $D \triangleq d/dt$. For example, we might have

$$\begin{bmatrix} D^2 + 2D + 2 & D + 1 \\ D + 1 & D^2 + 5D + 7 \end{bmatrix}$$

The sum and products of polynomial operators are perfectly meaningful (although the division of two polynomial operators leads to difficulties in interpretation). The following can be verified.

*The definition of the determinant of a square matrix and all the properties of determinants listed above are still valid when the matrix elements are polynomials in the operator $D \triangleq d/dt$.*

It is important to note that because polynomials in $D$ do not constitute a field, Cramer's rule (stated below) is not applicable to such matrices.

---

**2.3**    **Cramer's Rule**

Let us now consider the system of $n$ linear algebraic equations in $n$ unknowns

$$a_{11}x_1 + a_{12}x_2 + \cdots + a_{1n}x_n = b_1$$
$$a_{21}x_1 + a_{22}x_2 + \cdots + a_{2n}x_n = b_2$$
$$\cdots \cdots \cdots \cdots \cdots \cdots \cdots \cdots \cdots$$
$$a_{n1}x_1 + a_{n2}x_2 + \cdots + a_{nn}x_n = b_n$$

We may write this system in matrix form

(1)   $\mathbf{Ax} = \mathbf{b}$

where $\mathbf{x}$ and $\mathbf{b}$ are the column vectors with components $(x_1, x_2, \ldots, x_n)$ and $(b_1, b_2, \ldots, b_n)$, respectively.

If $\det \mathbf{A} \neq 0$, then the system of equations above has a unique solution

(2)   $$x_k = \frac{1}{\det \mathbf{A}} \begin{vmatrix} a_{11} & \cdots & a_{1,k-1} & b_1 & a_{1,k+1} & \cdots & a_{1n} \\ a_{21} & \cdots & a_{2,k-1} & b_2 & a_{2,k+1} & \cdots & a_{2n} \\ \cdots & & & & & & \cdots \\ a_{n1} & \cdots & a_{n,k-1} & b_n & a_{n,k+1} & \cdots & a_{nn} \end{vmatrix} \qquad k = 1, 2, \ldots, n$$

In other words, the numerator is the determinant of a new matrix obtained from $\mathbf{A}$ by replacing the $k$th column with the column vector $\mathbf{b}$. Using fact (1) for the evaluation of a determinant, we get

(3)   $$x_k = \frac{\displaystyle\sum_{i=1}^{n} b_i A_{ik}}{\det \mathbf{A}} \qquad \text{where } A_{ik} \overset{\Delta}{=} \text{cofactor of } a_{ik}$$

*Application*   Suppose $\det \mathbf{A} \neq 0$; consider then

$$\mathbf{AA}^{-1} = \mathbf{1}$$

as a set of $n$ systems of $n$ linear algebraic equations, with one system for each unknown column of $\mathbf{A}^{-1}$. Then Cramer's rule implies that if $c_{ij}$ is the element of the $i$th row and $j$th column of $\mathbf{A}^{-1}$, then

(4)   $$c_{ij} = \frac{A_{ji}}{\det \mathbf{A}} \qquad i, j = 1, 2, \ldots, n$$

where $A_{ji}$ is the cofactor of $a_{ji}$ in $\mathbf{A}$. Note that in (4) the subscripts of the cofactor $A_{ji}$ are those of $c_{ij}$ but in *reverse order*. An immediate consequence of (4) is the following corollary.

**COROLLARY**   Let $\mathbf{A}$ be an $n \times n$ nonsingular matrix. If $\mathbf{A}$ is a symmetric matrix (that is, $a_{ij} = a_{ji}$ for all $i$ and $j$), then $\mathbf{A}^{-1}$ is also a symmetric matrix.

**Example**   Let

$$\mathbf{A} = \begin{bmatrix} a_{11} & a_{12} \\ a_{21} & a_{22} \end{bmatrix}$$

Then,

$$\mathbf{A}^{-1} = \frac{1}{\det \mathbf{A}} \begin{bmatrix} a_{22} & -a_{12} \\ -a_{21} & a_{11} \end{bmatrix} = \begin{bmatrix} \dfrac{a_{22}}{a_{11}a_{22} - a_{12}a_{21}} & \dfrac{-a_{12}}{a_{11}a_{22} - a_{12}a_{21}} \\ \dfrac{-a_{21}}{a_{11}a_{22} - a_{12}a_{21}} & \dfrac{a_{11}}{a_{11}a_{22} - a_{12}a_{21}} \end{bmatrix}$$

The reader should verify that $\mathbf{AA}^{-1} = \mathbf{A}^{-1} \mathbf{A} = \mathbf{1}$.

---

**2.4**    **Determinant Inequalities**

In the study of linear resistive networks, an extension of the celebrated Hadamard inequality (clarified by O. Taussky) is very useful.

**THEOREM**    Let $\mathbf{A} = (a_{ij})$ be an $n \times n$ matrix with real elements.    Assume the following:

1.    Each diagonal element is larger than or equal to the sum of the absolute values of the other elements of the same row; that is,

(5)    $a_{ii} \geq \displaystyle\sum_{\substack{j=1 \\ j \neq i}}^{n} |a_{ij}| \qquad i = 1, 2, \ldots, n$

2.    The number of the inequalities in (5) that hold with an equal sign is at most $n - 1$.

3.    The matrix $\mathbf{A}$ cannot be transformed to the form

$$\begin{bmatrix} \mathbf{A}_{11} & \mathbf{A}_{12} \\ \mathbf{0} & \mathbf{A}_{22} \end{bmatrix}$$

by the *same* permutations of the rows and the columns, where $\mathbf{A}_{11}$ and $\mathbf{A}_{22}$ are square matrices.†    Then

$\det \mathbf{A} > 0$

**THEOREM**    Let $\mathbf{Z} = (z_{ij})$ be an $n \times n$ matrix with *complex* elements.    Assume the following:

1.    Each diagonal element is, in absolute value, larger than or equal to the sum of the absolute value of the other elements of the same row; that is,

---

† Condition 3, although often forgotten by many authors, is indispensable.    Without it the theorem is false.    From a circuit theory point of view, if the equations are written in the usual way, it is always satisfied for linear resistive networks (made of positive and negative resistors) whose graphs are connected and unhinged.

(6)   $|z_{ii}| \geq \sum\limits_{\substack{j=1 \\ j \neq i}}^{n} |z_{ij}|$     $i = 1, 2, \ldots, n$

2.  The number of inequalities in (6) that hold with an equal sign does not exceed $n - 1$.

3.  The matrix **Z** cannot be transformed to the form

$$\begin{bmatrix} \mathbf{Z}_{11} & \mathbf{Z}_{12} \\ \mathbf{0} & \mathbf{Z}_{22} \end{bmatrix}$$

by the *same* permutations of the rows and of the columns, where $\mathbf{Z}_{11}$ and $\mathbf{Z}_{22}$ are *square* matrices.† Then

$$\det \mathbf{Z} \neq 0$$

<br>

| **3** | **Linear Dependence and Rank** |
|---|---|

| **3.1** | **Linear Independent Vectors** |
|---|---|

A set of vectors $\mathbf{x}_1, \mathbf{x}_2, \ldots, \mathbf{x}_k$ is said to be **linearly dependent** if there exists a set of scalars $c_1, c_2, \ldots, c_k$ *not all zero*, such that

$$c_1\mathbf{x}_1 + c_2\mathbf{x}_2 + \cdots + c_k\mathbf{x}_k = \mathbf{0}$$

The vectors $\mathbf{x}_1, \mathbf{x}_2, \ldots, \mathbf{x}_k$ are also said to be **linearly dependent.** If a set of vectors is *not* linearly dependent, it is said to be **linearly independent.** In other words, the set of vectors $\mathbf{x}_1, \mathbf{x}_2, \ldots, \mathbf{x}_k$ is linearly independent if

$$c_1\mathbf{x}_1 + c_2\mathbf{x}_2 + \cdots + c_k\mathbf{x}_k = \mathbf{0}$$

implies

$$c_1 = c_2 = \cdots = c_k = 0.$$

If this is the case we say that the vectors $\mathbf{x}_1, \mathbf{x}_2, \ldots, \mathbf{x}_k$ are **linearly independent.**

| **3.2** | **Rank of a Matrix** |
|---|---|

Consider a rectangular matrix **A** with $m$ rows and $n$ columns. Any matrix obtained from **A** by deleting some rows and/or some columns is called a **submatrix** of **A**. The number $r$ is called the **rank** of **A** if (1) all $l \times l$ submatrices with $l > r$ have a determinant which is equal to zero, and if (2) at least one $r \times r$ submatrix has a determinant which is different from zero.

---

† Condition 3, though indispensable for the truth of the theorem, is always fulfilled when the network has a graph that is connected and unhinged and when the equations are written in the usual way.

Since the determinant of any matrix is equal to the determinant of its transpose, the rank of $\mathbf{A}$ is equal to the rank of $\mathbf{A}^T$.

The connection between the notion of linear independence and of rank is covered next. Let the vectors $\mathbf{x}_1, \mathbf{x}_2, \ldots, \mathbf{x}_k$ have the following components:

$$\mathbf{x}_1 = \begin{bmatrix} x_{11} \\ x_{21} \\ x_{31} \\ \vdots \\ x_{m1} \end{bmatrix} \quad \mathbf{x}_2 = \begin{bmatrix} x_{12} \\ x_{22} \\ x_{32} \\ \vdots \\ x_{m2} \end{bmatrix} \quad \cdots \quad \mathbf{x}_k = \begin{bmatrix} x_{1k} \\ x_{2k} \\ x_{3k} \\ \vdots \\ x_{mk} \end{bmatrix}$$

Let $k \leq m$: then the set of vectors $\mathbf{x}_1, \mathbf{x}_2, \ldots, \mathbf{x}_k$ is linearly independent if and only if the $m \times k$ matrix $\mathbf{X}$ has rank $k$, where

$$\mathbf{X} \triangleq \begin{bmatrix} x_{11} & x_{12} & \cdots & x_{1k} \\ x_{21} & x_{22} & \cdots & x_{2k} \\ \cdots\cdots\cdots\cdots\cdots \\ x_{m1} & x_{m2} & \cdots & x_{mk} \end{bmatrix}$$

We may think of the $n$ columns of the $m \times n$ matrix $\mathbf{A}$ as vectors (with $n$ components). If the rank of $\mathbf{A}$ is $r$, then (1) any subset made of $r + 1$ of its column vectors is a linearly dependent set of vectors, (2) any subset made of $r + 1$ of its row vectors is a linearly dependent set of vectors, and (3) there is at least one set of $r$ of its column vectors (or row vectors) which is a linearly independent set. A connection between the notions of linear independence, rank, and inverse of a matrix is given by the following theorem.

**THEOREM**   Let $\mathbf{A}$ be an $n \times n$ matrix. The following statements are equivalent:

1. Matrix $\mathbf{A}$ has an inverse; i.e., there is an $n \times n$ matrix, denoted by $\mathbf{A}^{-1}$, such that $\mathbf{AA}^{-1} = \mathbf{A}^{-1}\mathbf{A} = \mathbf{1}$.

2. $\det \mathbf{A} \neq 0$.

3. The rank of $\mathbf{A} = n$.

4. The $n$ columns of $\mathbf{A}$ form a set of $n$ linearly independent vectors.

5. The $n$ rows of $\mathbf{A}$ form a set of $n$ linearly independent vectors.

A useful theorem on the rank of a product of matrices is the following.

**THEOREM**   Let $\mathbf{A}$ and $\mathbf{B}$ be two *square* matrices of order $n$. Let $\rho(\mathbf{A})$ and $\rho(\mathbf{B})$ denote the rank of $\mathbf{A}$ and $\mathbf{B}$, respectively. Then

$$\rho(\mathbf{A}) + \rho(\mathbf{B}) - n \leq \rho(\mathbf{AB}) \leq \min \, [\rho(\mathbf{A}), \rho(\mathbf{B})]$$

In particular, if $\mathbf{A}$ is nonsingular,

$$\rho(\mathbf{AB}) = \rho(\mathbf{BA}) = \rho(\mathbf{B})$$

Observe that this theorem can be applied to the product of two rectangular matrices. Indeed, adding suitable rows or columns of zeros will make the matrices square and will affect neither their rank nor the rank of their product.

### 3.3   Linear Independent Equations

Consider now the following system of $m$ homogeneous linear algebraic equations in $n$ unknowns:

(7)   $a_{11}x_1 + a_{12}x_2 + \cdots + a_{1n}x_n = 0$

. . . . . . . . . . . . . . . . . . . . . . . . . . . .

$a_{m1}x_1 + a_{m2}x_2 + \cdots + a_{mn}x_n = 0$

These $m$ equations are said to be **linearly dependent** if there are $m$ constants $c_1, c_2, \ldots, c_m$, not all zero, such that the first equation multiplied by $c_1$, the second equation multiplied by $c_2, \ldots,$ and the $m$th equation multiplied by $c_m$ sum to zero *identically*. Referring to the definition of linearly independent vectors and considering the $n$ coefficients of each equation as defining a vector, we observe that the $m$ equations above are linearly dependent if and only if the $m$ row vectors of the coefficient matrix are linearly dependent.

Another way of expressing the idea of linear dependence is to observe that if the $m$ equations are linearly dependent, then some of the equations may be expressed as a linear combination of some of the others; in other words, some of the equations do not bring any information that is not yet contained in some of the others. Thus, we may conclude this discussion by stating the following theorem.

**THEOREM**   The system of Eq. (7) is linearly dependent if and only if the rank of the $m \times n$ coefficient matrix $(a_{ij})$ is smaller than $m$. Also, the system is a system of linearly independent equations if and only if the $m \times n$ matrix of coefficients has rank $m$; equivalently, this is true if and only if the $m$ row vectors of the coefficient matrix are linearly independent.

### 4   Positive Definite Matrices

Let $\mathbf{A}$ be a *symmetric* $n \times n$ matrix with real elements. Given any vector $\mathbf{x} = (x_1, x_2, \ldots, x_n)^T$ with real components, we can form the quadratic form

(8)   $\mathbf{x}^T \mathbf{A} \mathbf{x} = \displaystyle\sum_{i=1}^{n} \sum_{j=1}^{n} a_{ij} x_i x_j$

**Example 1**   Let $A = \begin{bmatrix} 1 & -2 \\ -2 & 1 \end{bmatrix}$

Then

$$\mathbf{x}^T A \mathbf{x} = x_1^2 + x_2^2 - 4x_1x_2$$

This quadratic form may be rewritten as

$$(x_1 - x_2)^2 - 2x_1x_2$$

Thus, for

$$\mathbf{x}_1 = \begin{bmatrix} 1 \\ 1 \end{bmatrix} \qquad \mathbf{x}_1^T A \mathbf{x}_1 = -2$$

for

$$\mathbf{x}_2 = \begin{bmatrix} 1 \\ -1 \end{bmatrix} \qquad \mathbf{x}_1^T A \mathbf{x}_2 = 6$$

and for

$$\mathbf{x}_3 = \begin{bmatrix} 1 \\ 2 + \sqrt{3} \end{bmatrix} \qquad \mathbf{x}_3^T A \mathbf{x}_3 = 0$$

Depending on what the vector $\mathbf{x}$ is, the quadratic form takes positive, zero, or negative values.

**Example 2**   Let $A = \begin{bmatrix} 2 & 1 \\ 1 & 1 \end{bmatrix}$

Then

$$\mathbf{x}^T A \mathbf{x} = 2x_1^2 + x_2^2 + 2x_1x_2$$
$$= (x_1 + x_2)^2 + x_1^2$$

Clearly, in this case, whenever $\mathbf{x} = (x_1, x_2)^T$ is not the zero vector, the quadratic form takes positive values. This suggests the following definition.

**DEFINITION**   The symmetric $n \times n$ matrix $A$ is said to be **positive definite** if $\mathbf{x} \neq \mathbf{0}$ implies $\mathbf{x}^T A \mathbf{x} > 0$. If $A$ is a positive definite matrix, the quadratic form $\mathbf{x}^T A \mathbf{x}$ is said to be a **positive definite quadratic form.**

It is immediately seen that a *diagonal* matrix with *positive* diagonal elements is a positive definite matrix; indeed, the corresponding quadratic form is

$$\mathbf{x}^T A \mathbf{x} = \sum_{i=1}^{n} a_{ii} x_i^2$$

There is a simple characterization of positive definite matrices. We call **leading principal minor of order** $k$ the determinant of the submatrix consisting of the *first* $k$ rows and the *first* $k$ columns. The characterization is given by the following theorem.

**THEOREM**   A symmetric $n \times n$ matrix is positive definite if and only if *all* its leading principal minors are positive.

Thus, in order to check whether or not a symmetric $n \times n$ matrix is positive definite, one has to calculate $n$ determinants of respective order $1, 2, \ldots, n - 1, n$.

**Example**   Let $A = \begin{bmatrix} 9 & 1 & 4 \\ 1 & 3 & 4 \\ 4 & 4 & 7 \end{bmatrix}$

The leading principal minors are

$$9 \qquad \det \begin{bmatrix} 9 & 1 \\ 1 & 3 \end{bmatrix} = 26 \qquad \text{and} \qquad \det \begin{bmatrix} 9 & 1 & 4 \\ 1 & 3 & 4 \\ 4 & 4 & 7 \end{bmatrix} = 4$$

Hence the given matrix is positive definite.

In some cases, Theorem 1 of Sec. 2.4 is useful in checking that all leading principal minors are positive.

# Appendix: Differential Equations

The purpose of this appendix is to summarize a number of basic facts concerning differential equations. We give no proofs because these are available in modern calculus texts. We hope that this collection of facts on differential equations will be useful for ready reference.

## 1 The Linear Equation of Order $n$

### 1.1 Definitions

We shall use throughout the notation $y^{(n)}$ for $d^n y/dt^n$. The differential equation

$$(1) \quad y^{(n)}(t) + a_1(t)y^{(n-1)}(t) + \cdots + a_{n-1}(t)y^{(1)}(t) + a_n(t)y(t) = b(t)$$

where the coefficients $a_i(\cdot)$ and the forcing function $b(\cdot)$ are *known* continuous functions, is called a **linear differential equation of order** $n$. If the coefficients $a_i(t)$ are all constants, the equation is said to be **linear with constant coefficients**. If one or more of the coefficients are functions of time, the equation is said to be **linear with time-varying coefficients**. If the forcing function $b(\cdot)$ is *identically* zero, Eq. (1) is said to be a **homogeneous linear differential equation of order** $n$, and it gives

$$(1H) \quad y^{(n)}(t) + a_1(t)y^{(n-1)}(t) + \cdots + a_{n-1}(t)y^{(1)}(t) + a_n(t)y = 0$$

A function $\psi$, which when substituted for $y$ in the differential equation (1), makes the left-hand side equal to $b(t)$ *for all* $t$, is called a **particular solution** of (1).

A function $\phi$, which when substituted for $y$ in the homogeneous equation (1H) makes the left-hand side equal to zero *for all* $t$, is called a **solution of the homogeneous equation** (1H).

**Example**  Consider the equation

$$y^{(2)}(t) + 3y^{(1)}(t) + 2y(t) = 1$$

It is easy to verify that $\psi_1(t) = 1 + \epsilon^{-t}$ and $\psi_2(t) = 1 + 3\epsilon^{-2t}$ are particular solutions of this equation.

For the homogeneous equation

$$y^{(2)}(t) + 3y^{(1)}(t) + 2y(t) = 0$$

it is easy to verify that, whatever the constants $c_1$ and $c_2$ may be, the function $\phi(t) = c_1\epsilon^{-t} + c_2\epsilon^{-2t}$ is a solution of the homogeneous equation.

These examples illustrate the fact that when no initial conditions are specified, a linear differential equation has infinitely many solutions.

Before considering what additional conditions are required to make the solutions unique, let us consider the relations between the solutions of (1) and (1$H$).

## 1.2   Properties Based on Linearity

The properties that we list below are all consequences of the fact that the left-hand side of Eqs. (1) and (1$H$) can be considered as the result of applying the differential operator

(2)   $L(D,t) \triangleq D^n + a_1(t)D^{n-1} + \cdots + a_{n-1}(t)D + a_n(t)$

to the function $y$.   (Here $D$ is a shorthand notation for $d/dt$.)   Thus, Eqs. (1) and (1$H$) may be written as, respectively,

(3)   $L(D,t)y(t) = b(t)$      and      $L(D,t)y(t) = 0$

The differential equations (1) and (1$H$) are justifiably called *linear* because $L(D,t)$ is a *linear operator*.   Indeed, it is homogeneous.   If $c$ is a constant, then

$L(D,t)[cy(t)] = cL(D,t)y(t)$

It is also additive; thus,

$L(D,t)[y_1(t) + y_2(t)] = L(D,t)y_1(t) + L(D,t)y_2(t)$

Now we state the properties of the linear differential equations which are based on linearity.

*Property 1*   If $\phi_1$ and $\phi_2$ are two solutions of the homogeneous equation (1$H$) and if $c_1$ and $c_2$ are any constants, then $c_1\phi_1 + c_2\phi_2$ is also a solution of the homogeneous equation (1$H$).

*Proof*   For all $t$, we have

$L(D,t)[c_1\phi_1(t) + c_2\phi_2(t)] = c_1L(D,t)\phi_1(t) + c_2L(D,t)\phi_2(t) = 0$

*Property 2*   If $\phi$ is any solution of the homogeneous equation (1$H$) and $\psi$ is any solution of the nonhomogeneous equation (1) (that is, $\psi$ is a particular solution), then $\phi + \psi$ is also a solution of the nonhomogeneous equation (1).

*Proof*   For all $t$, we have

$$L(D,t)[\phi(t) + \psi(t)] = L(D,t)\phi(t) + L(D,t)\psi(t) = 0 + b(t) = b(t)$$

*Property 3*   If $\psi_1$ and $\psi_2$ are solutions of the nonhomogeneous equation (1), then $\psi_1 - \psi_2$ is a solution of the homogeneous equation (1*H*).

*Proof*   For all $t$, we have

$$L(D,t)[\psi_1(t) - \psi_2(t)] = L(D,t)\psi_1(t) - L(D,t)\psi_2(t) = b(t) - b(t) = 0$$

*Property 4*   Suppose that in Eq. (1) the forcing function $b(\cdot)$ can be written as a linear combination of two other functions $g_1(\cdot)$ and $g_2(\cdot)$; that is $b(\cdot) = c_1 g_1(\cdot) + c_2 g_2(\cdot)$, where $c_1$ and $c_2$ are two *constants*. Let $\psi_1(\cdot)$ and $\psi_2(\cdot)$ be particular solutions of

$$L(D,t)\psi_1(t) = g_1(t)$$
$$L(D,t)\psi_2(t) = g_2(t)$$

respectively; then the function $c_1\psi_1(\cdot) + c_2\psi_2(\cdot)$ is a particular solution of (1); that is,

$$L(D,t)[c_1\psi_1(t) + c_2\psi_2(t)] = c_1 g_1(t) + c_2 g_2(t)$$

*Proof*   For all $t$, we have

$$\begin{aligned} L(D,t)[c_1\psi_1(t) + c_2\psi_2(t)] &= L(D,t)[c_1\psi_1(t)] + L(D,t)[c_2\psi_2(t)] \\ &= c_1 L(D,t)\psi_1(t) + c_2 L(D,t)\psi_2(t) \\ &= c_1 g_1(t) + c_2 g_2(t) = b(t) \end{aligned}$$

**Exercise**   Verify the four properties based on linearity on the following equations:

$$y^{(2)} + 4y^{(1)} + 3y = 8\epsilon^t \qquad \begin{cases} \psi_1(t) = \epsilon^t, \ \psi_2(t) = \epsilon^t + \epsilon^{-3t} \\ \phi_1(t) = \epsilon^{-t}, \ \phi_2(t) = \epsilon^{-3t} \end{cases}$$

$$t^2 y^{(2)}(t) - 2ty^{(1)}(t) + 2y(t) = 1 \qquad \begin{cases} \psi_1(t) = 1, \ \psi_2(t) = 1 + 2t^2 \\ \phi_1(t) = t, \ \phi_2(t) = t^2 \end{cases}$$

$$(t - 1)y^{(2)}(t) - ty^{(1)}(t) + y(t) = 1 \qquad \begin{cases} \psi_1(t) = 1, \ \psi_2(t) = 1 + 5t \\ \phi_1(t) = \epsilon^t, \ \phi_2(t) = t \end{cases}$$

## 1.3   Existence and Uniqueness

We know from experience that for any given forcing function $b(\cdot)$ and any given set of coefficients $a_1(\cdot), a_2(\cdot), \ldots, a_n(\cdot)$, Eq. (1) has an infinite number of solutions. Additional requirements must be imposed in order to make the solution unique. If we select *any time* $t_0$ and *any set* of $n$ real numbers $\alpha_0, \alpha_1, \alpha_2, \ldots, \alpha_{n-1}$, then there is *one and only one* solution of Eq. (1), call it $\phi$, such that

$$\phi(t_0) = \alpha_0$$

$$\phi^{(1)}(t_0) = \alpha_1$$

(4) $\qquad \phi^{(2)}(t_0) = \alpha_2$

$$\cdots\cdots\cdots\cdots\cdots$$

$$\phi^{(n-1)}(t_0) = \alpha_{n-1}$$

The equations in (4) are said to impose the initial conditions $\alpha_0$, $\alpha_1, \ldots, \alpha_{n-1}$ on the solution $\phi$. Equivalently, the solution $\phi$ is said to *satisfy the initial conditions* $\alpha_0, \alpha_1, \ldots, \alpha_{n-1}$ at $t_0$. Note that in order to obtain uniqueness the number of initial conditions to be specified [as in (4)] is equal to the *order n* of the differential equation.

It is possible to find systems of $n$ solutions of $(1H)$ $\phi_1, \phi_2, \ldots, \phi_n$ such that *any* solution $\phi$ of $(1H)$ can be written as

$$\phi = c_1\phi_1 + c_2\phi_2 + \cdots + c_n\phi_n$$

where the constants $c_1, c_2, \ldots c_n$ must be suitably chosen so that $\phi$ satisfies the initial conditions imposed on it. Such a system of solutions is called a **fundamental system of solutions.** The functions $\phi_1, \phi_2, \ldots, \phi_n$ of a fundamental system of solutions are said to be *linearly independent* because it is impossible to find $n$ scalars $c_1, c_2, \ldots, c_n$, which are not all equal to zero, such that

$$c_1\phi_1(t) + c_2\phi_2(t) + \cdots + c_n\phi_n(t) = 0 \qquad \text{for all } t$$

If $\psi_0$ is *any* particular solution of the nonhomogeneous equation (1) and if $\phi_1, \phi_2, \ldots, \phi_n$ is a fundamental system of solutions of Eq. $(1H)$, then *any solution* of Eq. (1), say, $\psi$, can be written as

$$\psi = c_1\phi_1 + c_2\phi_2 + \cdots + c_n\phi_n + \psi_0$$

where the constants $c_1, c_2, \ldots, c_n$ must be suitably chosen so that $\psi$ satisfies the prescribed initial conditions. The solution $\psi$ above is called the **general solution** of the linear differential equation (1). It is "general" in the sense that by particular choice of the constants $c_1, c_2, \ldots, c_n$ it can satisfy any set of initial conditions specified as in (4).

## 2    The Homogeneous Linear Equation with Constant Coefficients

Let $L(D) = D^n + a_1D^{n-1} + \cdots + a_n$, where $L(D)$ is a polynomial of degree $n$ with *constant* coefficients. The homogeneous differential equation can be written then as

(5H)   $L(D)y = 0$

The algebraic equation in the variable $s$

(6)   $L(s) = s^n + a_1 s^{n-1} + \cdots + a_n = 0$

is called the **characteristic equation** of the differential equation (5H).  The roots of the characteristic equation are called **characteristic roots.**

## 2.1   Distinct Characteristic Roots

If the characteristic equation has $n$ distinct roots $s_1, s_2, \ldots, s_n$, then the functions

(7)   $\phi_1(t) = \epsilon^{s_1 t}, \phi_2(t) = \epsilon^{s_2 t}, \ldots, \phi_n(t) = \epsilon^{s_n t}$

constitute a fundamental system of solutions of (5H).  Hence *any* solution of (5H) may be written in the form

(8)   $\phi(t) = c_1 \epsilon^{s_1 t} + c_2 \epsilon^{s_2 t} + \cdots + c_n \epsilon^{s_n t}$

where the constants $c_1, c_2, \ldots, c_n$ are appropriately chosen so that the solution satisfies the prescribed initial conditions.

## 2.2   Multiple Characteristic Roots

If the characteristic equation has $m$ distinct roots $s_1, s_2, \ldots, s_m$, where $m < n$, and if the multiplicities of these roots are $k_1, k_2, \ldots, k_m$, respectively, then

1.   $k_1 + k_2 + \cdots + k_m = n.$

2.   The $n$ functions

$\phi_1(t) = \epsilon^{s_1 t}, \phi_2(t) = t\epsilon^{s_1 t}, \ldots, \phi_{k_1}(t) = t^{k_1 - 1}\epsilon^{s_1 t}$

(9)   $\phi_{k_1+1}(t) = \epsilon^{s_2 t}, \phi_{k_1+2}(t) = t\epsilon^{s_2 t}, \ldots, \phi_{k_1+k_2}(t) = t^{k_2 - 1}\epsilon^{s_2 t}$

. . . . . . . . . . . . . . . . . . . . . . . . . . . . . . . . . . . .

$\phi_{n-k_m+1}(t) = \epsilon^{s_m t}, \phi_{n-k_m+2}(t) = t\epsilon^{s_m t}, \ldots, \phi_n(t) = t^{k_m - 1}\epsilon^{s_m t}$

constitute a fundamental set of solutions of Eq. (5H); hence any solution of Eq. (5H) can be written as a linear combination of the $n$ solutions listed above.  In other words, any solution of (5H) can be written in the form

(10)   $\phi(t) = p_1(t)\epsilon^{s_1 t} + p_2(t)\epsilon^{s_2 t} + \cdots + p_m(t)\epsilon^{s_m t}$

where $p_1, p_2, \ldots, p_m$ are polynomials in the variable $t$ of degree $k_1 - 1$, $k_2 - 1, \ldots, k_m - 1$, respectively.  The coefficients of these polynomials must be chosen so that the solution satisfies the initial conditions.

**Remark**   In applications, the coefficients of the characteristic equation are real numbers; consequently, if $s_1 \triangleq \alpha_1 + j\omega_1$ is a characteristic root, so is $s_2 \triangleq \alpha_1 - j\omega_1$. In such a case it is often convenient to use

$$\epsilon^{\alpha_1 t} \cos \omega_1 t \qquad \text{and} \qquad \epsilon^{\alpha_1 t} \sin \omega_1 t$$

as solutions rather than $\epsilon^{(\alpha_1 + j\omega_1)t}$ and $\epsilon^{(\alpha_1 - j\omega_1)t}$, which were suggested by (7).

---

## 3   Particular Solutions of $L(D)y(t) = b(t)$

*Case 1*   Consider the equation

(11)   $L(D)y = \epsilon^{\lambda t}$

    *a.*   If $L(\lambda) \neq 0$ (that is, $\lambda$ is *not* a characteristic root), then

(12)   $\psi(t) = \dfrac{1}{L(\lambda)} \epsilon^{\lambda t}$

        is a particular solution of Eq. (11). This is immediately verified by substitution in (11).

    *b.*   If $\lambda$ is a root of order $k$ of the characteristic equation $L(s) = 0$ and if $p(t)$ is an appropriate polynomial in $t$ of degree $k$, then

(13)   $\psi(t) = p(t)\epsilon^{\lambda t}$

        is a particular solution of Eq. (11); the coefficients of $p(t)$ can be obtained by substitution of (13) in (11).

*Case 2*   Consider the equation

(14)   $L(D)y = A \cos (\omega t + \phi)$

    where $A$, $\omega$, and $\phi$ are constants.
        If $L(j\omega) \neq 0$ and if $\theta$ is defined by

$$L(j\omega) = |L(j\omega)|\epsilon^{j\theta}$$

    then

(15a)   $\psi(t) = \dfrac{A}{|L(j\omega)|} \cos (\omega t + \phi - \theta)$

    is a particular solution of Eq. (14).
        There is an alternate way of writing the particular solution (15a). Put

(15b)   $\psi(t) = B \cos \omega t + C \sin \omega t$

    where the constants $B$ and $C$ are obtained by first, substituting (15b) into Eq. (14), second, writing the right-hand side of (14) in the form

$$A \cos \phi \cos \omega t - A \sin \phi \sin \omega t$$

and third, equating the coefficients of $\cos \omega t$ and $\sin \omega t$ of both sides of the resulting equation.

On the other hand, using well-known trigonometric identities, we can rewrite Eq. (15*b*) as follows:

(15*c*)   $\psi(t) = \sqrt{B^2 + C^2} \cos (\omega t + \alpha)$

where

$$\cos \alpha = \frac{B}{\sqrt{B^2 + C^2}} \qquad \text{and} \qquad \sin \alpha = \frac{-C}{\sqrt{B^2 + C^2}}$$

Note that (15*c*) is of the form (15*a*), hence

$$\frac{A}{|L(j\omega)|} = \sqrt{B^2 + C^2}$$

and $\alpha = \phi - \theta$.

If $j\omega$ is a root of order $k$ of the equation $L(s) = 0$ and if $p_1(t)$ and $p_2(t)$ are appropriate polynomials in $t$ of degrees $k$, then

$$p_1(t) \cos \omega t + p_2(t) \sin \omega t$$

is a particular solution of Eq. (14).   The coefficients of the polynomials $p_1$ and $p_2$ can be obtained by substituting the above expression into Eq. (14) and equating coefficients of like powers of $t$.

*Case 3*   Consider the equation

(16)   $L(D)y = f(t)$

where $f$ is a *polynomial* in $t$ of degree $k$.   If $L(0) \neq 0$, then there is a polynomial $p(t)$ of degree $k$ which is a particular solution of Eq. (16).   The polynomial $p$ can be found by substituting in (16) and equating coefficients of like powers of $t$.

If 0 is a root of order $m$ of $L(s) = 0$ and if $p$ is an appropriate polynomial of degree $k$, then

$$t^m p(t)$$

is a particular solution of Eq. (16).

---

**4**   **Nonlinear Differential Equations**

Given any lumped electric circuit and provided its elements have reasonably well-behaved characteristics, the equations of the circuit can be written as follows:

$$\dot{x}_1 = f_1(x_1, x_2, \ldots, x_n, t)$$
$$\dot{x}_2 = f_2(x_1, x_2, \ldots, x_n, t)$$
$$\cdots\cdots\cdots\cdots\cdots\cdots$$

$$\dot{x}_n = f_n(x_1, x_2, \ldots, x_n, t)$$

or, in vector form,

(17)    $\dot{\mathbf{x}}(t) = \mathbf{f}[\mathbf{x}(t), t]$

Such a system of differential equations is said to be in the **normal form.** Usually the variables $x_1, x_2, \ldots, x_n$ can be taken to be capacitor charges and inductor fluxes, or capacitor voltages and inductor currents (for example, see Chap. 5, Sec. 5 and Chap. 12, Sec. 3). The presence of the variable $t$ as a variable in the right-hand side is caused by two possibilities: the elements of the circuit are time-varying, or there are inputs that depend on $t$.

In the following we shall describe the main existence and uniqueness for $n = 2$; it is understood that it is still true for any positive integral value of $n$. Thus, we consider the following system of two ordinary differential equations in $x_1$ and $x_2$:

(18)
$$\dot{x}_1(t) = f_1[x_1(t), x_2(t), t]$$
$$\dot{x}_2(t) = f_2[x_1(t), x_2(t), t]$$

## 4.1    Interpretation of the Equation

The best way to get a feeling for the equations in (18) is to interpret them geometrically in the $x_1 x_2$ plane. The two functions

$$f_1(x_1, x_2, t)$$
$$f_2(x_1, x_2, t)$$

can be thought of as defining, for every point $(x_1, x_2)$ at any time $t$, the velocity $(\dot{x}_1, \dot{x}_2)$ of a particle. In other words, we can interpret the right-hand side of (18) as specifying a velocity field (see Fig. 1). The problem of "solving" the equations of (18) may be stated as follows: given an arbitrary point $(x_1^0, x_2^0)$ and an arbitrary initial time $t_0$ find the motion of a particle, subject to the conditions that (1) at time $t_0$ it starts at $(x_1^0, x_2^0)$, and (2) at every instant of time $t \geq t_0$, the velocity of the particle is equal to that prescribed by the field of velocities. Analytically, the motion of the particle can be described by two functions $\xi_1(\cdot)$ and $\xi_2(\cdot)$ subject to the conditions

(19)
$$\xi_1(t_0) = x_1^0$$
$$\xi_2(t_0) = x_2^0$$

(20)
$$\dot{\xi}_1(t) = f_1[\xi_1(t), \xi_2(t), t]$$
$$\dot{\xi}_2(t) = f_2[\xi_1(t), \xi_2(t), t]$$    for $t \geq t_0$

The pair of functions $(\xi_1, \xi_2)$ is called a **solution of the system of differential equations** (18). Because of (19) the solution $(\xi_1, \xi_2)$ is said to **satisfy the initial condition** $x_1^0, x_2^0$ **at time** $t_0$.

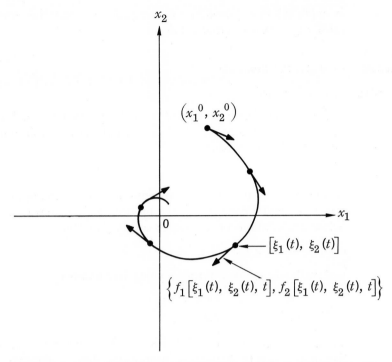

**Fig. 1**   Example of a field of velocities and the trajectory which starts from the point $(x_1{}^0,x_2{}^0)$.

Geometrically, the functions $\xi_1$ and $\xi_2$ may be thought of being the parametric equations of a curve in the $x_1x_2$ plane; thus

$$x_1 = \xi_1(t)$$
$$x_2 = \xi_2(t)$$    $t \geq t_0$

This curve is called a **trajectory,** or more precisely, the trajectory which starts from $(x_1{}^0,x_2{}^0)$ at time $t_0$.

**Remark**   If the functions $f_1$ and $f_2$ do not depend explicitly on $t$, then the field of velocities can be computed once and for all; it is no longer necessary to keep track of time when computing a velocity at a particular point in the $x_1x_2$ plane.

**4.2   Existence and Uniqueness**

So far we have imposed no restrictions on the functions $f_1$ and $f_2$. We wish now to state an often-quoted restriction—the *Lipschitz condition*—which guarantees the existence and uniqueness of the solution. The following example will show how a nice-looking equation may have several solutions satisfying the same initial condition.

---

**Example**   Consider the equation

(21)   $\dot{x}_1(t) = 2\sqrt{|x_1(t)|}$

(The absolute value sign is used to guarantee that we shall never take the square root of a negative number.)   Consider now solutions of Eq. (21) satisfying the initial condition

(22)   $x_1(0) = 0$

It is easy to verify that there are infinitely many solutions of (21) which satisfy the initial condition (22).   Indeed, the functions $\phi_1$ and $\phi_2$, defined below, are solutions of (21) which satisfy (22); thus,

$\phi_1(t) = 0$     for $t \geq 0$

and, with $c$ being an *arbitrary* nonnegative number,

$$\phi_2(t) = \begin{cases} 0 & \text{for } 0 \leq t \leq c \\ 2(t - c)^2 & \text{for } c \leq t \end{cases}$$

These solutions are shown on Fig. 2.   The reason for this infinitude of solutions is due to the *infinite* jump in the slope of the right-hand side of

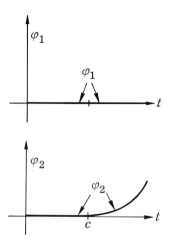

**Fig. 2**   Two solutions of $\dot{x} = 2\sqrt{|x|}$ subject to $x(0) = 0$; the curves $\phi_1$ and $\phi_2$ satisfy the same differential equation and the same initial conditions; observe that $c$ is an arbitrary nonnegative number.

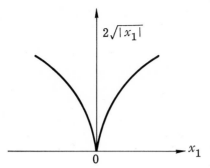

**Fig. 3**    The right-hand side of Eq. (21) plotted versus $x_1$; observe the jump in the slope at $x_1 = 0$.

Eq. (21) when plotted against $x_1$ (see Fig. 3). This suggests that to guarantee uniqueness, we must impose smoothness conditions on the functions $f_1$ and $f_2$; this is done by the Lipschitz condition.

The functions $f_1$ and $f_2$ are said to satisfy a **Lipschitz condition** in a domain $D$ of the $x_1x_2$ plane if there is a constant $k$ such that, for all $t$ under consideration and for all pairs of points $(x_1,x_2)$, $(\hat{x}_1,\hat{x}_2)$ in the domain $D$,

$$|f_1(x_1,x_2,t) - f_1(\hat{x}_1,\hat{x}_2,t)| + |f_2(x_1,x_2,t) - f_2(\hat{x}_1,\hat{x}_2,t)|$$
$$\leq k[|x_1 - \hat{x}_1| + |x_2 - \hat{x}_2|]$$

For instance, if the functions $\partial f_1/\partial x_1$, $\partial f_1/\partial x_2$, $\partial f_2/\partial x_1$, and $\partial f_2/\partial x_2$ are *continuous* functions in $D$, then there is a constant $k$ for which the Lipschitz condition is satisfied.

The key result may now be stated.

*Suppose $f_1$ and $f_2$ satisfy Lipschitz conditions.   Let $t_0$ be an arbitrary instant of time and $(x_1^0,x_2^0)$ be an arbitrary point in the domain $D$; then there is one and only one solution, say $[x_1(t),x_2(t)]$, of (18) which satisfies the initial condition*

$$x_1(t_0) = x_1^0$$
$$x_2(t_0) = x_2^0$$

Geometrically, it means that given any initial instant $t_0$ and any initial state $(x_1^0,x_2^0)$ in $D$, there is one and only one trajectory in $D$ that satisfies the differential equations (18). In circuit theory, these conditions are always satisfied, provided the characteristics of the elements are smooth. A detailed discussion of this question is the subject of more advanced courses.

# Index

Page references in **boldface** type indicate definitions.